Theoretical Advancement in Chromatography and Related Separation Techniques

NATO ASI Series

Advanced Science Institutes Series

A Series presenting the results of activities sponsored by the NATO Science Committee, which aims at the dissemination of advanced scientific and technological knowledge, with a view to strengthening links between scientific communities.

The Series is published by an international board of publishers in conjunction with the NATO Scientific Affairs Division

A Life Sciences	Plenum Publishing Corporation
B Physics	London and New York
C Mathematical	Kluwer Academic Publishers
and Physical Sciences	Dordrecht, Boston and London
D Behavioural and Social Sciences	
E Applied Sciences	
F Computer and Systems Sciences	Springer-Verlag
G Ecological Sciences	Berlin, Heidelberg, New York, London,
H Cell Biology	Paris and Tokyo
I Global Environmental Change	

NATO-PCO-DATA BASE

The electronic index to the NATO ASI Series provides full bibliographical references (with keywords and/or abstracts) to more than 30000 contributions from international scientists published in all sections of the NATO ASI Series.
Access to the NATO-PCO-DATA BASE is possible in two ways:

– via online FILE 128 (NATO-PCO-DATA BASE) hosted by ESRIN,
Via Galileo Galilei, I-00044 Frascati, Italy.

– via CD-ROM "NATO-PCO-DATA BASE" with user-friendly retrieval software in English, French and German (© WTV GmbH and DATAWARE Technologies Inc. 1989).

The CD-ROM can be ordered through any member of the Board of Publishers or through NATO-PCO, Overijse, Belgium.

Theoretical Advancement in Chromatography and Related Separation Techniques

edited by

Francesco Dondi

Dipartimento di Chimica,
Università di Ferrara,
Ferrara, Italy

and

Georges Guiochon

Chemistry Department,
The University of Tennessee,
Knoxville, TN, U.S.A.

Springer-Science+Business Media, B.V.

Proceedings of the NATO Advanced Study Institute on
Theoretical Advancement in Chromatography and Related Separation Techniques
Ferrara, Italy
August 18–30, 1991

Library of Congress Cataloging-in-Publication Data

Theoretical advancement in chromatography and related separation
  techniques / edited by Francesco Dondi and Georges Guiochon.
      p.   cm. -- (NATO ASI series. Series C, Mathematical and
  physical sciences ; v. 383)
    Includes index.
    ISBN 978-94-010-5189-7   ISBN 978-94-011-2686-1 (eBook)
    DOI 10.1007/978-94-011-2686-1
    1. Chromatography--Congresses.   I. Dondi, Francesco, 1943-
  II. Guiochon, Georges, 1931-   . III. Series: NATO ASI series.
  Series C, Mathematical and physical sciences ; no. 383.
  QD79.C4T44   1992
  543'.089--dc20   92-32221

Contents

PREFACE

Chromatography and all related separation techniques enjoy an immense popularity among chemists involved in pure and applied research, in Chemistry, Biochemistry, Clinical and the Environmental Sciences. This arises from the exceptional success of these techniques in solving one of the most important tasks of the experimental chemist, the physical separation of complex mixtures into its components, rapidly, completely, often rather inexpensively, even when aggressive compounds are involved, at either trace or large scale levels.

Although they came as an answer to the pressure of constantly renewed needs, the spectacular progress made in this area since World War II originated in theoretical work. Thus, came the most important advance in gas chromatography, the development of the open tubular columns of Golay. Thus, was originated the work in High Performance Liquid Chromatography. Thus, were invented the family of Field Flow Fractionation techniques. Thus, is evolving right now the field of preparative chromatography. All these breakthroughs originated in original theoretical work.

The key role played by theory in the advance of science was one of the main reasons to organize a meeting focussed on the theoretical aspects of chromatography and related separation techniques. Thanks to the generous support of grants provided by NATO and the Italian National Research Council (CNR), the Editors of these Proceedings were able to organize a NATO Advanced Study Institute in Ferrara (Italy) on August 18-30, 1991 at the Collegio di S.Spirito, as a part of the celebration of the Sixth Centenary of the foundation of the Universita' di Ferrara. The friendly atmosphere surrounding the fruitful and insightful scientific discussions made this meeting an unforgettable experience for the hundred participants. The work presented at this meeting is now available in book form to a much larger audience of chromatographers and scientists of other walks interested in the topic covered here.

The goal of those who organized this ASI and the subsequent book publication was to produce a coverage of the current status of this research area and place it in perspective with the general concepts of the fields of physical chemistry involved. We asked the lecturers and authors to strive at delivering and writing presentations at the graduate level, accessible to all those who have a good general background in the thermodynamics and mass transfer theory of phase equilibria. This book should, hopefully, be useful to the young scientists and engineers who want access to the current frontiers of research in chromatography and the other separation sciences.

In the first part of these proceedings, we placed the theory of chromatography, the basic thermodynamics of phase equilibria, and the molecular thermodynamics, which constitutes a bridge between classical thermodynamics and molecular properties. This allows a rational presentation of the recent developments in preparative chromatography and in other areas of this method. It also permits a unified exposition of this field and provides the necessary basis for the achievement of physico-chemical measurements by chromatography. The second part of this book is devoted to the separation of polymers and biopolymers, using hydrophobic interaction chromatography, field flow fractionation, classical or capillary zone electrophoresis.

The Editors are grateful to the members of the Scientific Organization Committee Gehrardt Findenegg, Eli Grushka and Michel Martin, to the lecturers-authors, the participants and auditors of the NATO ASI. They acknowledge the coworkers of the Laboratory of Analytical Chemistry of the University of Ferrara who devoted a substantial part of their time to the organization of the meeting, before, during and after the venue and contributed highly to make a great success of this venture.

Francesco Dondi, Ferrara, Italy
Georges Guiochon, Knoxville, TN, USA.

LECTURERS

Michel MARTIN
ESPCI - 10 Rue Vauquelin
75231 Paris Cedex 05
FRANCE

Patrick VALENTIN
Centre de Recherche Elf-Solaize Modélisation et Mathematiques BP22
69360 SAINT-SYMPHORIEN D'OZON
FRANCE

Claire VIDAL-MADJAR
L. P. C. B. - C. N. R. S.
B. P. 28
94320 THIAIS
FRANCE

Gerhard H. FINDENEGG
TU Berlin, Iwan-N-Stranski Inst. of Physical and Theoretical Chem.
Strasse des 17, Juni 122
W-1000 Berlin
GERMANY

N. A. KATSANOS
Physical Chemistry Laboratory, University of Patras
26110 PATRAS
GREECE

Eli GRUSHKA
Department of Inorganic and Analytical Chemistry
The Hebrew University of Jerusalem
ISRAEL

Fabrizio BRUNER
Istituto di Scienze Chimiche Università di Urbino
P.zza Rinascimento 6
61029 Urbino
ITALY

Gianpaolo CARTONI
Università di Roma I "La Sapienza", Dipartimento di Chimica
P.le Aldo Moro 5
00185 Roma
ITALY

x

Francesco DONDI
    Università di Ferrara, Dipartimento di Chimica
    via L. Borsari 46
    44100 Ferrara
    ITALY

Arnaldo LIBERTI
    Università di Roma I "La Sapienza", Dipartimento di Chimica
    P.le Aldo Moro 5
    00185 Roma
    ITALY

Piergiorgio RIGHETTI
    Dip. Scienze e Tecn. Biomediche, Sezione Chimica Organica
    Via Celoria 2
    20133 MILANO
    ITALY

Ervin KOVATS
    Ecole Polytech. Fed. de Lausanne, Departement de Chimie
    Laboratoire de Chimie-Technique
    CH-1015 LAUSANNE
    SWITZERLAND

Carl CRAMERS
    Eindohoven University of Technology
    P.O. Box 513
    56MB Eindohoven
    THE NETHERLANDS

Robert TIJSSEN
    Dept. AG/21 Koninklijke/Shell Laboratorium
    Badhuisweg 3
    1031 CM AMSTERDAM
    THE NETHERLANDS

J. R. CONDER
    University College of Swansea, Department of Chemical Engineering
    Singleton Park
    SWANSEA, SA2 8PP
    U. K.

Sadroddin GOLSHAN SHIRAZI
    Department of Chemistry
    Knoxville, Tennessee   37996/1600
    U. S. A.

Georges GUIOCHON
    The University of Tennessee, Department of Chemistry
    1413 Circle Drive
    Knoxville, TN 37996-1368
    U. S. A.
Csaba HORVATH
    Dept. Chemical Engeneering
    P.O. Box 2159, Yale Station
    NEW HAVEN CONNECTICUT 06520-2159
    U. S. A.

Daniel MARTIRE
    Department of Chemistry, Georgetown University
    WASHINGTON DC 20057-0001
    U. S. A.

Steve WILLIAMS
    Department of Chemistry, University of Utah
    Salt Lake City 84112 Utah
    U. S. A.

PARTICIPANTS

1) B. Tilquin
   Université Catholique de Louvain U.C.L. 72.30. - Unité CHAM
   Avenue E. Mounier, 72 B-1200 Louvain-en-Woluwe
   BELGIUM

2) John H. Luong
   National Research Council Canada Biotechnology Research Institute
   6100 Royalmount Avenue MONTREAL, Quebec H4P 2R2
   CANADA

3) André Lawrence
   National Research Council Canada Institute for Aerospace Research
   Ottawa, Ontario
   CANADA K1A OR6

4) Petr Gebauer
   Czechoslovak Academy of Sciences Institute of Analytical Chemistry
   Veverí 97 611 42 Brno
   CZECHOSLOVAKIA

5) Michal Roth
   Czechoslovak Academy of Sciences Institute of Analytical Chemistry
   Veverí 97 CS-61142 Brno
   CZECHOSLOVAKIA

6) Jorgen Mollerup
   Danmarks Tekniske Hojskole Instituttet for Kemiteknik
   Bygning 229 DK-2800 Lyngby
   DENMARK

7) Suresh Chandra Rastogi
   National Environment Research Institute
   Morkhoj Bygade 26H DK-2860 Soborg
   DENMARK

8) Philippe Cardot
   Facultè de Pharmacie
   Rue J. B. Clement 92296 Chatenay Malabry
   FRANCE

9) Estelle Georges
   C. N. R. S. L. P. C. B.
   2 rue H. Dunant 94320 Thiais
   FRANCE

10) Mauricio Hoyos
    ESPCI Laboratoire PMMH
    10 Rue Vauquelin 75231 Paris Cedex 05
    FRANCE

11) Marcela Marquez-Mendez
    Universitè de Compiegne Dept. Genie Biologique
    Compiegne
    FRANCE

12) James Renard
    C. N. R. S. L. P. C. B.
    2 rue H. Dunant 94320 Thiais
    FRANCE

13) Christoph Buttersack
    Technical University Institut für Zuckerindustrie
    Langer Kamp 5 D-3300 Braunschweig
    GERMANY

14) Ulrike Jegle
    Hewlett Packard GmbH
    Postfach 1280 7517 Waldbronn 2
    GERMANY

15) Wolfgang List
    Emil-v.-Behringwerke AG
    Postfach 11 40 Behring-Straße 76 D-3550 Marburg 1
    GERMANY

16) Joachim Walter
    Dr. Karl Thomae GmbH
    Birkendorfer Straße 65 7950 Biberach an der Riss 1
    GERMANY

17) John Kapolos
    University of Patras Physical Chemistry Laboratory
    P. O. Box 1045 26110 Patras
    GREECE

18) Georgia Kodelia
    National Hellenic Research FDN Inst. of Biol. Res. & Biotechnol.
    48, Vassileos Constantinou Avenue Athens 116 35
    GREECE

19) Athanasia Koliadima
    University of Patras Physical Chemistry Laboratory
    P. O. Box 1045 GR-26110 Patras
    GREECE

20) E. Papadopuolou Mourkidou
Aristotelian University Pesticide Science Laboratory
P.O. Box 173 GR-570 01 Thermi
GREECE

21) P. A. Siskos
Univ. of Athens, Chemistry Dept. Laboratory of Analytical Chemistry
Panepistimiopolis - Kouponia GR-15771 Athens
GREECE

22) Christos Vassilakos
University of Patras Physical Chemistry Laboratory
P. O. Box 1045 26110 Patras
GREECE

23) Attila Felinger
University of Veszprèm Dept. of Analytical Chemistry
P.O. Box 158 H-8201 Veszprèm
HUNGARY

24) Tamás Pap
University of Veszprém Institute of Analytical Chemistry
P.O. Box 158 H-8201 Veszprém
HUNGARY

25) Shulamit Levin
The Hebrew University of Jerusalem Pharmaceutical Chemistry
P.O. Box 12065 Jerusalem
ISRAEL

26) Corrado Bighi
Università di Ferrara Dipartimento di Chimica
Via L. Borsari 46, 44100 Ferrara
ITALY

27) Gabriella Blo
Università di Ferrara Dipartimento di Chimica
Via L. Borsari 46, 44100 Ferrara
ITALY

28) Giuliano Costantini
Ist. Guido Donegani
Via Fauser 4, 28100 Novara
ITALY

29) Silvia Guanziroli
Zambon Research SpA
Bresso (MILANO)
ITALY

30) Lorena Mella
Farmitalia Carlo Erba
Via Carlo Imbonati 24, 20159 MILANO
ITALY

31) Luisa Pasti
Università di Ferrara Dipartimento di Chimica
Via L. Borsari 46, 44100 Ferrara
ITALY

32) Paolo Pastore
Università degli Studi di Padova Dip. di Chim. Inorg. Metall. Anal.
Via Marzolo 1, 35131 Padova
ITALY

33) Maria Chiara Pietrogrande
Università di Ferrara Dipartimento di Chimica
Via L. Borsari 46, 44100 Ferrara
ITALY

34) Fernando Pulidori
Università di Ferrara Dipartimento di Chimica
Via L. Borsari 46, 44100 Ferrara
ITALY

35) Maurizio Remelli
Università di Ferrara Dipartimento di Chimica
Via L. Borsari 46, 44100 Ferrara
ITALY

36) Pierluigi Reschiglian
Università di Ferrara Dipartimento di Chimica
Via L. Borsari 46, 44100 Ferrara
ITALY

37) Sunil Bharati
GEOLAB NOR
P.O. Box 1581 Nidarvoll 7002 Trondheim
NORWAY

38) Valborg Holten
MATFORSK
Osloveien 1 N-1430 Aas
NORWAY

39) Rune Ringberg
Dyno Industries Dynochrom As
P. O. Box 213, 2001 Lillestrom
NORWAY

40) Ove Solesvik
DYNOCHROM A. S. Svellevn
P. O. Box 160, N-2001 Lillestrom
NORWAY

41) Robert Nowakowski
Polish Academy of Sciences Inst. of Physical Chemistry
Kasprzaka str. 44/52 PL-01-224 Warsaw
POLAND

42) Carlos Costa
Universidade do Porto Departamento de Engen. Química
Rua dos Bragas 4099 PORTO Codex
PORTUGAL

43) José Miguel Loureiro
Universidade do Porto Dept. de Engenharia Química
Rua dos Bragas 4099 PORTO Codex
PORTUGAL

44) Alirio Rodrigues
Universidade do Porto Dept. de Engenharia Química
Rua dos Bragas 4099 PORTO Codex
PURTUGAL

45) Roberto Rosal Garcia
Universidad de Oviedo Dept. de Ingegnieria Quimica
33071 Oviedo
SPAIN

46) Jan Ståhlberg
ASTRA Pharm. Prod. AB
S-15185 SÖDERTÄLJE
SWEDEN

47) K. S. Reddy
Ecole Polytechnique Fédérale Dépt. Chimie Laboratoire de Chimie-Technique
CH-1015 Lausanne
SWITZERLAND

48) Hugo Billiet
TU-Delf Dept. of Biochemical Engineering de Vries
van Heystplantsoen 2 2628 RZ DELFT
THE NETHERLANDS

49) R. T. Ghijsen
Free University of Amsterdam Dept. of General and Anal. Chem.
De Boelelaan 1083 1081 HV Amsterdam
THE NETHERLANDS

50) J. G. M. Janssen
Technische Universiteit Eindhoven
P.O. Box 513 5600 MB Eindhoven
THE NETHERLANDS

51) Johan C. Kraak
Universiteit van Amsterdam Lab. voor Analytische Scheikunde
Nieuwe Achtergracht 166 1018 WV Amsterdam
THE NETHERLANDS

52) Adil Denizli
Hacettepe University Chemical Engineering Department
06532 Beytepe / ANKARA
TURKEY

53) Bedia Erim
ITÜ-Fen-Edebiyat Fakültesi Kimya Bölümü
80626 Maslak Istanbul
TURKEY

54) A. Ersin Karagozler
Inonu University Department of Chemistry
44069 Malatya
TURKEY

55) M. S. Doulah
Polytechnic of Wales Dept. of Science and Chem. Eng.
Pontypridd, CF37 1DL
U. K.

56) Graham McCreath
University of Cambridge Dept. of Chemical Engineering
Pembroke Street Cambridge, CB2 3RA
U. K.

57) David Hearle
University College of London SERC Centre for Biochem. Engineer.
Torrington Place London WC1E 7JE
U. K.

58) John Noble
AEA Environment and Energy Biotechnology Dept.,
Building 353 Harwell Laboratory Oxfordshire, OX11 ORA
U. K.

59) Sally Prime
Oxford GlycoSystems Unit 4
Hitching Court, Blacklands Way Abingdon, Oxon. OX14 1RG
U. K.

60) Leonid Blumberg
Hewlett-Packard Company Avondale Division
Route 41 - P.O. Box 900 Avondale, PA 19311-0900
U. S. A.

61) Joe M. Davis
Southern Illinois University Chem. Biochem. Dept.
Carbondale
U. S. A.

62) Eric Dose
The University of Tennessee Department of Chemistry
1413 Circle Drive Knoxville, TN 37996-1368
U. S. A.

63) Ziad El Rassi
Oklahoma State University Department of Chemistry
College of Arts and Sciences
Stillwater, OK 74078-0477
U. S. A.

64) Joe P. Foley
Villanova University Department of Chemistry
Villanova, PA 19085-1699
U. S. A.

65) Martha Hilton
University of Missouri-Rolla Department of Chemistry
142 Schrenk Hall Rolla, MO 65401
U. S. A.

66) Steve Jacobson
Univ. of Tennessee Chemistry Dept.
611 Buehler Hall Knoxville TN 37996-1600
U. S. A.

67) Benjamin J. McCoy
Dept. of Chemical Engineering University of California
Davis, CA 95616
U. S. A.

68) Donald Poe
University of Minnesota Department of Chemistry
10 University Drive Duluth, MN 55812-2496
U. S. A.

69) Rebecca Riester
Georgetown University Department of Chemistry
Washington, DC 20057
U. S. A.

70) Karen B. Sentell
The University of Vermont Department of Chemistry
Cook Physical Science Building
Burlington, VT 05405-0125
U. S. A.

71) John Wheeler
Department of Chemistry
Mail Location 172 University of Cincinnati
Cincinnati, OH 45221-0172
U. S. A.

72) James A. Wilkins
OTSUKA America Pharmaceutical,Inc.
9900 Medical Center Drive Rockville, Maryland 20850
U. S. A.

THE IDEAL MODEL OF CHROMATOGRAPHY

S. GOLSHAN-SHIRAZI AND G. GUIOCHON
Department of Chemistry
University of Tennessee
Knoxville, TN, 37996-1501, USA
and
Division of Analytical Chemistry
Oak Ridge National Laboratory
Oak Ridge, TN, 37831-6120, USA

ABSTRACT. The ideal model of chromatography assumes that the column efficiency is infinite. Although actual columns have a finite efficiency, the ideal model has several attractive features which make it important to understand. First, it takes into exact account the effects of the thermodynamics of the phase equilibria on the band profiles. The influence of a non-linear isotherm or of competition between the components for interaction with the stationary phase appear very clearly. Secondly, the differences between the band profiles observed at high concentrations on real columns and the prediction of the ideal model are moderate and qualitatively predictable. Finally, analytical solutions of the ideal model exist, for convex or concave isotherms in the single component case, and for the competitive Langmuir isotherm in the case of a binary mixture. We review these solutions, in overloaded elution with a wide and a narrow injection pulse, and in displacement chromatography.

1. Introduction.

The general problem of chromatography is complex. When the concentrations of the sample or feed components in the mobile phase is not negligibly small and their kinetics of mass transfer, adsorption / desorption or of association / dissociation with the stationary phase is no longer extremely fast, a host of complex problems have to be investigated at the same time and it may be difficult to sort out those which control the column response to a perturbation of the mobile phase composition. As always when we deal simultaneously with thermodynamic and kinetic problems, we have two options for a first approximation, to neglect one or the other and to compare the results obtained with experimental observations.

If we neglect the influence of a finite concentration and assume that, at equilibrium, the ratio of the concentrations of every sample component in both phases remain constant, i.e., that the equilibrium isotherms between phases are linear, we have the model of linear chromatography. This model is quite satisfactory for most analytical applications of chromatography. We can then focus our attention on the influence of the kinetics of the various phenomena involved. These problems are discussed in the third chapter of these Proceedings [1].

Alternately, we may consider that the concentrations of the sample components in the system are high and that the equilibrium isotherms between mobile and stationary phases are non-linear. This isotherm behavior will have major consequences on the band profiles because it

1

F. Dondi and G. Guiochon (eds.),
Theoretical Advancement in Chromatography and Related Separation Techniques, 1–33.
© 1992 *Kluwer Academic Publishers.*

turns out that in the immense majority of cases, the mass transfer and retention kinetics are fast enough and do not mask the thermodynamic effects, hardly even blur them. Then, in order to have a model which is simple enough to handle, but which we can study in depth in order to try and understand what is the column response to simple perturbations, we may assume that the column efficiency is infinite. With this assumption, the axial dispersion will be zero, while the mass transfer kinetics will proceed at an infinite rate, the two phases being in constant equilibrium.

This ideal model of chromatography is the simplest possible model which accounts for non-linear effects in chromatography. However, this model gives instantaneous concentration changes, profiles with sharp angles, while we expect that in practice profiles will be smooth. The most important simplification brought by the ideal model, the least realistic one, however, is that features of the column response predicted by the ideal model take place instantaneously, while with real columns a certain time will be required before we can notice them. In practice, however, we are always surprised to see how accurate the prediction of the ideal model are. This is obviously because the typical columns used in most preparative separations have a high efficiency, exceeding most often several thousand plates. This is more than enough to leave barely eroded the key features that non-linear thermodynamics tries to impart to concentration profiles.

We discuss first the general properties of the ideal model, the equation which expresses its assumptions and its properties. Then we indicate the solutions when they exist, their main features and discuss successively the single component elution profile, the separation of a binary mixture in the elution mode, with either a wide or a narrow rectangular injection, and the solution of displacement chromatography.

2. Origin and Basis of the Ideal Model.

The mathematical basis of the ideal model were formulated by Wilson fifty years ago [2]. We may write the differential mass balance of a single component by simply stating that the amount of the component considered which accumulates in an infinitely thin slice of the column is equal to the difference between what enters and what exits the slice. Since we assume that there is no axial dispersion, the component may enter or exit the slice only by convection. Hence, the mass balance is written:

$$\frac{\partial C}{\partial t} + F\frac{\partial q}{\partial t} + u\frac{\partial C}{\partial z} = 0 \tag{1}$$

In this equation, C and q are the concentrations of the component in the mobile and stationary phases, respectively. We assume these concentrations to be continuous and differentiable functions of t and z, the time and position along the column, respectively. u is the mobile phase velocity and F is the phase ratio, $V_s/V_m = (1 - \epsilon)/\epsilon$, ϵ the total porosity of the packing, V_s and V_m the volumes of stationary and mobile phases. The same equation may be written for the various components of a mixture:

$$\frac{\partial C_i}{\partial t} + F\frac{\partial q_i}{\partial t} + u\frac{\partial C_i}{\partial z} = 0 \quad , \qquad i = 1, \cdots, k \tag{1a}$$

Equation 1 is a first order partial differential equation which has some very important properties. Before we discuss them, we need to state some complementary assumptions of the model and some additional equations.

As implicit in equation 1, we consider the chromatographic problem as unidimensional, as we assume that the column is radially homogeneous. This may not always be true, especially when the column length and diameter become comparable. We further neglect the compressibility of the mobile phase and assume that the partial molar volumes of the components in both phases are equal. While this is not strictly true in liquid chromatography, the effects of deviations from these assumptions are negligible. In gas chromatography, however, these assumptions are unacceptable. As a consequence, we must write a mass balance equation for the carrier gas, similar to equation 1 [3]. Furthermore, in gas chromatography u depends on C and we should replace the third term of the LHS of equation 1 by: $\frac{\partial(u\,C)}{\partial z}$. The sorption effect, due to the difference between the partial molar volumes of the solute and its vapor will be important for weakly retained components [4]. We also neglect the thermal effect due to the change in enthalpy associated with the phase transfer and the influence of the finite concentration of the solute on the mobile phase viscosity [3].

Since we have assumed constant equilibrium between the two phases, the system of k equations 1a must be completed by the k isotherm equations:

$$q_i = f(C_1, C_2, \cdots, C_i, \cdots) \tag{2}$$

There is no restriction on the isotherm we may use, as long as it is continuous, differentiable, and satisfies conditions of thermodynamic consistency. As we see later, there is a general solution of equation 1 in the single component case and this solution is valid for all concave or convex isotherms. In the multi-component case, the isotherms are competitive and the equations 1a are coupled by these isotherms. For the sake of simplicity and because there is often no analytical solution with other isotherms, we shall consider most often the Langmuir competitive isotherm [5], written as:

$$q_i = \frac{a_i C_i}{1 + \sum_1^k b_j C_j} \tag{3}$$

in the case of the component i in a mixture of k components.

A partial differential equation or a system of such equations cannot be solved without a complete set of initial and boundary conditions. Often this set of conditions has a critical importance, determining whether a closed form solution is possible or not. The typical initial condition is that the concentration of the sample components or of some critical component of the mobile phase is constant along the column:

$$C_i(0, z) = C_i^0 \tag{4}$$

In many cases, this constant concentration is 0. The boundary condition should be realistic. It may be tempting to use certain mathematical procedures requiring an outlet boundary condition; this would not be realistic and the consequences of this condition should be checked. In elution, the boundary condition is often the injection of a rectangular concentration plug, but it can be any profile:

$$C_i(t, 0) = \phi(t) \tag{5}$$

The Danckwerts condition [6], or an exponential decay of the input may be more realistic conditions.

Finally, we need one set of equations (mass balance equation, isotherm equation, initial and boundary conditions) for each compound in the system, meaning for each retained component of the sample and for each component of the mobile phase, except for the weak solvent. When the mobile phase is a mixture but the additives are much less strongly adsorbed than the sample components, we may neglect their adsorption, provided that we formulate the equilibrium isotherm accordingly [7], and consider only the mass balance equations of the sample components.

Wilson [2] could solve the ideal model in part, deriving the equation of the continuous or diffuse part of the profile. Some serious mathematical difficulties arise, however, because the other side of the profile is self-sharpening, as we will explain. These difficulties were discussed by DeVault [8] who made a rigorous study of the properties of the equation 1 and demonstrated the formation of concentration shocks or discontinuities in the case of pulse or step injections. Weiss described in detail the progressive change of the band profile during its migration in the case of either a Langmuir of a Freundlich isotherm [9]. In the late forties and fifties, Glueckauf made major experimental and theoretical contributions to the problem [10,11]. He showed how the band profile is related to the isotherm in the cases when the isotherm is convex down, convex up or sigmoidal.

Unfortunately, the method used by Glueckauf and his choice of variables resulted in unnecessary complications in the solutions and his results are difficult to understand and use today. Furthermore, the progress of the theory of hyperbolic systems of partial differential equations were slow. The tools needed for a rigorous approach of the problem appeared only in the fifties and were applied to the study of chromatographic problems only much later [12]. The shock theory was not applied in chromatography before the late 60's and was not available to Glueckauf.

3. Migration and Evolution of a Single Component Band Profile.

The response of the column to a concentration perturbation at the inlet (e.g., injection of a rectangular plug) can be studied using the theory of characteristics which applies to equation 1 [12]. We present here, as simply as possible, the most salient results. Equation 1 can be rewritten:

$$\frac{\partial C}{\partial t} + \frac{u}{1 + F\frac{dq}{dC}}\frac{\partial C}{\partial z} = 0 \qquad (6)$$

This equation describes the propagation of a concentration C to which is associated a velocity, u_z, given by:

$$u_z = \frac{u}{1 + F\frac{dq}{dC}} \qquad (7)$$

For a given column, during a given experiment, this velocity depends only on the concentration with which it is associated, through the isotherm, q. Thus, *each concentration propagates along the column at a constant velocity.* If we represent each concentration by a point in a three-dimension space, t, z, C, each concentration of the injected profile follows a trajectory parallel to the plane t, z, which is a straight line. The elution time of a concentration C is:

$$t_R(C) = t_p + \frac{L}{u}\left(1 + F\frac{dq}{dC}\right) \qquad (8)$$

assuming that the injection profile is a rectangular profile of width t_p and height C_0. The sample size is proportional to $t_p C_0$.

The velocity u_z, however, depends on the concentration. In the most frequent case of a convex upwards isotherm (e.g., the Langmuir isotherm), $\frac{d^2q}{dC^2}$ is negative, $\frac{dq}{dC}$ decreases and u_z increases monotonically with increasing concentration. Thus, high concentrations move faster than low concentrations. The rear profile spreads while the front sharpens, if it is not already vertical, as in a rectangular injection pulse. The high concentrations cannot pass the low concentrations, however. In one location, at a given time, there can be only one concentration [3,8,12]. High concentrations pile up and a front shock or concentration discontinuity forms. This shock is stable and propagates at a velocity which is given by:

$$U_s = \frac{u}{1 + F\frac{\Delta q}{\Delta C}} \qquad (9)$$

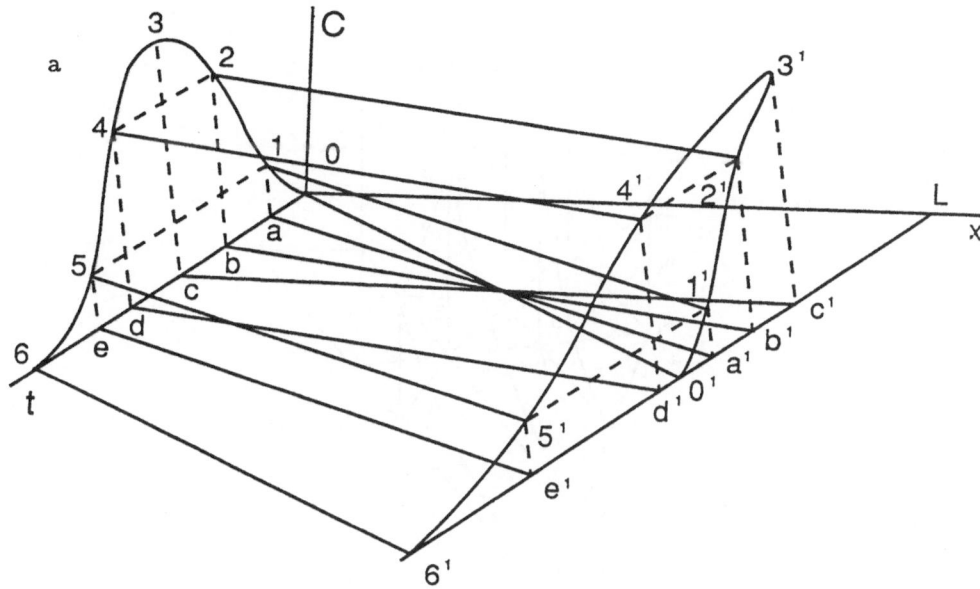

Figure 1. Velocity associated with a concentration, propagation of a band (or concentration signal) and formation of a shock. Reproduced from Ref. 17 with permission of the American Chemical Society.

Figure 1a. Why concentration shocks must form.

where Δq and ΔC are the concentration discontinuities in the stationary and mobile phases, respectively. While the velocity associated with the concentration is related to the slope of the isotherm itself, the velocity of the shock is related to the slope of the chord of the isotherm. The mathematical origin of the concentration shocks and their properties has been discussed by Courant and Friedrichs [13] and by Lax [14].

From a mathematical point of view, however, the introduction of discontinuities in band profiles is no simple matter. In the traditional theory of partial differential equations, a solution of equation 1a should be continuous. The concept of *weak* solutions was introduced by Lax as a generalization, to accept solutions that are not continuously differentiable [15]. A concentration profile including a concentration shock and diffuse boundaries is a *weak* solution of the equation 1a [3,12-15]. However, there are no unique weak solutions for equation 1a. Among the possible weak solutions, there is only one which is acceptable, because it prevents the crossing of the characteristics and makes physical sense. The rule of selection of the acceptable solution suggested by Oleinik [16] could be applied. Rouchon et al. [3] have shown how a condition formulated by Lax [14] permits also to define the correct weak solution of the physical problem.

The application of the shock theory to liquid chromatography has been illustrated by Lin et al. [17]. The mathematical shock has no physical sense. It is infinitely thin, thinner than a molecular diameter, which is impossible. The physical reality corresponding to the shock is the shock layer, a region of the plane t, z where the concentration varies very fast [18,19]. This concept will be discussed in the chapter on the equilibrium-dispersive model.

To summarize, if the equilibrium isotherm is convex upwards, the band profile has a diffuse rear whose profile is given by equation 8, while its front is a shock. If the isotherm is convex

6

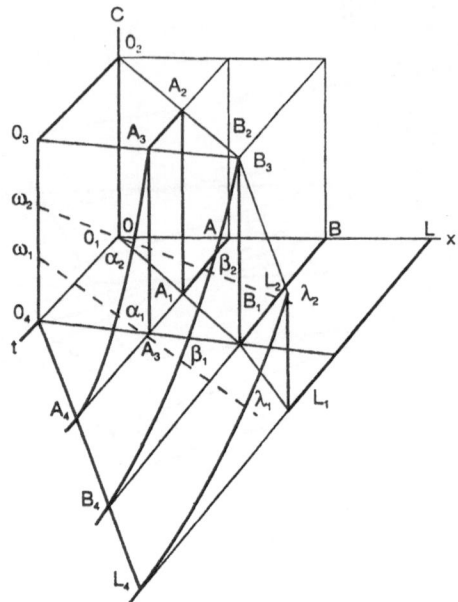

Figure 1b. Illustration of the concentration dependence of the propagation velocity of concentrations on a continuous profile and of the shock velocity.

downwards, the converse is true, the profile has a diffuse front and a rear shock. Figure 1 explains the migration velocity of a concentration, the progressive change of the band profiles, and the formation of a concentration discontinuity or shock.

Comparing equations 7 and 9, we see that for a convex upwards isotherm, the shock velocity is higher than the limit velocity of the compound at infinite dilution (itself given by the initial slope of the isotherm) since for a convex upwards isotherm we have $\frac{\Delta q}{\Delta C} < \left(\frac{dq}{dC}\right)_{C=0}$. On the other hand, we can consider the summit of the elution band as either the top of a shock of height $\Delta C = C_M$ or the top of the continuous rear profile. The two corresponding velocities are [12,17]:

$$u_z(C_M) = \frac{u}{1 + F \left(\frac{dq}{dC}\right)_{C=C_M}} \tag{10a}$$

$$U_s(C_M) = \frac{u}{1 + F \frac{q(C_M)}{C_M}} \tag{10b}$$

Obviously, for a convex upwards isotherm, the first velocity (eqn. 10a) is higher than the second (eqn. 10b), since $\frac{\Delta q}{\Delta C} < \frac{dq}{dC}$. Thus, the top of the diffuse part of the profile tends to go faster than the shock. It vanishes. The band spreads, broadens and in the same time gets shorter, which is necessary to conserve its area. Equation 9 cannot be used to calculate the retention time of the shock because the shock height decreases constantly during the elution and the shock slows down.

Finally, we note that the shock velocity is the velocity of the molecules actually injected in the column, on a concentration plateau (i.e., the velocity of an isotopically labelled tracer).

The velocity associated with a concentration is the velocity of the perturbation caused by the injection of a small pulse on a concentration plateau [19].

4. Retention Time of the Concentration Shock of a Single Component Band.

The retention of the front shock of a component band in the case of a Langmuir isotherm can be calculated directly [20]. We derive first a general relationship, which is valid for all isotherms, as long as they do not have an inflection point and that, accordingly, the position of the shock is unique. Then we explicit this equation in two important cases, the Langmuir isotherm and the bilangmuir isotherm.

4.1 - GENERAL SOLUTION.

We assume that the injection profile is rectangular, with a width t_p and a height C_0. The amount of sample injected is $n = C_0 t_p F_v = C_0 t_p \epsilon S u$, where S is the cross sectional area of the column, and $F_v = \epsilon S u$ is the mobile phase flow rate. The diffuse part of the profile is given by equation 8. The maximum concentration of the profile, C_M, is obtained by writing that the area under the diffuse profile is constant and equal to n/F_v [21]. By integration we obtain:

$$|q(C_M) - C_M \frac{dq}{dC}|_{C=C_M} = \frac{n}{F_v t_0 F} \tag{11}$$

Equation 11 is a simple algebraic equation which can be solved in closed form for some isotherm equation and is always easy to solve numerically. The solution is the maximum concentration of the band. Equation 8 gives the corresponding retention time of the shock.

4.2 - CASE OF THE LANGMUIR ISOTHERM.

For the Langmuir isotherm ($q = \frac{aF}{1+bC}$, a, b, numerical coefficients), the equations 7 and 9 giving the velocity associated with a concentration and the velocity of a shock, respectively, become:

$$u_z = \frac{u}{1 + \frac{Fa}{(1+bC)^2}} \tag{12a}$$

$$U_s = \frac{u}{1 + \frac{Fa}{1+bC}} \tag{12b}$$

Thus, the equation of the diffuse band rear (eqn. 8) can be written as:

$$t_R(C) = t_p + t_0(1 + \frac{aF}{(1 + bC)^2}) \tag{13}$$

We see that the band ends always for $C = 0$ and:

$$t_R(0) = t_p + t_0(1 + Fa) = t_p + t_0(1 + k_0') = t_{R,0} + t_p \tag{14}$$

The retention time of the band front or shock can be derived directly from equation 11 or can be calculated directly [12,20]. It is:

$$t_R = t_p + t_0 + (t_{R,0} - t_0)(1 - \sqrt{L_f})^2 \tag{15}$$

8

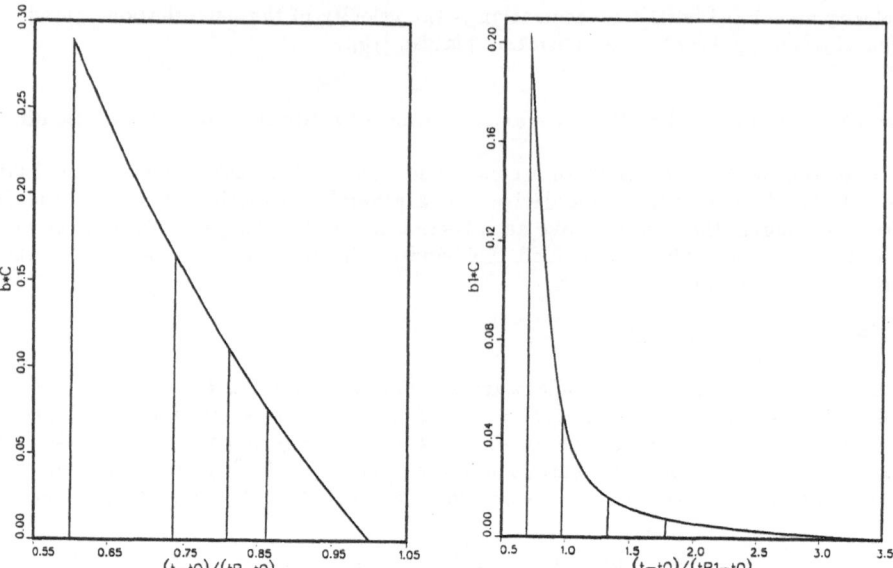

Figure 2. **Band profiles for a Langmuir isotherm, in reduced coordinates**. In this reference system, the profiles depend only on the sample size, given by reference to the column saturation capacity. For the four bands shown, the loading factors are: 0.5%, 1%, 2% and 5%, respectively.

Figure 3. **Band profiles for a bilangmuir isotherm, in reduced coordinates of the first Langmuir term**. The sample sizes are 0.5, 1, 2 and 5% of the column saturation capacity of the first term of the bilangmuir isotherm and 20, 40, 80 and 200% of the column saturation capacity of the second term (high energy sites).

where $L_f = \frac{nb}{\epsilon S L k_o'}$ is the loading factor or ratio of the sample size to the column saturation capacity. The maximum concentration of the band is:

$$C_M = \frac{\sqrt{L_f}}{b(1 - \sqrt{L_f})} \qquad (16)$$

From these equations we derive some important characteristics of the overloaded band profile when the isotherm is Langmuir:

- The band profile always ends for $t = t_{R,0} + t_p$. This is because there is no kinetic band spreading in the ideal model.

- The diffuse profile is not linear as sometimes assumed [22], but hyperbolic. When the loading factor tends towards unity, the retention time of the shock tends towards $t_p + t_0$ (eqn. 15) and the band height towards infinity (eqn. 16).

- If we adopt a simple system of reduced coordinates, replacing the time, t, with $(t - t_0)/(t_{R,0} - t_0) = k'/k_0'$ and the concentration, C, with $\Gamma = bC$, **the overloaded band profiles of all compounds whose adsorption is described with a Langmuir isotherm are identical and depend only on the loading factor** [20]. This result is illustrated in Figure 2, which shows in reduced coordinates a series of band profiles corresponding to increasing loading factors, for a compound which follows the Langmuir isotherm. This observation can be used to estimate rapidly the best parameters for the Langmuir adsorption model of a new compound.

4.3 - CASE OF THE BILANGMUIR ISOTHERM.

We have found that in many practical cases the experimental adsorption isotherm is well accounted for by the sum of two Langmuir isotherms [23-25]. The rationale behind such a model is the assumption that the adsorbent surface is heterogeneous but made of a patchwork of two homogeneous surfaces, e.g., a silica and an alkyl surface in the case of chemically bonded silica. This model has given also excellent results when applied to the adsorption of the enantiomers of the N-benzoyl derivatives of several amino acids on immobilized bovine serum albumin [25]. The isotherm is written:

$$q = \frac{a_1 C}{1 + b_1 C} + \frac{a_2 C}{1 + b_2 C} = \frac{(a_1 + a_2)C + (a_1 b_2 + a_2 b_1)C^2}{1 + (b_1 + b_2)C + b_1 b_2 C^2} \tag{17}$$

The numerical values of the four coefficients are such that $b_2 \gg b_1$ (the adsorption energy on the site 2 is much larger than on the site 1) and $a_2/b_2 \ll a_1/b_1$ (the column saturation capacity is much larger for the site 1 than for the site 2). With equation 17 as the isotherm equation, equation 8 becomes [20]:

$$t_R(C) = t_p + t_0 + Ft_0 \frac{a_1(1 + b_2 C)^2 + a_2(1 + b_1 C)^2}{(1 + (b_1 + b_2)C + b_1 b_2 C^2)^2} \tag{18}$$

The analytical retention time in this case is obviously:

$$t_R(0) = t_p + t_0[1 + F(a_1 + a_2)] \tag{19}$$

Equation 11 becomes [20]:

$$\left(\frac{b_1 C_M}{1 + b_1 C_M}\right)^2 + \frac{a_2 b_1}{a_1 b_2}\left(\frac{b_2 C_M}{1 + b_2 C_M}\right)^2 = L_{f,1} \tag{20}$$

where $L_{f,1}$ is the loading factor calculated from the total sample size and the column saturation capacity for the first site only. This equation must be solved numerically, or the retention time can be derived directly by numerical integration of the relationship:

$$\left| \int_{t_R}^{t_{R,0}+t_p} Cdt \right| = \frac{n}{F_v} \tag{20a}$$

using equation 18 to relate C and t. The band profile exhibits the characteristic tail often encountered in chromatograms obtained with polar compounds eluted in reversed phase chromatography. This confirms the attribution of the tail to "active sites" and provides a procedure for the investigation of these sites.

In Figure 3, we show the band profiles obtained for a bilangmuir isotherm, in the Langmuir reduced coordinates used for Figure 2. This time, however, the band tails much longer, until the retention time corresponding to the high adsorption energy site. In the caption, the loading factors are reported to the two column saturation capacities. The largest peak exceeds much the saturation capacity of the high energy site.

4.4 - ASYMPTOTIC SOLUTION.

When a band migrates along a column of infinite length, the profile tends towards a limit. It would be erroneous to believe that this limit is related in any way to a Gaussian profile.

Whatever smooth the injection profile can be, a shock will eventually appear, grow and control the band profile. In the ideal model, there is no kinetic band spreading. On the other hand, the band profile is sensitive to any deviation of the isotherm from linear behavior.

Mathematical considerations [21] show that the concentration is different from zero in the time interval between these two limits:

$$t = t_e = g'(0) z \tag{21a}$$

$$t = t_R = g'(0) z \pm \sqrt{|2mg''(0) z|} \tag{21b}$$

where the function $g(C)$ is related to the isotherm by:

$$g(C) = \frac{1}{u}(C + Fq) \tag{22}$$

Within the interval defined by equations 21a and 21b, the band profile is a right triangle lying on one side. Its equation is:

$$C(z, t) = \frac{1}{g''(0)} \left(\frac{t}{z} - g'(0) \right) \tag{23}$$

The band profile is zero for $t = t_e$ and has a shock for $t = t_R$. The time t_R is shorter than t_e if the isotherm is convex upward (sign - in eqn. 21b), longer if the isotherm is convex downwards.

The asymptotic band profile depends only on the initial slope and curvature of the isotherm.

4.5. COMPARISON WITH EXPERIMENTAL RESULTS.

We expect the ideal model to give band profiles in marked disagreement with experimental results. This is certainly so at low sample sizes. At high column loadings, however, with an efficient column, the result is not bad. In Figure 4a, experimental data obtained with phenol on a C18 chemically bonded column are compared to the profiles calculated with the ideal model for column loadings increasing from 2 to 11% [26]. In Figure 4b, the same comparison is made with data obtained with benzyl alcohol in normal phase chromatography for column loading factors increasing from 0.15 to 6are not vertical, but they are very steep. The difference in slope of the front from vertical is significant for the smallest peak, but it becomes nearly insignificant for the largest one. The rear part of the profile of the actual band is more strongly curved than the predicted profile and it tails longer.

These differences are essentially due to the influences of the axial dispersion and the finite rate of mass transfers in the column, influences which are neglected in the ideal model. An experimental band profile has a steep, but not vertical, front and a tailing rear. The long tail of the phenol band on octadecyl silica suggests, however, the possibility that a bilangmuir isotherm, with a second term (corresponding to the "active sites") having a small saturation capacity, could better explain the profile tail (see Figure 3) than a Langmuir isotherm.

5. Elution of a Two Components Band.

Wilson [2] and DeVault [8] both discussed the solution of the two component problem in the case of the elution of a pulse of a binary mixture. Glueckauf [27] was the first to give a comprehensive analysis of the separation process of a binary mixture when the adsorption behavior of the two components follows the competitive Langmuir isotherm [28]. In this case, a general solution can be derived directly, using the method of characteristics [12,29] or through the use of the h-transform [30,31]. Rhee and Amundson [32] made profound contributions to

11

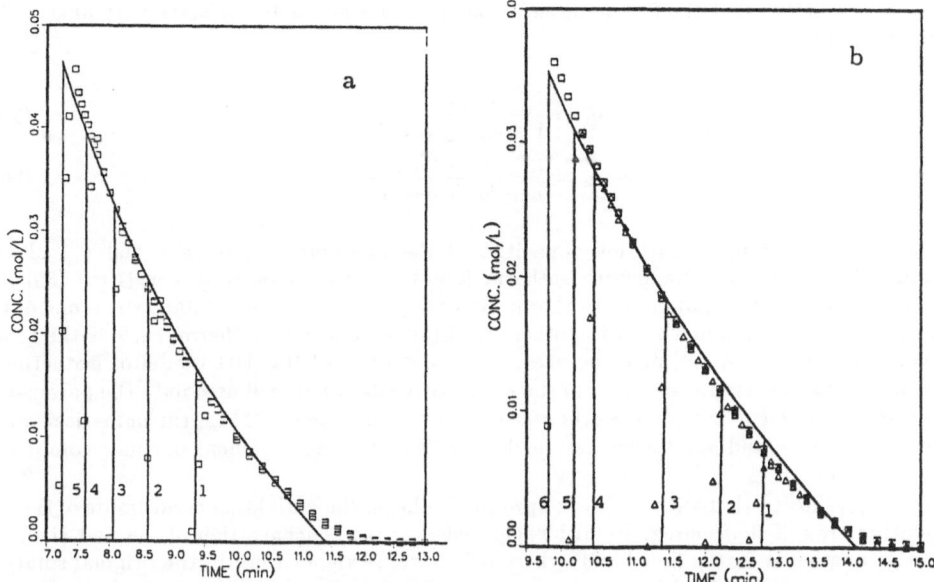

Figure 4. Comparison of experimental band profiles and profiles calculated with the ideal model.

Figure 4a. Phenol on C18 silica, eluted by a methanol/water (20:80) mixture. Flow rate 1 mL/min. Sample loading factors (size in mmole), 2.1(0.015); 4.3(0.03); 6.4(0.045); 8.5(0.06) and 10.7%(0.075). Reproduced from ref. 26, by permission of the American Chemical Society.

Figure 4b. Benzyl alcohol on silica, eluted by a THF/n-heptane (15:85) mixture. Flow rate: 1 mL/min. Sample loading factor (sample size): 0.47(0.00625); 0.95(0.0125); 1.9(0.025); 3.8(0.050); 4.6(0.060) and 5.7(0.075) millimole. Reproduced from ref. 26, by permission of the American Chemical Society.

the solution of the similar problem in frontal analysis. We discuss in this section the general principle of the solution. In the last three sections, we apply these principles to the study of the properties of the solution in the case of a wide rectangular injection (i.e., the succession of a positive and a negative step), in the case of a narrow pulse injection (elution) and in displacement chromatography.

By contrast to the single component problem which can be solved by a simple algebraic equation for all isotherms without an inflection point, the two-component problem can be solved in closed form only if the competitive isotherm is Langmuirian. This is an ideal case which is rarely met in practice. The increased complexity of the problem comes from the coupling between the mass balance equations of the components 1 and 2:

$$\frac{\partial C_1}{\partial t} + F\frac{\partial q_1}{\partial t} + u\frac{\partial C_1}{\partial z} = 0 \tag{24a}$$

$$\frac{\partial C_2}{\partial t} + F\frac{\partial q_2}{\partial t} + u\frac{\partial C_2}{\partial z} = 0 \tag{24b}$$

The coupling is through the competitive isotherms:

$$q_1 = f_1(C_1, C_2) \tag{25a}$$

$$q_2 = f_2(C_1, C_2) \tag{25b}$$

Equation 25 states that the two components compete for access to the stationary phase. The Langmuir isotherm is written:

$$q_1 = \frac{q_{s,1} b_1 C_1}{1 + b_1 C_1 + b_2 C_2} \tag{26a}$$

$$q_2 = \frac{q_{s,2} b_2 C_2}{1 + b_1 C_1 + b_2 C_2} \tag{26b}$$

In principle, the column saturation capacities of the two components, $q_{s,1}$ and $q_{s,2}$ should be equal. Otherwise, the Langmuir isotherm is not thermodynamically consistent [33]. In practice, the competitive Langmuir isotherm fits poorly the experimental data when one of the saturation capacities is larger than the other. In this case, another isotherm, such as the Levan and Vermeulen isotherm [34] must be used. The advantage of this last model of isotherms is that it permits a good representation of the competitive data but still uses only the parameters of the single component Langmuir isotherms of the two components. Thus, the numerical values of the coefficients b_i and $q_{s,i}$ needed can be derived from the measurement of single component adsorption data.

DeVault [8] showed that with a convex upwards isotherm the individual band fronts of the two components are self-sharpening. He also suggested that the concentration of the first (i.e., less retained) component in the first band, where it is pure, is higher than in the original solution, while the concentration of the second, slower moving component in the rear band, where it is pure, tends to be smaller than in the mixed band. Glueckauf [10,35] gave a comprehensive analysis of the problem and calculated the elution profiles of the two components. Coates and Glueckauf [11] measured the individual profiles of the two components and showed good agreement with the results of the calculation, except for the severe tailing of the second band, a phenomenon which plagued chromatographers for years. Unfortunately, the results of Glueckauf are very complex and difficult to understand. They are also incomplete as the shock theory was not available at the time. Progress in the theory of partial differential equations during the last forty years [36-40] has allowed the derivation of an exact solution based on the use of the method of characteristics and the shock theory [29].

Helfferich and Klein [30] have developed a theory of multi- component chromatography based on the concept of coherence and the competitive Langmuir isotherms [28]. They applied it to the study of separations by displacement chromatography, although it could be used in elution and frontal analysis as well. Rhee, Aris and Amundson [41] have studied the same problems using the theory of partial differential equations. They introduced the simple wave theory which applies to rectangular injections which are so wide that the elution profile still contains a segment of the injection plateau. Because of this plateau, the solution contains a constant state and considerable simplification ensues, which do not take place in the case of the elution of a narrow rectangular pulse. It always takes a finite time for the plateau to decay. During that time, all elution profiles corresponding to a rectangular injection are given by the simple wave solution [41-43].

The system of equations 24 is the classical system of reducible, quasi-linear, first-order partial differential equations of the ideal model of chromatography. Its properties have been studied in great detail. Noteworthy is the demonstration by Kvaalen, Neel and Tondeur [44] that this system is strictly hyperbolic, so the two bands are completely eluted in a finite time, but beyond the column hold-up time, $t_0 = \frac{L}{u_0}$. The contrast between the finite time required for complete elution of the band in the ideal model, and the infinitely long time required in linear chromatography illustrates the contrast between the hyperbolic properties of the system of equations 1a and the parabolicity property exhibited by diffusion.

As in the single-component case, there is a velocity associated with a given concentration of each component:

$$u_{z,1} = \frac{u_0}{1 + F\frac{Dq_1}{DC_1}} \tag{27a}$$

$$u_{z,2} = \frac{u_0}{1 + F\frac{Dq_2}{DC_2}} \tag{27b}$$

where the terms $\frac{Dq_i}{DC_i}$ ($i = 1, 2$) are the two directional derivatives:

$$\frac{Dq_1}{DC_1} = \frac{\partial q_1}{\partial C_1} + \frac{dC_2}{dC_1}\frac{\partial q_1}{\partial C_2} \tag{28a}$$

$$\frac{Dq_2}{DC_2} = \frac{\partial q_2}{\partial C_2} + \frac{dC_1}{dC_2}\frac{\partial q_2}{\partial C_1} \tag{28a}$$

Both velocities $u_{z,1}$ and $u_{z,2}$ depend on both concentrations. The coherence condition [30] results from the most important property of these directional derivatives: they are equal:

$$u_{z,1} = u_{z,2} \tag{29}$$

and so are the velocities associated with a given set of concentrations, C_1, C_2. Using this condition, we can construct characteristics and use them for a description of the band migration and separation. This approach will be used in the section dealing with elution.

The velocities of the shocks of the two components are:

$$U_{s,1} = \frac{u_0}{1 + F\frac{\Delta q_1}{\Delta C_1}} \tag{30a}$$

$$U_{s,2} = \frac{u_0}{1 + F\frac{\Delta q_2}{\Delta C_2}} \tag{30b}$$

Since the Langmuir isotherms are convex upwards, the velocities associated with a concentration (eqn. 27a and 27b) increase with increasing concentrations of either component. A stable shock appears on the front of the band. Alternately, the front shock of a rectangular injection is stable. Conversely, the rear shock of this injection is unstable and the rear part of the profile is diffuse.

Finally, we can represent the solutions of the system of equations 24 using the graph obtained with the hodograph transform [41,42]. The primary role of this transform is to give the integral curves of the two characteristic fields $\vec{C}(\lambda, t)$, i.e., the eigenvectors of the matrix associated to the system of equations 24a and 24b. Equations 27a and 27b and the coherence condition (eqn. 29) give the eigenvalue of this matrix. A more simple description of the hodograph transform may be given for the practitioner. The two concentrations C_1 and C_2 of the components of a binary mixture are defined everywhere in the column at any time (except at the shocks). The solutions are the concentration profiles along the column or the elution profiles (i.e., concentration profiles as a function of time at column exit $z = L$). Rather than looking for these profiles, we may search the column location and time where the two concentrations C_1 and C_2 are found simultaneously. The transform, giving the new functions $z(C_1, C_2)$ and $t(C_1, C_2)$ is possible provided that the Jacobian:

$$j = \frac{\partial(C_1, C_2)}{\partial(t, z)} = J^{-1} \tag{31}$$

has a finite value, different from 0. If the Jacobian matrix (eqn. 31) does not have its full rank (i.e., 2 for a binary mixture), the hodograph transform of the system of equations 24 is

14

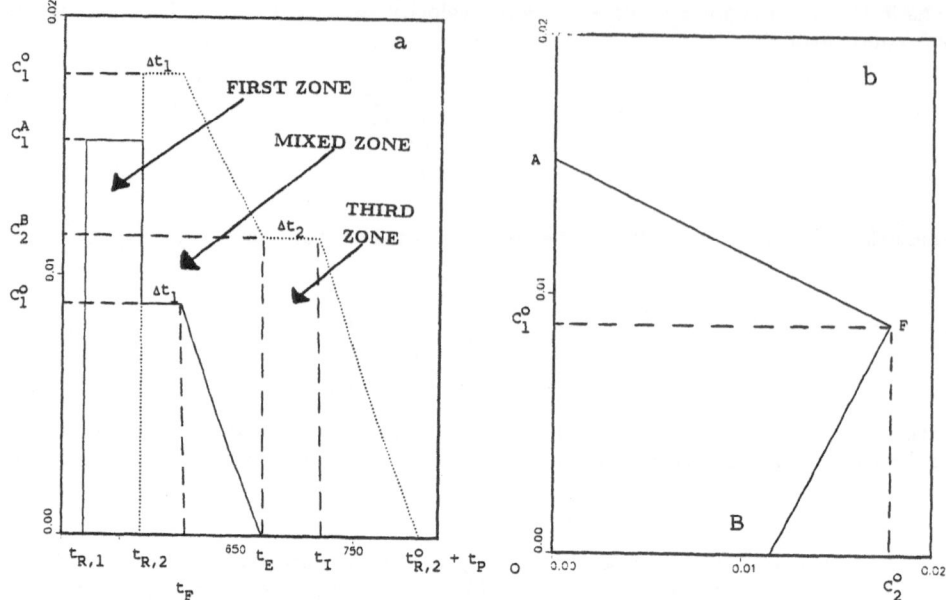

Figure 5. **Elution Profile of a Wide Rectangular Injection, showing the Characteristic Points of the Solution and Hodograph Transform** (see equations in Table I).

Figure 5a. Band profiles: solid line, first component; dotted line: second component. Bands of a 1/2 mixture, calculated under the same conditions as the Figures 6 to 11 below. Injection volume: 2.5 mL.

Figure 5b. Hodograph transform of the previous profile.

singular. This is the case whenever the solution has a simple state (e.g., C_1 and C_2 are constant in a certain (z, t) domain), such as when a rectangular injection is wide and the concentration plateau is not completely eroded. Then the hodograph transform of the solution is made of two lines which intersect at the point representing the constant state (i.e., coordinates C_1^0, C_2^0, concentration of the rectangular injection). In the case of a competitive Langmuir isotherm, the two lines are straight. From the coefficients of these straight lines, it is possible to derive the coefficients of the two single-component Langmuir isotherms, hence the competitive isotherms [42,43].

6. Elution of a Wide Rectangular Band.

The equations to be solved in this case are:

$$\frac{\partial C_1}{\partial t} + F\frac{\partial q_1}{\partial t} + u\frac{\partial C_1}{\partial z} = 0 \tag{1a}$$

$$\frac{\partial C_2}{\partial t} + F\frac{\partial q_2}{\partial t} + u\frac{\partial C_2}{\partial z} = 0 \tag{1b}$$

with the competitive Langmuir equation:

$$q_1 = \frac{a_1 C_1}{1 + b_1 C_1 + b_2 C_2} \tag{3a}$$

$$q_2 = \frac{a_2 C_2}{1 + b_1 C_1 + b_2 C_2} \tag{3b}$$

Obviously, in this case, we can have only partial separation. The solution is based on the simple wave theory [41,42]. We describe successively the solution for the first and the second components. We assume a rectangular injection of concentrations C_1^0 and C_2^0, and width t_p. The hodograph transform of the solution is made of the two characteristic lines through the constant state, which intersect in that point and intersect the coordinate axes in A and B (Figure 5b). During the elution of the profile, the composition of the eluent is represented by a point which moves along the two lines, from O $(0, 0)$ to A $(C_1^A, 0)$, to F (C_1^0, C_2^0), to B $(0, C_2^B)$ and to O. The velocity of the point is far from constant, however. The point figuring the eluent composition jumps instantly from O to A (first component shock), remains in A for a while, then jumps from A to F (second component shock), remains in F during the elution of the plateau residual. Then it moves at a finite velocity from F to B and O.

The concentrations C_1^A and C_2^B are derived from the roots of the equation 29 [29,41,42]. We have:

$$C_1^A = C_1^0 - r_2 C_2^0 \tag{32a}$$

$$C_2^B = C_2^0 - \frac{C_1^0}{r_1} \tag{32b}$$

where r_1 and r_2 are, respectively, the positive and the negative roots of the second degree equation:

$$\alpha b_1 C_2^0 r^2 - (\alpha - 1 + \alpha b_1 C_1^0 - b_2 C_2^0) r - b_2 C_1^0 = 0 \tag{32}$$

with $\alpha = a_2/a_1$. We note that C_1^A is larger than C_1^0 and that C_2^B is smaller than C_2^0.

The elution profile obtained upon the injection of a plug of a binary mixture is composed of three successive zones, a zone containing the first component pure, a mixed zone and a third zone containing the pure second component [29]. For a wide rectangular injection, we have a front shock of the pure first component, and the shock height is C_1^A. Hence the retention time is (see eqn. 12b) [29]:

$$t_{R,1} = t_0 \left[1 + \frac{F a_1}{1 + b_1 C_1^A} \right] \tag{33}$$

Since C_1^A is larger than C_1^0, the front shock of a binary mixture is eluted earlier than the front of a pure component, for the same value of C_1^0. Before the constant state is reached, a zone containing the pure first component is eluted. The first component concentration is higher than in the feed. This is the displacement effect.

When the second component appears, its concentration raises from 0 to C_2^0, while at the same time, the concentration of the first component falls from C_1^A to C_1^0. The two shocks are simultaneous. The second shock moves at the velocity associated with a concentration shock of amplitude C_2^0 in the presence of the concentration C_1^0 of the first component. The retention time of this second shock is (see eqn. 30b) [29]:

$$t_{R,2} = t_0 \left[1 + \frac{F a_2}{1 + b_1 C_1^0 + b_2 C_2^0} \right] \tag{34}$$

Consideration of the velocities associated with a couple of concentrations, C_1 and C_2 (eqns. 27), permits the calculation of the diffuse rear boundaries of the elution band. We find that there are two different arcs in the profile of the second component. To the segment FB correspond the rear, diffuse profile of the first component and the first arc of the rear profile of the second component. The equations of these profiles are [29]:

$$C_1 = \frac{1}{b_1 + b_2/(\alpha r_1)} \left[\sqrt{\frac{\gamma}{\alpha} \frac{t_{R,1}^0 - t_0}{t - t_p - t_0}} - 1 \right] \qquad \text{with} \qquad 0 < C_1 < C_1^0 \qquad (35a)$$

$$C_2 = \frac{1}{b_2 + \alpha b_1 r_1} \left[\sqrt{\gamma' \frac{t_{R,2}^0 - t_0}{t - t_p - t_0}} - 1 \right] \qquad \text{with} \qquad 0 < C_2 < C_2^0 \qquad (35b)$$

(with $\gamma = \frac{\alpha b_1 r_1 + b_2}{b_1 r_1 + b_2}$ and r_1 positive root of equation 32). When C_1 and C_2 reach the initial values C_1^0 and C_2^0, respectively, a concentration plateau appears on each of the component profiles. The retention time of the plateau concentrations (C_1^0, C_2^0) on a continuous profile can be derived by considering the velocity associated with a couple of concentrations (C_1, C_2), given by equation 27 (cf Eqn. 29, ref. 29). This retention time is given by (cf. eqn. 32, ref. 29):

$$t_F = t_p + t_0 \left[1 + \frac{F a_2}{\gamma(1 + b_1 C_1^0 + b_2 C_2^0)^2} \right] \qquad (36)$$

The retention time of the front of the second component band is given by equation 34, so the length of the plateaus at the concentrations $C_1 = C_1^0$ and $C_2 = C_2^0$ is given by:

$$\Delta t_1 = t_F - t_{R,2} = t_p - \frac{F a_2 t_0}{1 + b_1 C_1^0 + b_2 C_2^0} \left[1 - \frac{1}{\gamma(1 + b_1 C_1^0 + b_2 C_2^0)} \right] \qquad (37)$$

The first component profile ends for $C_1 = 0$ or, from equation 35a:

$$t_E = t_p + t_0 + \frac{\gamma}{\alpha}(t_{R,1}^0 - t_0) = t_p + t_0 \left(1 + \frac{\gamma F a_1}{\alpha} \right) \qquad (38)$$

Combination of equations 35b and 38 shows that when the concentration of the first component becomes equal to zero, the second component concentration becomes equal to C_2^B. Past that time, a zone of pure second component is eluted. The profile of that zone is made of a plateau and a diffuse rear boundary. The plateau concentration is given by point B (Fig. 5b). The profile of the diffuse boundary corresponds to the segment BO of the hodograph transform. Since along the segment BO, the second component is pure, this diffuse rear boundary is given by the same equation as we have derived in the case of a pure component (eqn. 13), which in the case of the second component becomes:

$$C_2 = \frac{1}{b_2} \left[\sqrt{\frac{t_{R,2}^0 - t_0}{t - t_p - t_9}} - 1 \right] \qquad (39)$$

It is remarkable, however, that the time when the concentration C_2^B is reached on this profile is:

$$t_I = t_p + t_0 \left[1 + F a_2 \left(\frac{\gamma}{\alpha} \right)^2 \right] = t_p + t_0 \left[1 + F a_1 \left(\frac{\gamma^2}{\alpha} \right) \right] \qquad (40)$$

By comparing equations 38 and 40, and since γ is always larger than unity, we see that t_E is always shorter than t_I. The reason for this difference is that the velocity associated with a certain concentration C_2 of the second component is not the same whether the second component is pure or is in the presence of the first component at a vanishing concentration (compare eqns.

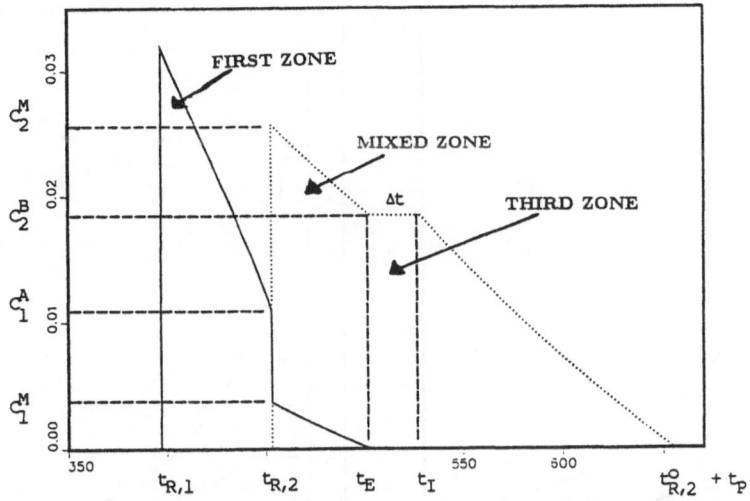

<u>Figure 6</u>. **Elution Profile of a Narrow Rectangular Injection, showing the Characteristic Points of the Solution** (see equations in Table II). Same conditions as Figure 5, except injection volume, 200 μL.

12a and 27). The concentration of the second component must remain constant and equal to C_2^B during the time interval t_E - t_I. The length of this plateau is given by:

$$\Delta t = t_I - t_E = \frac{\gamma F a_1}{\alpha}(\gamma - 1)t_0 \tag{40a}$$

This concentration plateau is at the origin of the tag-along effect.

These equations are valid for any rectangular injection, as long as the injection plateau has not been eroded. For a "narrow" injection, they are valid only during part of the band migration, but we should remember that any rectangular injection has the profile of a "wide" band during the first part of its migration, while any "wide" band would become "narrow" if it migrates along a long enough column. We show later how the elution profiles of a two components rectangular band changes during the migration and corresponds first to the profile of a wide rectangular band, to take later the shape of the profile of a narrow band. We show such profiles in the later figures. The angles and the sides of the profiles are noted by a symbol and the corresponding equations are listed in Table I. A wide injection is such that the plateau in the elution profile has a finite width. This means that the following condition must be met:

$$t_{R,2} < t_F \tag{41}$$

where $t_{R,2}$ is the retention time of the shock of the second component (eqn. 34) and t_F is the elution time of the end point of the rectangular injection (eqn. 36). This gives the minimum injection band width (see eqn. 33 in ref. 29):

$$t_p \geq \frac{F a_2 t_0 \left[1 + b_1 C_1^0 + b_2 C_2^0 - 1/\gamma\right]}{(1 + b_1 C_1^0 + b_2 C_2^0)^2} \tag{42}$$

7. Elution of a Narrow Rectangular Band.

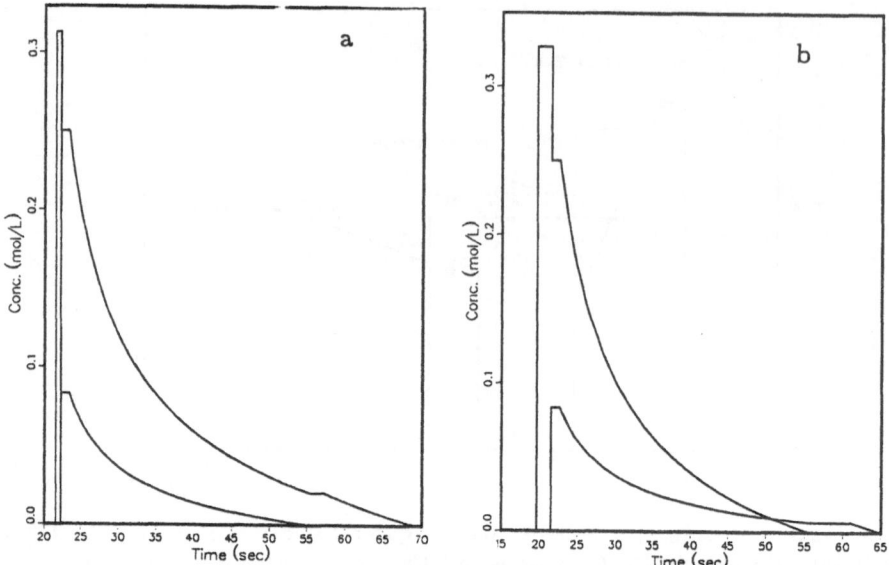

Figure 7. **Elution profile of a rectangular injection band at the end of a column of length shorter than z_I. The injection plateau is conserved.**

Conditions for the following figures: Phase ratio: $F = 0.25$; column diameter: 0.46 cm; Flow rate: $F_v = 1$ mL/min; Relative retention: $\alpha = 1.2$; retention factor: $k'_{0,1} = 6.0$; other isotherm coefficients: $b_1 = 6$; $b_2 = 7.2$; sample size: 66.7 μmole.

Figure 7a: $L = 0.9z_I = 0.87$ cm; 1:3 mixture. *Figure 7b*: $L = 0.9z_I = 0.81$ cm; 3:1 mixture.

The difference between a wide and a narrow injection band is that in the former case the condition given in equation 42 is satisfied, while it is no longer so in the latter case. The concentration plateau corresponding to the injection plug has disappeared, the height of the shocks decrease with increasing migration distance. Thus, the shock velocities are no longer constant and the solution is more complex. Nevertheless, it is possible to derive a closed form solution [29] for the equation of all the diffuse profiles, and for the retention time and the heights of the second shock (shock intensities of the concentrations of the two components). The only equations which cannot be given into closed form are those of the retention time and height of the front shock (first component). There are different successful approaches for obtaining this result, using either the theory of characteristics or the h-transform [29,31]. A typical example of the solution is shown in Figure 6.

We show the development of a typical solution in Figures 7 to 11. The solution includes three zones, as for the wide injection. A mixed zone lies between a zone of the pure first component and a zone of the pure second component. It is important to distinguish the features which are common to the solution of the wide and the narrow injection bands. The height of the second shock decreases as soon as the plateau disappears. The equations giving the rear, diffuse profiles of the two components behind this shock remain the same, however. When the sample size is reduced by contracting the injection width at constant concentrations, the retention time of the second component shock increases, but the rear profiles remain identical if they are corrected for the linear dependence of the retention times associated with a concentration on the duration of the injection, t_p.

The retention time of the second component shock is obtained by writing that the top of the

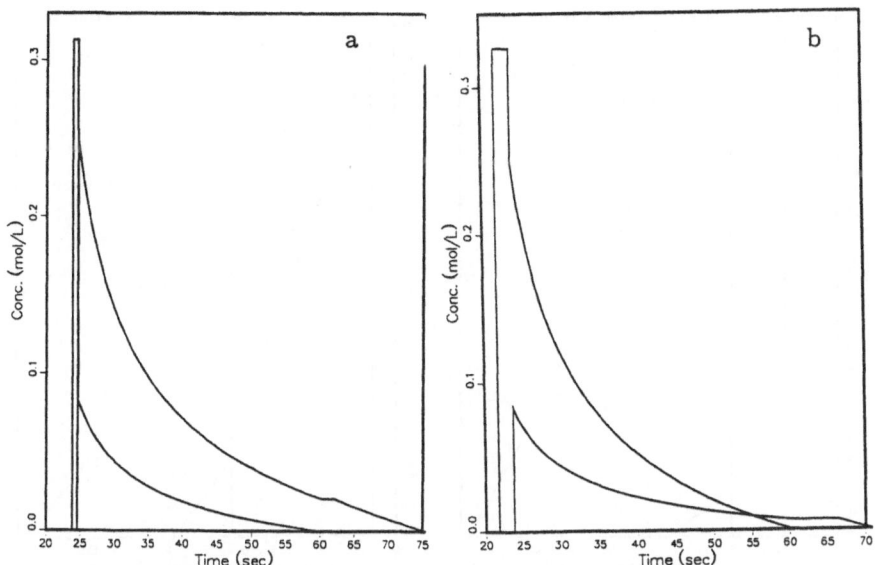

Figure 8. **Elution profile of a rectangular injection band at the end of a column of length equal to** z_I. The injection plateau has just disappeared. Same conditions as in Figure 7, except: *Figure 8a*: $L = z_I = 0.97$ cm; 1:3 mixture. *Figure 8b*: $L = z_I = 0.90$ cm; 3:1 mixture.

shock is also a point on the rear, diffuse profile of the band [29]. The retention time of this concentration is the same if calculated by integrating the equations with relate the velocity and the concentration of either a shock or a diffuse profile.

Finally, we obtain the equations in Table II. The location of the corresponding points are shown in Figure 6. Using these equations, we can calculate the concentrations profiles of the two components along a long column. The Figures 7 to 11 illustrate this variation. We show two series of figures corresponding to different relative compositions, 1:3 and 3:1. At the end of a very short column (Fig. 7), the injection plateau still subsists. This plateau disappears for a column length z_I (Eqn. 33 in ref. 29). The Figures 8 show the elution profiles at the end of a column of length z_I. For longer columns, the bands spread more and more and the dilution increases progressively (Fig. 9).

After a certain length, the two bands become resolved. The Figures 10 show the elution profile at that length, z_K (Eqn. 70 in ref. 29). The second component band has still a plateau at the concentration C_2^B.

Finally, this plateau disappears for a column of length z_L (Eqn. 73 in ref. 29). Figure 11 corresponds to that length. Beyond that length, the profile of the second component band is identical to a profile of the same amount of pure second component. The profile of the first component, on the other hand remains scarred by the interaction. Note that the lengths z_I and z_K calculated for the two mixtures studied are close, while z_L depends much on the mixture composition [29].

8. Displacement Chromatography.

In displacement chromatography, a rectangular band of sample is introduced in the column

20

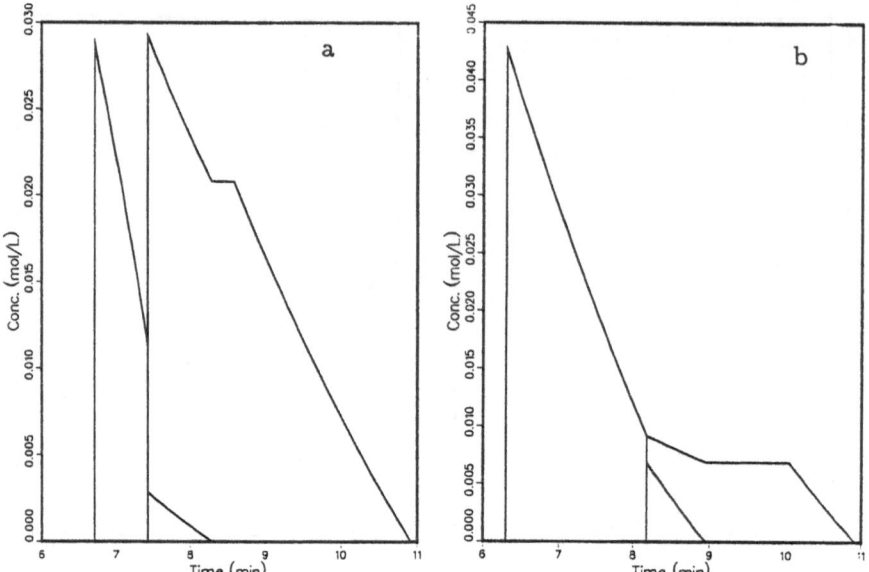

<u>Figure 9</u>. **Elution profile of a rectangular injection band at the end of a column of intermediate length**. The two bands overlap. Same conditions as in Figure 7, except: *Figure 9a*: $L = 10.$ cm; 1:3 mixture. *Figure 9b*: $L = 10.$ cm; 3:1 mixture.

which has been filled with a low or moderate strength eluent. Right after the end of the sample injection, the stream of mobile phase pumped into the column is replaced by a solution of the displacer, a compound which is more strongly adsorbed than any component of the mixture. The experimental conditions are such that the displacer breakthrough curve would be eluted before the sample bands, if the sample were to be eluted from the column with the displacer-free mobile phase. Depending on the experimental conditions, the individual band profiles are more or less complex. If the column is long enough, however, a steady state is reached and a succession of bands of the pure sample components is eluted, pushed by the displacer front.

The phenomenon of displacement chromatography has been studied in great detail with the ideal model of chromatography [27,30,35,45- 50]. The basic principles of displacement chromatography were presented by Tiselius [45], who realized that the adjacent bands of the sample components travel at the velocity of the displacer front and that, once stabilized, the plateau concentrations of the developed pure solute bands depend only upon the displacer concentration and are independent of the concentration of the corresponding solute in the feed. Glueckauf [27,35] made a major contribution to the theory of ideal displacement chromatography by presenting a procedure for the calculation of the band profiles when the sample components and the displacer follow the Langmuir isotherm model. Helfferich and Klein [30] presented a general theory of multi-component separations by using the so-called h- transform and the concept of coherence. In this approach, the set of concentrations C_i of the components of a system is replaced by a new set of variables, h_i, using this h-transform.

This "theory of interference" [30] was originally developed for stoichiometric ion exchange systems. However, adsorption processes can be viewed as equivalent to ion exchange processes by introducing a fictitious component [30]. Accordingly, an n- component adsorption system is equivalent to an (n+1)-component ion exchange system. Therefore the same approach may be applied to adsorption, provided that we may assume the validity of both the ideal model and

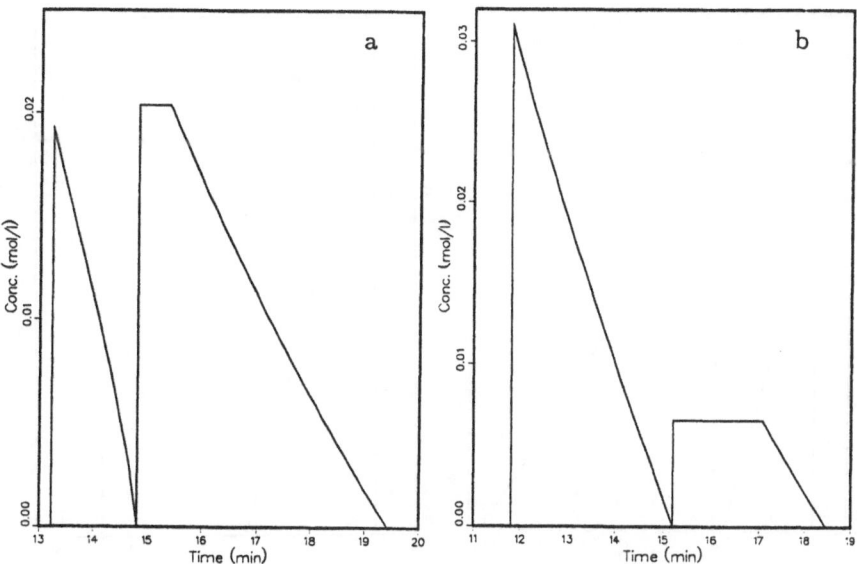

Figure 10. **Elution profile of a rectangular injection band at the end of a column of length equal to** z_k. The two bands are just resolved, but a plateau still exists at the top of the second component band. Same conditions as in Figure 7, except:
Figure 10a: $L = 17.6$ cm; 1:3 mixture. *Figure 10b*: $L = 16.75$ cm; 3:1 mixture.

the Langmuir competitive isotherm model. The band profiles obtained by the calculations using the h-transform were compared with experimental profiles for a four-component ion exchange sorption system [46]. The concentration profiles in the column effluent were also calculated from the h-transform method and compared with experimental results in displacement chromatography by Frenz et al. [47]. Other valuable contributions have been made by Basmadjian et al. [48], and Geldart et al. [49].

Rhee and Amundson [50] have fully analyzed the separation of multi- component mixtures using the ideal model of displacement chromatography. In this analysis, they also had to make the restrictive assumption of the validity of the Langmuir isotherms. They used the method of characteristics and studied the interactions between concentration shocks and centered simple waves [41]. Here we discuss first some general results on the steady state propagation of the band train, then the method of characteristics used by Rhee and Amundson [50].

8.1 - BASIC DISPLACEMENT THEORY: THE ISOTACHIC TRAIN.

As will appear clearly later, the only requirement to perform displacement is that the competitive isotherms of the mixture components and the displacer be convex upwards and do not intersect each other. Since the isotherms are convex, the front of the bands are self-sharpening. If an isotachic train is formed, all the fronts move at the same velocity. Each front separates two bands of pure compounds. Thus, the height of each component zone must be such that its front shock velocity is the same as the velocity of the displacer front, which we can control by pumping it at a constant velocity. The velocity of the displacer front depends on the flow rate, the displacer concentration, and its isotherm. Thus we have:

$$U_{s,1} = U_{s,2} = \cdots = U_{s,i} = \cdots = U_{s,D} \tag{43}$$

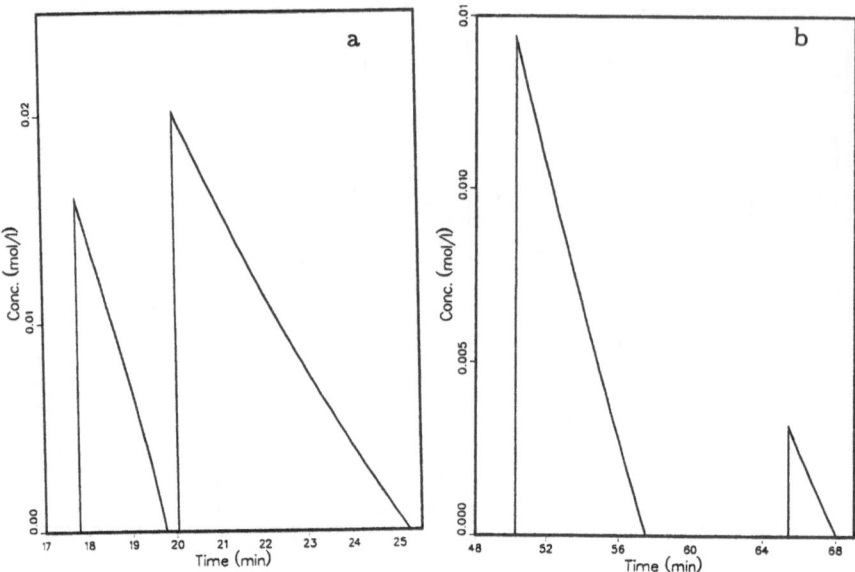

Figure 11. **Elution profile of a rectangular injection band at the end of a column of length equal to** z_l. The plateau at the top of the second component band has just disappeared. Same conditions as in Figure 7, except:
Figure 11a: $L = 22.97$ cm; 1:3 mixture. *Figure 11b*: $L = 62.2$ cm; 3:1 mixture.

The velocity of the displacer front is:

$$U_{s,D} = \frac{u_0}{1 + F\frac{\Delta q_D}{\Delta C_D}} = \frac{u_0}{1 + F\frac{q_D}{C_D}} \tag{44}$$

Most generally, the initial displacer concentration in the mobile phase, as well as the initial concentration of the analytes, is 0. Combination of equations 43 and 44 gives:

$$\frac{q_1}{C_1} = \frac{q_2}{C_2} = \cdots = \frac{q_i}{C_i} = \cdots = \frac{q_D}{C_D} \tag{45}$$

Equation 45 indicates that the concentrations of each of the successive bands is given by the intersection of the corresponding isotherm and the chord of the displacer isotherm. This line is called the operating line. By changing its slope, i.e., the displacer concentration, the chemist can change the heights of all the bands. It is not possible to adjust the heights independently, however. The band width are such that the area of each band is conserved, i.e., remains equal to the area of the injected profile.

Obviously, the chord must intersect the isotherm. If it does not, there is no displacement. The band of the corresponding (early eluting) component is eluted under conditions of overloaded elution. It is not exceptional that the less retained components of a mixture behave that way. If the isotherms of the compounds studied are Langmuir isotherms, the condition for a possible displacement of i by D can be written:

$$\frac{q_D}{C_D} = \frac{a_D}{1 + b_D C_D} \quad < \quad a_i \tag{46}$$

A successful displacement of i requires a displacer concentration such that:

$$C_D > \frac{1}{b_D}\left(\frac{a_D}{a_i} - 1\right)$$ (47)

These results can be obtained by a more general theory which describes the progressive change in the individual band profiles and shows how the overlaid component injection profiles are progressively transformed into a band train. In this process, it is frequent that the concentration of some of the feed components in their corresponding band of the isotachic train is higher than their concentration in the feed.

8.2 - THE THEORY OF CHARACTERISTICS.

This method is based on the mathematical theory of the systems of quasilinear partial differential equation and on the use of the characteristic method to solve these equations [50]. The h- transform is basically an equivalent theory, developed from a different point of view and more by definitions [30]. As expected the two methods give basically the same results. Both theories are valid only in the case of competitive Langmuir isotherm. This restriction should be kept in mind, because it might explain often the not so inconsequential disagreements observed between their predictions and observations reported in the literature.

The mass balance equations of the ideal model for an n-component mixture have been written above (eqns. 1a):

$$u\frac{\partial C_i}{\partial z} + \frac{\partial C_i}{\partial t} + F\frac{\partial q_i}{\partial t} = 0$$ (1a)

with:

$$q_i = f_i(C_1, C_2, \cdots, C_n)$$ (2)

These equations can be rewritten [50] in a dimensionless form by using the reduced variables $x = z/Z$ and $\tau = ut/Z$, where Z is a characteristic length:

$$\frac{\partial C_i}{\partial x} + \frac{\partial C_i}{\partial \tau} + F\frac{\partial q_i}{\partial \tau} = 0$$ (48)

We now assume that the isotherm of each component, i, is a given by the Langmuir competitive isotherm model:

$$q_i = \frac{a_i C_i}{1 + \Sigma_{j=1}^n b_j C_j}$$ (49)

and we number the mixture components in order of their increasing affinity for the stationary phases, so their rank is also their elution order. Now, we define n characteristic parameters $(\omega_1, \omega_2, \cdots, \omega_n)$ as the roots of the following nth-order algebraic equation:

$$\Sigma_{i=1}^n \frac{b_i q_i}{a_i - \omega} = 1$$ (50)

The definition of the characteristic parameters is analog to the H- function from which the h roots used in the h-transform [30] are obtained from:

$$\Sigma_{i=1}^n \frac{b_i C_i}{h(a_i/a_1) - 1} = 1$$ (51)

Thus, the characteristic parameters, ω_i, are analogous to the h roots, h_i. Neither equation 50 nor equation 51 can be solve explicitly when $n > 2$, but they can be calculated by simple numerical techniques. Equations 50 and 51 have n real, positive roots ω_k and h_k, respectively. The roots ω_k can be arranged in the following order:

$$0 \leq \omega_1 \leq a_1 \leq \omega_2 \leq a_2 \leq \cdots \leq \omega_j \leq a_j \leq \omega_{j+1} \leq a_{j+1} \leq \cdots \leq \omega_n \leq a_n \qquad (52)$$

Equation 50 represents the transformation of the concentration space (C_i or q_i) into the ω space (ω_i).

The solution of equation 48 in the (x, τ) plane consists of n waves. The advantage of the transformations just discussed is that the value of only one of the characteristic parameters, ω, changes when we move across each of these waves, the others remaining unchanged. By contrast, the value of several concentrations change across a wave, and even this number of parameters which change across a wave varies during the propagation of the band train.

If we identify a characteristic parameter, ω_k and the corresponding kth wave, and if the value of this parameter is equal to ω_k^l behind and to ω_k^r ahead of the wave, then, if $\omega_k^r < \omega_k^l$, the k-wave is a shock wave [50]. This shock wave propagates along a straight line in the (x, τ) plane, and the line slope is given by:

$$s_k = \frac{\Delta \tau}{\Delta x_k} = 1 + F \omega_k^l \Pi_{i=1}^m \frac{\omega_j^r}{a_j} \qquad (53)$$

where the superscripts l and r denote the left and right hand side of the shocks, respectively.

On the other hand, if $\omega_k^l > \omega_k^r$, the k-wave is a centered sample wave in the region where ω changes continuously [50]. So, a k-simple wave is represented by a family of straight characteristics (c_k), in the (x, τ) plane, and their slopes are given by:

$$\sigma_k = \frac{d\tau}{dx_k} = 1 + F \omega_k \Pi_{i=1}^n \frac{\omega_j}{a_j} \qquad (54)$$

The simple wave region is bound by the kth and $(k-1)$th constant states, on its left and right hand sides respectively. Thus, Equation 54 can be rewritten as:

$$\sigma_k = 1 + F \frac{\omega_k^2}{a_k} \Pi_{i=1}^{k-1} \frac{\omega_j^r}{a_j} \Pi_{i=k+1}^n \frac{\omega_j^l}{a_j} \qquad (55)$$

The only variable in Equation 55 is ω_k which varies in the following range: $\omega_k^l \leq \omega_k \leq \omega_k^r$. Thus, one can generate in this range a family of straight lines, each of which carries a fixed value of ω_k and thus a fixed set of solute concentrations.

Finally, if $\omega_k^l = \omega_k^r$, the kth wave does not exist and we have: $C_{i,k} = C_{i,k-1}$.

In practice, one or more component is absent from the mobile phase. As the separation progress, this situation becomes more and more frequent. Usually, the mobile phase ahead of the band train is free of all components and the displacer solution does not contain any mixture component. If some components are missing locally (and we hope that eventually the components will all separate, e.g., that they will be missing from other components' bands) the corresponding roots cannot be obtained directly from equation 50. However, this equation implies that for each missing component, j, there exists a characteristic parameter, whose value is equal to a_j. In a pure mobile phase, it results from Equation 50 that:

$$\omega_k = a_k \qquad \text{for} \quad k = 1, 2, \cdots, n \qquad (56a)$$

If a particular species, j, is absent of the solution, Equation 50 shows that one of the characteristic parameters should be equal to a_j, and it follows from Equation 52 that we have either:

$$\omega_j = a_j \qquad (56b)$$

or
$$\omega_{j+1} = a_j \tag{56c}$$

If the most strongly adsorbed species is absent from the solution, then, from equation 52 it is obvious that:
$$\omega_n = a_n \tag{56d}$$

The equations 56 are equivalent to the set of rules given by Helfferich and Klein [30] for evaluating the trivial roots of h.

8.3 - APPLICATION TO DISPLACEMENT CHROMATOGRAPHY.

In the initial state of displacement chromatography, no component is present, so the set of n characteristic parameters for the mobile phase are given by Equation 56a:
$$I[\omega_1^i, \omega_2^i, \cdots, \omega_n^i] = I[a_1, a_2, \cdots, a_n] \tag{57}$$

After the feed has been injected for a duration t_p, all the components are present, except the most strongly adsorbed one, the displacer. Thus, the characteristic parameter for the displacer is given by Equation 56d, while the characteristic parameters of the n-1 components of the feed can be determined by solving the Equation 50. Their values are denoted by $\omega_1^f, \omega_2^f, \cdots, \omega_{n-1}^f$. Thus, the set of n characteristic parameters for the feed is given by:
$$F[\omega_1^f, \omega_2^f, \cdots, \omega_{n-1}^f, \omega_n^f = a_n] \tag{58}$$

where $\omega_1^f < a_1$, $\omega_2^f < a_2$, $\cdots$, $\omega_{n-1}^f < a_{n-1}$. If we compare the Equations 57 and 58, we see that for all k, except the displacer, we have: $\omega_k^f < \omega_k^i$, which is the condition for a shock wave. So upon injection of the feed, n - 1 shock waves form and propagate from the origin. Obviously, for the displacer, $\omega_n^f = \omega_n^i$, so no shock is observed for the displacer when the feed is injected, which is rather obvious.

After the injection is completed, the displacer is introduced in the column as a continuous stream for the rest of the experiment. Since n - 1 species are absent from the displacer solution, the $n - 1$ values of their characteristic parameters, according to equation 56 are equal to $a_1, a_2, \cdots, a_{n-1}$. Another root of equation 50 can be obtained by solving it, which gives:
$$\omega^* = \frac{a_n}{1 + bC_d} \tag{59}$$

Obviously, the value of ω^* depends on the concentration of the displacer and can be varied from 0 to a_n. Therefore, the order of the set of ω_k is not fixed. It depends on the concentration of the displacer. If the displacer concentration is such that $a_{j-1} < \omega^* < a_j$, ω^* becomes the jth root of equation 50, so the ordering at the beginning of the displacement is:
$$D[\omega_1^d = a_1, \omega_2^d = a_2, \omega_{j-1}^d = a_{j-1}, \omega_j^d = \omega_*, \omega_{j+1}^d = a_j, \cdots \omega_n^i = a_{n-1}] \tag{60}$$

Comparing Equations 58 and 60 shows that, upon injection of the displacer, $j - 1$ simple waves (corresponding to the values of ω on the left side of ω_j^d in Equation 60) and $n - j$ shock waves (corresponding to the values of ω on the right side of ω_j^d in Equation 60) will develop. Starting from the point $(0, \tau_p = (t_p u)/Z)$. The jth wave could remain a shock if $\omega^* < \omega_j^f$ or become a simple wave if $\omega^* > \omega_j^f$. There are two sets of shock, the one which originates as a result of the feed injection, and the one which results from the displacer injection. They will be involved in a series of interactions which can be analyzed [50].

8.4 - CRITICAL VALUE OF THE DISPLACER CONCENTRATION.

Obviously, as ω^* gets smaller and smaller (i.e., as the displacer concentration gets larger and larger, eqn. 59), ω_j moves to the left hand side in Equation 60 and we get more and more shock waves. If we are interested in obtaining an isotachic train after the development is over, we want to obtain n shocks. This is possible only if the displacer concentration is large enough and ω^* becomes the first in the ordered characteristic set of Equation 60. This requires that $\omega^* < a_1$. Introducing this condition in equation 59 gives:

$$C_d > \frac{1}{b_n}(\frac{a_n}{a_1} - 1) \tag{61}$$

Hence the critical displacer concentration for successful displacement of a multi-component mixture depends only on the adsorption isotherm of the displacer (a_n and b_n) and on the slope of the less retained solute at infinite dilution (a_1). On the opposite, if the displacer concentration is so small that ω^* is the last element in the ordered characteristic set of displacement, in Equation 60, there is no shock and no isotachic train is possible. This happens when $\omega^* > a_{n-1}$. Combining this condition and Equation 59 gives:

$$C_d < \frac{1}{b_n}(\frac{a_n}{a_{n-1}} - 1) \tag{62}$$

More generally, if, as mentioned before, the displacer concentration is such that $a_{j-1} < \omega^* < a_j$, we have:

$$\frac{1}{b_n}(\frac{a_n}{a_{j-1}} - 1) > C_d > \frac{1}{b_n}(\frac{a_n}{a_j} - 1) \tag{63}$$

and the result is the successful formation of an isotachic train including $n - j$ bands, preceded by a series of $j - 1$ bands in a non-isotachic train. The jth wave is a shock if $\omega^* < \omega_j^f$, a simple wave if $\omega^* > \omega_j^f$.

8.5 - PLATEAU CONCENTRATIONS AND BAND WIDTH.

In the final stage of the development of the isotachic train, we get n plateaus and shock waves, one for each pure band j. Since all the other components are missing from that band:

$$\omega_j^{Pl,j} = \omega^*$$
$$\omega_k^{Pl,j} = a_k \qquad \text{for} \quad k \neq j \tag{64}$$

where $\omega_k^{Pl,j}$ is the value of ω for the component k in the band j when the isotachic train is formed and the band has a concentration plateau. Since at this final stage of the development ω^* is the smallest root of equation 50, while the other $(n - 1)$ roots are given for every pure component by a_k ($k \neq j$), we have:

$$\omega^* = \frac{a_1}{1 + b_1 C_1^P} = ... = \frac{a_j}{1 + b_j C_j^P} = ... = \frac{a_n}{1 + b_n C_n^P} \tag{65}$$

or:

$$C_j^P = \frac{1}{b_j}(\frac{a_j}{a_n} - 1) + \frac{a_j b_n}{a_n b_j} C_d \tag{66}$$

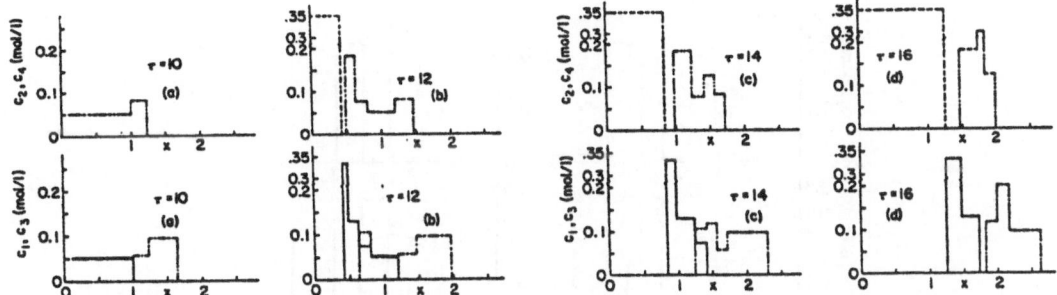

Figures 12a to 12d. **Separation of the components, A_1, A_2 and A_3, of a ternary mixture, displaced by a solution of A_4, as predicted by the ideal model** [50]. Reproduced with permission of the American Institute of Chemical Engineers.

Experimental conditions: $F = 1.5$, isotherm parameters $a_1 = b_1 = 5$, $a_2 = b_2 = 10$, $a_3 = b_3 = 15$, $a_4 = b_4 = 15$. Feed concentration: $C_1^0 = C_2^0 = C_3^0 = 0.05M$, $C_4 = 0.35M$. The displacement begins at $\tau = 10$.

where C_j^P is the plateau concentration of the component j in the isotachic train. Equation 66 shows that the concentration plateau of each solute under isotachic conditions is independent of the feed concentration and of its composition, but depends only on the displacer concentration and the isotherms of the displacer and the solute itself.

The concentration C_j^P can also be obtained from solving the equation:

$$b_j C_j = \left(\frac{a_j}{\omega_j} - 1\right) \Pi_{k=1}^n \frac{\frac{a_j}{\omega_k} - 1}{\frac{a_j}{a_k} - 1} \qquad j = 1, \cdots, n \tag{67}$$

Using Equation 66, and applying the conservation of mass of each component during the experiment, we can derive the time width of the band which is given by:

$$\Delta \tau_j = \tau_p \frac{C_j^f}{C_j^P} \tag{68}$$

Since the slope of the characteristic line is given by $s_j = \frac{\Delta \tau_j}{\Delta x_j}$, we obtain the bandwidth along the column:

$$\Delta x_j = \frac{\Delta \tau_j}{s_j} = \frac{\tau_p C_j^f}{s_j C_j^P} \tag{69}$$

The slope of the characteristics is given by the Equation 53:

$$s = s_1 = s_2 = \cdots = s_n = 1 + F \omega^* = 1 + \frac{F a_j}{1 + b_j C_j^P} \tag{70}$$

Since the paths of the n shocks are known, we can locate each shock using the Equations 68 to 70, at the final stage of the isotachic train, without going into a detailed analysis of the interactions between the bands.

In order to calculate the minimum bed length required for the formation of an isotachic train, all the interactions involved must be analyzed. Also, if the concentrations of the solutes are chosen so that ω^* is larger than ω_j for one or more components, the corresponding wave(s)

28

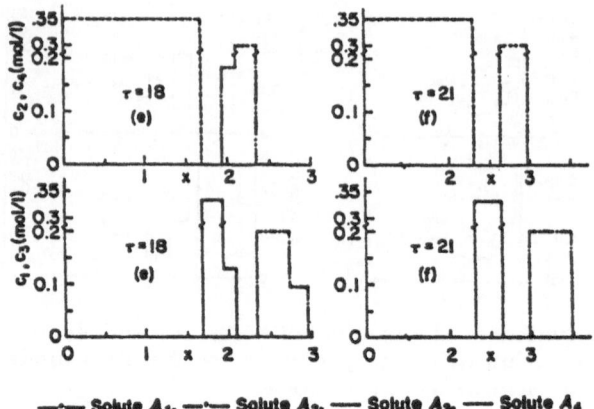

—·— Solute A_1, —··— Solute A_2, —— Solute A_3, —— Solute A_4

<u>Figures 12e and 12f</u>. **Separation of the components, A_1, A_2 and A_3, of a ternary mixture, displaced by a solution of A_4, as predicted by the ideal model** [50]. Reproduced with permission of the American Institute of Chemical Engineers.

of this(these) solute(s) is (are) simple wave(s), and the successive interactions between these simple waves and each of the shocks must be analyzed in order to obtain the solution [50]. We show in Figure 12 the separation of a ternary mixture (solutes A_1, A_2 and A_3), displaced by a solution of A_4, as predicted by the ideal model [50]. In this case, the displacer concentration is higher than the critical value, $\omega^* < a_1$, and the isotachic train is finally formed for all the component bands. For the sake of clarity, the band profiles of the first and third components are shown in the bottom chromatogram, the band profiles of the second component and the displacer in the top one. Although all the components are introduced at the concentration 0.05M, except the displacer, the bands in the isotachic train are more concentrated.

At $\tau = \frac{tu}{Z} = \tau_0 = 10$ (Fig. 12a), the feed mixture has just been introduced, so there is not yet a displacer front. The bands of components 1 and 2 exhibit already a plateau at an enriched concentration. In figure 12b ($\tau = 12$), the displacer has been introduced at the concentration C_4 = 0.35M and a narrow band of pure third component is formed. This band widens progressively in Figures 12c to 12f, where it becomes rectangular. Similarly, the bands of the second and third compound are separated, grow, and become narrower. At $\tau = 21$, the isocratic train forms and propagates unchanged at a constant velocity.

Acknowledgments.

This work has been supported in part by Grant CHE-8901382 of the National Science Foundation and by the cooperative agreement between the University of Tennessee and the Oak Ridge National Laboratory. We acknowledge support of our computational effort by the University of Tennessee Computing Center.

Literature Cited.

[1] Golshan-Shirazi, S., and Guiochon, G. (1992) NATO ASI Ser., Ferrara, 1991.
[2] Wilson, J.N. (1940) J. Amer. Chem. Soc., 62, 1583.
[3] Rouchon, P., Schonauer, M., Valentin, P., and Guiochon, G. (1987) Separat. Sci. Technol., 22, 1793.

 [4] Bosanquet, C.H., and Morgan, G.O. (1957) in Vapour Phase Chromatography, D.H. Desty Ed., Butterworths, London, UK, p. 35.
 [5] Langmuir, I. (1918) J. Amer. Chem. Soc., 40, 1361.
 [6] Dankwerts, P.V. (1953) Chem. Eng. Sci., 2, 1.
 [7] Kovats, E. sz, in The Science of Chromatography, F. Bruner Ed., Elsevier, 1985, p. 205.
 [8] DeVault, D. (1943) J. Amer. Chem. Soc., 65, 532.
 [9] Weiss, J. (1943) J. Chem. Soc., 297.
[10] Glueckauf, E. (1947) J. Chem. Soc., 1302.
[11] Coates, J.I., and Glueckauf, E. (1947) J. Amer. Chem. Soc., 69, 1309.
[12] Aris, R., and Amundson, N.R. (1973) Mathematical Methods in Chemical Engineering, Prentice Hall, Englewood Cliffs, NY, 1st Ed.
[13] Courant, R., and Friedrichs, K.O. (1948) Supersonic Flow and Shock Waves, Wiley-Interscience, New York, NY.
[14] Lax, P.D. (1957) Comm. Pure Appl. Math., 10, 537.
[15] Lax, P. (1971) in Contribution to Non-Linear Functional Analysis, E.H. Zarantonello Ed., University of Wisconsin, Madison, WI, 1971, 603.
[16] Oleinik, O.A. (1963) Amer. Math. Soc. Transl. Ser., 2, 33, 285.
[17] Lin, B., Golshan-Shirazi, S., Ma, Z., and Guiochon, G. (1988) Anal. Chem., 60, 2646.
[18] Rhee, H.-K., and Amundson, N.R. (1974) Chem. Eng. Sci., 29, 2049.
[19] Helfferich, F., and Peterson, D.L. (1964) J. Chem. Educ., 41, 410.
[20] Golshan-Shirazi, S., and Guiochon, G. (1988) Anal. Chem., 60, 2364.
[21] Golshan-Shirazi, S., and Guiochon, G. (1990) J. Phys. Chem., 94, 495.
[22] Knox, J.H., and Pyper, H.M. (1986) J. Chromatogr., 363, 1.
[23] Graham, D. (1953) J. Phys. Chem., 57, 665.
[24] Laub, R.J. (1986) ACS Symp. Ser., 297, 1.
[25] Jacobson, S., Golshan-Shirazi, S., and Guiochon, G. (1990) J. Amer. Chem. Soc., 112, 6492.
[26] Golshan-Shirazi, S., and Guiochon, G. (1989) Anal. Chem., 61, 462.
[27] Glueckauf, E. (1946) Proc. Roy. Soc. (London), A186, 35.
[28] Schwab, G.M. (1928) "Ergebnisse der Exacten Naturwissenschaften", Vol. 7, Julius Springer, Berlin, Germany, p. 276.
[29] Golshan-Shirazi, S., and Guiochon, G. (1989) J. Phys. Chem., 93, 4143.
[30] Helfferich, F., and Klein, G. (1970) "Multicomponent Chromatography. A Theory of Interference", M. Dekker, New York, NY.
[31] Golshan-Shirazi, S., and Guiochon, G. (1989) J. Chromatogr., 484, 125.
[32] Rhee, H.K., and Amundson, N.R. (1970) Chem. Eng. J., 1, 241.
[33] Kemball, C., Rideal, E.K., and Guggenheim, E.A. (1948) Trans. Faraday Soc., 44, 948.
[34] Levan, M.D., and Vermeulen, T. (1981) J. Phys. Chem., 85, 3247.
[35] Glueckauf, E. (1949) Disc. Faraday Soc., 7, 12.
[36] Shapiro, A.H. (1953) The Dynamics and Thermodynamics of Compressible Fluid Flow, Ronald Press, New York, NY.
[37] Jeffrey, A., and Tanuiti, T. (1964) Non-Linear Wave Propagation, Academic Press, New York, NY.
[38] Seinfeld, J.H., and Lapidus, L. (1964) Numerical Solutions of Ordinary Differential Equations, Academic Press, New York, NY.
[39] Whitman, G.B. (1971) Linear and Non-linear Waves, Wiley, New York, NY.
[40] Jeffrey, A. (1976) Quasilinear Hyperbolic Systems and Waves, Pitman, London, UK.
[41] Rhee, H.-K., Aris, R., and Amundson, N.R. (1970) Trans. Roy. Soc. (London), A267, 419.
[42] Ma, Z., Katti, A.M., Lin, B., and Guiochon, G. (1990) J. Phys. Chem., 94, 6911.
[43] Katti, A.M., Ma, Z., and Guiochon, G. (1990) AIChE J., 36, 1722.
[44] Kvaalen, E., Neel, L., and Tondeur, D. (1985) Chem. Eng. Sci., 40, 1191.
[45] Tiselius, A. (1943) Arkiv for Kemi, Mineral Geol., 16A, 1.

[46] Clifford, D. (1982) Ind. Eng. Chem. Fundamentals, 141, 21.
[47] Frenz, J., and Horvath, Cs. (1985) AIChE J., 31, 400.
[48] Basmadjian, D., and Coroyannakis, P. (1987) Chem. Eng. Sci., 42, 1723.
[49] Geldart, R.W., Qiming, Y., Wankat, P.C., and Wang, L.N.-H. (1986) Separat. Sci. and Technol., 21, 873.
[50] Rhee, H.-K., and Amundson, N.R. (1982) AIChE J., 28 (1982) 423.

TABLE I

EQUATIONS GIVING THE ELUTION PROFILE
OF A WIDE RECTANGULAR INJECTION PULSE[1]

First Zone: Pure First Component.
- Retention time of the First component:

$$t_{R,1} = t_0 \left[1 + \frac{F a_1}{1 + b_1 C_1^A} \right]$$

- Concentration:

$$C_1^A = C_1^0 - r_2 C_2^0$$

Mixed Zone.
- Retention time of the Second component:

$$t_{R,2} = t_0 \left[1 + \frac{F a_2}{1 + b_1 C_1^0 + b_2 C_2^0} \right]$$

- Composition of the Common Concentration Plateau: $C_1 = C_1^0$, $C_2 = C_2^0$.

- End Time of the Common Concentration Plateau:

$$t_F = t_p + t_0 \left[1 + \frac{F a_2}{\gamma (1 + b_1 C_1^0 + b_2 C_2^0)^2} \right]$$

- Duration of the Common Concentration Plateau:

$$\Delta t_1 = t_F - t_{R,2} = t_p - \frac{F a_2 t_0}{1 + b_1 C_1^0 + b_2 C_2^0} \left[1 - \frac{1}{\gamma (1 + b_1 C_1^0 + b_2 C_2^0)} \right]$$

- Diffuse Elution Profile of the First Component:

$$C_1 = \frac{1}{b_1 + b_2/(\alpha r_1)} \left[\sqrt{\frac{\gamma}{\alpha} \frac{t_{R,1}^0 - t_0}{t - t_p - t_0}} - 1 \right]$$

[1]Definitions: F, phase ratio ($F = (1 - \epsilon)/\epsilon$); ϵ, external porosity of the packing; t_0, hold-up time ($t_0 = L/u$, L, column length, u, mobile phase velocity); t_p, duration of the rectangular injection pulse; a_1, a_2, b_1, b_2, coefficients of the Langmuir isotherm; $\alpha = a_1/a_2$; $\gamma = \frac{\alpha b_1 r_1 + b_2}{b_1 r_1 + b_2}$; r_1, r_2, positive and negative roots of equation 32, respectively:

$$\alpha b_1 C_2^0 r^2 - (\alpha - 1 + \alpha b_1 C_1^0 - b_2 C_2^0) r - b_2 C_1^0 = 0 \qquad (32)$$

- Diffuse Elution Profile of the Second Component:

$$C_2 = \frac{1}{b_2 + \alpha b_1 r_1} \left[\sqrt{\gamma \frac{t^0_{R,2} - t_0}{t - t_p - t_0}} - 1 \right]$$

- End Time of the Mixed Zone:

$$t_E = t_p + t_0 + \frac{\gamma}{\alpha}(t^0_{R,1} - t_0) = t_p + t_0(1 + \frac{\gamma F a_1}{\alpha})$$

- Concentration of the Second Component at the End of the Mixed Zone:

$$C_2^B = \frac{\alpha - 1}{\alpha b_1 r_1 + b_2} = C_2^0 - \frac{C_1^0}{r_1}$$

Third Zone: Pure Second Component.
- Duration of the Concentration Plateau:

$$\Delta t_2 = t_I - t_E = F t_0 \frac{\gamma}{\alpha} \left[a_2 \frac{\gamma}{\alpha} - a_1 \right] = F a_1 \frac{\gamma(\gamma - 1)}{\alpha} t_0$$

- Beginning of the Diffuse Profile of the Pure Second Component:

$$t_I = t_p + t_0 \left[1 + \frac{F a_2}{(1 + b_2 C_2^B)^2} \right] = t_p + t_0 \left[1 + F a_2 \left(\frac{\gamma}{\alpha} \right)^2 \right]$$

- Elution Profile of the Pure Second Component:

$$C_2 = \frac{1}{b_2} \left[\sqrt{\frac{t^0_{R,2} - t_0}{t - t_p - t_0}} - 1 \right]$$

- End of the Profile:

$$t = t_p + t^0_{R,2} = t_p + t_0[1 + F a_2]$$

TABLE II

EQUATIONS GIVING THE ELUTION PROFILE
OF A NARROW RECTANGULAR INJECTION PULSE

First Zone: Pure First Component.
- Retention time of the First component[2]:
This retention time cannot be calculated in closed form. It is obtained as the lower boundary of the finite integral of the two profiles of the first component (in the first and second zones), such that this integral be equal to the area of the injected profile, i.e., corresponds to the mass of first component injected.

---

[2]Same symbols as in Table I.

- Concentration Profile of the First Component:

$$t = t_p + t_0 + (t^0_{R,1} - t_0) \left[\frac{1}{(1 + b_1 C_1)^2} - L_{f,2} \frac{\alpha - 1}{\alpha} \frac{1}{[(\alpha - 1)/\alpha + b_1 C_1]^2} \right]$$

- Concentration of the First Component on the Front Side of the Second Shock:

$$C_1^{A'} = \frac{1}{b_1} \frac{(1 - \alpha)/\alpha + \sqrt{L'_f}}{1 - \sqrt{L'_f}}$$

with:

$$L'_f = \left(1 + \frac{b_1 r_1}{b_2} \right) L_{f,2}$$

$L_{f,2}$ is the loading factor for the second component, $L_{f,2} = \frac{n_2 b_2}{\epsilon S L k'_2} = \frac{b_2 C_2^0 t_p}{t^0_{R,2} - t_0}$.

Mixed Zone.
- Retention Time of the Second Shock:

$$t_{R,2} = t_p + t_0 \left[1 + Fa_2 \gamma (1 - \sqrt{L'_f})^2 \right]$$

- Concentration of the First Component on the Rear Side of the Second Shock:

$$C_1^M = \frac{r_1}{b_2 + \alpha b_1 r_1} \frac{1 - \alpha + \alpha \sqrt{L'_f}}{1 - \sqrt{L'_f}}$$

- Concentration of the Second Component at the Top of the Second Shock:

$$C_2^M = \frac{1}{b_2 + \alpha b_1 r_1} \frac{\sqrt{L'_f}}{1 - \sqrt{L'_f}}$$

- Elution Profile of the First Component:

$$C_1 = \frac{1}{b_1 + b_2/(\alpha r_1)} \left[\sqrt{\frac{\gamma}{\alpha} \frac{t^0_{R,1} - t_0}{t - t_p - t_0}} - 1 \right]$$

- Elution Profile of the Second Component:

$$C_2 = \frac{1}{b_2 + \alpha b_1 r_1} \left[\sqrt{\gamma \frac{t^0_{R,2} - t_0}{t - t_p - t_0}} - 1 \right]$$

- Time the Mixed Zone ends:

$$t_E = t_p + t_0 + \frac{\gamma}{\alpha}(t^0_{R,1} - t_0) = t_p + t_0(1 + \frac{\gamma Fa_1}{\alpha})$$

- Concentration of the Second Component at the End of the Mixed Zone:

$$C_2^B = \frac{\alpha - 1}{\alpha b_1 r_1 + b_2} = C_2^0 - \frac{C_1^0}{r_1}$$

Third Zone: Pure Second Component.
- Duration of the Second Component Plateau:

$$\Delta t = t_I - t_E = F t_0 \frac{\gamma}{\alpha} \left[a_2 \frac{\gamma}{\alpha} - a_1 \right] = F a_1 \frac{\gamma(\gamma - 1)}{\alpha} t_0$$

- Time the diffuse Profile of the Second Component begins:

$$t_I = t_p + t_0 \left[1 + \frac{F a_2}{(1 + b_2 C_2^B)^2} \right] = t_p + t_0 \left[1 + F a_2 \left(\frac{\gamma}{\alpha} \right)^2 \right]$$

- Diffuse Profile of the Pure Second Component:

$$C_2 = \frac{1}{b_2} \left[\sqrt{\frac{t_{R,2}^0 - t_0}{t - t_p - t_0}} - 1 \right]$$

- End of the Elution Profile:

$$t = t_p + t_{R,2}^0 = t_p + t_0 [1 + F a_2]$$

THE EQUILIBRIUM-DISPERSIVE MODEL OF CHROMATOGRAPHY

S. GOLSHAN-SHIRAZI AND G. GUIOCHON
Department of Chemistry
University of Tennessee
Knoxville, TN, 37996-1501, USA
and
Division of Analytical Chemistry
Oak Ridge National Laboratory
Oak Ridge, TN, 37831-6120, USA

ABSTRACT. The equilibrium-dispersive model of chromatography assumes that the phase system is always very close to equilibrium and that the contributions of axial dispersion and of the finite rate of the mass transfer kinetics on the band profiles can be accounted for by the use of an apparent dispersion coefficient. Results of a similar study in linear chromatography have proven this assumption to be true for efficiencies exceeding 100 theoretical plates. There is no analytical solution of the equilibrium-dispersive model, except when the deviation of the isotherm from linear behavior and the loading factor are small. Otherwise, numerical solutions are necessary. The calculations procedures used, the approximations made and the errors are reviewed. Excellent agreement is observed between calculated and experimental band profiles, provided the equilibrium isotherms are known accurately and a good model is available to account for these data.

1. Introduction.

We have seen elsewhere [1] that the solution of the ideal model of chromatography gives a reasonable caricature of the elution profiles when the column efficiency is high and the column is seriously overloaded. Even in this case, however, the band profiles are smoother than predicted by the ideal model. The discontinuous parts of these profiles or shocks are replaced by steep boundaries. A more accurate model must be used to account for the experimental results. This model must include the various phenomena which are responsible for band broadening, the axial dispersion, the kinetics of mass transfer and the kinetics of adsorption/desorption.

These phenomena have been discussed in detail in our study of linear chromatography [2]. As we have seen, they can be included in a general rate model of linear chromatography for which solutions are available. The problem becomes much more complex in non-linear chromatography, when we need to discuss the elution band profiles of multi-component mixtures. When the equilibrium isotherm are not linear, we cannot assume than the profiles of these components are independent. Whereas approximate solutions are possible under some particular sets of restrictions in the single-component case, only numerical solution permit an extension of the discussion even to the binary case.

We discuss successively the single component and the multi- component problems. Although not particularly useful for investigating separation problems where, by definition, there should

F. Dondi and G. Guiochon (eds.),
Theoretical Advancement in Chromatography and Related Separation Techniques, 35–59.
© 1992 *Kluwer Academic Publishers*.

be at least two components, the study of the single component problem has two advantages. It is simple, and, under some simplifying assumptions, the problem has approximate analytical solutions from where concepts and general ideas regarding the multi-component problem can be derived.

2. The Single Component Problem.

2.1. THE APPARENT DISPERSION COEFFICIENT.

In linear chromatography, Glueckauf [3,4] suggested that the different contributions to band broadening are additive. Van Deemter et al. [5] made a detailed study of the properties of the analytical solution of the model of Lapidus and Amundson [6] which is at the crossing of the various approaches to studying linear and non-linear non-ideal chromatography.

The model of Lapidus and Amundson consider a set of two partial differential equations. The first one is the differential mass balance of the compound:

$$\frac{\partial C}{\partial t} + F\frac{\partial C_s}{\partial t} + u\frac{\partial C}{\partial z} = D_L\frac{\partial^2 C}{\partial t^2} \tag{1}$$

where C and C_s are the concentrations of solute in the mobile and the stationary phases, respectively; u is the mobile phase linear velocity; F is the phase ratio ($F = \frac{v_\star}{v_m} = \frac{1-\epsilon}{\epsilon}$, where ϵ is the total porosity of the column packing); and D_L is the coefficient of axial dispersion. The axial dispersion is the sum of the axial molecular diffusion and the eddy diffusion contribution.

The other partial differential equation relates the two functions in equation 1 and, thus, permits the solution of the problem. It is the kinetic equation. In the Lapidus and Amundson model a linear kinetic model is used:

$$\frac{\partial C_s}{\partial t} = k_a C - k_d C_s \tag{2a}$$

and k_a and k_d are the rate constants of adsorption and desorption, respectively. Another equation used in chromatography is the solid film linear driving force kinetic equation:

$$\frac{\partial C_s}{\partial t} = k_m(q - C_s) \tag{2b}$$

where k_m is the lumped mass transfer coefficient and q is the concentration of the compound studied in the stationary phase at equilibrium with the mobile phase. It is given by the equilibrium isotherm, $q = f(C)$. In linear chromatography, equation 2b is a particular case of equation 2a. This is no longer true in non- linear chromatography.

Van Deemter el al. [5] have shown that the solution of this model is equivalent to a Gaussian profile provided that the mass transfer kinetics is not very slow. We have shown [2] that this is true for a column efficiency which exceeds a hundred theoretical plates, a rather modest threshold given the current level of column performance. The variance of this Gaussian profile is the sum of the contributions of the axial dispersion and the mass transfer resistances, and the column height Equivalent to a Theoretical Plate (HETP) is given by:

$$H = \frac{2D_L}{u} + 2\left(\frac{k_0'}{1+k_0'}\right)^2 \frac{u}{k_0' k_m} \tag{3}$$

or:

$$\frac{1}{N_{apt}} = \frac{1}{N_{Disp}} + \frac{2}{N_m}\left(\frac{k_0'}{1+k_0'}\right)^2 \tag{4}$$

where $N_{Disp} = P_{e,z}/2 = \frac{Lu}{2D_L}$ is the number of mixing stages and $N_m = \frac{k_0' k_m L}{u}$ is the number of mass transfer stages.

According to Glueckauf [3,4], the lumped mass transfer coefficient k_m is related to the film mass transfer coefficient, k_f, and the pore diffusion coefficient, D_p, by:

$$\frac{F}{k_0' k_m} = \frac{d_p}{6k_f} + \frac{d_p^2}{60 D_p} \tag{5}$$

where d_p is the particle diameter. The Equations 3 and 5 are very important relationships. Combination of these equations show that in linear chromatography, for the general rate model as well as for the linear driving force models, the effects of the axial dispersion and the various contributions to mass transfer resistances are additive. This result constitutes the fundamental basis of the equilibrium-dispersive model of chromatography.

In the equilibrium-dispersive model, it is assumed that the mobile and the stationary phases are always in equilibrium and that the contributions of all the non-equilibrium effects can be lumped into an apparent dispersion coefficient. Accordingly, the equilibrium- dispersive model of chromatography is represented by a single partial differential equation:

$$\frac{\partial C}{\partial t} + F \frac{\partial q}{\partial t} + u \frac{\partial C}{\partial z} = D_a \frac{\partial^2 C}{\partial z^2} \tag{6}$$

and two algebraic equations, the isotherm equation, $q = f(C)$ and the equation relating the apparent dispersion term to the apparent column efficiency:

$$D_a = \frac{Hu}{2} = \frac{Lu}{2N_{apt}} \tag{7}$$

In linear chromatography, H or N_{apt} are related to the axial dispersion and to the coefficients of resistance to mass transfer through the Equations 3 to 5.

Giddings [7] has developed a non-equilibrium theory of chromatography and showed that the influence of the kinetics of radial mass transfers can be treated as a contribution to the axial dispersion.

Rhee and Amundson [8] have studied the effects of axial dispersion and mass transfer resistance in non-linear chromatography by using equations 1 and 2b. In the case of the breakthrough curve (Frontal Analysis for chromatographers, Riemann problem for mathematicians), and assuming the existence of the shock layers, they have shown that Equations 1 and 2b can be combined and reduced to a single equation. The contributions of the axial dispersion and the mass transfer resistances are additive, although there exists a coupling, second order term. This term, however, is negligible in most practical cases of interest in chromatography. However, instead of Equation 3, the additivity rule is given by:

$$H = \frac{2D_L}{u} + 2\left(\frac{k}{1+k}\right)^2 \frac{u}{k k_m} \tag{8}$$

where k is the slope of the isotherm chord, $k = F \frac{\Delta q}{\Delta C}$. For a Langmuir isotherm and for a step injection from 0 to C, it is given by $k = \frac{k_0'}{1+bC}$.

Equation 8 is extremely important. It proves that in non-linear chromatography, as well as in linear chromatography, and at least in the case of frontal analysis, assuming the solid film linear driving force model, the contributions of axial dispersion and of the mass transfer coefficient are additive. However, since k is concentration dependent, the apparent axial dispersion coefficient in non-linear chromatography, by contrast with linear chromatography, is concentration dependent. For linear chromatography $k = k_0'$ hence Equation 8 is reduced to Equation 3.

Equation 8 was derived for frontal analysis. In overloaded elution, we can apply neither the equation valid in linear chromatography nor the one derived for frontal analysis. We think that, in overloaded elution k in the previous equations should be replaced by the slope of the isotherm $k = Fdq/dC$. In the case of the Langmuir isotherm, k would be given by $k = \frac{k_0'}{(1+bC)^2}$. Accordingly, the situation in overloaded elution becomes more difficult than in the two previous cases, since C, hence k, varies all along the elution band profile.

In the equilibrium-dispersive model of chromatography, we assume that Equation 3 remains valid. Thus, we use Equation 6 assuming that the apparent dispersion coefficient D_a in Equation 6 is given by Equation 7 and assuming that the HETP is independent of the solute concentration and is the same in overloaded elution as measured in linear chromatography. As shown by the previous discussion this assumption is an approximation. However, as we showed elsewhere [9], Equation 3 is an excellent approximation as long as the column efficiency is greater than a few hundred theoretical plates. Thus, the use of Equation 3 is an excellent approximation.

In the case of an n-component mixture, we need to write n mass balance equations similar to Equation 6:

$$\frac{\partial C_i}{\partial t} + F\frac{\partial q_i}{\partial t} + u\frac{\partial C_i}{\partial z} = D_{a,i}\frac{\partial^2 C_i}{\partial t^2} \qquad \text{with} \qquad i = 1, 2, \cdots, n \qquad (9)$$

In order to solve the Equation 6 or the set of mass balance equations 9, we need to specify the initial and boundary conditions of the problem, as well as the isotherm equations which relate the concentration of each of the solutes in the stationary phase and the concentrations of the solutes in the mobile phase. In general, these adsorption isotherms are competitive, meaning the amount of component i adsorbed at equilibrium between phases from a solution with concentration C_i decreases usually with increasing concentration of the other components. Finally, the apparent dispersion coefficient, i.e., the column efficiency, must be known.

2.2. INITIAL AND BOUNDARY CONDITIONS.

In order to solve a partial differential equation, we need to specify the initial and boundary conditions. They depend on the mode of operation of the chromatographic column. The same calculation algorithm can be used for any kind of initial and boundary conditions. The difference between the profiles obtained in simulation is due to the different initial and boundary conditions used.

For example, in elution chromatography [10,11] the initial condition is that the column is filled with pure mobile phase, without any solute:

$$C_i(x, 0) = 0 \qquad (10a)$$

In displacement, as in elution, the column is filled with pure mobile phase at the beginning of the experiment. In some modes, e.g., in frontal analysis, or in the study of system peaks, the mobile phase contains a steady concentration of solutes or additives. In the study of system peaks, the concentration of additive in the mobile phase, in the whole column, at the beginning of the experiment is constant, as well as its concentration in the mobile phase at the column inlet. In frontal analysis, the concentration of the components studied is constant and finite along the column.

The boundary condition is given by the injection mode. In elution, we introduce a sample plug, during a certain time, t_p:

$$C_i(0, t) = C_i^0 \qquad 0 < t < t_p \qquad (10b)$$

Often this plug is rectangular, but with numerical solutions, it does not have to be so. After the injection is finished, pure mobile phase is pumped through the column:

$$C_i(0, t) = 0 \qquad t > t_p \tag{10c}$$

In Frontal analysis [12], the mobile phase pumped into the column has a constant concentration in the feed components. In displacement Chromatography [13], the (weak) mobile phase is replaced by a stream of mobile phase containing a finite concentration of the displacer. In the study of system peaks [14,15], the boundary conditions are the same as in elution. In gradient elution [16], the concentration of strong solvent or additive in the mobile phase increases monotonically.

2.3. THE ADSORPTION ISOTHERMS.

The isotherms relate the composition of the two phases at equilibrium. We need reliable equations to model experimental data, to predict equilibrium data when they are unknown and too difficult to measure, and to introduce them in numerical calculations in an orderly fashion.

2.3.1. The Single Component isotherms. The band exact profiles are very sensitive to the adsorption isotherms, hence we need accurate measurements and a correct model. The adsorption isotherm can be measured dynamically by Frontal Analysis, FA [17,18], by Elution by Characteristic Point, ECP [19], by Frontal Analysis by Characteristic Point, FACP [20], and by the retention time method, RTM [21], or by pulse technique which are more difficult to implement.

Often, experimental isotherm data for single component fit reasonably well to the Langmuir isotherm model [22]:

$$q = \frac{aC}{1 + bC} \tag{11}$$

In some cases, the bilangmuir isotherm model [23]:

$$q = \frac{a_1 C}{1 + b_1 C} + \frac{a_2 C}{1 + b_2 C} \tag{12}$$

permits a considerable improvement of the representation of the experimental data. This is the case, for example with enantiomers for which there exists at least two distinct adsorption sites on a chiral stationary phase, the chiral selective sites and the sites providing all the possible non-selective interactions [24]. This model is usually associated with a heterogeneous model of the surface, assuming the existence of two independent types of sites with different properties and adsorption energy. This is confirmed by a Scatchard plot in the shape of a hyperbola exhibiting two asymptotes each of which correspond to one of the Langmuir terms.

2.3.2. Competitive or Multicomponent Isotherms. In the case of multi-component systems, i.e., in all practical cases of separation, the solutes compete for access to the stationary phase. In other words, the adsorption isotherm of each solute, i, depends not only on C_i, but also on the concentrations of all the other solutes in the sample and the components of the mobile phase. The binary competitive isotherm can be measured either by frontal analysis [25] or by recording the elution profiles of a wide injection plug of solutes and applying the simple wave theory [26].

The competitive Langmuir isotherm is an extension of the single Langmuir isotherm:

$$q_i = \frac{a_i C_i}{1 + \Sigma_1^n b_j C_j} \tag{13}$$

If the system truly obeys the competitive Langmuir isotherm model, the parameters of the multicomponent isotherms should be the same as the parameters of the single component isotherm. This is not true in most practical cases. While the multicomponent isotherm data measured for a certain relative composition of a binary mixture can often be fitted well to a competitive Langmuir isotherm equation, the values of the parameters usually change with the relative composition of the mixture and they are not those obtained by measurement of the single component isotherm data [27].

The bilangmuir competitive isotherm is the sum of two Langmuir isotherm terms [23]:

$$q_i = \frac{a_{i,1}C_i}{1 + \Sigma_1^n b_{j,1}C_j} + \frac{a_{i,2}C_i}{1 + \Sigma_1^n b_{j,2}C_j} \tag{14}$$

where the subscripts 1 and 2 refer to the adsorption sites 1 and 2 respectively. This is the competitive form of the bilangmuir isotherm.

The Langmuir competitive isotherm model does not satisfy the Gibbs adsorption equation and, consequently, is not thermodynamically consistent, unless the column saturation capacities of the two solutes considered are the same [28]. The Levan-Vermeulen isotherm [29] is an attempt to correct the competitive Langmuir isotherm for this inconsistency in the case when the column saturation capacity of the two solutes are different. By assuming that each single component obeys the Langmuir isotherm when pure, and by using the Ideal Adsorbed Solution (IAS) theory [30] and the Gibbs adsorption equation, Levan and Vermeulen [29] derived a binary competitive isotherm in the form of a rapidly converging expansion series. When the column saturation capacities of two solutes are equal, the Levan-Vermeulen isotherm reduces to the Langmuir isotherm.

3. The Analytical Solution of the Equilibrium Dispersive Model.

There is no analytical solution of the equilibrium-dispersive model similar to the solution of the ideal model, for any classical isotherm. If we are interested by the phenomena appearing at the onset of column overloading, we can assume that the isotherm deviates little from linear behavior and replace any classical isotherm by its two-term expansion. In the case of the Langmuir isotherm, we obtain $q = aC(1 - bC)$, which will be referred to as the parabolic isotherm. The same equation would be valid for any isotherm, except for the relationship between the coefficients of the isotherm and of its expansion.

With the parabolic isotherm, we can rewrite equation 6 as:

$$\frac{\partial C}{\partial t} + \frac{\Lambda u C}{(1 + k_0')(1 + \Lambda C)}\frac{\partial C}{\partial \xi} = \frac{D_a}{(1 + k_0')(1 + \Lambda C)}\frac{\partial^2 C}{\partial \xi^2} \tag{15}$$

with:

$$\Lambda = -2b\frac{k_0'}{1 + k_0'} \tag{15a}$$

$$\xi = L\frac{t_{R,0} - t}{t_{R,0}} \tag{15b}$$

In order to obtain an analytical solution of Equation 15 Houghton [31] assumed that at low column loading (i.e., in the sample size range where $\Lambda C_{max} \leq 0.05$, where C_{max} is the maximum concentration of the band), the term ΛC can be ignored in the denominator of the second and third terms of equation 15. With this assumption, he derived an analytical solution of this equation:

$$\frac{\partial C}{\partial t} + \frac{\Lambda u C}{(1 + k'_0)} \frac{\partial C}{\partial \xi} = \frac{D_a}{(1 + k'_0)} \frac{\partial^2 C}{\partial \xi^2} \tag{16}$$

Because of the simplification made, however, Equation 16 is no longer a mass balance equation with the boundary conditions of chromatography. Mass is not conserved by equation 16. As a result the solution of Equation 16 does not have a constant area. There is an apparent mass loss for convex upwards isotherms (for which Λ is negative) and a mass gain for concave upwards isotherms (for which Λ is positive) [32,33].

Haarhoff and Van der Linde [34] have used exactly the same approach as Houghton until and including Equation 16. However, in the derivation of the solution they have replaced $D_a t$ by $D_a t_{R,0}$. Surprisingly, although this assumption gives an approximate solution of Equation 16, while Houghton has found an exact solution of the Equation 16, the approximate solution of Equation 16 derived by Haarhoff and Van der Linde [34] is a much better solution of the real mass balance equation (Equation 15) and their solution conserves mass. As we have shown previously [32], the reason for this dramatic improvement is that their assumption is equivalent to another, less drastic simplification of Equation 15. The Haarhoff and Van der Linde solution does not ignore ΛC with respect to 1 but incorporates the term $1 + \Lambda C$ in the coefficients of the second and third term of equation 15 and replaces it with its value derived from the solution of the Ideal model. The Ideal model solution gives $1 + \Lambda C = \frac{t}{t_{R,0}}$ [32]. The Haarhoff and Van der Linde [34] solution is given by:

$$X = |\frac{exp(-\tau^2/2)}{\sqrt{2\pi}[coth(m) + er f(\tau\sqrt{2})]}| \tag{17}$$

where:

$$m = N(\frac{k'_0}{1 + k'_0})^2 L_f \tag{17a}$$

$$X = |b|C(\frac{k'_0}{1 + k'_0})\sqrt{N} \tag{17b}$$

$$\tau = \frac{k'_0}{1 + k'_0}\sqrt{N}\frac{t - t_{R,0}}{t_{R,0} - t_0} \tag{17c}$$

where L_f is the loading factor which is the ratio of the amount of solute injected to the amount needed to saturate the column [21]:

$$L_f = \frac{n}{(1 - \epsilon)SLq_s} = \frac{nb}{\epsilon SLk'} = \frac{nb}{F_v(t_{R,0} - t_0)} \tag{18}$$

It is important to realize that it is only because of the assumption of a parabolic isotherm made at the beginning that this solution is valid only at low column loading. The range of validity of the solution depends, however, on the column efficiency. The results it gives are excellent when $bC_{max} \leq 0.05$. The solution can be used for $bC_{max} \leq 0.1$ with an error smaller than 5 % on the band width [32]. Increasing further the column loading increases the error significantly, since the parabolic isotherm deviates more and more from an actual isotherm (i.e., a Langmuir isotherm) with increasing sample sizes.

4. Numerical Solution of the Equilibrium- dispersive Model of Chromatography.

As we mentioned above, the only analytical solution of the equilibrium-dispersive model of chromatography that is available is the approximate solution of Haarhoff and Van der Linde for the single-component problem, assuming a parabolic isotherm and a low column loading. For any other problem, whether a single-component problem with any other isotherm at high column loading, or for multi-component problems, the solution can be obtained only by a numerical method.

Equation 6 can be written as:

$$\frac{\partial C}{\partial z} + \left[\frac{\partial[\frac{1}{u}(C + Fq)]}{\partial t}\right] = \frac{D_a}{u}\frac{\partial^2 C}{\partial t^2} \tag{19}$$

or, by replacing that $\frac{1}{u}(C + Fq)$ by G, we obtain:

$$\frac{\partial C}{\partial z} + \frac{\partial G(C)}{\partial t} = \frac{D_a}{u}\frac{\partial^2 C}{\partial t^2} \tag{20}$$

Various method of computation can and have been used to determine numerical solutions of this equation. They include finite difference methods, and orthogonal collocation on finite elements methods. In this work, however, we discuss only the finite difference methods.

The principle of the finite difference methods consist in dividing the space and time into a number of small, equal segments (of size h for space and τ for time) and in replacing each partial differential term in the equation by a finite difference term. Each first order term, for example, the first order term:

$$\frac{\partial C}{\partial z}$$

can be replaced by:

-   a forward finite difference: $\frac{C_{n+1}^j - C_n^j}{h}$,
-   a backward finite difference: $\frac{C_n^j - C_{n-1}^j}{h}$,
-   or a central finite difference: $\frac{C_{n+1}^j - C_{n-1}^j}{2h}$

Similarly, the second order term:

$$\frac{\partial^2 C}{\partial z^2}$$

can be replaced by:

-   a central finite difference: $\frac{C_{n+1}^j - 2C_n^j + C_{n-1}^j}{h^2}$.

Since there are many ways to combine together the various finite differences which may be used for each of the terms of the mass balance equation, there are many ways a partial differential equation can be approximated by a finite difference scheme. The choice can be more limited in practice, however, for two reasons. First, there is a condition of numerical stability to satisfy and, secondly, we need to control the amount of truncation error made.

Replacing a partial difference term with any of the possible finite difference term gives a truncation error. The error contribution is different for each way we replace the partial differential term by a finite difference. Thus, we need to evaluate these errors and consider which combination gives the numerical solution closest to the true solution of the system.

There are two different approaches for calculating numerical solutions of Equation 20. In the first approach, Equation 20 is solved directly. In the second approach, the second order term in the RHS of Equation 20 is ignored. Then, the space and time increments are chosen in such a way that the numerical or truncation error becomes equal to this neglected second order term and replaces it. We discuss successively the evaluation of the numerical errors, and the two approaches to a numerical solution.

4.1. ESTIMATION OF THE NUMERICAL ERRORS MADE DURING THE CALCULATION.

Let us assume that we have a continuous function, $\theta(x)$, and that we use a Taylor series expansion to calculate the first and second differentials, using a finite difference scheme involving the values of $\theta(x)$ at x_{i-1}, x_i and x_{i+1}. We let $\Theta_i = \theta(x_i)$, and write the Taylor expansions for θ_{i+1} and θ_{i-1},

$$\theta_{i+1} = \theta_i + \frac{\partial\theta}{\partial x}|_{x=x_i}\Delta x + \frac{\partial^2\theta}{\partial x^2}|_{x=x_i}\frac{\Delta x^2}{2!} + \frac{\partial^3\theta}{\partial x^3}|_{x=x_i}\frac{\Delta x^3}{3!} + \frac{\partial^4\theta}{\partial x^4}|_{x=x_i}\frac{\Delta x^4}{4!} + \cdots \tag{21}$$

$$\theta_{i-1} = \theta_i - \frac{\partial\theta}{\partial x}|_{x=x_i}\Delta x + \frac{\partial^2\theta}{\partial x^2}|_{x=x_i}\frac{\Delta x^2}{2!} - \frac{\partial^3\theta}{\partial x^3}|_{x=x_i}\frac{\Delta x^3}{3!} + \frac{\partial^4\theta}{\partial x^4}|_{x=x_i}\frac{\Delta x^4}{4!} + \cdots \tag{22}$$

with $\Delta x = x_i - x_{i-1} = x_{i+1} - x_i$. Rearranging equations 21 and 22 and dividing by Δx, we obtain the expressions of a forward finite difference:

$$\frac{\theta_{i+1} - \theta_i}{\Delta x} = \frac{\partial\theta}{\partial x}|_{x=x_i} + \frac{\Delta x}{2}\frac{\partial^2\theta}{\partial x^2}|_{x=x_i} + \cdots \tag{23}$$

and a backward finite difference:

$$\frac{\theta_i - \theta_{i-1}}{\Delta x} = \frac{\partial\theta}{\partial x}|_{x=x_i} - \frac{\Delta x}{2}\frac{\partial^2\theta}{\partial x^2}|_{x=x_i} + \cdots \tag{24}$$

Thus, replacing a partial differential term by a forward or a backward finite difference gives an error of the order $O(\Delta x)$, the coefficient being the second order partial differential, $\frac{\partial^2\theta}{\partial x^2}$. Alternately, we can subtract equation 22 from equation 21, divide by Δx, and obtain the expression of the central finite difference for the first order differential:

$$\frac{\theta_{i+1} - \theta_{i-1}}{2\Delta x} = \frac{\partial\theta}{\partial x}|_{x=x_i} + \frac{\Delta x^2}{3!}\frac{\partial^3\theta}{\partial x^3}|_{x=x_i} + \cdots \tag{25}$$

Thus, replacing a partial differential term by a central finite difference gives an error of the order $O(\Delta x^2)$, the coefficient being the third order partial differential. Finally, if we add equations 21 and 22, we rearrange and divide by Δx^2, we obtain the expression for the central finite difference for the second order differential:

$$\frac{\theta_{i+1} - 2\theta_i + \theta_{i-1}}{\Delta x^2} = \frac{\partial^2\theta}{\partial x^2}|_{x=x_i} + \frac{2\Delta x^2}{4!}\frac{\partial^4\theta}{\partial x^4}|_{x=x_i} + \cdots \tag{26}$$

Thus, replacing a second order partial differential term by a central finite difference gives an error of the order $O(\Delta x^2)$, the coefficient being the fourth order partial differential. In conclusion, replacing the first and second order partial differential terms in a partial differential equation by central finite difference terms give an error which is of the order of $O(\Delta x^2)$, a second order error which is negligible, while replacing the first order partial differential terms with a forward or a backward finite difference term gives a first order error $O(\Delta x)$, which is never negligible.

4.2. FIRST APPROACH: DIRECT SOLUTION OF EQUATION 20.

In this first approach, we try to calculate directly numerical solutions of Equation 20. In this case, a large number of combinations of terms can be written. However, we can reduce markedly the number of useful combinations by observing that (i) the solution should be stable; and (ii) the error term should be of the second order, $O(h^2 + \tau^2)$ in order to be small enough and negligible. Otherwise, if we use a scheme which gives a first order error term, a second order partial differential term equivalent to a numerical dispersion term will appear and it would be difficult to control or cancel.

For example, if we use a central finite difference for the first term in the RHS of Equation 20, a backward finite difference term in its second term, and a central finite difference term in its RHS, we obtain the following finite difference equation:

$$\frac{C_{n+1}^j - C_{n-1}^j}{2h} + \frac{G_n^{j+1} - G_n^j}{\tau} = \frac{D_a}{u} \frac{C_{n+1}^j - 2C_n^j - C_{n-1}^j}{h^2} \tag{27}$$

The error analysis of this calculation procedure shows that the error made in using this scheme is of the order of $O(h^2 + \tau)$. Thus, the scheme introduces an error term equivalent to a second order partial differential term. This procedure should not be used except possibly with very small time increments τ, which would make the computation time very long. In order to overcome this type of problems, Lax and Wendroff have suggested the addition to the axial dispersion term of an extra term, equivalent to but of opposite sign of the numerical dispersion term. This term compensates the first order error. In linear chromatography, it can be written as follows:

$$\frac{C_{n+1}^j - C_{n-1}^j}{2h} + \frac{G_n^{j+1} - G_n^j}{\tau} = \left(\frac{D_a}{u} + \frac{\tau u_{z,0}}{2} \right) \frac{C_{n+1}^j - 2C_n^j - C_{n-1}^j}{h^2} \tag{28}$$

with $G(C) = C/u_{z,0} = (1+k_0')C/u_0$ (see eqns. 19 and 20). This scheme is widely used in linear chromatography. It gives the exact solution of Equation 20. This is no longer true in non-linear chromatography, however, because, then, the numerical error generated by the second term in the RHS of Equation 28 is not equal to $\frac{\tau u_{z,0}}{2}$. Nevertheless, equation 28 was used by Lin et al. [35] for the calculation of band profiles in non-linear chromatography. They reported, however, serious difficulties in finding conditions giving numerical stability.

4.3. SECOND APPROACH: REPLACEMENT OF THE AXIAL DISPERSION BY NUMERICAL DISPERSION.

In this approach, the RHS of equation 20 is initially ignored, and the mass balance equation becomes:

$$\frac{\partial C}{\partial z} + \frac{\partial G(C)}{\partial t} = 0 \tag{29}$$

Any combination of forward, backward and center finite differences can be used for replacing the differential terms in Equation 29. However, our choice is limited for two reasons, first the need for numerical stability of the solution and secondly the need to control the order of the numerical error. For example, the stability analysis shows that the finite difference scheme which replaces the first term of equation 29 by a forward finite difference term and the second term by a center finite difference term:

$$\frac{C_{n+1}^j - C_n^j}{h} + \frac{G_n^{j+1} - G_n^{j-1}}{2\tau} = 0 \tag{30}$$

is unstable under any condition. It should not be considered any further. The second important thing to consider is the truncation error. For example, if we are interested in solving Equation 29, the Lax-Wendroff method [36] is one of the best choices [37].

4.3.1. The Lax-Wendroff Two-Step Scheme. The method proceeds as follows. If one has determined the value of C at the grid point (n,j), then one proceeds to the level n+1 in two steps. First, one determines C at the point whose coordinates are the form $(n + 1/2, j + 1/2)$, i.e., half-way between two points of the grid, by:

$$C_{n+1/2}^{j+1/2} = \frac{C_n^{j+1} + C_n^j}{2} - \frac{h}{2\tau}\left[G(C_n^{j+1}) - G(C_n^j)\right] \tag{31a}$$

and:

$$C_{n-1/2}^{j-1/2} = \frac{C_n^j + C_n^{j-1}}{2} - \frac{h}{2\tau}\left[G(C_n^j) - G(C_{n-1}^j)\right] \tag{32}$$

then, one determines C at the mesh points $(j, n + 1)$ from:

$$C_{n+1}^j = C_n^j - \frac{h}{\tau}\left[G(C_{n+1/2}^{j+1/2}) - G(C_{n-1/2}^{j-1/2})\right] \tag{33}$$

Since the difference term in the RHS of equation 33 is a central difference term, the computational error made in this calculation scheme is of the order of $O(\tau^2 + h^2)$. Thus, if we are interested in solving the ideal model (Eqn. 29), the Lax-Wendroff scheme appears to be one of the best choices [37]. Unfortunately, in the non-linear case, the Lax-Wendroff scheme has a strong tendency to oscillate near shocks and is not much used for this reason. In linear chromatography, by contrast, we have: $G(C) = \frac{1+k_0'}{u}\frac{\partial C}{\partial t} = \frac{1}{u_{z,0}}\frac{\partial C}{\partial t}$, and in this case, equation 31a can be written as:

$$\frac{C_{n+1}^j - C_{n-1}^j}{2h} + \frac{1}{u_{z,0}}\frac{C_n^{j+1} - C_n^j}{\tau} = \frac{\tau u_{z,0}}{2}\frac{C_{n+1}^j - 2C_n^j + C_{n-1}^j}{h^2} \tag{31b}$$

which is equivalent to equation 28, with $D_a = 0$, which is obvious since we are now writing the Lax-Wendroff equation for solving equation 29, which is equivalent to equation 20 with $D_a = 0$. However, we are now interested in solving the equilibrium- dispersive model (Eqn. 20), not the ideal model (Eqn. 29), so neither the Lax-Wendroff nor any similar scheme which gives a high order error is suitable for our purpose.

We need schemes that give a larger error, namely that give a first order error, $O(\tau + h)$. This means that the error should be equivalent to a second order partial difference term. Then, we can adjust the coefficient of this error term to replace the axial dispersion term that we have neglected. The time and space increments can be chosen for that purpose, with values such that the first order error term becomes equivalent to the second order term of Equation 20 which we have originally ignored. Some of the schemes used for solving Equation 29 and which give a first order error term are discussed now.

4.3.2. Scheme 1: The Forward-backward Differences. In this scheme, the first term of Equation 29 is replaced by a forward finite difference while the second term is replaced by a backward finite difference. We obtain the equation:

$$\frac{C_{n+1}^j - C_n^j}{h} + \frac{G_n^j - G_n^{j-1}}{\tau} = 0 \tag{34}$$

which can be solved easily for C_{n+1}^j:

$$C_{n+1}^j = C_n^j + \frac{h}{\tau}(G_n^j - G_n^{j-1}) \tag{35}$$

Equation 35 permits the calculation of the concentration at the new space position, $n + 1$, from the concentration at the previous space position, n (Godunov scheme). This scheme was introduced by Rouchon et al. [38] and is widely used. It is particularly attractive because of the fast rate of execution by modern computers [39].

4.3.3. Scheme 2: The Backward-forward Differences. In this scheme, the first term of Equation 29 is replaced by a backward finite difference, while the second term is replaced by a forward finite difference. We obtain the following finite difference equation:

$$\frac{C_n^j - C_{n-1}^j}{h} + \frac{G_n^{j+1} - G_n^j}{\tau} = 0 \tag{36}$$

which can be solved for G_n^{j+1}:

$$G_n^{j+1} = G_n^j - \frac{\tau}{h}(C_n^j - C_{n-1}^j) \tag{37}$$

As can be seen from Equation 37, in this method the value of $G = \frac{1}{u}(C + Fq)$ can be calculated at any time, $j + 1$, from the value at the previous time, j. However, the concentration C of the solute in the mobile phase at this time must be calculated from the value of $G = \frac{1}{u}(C + Fq)$ and this can be done only numerically, using some iteration method.

This scheme is identical to the Craig model if we chose the time and space increments such that $\frac{h}{\tau} = u$. It has been used by many authors, including Snyder et al. [40], Czok et al. [41]. It affords a good numerical solution of the gradient elution problems which cannot be solved by the scheme 1.

4.3.4. Scheme 3: The Forward-backward$_{n+1}$ Differences. In this scheme, the first term of Equation 29 is replaced by a forward finite difference, while the second term is replaced by a backward finite difference, but this time the backward finite difference is calculated at the space position $n + 1$. We obtain the following equation:

$$\frac{C_{n+1}^j - C_n^j}{h} + \frac{G_{n+1}^j - G_{n+1}^{j-1}}{\tau} = 0 \tag{38}$$

which gives:

$$\left(C + \frac{hG}{\tau}\right)_{n+1}^j = C_n^j + \frac{h}{\tau}G_{n+1}^{j-1} \tag{39}$$

In this method, the value of $C + \frac{h}{\tau}G = C + \frac{h}{\tau u}(C + Fq)$ can be calculated at any space position, $n + 1$, and for any specific time, j, from its values at the same space position, at the previous time, and at the same time, at the previous space position. However, the concentration, C, of the solute at this space must be calculated from $C + \frac{h}{\tau u}(C + Fq)$, which has to be done numerically by some iteration method. This implicit calculation scheme was used by Lin et al. [42].

4.3.5. Numerical Stability in Linear Chromatography. In order to obtain a stable solution and to manage the truncation error, we should evaluate the stability condition of the calculation procedure and determine the amount of truncation error done with each of these non-linear schemes. However, because of the enormous difficulty found in these theoretical studies when handling the non-linear case, we only evaluate here the stability condition and the truncation error made with the three schemes described above in the case of a linear isotherm.

If the isotherm is linear, G is equal to $C/u_{z,0}$ and Equation 29 can be written as:

$$\frac{\partial C}{\partial z} + \frac{1}{u_{z,0}} \frac{\partial C}{\partial t} = 0 \tag{40}$$

where $u_{z,0}$ is the velocity of solute in linear chromatography. This velocity is given by:

$$u_{z,0} = \frac{u}{1 + k_0'} \tag{40a}$$

The most widely used procedure for determining the stability of a calculation scheme using a finite difference approximation is the Neumann stability analysis [43]. In this analysis, an initial error is introduced as a finite Fourier series and one studies the growth or decay of this error during the calculation. The method applies only to initial value problems with a periodical initial condition, it neglects the influence of the boundary condition, and it is applied only to linear finite difference approximations with constant coefficients, i.e., to linear equations. This approach gives a necessary condition for the stability of a numerical procedure which in many cases is also a sufficient condition.

By using the Neumann stability analysis for the three calculation schemes discussed above, and assuming a linear isotherm, one obtains the following stability conditions:

- For scheme 1:
$$a = \frac{\tau u_{z,0}}{h} \geq 1 \tag{41}$$

- For scheme 2:
$$0 \leq a = \frac{\tau u_{z,0}}{h} \leq 1 \tag{42}$$

- For scheme 3:
$$a = \frac{\tau u_{z,0}}{h} \geq 0 \tag{43}$$

These conditions are frequently called the Courant-Friedrichs-Lewy (CFL) convergence conditions [44] and a is called the Courant number.

4.3.6. The Error Analysis in Linear Chromatography. The error analysis can be done by using the Taylor series expansions. For the concentrations at the four points of the grid which are the neighbors of the point n, j, we have:

$$C_{n+1}^j = \left[C + h\frac{\partial C}{\partial z} + \frac{h^2}{2}\frac{\partial^2 C}{\partial z^2} + ...\right]_n^j \tag{44}$$

$$C_{n-1}^j = \left[C - h\frac{\partial C}{\partial z} + \frac{h^2}{2}\frac{\partial^2 C}{\partial z^2} + ...\right]_n^j \tag{45}$$

$$C_n^{j+1} = \left[C + \tau\frac{\partial C}{\partial t} + \frac{\tau^2}{2}\frac{\partial^2 C}{\partial t^2} + ...\right]_n^j \tag{46}$$

$$C_n^{j-1} = \left[C - \tau\frac{\partial C}{\partial t} + \frac{\tau^2}{2}\frac{\partial^2 C}{\partial t^2} + ...\right]_n^j \tag{47}$$

By combining Equations 44 to 47 and Equations 34, 36 and 38, and assuming a linear isotherm, one obtains:

- For the Scheme 1:

$$\frac{\partial C}{\partial z} + \frac{1}{u_{z,0}}\frac{\partial C}{\partial t} = \frac{h}{2}(a-1)\frac{\partial^2 C}{\partial z^2} \tag{48}$$

- For the Scheme 2:

$$\frac{\partial C}{\partial z} + \frac{1}{u_{z,0}}\frac{\partial C}{\partial t} = \frac{h}{2}(1-a)\frac{\partial^2 C}{\partial z^2} \tag{49}$$

- For the Scheme 3:

$$\frac{\partial C}{\partial z} + \frac{1}{u_{z,0}}\frac{\partial C}{\partial t} = \frac{h}{2}(1+a)\frac{\partial^2 C}{\partial z^2} \tag{50}$$

In all these three equations, $u_{z,0}$ is the velocity of the solute given by Equation 40a, a is the Courant number and the order of the error made in all the three models is the first order, $O(h + \tau)$.

Comparing now the Equations 48 to 50 with the Equation 20 shows that, although we started calculating a numerical solution of the Equation 29, all the three models give an exact solution of Equation 20 in linear chromatography, provided that:

- In the first calculation scheme, the space increment, h, and the time increment, τ, are chosen such that $\frac{h}{2}(a-1) = D_a/u$. Since $D_a = Hu/2$, where H is the column HETP, we obtain for first scheme the condition:

$$h(a-1) = h(\frac{\tau u_{z,0}}{h} - 1) = H \tag{51}$$

- In the second scheme, the space increment, h, and the time increment, τ, are chosen such that $\frac{h}{2}(1-a) = D_a/u$ or:

$$h(1-a) = h(1 - \frac{\tau u_{z,0}}{h}) = H \tag{52}$$

- In the third scheme, the space increment, h, and the time increment, τ, are chosen such that $\frac{h}{2}(1+a) = D_a/u$ or:

$$h(1+a) = h(1 + \frac{\tau u_{z,0}}{h}) = H \tag{53}$$

4.3.7. Extension to Non-Linear Chromatography. The above analyses of the numerical stability of the solution and of the importance of the truncation errors made have been done for the linear case. They are not automatically valid for non-linear chromatography, but their extension raises very different questions.

Regarding the condition for numerical stability, it is very clear whether the numerical procedure is stable or not. We have used in non-linear chromatography the conditions given by Equations 41 to 43, which were derived assuming a linear isotherm for the three methods. In absolutely all cases in which numerical calculations were performed and the proper condition was satisfied, a stable solution was obtained. As the conditions are not very stringent, there is no reason in trying to adopt combinations of increments which would not meet the conditions set in Equations 41 to 43.

The more serious problem is in equating the numerical dispersion with the axial dispersion. The Equations 51 to 53 were derived for linear chromatography. Applying the proper condition in linear chromatography renders the numerical dispersion equal to the axial dispersion. Therefore, starting from Equation 29, we can obtain an exact solution of Equation 20. This is no longer true in non- linear chromatography, since Equations 51 to 53 do not held for a non-linear isotherm. As a first approximation, we may assume that in non-linear chromatography $u_{z,0}$ is replaced in equations 51 to 53 by the local velocity of the solute or velocity associated with the concentration C, u_z. Since the value of u_z for a convex upwards isotherm (e.g., the Langmuirian isotherm and the isotherms most often found in HPLC) increases with increasing concentration, we see that:

- For the first scheme:

$$h(a-1) = h(\frac{\tau u_z}{h} - 1) > H \tag{54}$$

- For the second scheme:

$$h(1-a) = h(1 - \frac{\tau u_z}{h}) < H \tag{55}$$

- For the fourth scheme:

$$h(1+a) = h(1 + \frac{\tau u_z}{h}) > H \tag{56}$$

These results mean that in non-linear chromatography, the schemes 1 and 3 overestimate and the scheme 2 underestimates the axial dispersion term.

Although the Equations 51 to 53 are an approximation in non-linear chromatography, we used and are using these conditions in numerous calculations of band profiles in non-linear chromatography. We use extensively the scheme 1 with $a = 2$ and $h = H$, which satisfies Equation 51 [10,11,39]. Czok et al. used the scheme 2 with different values of a [41]. Lin et al. used the scheme 3 with $a = 2$ and $h = H/3$ [42]. Unfortunately, all these three calculation schemes give only approximate solutions in non-linear chromatography. Nevertheless, the column efficiency is usually high in modern chromatography, the influence on the band profile of a finite column efficiency is small compared to the influence of a non-linear isotherm, especially at high column loading, which is the most important case for preparative chromatography.

Thus, the numerical solutions of the equilibrium-dispersive model calculated by this procedure are in good agreement with experimental results, despite the use of this approximation and of another approximation already discussed, that the column HETP is independent of the solute concentration. The reason for this agreement can be explained as follows. At low concentrations and especially on the diffuse boundary (usually the rear, tailing part of the profile) the solute concentration is low and $H_{nonlinear} \approx H_{linear}$, while the velocity associated with a concentration closely approximates $u_{z,0}$. Thus, the simulation gives very accurate results at low concentrations. At high concentrations, the deviation of the isotherm from linear behavior increases and the error made in approximating the axial dispersion increases. However, in this part of the profile, the non-linear, thermodynamic effect has the dominant role. Accordingly, the error made may be more important but it has a small effect on this part of the profiles. In the case of a single component, the scheme 1 is most often the best one, because no iteration is necessary and the computation time is much less than with the other two methods.

5. Extension of the Numerical Solution to the Multi-component Problem.

The numerical solution discussed in the previous sections for a single component can easily be extended to the case of a multi- component mixture. Then, we have n partial differential equations similar to the Equation 20:

$$\frac{\partial C_i}{\partial z} + \frac{\partial G_i(C_1, C_2, .., C_i, ..)}{\partial t} = \frac{D_{a,i}}{u} \frac{\partial^2 C_i}{\partial t^2} \tag{57}$$

with:

$$G_i = \frac{1}{u}[C_i + F q_i(C_1, C_2, .., C_i, ..)] \quad \text{with} \quad i = 1, 2, ..., n \tag{58}$$

The second approach discussed in the single component case consisted in neglecting the second order term and in trying to calculate numerical solutions of the ideal model, using the numerical dispersion (i.e., a first order error term) to simulate the neglected axial dispersion. The results are excellent with one component [12,16,24,26,39], and it is reasonable to use the same method in the calculation of solutions of the multi-component problem. If we use this approach, however, we find a third source of error besides the two errors discussed earlier in detail.

In this approach we try to compensate the axial dispersion term which is dropped from equation 57 by a numerical error term. However, we can use only fixed values of the time and space increments for all the components of the system. We see from Equations 51 to 53 that if we choose these values for one of the components, for the other components we have no freedom left. When the two increments are chosen, the Courant numbers (Eqn. 41) of all the components are determined. As a result, we can chose the value of H and, accordingly, of D_a derived from the truncation error for one single component. For all the other components H and D_a are fixed and there is no reason that they have their actual value in the practical chromatographic problem studied, nor even a realistic value [41]. Even in linear chromatography, where the Equations 51 to 53 are exact, if one simulates the band profiles of a series of compounds using the multi-component scheme (although there would not be any good reason to do that, as the equations 57 are not coupled when the isotherms are linear, for obvious thermodynamic reasons), one can obtain the exact band broadening for one component only, provided the space and time increments are adjusted according to the HETP of this component. For the other components, the band width would be related, following a dependence on the retention factor which has nothing to do with the prediction of any plate height equation.

For example, let us consider a two-component problem and use scheme 1 to solve it, with a Courant number equal to a_1 for the first component. In linear chromatography, the contribution of the numerical dispersion for this first component would be exactly equivalent to the true HETP, H_1, if the space and time increments are chosen according to equations 41 and 51, $h = \frac{H_1}{a_1 - 1}$ and $\tau = \frac{a_1 H_1}{u_{z,0,1}(a_1 - 1)}$. For these values of the time and space increments, the value of the Courant number for the second component is given by $a_2 = \frac{\tau u_{z,0,2}}{h} = \frac{a_1 u_{z,0,2}}{u_{z,0,1}} = a_1 \left(\frac{1+k'_{0,1}}{1+k'_{0,2}}\right)$ and the numerical dispersion for the first component would be equivalent to:

$$H_2 = h(a_2 - 1) = H_1 \frac{(a_2 - 1)}{(a_1 - 1)} = H_1 \frac{\frac{a_1(1+k'_{0,1})}{(1+k'_{0,2})} - 1}{a_1 - 1} \tag{59}$$

As an example, for $a_1 = 2$, $k'_{0,1} = 3$ and $k'_{0,2} = 4$, we have $H_2 = 0.2H_1$, quite an unreasonable change in chromatography.

Similarly, for the second scheme we find:

$$H_2 = h(1 - a_2) = H_1 \frac{(1 - a_2)}{(1 - a_1)} = H_1 \frac{1 - \frac{a_1(1 + k'_{0,1})}{(1 + k'_{0,2})}}{1 - a_1} \tag{60}$$

and for the third scheme:

$$H_2 = h(1 + a_2) = H_1 \frac{(1 + a_2)}{(1 + a_1)} = H_1 \frac{1 + \frac{a_1(1 + k'_{0,1})}{(1 + k'_{0,2})}}{1 + a_1} \tag{61}$$

So the band width calculated for the second component would correspond to a value of the HETP which decreases with increasing retention factor for the first and the third schemes and which increases with decreasing retention factor for the second scheme.

Thus, in multi-component simulations, besides the two errors which we have discussed in the study of the single component problem and which are found also in calculations of solutions of the multi- component problems, we have another error. As can be seen from the Equations 59 to 61, this error increases with the difference between the retention factors of the two components. Another interesting results from these equations is that this error can be decreased by selecting a lower value of the Courant number. The error would disappear with the schemes 2 and 3 and the numerical dispersion for the two solutes would become equal and correspond to the proper value of H if a would be close enough to 0. This can be achieved by the combination of a large space increment and a small time increment (Eqn. 42).

In this respect, both schemes 2 and 3 have the advantage that the Courant number can be chosen very small, since the condition of stability of the numerical procedure for these two schemes is $0 < a < 1$ and $a > 0$, respectively. For the first scheme, on the other hand, a should be greater than 1. This condition has important consequences in the calculation of solutions in the problem of gradient elution. In this case, the retention factor, k'_0, varies during the separation and its initial and final values are very different. When choosing the value of the Courant number, a, one should keep in mind that the value of a for both components should always satisfy the stability condition (Eqns. 41-43). Otherwise, a numerical instability may occur. This results in either dramatic oscillations of the concentrations, negative values of the concentrations or an interruption of the program run for overflow or underflow.

The band profiles generated with the first and the second calculation schemes for multi-component mixtures are usually quite comparable [41], although none of the two solutions is exact. These profiles are also close to those calculated with a finite element method [45]. the differences are significant only when the relative concentration of the second component to that of the first one is low and a strong tag-along effect takes place.

In practice, instead of choosing the value of $u_{z,0}$, hence the Courant number, for one component of a pair, the average value of $u_{z,0} = u(\frac{1}{t_{R,0,1}} + \frac{1}{t_{R,0,2}})$ is used for simulations.

In spite of some awkwardness in its formulation, the scheme 1 of the numerical integration of the ideal model (Eqn. 29) seems the most efficient way of calculating the band profiles of the equilibrium-dispersive model. The best alternate procedure is not another finite difference scheme but one using orthogonal collocation on finite elements [45]. This procedure is accurate but it requires a longer computation time than scheme 1. It is more mathematically involved than finite difference schemes, and its detailed explanation is not within the scope of this work. Although collocation methods have been developed for the rapid numerical calculation of solutions of partial differential equations, and are widely used for this reason in physics, mechanics and chemical engineering, their speed advantage over finite difference methods is important only in problems which have three spatial dimensions and a complicate geometry. Chromatography has only one such dimension, but we require a much denser definition of the time and space dimension than is encountered in other problems. This explains the difficulties and peculiarities of its solution.

6. Role of the Strong Solvent in Binary Mobile Phases.

In principle, for a chromatographic system having n components, including the sample and the mobile phase components, we should write n-1 mass balance equations. It is important to understand that, except for the main, weak solvent, and provided the equilibrium isotherm is defined accordingly [12,46], the mass balance equations of all the components of the mobile phase should be included in the set of mass balance equations of the problem. In liquid chromatography, the absolute adsorption isotherm is meaningless, the Gibbs excess isotherm is difficult enough to use with a single component. It becomes extremely complex for multi-component systems, and a theory of chromatography should be able to give answers to some of the relevant separation problems. The conventional adsorption isotherms depend on the position of the Gibbs surface, which we can choose to define [46]. Different choices are possible, e.g., that the weak solvent does not adsorb or that the whole mobile phase does not adsorb [46]. Obviously, this last choice is easier since in this case the mass balance equations for all the components of the mobile phase could be ignored, and only the mass balance equation of the mixture components have to be written and solved and in the isotherm equations we would need to include only the competition between these solutes, but could forget about the mobile phase components which would not need to be taken into account.

It has been shown [12] that, as long as the initial slope of the isotherm of the pure additive in the weak solvent is more than 5 times smaller than the initial slope of the isotherms of the pure solutes in the mobile phase, the two choices are equivalent. We may assume as well that only the weak solvent does not adsorb and that all the other constituents of the mobile phase adsorb and compete for adsorption (which is exactly what they do). Alternatively, we may assume that none of the components of the mobile phase adsorbs and use only the competitive isotherms of solutes, ignoring the competition between solutes and strong solvent or additives. With either assumption, the same results are obtained. The solution of the multi-component problem is much easier in the latter case. This assumption was used in most comparisons made between the band profiles recorded and those calculated. It would be extremely difficult to measure all the additive or strong solvent competitive isotherms. The only exception arises when the additive is nearly as strongly adsorbed as the solutes or more strongly adsorbed than them. Then the competition between solutes and strong solvent cannot be ignored. The correct and complete set of mass balance and competitive isotherms must be used [14,15].

7. Comparison of the Numerical Solution of the Equilibrium-dispersive Model with the Analytical Solution of the Ideal Model.

We have shown [32] that the degree of agreement between the solutions of the equilibrium-dispersive model and the Ideal model depends on the value of the effective loading factor:

$$m = N(\frac{k^{'}}{1 + k^{'}})^2 L_f \tag{62}$$

Thus, increasing the column efficiency or the loading factor increases m, hence, makes the solutions of the two models to become closer. In the case of a single component band, there is

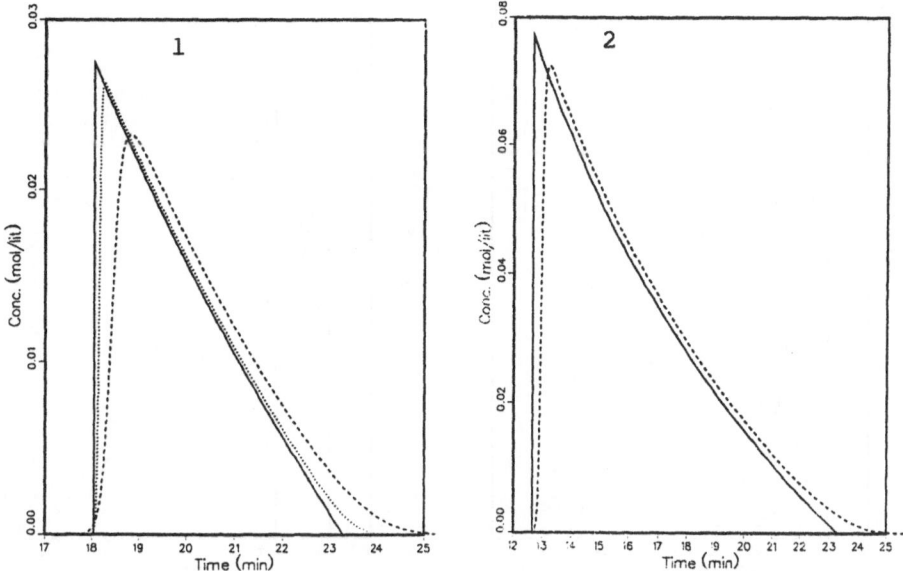

Figure 1. **Comparison of the solution of the Ideal model (solid line, $N = \infty$) with two numerical solutions of the equilibrium-dispersive model.** Dotted line, $N = 5000$ plates and dashed line, $N = 1000$ plates. Loading factor: $L_f = 2\%$.

Figure 2. **Comparison of the solution of the Ideal model (solid line, $N = \infty$) with a numerical solution of the equilibrium-dispersive model.** Dashed line, $N = 1000$ plates). $L_f = 10\%$.

Calculation conditions: Column length: $L = 25$cm; column i.d.: $d = 4.6$ mm; phase ratio: $F = 0.25$; volumetric flow rate: $F_v = 1$ mL/min; Langmuir isotherm parameters: $k'_0 = 6$, $q_s = 4$ mol/L.

a good agreement between the band profiles predicted by the two models when the value of m exceeds 50. In Figure 1, we compare the solutions of the two models for a single component, with a value of the loading factor of 2%. The solid line is a solution of the Ideal model [1], the dotted and dashed lines are the corresponding solutions of the equilibrium-dispersive model, with column efficiencies of 5000 ($m \approx 73$), and 1000 theoretical plates ($m \approx 15$), respectively. There is a good agreement between the profiles predicted by the two models in the case of a 5000 theoretical plate column, while, when the column efficiency is reduced to 1000 plates, there is a considerable difference between the results of the two models. In Figure 2, the same comparison is made, but with a loading factor increased to 10 %, so m is increased to 73 for $N = 1000$. For the sake of clarity, the line corresponding to an efficiency of 5000 plates is not shown; it was very close to the solution of the Ideal model. The dashed line corresponds again to an efficiency of 1000 theoretical plates, as in Figure 1. However, because of the large increase of the loading factor, the agreement between the two solutions has improved considerably.

Figure 3 shows the comparison of the individual band profiles predicted by the Ideal (solid line) and the equilibrium-dispersive Models (dotted line) for a 1:3 mixture. The loading factors for the first and second components are $L_{f,1} = 1$ and $L_{f,2} = 3$, respectively. For the equilibrium-dispersive model, the column has 5000 plates and the values of m for the first and the second components are 37 and 110, respectively. Despite significant differences between the solutions given by the two models, the main features of the elution bands are correctly predicted by the

54

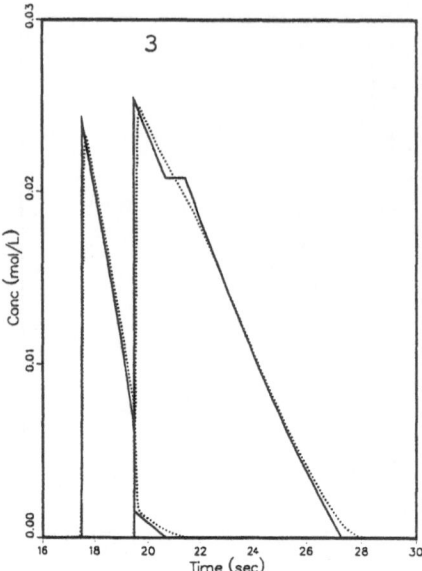 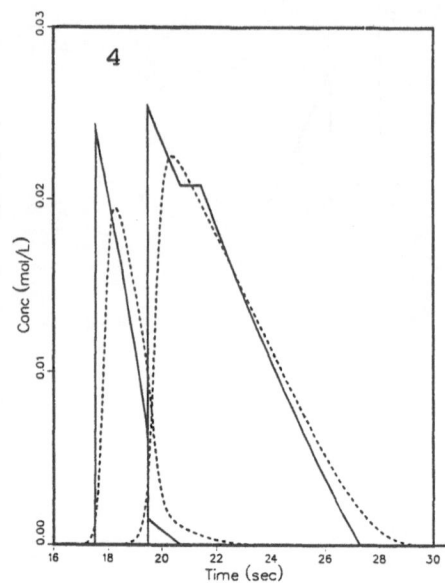

<u>Figure 3.</u> **Comparison of the solution of the Ideal model for a binary mixture (solid line, $N = \infty$) with a numerical solution of the equilibrium-dispersive model.** Dotted line, $N = 5000$ plates.

<u>Figure 4.</u> **Comparison of the solution of the Ideal model for a binary mixture (solid line, $N = \infty$) with a numerical solution of the equilibrium-dispersive model.** Dotted line, $N = 1000$ plates.

Calculation conditions: $L = 25$cm; $d = 4.6$ mm; $F = 0.25$; $F_v = 1$ mL/min; $\alpha = 1.2$. Competitive Langmuir isotherms, with: $k'_{0,1} = 6$, $k'_{0,2} = 7.2$, $q_{s,1} = q_{s,2} = 4$ mol/L. Composition of the mixture, 1:3; $L_{f,1} = 1\%$; $L_{f,2} = 3\%$.

ideal model. The main differences are observed (i) for the plateau predicted by the ideal model on the rear boundary of the second component, plateau which is eroded due to kinetic effects, (ii) for the front shock and (iii) for the rear shock in the profile of the first component. The later takes place in the same time as the front shock of the second component. All these shocks are replaced by shock layers [47].

In Figure 4 we compare the band profiles obtained for the same conditions as in Figure 3 with the Ideal model and with the equilibrium-dispersive model for a column efficiency of only 1000 plates. The values of m for the two components are now only 7.3 and 22, respectively. As expected from these numbers, the agreement between the solutions of the two models is now poor.

7. Comparison of Calculated Band Profiles with Experimental Results in the Case of Single Component Bands.

In order to simulate experimental band profiles, we need to know the experimental conditions. They include (i) the adsorption isotherm; (ii) the sample size; (iii) the mobile phase flow rate; (iv) the column dimensions; and (v) the HETP under linear conditions. The column HETP can

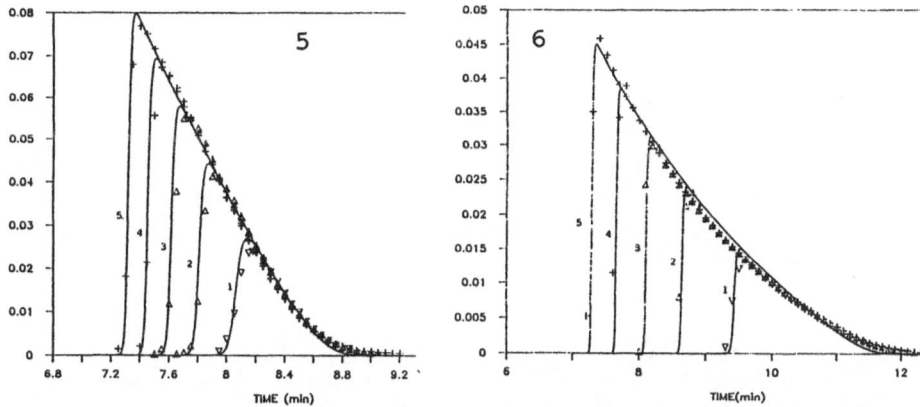

Figure 5. **Comparison of the calculated (solid line) and experimental (symbols) band profiles of acetophenone on silica.** Mobile phase: mixture of ethylacetate and n-hexane (2.5:97.5). $L = 25$ cm; $d = 4.6$ mm; $F_v = 2$ mL/min; $N = 5000$ plates; Sample sizes: 1, 0.025 mmole; 2: 0.05 mmole; 3: 0.075 mmole; 4: 0.1 mmole; 5: 0.125 mmole. (Figure 8 of reference 12, reproduced with permission of the American Chemical Society)

Figure 6. **Comparison of the calculated (solid line) and experimental (symbols) band profiles of phenol on octadecyl chemically bonded silica.** Mobile phase: mixture of methanol and water (20:80). $L = 25$ cm; $d = 4.6$ mm; $F_v = 1$ mL/min; $N = 5000$ plates; sample sizes: 1, 0.015 mmole; 2: 0.03 mmole; 3: 0.045 mmole; 4: 0.06 mmole; 5: 0.075 mmole. (Figure 10 of reference 12 reproduced with permission of the American Chemical Society)

be obtained easily by injecting a very small amount of sample and measuring the band width. The band profile under overloaded conditions is very sensitive to the adsorption isotherm. It must be measured accurately in order to achieve good agreement between theoretical and experimental profiles.

Figures 5 and 6 [12] shows the comparison of experimental results (symbols) with the results of simulation for normal and reversed phase chromatography (solid lines). Figure 5 corresponds to acetophenone eluted on silica with a (97.5:2.5) mixture of n-hexane and ethyl acetate. Figure 6 corresponds to phenol eluted on C18 chemically bonded silica with a (20:80) mixture of methanol and water. In both cases, there is a very good agreement between theory and experiments. However, it must be underlined that in each case, the isotherm was measured with the same column as used for the determination of the overloaded band profiles, and fitted to a Langmuir isotherm. The use of the ECP method should be avoided in the comparison between experimental and calculated band profiles, as it is easy to fall into a circular argument.

In most cases, the experimental data obtained in the determination of single component isotherms can be fitted correctly to the Langmuir equation ($q = \frac{aC}{1+bC}$). In some cases, the results may be improved by using a bilangmuir isotherm instead. In some cases, such as in the separation of optical isomers, we know that two types of sites at least exist, one type of sites being selective and the other non-selective [24]. In such cases, the use of a bilangmuir isotherm ($q = \frac{a_1 C}{1+b_1 C} + \frac{a_2 C}{1+b_2 C}$) permits the achievement of excellent simulations of the experimental bands.

Figure 7 compares the experimental band profiles of various samples of increasing sizes of N-benzoyl- D- and L-phenylalanine on a Resolvosil-BSA-7 column [24]. The stationary phase is immobilized bovine serum albumin and the mobile phase is a 0.1 M buffer aqueous solution at

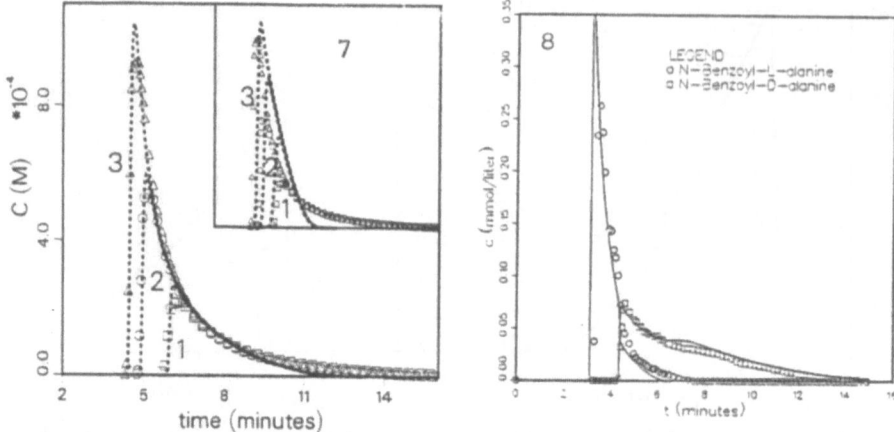

Figure 7. **Comparison of the calculated (solid line) and experimental (symbols) band profiles of N-benzoyl-L- Phenylalanine on a Resolvosil-BSA-7 Column.** Mobile phase: 0.1 M buffer aqueous solution (pH = 6.8) with 7% of 1-propanol (v/v). $L = 15$ cm; $d = 4$ mm; $F_v = 1$ mL/min; $N = 700$; sample sizes: 1, 0.494 μmole; 2: 0,989 μmole; 3: 1.48 μmole. Bilangmuir isotherm: selective site: $a_1 = 20.1$, $q_{s,1} = 0.000556$ mole/L; non-selective site: $a_2 = 7.09$, $q_{s,2} = 0.217$ mole/L (Figure 2 of reference 49, reproduced with permission of the American Institute of Chemical Engineers)

Figure 8. **Comparison of the calculated (solid line) and the experimental (symbols) individual band profiles for a racemic mixture of N-benzoyl-L-(1) and D-(2) Phenylalanine on a Resolvosil-BSA-7 Column.** Mobile phase: 10 mM buffer aqueous solution (pH = 6.7) with 3 % of 1-propanol (v/v). $L = 15$ cm, $d = 4$ mm; $F_v = 1$ mL/min; $N = 700$; sample sizes: 0.26 μ mole for each isomer. Binary competitive Langmuir isotherm: selective site: $a_{1,L} = 14.16$, $a_{1,D} = 35.09$, $q_{s,1,L} = 0.0019$ mole/L, $q_{s,1,D} = 0.0020$ mole/L; non-selective site: $a_2 = 4.41$, $q_{s,2} = 0.01995$ mole/L (Figure 5 of reference 24, reproduced with permission of the American Chemical Society)

pH = 6.8, with 7% of 1-propanol (v/v). The single component isotherms were measured on the same column by frontal analysis and fitted to a bilangmuir isotherm. The choice of this model is justified by the existence of two simultaneous retention mechanisms on this column, an achiral mechanism and a chiral selective one. The agreement between the calculated and experimental profiles is very good. In the insert, a comparison is made with the profiles calculated using a single-site Langmuir isotherm. As expected, the agreement is not good. These results demonstrate that the choice of the proper isotherm model is as crucial for accurate prediction of band multi-component band profiles as the determination of accurate experimental adsorption data.

8. Comparison of Calculated Band Profiles with Experimental Results in the Case of Multi-Component Bands.

As mentioned earlier, the extension of the numerical solution from the single-component problem to the multi-component problem is straightforward. However, in this case we need to use competitive isotherms instead of single-component isotherms. The quest for these isotherms has been only moderately successful so far.

In the case of multi-component systems and when the column is overloaded to the degree that the bands of two components overlap, the determination of the individual band profiles from

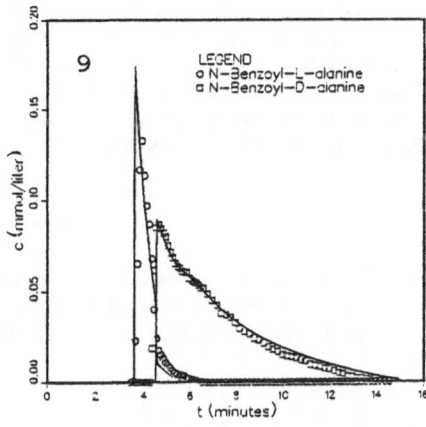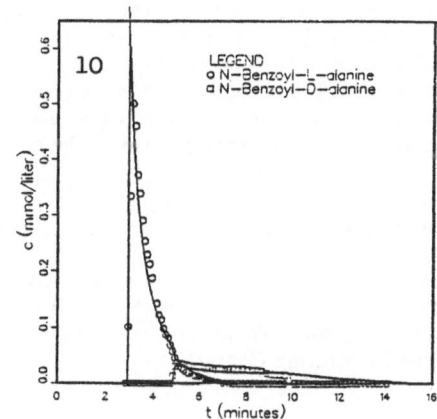

Figure 9. **Same comparison as in Figure 8, except 1:3 mixture, sample size: 0.105** μ**mole of L and 0.392** μ**mole of D isomer.** (Figure 6 of reference 24, reproduced with permission of the American Chemical Society)

Figure 10. **Same comparison as in Figure 8, except 3:1 mixture, sample size: 0.491** μ **mole of L and 0.165** μ**mole of D isomer.** (Figure 7 of reference 24, reproduced with permission of the American Chemical Society)

the detector response is not possible in general. It is exceptional that two compounds whose interference on a column is not accidental have UV spectra which differ enough to permit the accurate detection of one in the presence of the other. In order to determine the individual profiles, fractions must be collected at sufficiently close intervals and analyzed in order to find the composition of the eluent as a function of time [48,49].

Advanced knowledge of the competitive isotherm data is of critical importance for the prediction of the individual band profiles. Unlike the single component case in which the Langmuir model is often a reasonable first approximation, the competitive isotherms follow rarely the Langmuir isotherm model. The only exception we found is in the separation of some optical isomers. This is explained by the combination of two effects. First, the column saturation capacities of the two components are exactly the same, so the Langmuir model is sound. More importantly, the column saturation capacity for the active sites is low. As a result, low concentrations of solutes only are necessary to overload the column (orders of magnitude lower than in other cases). Because adsorption is measured in a low range of concentrations, the liquid and adsorbed phases are close to ideal, and the conditions required for the validity of the Langmuir isotherm are nearly fulfilled. As a result, in most cases of separation of optical isomers we have studied so far, the band profile of the components can be accounted well by using a bilangmuir competitive isotherm model, whose parameters are those obtained by single component isotherm measurement [24,49].

In Figures 8 to 10, we compare a series of experimental individual band profiles and the profiles calculated using the scheme 2. The components are N-benzoyl D- and L-alanine. The single component isotherms of each optical isomer was measured on the same column. The experimental data were well fitted by a bilangmuir isotherm. A bilangmuir competitive isotherm using these coefficients was used for the calculation of the band profiles. In all the Figures, the solid lines are calculation results and the symbols (O for the N- isomer, for the D-isomer) are the experimental data. There is a very good agreement in all cases between the calculated and the experimental profiles. Figure 8 compares the bands of the two enantiomers for a racemic mixture. The displacement and the tag- along effects can be seen in these bands. Figure 9 compares the bands of the isomers in a 1:3 mixture. As predicted by theory, the intensity of the displacement of the first component by the second one increases with increasing the relative concentration of the second component (compare Figures 8 and 9). Figure 10 compares the bands of the two isomers in a 3:1 mixture. As predicted by theory, the intensity of the tag-along effect increases with increasing ratio of the concentrations of the first and second components.

Acknowledgments.

This work has been supported in part by Grant CHE-8901382 of the National Science Foundation and by the cooperative agreement between the University of Tennessee and the Oak Ridge National Laboratory. We acknowledge support of our computational effort by the University of Tennessee Computing Center.

Literature Cited.

[1] Golshan-Shirazi, S., and Guiochon, G. (1992) NATO Advance Study Institute Ser., p.
[2] Golshan-Shirazi, S., and Guiochon, G. (1992) NATO Advance Study Institute Ser., p.
[3] Glueckauf, E. (1955) in Ion- Exchange and its Applications, Metcalfe and Cooper, London, UK, 1955, P.34.
[4] Glueckauf, E. (1955) Trans. Faraday Soc., 51, 1540.
[5] Van Deemter, J.J., Zuiderweg, F.J., and Klinkenberg, A. (1956) Chem. Eng. Sci., 5, 271.
[6] Lapidus, L., and Amundson, N. L. (1952) J. Phys. Chem., 56, 984.
[7] Giddings, J. C. (1965) Dynamics of Chromatography, M. Dekker, New York, NY.
[8] Rhee, H.-K., and Amundson, N.R. (1972) Chem. Eng. Sci., 27, 199.
[9] Golshan-Shirazi, S., and Guiochon, G., In preparation.
[10] Guiochon, G., Golshan-Shirazi, S., and Jaulmes, A. (1988) Anal. Chem., 60, 1856.
[11] Guiochon, G., and Ghodbane, S. (1988) J. Phys. Chem., 92, 3682.
[12] Golshan-Shirazi, S., and Guiochon, G. (1988) Anal. Chem., 60, 2364.
[13] Katti, A.M., and Guiochon, G. (1988) J. Chromatogr., 449, 25.
[14] Golshan-Shirazi, S., and Guiochon, G. (1989) J. Chromatogr., 461, 1.
[15] Golshan-Shirazi, S., and Guiochon, G. (1989) Anal. Chem., 61, 2373.
[16] El Fallah, M.Z., and Guiochon, G. (1991) Anal. Chem., 63, 859.
[17] Schay, G., and Szekely, G. (1954) Acta Chim. Hung., 5, 167.
[18] James, D.H., and Phillips, C.S.G. (1954) J. Chem. Soc., 1066.
[19] Cremer, E., and Huber, J.F.K. (1961) Angew. Chem., 73, 461.
[20] Glueckauf, E. (1955) Trans. Faraday Soc., 51 1540.
[21] Golshan-Shirazi, S., and Guiochon, G. (1989) Anal. Chem., 61, 462.
[22] Golshan-Shirazi, S., and Guiochon, G. (1988) Anal. Chem., 60, 2360.
[23] Graham, D. (1953) J. Phys. Chem., 57, 665.
[24] Jacobson, S., Golshan-Shirazi, S., and Guiochon, G. (1990) J. Amer. Chem. Soc., 112, 6492.
[25] Jacobson, J.M., Frenz J.H., and Horvath, Cs. (1987) Ind. Eng. Chem. Res., 26, 43.
[26] Ma, Z., Katti, A.M., Lin, B., and Guiochon, G. (1990) J. Phys. Chem., 94, 6911.
[27] Zhu, J., Katti, A.M., and Guiochon, G. (1991) J. Chromatogr., 552, 71.
[28] Kembell, C., Rideal, E.K., and Guggenheim, E.A. (1984) Trans. Faraday Soc., 44, 948.
[29] Levan, M.D., and Vermeulen, T. (1981) J. Phys. Chem., 85 3247.
[30] Myers, A.L., and Prausnitz, J.M. (1965) AIChE J., 11, 121.
[31] Houghton, G. (1963) J. Phys. Chem., 67, 84.
[32] Golshan-Shirazi, S., and Guiochon, G. (1990) J. Chromatogr., 506, 495.

[33] Lucy, C.A., Wade, J.L., and Carr, P.W. (1989) J. Chromatogr., 484, 61.

[34] Haarhoff, P.C., and Van der Linde, H.J. (1966) Anal. Chem., 38, 573.

[35] Lin, B., Ma, Z., and Guiochon, G. (1989) Separat. Sci. Technol., 24, 809.

[36] Lax, P.D., and Wendroff, B. (1960) Comm. Pure. Appl. Math., 13, 217.

[37] Ames, W.F. (1977) Numerical Methods for Partial Differential Equations, Academic Press, New York, NY.

[38] Rouchon, P., Schonauer, M., Valentin, P., and Guiochon, G. (1987) Separat. Sci. and Technol., 22, 1793.

[39] Czok, M., and Guiochon, G. (1990) Comput. Chem. Eng., 14, 1435.

[40] Eble, J.E., Grob, R.L., Antle, P.E., and Snyder, L.R. (1987) J. Chromatogr., 405 1.

[41] Czok, M., and Guiochon, G. (1990) Anal. Chem., 62, 189.

[42] Lin, B., Golshan-shirazi, S., and Guiochon, G. (1989) J. Phys. Chem., 93, 3363.

[43] Lapidus, L., and Pinder, G.F. (1982) Numerical Solutions of Partial Differential Equations in Science and Engineering, Wiley, New York, NY, p. 491.

[44] Courant, R., Friedrichs, K., and Lewy, H. (1928) Math. Ann., 100, 32.

[45] Ma, Z., and Guiochon, G. (1991) Comput. Chem. Eng., 15, 415.

[46] Kovats, E. sz, (1985) The Science of Chromatography, F. Bruner, Ed., Elsevier, Amsterdam, The Netherlands, p. 205.

[47] Lin, B.C., Golshan-Shirazi, S., Ma, Z., and Guiochon, G. (1988) Anal. Chem., 60, 2647.

[48] Katti, A.M., and Guiochon, G. (1990) J. Chromatogr., 499, 21.

[49] Jacobson, S., Golshan-Shirazi, S., and Guiochon, G. (1991) AIChE J., 37, 836.

REVIEW OF THE VARIOUS MODELS OF LINEAR CHROMATOGRAPHY, AND OF THEIR SOLUTIONS.

S. GOLSHAN-SHIRAZI AND G. GUIOCHON
Department of Chemistry
University of Tennessee
Knoxville, TN, 37996-1501, USA
and
Division of Analytical Chemistry
Oak Ridge National Laboratory
Oak Ridge, TN, 37831-6120, USA

ABSTRACT. The various models of linear chromatography are compared. Those include the discrete and continuous plate models, the stocchastic model, the equilibrium-dispersive model of chromatography, the lumped kinetic models, and the general rate model. We show that all these models are equivalent and can be correctly approximated by a Gaussian profile, even when the column efficiency is as low as 100 theoretical plates. The sources of band asymmetry and tailing in linear chromatography are also discussed. We show that the existence of active sites on the adsorbent surface is probably the main reason of band asymmetry, in most cases.

1. Introduction.

In most analytical applications of chromatography, a small amount of sample is injected into the column. The concentrations of all the analytes in the mobile phase are very low. By contrast with the preparative applications of chromatography, these concentrations are usually so low that we may consider as linear their equilibrium isotherms between the two phases of the chromatographic system. Let us note here that the assumption of a linear isotherm is nearly always an approximate one. It may be reasonable, but the cases where the isotherm is truly linear are exceptional. In most cases, the effects of a non-linear isotherm (e.g., the peak asymmetry, the dependence of the retention time on the sample size) may be smaller than the precision of the experiments, or simply smaller than what we are ready to tolerate in order to benefit from a simple model.

This assumption of a linear behavior of the equilibrium isotherm has two important consequences. First, the different analytes behave independently from each other. There is no competition between them for access to the stationary phases. Thus, the band profiles of the various components of a mixture are independent of the presence of the other solutes. Each band profile is the same as if the corresponding solute were alone. Therefore and again by contrast with non-linear chromatography, in linear chromatography there is a single problem to solve, the calculation of the band profile of a single compound. Secondly, the assumption of a linear isotherm makes the mathematics of describing the migration of these independent, individual band profiles simple. As we show in this review, an analytical solution or, at least, a closed form solution in the Laplace domain can be obtained with any model of linear chromatography.

61

F. Dondi and G. Guiochon (eds.),
Theoretical Advancement in Chromatography and Related Separation Techniques, 61–92.
© 1992 *Kluwer Academic Publishers.*

The different models which have been used to describe linear chromatography and to predict the elution profiles can be classed in three broad types of models. The plate or "tank-in series" models, the statistical models, and the solutions of the partial differential equations which state the mass balance of the studied compound and its mass transfer kinetics along the column.

2. The Plate Models.

In the plate models, we assume that the column consists of a number of identical equilibrium stages, or theoretical plates, placed in series. Furthermore, we assume that, in each of these plates the mobile and the stationary phases are in equilibrium. Thus, plate models are by nature approximate models since the equilibrium assumption requires a mixing mechanism which is clearly absent from the physical system.

There are two categories of plate models. The first one is the discrete stage distribution or Craig model [1], where a finite volume of eluent is equilibrated step by step, within one theoretical plate in the column after another. In the second plate model, there is continuous flow of the mobile phase through the series of stages; this is the Martin and Synge model [2]. The difference between these two types of plate models has been discussed by Klinkenberg et al [3].

2.1 - THE CRAIG PLATE MODEL.

In the Craig model [1] we have a series of N_c discrete stages numbered in the direction of flow, with a rank $l = 0, 1, 2, \cdots, N_c - 1$. Each one of the N_c stages contains a volume v_m of mobile phase and a volume v_s of stationary phase. We also assume that initially the first stage (stage number $l = 0$) contains all the solute used and that the other stages are entirely free of solute. Similarly, all further portions of the mobile phase introduced in the first stage ($l = 0$) are pure mobile phase. The distribution coefficient of the solute is $a = C_s/C_m$.

After equilibrium is reached in the first stage, the mobile phase is moved from one stage to the next, new mobile phase is added to the initial stage ($l = 0$) and mobile phase withdrawn from the last stage ($l = N_c - 1$). Then, after equilibrium is reached in all the stages containing some solute, the mobile phase is moved again from one stage to the next and the processed repeated. At each stage, at each step, the fraction of solute $R = \frac{v_m C_m}{v_m C_m + v_s C_s} = \frac{1}{1+Fa} = \frac{1}{1+k_o'}$ moves on from any stage to the next, and the fraction $1 - R$ remains in this stage. After r such operations have been performed, the composition of the first $r + 1$ stages is given by the terms of the binomial distribution of the expression:

$$[(1 - R) + R]^r \tag{1}$$

Thus, the probability of finding a molecule in the stage of rank l (i.e., the (l+1)th stage) after r operations have been completed (with $l < r$) is given by:

$$P_{l,r} = \frac{r!}{l!(r - l)!} R^l (1 - R)^{r-l} \tag{2}$$

After these r operations have been made, the mean location of the amount of solute introduced initially is the stage number μ_s, with:

$$\mu_s = rR \tag{3}$$

and the variance of the distribution of the solute is:

$$\sigma_s^2 = rR(1 - R) = \mu_s(1 - R) \tag{4}$$

If r is sufficiently large, the binomial distribution (eqn. 1) can be considered as identical to a Gaussian distribution with the same average, μ_s, and standard deviation, σ_s, according to the central limit theorem of statistics. The distribution of the solute in the Craig model becomes:

$$f(l) = \frac{1}{\sigma_s\sqrt{2\pi}} \; exp\left[-\frac{(l - \mu_s)^2}{2\sigma_s^2}\right] \tag{5}$$

The maximum concentration of the solute is eluted from the column when it leaves the last stage (N_cth stage, of rank $l = N_c - 1$), i.e., only when the mean location of the solute distribution is in this stage, so the number of operation needed to elute the band maximum is given by:

$$\mu_s = rR = N_c \tag{6}$$

The total volume of mobile phase which has passed through the last stage during the entire process is $V_R = rv_m$. From equation 6, we obtain:

$$V_R = rv_m = \frac{N_c}{R}v_m \tag{7}$$

We note that $N_c v_m$ is equal to the total volume of mobile phase, V_m, contained in the column. By substituting V_m and inserting the value of R in equation 7, we obtain:

$$V_R = V_m(1 + k_0') \tag{8}$$

Since the mobile phase move at a velocity which is $1/R$ times faster than the solute, the standard deviation of the distribution at the end of the column, expressed in volume units, is given by:

$$\sigma_v = \frac{1}{R}\sigma_s v_m = \frac{v_m\sqrt{rR(1 - R)}}{R} \tag{9a}$$

Combining equations 7 and 9a gives:

$$\sigma_v = V_R\sqrt{\frac{1 - R}{N_c}} \tag{9b}$$

Finally, by inserting the value of R in equation 9b, we obtain the number of stages in the Craig model corresponding to a certain Gaussian distribution of the solute in a column, which is given by:

$$N_c = \frac{V_R^2}{\sigma_v^2}\frac{k_0'}{1 + k_0'} \tag{10}$$

It is important to understand that the distributions given by equations 2 and 5 are not elution profiles but rather the spatial profile of the solute concentration along the column after a given number of transfers, r, has been performed and for the stages with a rank l between 0 and $r - 1$. The elution profile is determined by the amount of solute in the mobile phase which leaves the last or N_cth stage (rank $N_c - 1$) during each operation. This amount is equal to $R * P_{l=N_c-1,r}$. Unlike in Equation 2, here l is constant and equal to $N_c - 1$, since we consider the last stage, and r varies from $N_c - 1$ (corresponding to the hold-up volume) to infinity.

$$f(r) = R\frac{r!}{(N_c - 1)!(r - (N_c - 1))!}R^{(N_c-1)}(1 - R)^{r-(N_c-1)} \tag{11}$$

2.2 - THE MARTIN AND SYNGE PLATE MODEL.

In the Martin and Synge continuous plate model [2], it is assumed that the column is a series of continuous flow mixers. By contrast with the Craig model, the mobile phase flows continuously, but in each mixer the volumes of the mobile phase, v_m, and of the stationary phase, v_s, remain constant. It is assumed also that, at the beginning of the run, only the first plate (rank $l = 0$) is loaded with solute and that the other plates are free of solute. Said [4] extended the continuous plate theory model to the case when the solute is initially distributed on several plates according to a certain distribution function. The mass balance for the mixer of rank l, when a volume dv of mobile phase is moved through the series of plates is given by:

Amount of solute entering the mixer l =
Amount of solute exiting from the mixer l +
Accumulation due to concentration change,

or:

$$C_{m,l-1}dv = C_{m,l}dv + v_m dC_{m,l} + v_s dC_{s,l} \tag{12}$$

Since we assume that complete equilibrium is reached in each plate, we have:

$$dC_{s,l} = adC_{s,m} \tag{13}$$

where a is the slope of the adsorption isotherm. By combining Equations 12 and 13, we obtain a linear, first order, differential equation. Its solution is obtained by assuming that at the beginning of the experiment, the solute exist only in the first plate, where its concentration is $C_{m,0}$ in the mobile phase and $C_{s,0}$ in the stationary phase. Then one obtains:

$$\frac{C_l}{C_0} = \frac{e^{-x}x^l}{l!} \tag{14}$$

This equation is a Poisson distribution function, where the variable is the plate rank, l and x is a constant equal to:

$$x = \frac{V}{v_m + av_s} = \frac{F_v t}{v_m + av_s} = \frac{t}{\tau} \tag{15}$$

where $\tau = (v_m + av_s)/F_v$, V is the total volume of mobile phase passed through the plate l, $v_m + av_s$ is the equivalent volume of each plate, and F_v is the volumetric flow rate of the mobile phase. For large values of x, this Poisson distribution is also approximated by the Gaussian distribution in Equation 5. Then, according the laws of statistics, the mean of the distribution in Equation 14 is given by $\mu_s = x$ and its variance is given by $\sigma_s^2 = x$.

The maximum concentration of solute leaves the last or Nth stage only when the mean value of the solute distribution is located in this plate, so the number of operation needed to elute it should satisfy the following relationship:

$$\mu_s = x = N \tag{16}$$

or:

$$\frac{\mu_s^2}{\sigma_s^2} = x = N \tag{17}$$

Comparison between Equations 10 and 17 shows that the peak dispersion in the Craig model is less than in the Martin-Synge model if we use the same number of stages. In order for the two results to become equal, the number of stages in the Craig model should be less than in the Martin-Synge model. We should have:

$$N_c = \frac{k_0'}{1 + k_0'} N \tag{18}$$

Again, Equation 14 is not the elution profile of the peak, but the distribution profile of the solute along the column. The elution profile is the distribution of the amount of the solute which leaves the last stage (rank $(N-1)$). This is given by $\frac{1}{\tau} * \frac{C_{(N-1)}}{C_0}$. We notice that, unlike in Equation 14, the number l is now constant and equal to $(N-1)$, while x is a variable. Finally, the elution profile in the case of the continuous plate model is given by:

$$f(t) = \frac{1}{\tau} e^{-\frac{t}{\tau}} \left[\frac{t}{\tau} \right]^{(N-1)} \frac{1}{(N-1)!} \tag{19}$$

Equation 19 with a fixed value of N and a variable value of t is not a Poisson distribution function but a gamma density function [5] with a first moment given by:

$$\mu = \tau N \tag{20}$$

and a variance equal to:

$$\sigma^2 = \tau^2 N \tag{21}$$

. Since the first moment of the distribution is equal to the retention time, we obtain from Equation 20:

$$\tau = \frac{t_R}{N} \tag{22}$$

Combining equations 19 and 22 gives:

$$f(t) = \frac{N}{t_R} e^{-\frac{Nt}{t_R}} \left(\frac{Nt}{t_R} \right)^{(N-1)} \frac{1}{(N-1)!} \tag{23}$$

For large values of the plate number, N, the gamma density function approaches the Gaussian function [6].

Figure 1 compares three band profiles, the Gaussian profile (Equation 5, solid line), the solution of the continuous plate model (Equation 23) for a hypothetical column having only 100 theoretical plates (dotted line), and the solution of the Craig model (Equation 11) for the same column (dashed line). For the Craig model, the number of stages has been calculated according to Equation 18. As can be seen in the Figure, the difference between the three profiles is entirely negligible. It is insignificant for columns having an efficiency in excess of 100 theoretical plates. Even for 25 stages (Figure 1b), the difference is nearly insignificant. The profiles calculated from the Martin and Synge and from the Craig concentration profiles cannot be distinguished, even for 25 stages (Figure 1b).

3. The Statistical Approach.

In this second approach for the modeling of linear chromatography, a "microscopic statistical" method is used to derive the probability density function at l and t of a single molecule of solute. The "random walk" approach [7] is used to calculate the profile of the chromatographic band in a simple way.

66

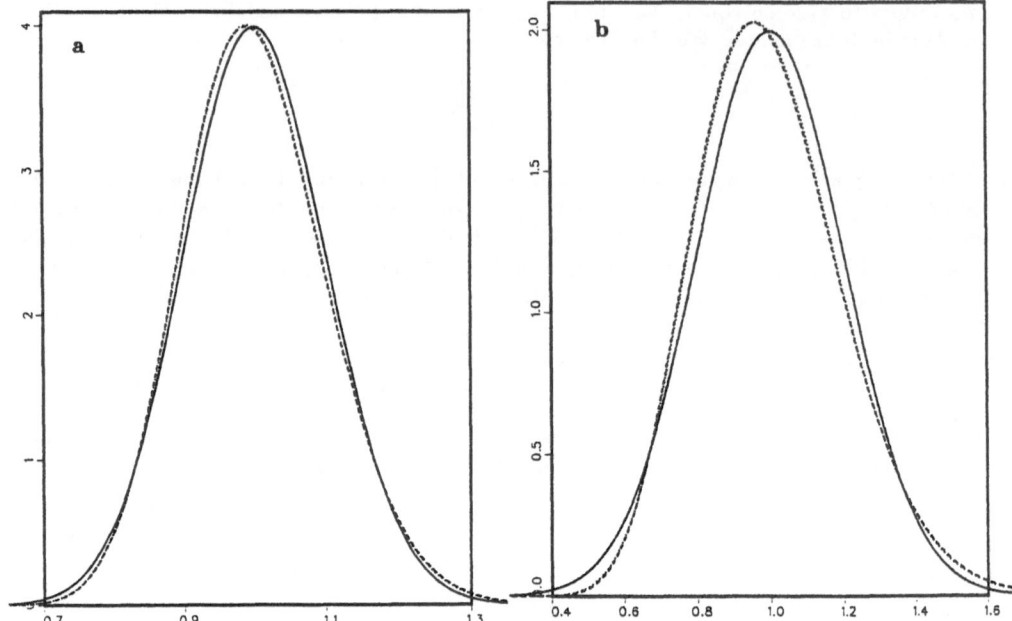

Figure 1. **Comparison of the Chromatograms given by the two Plate Models of Chromatography.** The number of stages in the Craig model is calculated using Equation 18. *Solid line*: Gaussian profile. *Dotted line*: Continuous (Martin and Synge) plate model. *Dashed line*: Discrete or Craig plate model.

Figure 1a- Plate number: $N = 100$.   *Figure 1b-* Plate number: $N = 25$.

In 1955, Giddings and Eyring introduced another probabilistic approach, the stocchastic model, for the description of the molecular migration in chromatography [8]. Their molecular dynamic approach is based on the statistical ideas and treat chromatographic process as a Poisson distribution process. They considered the random migration of a single solute molecule down a chromatographic column. They derived an expression for the elution profile, or distribution of the residence times of a molecule in the column, assuming random adsorption/desorption processes, with a single type of sites on the stationary phase and for an impulse injection. They ignored the axial dispersion and the mass transfer kinetics. In this case the probability density function is given by:

$$P = \frac{X}{2(t - t_0)} I_1(X) \; exp[-k_d(t - t_0) - k_a t_0] \tag{24}$$

where k_a and k_d are the first-order rate constants for adsorption and desorption, respectively, I_1 is the Bessel function and X is given by:

$$X = \sqrt{4 k_a k_d t_0 (t - t_0)} \tag{25}$$

For large values of X, the first term of the asymptotic expansion of the Bessel function, $I_1(X)$ can be used, and the final result is:

$$P = \frac{k_a k_d t_0}{2\sqrt{\pi}(t - t_0)^{0.75}} \; exp[-(\sqrt{k_d(t - t_0)} - \sqrt{k_a t_0})^2] \tag{26}$$

McQuarrie [9], has extended the stocchastic theory to the case of multiple adsorption sites, and to a column with one single type of sites, but with various input distribution functions, by means of the theory of the Laplace transform.

Dondi et al. [10] have applied the "Characteristic Function Method" to the stocchastic theory of chromatography, and obtained basically the same result. The method has potential to account for axial dispersion and the resistances to mass transfer.

4. The Solutions of the Mass Balance Equation.

The third of the approaches which have been extensively used to calculate the chromatographic response to a given input function is the transport approach, which starts with an equation of motion. In this method, we search for the mathematical solution of the set of partial differential equations describing the chromatographic process, or rather the differential mass balance of the solute in a slice of column and its kinetics of mass transfer in the column. Various mathematical models have been developed to describe the chromatographic process. The most important of these models are the equilibrium-dispersive model, the lumped kinetic model and the general rate model of chromatography. We discuss here these three models.

The ideal model of chromatography, which has a great importance in non-linear chromatography has little interest in linear chromatography. Along an infinitely efficient column, with a linear isotherm, the injection profile travels unaltered and the elution profile is the same as the injection profile. We also note that, because of the profound difference in the formulation of the two models, the solutions of the mass balance equation of chromatography for the ideal, non-linear and the non-ideal, linear models rely on entirely different mathematical techniques.

4.1 - THE EQUILIBRIUM-DISPERSIVE MODEL OF CHROMATOGRAPHY.

In the equilibrium-dispersive model, we assume that the mobile and the stationary phases are constantly in equilibrium. We recognize, however, that there are non-equilibrium effects (e.g., axial dispersion, mass transfer resistances, finite kinetics of adsorption/desorption). We assume that their contributions can be lumped together in an apparent dispersion coefficient. This coefficient is related to the experimental parameters by:

$$D_a = \frac{Hu}{2} = \frac{Lu}{2N} \tag{27}$$

where H is the column height equivalent to a theoretical plate (H.E.T.P), N is the column plate number or efficiency, u is the mobile phase linear velocity, L is the column length, and D_a is the apparent axial dispersion coefficient. From eqn. 27, we derive that $N = \frac{P_{e,z}}{2}$, where $P_{e,z}$ is the axial Peclet number (uL/D_a).

With these assumptions, the differential mass balance of the solute in a slice of column is given by:

$$\frac{\partial C}{\partial t} + F\frac{\partial q}{\partial t} + u\frac{\partial C}{\partial z} = D_a\frac{\partial^2 C}{\partial z^2} \tag{28}$$

where $F = \frac{V_s}{V_m} = \frac{1-\epsilon}{\epsilon}$ is the phase ratio, ϵ is the total porosity or void volume fraction of the packing, C is the solute concentration in the mobile phase, q is the solute concentration in the stationary phase in equilibrium with the mobile phase at concentration C, and u is the mobile phase linear velocity. Since we assume that the isotherm is linear, $q = aC$, and $k_0' = Fa$. Equation 28 can be rewritten as:

$$(1 + k_0')\frac{\partial C}{\partial t} + u\frac{\partial C}{\partial z} = D_a\frac{\partial^2 C}{\partial z^2} \tag{29}$$

Equation 29 can be written in dimensionless form:

$$\frac{\partial C_d}{\partial t_d} + \frac{\partial C_d}{\partial z_d} = \frac{1}{2N} \frac{\partial^2 C_d}{\partial z_d^2} \tag{30}$$

with the following dimensionless variables:

$$z_d = \frac{z}{L} \tag{30a}$$

$$t_d = \frac{ut}{L(1+k_0')} = \frac{tu_{z,0}}{L} = \frac{t}{t_R} \tag{30b}$$

$$C_d = \frac{Ct_R}{A_p} = C\frac{F_v t_R}{n} = C\frac{\epsilon SL(1+k_0')}{n} \tag{30c}$$

where t_R is the retention time of the band, $A_p = n/F_v$ is the area of the injected pulse, F_v is the volume flow rate of mobile phase, n is the amount (mole) injected, and S is the column cross-sectional area.

The solution of Equation 30 depends on the choice of the boundary conditions. The problem of the selection of appropriate boundary conditions for the solution of a partial differential equation is mathematically subtle and full of pitfalls. Slightly different boundary conditions may result in very different solutions for a given equation.

4.2.1) Open-Open Boundary Solution. A closed-form solution has been derived by Lapidus and Amundson [11], Levenspiel and Smith [12], Carberry and Bretton [13], Reilley et al. [14], and von Wicke [15], with an "open-open" condition, i.e., assuming that the column stretches to infinity in both directions ($z \to -\infty$, $\frac{\partial C}{\partial z} = 0$; $z \to \infty$, $\frac{\partial C}{\partial z} = 0$), and a Dirac δ pulse of solute injected at $z = 0$. The solution is given by:

$$C_d(z_d, t_d) = \sqrt{\frac{N}{2\pi t_d}} e^{-\frac{N}{2t_d}(z_d - t_d)^2} \tag{31}$$

The equation 31 gives the concentration profile of the solute along the column. The time, t_d, is constant and the position, z_d, is variable. Then, we obtain a Gaussian profile. On the contrary, the elution profile, which is given by writing $z_d = 1$ in equation 31 and taking t_d as the variable, is not Gaussian. It is obtained by making $z_d = 1$ in Equation 31:

$$C_d = \sqrt{\frac{N}{2\pi t_d}} e^{-\frac{N}{2t_d}(1 - t_d)^2} \tag{32}$$

The dimensionless plot of C_d versus t_d depends only on the plate number, N. The first and second moment of the profile are given [12] by:

$$\bar{t}_d = 1 + \frac{1}{N} \tag{33a}$$

$$\sigma_d^2 = \frac{1}{N} + \frac{2}{N^2} \tag{33b}$$

Equation 32 is not a Gaussian profile. However, if we use the approximation $\frac{N}{t_d} \simeq N = \frac{1}{\sigma_d^2}$ (see eqn. 33b), we obtain a Gaussian profile:

$$C_d = \frac{1}{\sigma_d \sqrt{2\pi}} e^{-\frac{(1-t_d)^2}{2\sigma_d^2}} \tag{34}$$

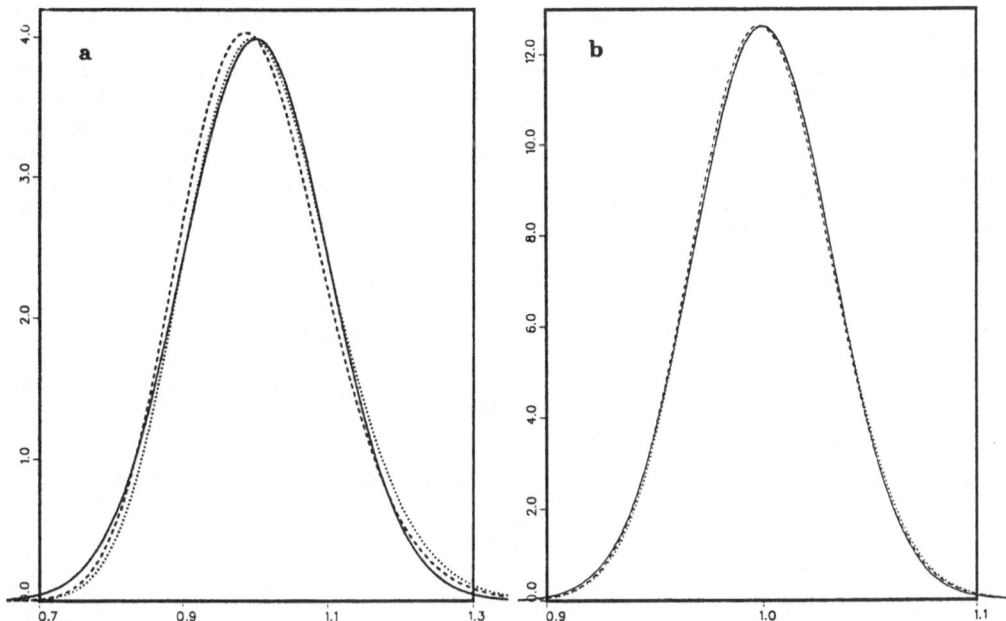

Figure 2a. **Comparison of the Chromatogram given by the Equilibrium-dispersive Model of Chromatography with a Gaussian profile.** Dimensionless plot of $C_d = Ct_R/A_p$ versus $t_d = t/t_R$. *Solid line:* Gaussian profile with $N = 100$ theoretical plates. *Dotted line:* equilibrium-dispersive model with an "open-open" boundary condition and $N_{Disp} = \frac{Lu}{2D_a} = 100$. *Dashed line:* equilibrium-dispersive model with a δ pulse boundary condition and $N_{Disp} = \frac{Lu}{2D_a} = 100$.

Figure 2b. Same as Figure 2a, except $N = N_{Disp} = 1000$ theoretical plates.

Unless σ_d is very small (i.e., unless N is sufficiently large), the elution profile is not Gaussian (we show later that the profile is nearly Gaussian for $N = 100$).

When the boundary condition is a δ function at $z = 0$, the solution is different. It can be obtained by the inverse Laplace transform [16,17]:

$$C_d = \frac{1}{t_d} \sqrt{\frac{N}{2\pi t_d}} e^{-\frac{N}{2t_d}(1-t_d)^2} \tag{35}$$

The first and second moment of this solution are now:

$$\bar{t}_d = 1 \tag{36a}$$

$$\sigma_d^2 = \frac{1}{N} \tag{36b}$$

Comparing equations 32 and 35 shows that the ratio of the two solutions is $1/t_d$:

$$C_d(\text{pulse boundary condition}) = \frac{1}{t_d} C_d(\text{open-open boundary condition})$$

In Figures 2a and 2b, we compare the Gaussian profile and the exact solutions of the equilibrium-dispersive model with the two different boundary conditions, for columns having 100 (Fig. 2a) and 1,000 (Fig. 2b) theoretical plates, respectively. On each Figure, the dotted lines are the band profiles calculated as the solution of the equilibrium-dispersive model with the open-open boundary condition (Eqn. 32), the dashed line is the solution of the same model with the δ pulse boundary condition (Eqn. 35), and the solid line is the Gaussian approximation (Eqn. 34). In Figure 2a, for $N = 100$, there is a slight difference between the two band profiles, especially when the time is far from the retention time of the band maximum, t_R. In Figure 2b, for $N = 1000$, the difference between the two solutions is negligible.

4.3.2) Open-Closed Boundary Condition. Another analytical solution can be obtained by assuming a semi- infinite column which stretches from $z = 0$ to infinity in the positive direction. This is the "close-open" boundary condition, which corresponds to:

$$\text{At } z = 0, \quad n\delta(t) = (uC - D_a\frac{\partial C}{\partial z})|_{z\to 0}\epsilon S \quad \text{for} \quad 0 < t \leq t_p \tag{37a}$$

$$C = 0 \qquad\qquad t > t_p$$

This boundary condition is equivalent to the Danckwerts [18] boundary condition:

$$uC_0 = uC - D_a\frac{\partial C}{\partial z}|_{z\to 0} \quad \text{for} \quad 0 < t \leq t_p \tag{37b}$$

$$0 = uC - D_a\frac{\partial C}{\partial z}|_{z\to 0} \quad \text{for} \quad t > t_p \tag{37c}$$

$$\text{and:} \quad z \to \infty \quad C \to 0 \tag{37d}$$

In dimensionless form, these equations become:

$$\text{At } z = 0, \quad \delta(t_d) = C_d - \frac{1}{2N}\frac{\partial C_d}{\partial z_d}|_{z_d\to 0} \tag{37e}$$

$$\text{and for} \quad z_d \to \infty \quad C_d \to 0 \tag{37f}$$

With this last boundary condition, the solution of equation 30 has been derived by Villermaux and Van Swaaig [19], using the inverse Laplace transform:

$$C_d = \sqrt{\frac{2N}{\pi t_d}}e^{-\frac{N}{2t_d}(1-t_d)^2} - 2Ne^{2N}erfc\left[\frac{1+t_d}{2}\sqrt{\frac{2N}{t_d}}\right] \tag{38}$$

where erfc is the complementary error function. The first and second moments of the band profile are:

$$\bar{t}_d = 1 + \frac{1}{2N} \tag{39a}$$

$$\sigma_d^2 = \frac{1}{N} + \frac{3}{4N^2} \tag{39b}$$

4.3.3) Closed-Closed Boundary Condition. For a finite column of length L, the boundaries are closed for dispersion at the column inlet and outlet. In dimensionless form, this "closed-closed" boundary condition is written:

$$\text{At column inlet:} \quad \delta(t_d) = C_d - \frac{1}{2N}\frac{\partial C_d}{\partial z_d}|_{z_d\to 0} \tag{40a}$$

$$\text{At column outlet} \quad \frac{\partial C_d}{\partial z_d}|_{z_d\to 1} = 0 \tag{40b}$$

The solution of equation 30 for these boundary conditions has been derived by Otake and Kunigita [20]:

$$C_d = exp\left[N(1 - \frac{t_d}{2})\right] \sum_{n=1}^{\infty} \frac{[2N sin\delta n + 2\delta n cos\delta n]\delta n}{\delta n^2 + N^2 + 2N} e^{-\frac{\delta n^2 t_d}{2N}} \tag{41a}$$

where δn is the nth root of the transcendental equation:

$$cot\delta n = \frac{\delta n}{2N} - \frac{N}{2\delta n} \tag{41b}$$

The first and second moment of the concentration distribution given by equation 41a are:

$$\bar{t}_d = 1 \tag{42a}$$

$$\sigma_d^2 = \frac{1}{N} - \frac{1}{2N^2}(1 - e^{-2N}) \tag{42b}$$

The differences between the four solutions derived for equation 30 are very close, unless the plate number is small.

4.3 - SOLUTION OF THE LUMPED KINETIC MODEL.

In the Lumped Kinetic Model of chromatography, the contributions of all the mechanisms which are involved in the band broadening due to a slow kinetics are lumped in a single coefficient. Depending on the main cause of sluggishness in reaching equilibrium in the column, we can distinguish several kinetic models.

4.3.1) The Reaction-dispersive Model. In this model we assume that the contributions of the mass transfer resistances are negligible, but that the kinetics of adsorption/desorption is slow. So the behavior of the chromatographic system is described by a mass balance equation:

$$\frac{\partial C}{\partial t} + F\frac{\partial C_s}{\partial t} + u\frac{\partial C}{\partial z} = D_L\frac{\partial^2 C}{\partial z^2} \tag{43}$$

where C_s is the concentration of the solute in the stationary phase and D_L accounts for the axial dispersion. If we assume now that the kinetics of adsorption/desorption is of the first order, we have the kinetic equation:

$$\frac{\partial C_s}{\partial t} = k_a C - k_d C_s \tag{44}$$

where k_a and k_d are the rate constant for the adsorption and the desorption, respectively.

Lapidus and Amundson [11] have derived the analytical solution of this reaction-dispersive model of chromatography where the reaction kinetics is linear (Equations 43 and 44), for any injection profile and for a column of infinite length. The solution derived by Giddings and Eyring [8] (Equation 24 or 26) is a special case of the solution derived by Lapidus and Amundson [11], although these two solutions have been derived from two completely different approaches. In this latter solution, the contribution of the axial dispersion is also included, besides that of the adsorption/desorption kinetics, while the axial dispersion is ignored in the treatment of Giddings and Eyring [8].

4.3.2) The Transport-dispersive Model. In this model we assume that the kinetics of adsorption/desorption is infinitely fast but that the mass transfer kinetics is not. More specifically, the mass transfer kinetics of the solute to the surface of the adsorbent is given by either the liquid film linear driving force model or by the solid film linear driving force model. In the former case and instead of Equation 44 we have for kinetic equation:

$$\frac{\partial C_s}{\partial t} = k_m'(C - C^*) \tag{45}$$

where C^* is the solute concentration in the mobile phase which is in equilibrium with the solid phase concentration C_s. Thus, $C^* = C_s/a$ where a is the slope of the linear isotherm ($a = k_0'/F$, F, phase ratio), and k_m' is the apparent mass transfer coefficient.

In the latter case (Solid film linear driving force model of mass transfer kinetics), we have:

$$\frac{\partial C_s}{\partial t} = k_m(q - C_s) \tag{46}$$

where q is the solute concentration in the stationary phase which is in equilibrium with the mobile phase concentration C. Thus, $(q = aC)$ and k_m is the apparent mass transfer coefficient.

In linear chromatography, the solid film driving force model (Equation 46), and the liquid film driving force model (Equation 45) are special cases of the first order linear kinetics (Equation 44), since this last equation (Eqn. 44) can be rewritten as:

$$\frac{\partial C_s}{\partial t} = k_d(\frac{k_a}{k_d}C - C_s) = k_d(\frac{k_a}{k_d a}q - C_s) \tag{47}$$

The Equation 47 is equivalent to the Equation 46 provided that $k_d = k_m$ and that $k_a = ak_m$. So, in linear chromatography, the solid film driving force model is a special case of the linear kinetic model, with $k_d = k_m$ and $k_a = ak_m$.

Equation 45 can be rewritten as:

$$\frac{\partial C_s}{\partial t} = \frac{k_m'}{a}(q - C_s) \tag{48}$$

Thus, in linear chromatography the liquid film driving force model (Equation 45) is equivalent to the solid film driving force model (Equation 46) with $k_m' = ak_m$. These two models are special cases of the linear kinetic model, with $k_d = k_m'/a$ and $k_a = k_m'$. Thus, the solution of the kinetic model derived by Lapidus and Amundson [11], which was obtained for a first order linear kinetic model, is directly applicable to both the solid film and the liquid film driving force models. We supply now these equivalent solutions.

The solution of the solid film driving model is the same as the analytical solution of the linear kinetic model derived by Lapidus and Amundson [11], provided that k_a and k_d be replaced by ak_m and k_m, respectively. The solution of the liquid film driving force model is the same as the analytical solution of the linear kinetic model derived by Lapidus and Amundson [11], provided that k_a and k_d be replaced by k_m' and k_m'/a, respectively. For example, the analytical solution of Lapidus and Amundson [11] for an impulse injection, for the solid film driving force model can be written [21]:

$$C = C_0\Big[\frac{zt_p}{2t\sqrt{\pi D_L t}}e^{-\frac{(z-ut)^2}{4D_L t}-k_m k_0't} + \int_0^t \frac{zt_p}{2t'\sqrt{\pi D_L t'}}e^{-\frac{(z-ut')^2}{4D_L t'}}F(t')dt'\Big] \tag{49}$$

with:

$$F(t') = \sqrt{\frac{k_m^2 k_0' t'}{t - t'}}e^{-k_m[t+(k_0'-1)t']}I_1(2\sqrt{k_m^2 k_0' t'(t - t')}) \tag{50}$$

Equations 49 and 50 supply the concentration profile along the column (t constant, z variable). The elution profile can be obtained by letting $z = L$, which, after some rearrangements, can be written under dimensionless form as:

$$C_d = \frac{C t_R}{A_p} = (1 + k_0')\sqrt{\frac{N_{Disp}}{2\pi}}$$

$$\left[(\frac{1}{t_{0,d}})^{1.5} e^{-[N_m t_{0,d} + \frac{N_{Disp}(1 - t_{0,d})^2}{2 t_{0,d}}]} + \int_0^{t_d} (\frac{1}{t_{0,d}'})^{1.5} e^{-\frac{N_{Disp}(1 - t_{0,d}')^2}{2 t_{0,d}'}} F(t_{0,d}') dt_{0,d}' \right] \quad (51)$$

with:

$$F(t_{0,d}') = N_m \sqrt{\frac{t_{0,d}'}{k_0'(t_{0,d} - t_{0,d}')}} e^{-N_m[t_{0,d}' + \frac{t_{0,d} - t_{0,d}'}{k'}]} I_1(2 N_m \sqrt{t_{0,d}'(t_{0,d} - t_{0,d}')/k_0'}) \quad (52)$$

In these equation, I_1 is the First order Bessel Function, N_{Disp} and N_m are the numbers of dispersion units and of mass transfer units, respectively. The other parameters are dimensionless quantities:

$$t_{0,d} = \frac{t}{t_0} = t_d(1 + k_0') \quad (52a)$$

$$t_{0,d}' = \frac{t'}{t_0} = t_d'(1 + k_0') \quad (52b)$$

$$t_d = \frac{t}{t_R} \quad (52c)$$

$$t_d' = \frac{t'}{t_R} \quad (52d)$$

The number of Dispersion units or mixing stages is given by:

$$N_{Disp} = \frac{P_{e,z}}{2} = \frac{Lu}{2D_L} = \frac{L^2}{2D_L t_0} \quad (53)$$

where $P_{e,z}$ is the axial dispersion number or Peclet number. The number of mass transfer units is given by:

$$N_m = \frac{k_m k_0' L}{u} = k_m k_0' t_0 \quad (54)$$

Van Deemter et al. [21] have shown that, if the mass transfer kinetics is not very slow, the Equations 49 and 50 can be reduced to a Gaussian profile:

$$C = \frac{A_p}{(1 + k_0')\sqrt{2\pi(\sigma_1^2 + \sigma_2^2)}} e^{-\frac{(\frac{z}{u} - \frac{t}{1+k_0'})^2}{2(\sigma_1^2 + \sigma_2^2)}} = \frac{A_p}{(1 + k_0')\sqrt{2\pi(\sigma_1^2 + \sigma_2^2)}} e^{-\frac{((1 + k_0')z - ut)^2}{2u^2(1 + k_0')^2(\sigma_1^2 + \sigma_2^2)}} \quad (55)$$

with:

$$\sigma_1^2 = 2\frac{D_L z}{u^3} \quad (56a)$$

$$\sigma_2^2 = \frac{2 k_0' z}{(1 + k_0')^2 k_m u} \quad (56b)$$

From the concentration profile along the column, given by Equation 55, we can derive the corresponding elution profile by inserting $z = L$, and, using Equations 53 and 54, we can write it under a dimensionless form:

$$C_d = \frac{C * t_R}{A_p} = \frac{1}{\sqrt{2\pi[\frac{1}{N_{Disp}} + \frac{2}{N_m}(\frac{k_0'}{1+k_0'})^2]}} e^{-\frac{(1-t_d)^2}{2[\frac{1}{N_{Disp}} + \frac{2}{N_m}(\frac{k_0'}{1+k_0'})^2]}} \tag{57}$$

Comparing Equations 34 and 57 shows that:

$$\frac{1}{N_{apt}} = \frac{1}{N_{Disp}} + \frac{2}{N_m}\left(\frac{k_0'}{1+k_0'}\right)^2 \tag{58}$$

or:

$$H = \frac{2D_L}{u} + 2\left(\frac{k_0'}{1+k_0'}\right)^2 \frac{u}{k_0' k_m} \tag{59a}$$

According to Glueckauf [22,23], the lumped mass transfer coefficient, k_m is related to the film mass transfer coefficient and to the pore diffusion by the equation:

$$\frac{1}{k_0' k_m} = \frac{d_p}{6Fk_f} + \frac{d_p^2}{60FD_p} \tag{59b}$$

where k_f is the mass transfer coefficient for the film around the particles, D_p is the pore diffusion coefficient, $F = \frac{1-\epsilon_e}{\epsilon_e}$, ϵ_e is the external porosity, and d_p is the diameter of the particle. Equations 59a and 59b are very important relationships in linear chromatography. Combined, they constitute the famous **"Van Deemter Equation"**, which shows that the effects of the axial dispersion and of the mass transfer resistances are additive. This is the basis tenet of the equilibrium-dispersive Model of chromatography.

In Figure 3, we compare the band profiles obtained as the exact solution of the Equations 51 and 52 (dotted line) and a Gaussian profile (Equations 34 or 57). For the calculation of the exact solution of Equations 51 and 52, we have selected the following values: $N_{Disp} = 200$ and $(\frac{1+k_0'}{k_0'})^2 N_m = 400$. According to Equation 58, N_{apt} is then equal to 100. Thus, for the Gaussian profile a value of $N = 100$ is used. As can be seen in the Figure 3a, there is a good agreement between the two band profiles, although the column efficiency is only 100 theoretical plates. The difference between the exact solution of the lumped kinetic model and a Gaussian profile becomes rapidly insignificant for higher efficiencies. Even in Figure 3b, in the case when $N = 25$, there is a rather small difference.

5. The General Rate Model of Chromatography.

Chromatography involves an intricate combination of complex phenomena. In all the models we have considered until now, we have made considerable simplifications. Unfortunately, there are many steps in the migration of the solute molecules along the columns and it is often impossible to identify a single one as rate controlling.

The General Rate Model attempts to consider simultaneously all the contributions to the mass transfer kinetics. The first problem is to make a complete census of them. The general rate models consider usually the axial dispersion as defined above, the external film mass transfer

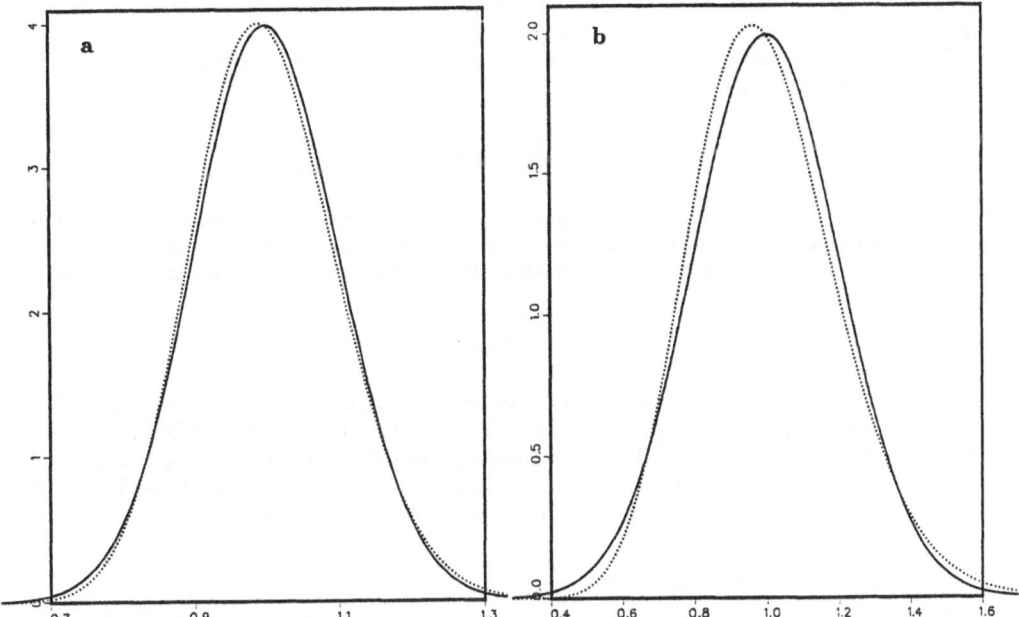

Figure 3. **Comparison of the Chromatogram given by the Lapidus and Amundson Model of Chromatography with a Gaussian Profile.** Dimensionless plot of C_d versus t_d.

Figure 3a: Solid line: Gaussian profile with $N = 100$. *Dotted line:* Lapidus and Amundson model with $N_{app} = 100$.

Figure 3b: Same as Figure 4a, with $N_{app} = N = 25$.

resistance, the intraparticle diffusion and the rate of adsorption/desorption. Since we consider now separately what is happening to the solute inside and outside the particle, we need to write two mass balance equations, one for the mobile phase flowing between particles, one for the stagnant mobile phase inside the particle.

In this model the mass balance equation in the mobile phase can be written as:

$$u\frac{\partial C}{\partial z} + \frac{\partial C}{\partial t} + F\frac{\partial \bar{q}}{\partial t} = D_L\frac{\partial^2 C}{\partial z^2} \tag{60}$$

where u is the linear velocity, $F = \frac{1-\epsilon_e}{\epsilon_e}$, and D_L is the axial dispersion coefficient, which is the sum of the molecular and eddy diffusion coefficients. $\bar{q}$ is the value of the stationary phase concentration, q, averaged over the entire particle. Calculated for a spherical particle it is:

$$\bar{q} = \frac{3}{R_p^3}\int_0^{R_p} r^2 q\,dr \tag{61}$$

The term $\frac{\partial \bar{q}}{\partial t}$ in Equation 60 is the rate of adsorption averaged over the particle. For a spherical particle it is given by $\frac{\partial \bar{q}}{\partial t} = 3/R_p M_F$ where M_F is the mass flux of solute from the bulk solution to the external surface of the particle. The boundary condition at the outer surface of the particle is given by:

$$M_F = D_p\frac{\partial C_p}{\partial r}\big|_{r=R_p} = k_f(C - C_p|_{r=R_p}) \tag{62}$$

where k_f is the external mass transfer coefficient and C_p is the concentration of the solute within the pores inside the particle, a function of r.

The differential mass balance of the solute inside the pores of the adsorbent particle is given by another partial differential equation:

$$\epsilon_p \frac{\partial C_p}{\partial t} + (1 - \epsilon_p)\frac{\partial C_s}{\partial t} = D_p\left(\frac{\partial^2 C_p}{\partial r^2} + \frac{2}{r}\frac{\partial C_p}{\partial r}\right) \tag{63}$$

where ϵ_p is the internal porosity of the particle, D_p is the diffusion coefficient of the solute in the particle pores, C_p is the concentration of the solute inside the pores, and C_s is the solute concentration at the stationary phase. Further, due to the symmetry condition, we have:

$$\frac{\partial C_p}{\partial r}\Big|_{r=0} = 0 \tag{64}$$

The solute concentrations in the stationary phase, C_s, and in the mobile phase, C_p, are related. If we assume that the kinetics of adsorption/desorption is infinitely fast, these concentrations are related through the (linear) adsorption isotherm equation, $C_s = K_a C_p$. If the kinetics of adsorption/desorption is too slow, then they are related through the kinetic equation of the adsorption/desorption. Assuming a first-order slow adsorption/desorption kinetics we have:

$$\frac{\partial C_s}{\partial t} = k_{ads}(C_p - C_p^*) = k_{ads}\left(C_p - \frac{C_s}{K_a}\right) \tag{65}$$

where k_{ads} and K_a are the adsorption rate constant and the adsorption equilibrium constant, respectively. Finally, for a pulse injection, the initial and boundary conditions are:

$$C(z,0) = 0 \tag{66a}$$
$$C_p(r,z,0) = 0 \tag{66b}$$
$$C(0,t) = C_0 \quad \text{for} \quad 0 \le t \le t_p \tag{66c}$$
$$C(0,t) = 0 \quad \text{for} \quad t \ge t_p \tag{66d}$$

For the general rate model, which accounts for the axial dispersion, the film mass transfer resistance, the pore diffusion and a first-order, slow kinetics of adsorption/desorption (Equations 60 to 66), a closed form analytical solution is impossible to derive in the time domain.

Kucera [24] and Kubin [25] have derived the solution in the Laplace domain. From that solution, they have derived the expressions for the first five statistical moments. For a linear isotherm, this model has been studied extensively in the literature, as well as its extension using a macro-micropore diffusion model with external film mass transfer resistance [26]. All these studies use the Laplace domain solution and moment analysis.

5.1 - ANALYTICAL SOLUTION OF THE GENERAL RATE MODEL IN A PARTICULAR CASE.

If we assume that in the same time the axial dispersion can be neglected (i.e., $D_L = 0$ in Equation 60), and the kinetics of adsorption/desorption can be considered as infinitely fast, instead of Equation 65 we have:

$$C_s = K_a C_p \tag{67}$$

Now, we have a simplified general rate model which considers only film mass transfer and pore diffusion. The advantage of this model is that it is possible to calculate the inverse Laplace transform of its solution in the Laplace domain, provided that we may assume that the injection is periodical. Carta [27] has obtained an analytical solution for a periodic injection in the form of a rapidly convergent infinite series. The solution in the time domain is derived by applying the residue theorem to the inversion integral. Although the solution is strictly valid only for periodical injections, it can be used as a very good approximation to calculate the response of the column to a single injection by choosing the injection period long enough, so that the effluent concentration has reduced to zero at the end of first cycle. The elution profile, at the end of a column of length L, for an injection pulse of width t_p, and a period t_C can be written as:

$$\frac{C}{C_0} = \frac{t_p}{t_C + t_p} +$$

$$\frac{2}{\pi} \sum_1^\infty \left[\frac{1}{k} \; exp(-a_k N_f) \; sin\left(\frac{\pi k t_p}{t_C + t_p}\right) \; cos\left(\frac{2\pi k(t - t_0 - t_p/2)}{t_C + t_p} - N_f b_k\right)\right] \quad (68)$$

where the coefficients a_k and b_k are given by:

$$a_k = \frac{(\gamma_k - \lambda_k)[\gamma_k - (1 - 5N_f/N_p)\lambda_k] + \eta_k^2}{[\gamma_k - (1 - 5N_f/N_p)\lambda_k]^2 + \eta_k^2} \quad (69)$$

and:

$$b_k = \frac{(\gamma_k - \lambda_k)\eta_k + [\gamma_k - (1 - 5N_f/N_p)\lambda_k]\eta_k}{[\gamma_k - (1 - 5N_f/N_p)\lambda_k]^2 + \eta_k^2} \quad (70)$$

and where:

$$\gamma_k = \sqrt{k/2r}(sinh\sqrt{2k/r} + sin\sqrt{2k/r}) \quad (71)$$

$$\eta_k = \sqrt{k/2r}(sinh\sqrt{2k/r} - sin\sqrt{2k/r}) \quad (72)$$

$$\lambda_k = (cosh\sqrt{2k/r} - cos\sqrt{2k/r}) \quad (73)$$

$$r = \frac{(t_C + t_p)N_p}{30k_1\pi t_0} \quad (74)$$

$$k_1 = F[\epsilon_p + (1 - \epsilon_p)K_a] \quad (75)$$

N_f is the number of mass transfer units corresponding to diffusion across the film around the particle:

$$N_f = \frac{6Fk_f L}{d_p u} = \frac{6Fk_f t_0}{d_p} \quad (76)$$

and N_p is the number of mass transfer units in the pores which is given by:

$$N_p = \frac{60FD_p L}{d_p^2 u} = \frac{60FD_p t_0}{d_p^2} \quad (77)$$

As shown by Glueckauf [22,23], the effects of the different phenomena contributing to band broadening are additive. Since in this simplified model the axial dispersion is neglected and the kinetics of adsorption/desorption is also ignored, the apparent number of transfer units can be evaluated through Eqs. 59a and 59b, and we obtain:

$$\frac{1}{N_{app}} = 2(\frac{k_1}{1 + k_1})^2(\frac{1}{N_f} + \frac{1}{N_p}) \quad (78)$$

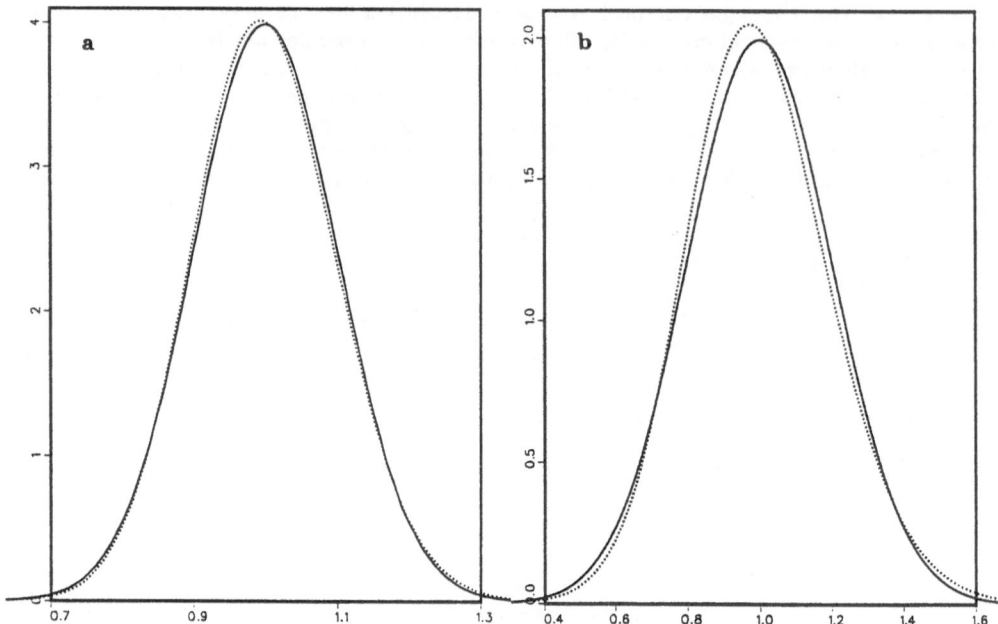

Figure 4. **Comparison between the Chromatogram given by the Film Mass Transfer - Pore Diffusion Model of Chromatography and a Gaussian Profile.** Dimensionless plot of C_d versus t_d.

Figure 4a: Solid line: Gaussian profile with $N = 100$. *Dotted line:* Carta's solution [27] with $N_{app} = 100$.

Figure 4b: Same as Figure 4a, with $N_{app} = 25$.

In Figure 4, we compare the solution derived by Carta [27] and calculated from Equations 68 to 75 (dotted line) with a Gaussian profile (Eqn. 34, solid line) for column efficiencies $N = 100$ (Figure 4a) and $N = 25$ (Figure 4b). To calculate the Carta solution, we have chosen for Figure 4a the following values of the expressions: $\left(\frac{k_1+1}{k_1}\right)^2 N_f = 400$ and $\left(\frac{k_1+1}{k_1}\right)^2 N_p = 400$. With these values, and according to equation 78, N_{apt} becomes equal to 100. The Figure 4a shows clearly that, although the column efficiency is only 100 plates, there is a very good agreement between the two band profiles. Even with 25 plates (Figure 4b), the difference would be difficult to establish through experimental measurements. This demonstrates that this solution, as well as the Lapidus and Amundson solution, can be approximated very well with a Gaussian band profile for any column efficiency of practical interest.

5.2 - INVERSE LAPLACE TRANSFORM OF THE GENERAL RATE MODEL SOLUTION.

Rosen [28] was the first to find and calculate a solution for the inverse Laplace transform of the solution of the general rate model in the Laplace domain. He obtained it in the form of an infinite integral, for the case of a breakthrough curve (step input), and he used contour integration for the integration, assuming (i) that axial dispersion can be neglected (i.e., $D_L = 0$ in Equation 60) and (ii) that the kinetics of adsorption/desorption is infinitely fast (i.e., using eqn. 67 instead of eqn. 65). Hence, he considered in his solution only the effects of intraparticle diffusion and of the external film resistance. Thus, Rosen's model is equivalent to Carta's [27].

Pellett [29] and Rasmuson et al. [30] extended the solution of Rosen by including the axial dispersion, but still assuming an infinitely fast kinetics of adsorption/desorption. Later, Rasmuson et al. [31] extended their solution to the case of the general rate model (Equations 60 to 66), which include the axial dispersion, the film mass transfer resistance, the pore diffusion, and a first- order slow kinetics of adsorption/desorption, for a breakthrough curve.

All the solutions discussed above are given in the form of infinite integrals. In the case of a pulse injection, a similar analytical solution has not been derived yet, except for the Carta's solution of Rosen's model. However, the numerical evaluation of the inverse Laplace transform is possible. It has been calculated in the case of the general rate model (i.e., Equations 60 to 66) by Lenhoff [32]. The numerical integral is given by:

$$\theta(\tau)|_{\xi=1} = \frac{1}{\pi} \int_0^\infty \mathcal{R}(\bar{\theta}(i\omega)|_{\xi=1} e^{i\omega\tau}) d\omega \tag{79}$$

and

$$\bar{\theta}(s)|_{\xi=1} = \frac{be^{2N_{Disp}m_2}}{m_1^2 - m_2^2 e^{-2N_{Disp}b}} \tag{80}$$

where:

$$N_{Disp} = \frac{Pe,z}{2} = \frac{uL}{2D_L} \tag{80a}$$

$$m_1 = \frac{1+b}{2} \tag{80b}$$

$$m_2 = \frac{1-b}{2} \tag{80c}$$

$$b = \sqrt{1 - 4\frac{\phi}{4N_{Disp}^2}} \tag{81}$$

$$\phi = -2sN_{Disp} - \frac{2N_{Disp}N_f}{1 + \frac{5N_f}{N_p(w\coth w - 1)}} \tag{82}$$

$$w = \sqrt{15Fs/N_p(\epsilon_p + \frac{D}{s + \frac{D}{K_a(1-\epsilon)}})} \tag{83}$$

N_f and N_p are defined in Equations 76 and 77 and $D = \frac{k_{ads}d_p}{u}$ is Dankohler number.

In Figure 5, we compare a Gaussian profile (eqn. 34, solid line) corresponding to a column efficiency equal to 100 (Figure 5a) and 25 (Figure 5b) theoretical plates with the band profile calculated using the numerical inversion of the Laplace transform of the general rate model (Equations 79 to 83) (dotted line). For the solution of the general rate model we have chosen the values of N_{Disp}, N_f and N_p so that $N_{app} = 100$. As we can see, the two band profiles are still very close for a column efficiency of 25 theoretical plates.

5.3 - MOMENT ANALYSIS AND PLATE HEIGHT EQUATIONS.

As mentioned earlier, a closed form, analytical solution of the Inverse Laplace transform has not been derived yet in the case of the general rate model for a pulse injection. However, Kucera [24] and Kubin [25] have derived the first five moments of the solution. These moments can be easily calculated from their solution of the general rate model giving the band profile in the Laplace domain (Equations 60 to 66).

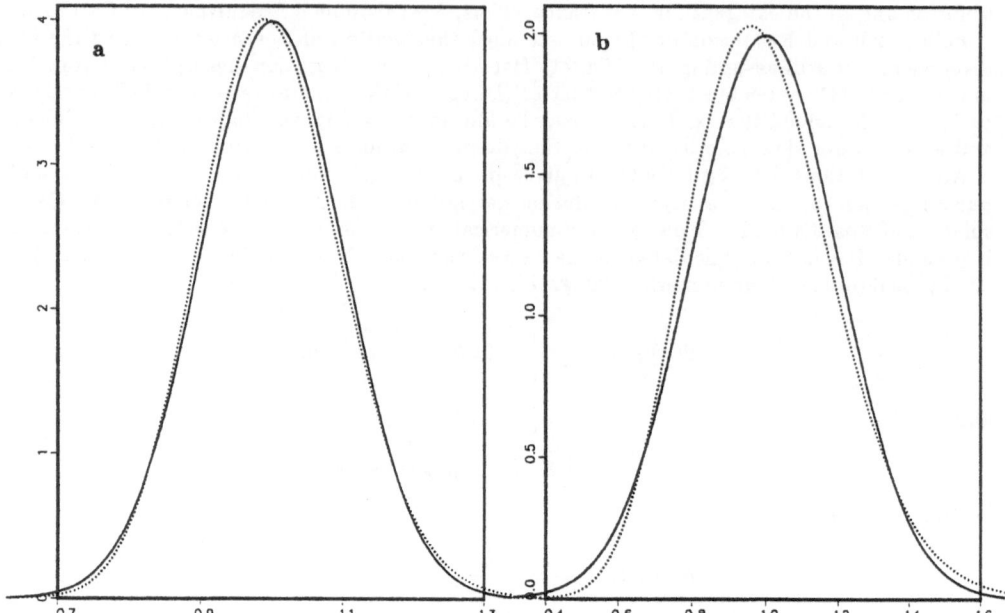

Figure 5. **Comparison between the Chromatogram given by the General Rate Model of Chromatography and a Gaussian Profile.** Dimensionless plot of C_d versus t_d.

Figure 5a: Solid line: Gaussian profile with $N = 100$. *Dotted line*: General rate model solution with $N_{app} = 100$.

Figure 5b: Same as Figure 5a, with $N_{app} = 25$.

5.3.1 - Moments of a Chromatographic Peak. The nth moment of the band profile at the exit of a chromatographic bed of length $z = L$ is given by

$$\mathcal{M}_n = \int_0^\infty C(t, \, z = L) \, t^n \, dt \tag{84}$$

The nth absolute or normalized moment is given by:

$$\mu_n = \frac{\mathcal{M}_n}{\mathcal{M}_0} = \frac{\int_0^\infty C(t, \, L) \, t^n \, dt}{\int_0^\infty C(t, \, L) \, dt} \tag{85}$$

and the nth central moment is given by:

$$\bar{\mu}_n = \frac{\int_0^\infty C(t, \, L) \, (t - \mu_1)^n \, dt}{\int_0^\infty C(t, \, L) \, dt} \tag{86}$$

The moments of the response of the chromatographic column to a pulse injection can be derived from the analytical solution in the Laplace domain by using the theorem of Van der Lann [33]:

$$\mathcal{M}_n = (-1)^{n-1} lim_{s \to 0} \frac{\partial^n \bar{C}(s, \, L)}{\partial s^n} \tag{87}$$

The first moment of the general rate model solution (Equations 60 to 66) is given by:

$$t_R = \mu_1 = \frac{L}{u}(1 + k_1) + \frac{t_p}{2} \tag{88}$$

where k_1 is given by:

$$k_1 = F(\epsilon_p + (1 - \epsilon_p)K_a) \tag{89}$$

The second moment is given by:

$$\sigma^2 = \bar{\mu}_2 = \mu_2 - \mu_1^2 = 2\frac{L}{u}[\frac{D_L}{u^2}(1 + k_1)^2 + \frac{k_1^2}{F}(\frac{d_p^2}{60D_p} + \frac{d_p}{6k_f} + \frac{2Fu(1 - \epsilon_e)^2 K_a^2}{k_{ads}}] + \frac{t_p^2}{12} \tag{90}$$

5.3.2 - Plate Height Equation in the General Rate Model. From the moment analysis of the solution of the general rate model in the Laplace domain one can obtain the expression for the column HETP. By definition of height equivalent to a theoretical plate, we have:

$$H = \frac{L}{N} = \frac{\sigma^2}{t_R^2}L = \frac{\bar{\mu}_2}{\mu^2}L \tag{91}$$

If the width of the injection pulse is negligible, we obtain from Equations 88 and 90:

$$H = \frac{2D_L}{u} + 2(\frac{k_1}{1 + k_1})^2 \left[\frac{ud_p^2}{60FD_p} + \frac{ud_p}{6Fk_f} + (\frac{k_p}{1 + k_p})^2 \frac{u}{Fk_{ads}} \right] \tag{92}$$

with:

$$k_p = \frac{1 - \epsilon_p}{\epsilon_p}K_a \tag{92a}$$

or:

$$\frac{1}{N_{app}} = \frac{1}{N_{Disp}} + 2\left(\frac{k_1}{1 + k_1}\right)^2 \left[\frac{1}{N_p} + \frac{1}{N_f} + \left(\frac{k_p}{1 + k_p}\right)^2 \frac{1}{N_{rea}} \right] \tag{93}$$

where N_{Disp} is given by equation 53, N_f and N_p by equations 76 and 77, and $N_{rea} = \frac{Fk_{ads}L}{u}$. Equation 93 is used for the calculation of N_{app} when the profile predicted by a particular model is compared to a Gaussian profile. This equation is general. It includes the numbers of dispersion units, film mass transfer units, pore mass transfer units, and adsorption/desorption units. Equations 58 and 78 are particular cases of this general equation.

Equation 92 shows that the column HETP is the sum of the contributions of the axial dispersion (molecular diffusion and eddy diffusion), the film mass transfer resistance, the pore diffusion and a slow kinetics of adsorption/desorption. By comparing Equations 92 and 59a, we obtain:

$$\frac{F}{k_0'k_m} = \frac{d_p}{6k_f} + \frac{d_p^2}{60D_p} + (\frac{k_p}{1 + k_p})^2 \frac{1}{k_{ads}} \tag{94}$$

Equation 94 shows that the lumped mass transfer coefficient in the solid film driving force model is related to the film mass transfer coefficient, the intraparticle diffusion and the adsorption rate constant. It is a more complete form of equation 59b, an equation to which it reduces when the kinetic of adsorption/desorption is infinitely fast.

Equation 92 can be rewritten under a dimensionless form:

$$h = \frac{2}{\nu}\frac{D_L}{D_m} + 2(\frac{k_1}{1 + k_1})^2 \left[\frac{\nu D_m}{60FD_p} + \frac{\nu}{6FSh} + (\frac{k_p}{1 + k_p})^2 \frac{1}{FD} \right] \tag{95}$$

where $h = H/d_p$ is the reduced plate height, $\nu = Pe_p = \frac{ud_p}{D_m}$ is the reduced velocity, $Sh = \frac{k_f d_p}{D_m}$ is the Sherwood number and $D = \frac{k_{ads}d_p}{u}$ is the Dankohler number.

Equations 92 and 95 are the general equations for plate height and reduced plate height equations. All the plate height equations which have been reported and discussed in the literature can be derived from these two equations, themselves stemming from the general rate model of chromatography, the most sophisticated and comprehensive model of chromatography so far. The differences between the various expressions reported in the literature for the dependence of the column HETP on the experimental parameters result from the use of different expressions for the axial dispersion D_L and for the mass transfer coefficient, k_f. For example, Van Deemter et al. [21] have assumed that the axial dispersion coefficient is given by:

$$D_L = \gamma D_m + \lambda u d_p \tag{96}$$

The first term of equation 96 accounts for the molecular diffusion and the second term for the eddy diffusion. Inserting Equation 96 in Equation 92, we obtain:

$$H = A + \frac{B}{u} + Cu \tag{97}$$

Equation 97 is the classical Van Deemter equation.

5.3.3 - Coupling of Molecular and Eddy Diffusion. Giddings [7] has argued that the Van Deemter Equation (Eqn. 97) is too simplistic, because it ignores the coupling which exists between the flow velocity and the radial diffusion in the void space of the packing, around the particles. The engine for this coupling is the eddy diffusion and the turbulence it causes ("rugged flow" [34]). Since both the eddy diffusion and the radial diffusion are simultaneously responsible for the transfer of molecules between different flow paths of unequal velocity, the parameter A in the Van Deemter equation should depend on the velocity. Giddings proposed to replace the term A by a term $\frac{a}{1+bu^{-1}}$. Huber et al. [35] have proposed to replace the Equation 96 by:

$$D_L = \gamma D_m + \frac{2\lambda_1 d_p u}{1 + \lambda_2 (D_m/ud_p)^{0.5}} \tag{98}$$

and also to assume that the mass transfer coefficient in the mobile phase (i.e., k_f in Equation 92) is velocity dependent, by writing that k_f is proportional to $u^{0.5}$.

In order to include the coupling between the rugged laminar flow in a porous medium and the molecular diffusion, Horvath and Lin [36] used a model in which each particle is supposed to be surrounded by a stagnant film of thickness δ. Axial dispersion occurs only in the fluid outside this stagnant film, whose thickness decreases with increasing velocity. In order to obtain an expression for δ, they used the Pfeffer and Happel "free - surface" cell model [37] for the mass transfer in a bed of spherical particles. According to the Pfeffer equation, at high values of the reduced velocity the Sherwood number, and therefore the film mass transfer coefficient, is proportional to $\nu^{0.33}$.

$$Sh = k_f \frac{d_p}{D_m} = \Omega \nu^{0.33} \tag{99}$$

Horvath and Lin assume δ to be equal to the thickness of the Nernst diffusion layer, D_m/k_f. Then:

$$\delta = \frac{D_m}{k_f} \tag{100a}$$

and using equation 99, we obtain:

$$\delta = \frac{D_m}{k_f} = \frac{d_p}{\Omega Pe^{0.33}} \tag{100b}$$

Introducing this value of the stagnant film thickness, they derived a value of the axial dispersion coefficient:

$$D_L = D_m + \frac{\lambda d_p u}{1 + \Omega \nu^{-0.33}} \tag{101}$$

Inserting Equations 99 and 101 in Equation 95 gives the Horvath and Lin equation [36].

Arnold et al. [38] argued that Equation 99 is valid only at large values of the reduced velocity. According to their work, the Horvath and Lin equation is valid only for $\nu > 50$. When $\nu < 50$, according to Pfeffer et al. [37], the Sherwood number is constant and so is the film mass transfer coefficient. Furthermore, since k_f is independent of the reduced velocity at low values of the reduced velocity, according to Equation 100a, the stagnant boundary layer thickness, δ, is nearly constant and independent of the velocity. Under these conditions, Equation 101 reduces to Equation 96. Thus, at low reduced velocities the Van Deemter equation is more accurate than the Horvath and Lin Equation. At high reduced velocities, Horvath and Lin [36] have showed that their equation reduces to the Knox [39] empirical equation:

$$h = a\nu^{0.33} + \frac{b}{\nu} + c\nu \tag{102}$$

where a, b and c are constant parameters.

5.3.4 - The Golay Plate Height Equation. By integration of the mass balance equation, Golay [40] derived a plate height equation for open tubular columns in gas liquid chromatography. The column is assumed to be a linear, cylindrical tube with a constant circular cross-section. The mobile phase is a non-compressible gas. The stationary phase is a liquid film of constant thickness coating the inner wall of the column. The equation applies also if the column is coiled (provided the coil diameter is much larger than the column inner diameter) and that a correction can be applied to take the effect of the gas compressibility *into account, if needed*.

The *equation is written*:

$$H = \frac{2D_m}{u} + \frac{1 + 6k_0' + 11k_0'^2}{96(1 + k_0')^2} \frac{u d_p^2}{D_m} + \frac{k_0'^3}{24(1 + k_0')^2} \frac{u d_p^2}{F^2 a^2 D_l} \tag{103a}$$

where k_0' is the retention factor, d_p is the column diameter, D_m and D_l are the molecular diffusivity of the solute in the gas and liquid phases, respectively. a is the partition coefficient, or ratio of the equilibrium concentrations of the solute in the two phases, and F is the ratio of the cross-section areas of the liquid (stationary) and gas (mobile) phases. The second part of the last term is often replaced by $u d_f^2/k_0'^2 D_l$. The equation is now classically written:

$$H = \frac{2D_m}{u} + \frac{1 + 6k_0' + 11k_0'^2}{96(1 + k_0')^2} \frac{u d_p^2}{D_m} + \frac{k_0'}{24(1 + k_0')^2} \frac{u d_f^2}{D_l} \tag{103b}$$

The first term of the RHS of equations 103a and 103b accounts for static, axial, molecular diffusion. There is no eddy diffusion or "A" term in the equations 103a and 103b, in contrast to equation 97, as expected in the case of a cylindrical open tube, where the flow is ideally laminar. The second term in these equations accounts for the dynamic diffusion related to the non-homogeneity of the gas phase (because of the Poiseuille radial flow profile) and its coupling with the retention. This term is absent in other plate height equations, although it could be incorporated in the "C" term of equation 97. Originally [40], it was anticipated that this term would be small compared to the last term, the resistance to mass transfer in the stationary phase, or "C" term of equation 97. However, it has been possible to prepare and use

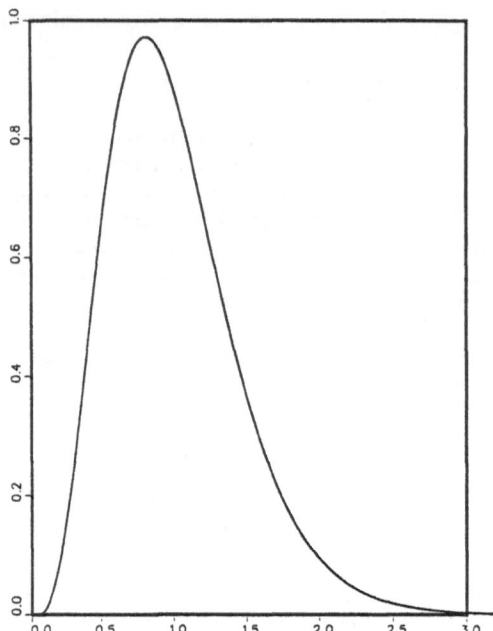

Figure 6. **Effect of a slow rate of the adsorption/desorption kinetics on the shape of the band profile and its asymmetry.** Diensionless plot of C_d versus t_d. Chromatogram calculated with the Lapidus and Amundson model, with $N_{app} = 5$.

columns with very thin films of liquid phase. With most modern open tubular columns for gas chromatography, the second term is more important than the third one [41].

6. Sources of Band Asymmetry and Tailing in Linear Chromatography.

As we discussed above, even when the column efficiency is only 100 theoretical plates, the elution profile predicted by any model of chromatography can be represented more than reasonably well by a Gaussian profile. When the efficiency is in the 1,000 theoretical plates or higher, as typical of modern analytical chromatography, there is no possibility to ever be able to show a difference between the actual band profile and a Gaussian curve. Nevertheless, it is not uncommon to observe that an actual band profile is unsymmetrical. A complete theory of linear chromatography should be able to justify this observation and offers reasons to explain the apparent contradiction and possible remedies in serious cases.

If the mass transfer kinetics becomes extremely slow, the peak exhibits a significant degree of asymmetry and tailing. In Figure 6 we show a band profile simulated using the Lapidus and Amundson solution in a case when the apparent column efficiency is only 5 theoretical plates. We see that a slow mass transfer kinetics could be responsible for considerable band broadening and for a tailing peak, but chromatograms similar to the one in Figure 6 are observed only in unusual cases, such as in affinity chromatography, where the mass transfer or adsorption/desorption kinetics is very slow. The tailings which are observed in most practical cases of analytical chromatography cannot be explained by a slow kinetics alone.

One model which is widely used for the evaluation of band asymmetry is the exponentially modified Gaussian (EMG) function [42-44]. It is the convolution of the unit area Gaussian

Function and of an exponential decay of unit area. By definition, the convolution of two functions $\phi_1(t)$ and $\phi_2(t)$ is the following function $\Phi(t)$:

$$\Phi(t) = \int_0^t \phi_1(t - t')\phi_2(t')dt' \tag{104}$$

The Gaussian profile is written:

$$G(t) = \frac{1}{\sigma\sqrt{2\pi}}e^{-\frac{t_R^2}{2\sigma^2}(\frac{t}{t_R}-1)^2} \qquad \text{for} \quad t > 0 \tag{105a}$$

$$G(t) = 0 \qquad \text{for} \quad t < 0 \tag{105b}$$

The exponential decay is written:

$$H(t) = \frac{1}{2\tau}e^{-t/\tau} \qquad \text{for} \quad t > 0 \tag{106a}$$

$$H(t) = 0 \qquad \text{for} \quad t < 0 \tag{106b}$$

According to equation 104, the convolution of these two functions is written:

$$C = \frac{1}{\tau\sigma\sqrt{2\pi}} \int_0^t exp\left(-\frac{t_R^2}{2\sigma^2}(\frac{t - t'}{t_R} - 1) \right)e^{-\frac{t'}{\tau}}dt' \tag{107a}$$

or:

$$C = \frac{1}{2\tau}e^{[0.5(\sigma/\tau)^2 - \frac{t-t_R}{\tau}]}(er\!f[\frac{1}{\sqrt{2}}(\frac{t_R}{\sigma} + \frac{\sigma}{\tau})] + er\!f[\frac{1}{\sqrt{2}}(\frac{t - t_R}{\sigma} - \frac{\sigma}{\tau})]) \tag{107b}$$

The first error function approaches unity when N exceeds 5. Such a model could only be justified if extra-column sources of band broadening are at work and if they are important. Such sources may include a tailing injection profile. As an example, we show in Figure 7 a comparison between a Gaussian band profile calculated using Equation 34 (solid line) and a band calculated with an EMG function (Equation 107b), for $\frac{\tau}{t_R} = 0.02$ (dotted line), both profiles corresponding to a column efficiency of $N = 5000$ theoretical plates. We see that, as expected, the EMG function can produce significant band asymmetry and tailing, even at high efficiency.

Peak tailing can also be explained by the heterogeneity of the surface of the stationary phase. For the sake of simplicity in this illustration, assume the existence of two different adsorption sites on the surface (e.g., alkyl groups bonded to a silica surface and residual silanols). These sites include a first type of sites, or ordinary sites, covering most of the surface and a very small proportion of highly active sites. On such a surface model, peak tailing could happen for one of two different reasons. First, the isotherm behavior may be linear for the ordinary sites and non- linear for the high energy sites. Second, the kinetics of adsorption/desorption on the high energy sites may be too slow. We discuss now these two options.

In the first case, since the active sites occupy a very small fraction of the surface, and since the interaction between active sites and solute molecules is very strong, the active sites will become saturated for a rather low concentration of the solute, a much smaller one than what would be needed to saturate the ordinary types of sites. In other words, although the isotherm for the main adsorption sites could be considered as linear, the same is no longer true for the high active site isotherm. It is not linear, and we should use a non-linear isotherm instead, e.g., a Langmuir isotherm.

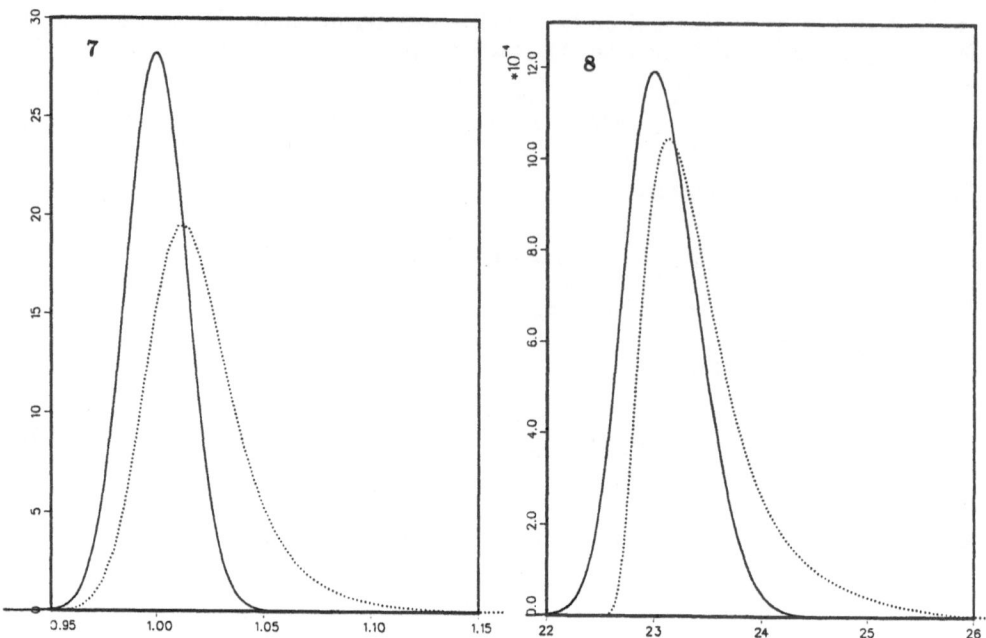

Figure 7. **Comparison between the Exponentially Modified Gaussian (EMG) and a Gaussian Profile.** Dimensionless plot of C_d versus t_d. *Solid line*: Gaussian profile with $N = 5000$. *Dotted line*: EMG function with $N = 5000$ and $\tau/t_R = 0.02$.

Figure 8. **Comparison between the Band Profiles Calculated with a One-site and a Two-sites Model, using a Langmuir Isotherm.** Dimensionless plot of C_d versus t_d. *Solid line:* one-site Langmuir isotherm model $q = \frac{24C}{1+6C}$. *Dotted line*: two-sites bilangmuir isotherm: $q = \frac{24C}{1+6C} + \frac{2.4C}{1+600C}$. In both cases, $N = 5000$.

In Figure 8, we compare two band profiles corresponding to the same amount of solute and calculated as numerical solutions of the equilibrium-dispersive model for a non-linear isotherm [45]. These profiles were calculated using a one-site (solid line) and a two- sites isotherm model (dotted line), respectively. For both sites, we chose a Langmuir isotherm. For the single-site model, this isotherm is $q = 24C/(1 + 6C)$. For the two-sites model, the isotherm is $q = 24C/(1 + 6C) + 2.4C/(1 + 600C)$. This combination of two Langmuir terms indicates that the fraction of the surface occupied by the active sites is 1000 times smaller than the fraction occupied by the ordinary sites, or practically 0.1% of the total surface area of the adsorbent. On the other hand, the solute has a 100 times stronger affinity for the active-site surface than for the bulk of the surface. As a result, the presence of the active sites increases by only 10% the retention factor at infinite dilution. Although a Langmuir isotherm and not a linear isotherm is used for the single-site surface, the band profile obtained is Gaussian, because the sample amount is low enough and the value of $b_1 C$ is small. For the two-sites surface, however, since the term $b_2 C$ is large for the active sites, the isotherm behaves non-linearly and the band profile exhibits significant asymmetry and tailing.

Obviously, if a linear isotherm had been used for both sites, the band profile would have been Gaussian, regardless of whether a single-site or a two-sites model is chosen. Thus, is a hundred-fold smaller sample were used, the calculated band profile with the two-sites model would be

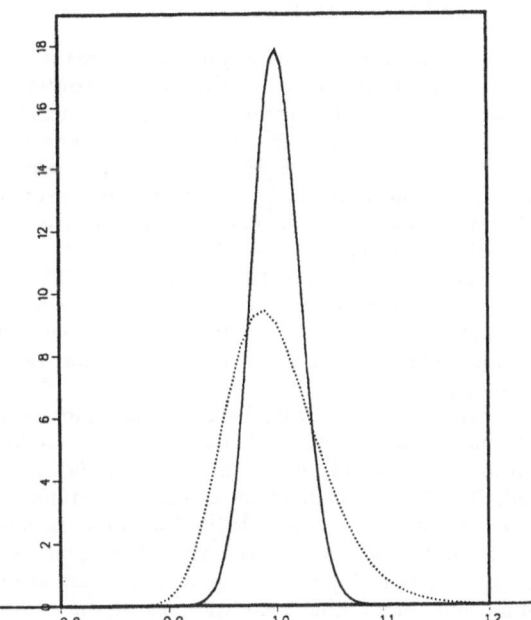

Figure 9. **Comparison between the Band Profiles Calculated with a One site and a Two-sites Isotherm Models, using Linear isotherms.** Dimensionless plot of C_d versus t_d. *Solid line*: one-site model, fast kinetics of adsorption/desorption ($N_{app} = 2000$, $D = \infty$). *Dotted line*: Two-sites model, fast kinetics of adsorption/desorption on the ordinary sites, slow kinetics on the active sites ($D_1 = \infty$, $D_2 = 20$).

Gaussian.

If, because of the small sample size used and the numerical values of the parameters, the isotherm for a two-sites surface behaves linearly, the peak tailing can still be explained by assuming that the kinetics of adsorption/desorption (A/D) on the ordinary sites is fast, but that it is slow on the active sites. Such an assumption is reasonable because we know that when a molecule is adsorbed on an active site, its desorption from this site is slow.

We compare in Figure 9 two profiles. They were both calculated as numerical solutions of the equilibrium-dispersive model, using a linear isotherm. The first profile (solid line) is calculated with a single-site isotherm ($q = 26.4C$) and an infinitely fast A/D kinetics. The second profile (dashed line) uses a two-sites isotherm model ($q = 24C + 2.4C$), which is identical to the single-site isotherm, and assumes an infinitely fast A/D kinetics on the ordinary sites but a slow A/D kinetics on the active sites. The inverse Laplace transform of the general rate model given by Lenhoff [32] (eqns. 79-83) is used for the simulation. In the case of a surface with two types of adsorption sites, equation 83 is modified to take into account the kinetics of adsorption/desorption on these two site types.

As expected, the profile obtained with the two-site model and a slow A/D kinetics for the active sites exhibits a very significant degree of tailing. Situations of this type are expected to arise in reversed phase chromatography using chemically bonded silica, as there is always some unreacted silanol groups which may behave as active site towards strongly polar molecules. Obviously, the band profiles obtained with either of the two isotherm models are identical when the A/D kinetics is very fast on both sites of the second isotherm.

7. Extension of these Various Models to Non- Linear Chromatography.

Most of the models discussed above have been applied to preparative chromatography as well. In preparative chromatography, a large amount of an usually concentrated solute mixture is injected. Because the concentration of the bands is too large during all or most of their migration, we cannot consider the isotherm as linear any longer. The non-linear behavior of the isotherm has several important consequences. First and foremost, all the concentrations no longer move at the same velocity. Depending on the direction of curvature of the isotherm, the high concentrations move faster or slower than the low ones. This causes profound deformation of the band profiles during their migration.

Secondly, non-linear isotherms are also competitive isotherms. The isotherms should be written as $q_i = f(C_1, C_2, ..)$. There is competition between the different components of the mixture for access to the stationary phase. As a result and unlike linear chromatography, the elution profiles of these different components are not independent, but they are affected by the presence of the other components.

Finally, the mathematical aspects of the theory become much more complex. For a single solute, the complexity of the mathematics with a non-linear isotherm is already such that there is no analytical, closed form solution available, except for two simplified models, the ideal model and the Thomas model [46]. The ideal model, which assumes an infinite column efficiency, is discussed in detail elsewhere in these Proceedings [47]. The Thomas model, which assumes a Langmuir slow A/D kinetics but which ignores the axial dispersion and the mass transfer resistance, has been solved by Goldstein [48], and this general solution has been simplified for a pulse injection by Wade et al. [49]. In all other cases, the problem must be solved numerically.

In the multi-component case, we need to use one differential mass balance equation for each of the components of the mixture (and each additive in the mobile phase if they are significantly adsorbed). Furthermore, these partial differential equations are coupled through the competitive isotherms, which renders the problem most complex. There is no analytical solution in the case of the Thomas model for two components. Even the ideal model cannot be solved completely in closed form for a binary mixture, in the case of the competitive Langmuir isotherm, the simplest non- linear isotherm [47,50]. The equilibrium-dispersive model of chromatography has been applied successfully in non-linear chromatography, but numerical solutions have to be calculated [45,51]. The Reaction-dispersive [52] and the Transport-Dispersive [53,54] models of chromatography also have to be solved numerically for non-linear isotherms. The general rate model of chromatography, and its extension which includes reaction in the mobile and stationary phases is solved numerically with non-linear isotherms [55,56].

The various lumped kinetic models of chromatography using a Langmuir isotherm have been compared [57]. It was shown that, like in linear chromatography, experimental data could be fitted well to each of the different models. However, unlike in linear chromatography, the apparent number of theoretical plates derived from this fitting is concentration dependent. This means that the Equations 92 or 93 are no longer valid in non-linear chromatography. That result contribute to make more difficult the analysis of non-linear chromatography. However, it was shown also that in overloaded elution the Equations 92 and 93 are a good approximation as long as the A/D or mass transfer kinetics are not too slow [58].

Finally, it has been shown that, in non-linear chromatography, the extra-column contributions to band broadening cannot be modeled by shift-invariant convolution as in linear chromatography [59]. The use of the more complex shift-variant convolution is required. Among other consequences, this fact removes the property of variance additivity which is so familiar and useful in linear chromatography. This result also demonstrates that the convolution of the solution of the ideal model by a Gaussian does not constitute a good model of non-linear chromatography. The former gives the band profile controlled by the thermodynamics of phase equilibria, in the case of a column of infinite efficiency. The second accounts well for the axial dispersion due to axial diffusion and resistances to mass transfer in the case of a column of high efficiency. The convolution should be shift-variant, however. As a consequence, the suggestion by Knox and Pyper [60] to account for the finite column efficiency in the calculation of the band width by adding the separate contributions to band broadening due to kinetics and thermodynamics cannot be accepted, except at very low column loading.

Acknowledgments.

This work was supported in part by grant CHE-8901382 from the National Science Foundation and by the cooperative agreement between the University of Tennessee and the Oak Ridge National Laboratory. We acknowledge the continuous support of our computational effort by the University of Tennessee Computing Center.

Literature Cited.

[1] Craig, L.C. (1944) J. Biol. Chem., 155, 519.
[2] Martin, A.J.P., and Synge, R.L.M. (1941) Biochem. J., 35 1358.
[3] Klinkenberg, A., and Sjenitzer, F. (1956) Chem. Eng. Sci., 5, 258.
[4] Said, A.S. (1956) AIChE J., 2, 477.
[5] Kendall, M., and Stewart, A. (1977) The Advanced Theory of Statistics, 4th edition, Chas. Griffen, London, UK.
[6] Smit, H.C., Smit, J.C., and de Jager, E.M. (1986) Chromatographia, 23, 229.
[7] Giddings, J.C. (1965) Dynamics of Chromatography, M. Dekker, New York, NY.
[8] Giddings, J.C., and Eyring, H. (1955) J. Phys. Chem., 59, 416.
[9] McQuarrie, D.A. (1963) J. Chem. Phys., 38, 437.
[10] Dondi, F., and Remelli, M. (1986) J. Phys. Chem., 90, 1885.
[11] Lapidus, L., and Amundson, N.L. (1952) J. Phys. Chem., 56, 984.
[12] Levenspiel, O., and Smith, W.K., (1957) Chem. Eng. Sci., 6, 227.
[13] Carberry, J.J., and Bretton, R.H. (1958) AICHE J., 4, 367.
[14] Reilley, C.N., Hildebrand, G.P., and Ashley, J.W. (1962) Anal. Chem., 34, 1198.
[15] Von Wicke, E. (1965) Ber. Bunsen-Ges. Phys. Chem., 69, 761.
[16] Aris, R. (1959) Chem. Eng. Sci., 9, 266.
[17] Gibilaro, L.G. (1978) Chem. Eng. Sci., 33, 487.
[18] Danckwerts, P.W. (1953) Chem. Eng. Sci., 2, 1.
[19] Villermaux, J, and Van Swaaj, W.P.M. (1969) Chem. Eng. Sci., 24, 1097.
[20] Otake, T., and Kunigita, E. (1958) Kagaku Kogaku, 22, 144
[21] Van Deemter, J.J., Zuiderweg, F.J., and Klinkenberg, A. (1956) Chem. Eng. Sci., 5, 271.
[22] Glueckauf, E. (1955) Ion- Exchange and its Applications, Metcalfe and Cooper, London, UK, P.34.
[23] Glueckauf, E. (1955) Trans. Faraday Soc., 51, 1540.
[24] Kucera, E. (1965) J. Chromatogr., 19, 237.
[25] Kubin, M. (1965) Collect. Czech Chem. Commun., 30, 2900.
[26] Haynes, H.W., and Sarma, P.N. (1973) AIChE J., 19, 1043.
[27] Carta, G. (1988) Chem. Eng. Sci., 43, 2877.
[28] Rosen, J.B. (1952) J. Chem. Phys., 20, 387
[29] Pellet, G.L. (1966) Tappi, 49, 75.
[30] Rasmuson, A., and Neretnieks, I. (1980) AIChE. J., 26, 686.
[31] Rasmuson, A. (1981) AIChE. J., 27, 1032.
[32] Lenhoff, A.M. (1987) J. Chromatogr., 384, 285.
[33] Van der Laan, E. Th. (1958) Chem. Eng. Sci., 7, 187.

90

[34] Guiochon, G. (1966) in Chromatographic Reviews, M. Lederer Ed., Elsevier, Amsterdam The Netherlands, p. 1.
[35] Huber, J.F.K. (1973) Ber. Bunsen-Ges. Phys. Chem., 77, 179.
[36] Horvath, Cs., and Lin, H.J. (1978) J. Chromatogr., 149, 43.
[37] Pfeffer, R., and Happel, J. (1964) AIChE J., 10, 605.
[38] Arnold, F.H., Blanch, H.W., and Wilke, C.R. (1985) J. Chromatogr., 330, 159.
[39] Knox, J.H. (1977) J. Chromatogr. Sci., 15, 352.
[40] Golay, M.J.E. (1958) Gas Chromatography, Desty, D.H., Ed., Butterworths, London, 1959 36.
[41] Guiochon, G., and Guillemin, G. (1988) Quantitative Gas Chromatography, Elsevier, Amsterdam, The Netherlands.
[42] Foley, J.P., and Dorsey, J. G. (1984) J. Chromatogr. Sci., 22, 40.
[43] Grushka, E. (1972) Anal. Chem., 44, 1733.
[44] Hanggi, D., and Carr, P.W. (1985) Anal. Chem., 57, 2394.
[45] Guiochon, G., Golshan-Shirazi, S., and Jaulmes, A. (1988) Anal Chem., 60, 1856.
[46] Thomas, H.C. (1944) J. Amer. Chem. Soc., 66, 1664.
[47] Golshan-Shirazi, S., and Guiochon, G. (1992) NATO Advance Study Institute Ser., Ser. E.
[48] Goldstein, S., (1953) Proc. Roy. Soc. London, A219, 151.
[49] Wade, J.L., Bergold, A.F., and Carr, P.W. (1987) Anal. Chem., 59, 1286.
[50] Golshan-Shirazi, S., and Guiochon, G. (1989) J. Phys. Chem., 93, 4243.
[51] Guiochon, G., and Ghodbane., S. (1988) J. Phys. Chem., 92, 3682.
[52] Golshan-Shirazi, S., and Guiochon, G., (1991) J. Phys. Chem., 95, 6390.
[53] Lin, B., Golshan-Shirazi, S., and Guiochon, G. (1989) J. Phys. Chem., 93, 3363.
[54] Golshan-shirazi, S., Lin, B., and Guiochon, G. (1989) J. Phys. Chem., 93, 6871.
[55] Yu, Q., and Wang, N.-H.L. (1989) Comput. Chem. Eng., 13, 915.
[56] Whitley, R.D., Van Cott, K.E., Berninger, J.A., and Wang, N.-H.L. (1991) AIChE J., 37 555.
[57] Golshan-Shirazi, S., and Guiochon, G., J. Chromatogr., In Press.
[58] Golshan-Shirazi, S., and Guiochon, G., In preparation.
[59] Dose, E., and Guiochon, G. (1990) Anal. Chem., 62, 1723.
[60] Knox, J.H., and Pyper, H.M. (1986) J. Chromatogr., 363, 1.

Glossary of Symbols.

A Constant in the Van Deemter equation (Eqn. 97)
A_p Area of the injected pulse ($A_p = n/F_v$)
a Distribution coefficient (ratio q/C_m)
   Also, constant in the dimensionless plate height equation (eqn. 102)
B Constant in the Van Deemter equation (Eqn. 97)
C Constant in the Van Deemter equation (Eqn. 97)
C Mobile phase concentration
C^* Solute mobile phase concentration in equilibrium with the solid phase concentration C_s ($C^* = C_s/a$)
C_d Dimensionless concentration (eqn. 30c)
C_m Concentration of the solute in the mobile phase
C_p Concentration of the solute inside the pores (eqn. 63)
C_0 Initial concentration
C_s Concentration of the solute in the stationary phase
D Dankohler number ($D = k_a d_p/u$)
D_a Apparent axial dispersion coefficient.
D_L Axial dispersion coefficient (eqn. 43)

D_l Molecular diffusivity in the coated liquid phase (open tubular column in GC, eqn. 103b)

D_m Molecular diffusivity in the mobile phase

D_p Diffusion coefficient of the solute in the particle pores (eqn. 59b)

d_p Particle diameter

F $= (1-\epsilon)/\epsilon$ in eqns. 28 and 43, and $(1-\epsilon_e)/\epsilon_e$ in eqns. 59 to 95.

F_v Flow rate

$f(l)$ Distribution of the solute in the Craig model (eqn. 5)

$f(r)$ Elution profile in the Craig model (eqn. 11)

$f(t)$ Elution profile

H Height equivalent to a theoretical plate (H.E.T.P)

h Reduced height equivalent to a theoretical plate (eqn. 95)

I_1 First order Bessel function (eqn. 24)

K_a Adsorption equilibrium constant

k_0' Retention factor ($k_0' = Fa$)

k_a first-order rate constant for adsorption (eqn. 24)

k_{ads} Adsorption rate constant (eqn. 65)

k_d first-order rate constant for desorption (eqn. 24)

k_m, k_m' Apparent mass transfer coefficient (eqn. 45 and 46)

k_f Film mass transfer coefficient (eqn. 59b)

L Column length

l Rank of a stage in the Craig model (eqn. 2)

$\mathcal{M}_n$ nth moment of a distribution

M_F Mass flux of solute from the bulk solution to the external surface of the particle (eqn. 62)

N_c Total number of stages in the Craig model (eqn. 6)

$_{pp} = N$ Column plate number or efficiency (eqn. 27)

N_{Disp} number of dispersion units in the column (eqn. 53)

N_f Number of mass transfer units corresponding to diffusion across the film around the particle (eqn. 76)

N_m number of mass transfer units in the column (eqn. 54)

N_p Number of mass transfert units due to the diffusion kinetics in the pores (eqn. 77)

N_{rea} Number of mass transfer units corresponding to the rate of adsorption/desorption ($N_{rea} = \frac{Fk_{ads}L}{u}$)

n Amount (mole) of solute injected

P Probability density function (eqn. 24)

$P_{e,z}$ Axial Peclet number

$P_{l,r}$ Probability of finding a molecule in the stage number l after r operations have been completed with the Craig model (eqn. 2)

q Concentration in the stationary phase in equilibrium with the mobile phase ($q = aC$)

R Fraction of solute in the mobile phase ($R = \frac{v_m C_m}{v_m C_m + v_s C_s} = \frac{1}{1+Fa} = \frac{1}{1+k_0'}$ (eqn. 1)

R_p Particle radius (eqn. 61)

r Number of operations performed during the operation of a Craig machine (eqn. 1)

S Column cross-sectional area

Sh Sherwood number ($Sh = k_f d_p/D_m$)

t Time

t_c Cycle time (eqn. 68)

t_0 Dead time or hold-up time of the column ($t_0 = V_m/F_v = L/u$)

t_d Dimensionless time (eqn. 30b)

$\bar{t}_d$ First moment of a peak, in dimensionless form (eqn. 33a)

t_p Width of the injected sample (concentration, C_0)

t_R Retention time

u Mobile phase linear velocity

$u_{z,0}$ Velocity of the peak of solute.

V_m Total volume of mobile phase contained in the column (eqn. 8)

V_R Total volume of mobile phase passed through the last stage of a Craig model during the entire process ($V_R = r v_m$, eqn. 7)

v_m Volume of mobile phase in a stage of the Craig model

v_s Volume of stationary phase in a stage of the Craig model. Also volume of stationary phase in the column.

X Reduced variable (eqn. 25)

x Reduced volume (eqn. 15)

z Distance along the column

z_d Dimensionless distance ($z_d = z/L$)

Greek Letters.

γ Tortuosity of the packing (eqn. 94)

δ Thickness of the Nernst diffusion layer

$\delta(t)$ Dirac pulse injection

$\delta(t_d)$ Dimensionless Dirac pulse injection

ϵ Total porosity of the column, or void volume fraction of the column or packing (eqn. 28)

ϵ_e External porosity of the packing (eqn. 59)

ϵ_p Internal porosity of the particle (eqn. 63)

μ First moment of a distribution

μ_s Mean location of the amount of solute (eqn. 3)

μ_n nth absolute or normalized moment (eqn. 85)

$\bar{\mu}_n$ nth centered moment (eqn. 86)

ν Reduced velocity or particle Peclet number

σ Standard deviation of a distribution

σ^2 Variance of a distribution

σ_d^2 Variance of the chromatographic peak in dimensionless form (eqn. 33b)

σ_s^2 Variance of the distribution of the solute (eqn. 4)

σ_t^2 Variance of a chromatographic distribution in time

σ_v Variance of a chromatographic distribution in volume (eqn. 9)

τ Reduced time (eqn. 15)

THEORY OF ZONOIDS: I A MATHEMATICAL SUMMARY

Patrick R. VALENTIN
Elf- France
Elf- Solaize Research Center
69360 St SYMPHORIEN D'OZON BP 22
FRANCE
e-mail elfcnrs@cism51.fr

ABSTRACT. Requirements to be fulfilled by a meaningful measure of the **separation state** of a physico-chemical system are set up and it is shown that the separation state can be represented by a geometric entity called a **zonoid**. A brief account is given of zonoid mathematical properties, necessary for chromatographic applications. Several new concepts are introduced such as regular selectivity and a classification of transformation processes in four classes: separation, mixing, null and sepmix processes, with the last two being new types of processes. Zonoid volume gives a useful (although degraded) measure of separation. For problems of evolution modelled by scalar partial differential equations in form of conservation laws, theorems on non increase of the separation state are given and the fundamental importance of the increase in the minimum volume enclosing a given quantity of species is stressed. Evolution equation for the zonoid itself as well as its volume show that the dimension of the Euclidean space where a process occurs has a deep impact on separation.

1. Introduction

These two papers describe a brief account of a new geometric approach to Separation Engineering, called **Zonoid Theory** and show its potential by applying it to chromatography. A small part of this theory, has been presented elsewhere [1,2]. For the sake of completeness, relevant parts of the backbone of the theory are recalled and definitions and theorems given often without proofs or validity statements. Since this paper is written primarily for chemists and not for mathematicians, most of the standard definitions and results used are recalled in order to be self-consistent. A full account of theory would involve a background in measure theory although we shall limit reference to it as much as possible.

Previous separation theories [3,4] failed mainly about two points: to begin with, they did not take into full account the fact that, in most physico-chemical systems, composition is a *continuous* function of space and time. In the second place, these theories concentrated upon the separation of the target species, neglecting the fact that this goal can be achieved only through admixing the system with some separating agent, e.g. energy. This mixing process is an essential part of the overall process and of its thermodynamic consistency. It should be taken properly into account from within the theory. The aim of this paper is to show that zonoid theory removes such limitations.

93

F. Dondi and G. Guiochon (eds.),
Theoretical Advancement in Chromatography and Related Separation Techniques, 93–135.
© 1992 *Kluwer Academic Publishers.*

The ability to cope with multicomponent *differential* families, either in time or in several space variables, is the core of the separation theory presented here. On practical grounds a discretization of the outlet flow into "cuts" is unescapable, but it must be done thoughtfully and on a clearly defined basis, after taking due account of the separation really produced by the column.Indeed, zonoid theory gives us a safe method with which to optimize the trapping procedure, thus minimizing the loss of separation through the mixing induced by trapping. In this manner, the separation really produced by the column is actually taken into account. Further work will address to this problem.

In contrast with previous theories, a salient feature of the present theory is that the very nature of a process can be established by the sheer observation of the conservative species evolution. It is model-independent. It is also independent of the physical nature of the conservative species themselves, and therefore, it includes e.g. energy. In this respect, the abstract mathematical presentation is intended to ease applications to other fields of separation science than chromatography.

This paper consists of two parts: the first one, covered by sections 2 to 5, introduces concepts, with some axiomatic justification, but without proving theorems or making precise conditions of validity. The second part, covered by section 6, presents an abstract application of the theory to the conservation laws. There, proofs are sketched in order to underline how to use the tools and also because the subject is completely new. Concepts and results of the zonoid theory are given in a somewhat restricted frame, suited to chromatography in one or several space dimensions. Although strong links exist between entropy and zonoid theory we reserve them for further study. To help the reader, a short physical motivation is inserted at the beginning of each section and subsection. References to part II (noted with a prefix II, e.g. II-§5-2) are also included to help the reader to grasp these abstract concepts on drawings and practical examples. Grouped references [14-25] forming together a short bibliography on the mathematical properties of zonoids and related topics have been added for those intending to plunge deeper into the theory.

In such a borderline matter, conflicts between the traditional notations in mathematics and in physico-chemistry are not easy to resolve, e.g. x designates usually a point in the former, a molar fraction in the latter, u is a "universal" unknown quantity in the former, a velocity in the latter. We have chosen to stick rather to the mathematical notations in Part I of this paper, to the physical ones in Part II.

2. The properties required for a *Separation state*

Since no acceptable criterion for separation exists at present, we begin by listing what properties such a hypothetical criterion should satisfy and then exhibit an entity endowed with them.

The SEPARATION STATE of a Physico-Chemical SYSTEM in a given state must be

1. **defined** for any type of system and must pertain to any finite number of conservative species, including energy
2. **invariant** by any permutation of the relative positions of subsystems
3. **additive** for any disjoint union of systems
   extensive for spatial similarity between systems
4. continuous in regard to the adjunction of a "small system"
5. endowed with **simple transformation laws** for any change in the way we observe and analyze the system:
   - nature of discretization
   - nature of conservative species

COMPARISON of SEPARATION STATES must have the following properties:

6. "separation state 1 is greater than separation state 2" can be defined as a partial order **relationship**

7. partial order defined in 6. must be compatible with the addition defined in 3.

8. **separation state is lower** when homogenization has occurred in a part of the system. The state of null separation, relative to a wholly homogeneous system, is a lower bound on the set of separation states of a system

9. **separation state is greater** when a homogeneous part of a system has been transformed into a non homogeneous part. The state of total separation, relative to a system composed of n species isolated pure in n regions, is an upper bound on the set of separation states of a system.

It should be noted that these requirements and specially 3. pertain to the definition of an extensive entity. We shall use rather the wording separation *content* than state when alluding specifically to that. Requirements 1 to 5 define the general mathematical properties useful in handling any criterion of physical interest, requirements 6 to 9 define those specific properties qualifying it for a "separation criterion". It will be clear in §5 that this set of requirements does not have a solution in the range of the usual classical tools, e.g. numbers or functions and that we must introduce new ones, what we do now, starting back at the beginning, i.e. defining the state of a system. We leave to the reader to verify which of the above requirements are meaningful and stand true at each step. Let us just mention here that requirements 3 and 8 imply that the sum of null separation states is not generally a null separation state: although it is the separation state of a system composed of homogeneous parts, the system is not homogeneous as a whole.

3. Systems, linear space E and differential families

Let us consider a multicomponent physico-chemical system $\mathscr{Y}$ with properties varying continuously as a function of a parameter. e.g. the time. This will be the case if we are looking at the chromatogram produced at the outlet of a chromatographic column. Then the state in consideration will be entirely characterized by the mass or molar flow-rate of each species, that is the solutes and the carrier, in course of time.

A formalization of this concept is given in the following

Definition Given a vector space E of finite dimension n, state Σ of a system $\mathscr{Y}$ is a 1-differential form on real line R, taking its values in the positive orthant E^+, and such that coefficient **F** of Σ is a Lebesgue integrable vector valued function, i.e. $F^i \in L^1(\Delta t)$:

$$\Sigma = \mathbf{F}(t) \, dt, \, t \in \Delta t \subset R$$

E is called the space of conservative species quantities and vector $\mathbf{N} \in E$ is called a quantity vector. E is referred to as an n-D space. That n is finite for physico-chemical systems results from the phase Gibbs law. In accordance with tensor notations, contravariant coordinates are indexed in upper position. Since an Euclidean norm would have no physical justification, E is given norm L_1,

$$\|\mathbf{N}\|_1 = \sum_{i=1}^{n} \left| N^i \right|$$

The unit ball B^1, for this norm, i.e. the set of points in E such that $\|\mathbf{N}\|_1 \leq 1$ is a cube centred on O, edges of which are translates of the segments [-1,1] on the axes.

Consequently, in the sequel, we shall refrain from using properties involving the Euclidean norm in E, such as scalar product or orthogonality of vectors.

A **problem of evolution** arises when F also depends on another parameter, say x, i.e. we consider the evolution of the state of a system as a function of x. We note the x-evolution problem:

$$\Sigma(x) = F(x, t)\, dt, \ t \in \Delta t\, (x) \subset R, \ x \in L \subset R$$

Clearly in the above definitions we may take any convenient parameters instead of t and x, as they are dictated by the nature of the problem, e.g. exchange t and x to get a t-evolution problem. In the context of Part I, we shall consider t-evolution problems. We also may extend x into a multivalued parameter. For chromatography in several space dimensions we take, e.g. $x = (x_1, x_2, x_3)$. There, we usually have a state $\Sigma(\beta) = \{F(x,t)\, dt\}$, $t \in \Delta t$, $x \in \Omega(\beta) \subset R^3$ - more generally, the Euclidean space R^m -, and the evolution is followed along a family $\Omega(\beta)$, depending continuously on a parameter β, of sets in R^3.

We shall deal in the sequel with the two different vector spaces E and R^m we just introduced and we must use consistent notations: at odds with standard mathematical notations, m will denote the dimension of the Euclidean space and n the dimension of E.

In Part II, we consider often n=3, and diffuse, smooth states, i.e.

$$F(x,t) = (F^1(x,t),\ F^2(x,t),\ F^3(x,t)),\ t \in \Delta t,\ F^i \text{ of class } C^2\, (\Delta t) \cap C^1\, (L)$$

but we shall also consider piecewise continuous states, therefore allowing F to possess discontinuities. All these definitions can be extended to an unbounded Ω if F is *locally integrable,* i.e. integrable on any compact subset, what we denote by $F^i \in L^1_{loc}(\Omega)$.

Formally and for better physical grasp, 1-differential form Σ may be thought of as a set of "infinitesimal" quantity vectors "dN" in E, this set being endowed with invariance property given by eqn. 1. Let us write it:

$$\mathcal{F} = \{dN(t)|\, t \in \Delta t\} = \{F(t)dt\}_{t \in \Delta t}$$

which is called a **1-differential** (in short, 1-d) **family** and generically noted as $\mathcal{F}$. System state and families or 1-differential forms are therefore the same concept. A problem of evolution would therefore involve a 1-d family depending on an evolution parameter, e.g. x.

The concept of a family of vectors is central to zonoid theory since family contains exactly the information needed to characterize the separation present in the system.

The family taken at some fixed value of x will be denoted $\mathcal{F}(x)$. When no ambiguity can occur on the parameter t and its domain Δt, we may denote $\mathcal{F}$ simply by $F(t)$, the n-tuple of the coefficients of $\mathcal{F}$. In the sequel, we shall consider mainly families with two components (n=2) in the following special cases: $\mathcal{F} = \{(1, u(\alpha))\, d\alpha\}_{\alpha \in \Omega}$ where α may be a real or a m-tuple. For m=1, we may always achieve this form of $\mathcal{F}$ by changing the variable in the family (see § 3.1), but this need not hold for m > 1. Physically, the first species is homogeneously distributed on the domain Ω and it may be a separation agent or the volume itself in the Euclidean space R^m.

It is important to realize that the functions commonly used for defining the composition, i.e. concentrations, molar fractions... do not contain all the information needed to characterize separation present in the system because they do not tell us how much quantity of a given composition is present in the system. Furthermore, they have not the invariance property exhibited by families (§ 3.1). Although we could have done all the following theory using e.g. molar fractions *together with* a quantity function, the latter taken in account using the lever rule, we have chosen to stick to the vector theory, much simpler and natural. Composition functions will rather be useful for characterizing the **type** - or selectivity - of a separation state (see § 4.4).

Let Δ_2 be the standard simplex of E, that is, $\Delta_2 = \{ X \in E^+ \mid X^1 + X^2 + X^3 = 1 \}$

Calling $F^0 = F^1 + F^2 + F^3$, $\| F(x,t) \| > 0$ on $L \times \Delta t$, we associate the composition function X, point valued in Δ_2, to F

$$X(x,t) = (\frac{F^1}{F^0}(x,t), \frac{F^2}{F^0}(x,t), \frac{F^3}{F^0}(x,t)), \ t \in \Delta t, \ z \in L$$

3.1   INVARIANCE OF FAMILIES BY A CHANGE OF VARIABLE

The conservation of moles or quantity implies that families must be endowed with some property of mathematical invariance, which we express in the following way.

Changing variables in families follows the chain rule; more precisely let $t = f(\tau)$, f, continuous, derivable, monotonic on segment Δt, $\tau = f^{-1}(t)$

$$\{dN(t)\}_{t \in \Delta t} = \{F(t)dt\}_{t \in \Delta t} = \left\{ F \circ f(\tau) \left| \frac{df}{d\tau}(\tau) \right| d\tau \right\}_{\tau \in f^{-1}(\Delta t)} \tag{1}$$

This is really a property of differential forms. Note that the inclusion of dt in the notation of differential families allows for automatic use of the chain rule and is the reason for imbedding dt in the definition of the family. Invariance property is easily generalized for families depending on several parameters, using the Jacobian of the change of variables. Finally, in case f would be decreasing, absolute values in eqn. 1 are taken to preserve positivity. See II- § 2.3 for an example.

Definition Two families which can be set to correspond by changing the variable are called equivalent.

The following families are equivalent, for $u^1 > 0$:

$$\left\{ u^1(t)\, dt, u^2(t)\, dt \right\}_{t \in [0, \Delta t]} = \left\{ dV, \frac{u^2(t(V))}{u^1(t(V))}\, dV \right\}_{V \in [0, V_o]} \tag{2}$$

since we may pass, in eqn. 2, from the first to the second one by changing the variable t into V, through $V(t) = \int_0^t u^1(t')\, dt'$, $V_o = V(\Delta t)$ provided $u^1 > 0$. This result can be extended to the case when u is bounded from below.

3.2   THE INTEGRATION OF FAMILIES

A fundamental operation on families allows to know what is the species content of a specified part of a system. This corresponds mathematically to the concept of integration.

A region ω is any interval or reunion of disjoint intervals, $\omega \subset \Delta t$ and generates a *quantity vector* by

$$N(\omega) = \int_{\omega} F(t)dt = \int_{R} \chi_{\omega} F(t)dt, \quad N(\omega) \in E \tag{3}$$

where the characteristic function, $\chi_{\omega} : R \rightarrow \{0,1\}$ (or discrete sampling function) is defined by

$$\chi_{\omega}(t) = \begin{cases} 1 \text{ if } t \in \omega \\ 0 \text{ if } t \notin \omega \end{cases}$$

$$N_o = \int_0^{\Delta t} F(s) \, ds \text{ is called the sum (or distal) vector of family } \mathcal{F}$$

By the definition of a family, N_0 is finite. In an evolution problem, a system is isolated if and only if $N_0(z)$ is fixed. In fact, eqn. 3 holds for m-d families with ω a measurable set in R^m.

At this point some definitions and notations relative to level sets of a scalar function are needed (see e.g. [6], p 52) and we shall connect them in an essential way with families:

Definition A function $u(x)$, $x \in \Omega \subset R^m$ is of bounded variation on Ω if ∇u, the generalized gradient of u is a R^m vector-valued family, i.e. it has a finite distal vector in R^m, or, equivalently, if

$$TV_{\Omega}(u) = \|\nabla u(x)\|(\Omega) = \int_{\Omega} \|\nabla u(x)\| \, dx < +\infty$$

with $\|\nabla u(x)\| = \sqrt{\sum_{j=1}^{m} \frac{\partial u}{\partial x_j}^2}$, the Euclidean norm on R^m. Let $BV(\Omega)$ be the space of the functions of bounded variation on Ω. In fact, a more general definition would say that the gradient of u is a measure and we would write $TV_{\Omega}(u) = \|\nabla u(x)\|(\Omega) < +\infty$.

Call ω_{τ} the **level set** of $u(x)$, $x \in \Omega \subset R^m$, the set of points in Ω such that u assumes at least the value τ, that is $\omega_{\tau} = \{x \in \Omega \mid u(x) \geq \tau \}$. Note that ω_{τ} can be the union of many unconnected subsets of Ω. We consider specially in the following, u continuous, or u $\in BV(\Omega)$.

$$V(\tau) = |\omega_{\tau}| = \mathcal{L}_m(\omega_{\tau}), \text{ the m-dimensional Lebesgue measure of } \omega_{\tau},$$

$$Q(\tau) = \int_{\omega_{\tau}} u(x) \, dx = \int_{\Omega} u(x)\chi_{\omega_{\tau}}(x) \, dx \text{ the enclosed content of species in } \omega_{\tau}$$

In fact, the pair $(V(\tau), Q(\tau))$ is obtained by reducing and integrating the m-d family : $\mathcal{F}(u)$ = $\{(1,u(x))\}_{x \in \Omega}$. Clearly $V(\tau)$, $Q(\tau)$ are right continuous, non increasing functions of τ, and

$$V(\tau) \in [0, V_0=V(-\infty) \,], \quad Q(\tau) \in [0, Q_0= Q(-\infty)].$$

Call $\partial\omega_\tau = \{x \in \mathbb{R}^m | u(x) = \tau \}$, the **level surface** of u, that is also, the boundary of ω_τ, $n(x)$ the normal outer unit vector at x on ω_τ. Whenever the gradient of u exists, it is collinear to n and points to the opposite direction:

$$\nabla u(x) = - \|\nabla u(x)\| n(x) \tag{4}$$

Let $\mathcal{H}_{m-1}$ be the (Hausdorff) measure of surface in $\mathbb{R}^m$. $A(\tau) = H_{m-1}(\partial\omega_\tau)$ is the perimeter of ω_τ. It is finite since u is of bounded variation and we have

$$A(\tau) = \int_{\partial\omega_\tau} d\sigma = \|\nabla\chi_{\omega_\tau}\| (\Omega) \tag{5}$$

An important formula, given by Fleming and Rishel [9] holds for $u \in L^1_{loc}(\Omega)$ (that is under very general circumstances) and links the variation of u and the perimeter of level sets. More precisely, for any subset G in Ω and $A(\tau | G) = A(\partial(\omega_\tau \cap G))$ denote the perimeter of the restriction of ω_τ to G,

$$TV_G(u) = \|\nabla u(x)\|(G) = \int_{-\infty}^{+\infty} A(\tau | G) \, d\tau \tag{6}$$

In essence, this formula says that the variation of u is the area of the developed projection of the graph of u onto a vertical cylinder. It is applicable to discontinuous functions. If u is smooth, the gradient exists in the classical sense, the central term in eqn. 6 is just the Lebesgue integral of $\|\nabla u(x)\|$.

Turning to the volume of level sets of u (vertical projections onto the ground space), we need a more general variant of eqn. 6: by multiplying the gradient with an integrable function f, we get (Federer [10], p. 248):

$$\int_G f(x) \|\nabla u(x)\| \, dx = \int_{-\infty}^{+\infty} \int_{\partial(\omega_\tau \cap G)} f(x) \, d\sigma \, d\tau \tag{7}$$

Although the above formula holds for a certain class of continuous functions, they hold true for a class of functions which are only piecewise continuous of BV provided these are solutions of conservation laws, see §6 and left-hand member written in measure as in eqn. 5. Let us apply this with $f(x) = \dfrac{1}{\|\nabla u(x)\|}$, we get, for any $G \subset \Omega$

$$\mathcal{L}_m(G) = \int_G dx = \int_{-\infty}^{+\infty} \int_{\partial(\omega_\tau \cap G)} \frac{1}{\|\nabla u(x)\|} d\sigma \, d\tau \qquad (8)$$

From which, taking $G = \omega_\tau$, noting that $\partial(\omega_\tau \cap G) = \partial\omega_\tau$, we get the derivative of the volume V of level sets with respect to the level τ:

$$\frac{dV}{d\tau}(\tau) = - \int_{\partial\omega_\tau} \frac{1}{\|\nabla u(x)\|} d\sigma \qquad (9)$$

whereas, for m = 1, we get (see also Duff [8], Federer [10])

$$\frac{dV}{d\tau}(\tau) = \sum_{j=1}^{p(\tau)} \frac{1}{|\nabla u(\alpha_j(\tau))|} \qquad (10)$$

with p, the number of points, with coordinates α_j, composing $\partial\omega_\tau$. Linear interpolation allows to treat the case when u has 'flat parts' or plateaus. Then, for some τ, $\partial\omega_\tau$ is such that $|\partial\omega_\tau| > 0$. Lett $\omega^*_{\tau_i}$ be the x-domain of the plateau, then $V(\tau^-) - V(\tau^+) = |\omega^*_\tau|$, with $V(\tau^+)$ and $V(\tau^-)$, the right and left limits of V at τ.

3.3   THE REDUCED REPRESENTATION OF A FAMILY

The reduced representation of a family contains only features relevant to the separation state representation. Reduction is a fundamental operation preserving the state of separation, which involves integration upon sets of same local composition. We may consider this as an application of a mixing convention to a family to obtain an equivalent family,

Definition A family $\mathcal{F}$ is said **reduced**, and noted $\mathcal{F}^*$, if all of its collinear vectors in E^+ are have been added together.

By the theorem of implicit functions we need to reduce a general m-d family with values in E, only in the case when dim E -1 $\leq$ m. If this arises, we will have to remix all elements of same composition in the family, that is integrate the family over (m-n +1)-D surfaces.
    From a slight extension of eqn. 2 in § 3.1, any m- differential family with values in R^2, whose one component is positive is equivalent to a family of the form:

$$\mathcal{F} = \left\{ F(u(x))dx \right\}_{x \in \Omega} \quad , F(u) = (1, u(x)), \, u: \mathbb{R}^m \to \mathbb{R}, u \text{ continuous.}$$

According to this, we consider the reduction of m- differential families of this simple type.

Reduction formulas are easily obtained using either eqn 9 or 10. Result depends on m and on n. The case of dim E -1 = m is a special case because we integrate over 0-D surfaces, i.e. sum values taken at points.

Let dim E = n = 2 and take in turn:

m=1: u: $\mathbb{R} \to \mathbb{R}$,

$$\mathcal{F}^* = \left\{ F(\tau) \sum_{j=1}^{p(\tau)} \frac{1}{\left|\nabla u(\alpha_j(\tau))\right|} \, d\tau \right\}_{\tau \in [u_{min}, u_{max}]} \tag{11}$$

where the α_j are the solutions of:

$$u(\alpha_1) = \ldots = u(\alpha_p) = \tau$$

m ≥ 2: u: $\mathbb{R}^m \to \mathbb{R}$

$$\mathcal{F}^* = \left\{ F(\tau) \int_{\partial \omega_\tau} \frac{1}{\|\nabla u(x)\|} \, d\sigma \, d\tau \right\}_{\tau \in [u_{min}, u_{max}]} \tag{12}$$

For an example see II-eqn. 7.

The mixing convention avoids trivial complications, and is quite natural since we are looking at separation. Note that many physically different systems can map into same reduced state. The separation state of a system with p homogeneous phases such as a thermodynamic system of given volume, at the equilibrium, reduces to a family of p vectors, irrespective of the arrangement of these phases in the Euclidean space.

3.4   THE APPROXIMATION OF A FAMILY

To compute separation states, we need, in most cases, to approximate continuous families by discrete families and frequently, we would like to do this "in the best way".

A partition of Δt into p intervals (or sub-regions) $\omega_1, \ldots \omega_p$ defines a discrete sampling (approximation) of system state by the (discrete) family of quantity vectors,

$$\{N_j\}_J = \{N_j \in E \mid j \in J = \{1, 2, \ldots, p\}\} = \{N_1, N_2, \ldots, N_p\} \tag{13}$$

If needed, the N_j can be looked upon as columns of the non negative quantity matrix [N], whose entry N^i_j represents the quantity of species i in region j. This extends naturally to a m-d family defined on $\Omega \subset \mathbb{R}^m$, with $\omega_1, \ldots \omega_p$ a partition of Ω.

4. Zonotopes and zonoids

We introduce now the central geometric tool, which will turn out to represent the separation state. We will need both a tool for discrete systems and a somewhat more elaborate one for continuous systems such as a chromatographic column.

Let us first define a very simple geometric operation on sets of E, taking advantage of the vector space structure,

Definition. The Minkowski sum of the sets A and B, is the set

$$A + B = \{M + N \mid M \in A, N \in B, A, B \subset E\} \tag{14}$$

with E considered as an affine or point space.

A sketch of the Minkowski addition of two sets in the plane is given on fig 1. The reader may easily verify that to obtain the sum of two sets A and B one needs just to slide origin 0 of translates of A along boundary of B and retain all points covered by these translates.

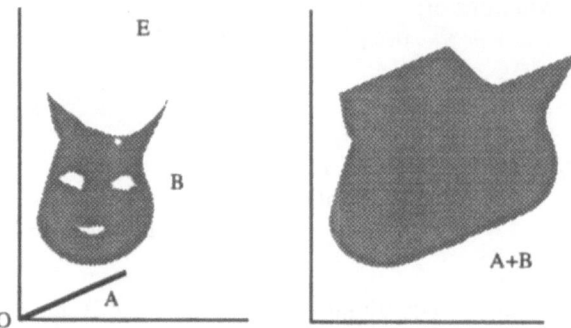

Figure 1 The Minkowski cat or the sum of a cat's head and a segment.

To a quantity vector **N**, we associate, in a straightforward manner, a segment noted as

[0,N] or N for short, with $N = \{\lambda N \mid \lambda \in [0,1], N \in E\}$. Here E is considered a vector space.

Closely related to the Minkowski sum is the Hausdorff distance between two sets A, B

$$d(A,B) = \text{Min} \{ r \mid A + r B^1 \supset B, B + r B^1 \supset A \}$$

given B^1 the unit ball of norm of E.

Definition (Coxeter [12,13]) A zonotope Z is the Minkowski sum of a finite set of segments of E:

$$Z = \sum_{j=1}^{p} [0, N_j]$$

A zonotope is a special type of polytope, recalling that a polytope is a general m-D version of a polyhedron. If A and B reduce to a point, A+B is also a point, the extremity of vector OA + OB. Less trivial examples of zonotopes (zonohedra if n=3) are the parallelogram, plane polygon with central symmetry, cube, parallelepiped. See §5-3, Fig 3 for a drawing of some zonohedra.

Loosely speaking, a zonoid is the equivalent of a zonotope, but with, almost everywhere, a curved and smooth boundary. It is the Minkowski sum of a continuous family of "infinitesimal" segments. More precisely, we have:

Call **K** the set of compact zonotopes in E. **K** is a metric space using the Hausdorff distance induced by norm L_1 on E.

Theorem 1 A zonoid Z, is the limit in the sense of the Hausdorff distance of a convergent sequence of zonotopes in **K**.

Example of zonoids are the circle, closed plane curve with central symmetry, sphere, ellipsoid. For a 2D zonoid, see fig 6, part II, for a 3D zonoid see fig 2 part II. These zonoids are generated by families of flow-rates going out from a chromatographic column.

Let $\mathbf{\mathcal{Z}}$ be the closed set of n-dimensional zonoids. In fact, $\mathbf{\mathcal{Z}}$ is the closure of **K**. Hausdorff metric on $\mathbf{\mathcal{Z}}$ allows to calculate the distance between two zonoids. We have $d(Z_1, Z_2) = 0$ if and only if $Z_1 = Z_2$.

4.1 LINEAR TRANSFORMATIONS AND PROJECTIONS

Neglecting the presence of any species in a system, e.g. the carrier in chromatography, or using new species which are linear combination of the old ones, will be represented by a projection or a linear transformation of the zonoid (see §2, requirement 5).

Any partition of a base of E in two subsets and their spanned subspaces V, $V^\perp$, defines a canonical projection in E, that is a projection parallel to $V^\perp$ onto V. In a canonical projection of Z onto V = linear span (e_1, e_2), we will use the convention that the projected object takes on the indexes of the plane, e.g. we denote it by Z_{12}.

Theorem 2 The image of a zonoid $Z\{\mathbf{\mathcal{F}}\}$ by a linear transformation T (e.g. projection), $T(Z)$ is the zonoid generated by the transformed family $T\mathbf{\mathcal{F}}$ i.e

$$T(Z\{\mathbf{\mathcal{F}}\}) = Z\{T(\mathbf{\mathcal{F}})\} \tag{15}$$

Corollary: any zonotope built on p vectors of E, dimE = n, is the projection of a cube from R^p into E.

4.2 THE SEPARATION STATE

Definition The separation content Z, of a state Σ of system $\mathbf{\mathcal{Y}}$ is
for a discrete system state with p vectors, the zonotope

$$Z\{N_1, N_2, \ldots, N_p\} = \left\{ \mathbf{M} \in E \ \middle| \ \mathbf{M} = \sum_{j=1}^{p} \lambda^j N_j, \lambda^j \in [0,1], j = (1, \ldots, p) \right\} \tag{16}$$

for a diffuse system, the zonoid

$$Z\{F(t)dt\}_{\Delta t} = \left\{ \mathbf{M} \in E \ \middle| \ \mathbf{M} = \int_{\Delta t} \lambda(t)F(t)dt, \lambda \in [0,1] \right\} \tag{17}$$

$\lambda(t)$ is called the sampling function and is a measurable function of t on Δt.

If Σ is a discrete (diffuse) system state, its separation content $Z(\Sigma)$ is a zonotope (resp. zonoid) said to be generated by the discrete (resp. differential) family of quantity vectors. In both cases, $Z(\Sigma)$ is defined as the set of all mixtures feasible by sampling from Σ

The Liapunov convexity Theorem asserts that Z is convex and closed. The Krein-Milman Theorem makes it possible to prove that, without any loss in the possible mixtures, we may restrict the sampling functions in eqn. 17, to take only values 0 or 1, that is to be the characteristic function of some set ω. Therefore eqn. 17 can be restated as:

Theorem 3 Z is the range of the vector measure μ, whose density is $F(t)$ relative to the Lebesgue measure on Δt.

From theorem 3 we get the **fundamental property of zonoids**

Theorem 4 Given a vector $M \in E$ and a family $\mathcal{F}$, the following properties are equivalent:

    (i) M is feasible by a sampling from the family $\mathcal{F}$

    (ii) M belongs to the zonoid $Z\{\mathcal{F}\}$

4.3 THE SUPPORT FUNCTION OF A ZONOID

Support function, a classical tool in convexity theory, is very useful in comparing zonoids, e.g. in problems of evolution modelled by partial differential equations: given a family of parallel hyperplanes of direction η, the support function of a convex set K gives the position of the extreme member of the family still touching K. (see fig 2 for a 2-D example).

Call E^*, the dual space of E, that is the vector space of linear forms over E, $\dim E = \dim E^*$ = n. Define the duality product as $<M,\eta> = \sum_{k=1}^{n} M^k \eta_k$, $M \in E$, $\eta \in E^*$

Definition. The support function of a convex set K is the homogeneous function ψ: $E^* \to R$

$$\psi(\eta \mid K) = \text{Sup} \{<M,\eta> \mid \eta \in E^*, M \in K\}$$

Support function is a convenient tool to test for inclusion of convex sets, even unbounded, by the equivalence:

$$\psi(\eta \mid K_1) \geq \psi(\eta \mid K_2) \Leftrightarrow K_1 \supset K_2$$

Zonoids have very simple support functions (Bolker [18])

$$\psi(\eta \mid Z) = \{ \int_{\omega} <u(x),\eta> dx \mid \omega = \{x \in \Omega \mid <u(x),\eta> > 0\}\} \tag{18}$$

At the same time we also get a point M on boundary of zonoid since tangent plane in direction η touches Z at

$$M = \int_\omega \mathbf{u}(x)\, dx, \quad \omega = \{x \in \Omega \mid <\mathbf{u}(x),\eta> > 0\} \tag{19}$$

This formula tells us that M is obtained simply by summing up every element of the family which points into the positive half space defined by η. More generally, the tangent plane supports Z along a **face** which is the Minkowski sum of the subfamily defined by $\{x \mid <\mathbf{u}(x),\eta> = 0\}$.

Let us introduce now a function H which will be used extensively in section 6 of this paper.

Definition $H(\tau \mid Z) = \displaystyle\int_{\mathbf{u}(x)\, -\, \tau\, \geq\, 0} (\mathbf{u}(x) - \tau)\, dx \geq 0$ \hfill (20)

In the sequel, $H(\tau \mid Z)$ will also be denoted $H(\tau)$ in case of no ambiguity on Z. The link of H with the 2-D zonoid support function takes place through the following:

Proposition Calling $\eta = \eta_2 (-\tau, 1)$, τ, the slope of support line to Z, then $H(\tau \mid Z)$ is the intercept of cone $\psi(\eta \mid Z)$ with straight line $\eta_2 = 1$.

4.4 EXTERIOR PRODUCT OF VECTORS

Calculation of volume of zonotopes and zonoids require some definitions relative to exterior products of elements of space E. Exterior product generalizes determinants which, in turn, are known to give the volume of a parallelepiped, the simplest sort of zonotope.

Define the exterior (or cross) product of two vectors $N_1, N_2 \in E = R^2$ as

$$N_1 \wedge N_2 = \mathrm{Det}\,[N_1, N_2] = N_1^1 N_2^2 - N_1^2 N_2^1 \tag{21}$$

and for n=3, $N_1, N_2 \in E = R^3$ is a vector (in a 3-D space noted $\Lambda^2 E$)

$$(N_1 \wedge N_2)^k = = N_1^i N_2^j - N_1^j N_2^i, \ 1 \leq i < j \leq 3, \ i, j \neq k, \ k = 1,2,3 \tag{22}$$

If $N_1, N_2, N_3 \in E = R^3$, exterior (or mixed) product is

$$N_1 \wedge N_2 \wedge N_3 = \mathrm{Det}\,[N_1, N_2, N_3] \tag{23}$$

4.5 REGULAR SELECTIVITY OF A SEPARATION STATE

The concept of selectivity is important both because of its physical meaning and of its deep link with the linear dependencies between vectors of a family and the structure of zonoid boundaries. The physical meaning of Theorem 6 is that, given any vector in a family enjoying the regular selectivity property, no vector of the same composition can be made from the other vectors of the family through mixing. In this sense any vector of a regularly selective family is extreme.

Definition A family of p vectors in 3-D space E has a 3-regular selectivity property iff its elements can be linearly ordered in such a manner that

$$N_{j_1} \wedge N_{j_2} \wedge N_{j_3} \geq 0 \, , \, \forall \, 1 \leq j_1 < j_2 < j_3 \leq p \tag{24}$$

Similarly, a 1-differential family has the 3-regular selectivity property iff a regular parameter t exists such that

$$F(s) \wedge F(t) \wedge F(u) \geq 0 \, , \, \forall \, 0 \leq s < t < u \leq \Delta t \tag{25}$$

Theorem 5 If F is twice differentiable a local requirement for 3-regular selectivity is that the Wronskian determinant W(F) be non negative

$$W(F) = \, F(t) \wedge F'(t) \wedge F''(t) \geq 0 \, , \, \forall \, 0 \leq t \leq \Delta t \tag{26}$$

Wronskian matrix contains all the information relative to selectivity since condition of 2-selectivity of separation between two species,

$$\mathrm{proj}_{ij}F(t) \wedge \mathrm{proj}_{ij}F'(t) \geq 0$$

can be looked at directly on corresponding minor on the Wronskian matrix W(F).

Definition The points of 3-selectivity reversal are the roots of equation W(F(t)) = 0, such that W(F(t)) changes its sign.

Definition A discrete family has the same selectivity as a differential family if both of them can be ordered in such a way that determinants in eqn. 24 follow the same sequence of signs.

Since, from above definitions, selectivity is conserved for a family $\lambda(t)$ $\mathcal{F}$, $\lambda > 0$, selectivity does not depend on quantities, but only on composition of vectors of the family. In the sequel, we shall always suppose that, in case of regular selectivity, the parameter of the family has been chosen to ensure non negativity of exterior products, eqns. 24, 25. Note however that the selectivity is not conserved in more general linear operations such as projections.

Geometrically, selectivity can be monitored directly from the graph γ of x(t) in simplex Δ_2 (and, more generally, in any plane not passing by the origin -such a diagram is known as the first projective diagram [16,23]) as shown by

Theorem 6 Selectivity is 3-regular iff γ is convex
Selectivity between species i, j, i ≠ j is 2-regular iff one of the equivalent properties hold for any straight line KQ, K a vertex of Δ_2 (extremity of e_k), k ≠ i,j, Q on γ :

(i) KQ rotates uniformly when Q runs uniformly along γ

(ii) no line KQ has another point on γ than Q

Definition A 1-differential family of n species has a totally regular selectivity iff it has a r-regular selectivity for any $1 \leq r \leq n$.

For n=3, study of selectivity involves the Wronskian matrix [F,F',F''] and its 2×2 minors. There are rules (Karlin, [5]) to build or recognize families enjoying totally regular selectivity [1]. The above definitions generalize themselves to a constant arbitrary sign in inequations 24, 25.

4.6 BOUNDARY OF 3-D ZONOIDS

Zonohedra are named from the fact that for each vector N_j, of the generating family, there is a sequence of 2-faces whose adjacent elements have a common edge which is a translate of N_j. From this, it is easy to infer that the zonohedron generated by family with p vectors has p(p-1) faces in general.Similarly, a particular set of lines, called zonal lines or zones, covers the boundary of a zonoid and motivates the name of this body. Call ∂Z the boundary of a 3D zonoid Z. Structure of ∂Z is strongly linked with the topology of curve γ in Δ_2.

Lemma 6 Every tangent plane to Z is spanned by at least two vectors $F(t_1)$, $F(t_2)$, t_1, t_2 $\in [0, \Delta t]$

Definition A zone $L(t_1)$ is the set of points M on ∂Z where the tangent plane T_M is parallel to $F(t_1)$.

Theorem 7 $L(t_1)$ is a closed simple curve and the set of tangent planes envelopes a convex cylinder whose generatrix is parallel to $F(t_1)$.

To each intersection point of a straight line with curve γ in the projective diagram corresponds a zone which passes through M. Therefore the number of zones which cross at any particular point on ∂Z is equal to the number of points of intersection of some straight line with γ. When γ is convex, there are exactly two zones crossing at a point (except at the origin or the distal point). This holds in fact for any $F(t_1)$ such that, seen from $X(t_1)$, γ is star-shaped. When γ is not convex, many zones can cross and even tangent each other at a point.

When γ is not smooth, ∂Z is composed of patches which paste together along lines of discontinuity of the tangent plane.

Theorem 8 If the family $\mathcal{F}$ is 3-regular selective, the cumulate curve splits $\partial Z(\mathcal{F})$ into two pieces, ∂Z^+, ∂Z^-. ∂Z^+, called the "upper" (or "positive") part of ∂Z, gets, as a 2-D surface, a two parameter expression:

$$\partial Z^+ : \quad M(t_1, t_2) = \int_0^{t_1} F(\tau)d\tau + \int_{t_2}^{\Delta t} F(\tau)d\tau \quad , 0 \le t_1 \le t_2 \tag{27}$$

or

$$\partial Z^+ : \quad M(t_1, t_2) = N_0 + N(t_1) - N(t_2) \quad , 0 \le t_1 \le t_2 \le \Delta t$$

For the "lower" part, ∂Z^- we get

$$M(t_1, t_2) = \int_{t_1}^{t_2} F(\tau)d\tau \ , \ 0 \le t_1 \le t_2 \tag{28}$$

Clearly the points M with same arguments in eqns. 27, 28 are antipodal (i.e. they sum to N_0), a result in line with the central symmetry of Z.

In the general case we get a formula for any point on ∂Z^+

$$\left| \quad \partial Z^+ \ : \ M(t_1, t_2) = \int_{\omega^+(t_1, t_2)} F(\tau)d\tau \ , \ 0 \le t_1 < t_2 \le \Delta t \right. \tag{29}$$

$$\omega^+(t_1, t_2) = \{\tau \in [0, \Delta t] \mid F(t_1) \wedge F(t_2) \wedge F(\tau) \ge 0\}$$

where the boundary of the integration domain is the set of solutions of equation $F(t_1) \wedge F(t_2) \wedge F(\tau) = 0$.

At this point let us give four easy Lemmas which will be useful in the study of evolution problems (§ 6). They also show the meaning of a function H which was previously introduced (in place of Q used here) by C. Bandle for technical reasons [6, p. 212] in studying some parabolic partial differential equations of reaction-diffusion.

Lemma 8 Let $u(x)$ be a positive Lebesgue integrable continuous function, Ω an open bounded set, $\Omega \subset R^m$, $\mathcal{B}(\Omega)$ the set of Borel sets on Ω. Suppose further that all τ-level sets of u (on which $u = \tau$) are m-Lebesgue negligible. Then ∂Z^+ is pointwise generated by three equivalent optimization problems on $\mathcal{B}(\Omega)$:

(i) $\quad V^-(Q) = \text{Min} \left\{ |\omega| \ \Big| \ \int_\omega u(x) \, dx \ = Q, \omega \in \mathcal{B}(\Omega) \right\}$

(ii) $\quad Q^+(V) = \text{Max} \left\{ \int_\omega u(x) \, dx \ \mid \ |\omega| = V, \omega \in \mathcal{B}(\Omega) \right\}$

(iii) $\quad H^+(\tau) = \text{Max} \left\{ \int_\omega (u(x) - \tau) \, dx \ \mid \tau = \text{const}, \omega \in \mathcal{B}(\Omega) \right\}$

Solution of this problem is

(i) $\quad V^-(Q) = \int_{\omega_\tau} dx$ for τ such that $Q = \int_{\omega_\tau} u(x) \, dx$

(ii) $\quad Q^+(V) = \int_{\omega_\tau} u(x) \, dx$ for τ such that $V = \int_{\omega_\tau} dx \tag{30}$

(iii) $\quad H^+(\tau) = \int_{\omega_\tau} (u(x) - \tau) \, dx$

The hypothesis made upon level sets of u means that u has no plateau but is not essential and can be dispensed with by filling in linearly the possible discontinuities of V⁻ and Q⁺. In the latter case we loose the unicity of the optimum ω, but this is of no consequence to the separation theory in view of requirement 2 in §2. Note however that, even in this case, H⁺ remains a bounded *continuous* function of τ.

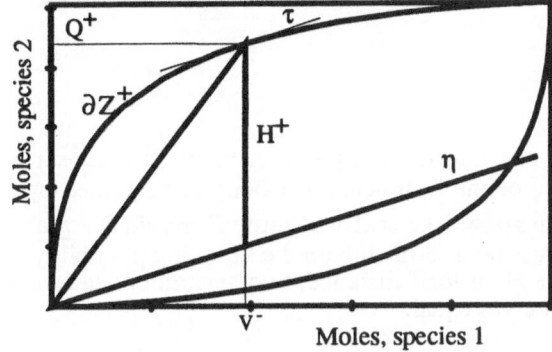

Figure 2 A graphical interpretation of the function H⁺ and related quantities in Lemma 8

In view of the importance of the function H, already defined by eqn. 20, let us give its physical meaning: $H^+(\tau)$ give the maximum excess species quantity above the given density level τ.

Lemma 9 on the 2D zonoid boundary for a class of discontinuous functions:
For $\mathcal{F} = \{(1,u(x))\,dx\}$, $x \in \Omega \subset R^m$, u a piecewise smooth function, $Z(\mathcal{F})$, at point M $\in \partial Z(\mathcal{F})$, is either smooth (a), has a corner (b), has a linear edge (c). Thus we get

(a) iff $\partial\omega_\tau^o = \{x : u(x) = \tau\}$ is of positive (m-1)-Lebesgue measure, zero m-Lebesgue measure and u is continuous on part of $\partial\omega_\tau$

(b) $\partial\omega_\tau^o$ is of zero (m-1) measure, u is discontinuous on the whole of $\partial\omega_\tau$

(c) $\partial\omega_\tau^o$ is of positive m-Lebesgue measure, u has a plateau at $u = \tau$

Lemma 10 In problems of t-evolution, under very mild conditions, for u being continuous, the time derivatives of the functions $Q^+(V,t)$, $V^-(Q,t)$ and $H^+(\tau,t)$ of Lemma 8, exist, are continuous and we have

$$\frac{\partial Q^+(V,t)}{\partial t} = -\tau\frac{\partial V^-(Q,t)}{\partial t} = \frac{\partial H^+}{\partial t}(\tau,t) = \int_{\omega_\tau}\frac{\partial u(x,t)}{\partial t}\,dx \qquad (31)$$

This result does not extend, in general, to discontinuous functions. However it holds if we impose additional restrictions to the discontinuities, such as the so-called Rankine-Hugoniot condition, see §6.1.

Lemma 11 The solution of Lemma 8 for $\mathcal{F}_1 = \{(1,u(x))\,dx\}$, $x \in \Omega \subset R^{m\prime}$, applies to $\mathcal{F}_2 = \{(1,f(u(x)))\,dx\}$, $x \in \Omega \subset R^m$, with f continuous, increasing.

Since $\omega_{f(\tau)}(f(u)) = \omega_\tau(u)$, we just need to insert f(u) in place of u in eqn. 30 to get the boundary of $Z(\mathcal{F}_2)$ from the boundary of $Z(\mathcal{F}_1)$ by a nonlinear affinity on the Q axis

From now, for the sake of simplicity, let us drop upper indexes in V,Q,H, being understood that these functions refer to the extremal quantities of Lemma 8. Also, we denote $\frac{\partial H}{\partial t}(\tau,t)$ by H_t.

4.7   THE n-VOLUME OF A ZONOID

Let Ξ, or, more specifically, $\Xi_{ijk...}$, if one considers separation between a subset of species labelled i, j, k..., be the volume of the n-dimensional body Z generated by a family. Ξ is called the **n-volume of separation**. The scalar quantity Ξ provides a global information on the separation state but gives no information on the sampling possibility of a particular mixture. In contrast with the Hausdorff distance, two separation states can have equal n- volume without being themselves equal.

The 2-volume of the 2-D zonotope generated by family $\{N_1, ..., N_p\}$, is given by

$$\Xi = \sum_{1 \le j_1 < j_2 \le p} \left| N_{j_1} \wedge N_{j_2} \right| \tag{32}$$

which comes from the fact that Z can be paved by a tight assembly of all possible parallelograms built on pairs of vectors. Reordering the family by increasing slope order, we may drop the absolute values.

Passing to the limit for 2-D zonoids, we integrate the differential 2-volume element:,

$d\Xi = \left| F(s) \wedge F(t) \right| ds \, dt$ and get the following theorem:

Theorem 9 The 2-volume of separation of the 1-differential family F is given by

$$\Xi = \iint_{0 \le s < t \le \Delta t} \left| F(s) \wedge F(t) \right| ds \, dt \tag{32a}$$

If we reparametrize the family in such a manner that $F(s) \wedge F(t) \ge 0$, $0 \le s < t \le \Delta t$, we are in the case of 2-regular selectivity and, dropping the absolute values, integrating and using multilinearity of exterior product, we get:

$$\Xi = \int_{0 \le t \le \Delta t} N(t) \wedge F(t) \, dt \tag{33}$$

with the cumulated vector N(t) defined as

$$N(t) = \int_0^t F(s) \, ds$$

Note that eqn. 33 is a classical expression for signed area (seen from O) generated by a plane curve given by parametric equations.

These formulas readily extend to zonotopes and zonoids in spaces of higher dimensions:

The 3-volume of a 3-D zonotope is (from Shephard, [20]),

$$\Xi = \sum_{1 \le j_1 < j_2 < j_3 \le p} \left| N_{j_1} \wedge N_{j_2} \wedge N_{j_3} \right| \tag{34}$$

For a diffuse family generating a bounded 3-D zonoid, equivalent of eqn. 34 is

$$\Xi = \iiint_{0 \le r < s < t \le \Delta t} \left| F(r) \wedge F(s) \wedge F(t) \right| dr \, ds \, dt \tag{35}$$

In general, letting Δ_3 be the 3-D simplex defined by $0 \le r < s < t \le \Delta t$, and Δ^-_3 the subset of Δ_3 on which $F(r) \wedge F(s) \wedge F(t) < 0$, we have

$$\Xi = \iiint_{\Delta_3} F(r) \wedge F(s) \wedge F(t) \ dr \, ds \, dt - 2 \iiint_{\Delta_3^-} F(r) \wedge F(s) \wedge F(t) \ dr \, ds \, dt \tag{36}$$

In the case of 3-regular selectivity, Δ^-_3 is void and upon integration in the above equation we get the expression (which is, more generally, a lower bound for the right-hand member in eqn. 35)

$$\Xi = N(\Delta t) \wedge \int_{0 \le s \le \Delta t} N(s) \wedge F(s) \ ds \tag{37}$$

A further integration of eqn. 37, in the case of totally regular selectivity shows that, in this case, the 3-volume of a 3-D zonoid Z is simply a linear combination of the 2-volumes of the canonical projections of Z:

$$\Xi = N_o^1 \, \Xi_{23} - N_o^2 \, \Xi_{13} + N_o^3 \, \Xi_{12} \tag{38}$$

See II-2.3.3 for a partial relaxation of these hypotheses.

In problems involving several space dimensions, and specially in problems of evolution, these formulas do not apply directly, but only after reduction of the family. However the 2-volume of zonoid generated by 1-d families obtained through reduction from m-d families may be calculated as follows: consider the family $\{dx, u(x) \, dx\}_{x \in \Omega}$, $\Omega \subset R^m$.

Following notations and material of §3.2., let $u_{min} \le \tau \le u_{max}$, ω_τ the set of level τ of $u(x)$, $V(\tau)$ the volume of ω_τ, $Q(\tau)$ the species quantity in ω_τ. Then, from eqn. 33,

$$\Xi = \int_{u_{min}}^{u_{max}} \left| \begin{matrix} \dfrac{dV(\tau)}{d\tau} & V(\tau) \\ \dfrac{dQ(\tau)}{d\tau} & Q(\tau) \end{matrix} \right| d\tau \tag{39}$$

and upon development and integration by parts:

$$\Xi = 2 \int_{u_{min}}^{u_{max}} Q(\tau) \frac{dV(\tau)}{d\tau} \, d\tau - V_o Q_o \tag{40}$$

We have also, after a change of variable, eqn. 41, which holds even if u has plateaus :

$$\Xi = 2 \int_0^{V_o} Q(V) \, dV - V_o Q_o \tag{41}$$

Finally, expanding eqn. 39 and using the definition of H, we get the last equation of the set of equivalent expressions 40-42:

$$\Xi = 2 \int_{u_{min}}^{u_{max}} H(\tau) \frac{dV(\tau)}{d\tau} \, d\tau \tag{42}$$

If the function u has p plateaus, $V(\tau)$ has p discontinuities at τ_i, numbering them by decreasing values, noting that, from Lemma 9, $H(\tau)$ is constant, we may compose eqns. 32, 42, in the following way:

$$\Xi = 2 \sum_{i=1}^{p} \int_{\tau_i^-}^{\tau_{i+1}^+} H(\tau) \frac{dV(\tau)}{d\tau} \, d\tau + \sum_{i=1}^{p} H(\tau_i) \left(V(\tau_i^-) - V(\tau_i^+) \right) \tag{42a}$$

where the left sum is taken on domains where u is not constant, that is the segment $[u_{min}, u_{max}]$ deprived of the points τ_i, the right sum on plateaus.

Coming back to the main case, inserting eqn. 9 and expression for $Q(\tau)$ in eqn. 40, we get

$$\Xi = 2 \int_{u_{max}}^{u_{min}} \left(\int_{\omega_\tau} u(x) \, dx \int_{\partial \omega_\tau} \frac{1}{\|\nabla u(x)\|} \, d\sigma \right) d\tau - V(u_{min})Q(u_{min}) \tag{43}$$

The above formula allows to calculate directly the volume of 2-D zonoid generated by the family $\{dx, u(x) \, dx\}_{x \in \Omega}$ in function of volume integrals on level sets and of surface integrals on their boundary. If u has plateaus, eqn. 43 does not hold and we must apply eqn. 42a.

An important application arises in connection with t-evolution problems. A straightforward generalization of eqn. 42 gives

$$\Xi(t) = 2 \int_{u_{min}(t)}^{u_{max}(t)} Q(\tau,t) \frac{\partial V(\tau,t)}{\partial \tau} d\tau - V_o Q_o \qquad (44)$$

from which, using eqn. 32, after some tricks we derive,

$$\frac{d\Xi}{dt}(t) = 2 \int_{u_{min}(t)}^{u_{max}(t)} \frac{\partial H(\tau,t)}{\partial t} \frac{\partial V(\tau,t)}{\partial \tau} d\tau \qquad (45)$$

an expression which will prove to be very useful. Again using eqn. 32, 9, we get the final result, holding for the case $m \geq 2$:

$$\frac{d\Xi}{dt}(t) = 2 \int_{u_{max}(t)}^{u_{min}(t)} \int_{\omega_\tau} \frac{\partial u(x,t)}{\partial t} dx \int_{\partial\omega_\tau} \frac{1}{\|\nabla u(x)\|} d\sigma \, d\tau \qquad (46)$$

When $m = 1$, a slightly different result holds

$$\frac{d\Xi}{dt}(t) = 2 \int_{u_{max}(t)}^{u_{min}(t)} \int_{\omega_\tau} \frac{\partial u(x,t)}{\partial t} dx \sum_{j=1}^{p(\tau)} \frac{1}{|\nabla u(\alpha_j(\tau,t))|} d\tau \qquad (47)$$

where p is the number of extremities of segments composing ω_τ.

When the system is unbounded, e.g. when $\Omega = R^m$, by hypothesis, $Q_o = \int_{R^m} u(x) \, dx$

remains bounded but V_o is unbounded and so is the zonoid and Ξ. In this case, the 2-volume looses its interest. This problem can be managed with, in some limited situations, by passing to the limit as $\Omega \to R^m$, in the loss of 2-volume in place of the 2-volume itself:

Definition The **loss of 2-volume** is the 2-volume of the complement of the 2-D zonoid to the rectangle of total separation, that is, if it exists, the quantity

$$\zeta = Lim_{\Omega \to R^m} (V_o(\Omega) \, Q_o(\Omega) - \Xi(\Omega))$$

We have, even for Ξ unbounded:

$$\frac{d\Xi}{dt}(t) = -\frac{d\zeta}{dt}(t)$$

See in II-§2.3.1 the case of loss of 2-volume of separation between a solute and carrier in uniaxial chromatography.

5 A geometric comparison of separation states

Requirement 6 in section 2 states that we must be able to compare separation states. Although inclusion of zonoids seems to be a good candidate for such an order, it turns out that it is not sufficiently strong for our purpose: inclusion tests that we may produce any M_1 from family $\mathcal{F}$ by mixing from family $\mathcal{F}'$ but does not tests that we may produce all the vectors of family $\mathcal{F}$ taken all together. Therefore we must introduce a stronger order called existence order which will exactly specify this. Physical meaning of existence order is therefore that the whole system state Σ' can be made by sampling of system Σ. Another natural requirement is that we compare separation states of a physico-chemical system only if they have, globally, the same species content, that is, if the families $\mathcal{F}$ and $\mathcal{F}'$ have the same sum vector. This occurs, e.g., if the system is isolated.

Definition A discrete system state Σ represented by family $\mathcal{F} = \{N_j \mid j \in J\}$ is said to contain a greater separation than system state Σ', or $\mathcal{F}' = \{M_k \mid k \in K\}$ iff family $\mathcal{F}'$ can be made by sampling from family $\mathcal{F}$, that is iff a $J \times K$ matrix $[\mu]$ exists such as

$$M_k = \sum_{j \in J} \mu_k^j N_j , \, k \in K \tag{49}$$

$$0 \le \mu_k^j \le 1, \, j \in J, \, k \in K , \quad \sum_{k \in K} \mu_k^j = 1 , \, j \in J \tag{49a}$$

Clearly this implies $N_0 = \sum_{k \in K} M_k = \sum_{j \in J} N_j$

For diffuse systems represented by 1-differential families, we let the

Definition A system state (family) $\mathcal{F}_2 = \{F_2(\tau) \, d\tau\}_{\tau \in [\tau_1, \tau_2]} = I_2 \subset R$
has a greater separation content than
system state (family) $\mathcal{F}_1 = \{F_1(t) \, dt\}_{t \in [t_1, t_2]} = I_1 \subset R$
iff a function λ, called a sampling kernel, $\lambda : I_1 \times I_2 \to [0,1]$ exists such that

$$F_1(t) = \int_{I_2} \lambda(t,\tau) \, F_2(\tau) d\tau , \, t \in I_1 \tag{50}$$

$$\int_{I_1} \lambda(t,\tau) \, dt = 1 , \, \tau \in I_2 \tag{50a}$$

Both eqn. 50 and 50a imply, through switching the order of integration, that

$$\int_{I_1} \mathbf{F}_1(t) \ dt = \int_{I_2} \mathbf{F}_2(\tau)d\tau = \mathbf{N}_o$$

Sampling kernels generalize sampling functions defined in eqn. 17. and give a constructive way to obtain $\mathscr{F}_1$ from $\mathscr{F}_2$.

Sampling kernels generate a partial order relation, called *existence*, on separation states or zonoids, i.e. on $\mathbf{Z}$: one writes

$$Z(\Sigma') \prec Z(\Sigma) \tag{51}$$

and reads eqn. 51 as: separation state Σ' *exists* in separation state Σ. We have

$$Z(\Sigma') \prec Z(\Sigma) \text{ and } Z(\Sigma) \prec Z(\Sigma') \Rightarrow \Sigma = \Sigma'$$

By a theorem from Blackwell [22], if dim $E = 2$, the existence order is equivalent to the inclusion of zonoids. However, in general, it is a stronger order as shown by the counter-example in §5.1. Note further that in the case of $n = 1$, the existence reduces trivially to the law of conservation of quantity.

Since the subtraction has no meaning for zonoids we cannot simplify the equalities or inequalities as if they were scalar. However we have a simplification property, which expresses the compatibility between the Minkowski sum and the existence order:

$$\text{If } Z' = Z'_1 + Z_2, Z = Z_1 + Z_2, \text{ then } Z' \prec Z \Leftrightarrow Z'_1 \prec Z_1 \tag{52}$$

Proposition The set of zonoids $\mathbf{Z}$ is a convex cone
$$Z(\alpha\mathscr{F}) = \alpha \, Z(\mathscr{F}), Z(\alpha\mathscr{F}_1 + \beta\mathscr{F}_2) \prec \alpha \, Z(\mathscr{F}_1) + \beta \, Z(\mathscr{F}_2), \alpha, \beta \geq 0 \tag{53}$$

5.1   EXISTENCE: COUNTER-EXAMPLE AND EXAMPLE FOR N =3

A counter-example (from D. Girard, personal communication) shows that the existence order turns out, for $n \geq 3$, to be a stronger order than the inclusion of zonoids:
Let $N_1, N_2, N_3 \in E$, dim $E \geq 3$, be linearly independent vectors. We consider families

$$\mathscr{F}_1 = \{N_1, N_2, N_3, N_4 = N_1+N_2+N_3\}$$
$$\mathscr{F}_2 = \{M_1 = N_1+N_2, M_2 = N_1+N_3, M_3 = N_2+N_3\}$$

The zonohedra generated by these families are represented on fig 3. As one may see on fig 3c, inclusion $Z(\mathscr{F}_2) \subset Z(\mathscr{F}_1)$ holds, (test on six vertices of $Z(\mathscr{F}_2)$) but $Z(\mathscr{F}_2)$ does not exist in $Z(\mathscr{F}_1)$: there is only one way to sample M_1 from $\mathscr{F}_1$ and, from the residual family, $\{N_3, N_1+N_2+N_3\}$, it is not possible to sample M_2 or M_3 and, a fortiori, both.

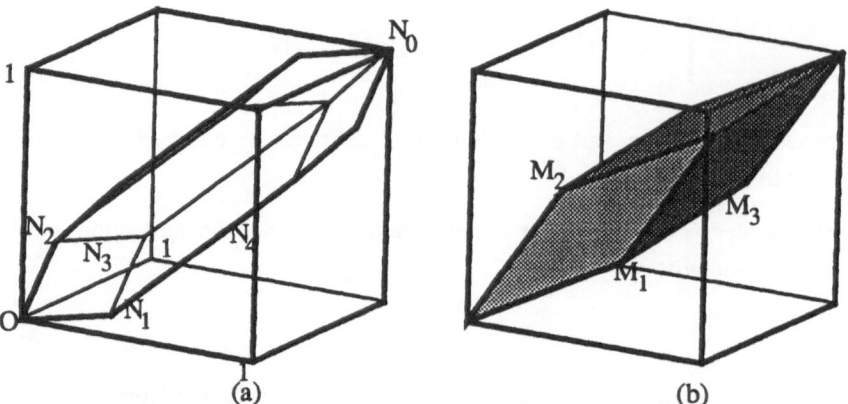

Figure 3 Zonohedra generated by families $\mathcal{F}_1$, (a), $\mathcal{F}_2$, (b), in the counter-example.
$N_1 = (1/2, 1/2, 0)$, $N_2 = (1/2, 0, 1/2)$, $N_3 = (0, 1/2, 1/2)$

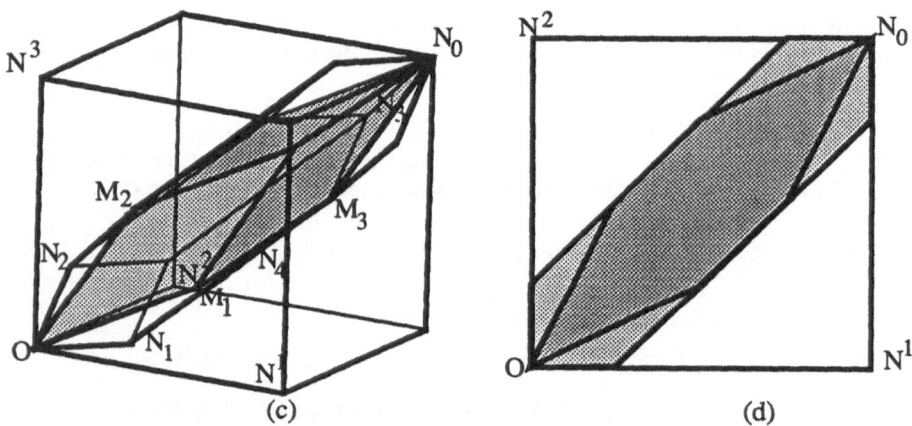

Figure 3c, Inclusion of the zonohedron $Z(\mathcal{F}_2)$ into $Z(\mathcal{F}_1)$
        3d, Projection on the plane of species 1,2

However, if we suppress any one of the three species, that is if we project Z onto one of the coordinate planes, then inclusion becomes equivalent to existence.

An example of existence: the three-component family

$$\left(\frac{1}{\sqrt{2\pi}\,\sigma} e^{-\frac{1}{2}\left(\frac{t}{\sigma}\right)^2} dt \,,\; \frac{1}{\sqrt{2\pi}\,\sigma} e^{-\frac{1}{2}\left(\frac{\delta-t}{\sigma}\right)^2} dt,\, 1 dt \right)_{t \in (-\infty,+\infty)}$$

exists in the family $\left(\dfrac{1}{\sqrt{2\pi}\,\sigma'} e^{-\frac{1}{2}\left(\frac{t}{\sigma'}\right)^2} dt \,,\; \dfrac{1}{\sqrt{2\pi}\,\sigma'} e^{-\frac{1}{2}\left(\frac{\delta-t}{\sigma'}\right)^2} dt,\, 1 dt \right)_{t \in (-\infty,+\infty)}$

provided $0 < \sigma < \sigma'$.

From the classical identity of variance additivity, see [1],

$$\frac{1}{\sqrt{2\pi}\ \sqrt{\sigma'^2 + \sigma^2}}\ e^{-\frac{\tau^2}{2(\sigma'^2 + \sigma^2)}} = \int_{-\infty}^{+\infty} \frac{e^{-\frac{w^2}{2\sigma^2}}}{\sqrt{2\pi}\ \sigma}\ \frac{e^{-\frac{(w-\tau)^2}{2\sigma'^2}}}{\sqrt{2\pi}\ \sigma'}\ dw$$

and calling $\sigma'' = \sqrt{\sigma'^2 - \sigma^2}$, we get

$$\lambda(t,t') = \frac{1}{\sqrt{2\pi}\ \sigma''}\ e^{-\frac{1}{2}\left(\frac{t-t'}{\sigma''}\right)^2} > 0, \int_{-\infty}^{+\infty} \lambda(t,t')\ dt = 1, \int_{-\infty}^{+\infty} \lambda(t,t')\ dt' = 1 \qquad (54)$$

The above kernel is therefore a mixing kernel. This process describes diffusion of "spots" of two isotopic species put at distance δ and diffusing along an uniaxial homogeneous medium.

5.2   A CLASSIFICATION OF TRANSFORMATIONS

Definition A transformation (or process) T, $\mathcal{F} \rightarrow \mathcal{F}' = T\mathcal{F}$, occurring in an isolated physico-chemical system is qualified as

  a pure separation iff $Z(\mathcal{F}) \prec Z(T\mathcal{F})$
        i.e. separation state increases
  a pure mixing iff $Z(T\mathcal{F}) \prec Z(\mathcal{F})$
        i.e. separation state decreases
  a **null** iff $Z(\mathcal{F}) = Z(T\mathcal{F})$
        i.e. separation state is invariant
  a **sepmix** in any other cases,
        i.e. separation contents are incomparable

In the above definition, increase and decrease are taken in the strict sense.

The counter-example in § 5.1 gives us a non trivial sepmix which mimics a separation process. A null process can be different from a trivial identity process, i.e. where spatial distribution functions of conservative species are invariant: i.e. **a null process may appear to do something** in the euclidean space, but, with respect to the separation state, it does nothing, i.e. the reduced family is an invariant of the process, see Theorem 11, §6).

5.3   ZONOIDS AND SEPARATION STATES

We are now in a position to state the basic theorem which shows that all properties required in Section 2 are indeed fulfilled by zonoids:

Theorem 10
  The *SEPARATION STATE* of a *Physico-Chemical SYSTEM* in a defined state $\mathcal{F}$, where $\mathcal{F}$ is a family, is the zonoid $Z(\mathcal{F})$. $Z(\mathcal{F})$ fulfills the requirements, given in Section 2, of being:

A. general properties
1. always **defined** (the set of zonoids $\mathbf{Z}$ is complete)
2. **invariant** (as an image of the reduced family $\mathcal{F}^*$)
3. **additive** (sum of Minkowski)
   extensive (similarity between zonoids)
4. **continuous** (Hausdorff distance between zonoids: $\mathbf{Z}$ is a metric space)
5. **linear transformation laws** (stability of zonoids by linear transformations in E)

   B. characteristic properties

6. **partially ordered**: existence ordering between zonotopes
7. Minkowski sum and existence order are compatible:

$$Z(\mathcal{F}_1) \prec Z(\mathcal{F}_2), Z(\mathcal{F}_3) \prec Z(\mathcal{F}_4) \Rightarrow Z(\mathcal{F}_1) + Z(\mathcal{F}_3) \prec Z(\mathcal{F}_2) + Z(\mathcal{F}_4)$$

8. **separation state decreases**, relative to the existence order, when any homogenization occurs in some part of the system
   The greatest lower bound is the distal segment N_0 in E.
9. **separation state increases**, relative to the existence order, when any homogeneous part of the system is made unhomogeneous.
   The least upper bound is the parallelepiped with diagonal N_0.

6 Zonoids and Partial differential Equations

In many problems of evolution, such as those involving transport, diffusion or transport-diffusive phenomena, the system state is a solution of a set of one or several partial differential equations describing local species balance and this solution is not explicitly known in most cases. A somewhat general form of these laws is

$$\frac{\partial u}{\partial t} + \text{div } J(u, \nabla u) = 0$$

with $u(x,t)$ the density of a species and J the flux of this species. The divergence operator is defined to be div $v(x,t) = \sum_{j=1}^{m} \frac{\partial v^j}{\partial x^j}$ for $v = (v^1, ..., v^m)$.

How J depends on u, on ∇u, or on both strongly influence the evolution. A convenient physical hypothesis is that $J(0, 0) = 0$ and that J can be split into two terms, one, the diffusion term, depending linearly on ∇u with coefficients depending, possibly nonlinearly, on u and the remaining transport or convection term, depending only on u. Further precisions on the splitting between these terms will be given in § 6.3.

A natural idea in the zonoid theory is to associate, to each value of the evolution parameter, say t, not only the solution $u(.,t)$, but also the zonoid $Z(t) = Z(\mathcal{F}(t))$ generated by the family $\mathcal{F}(t) = \{(1, u(x,t)) \, dx\}_{x \in \Omega}$. The physical motivation behind this is that there exists most often another conservation law for the volume, considered as a species (whose density is one). Flux of this species is the speed v and conservation law states that div $v = 0$. Therefore we wish to compare zonoids $Z(t)$ or give laws of evolution for the separation state, as a function of t. More precisely, we wish to assess the nature of transport,

diffusion or transport-diffusion in the light of the classification of processes given in §
5.2. The matter is complicated however by the fact that evolution laws for separation relate
not only to the very nature of balance equations but also to the boundary conditions. Since
it is hopeless to solve the general balance equations we shall use "a priori estimates", that
allow to deduce these evolution laws from the very shape of the balance equations without
the need to calculate the solution in each particular case. All the proofs work along the
same line, that is to seek for estimates for the function H_t in the course of evolution.

6.1 ZONOID EVOLUTION FOR AN IDEAL TRANSPORT EQUATION

This section is intended to provide the general material needed for a study of separation
evolution in nonlinear equilibrium multidimensional chromatography (part II). Physically,
the following Problem I models the evolution, under an ideal transport, of a conservative
extensive quantity Q whose volume density is u. The qualificative 'ideal' refers to two
hypotheses: the absence of dissipative phenomena, therefore, J depends only on u, and
the instantaneous equilibrium between phases in any small control volume. We have
chosen to skip up the fundamental topic of entropies, partly because of lack of space,
partly because zonoids will suffice for the present needs.

Problem Let $u(x,t)$, $u: \Omega \times I \to R^+$, be the density, $J = w(u)$, $w : R^+ \to R^m$ be a flux
function, with w a smooth nonlinear vector function of u. We consider a first order
quasilinear scalar equation written in conservative form:

$$
\text{Problem I} \quad \left\{
\begin{array}{l}
\dfrac{\partial u}{\partial t} + \displaystyle\sum_{j=1}^{m} \dfrac{\partial}{\partial x^j} J^j(u) = 0 \,, x \in \Omega \subset R^m, t \in I \subset R^+ \\[3mm]
u(x,0) = u_o(x) \geq 0 \quad , x \in \Omega
\end{array}
\right.
$$

Ω, an open bounded domain, $I = [0, T]$. u_o, the initial data are supposed of compact
support, i.e. to be zero outside of a compact set. Since speed of propagation is finite, Ω
may be chosen so that support $u(.,t) \subset \Omega$ for $t < T$ and we may extend u_o with 0 to get a
Cauchy problem (i.e. with initial data given on R). Furthermore, let us take $u_o \in L^1(\Omega)$,
i.e. $Q_o = \int_\Omega u(x,0)\, dx$ finite, i.e. and, to simplify matters, piecewise continuous.

From the theory of hyperbolic equations, it is well known that the solution $u(x,t)$ of
Problem I is locally
a) either constant along characteristic lines which are, in general, the integral lines, $x(t)$, of
the following system of m ordinary differential equations

$$
\frac{dx}{dt}(t) = \frac{dJ}{du}(x(t),t, u(x(t),t)) = v(u)
$$

In our case, these lines reduce to straight lines of known direction $v(u)$.
b) or, if discontinuous between u^+ and u^-, the discontinuity follows a curve $x(t)$, whose
tangent direction must satisfy the Rankine-Hugoniot condition:

$$
\frac{dx}{dt}(t) = \frac{J(u^+) - J(u^-)}{u^+ - u^-} = v(u^+, u^-)
$$

Discontinuities are labelled as follows: u^+ (resp. u^-) is the one-sided inner (resp. outer) limit at a discontinuity point of u, $n(x)$ is the outer normal vector at this point. Inner is taken here, relative to the level surface, (m-1)D surface of discontinuity of u or any level surface which crosses the discontinuity.

However we can place such discontinuities in a solution in many ways and we need a selection rule to pick up those ones which are "physically meaningful", in fact, those which will ensure both existence and unicity of solutions in the class of functions to which belongs the initial data.

The condition E (Oleinik, Krushkov [12,13]) states that, for any $u^- \le \tau \le u^+$, an admissible, or so called entropic, shock with speed $v(u^+,u^-)$ must satisfy:

$$\text{sgn}(u^+ - \tau)\left\{ (u^+ - \tau)(v(u^+,u^-), n) - ((J(u^+) - J(\tau)), n)\right\} \le 0$$

It is well known that this shock selection rule has the above required properties : if condition E holds, solution of problem (I) exists, is unique and depends continuously on initial data, i.e. problem (I) is well posed. It is also a classical result that if u_0 is non negative for a.e. x in Ω, u will remain so.

Condition Z, or rule Z, for the shock selection states:
the rate of separation decrease must be maximum at each instant, i.e. no other choice of discontinuities can increase this rate.

Therefore, the zonoid itself at any given time must be minimum, in the sense of inclusion compared to zonoids associated with solutions which would satisfy only the Rankine-Hugoniot conditions. In this respect, condition Z is of the same type as the entropy rate of admissibility criterion given by Dafermos [11] which he showed to be equivalent to condition E.

Theorem 11 The class $Z(\mathcal{F}(t))$ of zonoids associated with a positive entropic solution u of (I) is

invariant if u is a classical solution, $\quad Z(\mathcal{F}(t)) = Z(\mathcal{F}(0))$

decreasing in t if u is a weak solution , $Z(\mathcal{F}(t_1)) \supset Z(\mathcal{F}(t_2)) \ \forall \ t_1 < t_2$

$$Z(\mathcal{F}(t_1)) \succ Z(\mathcal{F}(t_2)) \ \forall \ t_1 < t_2$$

and conversely, if the above conditions hold, the solution is entropic.

Therefore the Theorem 11 states that the Oleinik rule E is equivalent to the rule Z.

Corollary 11 a In the course of time we have, under the conditions of theorem 11, the following equivalent principles of dilution:
(i) the minimum volume of a given quantity Q of species does not decrease
(ii) the maximum content of species in a given volume V does not increase

(iii) the maximum excess species content above a given density level τ of species does not increase.

Proof Write the condition Z as in lemma 8:

$$\frac{\partial H}{\partial t}(\tau,t) = \frac{\partial Q(\tau,t)}{\partial t} - \tau \frac{\partial V(\tau,t)}{\partial t} \le 0$$

with

$$\frac{\partial V(\tau,t)}{\partial t} = \int_{\partial \omega_\tau} (v, n)\, d\sigma$$

At point x on $\partial\omega_\tau$, v is given by one of the above equations, depending if $u = \tau$ or, if $u^+ > \tau$. From the property of finite speed of propagation, Ω can be chosen such that $\partial\omega_\tau$ never intersects $\partial\Omega$. Let us split $\partial\omega_\tau : u \geq \tau$ into $\partial\omega_\tau^0$ u(x) = τ and $\partial\omega_\tau^+ = \partial\omega_\tau \backslash \partial\omega_\tau^0$. Consider now the three cases of lemma 9:

(a) $\partial Z^+(t)$ is smooth at τ. Then, using the Gauss divergence formula, which holds for almost all t since $u \in BV(\Omega)$,

$$\frac{\partial Q(\tau,t)}{\partial t} = \int_{\omega_\tau} \frac{\partial u}{\partial t} dx = - \int_{\partial\omega_\tau^+} (J(u^+)-J(\tau),n) \, d\sigma$$

Although the time derivative under the integral has no meaning in general because u may be discontinuous on surfaces in ω_τ, it is easy to show that it makes a good sense here; indeed, by excluding these surfaces from ω_τ, we do not change the value of the integral in view of the Rankine-Hugoniot condition and the above expression is meaningful. From this, the final form of condition Z follows

$$\frac{\partial H}{\partial t}(\tau,t) = \int_{\partial\omega_\tau} \left((u^+ - \tau)\, v(u^+,u^-) -(J(u^+) - J(\tau)),\, n \right) d\sigma \leq 0 \qquad (55)$$

and we recognize that it is just the integral of condition E along $\partial\omega_\tau$.

(b) u is discontinuous on the whole of $\partial\omega_\tau$ and, on $\partial\omega_\tau$: $\tau_{max} = $ Min u^+, $\tau_{min} = $ Max u^-, $\tau_{max} > \tau_{min}$. For $\tau_{min} < \tau < \tau_{max}$, point $(V(\tau), Q(\tau))$ is stationary (a corner point on ∂Z^+) and Condition Z remains identically true.

(c) u has a the plateau at level τ. If $\tau > 0$, by applying the preceding argument for $\tau - \delta\tau$ and $\tau + \delta\tau$, $\delta\tau > 0$, and passing to the limit we get the inclusion property for the limit points A, B and hence, by convexity, for the segment AB $\in \partial Z^+$. If the plateau is at $\tau = 0$, same argument with one sided limit if necessary (i.e. if there is a shock with lower limit $u^- = 0$), show that upper point of ∂Z^+ can be completed by an horizontal segment.

Therefore, if condition E holds for any weak solution, then condition Z holds for the associated zonoid. Conversely, if u contains a nonentropic discontinuity, which means that for τ in some interval, the reverse inequality holds in ineqn. 55, then we may always find a smooth function u' which coincides with u on this part of the shock. From eqn. 55, for the evolution of the zonoid associated with u', the reverse inclusion holds and condition Z is violated. $\qquad\square$

The above Theorem can be extended to functions of bounded variation, $u_0 \in BV(\Omega)$, that is $u_0 \in L^1(\Omega) \cap BV(\Omega)$. Indeed, observe that eqn. 55 contains no derivative of u and makes perfectly good sense if $u \in L^1(\Omega) \cap BV(\Omega)$. However the example in II-§3.2 shows that, since, even in the solution of simple physical models, one may find Dirac measures, a more general solution space than $L^1(\Omega) \cap BV(\Omega)$ is needed.

122

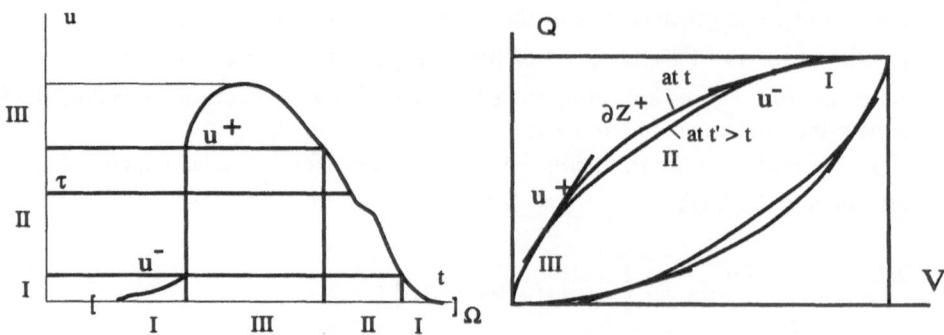

Figure 4 left: Species distribution with different cases for level sets: in part I, III propagation is locally reversible. Mixing due to shock wave localizes in smooth part II of the system, **opposite the shock**.

right: Schematic evolution of 2D zonoid (V, volume, Q, species quantity).

From the above proof it is easily inferred that the volume of plateaus cannot increase in the course of evolution. Therefore also, a plateau cannot appear in a domain where u was previously smooth.

Finally, we remark that we have introduced a new invariant or, more generally, a new inequality, for the problem of evolution: if there is no shock wave, the zonoid, and the separation state, are **invariant** in the course of evolution. This was, of course, the motivation to introduce the *null* process in our classification, §5.2. The fundamental reason for this invariance is that level surfaces propagate without deformation along characteristic lines. However, since their speed depends on u, they shift their relative position.

6.2 ZONOID EVOLUTION FOR A DIFFUSION EQUATION

We shall now prove that nonlinear diffusion in several space dimensions of a species in an immobile homogeneous medium is a mixing process. Here we have no transport terms in the flux, which can be written $J(u,\nabla u) = - D(u) \nabla u$, that is the flux depends linearly on the gradient.

Problem We consider a second order quasilinear parabolic equation in divergence form :

Problem II
$$\begin{cases} \dfrac{\partial u}{\partial t} - \text{div } ([D(u)] \nabla u) = 0 \\[2mm] u(x,0) = u_o(x) \geq 0 \\[2mm] ([D(u)] \nabla u, n) = 0 \text{ on } \partial\Omega \times [0,\infty] \end{cases}$$

Here, Ω is an open, bounded domain in R^m, $m \geq 2$, with smooth boundary $\partial\Omega$ and n the outer unit normal vector to $\partial\Omega$., (.,.) is the scalar product in R^m, $D(v)$ is a m×m matrix strictly positive definite with bounded elements $D_{jk}(v)$, $j,k = 1,\ldots,m$,

$$\nu \sum_{j=1}^{m} \xi_j^2 \leq \sum_{j,k=1}^{m} D_{jk}(v)\, \xi_j \xi_k \leq \mu \sum_{j=1}^{m} \xi_j^2, \quad \forall\, \xi \in R^m, \forall\, v \in R^+ \tag{56}$$

where ν, μ are strictly positive constants.

The solutions space is defined as $\left\{ u \in C^0(\Omega \cup \partial\Omega) \cap C^1(\Omega), \dfrac{\partial u}{\partial n} = (\nabla u, n) \in C^0(\Omega \cup \partial\Omega) \right\}$.

Now comes a formal statement of

Theorem 12 For any classical solution $u(x,t)$ of problem II we have

(i) $Z(t)$ is strictly decreasing, relative to inclusion : $Z(t_2) \subset Z(t_1)$, $t_2 > t_1$

(ii) $Z(t)$ is strictly decreasing, relative to existence : $Z(t_2) \prec Z(t_1)$, $t_2 > t_1$

Proof It is well known that when matrix $D_{jk}(u)$ is strictly positive definite, provided mild conditions hold on domain, coefficients, initial and boundary conditions, then a unique classical solution u exists, which is smooth, bounded, together with its gradient. Since u is bounded, by translation we may set u to be non negative.

Since Problem (II) is a problem of evolution in time, for a $\tau \in [u_{min}(t), u_{max}(t)]$, let $\omega_\tau(t)$ be the level set at fixed t, $\partial\omega_\tau(t)$ its boundary. We split $\partial\omega_\tau(t)$ in two parts, $\partial\omega_\tau(t) = \partial\omega_\tau{}^i(t) \cup \partial\Omega_\tau(t)$ with

$$\partial\omega_\tau{}^i(t) = \partial\omega_\tau(t) \cap \Omega \text{ (on which } u(x,t) = \tau)$$
$$\partial\Omega_\tau(t) = \omega_\tau(t) \cap \partial\Omega \text{ (on which } u(x,t) \geq \tau)$$

Using Lemma 10 and eqn. 31 we want to show that the following expression is negative

$$\frac{\partial H}{\partial t}(\tau,t) = \int_{\omega_\tau} \frac{\partial u(x,t)}{\partial t}\, dx < 0$$

Using the equation of Problem (II) then the Gauss divergence Theorem we get

$$\frac{\partial H}{\partial t}(\tau,t) = -\int_{\partial\omega_\tau^i(t)} ([D(u)]\, \nabla u,\, n)\, d\sigma + \int_{\partial\Omega_\tau(t)} ([D(u)]\, \nabla u,\, n)\, d\sigma$$

In view of the homogeneous Neumann conditions last integral vanishes and, recalling eqn. 4, we get

$$\frac{\partial H}{\partial t}(\tau,t) = -\int_{\partial\omega_\tau^i(t)} \frac{([D(u)]\, \nabla u,\, \nabla u)}{\|\nabla u\|}\, d\sigma < 0 \tag{57}$$

which is negative since matrix D is positive definite. In fact, from eqn. 56, it is bounded away from 0. Since eqn. 57 holds for all τ, the support function $\psi(\eta \,|\, Z(t))$ is a decreasing function of time. Taking $\tau \leq u_{min}(t)$ in $V(\tau)$ and $Q(\tau)$, we see that V_0 and Q_0 are constant, i.e. all zonoids have the same distal point: therefore zonoid is a decreasing function of time as well, in the sense of inclusion. This closes part (i) of the proof. The Blackwell Theorem, quoted in §5, yields equivalence between (i) and (ii). ☐

Observe that in eqn. 57 requirements on smoothness of u are low since it contains no second derivatives of u. Therefore one may think that this theorem holds for generalized solution of Problem II.

The maximum principle is imbedded in Theorem 12 since it comes up directly from inclusion of the zonoid supporting cones at O, i.e. from evaluation of Limit$_{V \to 0}$ $\partial Q(V,t)/\partial V$.

Some equivalent forms of Theorem 12 are of direct physical interpretation:

Corollary 12 a In the course of time we have, under the conditions of Theorem 12, the following equivalent principles of dilution:
(i) the minimum volume of a given quantity Q of species increases
(ii) the maximum content of species in a given volume V decreases

(iii) the maximum excess species content above a given density level τ of species decreases.

To each of these statements we may associate an equation analog to eqn. 57 - which holds for (iii).

6.3 ZONOID EVOLUTION FOR A TRANSPORT-DIFFUSION EQUATION

Since Problem I can be considered to be an idealization of the real physical situation where diffusive effects are always present, we deal now with the general case. The results will be used in Part II to show that the nonlinear transport and diffusion of a single species in multidimensional chromatography is a solute-carrier mixing process.

Consider a model for the convection-diffusion of a conservative species in a volume Ω in m-D space (e.g. m = 2, 3) in the creeping flow assumption, that is when a flow field v, preserving volumes in R^m, is given such that:

$$
\text{Problem III} \quad
\begin{cases}
\text{div } v(x,u) = 0 \\[4pt]
\dfrac{\partial u}{\partial t} - \text{div } (J_t(x, u) + J_d(x, u, \nabla u)) = 0 \\[4pt]
u(x,0) = u_o(x) \geq 0 \\[4pt]
(J_t(x, u) + J_d(x, u, \nabla u), n) = 0 \text{ on } \partial \Omega \times [0,\infty]
\end{cases}
$$

The transport term, $J_t = (J^1, J^2, \ldots, J^m)$, a smooth function of (x, u), is such that

$$J_t(x,u) = u \, v(x,u)$$

The diffusion term, is given by,

$$J_d^j(x, u, \nabla u) = -\sum_{k=1}^{m} D_{jk}(x,u) \frac{\partial u}{\partial x^k}, \quad j = 1, \ldots, m$$

with the Fickian diffusion matrix $D_{jk}(x,u)$, $j,k = 1,2,\ldots m$, positive definite.

Theorem 13 For any classical solution $u(x,t)$ of problem III we have

    (i) $Z(t)$ is strictly decreasing, relative to inclusion : $Z(t_2) \subset Z(t_1)$, $t_2 > t_1$

    (ii) $Z(t)$ is strictly decreasing, relative to existence : $Z(t_2) \prec Z(t_1)$, $t_2 > t_1$

Proof results from slight modifications in proof of Theorem 12. In fact we get an additional term in H_t, eqn. 57, the shape of which is

$$\tau \int_{\partial\omega_\tau} (\, v(x, \tau), n) \, d\sigma = \tau \int_{\omega_\tau} \text{div } v(x, u) \, dx = 0$$

which vanishes, provided div $v(x,u) = 0$. Note that this happens to be the case when v depends only in u.   □

There are two interesting connections between problem I and Problem III:

1 In the well known method of artificial viscosity, one considers the limit of problem III when diffusion tends to zero, usually by replacing D by ε D with $\varepsilon > 0$, $\varepsilon \to 0$. It is a classical result that then, u_ε, the solution of problem III, tends to the entropic solution u of problem I. If we call Z_ε (resp. Z) the zonoid associated to u_ε (resp. u), we have that $Z_\varepsilon \to$ Z from below; indeed, it is clear from eqn. 57, written with ε D, at a given time, that $\dfrac{\partial H_\varepsilon}{\partial t}$

is a linear, decreasing function of ε, hence Z_ε itself is a decreasing function of ε, in the sense of the inclusion. Therefore, the zonoid satisfying condition Z in Problem I can be obtained as a limit, in two different ways: from above, by taking the least upper bound from all the weak solutions of problem I, from below, by taking the greatest lower bound from all the diffusive solutions of problem III.

2 We note that the integrands on level surfaces in the expression of $\dfrac{\partial H}{\partial t}$ in eqns. 55, 57 look very different. However one may show that, a direct link between these results comes from the fact that, along level surfaces,

$$\text{Lim}_{\varepsilon \to 0} \left(-\varepsilon \, [D(\tau)] \, \nabla u_\varepsilon, n_\varepsilon \right) = ((u^+ - \tau) \, v(u^+, u^-) - (J(u^+) - J(\tau)), n)$$

In fact, in many physical situations $\partial\omega_\tau$ is sufficiently smooth and we may drop the scalar product in the above formula to get the corresponding vector identity.

A first generalization of Theorems 11, 12, 13 allows the flux J to depend smoothly on (x,t) as well as on u. Smoothness requirement on J is not essential and can be relaxed using more sophisticated tools in the proofs (the so-called weak formulation of the problem).
A second generalization allows to replace the density u by any smooth, increasing function $f(u)$. Clearly u and $f(u)$ have same level sets and, using the Lemma 11, proofs of theorems 11, 12, 13 remain in force, for the zonoids $Z_{f(u)}$ associated to the solutions of the equation:

$$\frac{\partial f(u)}{\partial t} + \text{div } (J_t(x,u) + J_d(x,u,\nabla u)) = 0 \tag{57a}$$

The case of chromatography in Section II-§3 will provide us with a case of this type, in which u will be the concentration and $\tilde{q} = f(u)$, the total accumulated solute density.

More generally, in Problem III, we may consider a general flow, that is

$$\text{div } v(x,u) = r(x,u) \neq 0$$

then we get a term involving r(x,u), the rate of volume creation inside the level set:

$$\frac{\partial H}{\partial t}(\tau,t) - \tau \int_{\omega_\tau} r(x, u) \, dx = - \int_{\partial \omega_\tau} \frac{([D(u)] \nabla u, \nabla u)}{\|\nabla u\|} \, d\sigma < 0 \tag{58}$$

Thus we arrive at the general conjecture that the elementary effects, be they diffusion or transport must be mixing effects for a single species if we wish to have a unique solution depending continuously on the initial and boundary conditions.

The case of several species is of course the ultimate goal of our investigation and we have not even scratched it. The reason is that in the mathematical theory of nonlinear conservation laws, there are specific difficulties which have prevented definitive results from being obtained at present: in the presence of shock waves, the principle of maximum needs not hold. Consequently, inclusion cannot be hoped in this case. This fact could well be linked with the special properties of 2-D zonoids, which do not extend into higher dimensional spaces, e.g. the fact that the intersection of two 2-D zonoids with same sum is a 2-D zonoid does not extend to 3-D zonoids. An interesting and encouraging remark is that nonlinear effects have not affected the simplicity of the results given in Theorems 11, 12, 13. Therefore we may be confident that zonoids will prove useful for systems of conservation laws.

6.4 EQUATIONS OF 2-VOLUME DECREASE

As already mentioned, Ξ, the 2-volume of zonoid is a natural although degraded index for separation. We first evaluate Ξ, its rate of time variation and then apply this to a problem in the "design of the best system", which gives an insight on either what are the best initial conditions or what are the best boundary conditions.

6.4.1. The rate of 2-volume decrease For the problem I, inserting eqn. 55 into eqn. 46 we get the immediate result:

$$\frac{d\Xi}{dt}(t) = 2 \int\limits_{\substack{\partial\omega_\tau^+ \\ u^+(t)}}^{u^-(t)} ((u^+ - \tau)\, v(u^+,u^-) - (J(u^+) - J(\tau)), n)\, d\sigma \quad \int\limits_{\partial\omega_\tau^0} \frac{1}{\|\nabla u(x)\|}\, d\sigma \quad d\tau \le 0 \,(59)$$

in the case of a single shock and no plateau. Therefore the only cause in 2-volume loss is the shock. However the loss depends not only on the shock strength, but also on ∇u values, taken along part of the level surface where u is smooth. For any number of shocks, one just need to extend the integration range in eqn. 56 to cover all the shocks. If u has plateaus, for some values τ_i of u, we take them into account by adding to eqn. 59, a sum of terms like

$$\frac{\partial}{\partial t}\big(H(\tau_i,t)\, |\omega_{\tau_i}^*(t)|\big) = H(\tau_i,t)\, \frac{\partial}{\partial t}\big(|\omega_{\tau_i}^*(t)|\big) = H(\tau_i,t) \int\limits_{\partial\omega_{\tau_i}^*(t)} (v,n)\, d\sigma \qquad (59a)$$

This comes from eqn. 42a, noting that $H(\tau_i,t)$ is constant on a neighborhood of t - condition Z vanishes on the limit $u \to \tau_i$. Here, v designates, either the characteristic speed, or the shock speed, according to the case, n, the outer normal to $\partial\omega^*_{\tau_i}$.

The case of m=1 is simpler and can be further simplified into an integral extending over the x_0-domain, opposite the shock, where u is smooth, see domain II on fig. 4:

$$\frac{d\Xi}{dt}(t) = 2 \int\limits_{u^+(t)}^{u^-(t)} [(u^+ - \tau)\, v(u^+,u^-) - (J(u^+) - J(\tau))]\, \frac{\partial x}{\partial u}(\partial\omega_\tau^0, t)\, d\tau \le 0 \qquad (60)$$

and the above extension hold true, eqn. 59a becoming $H(\tau_i,t)\,\big(v(\tau_i^-) - v(\tau_i^+)\big)$. Recall that in this case, the τ_i are excluded from the integration domain in eqns. 59, 60.

The case of Problems II, III is solved by the following:

Theorem 14 For u(x,t), a solution of Problem II or III, the rate of 2-volume decrease in course of time is given by
for $m \ge 2$

$$\frac{d\Xi}{dt}(t) = 2 \int_{u_{max}(t)}^{u_{min}(t)} \int_{\partial\omega_\tau} \frac{([D(u)]\,\nabla u, \nabla u)}{\|\nabla u\|}\, d\sigma \int_{\partial\omega_\tau} \frac{1}{\|\nabla u\|}\, d\sigma\, d\tau \qquad (61)$$

for m = 1

$$\frac{d\Xi}{dt}(t) = 2 \int_{u_{max}(t)}^{u_{min}(t)} D(\tau) \sum_{j=1}^{p(\tau)} \left|\nabla u(\alpha_j(\tau,t))\right| \sum_{j=1}^{p(\tau)} \frac{1}{\left|\nabla u(\alpha_j(\tau,t))\right|}\, d\tau \qquad (62)$$

Proof Inserting eqn. 57 into eqns. 46 or 47, we get the required results.

Remarks In the case m= 1, the finite sums are taken at points where u = τ, as in eqn. 11.

In eqns. 61, 62, dependence on t is implicit through u(x,t) and ω_τ which depends also on u(x,t). This also explains why Problems II and III may give the same formal expression. However, if we can have any *a priori* information on level sets (e.g. isoperimetric inequality or symmetry) we will be able to get a corresponding a priori bound on Ξ. Similarly, inserting bounds given by eqn. 56, we may derive bounds in eqn. 61, 62 and also proceed as in the following corollary:

Corollary 14 In problems II, III, if [D(u)] = D(u) [I], with [I], the m×m identity matrix, then

$$\frac{d\Xi}{dt}(t) = 2 \int_{u_{max}(t)}^{u_{min}(t)} D(\tau) \int_{\partial\omega_\tau} \|\nabla u\|\, d\sigma \int_{\partial\omega_\tau} \frac{1}{\|\nabla u\|}\, d\sigma\, d\tau \qquad (63)$$

Corollary 14 holds for diffusion or transport-diffusion in an isotropic medium. If furthermore diffusion is constant, taking D out of the integral, we have the following quite remarkable result :

the diffusion coefficient and the geometric features of level surfaces of the solution u contribute independently to the rate of decrease of 2-volume.

Recalling concepts introduced in § 3.2, we see that, in eqn. 63, integral terms relate, the former, to the vertical variation of u along the boundary of level sets, the latter, to the corresponding horizontal variation.

6.4.2 A problem of comparison A theoretical problem arises quite naturally in term of zonoids. It has been looked at in recent studies [10,11] and has potentially important practical applications in design of physico-chemical systems.

Problem IV Given a zonoid Z_0, from all solutions u of Problems II, III such that $Z_{u_0(x)}$ = Z_0, find those which minimize some criterion ϕ:

$$\text{Min} \quad \phi(u(x,t)) \text{ at a fixed t}$$
$$u_o(x),\ \Omega$$
$$\begin{vmatrix} Z_{u_0(x)} = Z_0 \\ |\Omega| = V_0 \end{vmatrix}$$

A first problem allows variation of the initial data, at Ω fixed. A more general one allows furthermore shape variation of Ω, under the constraint $|\Omega|$ = V_0. In the case of interest to us,, $\phi = -\dfrac{d\Xi}{dt}$, the solution of this design problem is given by the following:

Theorem 15 For any classical solution of Problem II,III with matrix $[D(u)] = D(u)\ [I]$, the following a priori inequality holds

$$2 \int_{u_{min}}^{u_{max}} \left(\int_{\partial\omega_\tau} d\sigma \right)^2 D(\tau)\, d\tau \ \leq\ \left| \frac{d\Xi(t)}{dt} \right| \tag{64}$$

with equality iff $\|\nabla u\|$ = const on $\partial\omega_\tau$ for $\tau \in [u_{min}, u_{max}]$.

Proof comes directly from inserting, into in eqn. 63, the Cauchy-Schwarz inequality

$$\left| \int_\Omega f(x)\ g(x)\ dx \right|^2 \leq \int_\Omega f(x)^2\ dx \int_\Omega g(x)^2\ dx \text{ with } f = \sqrt{\|\nabla u\|},\ g = \frac{1}{\sqrt{\|\nabla u\|}}.$$

However one cannot expect that the equality case in 15 applies for Problem III to other than exceptional cases since nonlinear, as well as any anisotropic, effects tend to break any existing symmetry of the gradient.

Theorem 16 For a classical solution of Problem II, in a space of dimension m ≥ 2, B_m, the volume of the unit ball, with constant diffusion, the following a priori inequality holds,

$$2\, D\, m^2\, B_m^{\frac{2}{m}} \int_{u_{min}}^{u_{max}} |\omega_\tau|^{\frac{2(m-1)}{m}}\, d\tau \ \leq\ \left| \frac{d\Xi(t)}{dt} \right| \tag{65}$$

Proof is easy: recalling the isoperimetric inequality between A, the perimeter of a set and V, the volume of the set [6],

$$A^2 \geq m^2\, B_m^{\frac{2}{m}}\, V^{\frac{2(m-1)}{m}} \text{ with } B_m = \frac{2\, \pi^{\frac{m}{2}}}{m\, \Gamma(\frac{m}{2})}$$

we take $A = \int_{\partial\omega_\tau} d\sigma$, $V = |\omega_\tau|$ and insert this in eqn. 64 to get eqn. 65.

We get the equality case in the isoperimetric inequality when the level sets are balls, an hypothesis already implied by the case of equality in 15, stating that grad u should then be constant on level surfaces of u. Therefore u has a spherical symmetry if equality holds. In this case, letting $u(x,t) = u(r,t)$, $r = \|x\|$, we get another form of this result $(m\geq 2)$

$$2 D (m-1) B_{m-1}{}^2 \int_{r(u_{min})}^{r(u_{max})} r^{2m-3} u(r,t)\, dr = \left|\frac{d\Xi(t)}{dt}\right|$$

The type of functional dependance of the rate of zonoid 2-volume decrease on quantities derived from u, is strongly dependent on the dimension of the Euclidean space:

When $m = 1$, eqn. 65 becomes

$$8 D \left(u_{max}(t) - u_{min}(t)\right) = 4 D\, TV_\Omega(u(.,t)) \leq \left|\frac{d\Xi(t)}{dt}\right| \tag{66}$$

and the lower bound of the rate of decrease of 2-volume depends only on the total variation of u.

When $m = 2$, we get a lower bound for the rate of decrease of 2-volume, which depends only on $\mathcal{V}$, the volume enclosed between the plane at u_{min} and the graph of u on Ω,

$$8 \pi D \int_{u_{min}}^{u_{max}} |\omega_\tau|\, d\tau = 8 \pi D\, \mathcal{V} \leq \left|\frac{d\Xi(t)}{dt}\right| \tag{67}$$

Physical meaning: in all systems isolated with isotropic diffusion represented by a given zonoid at the initial state, the system which has the smallest rate in separation decrease (counted either as Hausdorff distance between zonoids, as pointwise on ∂Z^+ or as 2-volume) is of radial symmetry. Of course, a necessary and sufficient condition to that is that Ω and $u_0(x)$ be themselves of radial symmetry.

6.5 THE EQUATIONS OF ZONOID EVOLUTION

We may derive interesting results by establishing the evolution equations of the zonoid boundary, e.g. for ∂Z^+. Such a method was initiated by C. Bandle in a reaction-diffusion problem [6]. These equations, with Q as the unknown quantity and V,t as the variables define a time evolution problem in one space dimension, and it turns out that they define in effect a Cauchy problem of the same PDE type as the starting equations.

From eqn. 31 in Lemma 10, we get

$$\frac{\partial H}{\partial t}(\tau,t) = \frac{\partial Q}{\partial t}(V, t)\,,\, \tau = \frac{\partial Q}{\partial V}(V, t) > 0,\, \frac{\partial \tau}{\partial V}(V,t) = \frac{\partial^2 Q}{\partial V^2}(V, t) \leq 0$$

and deal now in turn with the three above defined problems.

Problem I: Inserting theses values into the corresponding equations for H, we get again a hyperbolic problem, but with different initial and boundary conditions: from eqns. 55, 9 and Lemma 8,

$$\frac{\partial Q}{\partial t} - \frac{1}{\tau} \int_{\partial \omega_\tau} ((u^+ - \tau) \, v(u^+, u^-) - (J(u^+) - J(\tau)), \, \mathbf{n}) \, d\sigma \frac{\partial Q}{\partial V} = 0 \tag{68}$$

$$Q(V,0) = 0, \, V < 0; \quad Q(V,0) = \partial Z^+, \, 0 \le V \le V_0; \quad Q(V,0) = Q_0, \, V > V_0$$

that is a hyperbolic equation since the second coefficient in eqn. 68 is non negative. Initial data define a Cauchy problem on R.
This shows that Q at a given V propagates with a positive speed along a characteristic line and we construct the zonoid boundary at any time if we know the position and the strength of shocks. The edges corresponding to the u plateaus propagate as edges with same slope but a decreasing length.

Problem II: from eqn. 59, we get the Dirichlet problem:

$$\frac{\partial Q}{\partial t} - \int_{\partial \omega_\tau} \frac{([D(u)] \nabla u, \nabla u)}{\|\nabla u\|} \, d\sigma \int_{\partial \omega_\tau} \frac{1}{\|\nabla u\|} \, d\sigma \frac{\partial^2 Q}{\partial V^2} = 0 \tag{69}$$

with $Q(V,0) = \partial Z^+, \, 0 \le V \le V_0, \, Q(0,t) = 0, \, Q(V_0,t) = Q_0$.

Problem III: from eqn. 60, in the general case,

$$\frac{\partial Q}{\partial t} - \int_{\partial \omega_\tau} (\, v(x, \tau), \, \mathbf{n}) d\sigma \frac{\partial Q}{\partial V} - \int_{\partial \omega_\tau} \frac{([D(u)] \nabla u, \nabla u)}{\|\nabla u\|} \, d\sigma \int_{\partial \omega_\tau} \frac{1}{\|\nabla u\|} \, d\sigma \frac{\partial^2 Q}{\partial V^2} = 0 \tag{70}$$

we get a parabolic problem with $Q(V,0) = \partial Z^+, \, 0 \le V \le V_0, \, Q(0,t) = 0, \, Q(+\infty,t) = Q_0$.

By an appropriate change in the initial conditions and in the sign of coefficients in the above equations we would get the evolution equations of ∂Z^-.

All these equations become very simple if we take into account the rate of 2-volume decrease, e.g. eqn. 69 becomes

$$\frac{\partial Q}{\partial t} + \frac{1}{2} \frac{\partial^2 \Xi}{\partial t \partial \tau} \frac{\partial^2 Q}{\partial V^2} = 0 \tag{71}$$

Any of the minorations we have done in the preceding subsection may be applied here also. For example, in the case of 15, noting that all terms are negative in eqn. 71, we get a semilinear inequation describing the evolution of zonoid boundary:

$$\frac{\partial Q}{\partial t} - D \, m^2 \, B_m^{\frac{2}{m}} \, V^{\frac{2(m-1)}{m}} \, \frac{\partial^2 Q}{\partial V^2} \leq 0 \tag{72}$$

with the equality case for a solution with radial symmetry.

The stationary solution in these equations can also be used to study the long time approximation of the separation state in problems I to III. Classical a priori inequalities applied to the above equations allows to derive inequalities upon volume and species content of the level sets of the solution.

7. Conclusions

We have presented here a justification for the use of zonoids as separation states in systems of laws of conservation as well as a summary of elementary mathematical properties of zonoids. Section 6 illustrate the capacity of this theory to produce unanticipated and basically simple results such that the transport is a null process, as well as results which comfort the intuition such that diffusion is a mixing process. The importance of the excess species quantity function, $H(\tau,t)$ and of its time derivative H_t is also pointed out. Let us just show how it also roots into the thermodynamic of irreversible processes [26]; by integration of eqn. 57, using Federer formula, eqn. 7, with an appropriate $f(x)$, we get

$$-\int_{u_{min}}^{u_{max}} \frac{\partial H}{\partial t}(\tau,t) \, d\tau = \int_{\Omega} \left(-[D(u)] \, \nabla u, -\nabla u\right) dx = \int_{\Omega} (\mathbf{J}, \mathbf{X}) \, dx = \sigma[S] > 0$$

in which, letting $\mathbf{J} = -[D(u)] \, \nabla u$ be the flux, $\mathbf{X} = -\nabla u$, the driving force, we recognize $\sigma[S]$, the rate of entropy production in volume Ω, in the notations of the irreversible thermodynamic. The extension of this formula to multicomponent processes will be carried out in a further work.

The unification power of the methodology is underlined by the fact that we are able to deal with three different topics, the zonoid evolution, the equations for the zonoid boundary and the equation for the rate of 2-volume decrease, and that, independently of the type of the conservation law under consideration.

Part II of these two papers describe applications of the theory to assess the real status of chromatography as a separation process.

ACKNOWLEDGMENT

Part of this work was made at time when the author was lecturer with Professor G. Guiochon at Ecole Polytechnique and greatly influenced by many discussions with him. I gratefully acknowledge also discussions with L. Tartar (now at Carnegie Mellon University), M. Schœnauer, of the Applied Mathematics Department of this school. The continuous interest of P.J. Laurent, B. Lacolle from IMAG (University Joseph Fourier, Grenoble) was equally important for completion of this work. I also thank W. Jäger (Univ. Heidelberg) and E. Canon. Figure 2 was created using a program for computer manipulation of zonohedra created by N. Szafran (LMC, IMAG, Univ. Joseph Fourier, Grenoble). This work was supported by Société Nationale Elf-Aquitaine.

SYMBOLS

Latin Alphabet

E	vector space of quantities	(moles)
dx	the volume element, $dx = dx_1 dx_2 \ldots dx_m$	
e	base vector of E	
F	flow-rate vector	(moles/sec)
H	excess quantity function	
H_t	$\dfrac{\partial H}{\partial t}$	
$\mathcal{H}_{m-1}$	(m-1)-dimensional Hausdorff (surface) measure	
I	interval of natural numbers, $\{1,\ldots, n\}$	
i	chemical species	
J	interval of natural numbers	
J	flux of a conservative species	
$\mathcal{L}_m$	m-dimensional Lebesgue (volume) measure	
M, N	quantity vector	
N_o	overall (or sum) quantity vector or distal point	
n	dimension of the space E number of species	
O	origin	
p	number of regions or of cuts	
Q	Content of species	
R	vector space of real numbers $(-\infty, +\infty)$	
R^3	usual Euclidean space	
t	current parameter on a curve	
u	a function	
X	molar fraction vector	
x	spatial coordinate in R or R^3	
Z	zonoid, convex set of mixtures	

Greek Alphabet

α	a real parameter
Δ_{n-1}	n-1 dimensional simplex (of molar fractions in E, dim E = n)
$\Gamma(m)$	the Euler gamma function : $\Gamma(m+1) = m\,\Gamma(m)$, $\Gamma(1) = 1$, $\Gamma(1/2) = \sqrt{\pi}$
∂	operator taking boundary of an open set
Ξ	n-volume of Z, $(Z \subset E, \dim E = n)$ $(\text{moles})^n$
ξ	extent of separation
Σ	system state
τ	reduced time, parameter on a curve

Indices

i, k	chemical or conservative species (upper index), row of a matrix

134

j	region (lower index), column of a matrix
+	positive orthant (upper index)

Signs

[ ]	matrix
{...}	a set of elements
$\in$	belongs to
| |	the volume of a set
‖ ‖	the Euclidean norm
‖ ‖$_1$	the L^1 norm
o	composition of functions: f o g (x) = f(g(x))
+	sum or Minkowski sum
$\wedge$	exterior product of vectors
n-D	n dimensional (space)
1-d	1-differential (family)

General notations

C$^k(\Omega)$	the class of functions on Ω, whose kth derivatives exist and are continuous.
iff	if and only if
loc	local, true in any bounded open set
sgn(x)	-1 if x < 0, 1 if x > 0, undefined if x = 0

REFERENCES

[1] Valentin, P., Is chromatography a separation process, The zonoid answer, J. Chromatog., **556**, 25-80 (1991)

[2] Valentin, P., *Design and optimisation of preparative chromatographic separations*, in *Percolation Process and Applications*, Rodrigues and Tondeur (ed) NATO-ASI Series E33, Sijthoff and Noordhoff, Netherlands, pp 141-195, 1980

[3] Rony, P. R., The extent of separation, a universal separation index, *Separation Science*, **3**, 239, (1968)

[4] Rony, P. R., The extent of separation: application to elution chromatography, *Separation Science*, **3** 357, 1968

[5] Karlin, S., *Total Positivity*, Stanford University Press 1968

[6] Bandle, C., Isoperimetric Inequalities and applications, Pitman, New-York, 1980

[7] Vol'pert A.I., Hudjaev S.I., Analysis in classes of discontinuous functions,and equations of mathematical physics, Martinus Nijhhoff Publishers, Dordrecht, 1985

[8] Duff, G.F.D., Integral Inequalities for Equimeasurable Rearrangements, Can. J. Math., **22**, 408-430, (1970)

[9] Fleming, W.H.,Rishel R., An Integral formula for total gradient variation, Arch. Math., **11**, 218-222, (1960)

[10] Federer, H., *Geometric Measure Theory*, Springer Verlag, Berlin (1969)

[11] Dafermos, C.M., The entropy rate admissibility criterion for solutions of hyperbolic conservation laws, J. Diff. Equ.,**14**, 202-212 (1973)

[12] Krushkov, S.N., First order quasilinear equations in several independent variables, Math. USSR Sb., **10**, 217-273 (1970)

[13] Oleinik, O. A., Uniqueness and Stability of the general solution of the Cauchy problem for a quasi-linear equation, Amer. Math. Soc. Transl. Ser., 2, **33**, 285-290 (1963)

[14] Alvino, A., Lions P.L., Trombetti, G., On optimization problems with prescribed rearrangements, Nonlinear Analysis, TMA., **13** 185-220 (1989)

[15] Ferone, V., Posterario, M. R., Maximization on classes of functions with fixed rearrangement, Diff. and Integral Equ., **4**, 707-718 (1991)

[16] Coxeter, H.S.M., The classification of zonohedra by means of projective diagrams, J. Math. Pures Appl., (9) **41**, 137-156,1962

[17] Coxeter, H.S.M., *Regular Polytopes*, 2nd ed., Mac Millan, New-York, 1963

[18] Bolker, E.D., A class of convex bodies, Trans. Amer. Math. Soc., **145**, 323-345,1969

[19] McMullen, P., On zonotopes, Trans. Amer. Math. Soc., **159** 91-110 1971

[20] Shephard, G.C., Combinatorial properties of associated zonotopes, Canad. J. Math., **26** 302-321, 1974

[21] Schneider, R., Weil, W., Zonoids and related topics, in: *Applications of Convexity*, Birkhauser Verlag, Basel, pp 296-317 (1982)

[22] Blackwell, D., Comparison of experiments, in *Proceedings of the second Berkeley Symposium on Mathematical Statistics and Probability*, [1950, Berkeley], pp. 93-102 Berkeley, University Press, Los Angeles, (1951)

[23] Szafran, N., Zonoèdres: de la Géométrie Algorithmique à la Théorie de la Séparation, Thèse de Doctorat en Mathématiques Appliquées, University J. Fourier, Grenoble, 25 Oct. 91.

[24] J. Mossino, J.M. Rakotoson, Isoperimetric Inequalities in Parabolic Equations, Ann. Scuola Norm. Sup. Pisa, ser. IV, **XIII**, pp 56-71 (1986).

[25] J.M. Rakotoson, Réarrangement relatif: propriétés et applications aux équations aux dérivées partielles, Thèse Doctorat en Sciences Mathématiques, Univ. Paris Sud, Orsay (1987).

[26] Glansdorff, P., Prigogine, I., Structure, Stabilité et Fluctuations, Masson, Paris, 1971.

THEORY OF ZONOIDS: II APPLICATION TO CHROMATOGRAPHIC PROCESSES

Patrick R. VALENTIN
Société Nationale Elf- Aquitaine
Elf- Solaize Research Center
69360 St SYMPHORIEN D'OZON BP 22
FRANCE
e-mail elfcnrs@cism51.fr

ABSTRACT.Following the methodology set up in paper I, it is shown that linear diffusional chromatography is not a separation, but a sepmix process. Let N be the number of theoretical plates, then the **loss** in 2-volume of separation, between a solute and the carrier increases along the column approximately according to $\sqrt{N}$, but **gain** in 2-volume of separation between two solutes increases as $\sqrt{N}$. The 3-volume of separation between two solutes and carrier shows two distinct phases when N increases: it first increases and then decreases. Nonlinear multidimensional chromatography of a single solute, in the absence of diffusion, is proved, using new "a priori estimates", to be a null process if no shock waves arise in it, a mixing process if shock waves arise in the column. In the presence of diffusion, chromatography is proved, using the same tools, to be a mixing process. The Rony "extent of separation" fits into present theory but reflects only part of separation produced by the column. A few examples are given of how to set up practical problems with zonoids.

1. Introduction

This second paper applies the **zonoid theory** to chromatography. It is a synthesis of three different papers: the first [1] deals with linear diffusional multicomponent column chromatography, while the second [2] deals with the nonlinear transport of a single species in several space dimensions, and the third [3] deals with non linear diffusion. These last two are more mathematical in nature and do not show practical applications.

The separation of two species by chromatography is basically, a ternary process, since it necessarily involves some spending of a third species called carrier : space E of quantities is three-dimensional, dimE = 3. Furthermore the column effluent composition is a *continuous* function of time. These essential features where not taken in account by previous separation theories [3,4].

The two main sections of this paper illustrate two different and complementary uses of the theory. One assesses the separation characteristics of a column by observing the outlet family in reference to the inlet family, experimentally or by an explicit model giving the outlet composition in function of time. The other one starts from the partial differential equations modelling the local species balance to assess the separation class of the process directly from the equations. It is more powerful but requires more sophisticated tools,.

137

F. Dondi and G. Guiochon (eds.),
Theoretical Advancement in Chromatography and Related Separation Techniques, 137–172.
© 1992 *Kluwer Academic Publishers.*

In section 2, we choose the first possibility, using a simple explicit model of linear chromatography with dispersion terms. We shall show, in a rather detailed way, that evolution of the solute-solute-eluent ternary separation in the column can be computed as a function of z, without the need to resort to any (arbitrary) "cut point". The interplay between the solute dilution process in the eluent and the separation of solutes is clarified.

In section 3 we use a quite different method known as a method of "a priori estimates" in the case of the non linear model of propagation with or without dispersion. As indicated by the name, this method does not require the explicit solution of the model.

This work is a first step towards a deeper understanding of how the separation evolution is governed by the partial differential equations of propagation themselves, together with their initial and boundary conditions and the related thermodynamic constraints.

We sketch here, without bothering about proofs, except where they have a special interest, the methodology. Proofs can be found, or will be published, in the quoted or announced papers. References to elements, theorems, sections... of part I will be prefixed with I, e.g. Theorem I-11.

2. Linear Column Chromatography

2.1 FAMILY OF FLOW-RATES

The system state is obtained from observation of flow-rates given by a detector located at z as a function of time, during time interval Δt. Passing to space of quantities E is straightforward since $dN(z,t) = F(z,t)dt$ formally represents the infinitesimal quantity which would be collected in the mobile phase between t and t + dt at z in the column. It is important to note, as emphasized in part I, that $C(z,t)$, the vector of concentrations, gives the composition but does not contain the information upon the *quantity* which would be necessary to build the zonoid.

A formal distinction here is that species indicates any chemical components injected in column, solute, only a species one wishes to analyze or separate. We make the convention of numbering solutes by order of increasing retention times, carrier being put in the last position. A natural base for E is formed on the unit quantity of each pure species, e.g. axes will be labelled in moles of species 1, 2, 3, carrier. In this base, a vector of components (N^1, N^2, N^3) is associated with any mixture of carrier and solutes. Note the upper species label.

2.1.1 The Gaussian Model, Isovariance Assumption The following model, although a simplified one is sufficient for the example we want to discuss. Furthermore all complicated models tend to be more and more similar to it when the number of plates becomes high. Let H be the **height equivalent to a theoretical plate**; u, the speed of the carrier; σ_i, the standard deviation of the Gaussian distribution of the flow rate.

Since $z = N \, H$, $\bar{t}_i$, at location z, is given as usual by

$$\bar{t}_i = \frac{z}{u}(1 + k^i) \approx \frac{(1 + k^i) N H}{u} \tag{1}$$

$$\sigma_i = \frac{\bar{t}_i}{\sqrt{N}} \tag{2}$$

For the sake of computational simplicity, and owing to the fact that interest in chromatography focuses mainly on difficult separations, in which the difference between retention times is much smaller than the retention times themselves, we may assume that σ_i, in eqn. 2 depends not on i, but on some mean retention time for the group of solutes. Physically, this means that although the two peaks translate at different speeds, they broaden at the same rate, depending on z or N but not on the species. Injected quantity of the species acts on the peak only through a vertical affinity. We shall set (although other mean values could have be chosen)

$$\bar{t} = \frac{1}{2}(\bar{t}_1 + \bar{t}_2)$$

Therefore, in the isovariance assumption we have

$$\bar{\sigma} = \frac{\bar{t}}{\sqrt{N}} = \frac{(\bar{t}_1 + \bar{t}_2)}{2\sqrt{N}} \tag{3}$$

With these assumptions, and considering the carrier flow to stay constant, we get the 1-differential flow-rate family $\{F(N,t)\,dt\}$, the three components of which are given in the first column of Table 1.

Table 1 Flow rate family and cumulated vectors

$\{F(N, t)\,dt\}_{\Delta t}$	$N(N, t)$
$N_0^1 \dfrac{\sqrt{N}}{\sqrt{2\pi}\,\bar{t}}\, e^{-\frac{1}{2}N\left(\frac{\bar{t}_1 - t}{\bar{t}}\right)^2}\, dt$	$N_0^1\,\mathrm{erf}\left(\sqrt{N}\,\frac{\bar{t}_1 - t}{\bar{t}}\right)$
$N_0^2 \dfrac{\sqrt{N}}{\sqrt{2\pi}\,\bar{t}}\, e^{-\frac{1}{2}N\left(\frac{\bar{t}_2 - t}{\bar{t}}\right)^2}\, dt$	$N_0^2\,\mathrm{erf}\left(\sqrt{N}\,\frac{\bar{t}_2 - t}{\bar{t}}\right)$
$F_0^3\, dt$	$F_0^3\, t$

In second column of Table 1, N is the quantity (cumulated starting from the onset of the cycle). The following definition has been used

$$\mathrm{erf}\,(u) = \frac{1}{\sqrt{2\pi}} \int_{-\infty}^{u} e^{-\frac{t^2}{2}}\, dt \tag{4}$$

Although the present definition of erf slightly differs from the customary one, it is more natural in our case.

It remains only to define Δt, the interval of variation of parameter t. In principle, due to the nature of diffusion, cycle or interval of time collection $\Delta t(z)$ (and therefore carrier quantity and zonoids) is unbounded. To avoid mathematical complications, it will be often considered that essentially all of the injected feed is recovered over a finite time interval called a cycle. An interval Δt of width $6\,\sigma + t_2 - t_1$, centered on mid-point between peaks maxima would be a good assumption in the present case:

$$\Delta t = \left[\bar{t}_1 - \frac{3\bar{t}}{\sqrt{N}} , \bar{t}_2 + \frac{3\bar{t}}{\sqrt{N}} \right] \tag{5}$$

It is useful to get a simpler expression of this family via the invariance property (I-§3.1). Let us define the reduced time τ:

$$\tau = \sqrt{N}\,\frac{t}{\bar{t}} , \quad \bar{\tau}_i = \sqrt{N}\,\frac{\bar{t}_i}{\bar{t}}$$

and the reduced translation parameter

$$\delta_{12} = \bar{\tau}_2 - \bar{\tau}_1 = \sqrt{N}\,\frac{\bar{t}_2 - \bar{t}_1}{\bar{t}} \approx 2\,\frac{\alpha - 1}{\alpha + 1}\,\sqrt{N} \quad \text{if } k^1 \gg 1 \tag{6}$$

noting that $d\tau = \frac{\sqrt{N}}{\bar{t}}\,dt$ we get the new expression of the family

$$\mathcal{F} = \left\{ N_0^1 \frac{1}{\sqrt{2\pi}}\, e^{-\frac{1}{2}(\bar{\tau}_1 - \tau)^2}\, d\tau,\ N_0^2 \frac{1}{\sqrt{2\pi}}\, e^{-\frac{1}{2}(\bar{\tau}_2 - \tau)^2}\, d\tau,\ F_0^3 \frac{\bar{t}}{\sqrt{N}}\, d\tau \right\}\ \tau \in \sqrt{N}\,\frac{[\Delta t]}{\bar{t}} \tag{7}$$

and of the cumulated vector

$$\left\{ N_0^1 \operatorname{erf}(\bar{\tau}_1 - \tau),\ N_0^2 \operatorname{erf}(\bar{\tau}_2 - \tau),\ F_0^3 \frac{\bar{t}}{\sqrt{N}}\,\tau \right\}\ \tau \in \sqrt{N}\,\frac{[\Delta t]}{\bar{t}} \tag{8}$$

We note that, in the case of binary separations, we could further use the change of variables described in I-§3 to get an equivalent family of the type $\{(1,u(\alpha))\,d\alpha\}$.

2.1.2 *x- Evolution Problem* A cycle transforms a system state Σ into a system state Σ'. From this point of view, the inlet (resp. outlet) family can also be called the **initial** (resp. **final**) **family**.

For theoretical and design purposes, evolution of the 3-D zonoid has to be considered, together with all of its canonical projections, i.e. projections on coordinate planes (1,2), (2,3), (2,3). These tell us of the binary separation evolution between the solutes or between a solute and the carrier.

In elution chromatography, the reduced inlet family is discrete and consists of two vectors:

$$\mathcal{F}(0) = \left\{ \begin{bmatrix} N_o^1 \\ N_o^2 \\ 0 \end{bmatrix}, \begin{bmatrix} 0 \\ 0 \\ F_o^3 \Delta t \end{bmatrix} \right\}$$

the first one is the quantity of feed, the second one is quantity of pure carrier taken in by the column during the cycle time, with the assumption of constant carrier flow, $F^3(z,t) = F^3_o = const.$. In contrast, the reduced outlet family produced by the column is a 1-d family noted $\mathcal{F}(L) = \{F(L, t)dt\}_{t \in \Delta t}$. Overall species balance imposes that these two families have same sum vector. The natural parameter t of the family is time and the evolution problem has parameter z, abscissa in the column of bounded length L, so that we may put $0 \le z \le L$.

2.2 GRAPHICAL REPRESENTATIONS

2.2.1 The graph Γ of cumulated quantities The graph of cumulate family, $N(z,t)$ of table I is shown for z = 25 on Fig 1. It corresponds to the first peak in chromatogram of fig 4.

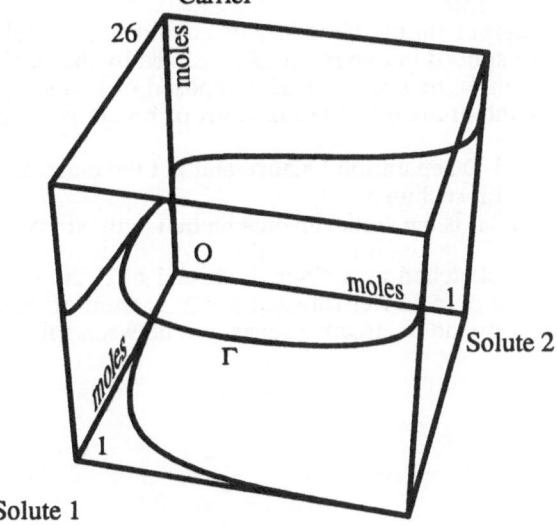

Fig 1 Graph of the cumulated quantity vector function $N(t)$ with its canonical projections z = 25 cm, $F^3_o = 0.2$ moles /sec, HETP = 0.1 cm, u = 5 cm/sec, $k^1 = 20$, $k^2 = 22$, $\alpha = 1.10$, Injected quantity per cycle: (1, 1, 26) moles, Carrier flow-rate : 0.2 mole/s, Cycle time $\Delta t = 130$ seconds.

Although the primary ingredient for the zonoid construction is curve $\Gamma = N(t)$, it is very difficult to infer from it the shape of the zonoid. More direct tools are needed. A direct examination of Γ and of its projections on the coordinate planes reveals that:

1. the vertical projection of Γ on the ground plane (1,2) is convex. Therefore the canonical projection of Γ on the ground plane shows the projection of the 3-D zonoid boundary.

2. the projections on planes (1,3) and (2,3) are not convex. This topic will be discussed further in Section 2.3.

There is clearly a strong difference between the type of separation of solute 1 from solute 2 and between separation of carrier from solute 1 or solute 2. We shall study meaning and significance of convexity under the heading "selectivity" at the end of this section. A quantitative study of the growth or decay of separation represented by a 3-D zonoid together with its projections onto ground planes will be given at the end of the present section.

2.2.2 Initial and Final Zonoid It must be mentioned that each component of the 1-differential family is proportionate to the corresponding coordinate of N_0; that is, the zonoid obtained from injected vector N_0 deduces from the zonoid calculated for the mixture (1,1,1) by the product of affinities N_0^i along axes e_i (i=1,2,3). Moreover, the canonical projection (1,2) does not depend on the dilution of the injection in the carrier, as long as the injection time remains short.

Now let's get a visual grasp of the 3-D zonoid itself; that is let's to draw its boundary. Although different zonoid boundary shapings are possible, the simplest choice is to cover Z by zonal lines. Let us postpone construction of such lines until the next paragraph and suppose we have constructed a set of zonal lines.

Looking at fig 2a, we see different curves on the boundary of zonoid: the cumulate curve has disappeared. In fact it has been spliced in two replicas translated by the vertical vector representing pure carrier. Each zonal line such as L is composed of two smooth parts which join angularly at the origin and at the distal point. A set of regularly spaced zonal lines are drawn.

We are now in a position to 'see' the 3-D separation balance and get the class of the process, using classification defined in part I, section 5.

The initial or injected zonoid Z_0 (figure 2a) is the vertical parallelogram built on vectors (1, 1, 0) and (0, 0, 26).

The final or outlet zonoid, Z may be sketched as a slanted ovoidal box: the quasi-vertical part of Z correspond to the almost pure carrier (present partly in front of, partly behind peaks), the top and the bottom correspond to effective separation between solutes.

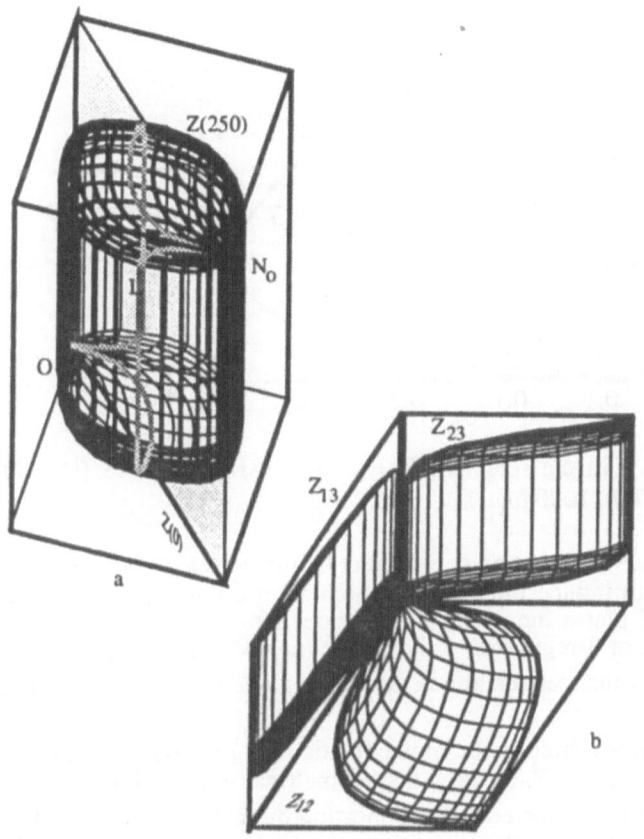

Fig 2a Injected and recovered 3-D zonoids, z= 0, 25 cm, $N^1 = 1$, $N^2 = 1$, $N^3 = 26$ moles.
  2b Projected 2-D zonoids representing binary separation

Clearly zonoids Z_0 and Z intersect - in the sense that they have common points although none is contained in the other and this is a sufficient condition for the process to be a sepmix process. Indeed, from the fundamental property of zonoids, a mixture of solutes and carrier, M is feasible by mixing from effluent of the column iff it belongs to Z. In fact, mixtures $M \in Z$, $M \notin Z_0$, feasible from the final separation state are not feasible from the initial state because they have not the initial composition in solutes and conversely, mixtures $M \notin Z$, $M \in Z_0$ feasible from the initial separation state are not feasible from the final state because they are either too rich or too poor in carrier.

Finally, one may look at canonical projections of Z on coordinate planes, Z_{12}, Z_{13}, Z_{23}, which are 2-D zonoids and represent states of separation between species 12, 13 and 23 (fig 2b). We have represented also the projection of zonal lines. The differential family has been approximated by a discrete family with 50 vectors using a computer program developed by N. Szafran [9].

2.2.3 A Study of the 3-selectivity To see the influence the abscissa in the column has on selectivity, let us draw the curves γ of the molar fraction family at different values of N (fig. 3)

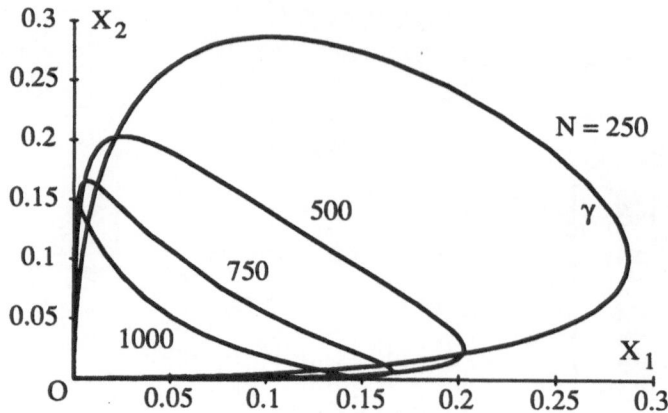

Fig 3 Curves γ in Δ_2, in the selective (N= 250), nearly selective (N= 500) and non selective cases (N= 750, 1000 plates).

We note that these curves are smooth, except at O, the vertex of Δ_2 which corresponds to pure carrier. This will induce lines along which zonoid boundary is not smooth. Also, when the number of plates increases above about 500, a change in convexity occurs, which marks the loss of 3-regular selectivity. We must skip a more precise discussion on how the shape of γ influences the zones. It can be made using the material given under Theorem I-7 or in [9].

Although we have drawn directly the molar fraction simplex Δ_2, in its plane, it is useful, see [1], to embed it in space E, to show the geometrical relationship between family and curve γ: γ is the intercept of the cone Γ, supporting the family, with Δ_2. This embedding also makes much more intuitive the projective rules for construction of zonoid boundary. It also makes clear, that we could have used, in place of Δ_2, the plane P of solutes concentration C_1, C_2: the corresponding curves, intercept of Γ with P, are known under the name of hodograph and the whole study of selectivity could then be done in term of the concentrations. Examples of hodographs are given in the paper from Guiochon and coll. [13].

From Theorem I-5, and after some algebraic simplifications (see [1] for details), the selectivity of the outlet family is 3-regular iff $W(F) \geq 0$, with

$$W(F) = \text{Det} \begin{bmatrix} 1 & -(t-\bar{t}_1) & N\left(\frac{(t-\bar{t}_1)}{\bar{t}}\right)^2 -1 \\ 1 & -(t-\bar{t}_2) & N\left(\frac{(t-\bar{t}_2)}{\bar{t}}\right)^2 -1 \\ 1 & 0 & 0 \end{bmatrix} \tag{9}$$

which reduces, after further simplifications, to the first entry in Table 2. The critical number of plates N_c and time t_c, at the 3-selectivity reversal, are also given in this table.

Table 2. Wronskian and critical parameters

W(F)	N_c	t_c
$N\left(\dfrac{(t-\bar{t}_1)(t-\bar{t}_2)}{\bar{t}^2}\right)+1$	$\left(\dfrac{\bar{t}_2+\bar{t}_1}{\bar{t}_2-\bar{t}_1}\right)^2$	$\dfrac{\bar{t}_2+\bar{t}_1}{2}$

Therefore, the condition for 3-regular selectivity inversion, $N = N_c$, is just the condition for the existence of a "double" inflexion point (i.e. a point where the tangent has a contact of order 4) on γ and this point happens at parameter t_c. One has also the property

If the inflexion point on the chromatogram of each one of the two solutes appear at the same time t at location x in the column, then 3-selectivity reversal occurs at (x,t).

In the conditions in fig 1, we get $N_c = 484$ plates, $t_c = 110$ sec. Clearly, the separation is far from complete at the critical number of plates.

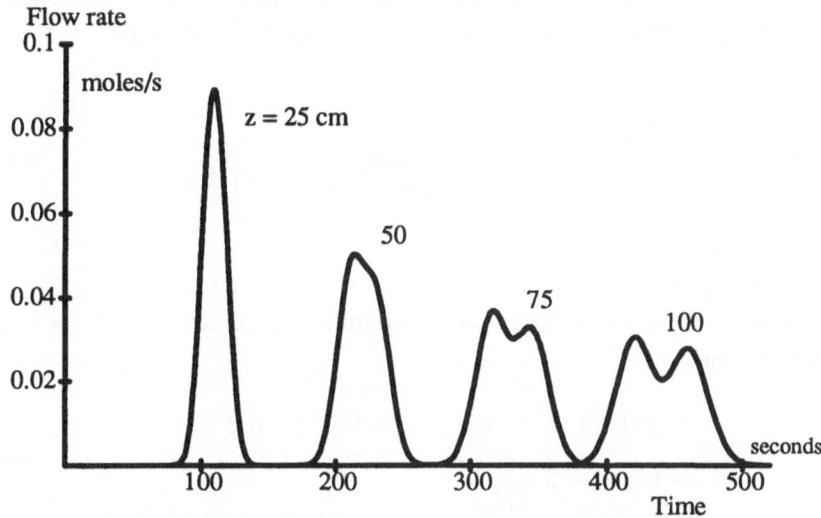

Fig 4 Total flow rate of solutes in case of 3-regular selectivity (25 cm), just after selectivity inversion (50 cm), non 3-regular selectivity (75 cm)

Even in the simple model presented here, one finds rather complex features for selectivity, including type changes along column:

1. 3-regular selectivity or not
2. absence of 2-selectivity between carrier and any solute
3. 2-selectivity between the two solutes

Combined with the above comment, this gives a hint as to how intricate and rich the study of separation can be for even a moderate number of species.

2.3. IS CHROMATOGRAPHY A SEPARATION PROCESS ?

Quantitative assessment of separation evolution in chromatography involves evolution of a zonoid "passing through" the abscissa z in the column, as well as its canonical projections as a function of z.

Although we are interested, ultimately, only in the separation between solutes. we must consider in turn the 3 canonical projections of $Z(z)$, then the 3-D zonoid itself.

2.3.1 The solute-carrier Separation ; The condition for 2-selectivity of separation of a solute i and the carrier is derived from I-§4.5: using subdeterminant built on lines i, 3 and columns 1,2 of the Wronskian matrix eqn. 5, we get $t_i - t \geq 0$, which means that the 2-selectivity reversal between solute and carrier occurs at the maximum of the solute peak. Since we are considering only two species we can apply the mixing convention. Reduction of family $\mathcal{F}$ in the eqn. 7 is achieved using formula I-11:

$$F^*(\tau) = F(\tau)\left(1 - \frac{\dfrac{dF^1}{d\tau}(\tau)}{\dfrac{dF^1}{d\tau}(\tau')}\right), \ \tau \in [0, \tau_{max}] \tag{10}$$

calling τ_{max} the reduced retention time of peak maximum. Since $F^1(\tau)$ is symmetrical around τ_{max}, $t' = 2\, t_{max} - t$, therefore eqn. 10 reduces to an explicit result: $F^* = 2\, F$ and finally, we get

$$\left\{2\, N_0^i \frac{1}{\sqrt{2\pi}}\, e^{-\frac{1}{2}\left(\bar{\tau}_i - \tau\right)^2} d\tau, \ 2F_0^3 \frac{\bar{i}}{\sqrt{N}} d\tau\right\} \ \tau \in \frac{1}{2}\sqrt{N} \frac{[\Delta t]}{\bar{t}} \tag{11}$$

Evolution of the separation state

First, to perform a **geometric** study, let's compare families of curves $\Gamma_{i3}(N)$, $i = 1, 2$. From eqn. 10, equation of Γ_{i3} is

$$N^i(\tau_i) = 2\, N_0^i \operatorname{erf}(\tau_i), \quad i = 1 \text{ or } 2, \quad 0 < \tau_i < \infty$$
$$N^3(\tau_i) = 2\, F_0^3 \frac{\bar{i}}{\sqrt{N}}\, \tau_i \tag{12}$$

That is, all curves Γ_{i3} map onto the graph of $\operatorname{erf}(x)$ over R^- by a linear application whose matrix $[A]$ is diagonal, with elements $\left(\dfrac{1}{2\, N^i_0}, \dfrac{1}{2\, F^3_0 \dfrac{t_i}{\sqrt{N}}}\right)$. Then curves Γ_{i3} do not intersect, except at the origin; note that this rules out these curves having exactly the same sum vector) although they have the same horizontal asymptote and lay above one another for increasing values of N. Decrease of the 2-D zonoid Z_{13} is clearly visible on fig. 5.

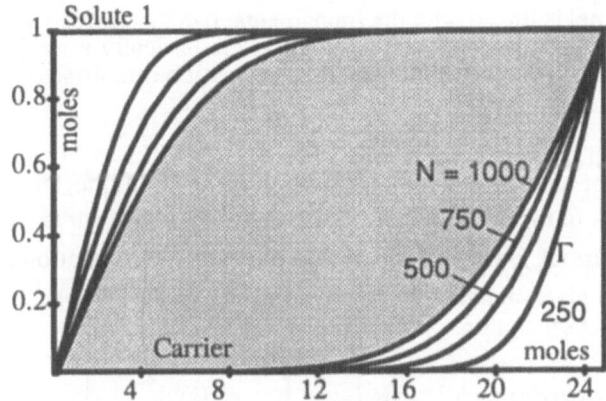

Figure 5 Evolution of 2-D zonoid (solute 1, carrier) (z= 25, 50, 75, 100 cm).
relative loss area ratios are 1, 1.41, 1.73, 2

We have thus established inclusion (recall $proj_{i3} Z = Z_{i3}$)

$$Z_{i3}(z') \supset Z_{i3}(z) \text{ iff } z' < z, i = 1,2$$

which amounts here to existence of canonical projections of $Z(z)$ in $Z(z')$ on planes (1,3) or (2,3).

$$Z_{i3}(z) \prec Z_{i3}(z') \text{ iff } z' < z, i = 1,2$$

This leads us to important and simple conclusions:

Linear diffusional chromatography is a solute-carrier mixing process.

Second, to perform an **analytical** study, we give a direct proof that solute propagation is a mixing process between solute and carrier by exhibiting a sampling kernel λ deduced from the one given in paper I, § 3. λ allows to get the family of flow-rate vectors passing through the $N + \Delta N$-th plate from the one passing through the N-th plate.

Proof: Referring at Table 1 for expression of the families, we see that, up to translations of time scale to get centred distributions, i.e. $w = t - \bar{t}_1$, $\tau = t' - \bar{t}_2$, we may use the coordinates 1, 3 of the kernel λ taken from example in I-§5.1. Therefore

$$\lambda(t,t') = \frac{e^{-\frac{(t - t' + \bar{t}_2 - \bar{t}_1)^2}{2\sigma'^2}}}{\sqrt{2\pi}\,\sigma'} \text{ with } \sigma' = \sqrt{\sigma''^2 - \sigma^2} = \frac{(1 + k^1)\sqrt{\Delta N}\,H}{u}$$

and the family $\mathcal{F}(N + \Delta N)$ results from the family $\mathcal{F}(N)$ through a mixing process associated with mixing kernel

$$\lambda(t,t') = \frac{1}{\sqrt{2\pi}}\sqrt{\frac{N + \Delta N}{\Delta N}}\,\frac{1}{\bar{t}_2}\,e^{-\frac{N + \Delta N}{\Delta N}\frac{(t - t' + \bar{t}_2 - \bar{t}_1)^2}{2\bar{t}_2^2}}$$

148

Of course existence of a mixing kernel is linked with the fundamental fact that diffusion, as represented here by the increasing variance of a Gaussian curve, is basically a mixing process, a fact that can be traced back to the diffusion equation. This will be emphasized in Section 3.

2-volume of solute-carrier separation

For the unbounded (rigorous) case of an infinite column, carrier quantity is infinite, and so is the 2-volume. However, the area in the rectangle of total separation, under curve Γ_{i3}, stays finite and we may use the **loss of 2-volume** ζ_{i3} in the place of the separation 2-volume. According to the definition given in part I-§4.7, ζ_{i3}, is the area of the complement of 2-D zonoid to the rectangle of total separation.

Let ζ_o, be the area under the graph of erf(x) over R^-, i.e. $\zeta_o = \int_{-\infty}^{0} \text{erf}(s)\, ds = \dfrac{1}{\sqrt{2\pi}}$.

Through transformation T^{-1}, ζ will be multiplied by det $[A]^{-1}$, so that we get a very simple expression:

$$\zeta_{i3} = 8\, N^i_o\, F^3\, \frac{\bar{t_i}}{\sqrt{N}}\, \zeta_o = \frac{8}{\sqrt{2\pi}} N^i_o\, F^3\, \sigma_i \tag{13}$$

At any given retention time of solute i, ζ_{i3} is proportional to $\dfrac{\bar{t_i}}{\sqrt{N}} = \sigma_i$, i.e. the 2-volume loss is an increasing function of the peak variance.

Eqn. 13 may be given other forms:

$$\zeta_{i3} = 8\, \zeta\, N^i_o\, F^3\, \sqrt{N}\ \text{HETP}\, \frac{1+k^i}{u}$$

Loss of 2-volume varies for a given z, as $\sqrt{\text{HETP}}$, or, for a given HETP, as $\sqrt{N}$.

The conclusion may be drawn over in still another way: although solute-carrier separation decreases when z increases, (i.e., zonoids become smaller and smaller the longer the column is), a given increment of length has less and less influence on this decrease. This is to be related to Fick law which states that the flux is proportional to the gradient of concentration. Indeed, the nonlinear effect of a **constant increment** of column length on the mixing is clearly visible on fig. 5.

These conclusions all fit quite well into intuitive, qualitative thinking about what the effect of the above parameters upon separation should be.

2.3.2 Solute 1-solute 2 Separation; Let us now consider the 1-2 solute separation, that is the canonical projection on the plane (1,2) of the 3-D zonoid. We must first construct proj $_{12}$ (Z) and then study its variation with N.

Study of 2-selectivity : taking the leading minor in W(F), we see that the 2-selectivity condition is always satisfied. The family of flow-rate vector possesses therefore the 2-regular selectivity property. Then, the zonoid construction from the chromatogram becomes very simple: integrate F from 0 to t, to get N(t). Tangent to Z at point N(t) is F(t).

Physically, 2-selectivity for species 1, 2, means that in the flow passing through section located at z, increases in purity of species 2, the more retained species, as the time goes on.

Construction of a 2-D zonoid

Projection Γ_{12} of the cumulate curve N(t) is a convex curve. Therefore, from eqn. 7, the half-boundary of Z_{12} has parametric equation,

$$\begin{cases} N^1(N, \tau) = N_o^1 \operatorname{erf}(\bar{\tau}_1 - \tau) \\ N^2(N, \tau) = N_o^2 \operatorname{erf}(\bar{\tau}_2 - \tau) \end{cases} \tau \in [0, \sqrt{N}\frac{\Delta t}{t}] \tag{14}$$

Evolution of the separation state

A quantitative study of growth of zonoid Z_{12} with z, covers both relative positions of Z and area growth. Introducing further the reduced translation parameter δ_{21} of eqn. 5 and translating the τ-scale by τ_1, we see that equation of Γ_{12} depends on the single parameter $\delta = \delta_{12}$.

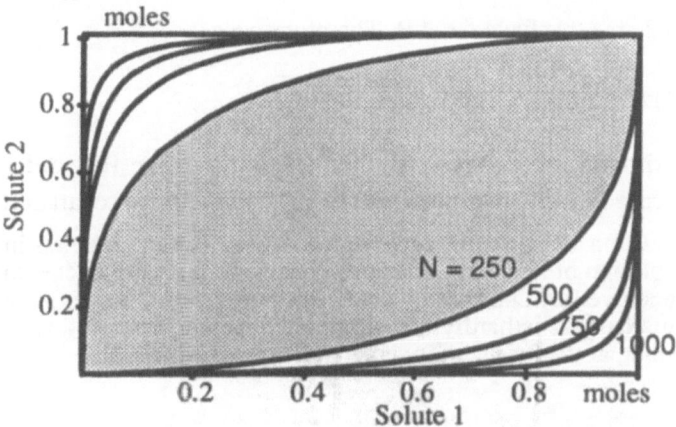

Figure 6   Growth of the separation, zonoid Z_{12} (L = 25, 50, 75, 100 cm).

The inclusion property follows then easily: at any given τ, increasing δ makes the second peak recede from the first one, thus $N^2(\tau)$ decreases. In this process, any point on ∂Z^- moves towards axis 1 on the vertical above $N^1(\tau)$ which proves inclusion. Note that variation of slope of the tangent itself in this process is not uniform: it decreases if $\tau < \tau_1 + \delta$, and it increases if $\tau > \tau_1 + \delta$

2-volume of separation

Calculation of the law of increase in separation 2-volume with δ is less easy but gives a simple result, both as a first order expansion of Ξ and in general.

$$\Xi_{12} = 2 \operatorname{erf}\left(\frac{\delta}{\sqrt{2}}\right) - 1 \tag{15}$$

See [1], Annex I for a proof using convolution and Fourier transform. A first order development of eqn. 15 gives

$$\Xi_{12} \approx \frac{\delta}{\sqrt{\pi}} = \frac{1}{\sqrt{\pi}}\sqrt{N}\,(\alpha - 1) \quad, \delta \le \alpha + 1$$

where the last inequality amounts to $N \le N_c$. Therefore, at incipient separation, 2-volume increases as only as $\sqrt{N}$, or, conversely, rate of growth of 2-volume is infinite at $N = 0$, and decreases as $1/\sqrt{N}$ since

$$\frac{d\Xi_{12}}{dN} = -\frac{1}{\sqrt{\pi}}\frac{1}{2\sqrt{N}}\,(\alpha - 1)$$

This relative inefficiency of theoretical plates in the process can be traced back to the fact that, at each instant, separation is produced only in part of the column since the width of the signal is approximately proportionate to $\sqrt{N}$. Quite remarkable also is the grouping of α and N into a single factor, which is indeed linked with the resolution R. Let be R = $R_{12} = 2\frac{\bar{t}_2 - \bar{t}_1}{w_1 + w_2}$, w_i, the length between intercept of inflexion tangents with base line. For two Gaussian peaks of equal area, we have $\delta = 4\,R$. Therefore we get

$$\Xi_{12} \approx \frac{4\,R}{\sqrt{\pi}}$$

However, use of these approximations is severely limited since the above first order expansion is valid only in the case of difficult separation ($\alpha \approx 1$) and incipient separation ($\sqrt{N}\,(\alpha - 1) < 1$) [1]. The condition of incipient separation, is usually not satisfied in practice since one aims at a resolution of at least one between peaks. But, R is restricted to being lower than about 5 if the above formula has any chance of being valid -i.e. giving a value lower than 1 (and, in this case, it is hardly measurable from the chromatogram). Finally it is not difficult to see that, at the 3-selectivity inversion point, N_c, R = 0. 5, which means a rather low resolution.

Therefore we may state:
    For difficult separation, at the incipient stage, the rate of 2-volume production for
    solutes 1, 2 per unit of column length is proportional to $\sqrt{N}$.

The study of canonical projections of Z leads us therefore to a remarkably simple conclusion concerning the binary separations:

    Linear chromatography with diffusion (Gaussian isovariant model) is a separation
    process between two solutes, a mixing process between a solute and carrier.

Since linear diffusional chromatography is neither a pure separation, nor a pure mixing process, it must be a **sepmix process**. Note that, although gained from study of binary separations, this conclusion holds for ternary separation since, clearly, it is a sufficient condition for a process to be sepmix to exhibit two 2-D projections in which that initial and final zonoids are not ordered in the same manner.

A generalization to n-solutes separation

Results on selectivity for two solutes are generalized to the **separation of any number of solutes**: using Polya frequency functions (PF). The general theorem, see [1], holding for any number of solutes, to the exclusion of the carrier, is :

Theorem The Gaussian isovariant model of chromatography generates 1-differential families endowed with a totally regular selectivity between solutes.

The same result has been obtained in [1] for the Poisson model of chromatography.

2.3.3 Separation between solutes and carrier An overview of the characteristic features of the chromatographic separation is gained by a 3-D representation of the zonoid rather than by the 2-D projections, as done above. Let's draw qualitative "visual" conclusions.

3-D zonoid comparison

The influence of the abscissa z on the separation state, for given operating conditions, is apparent in fig. 7. The cases a,b,c,d are the same as those figured in fig. 3. Clearly the loss of selectivity can be looked at, just by observing the zone arrangement on the zonoid surface on fig 7c, 7d, and, at the nascent state, on fig. 7b, near the distal point.

The injected zonoid would be a vertical parallelogram, (see fig 2a) since the solute mixture and the pure carrier are available separately at the column inlet.. Therefore, clearly, if solutes are to be separated to some extent, the resulting zonoid must not project onto a segment, hence no inclusion property between zonoids can hold: the process must be a sepmix.

All these figures have been drawn using the same quantity of carrier, by choosing a cycle time sufficient to recover essentially all of the injected solutes. Clearly, when this goal is achieved, it is no more useful to increase the time cycle: increasing Δt by δt would add a vector $(\varepsilon_1, \varepsilon_2, F^3 \delta t)$ with small ε_1, ε_2 to the initial and final families, or, equivalently, expand all zonoids with inserting, to a close approximation, a vertical cylinder of length $F^3 \delta t$ without adding any really interesting information. 3- volume of Z would then increase by $\Xi_{12} F^3 \delta t$.

When z increases, the general deformation trends are as follows

1. Z becomes thinner (maximum vertical width decreases): for a given carrier quantity, this means a more homogeneous medium since a given mixture of solutes 1, 2 can be made with an ever smaller fork of quantities of carrier.

2. Z becomes larger in horizontal projection (smallest vertical cylinder containing Z grows)

152

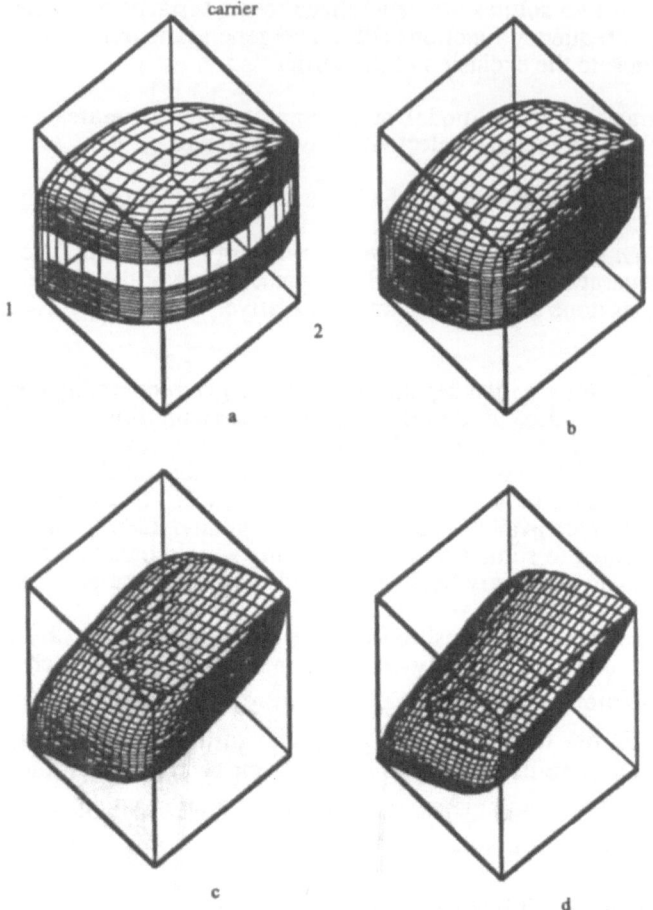

Fig. 7 Evolution of 3-D zonoid along the column ($z = 25, 50, 75, 100$)
       Conditions are those from Fig. 1.

<u>3-Volume evolution</u>
A noteworthy expression for the solutes-carrier separation 3-volume is given under the

Theorem For a family $\mathcal{F}$ given by the Gaussian isovariant model, see Table 1, when $N < N_c$, i.e. when 3-regular selectivity holds,

$$\Xi = N_o^1 N_o^2 N_o^3 \left\{ \xi_{12} - 2 \frac{\left(\bar{t}_2 - \bar{t}_1 \right)}{\Delta t} \right\} \tag{16}$$

$N_o^1 N_o^2 N_o^3$ is the cube of total separation. $N_o^1 N_o^2 N_o^3 \xi_{12}$ is the 3-volume of a vertical cylinder with base Z_{12}, height $N^3{}_o = F^3 \Delta t$ which would be obtained if the carrier were not an integral part in the process and could be spared (of course an hypothetical case). Note that this 3-volume is unbounded when $\Delta t = R$.

$N_o^1 N_o^2 F^3 2 \left(\bar{t}_2 - \bar{t}_1 \right)$ has the dimension of a 3-volume and is bounded. Formally, this term represents 3-volume of total separation between pure solutes and quantity of carrier which flows between the peaks retention times. It neither depends on Δt, (and thus it is finite), nor directly on the process efficiency. From the very shape of eqn. 14, its second term can be appropriately called **loss of 3-volume**.

Therefore, yield of 3-volume, for the Gaussian isovariant model, under the condition $N < N_c$ is given by

$$\xi = \left| 2\,\text{erf}\left(\sqrt{\frac{N}{2}}\,\frac{\bar{t}_2 - \bar{t}_1}{\bar{t}}\right) - 1 - 2\frac{(\bar{t}_2 - \bar{t}_1)}{\Delta t} \right| \tag{17}$$

where t_i can be evaluated using Equ (1).

$$\xi = \left| 2\,\text{erf}\left(\sqrt{\frac{N}{2}}(\alpha - 1)\right) - 1 - 2\frac{(\alpha - 1)}{\Delta t}\,\frac{(1 + k^1)\,N\,H}{u} \right| \tag{18}$$

The minimum collection time Δt itself can be evaluated for a given recovery ratio of solutes in function of N and, using eqn. 5, we get finally

$$\xi = \left| 2\,\text{erf}\left(\sqrt{\frac{N}{2}}\,\frac{\bar{t}_2 - \bar{t}_1}{\bar{t}}\right) - 1 - \frac{2\,(\alpha\text{-}1)}{(\alpha\text{-}1) + \dfrac{6}{\sqrt{N}}} \right| \tag{19}$$

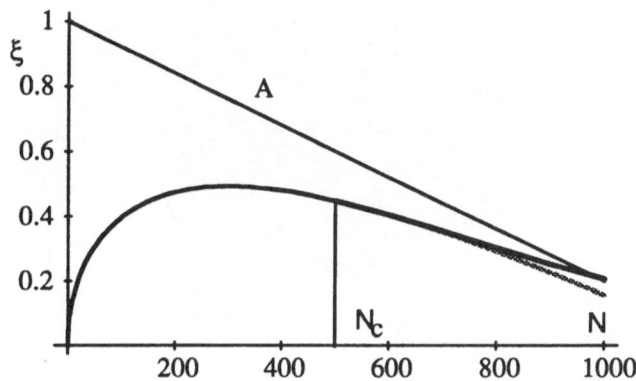

Figure 8 3-volume yield vs number of plates, conditions of fig. 1, $\Delta t = 100$ sec.
In black, the exact solution, in grey the approximate eqn. 18, with A the asymptote.

Specially interesting is the rapid (in $\sqrt{N}$) increase in 3-volume creation at the outset, followed by a step-up and a decrease. The mean rate of 3-volume creation per plate, $\dfrac{\xi}{N}$, goes to zero at a certain number of plates. Clearly, existence of such a maximum on ξ is a very important feature for optimization: it strongly indicates that one should consider some way to enhance efficiency of the process beyond this point. It also indicates a competition between two processes, the first of which, translation of peaks, creates separation, and the second one, dilution in the carrier, destroys separation.

154

The eqn. 16 ceases to stand valid for $N > N_c$ - in fact, ξ goes eventually negative, an impossibility - and curve in grey on fig. 8 branches, at N_c, to the to curve representing the rigorous solution. It is clear that after N_c, ξ in bounded from below by eqn. 16 and, more precisely, applying elementary property of absolute value,

2.3.4 Evolution of 3-Volume for 3 solutes In the special case of a totally regular selectivity, using formula I-38, we get the following expression of the separation 3-volume between 3 solutes, for the Gaussian isovariant model :

$$\xi = 2 \left\{ \text{erf}\left(\frac{\delta_{23}}{\sqrt{2}}\right) - \text{erf}\left(\frac{\delta_{13}}{\sqrt{2}}\right) + \text{erf}\left(\frac{\delta_{12}}{\sqrt{2}}\right) \right\} - 1 \qquad (20)$$

where δ_{ij}, given by eqn. 10 is the translation parameter between peaks of solutes i and j. Since $\delta_{13} = \delta_{12} + \delta_{23}$, ξ is a symmetric function of two independent variables δ_{12}, δ_{23}.

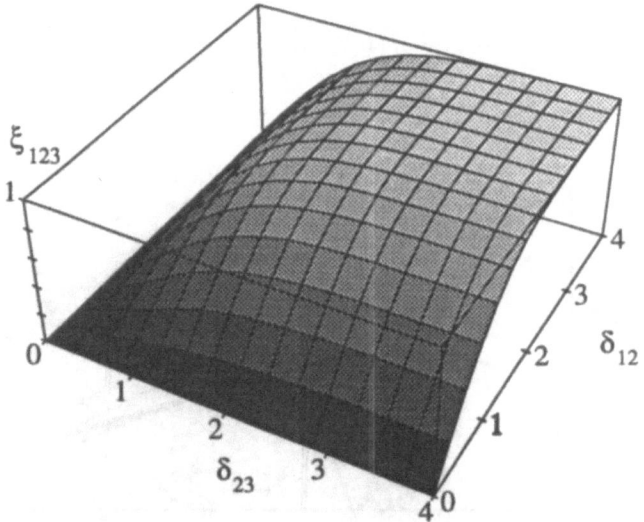

Fig 9 The 3-volume yield between three solutes vs reduced translation parameters δ_{ij}

3. Non linear Chromatography in One or Several Space Dimensions

The application of zonoid theory to chromatography is not limited to the linear case. We now prove that the nonlinear equilibrium chromatography without diffusion of a single solute is a null process if no shock waves arise in the column, i.e. for a continuous solution, a mixing process if shock waves arise, that is for the so-called weak solution.

3.1 EVOLUTION PROBLEMS IN 1-D CHROMATOGRAPHY

3.1.1 The general ideal model. The ideal model of chromatography can be found in the papers by Golshan-Shirazi and Guiochon in the present book [11,12] or in the book by Rhee, Aris and Amundson [13].

In column chromatography, the equation of solute conservation is:

$$\frac{\partial \tilde{q}(C)}{\partial t} + \frac{\partial J(C)}{\partial x} = 0 \tag{21}$$

and the initial and boundary conditions are

$$C(x,0) = C_0(x) \geq 0, \, 0 \leq x \leq L$$
$$C(0,t) = C_i(t), \, t \geq 0$$

where the total accumulated linear solute density, $\tilde{q}$, and its associated flux, J, depend on the concentration $C(x,t)$ in the mobile phase through

$$\tilde{q}(C) = \left(C + \frac{(1-\varepsilon)}{\varepsilon} q(C)\right), \, J(C) = u \, C \tag{22}$$

with $\varepsilon \in \,]0,1[$, the mobile phase porosity, $u > 0$, its speed. The isotherm $q(C)$ describes the equilibrium concentration in the stationary phase in function of C and is a smooth increasing function, $q(0) = 0$, $\frac{dq}{dC}(C) > 0$. $\tilde{q}$ is therefore a smooth increasing, therefore invertible, function of the concentration. Let $C(\tilde{q})$ be the inverse function.

Eqn. 21 is a special case of eqn. I-57a. In fact, eqn. 21 holds in the multicomponent case. However, the boundary conditions differ slightly from the Cauchy problem but it is easily shown that these conditions can be recast in the form of a Cauchy problem.

3.1.2 Different forms for evolution problems To the problem given by eqn. 21 in a symmetrical form, we may associate two different evolution problems: a t-evolution problem or a x-evolution problem, both problems being of type I.

The methodology given in part I needs first to identify the species and, consequently, the families from which to construct the 2-D separation state. To this end, we note that another conservation law is present, but used in a somewhat disguised manner. In the present case where the carrier speed u is assumed to be constant, we have $F_0 = u \, C_0$, which is a flux, if we take the column to be of unit area, clearly an unimportant assumption. Then eqn. 21 can be completed with

$$\frac{\partial C^o}{\partial t} + \frac{\partial F^o}{\partial x} = 0$$
$$C^o(x,0) = C^o(0,t) = C^o = \text{const} \geq 0, \, 0 \leq x \leq L, \, t \geq 0$$

which expresses the conservation of the carrier, indexed here with the superscript o. One may also use the conservation of the total quantity, mass etc, as is the most convenient. At low solute concentration, the above equation is uncoupled and has the obvious solution $C^o(x,t) = C^o = \text{const}$, but more realistic situations involving the sorption effect [15] could be accounted for.

Therefore we obtain different expressions, according to the variables chosen to represent the system state, namely the flow-rates or total accumulated solute density:

$$\left\{\begin{bmatrix} F^o \\ F(x,t) \end{bmatrix} dt\right\}_{t \in [-\infty, +\infty]} , \, \left\{\begin{bmatrix} C^o \\ \tilde{q}(x,t) \end{bmatrix} dx\right\}_{x \in [0,L]} \tag{23}$$

In fact, there is a natural choice of the family in connection with the type of evolution problem: the flow-family comes into play for x-evolution problems, the accumulated-family, for the t-evolution problems.

On physical grounds, the above remarks are quite significant: flow and total accumulated quantity have direct physical meaning, since the flow going out can be collected or the column can be cut into pieces. Therefore they will generate families. In contrast, the quantity accumulated in the mobile phase cannot be recovered as such, in the implicit hypothesis made of an homogeneous medium in the column. This confirms that concentration, although a useful and traditional variable in chromatography, is not a conservative quantity, and does not generate a family. We shall keep on the tradition, being aware, however, that in some situations, such when solution contains discontinuities, one must come back to the conservative form. Therefore the usefulness of concentration comes only from the fact that is is an intermediate variable in term of which both q and F can be expressed or, on experimental grounds, from the fact that it may be the monitored quantity.

Theorems in I-§6 have been proved for t-evolution problems, a traditional way in mathematics, inherited from the early ages of fluid mechanics. We must notice the fact that chromatography is usually considered as a x-evolution problem, (i.e. one records the chromatogram at the column outlet, see section 2 for an example).

We finally arrive at two equivalent Cauchy problems, taking the convention the the first partial derivative indicates the type of evolution problem:

t- evolution problem

$$
\left\{
\begin{array}{l}
\dfrac{\partial \tilde{q}(C)}{\partial t} + \dfrac{\partial J(C)}{\partial x} = 0 \\[2ex]
J(C(x,0)) = J_0(x) \geq 0 \ , \ -\infty < x < +\infty
\end{array}
\right.
\tag{24}
$$

x- evolution problem

$$
\left\{
\begin{array}{l}
\dfrac{\partial J(C)}{\partial x} + \dfrac{\partial \tilde{q}(C)}{\partial t} = 0 \\[2ex]
\tilde{q}(C(0,t)) = \tilde{q}_0(t) \geq 0 \ , \ -\infty < t < +\infty
\end{array}
\right.
\tag{25}
$$

It is, however, highly desirable to prove that the separation class of a process be not changed when we pass from a t- to an x-evolution. It is not difficult to prove this for column chromatography, an exchange of space and time, together with the flux and accumulation, take us from eqn. 24 to eqn. 25 and these problems are equivalent because both J(C) and $\tilde{q}(C)$ are increasing. Therefore, applying theorems given in part I, we get:

t-evolution and x-evolution problems in column chromatography are solute-carrier null or mixing processes.

However in general (m > 1), the application is not quite straightforward and we must recourse to different methods of proof.

3.2 ZONOID EVOLUTION FOR LANGMUIR 1-D CHROMATOGRAPHY

Let us apply the zonoid theory to the x-evolution problem in the case of frontal or elution column chromatography, with the Langmuir isotherm,

$$
q = N^a \frac{KC}{1+KC} \ \text{from which,} \ \tilde{q}(C) = C + a \frac{KC}{1+KC} = C\left(1 + k_o' \frac{1}{1+KC}\right)
$$

Therefore, making a global flux balance, using eqn. 27, we get finally (compare with [12], eqn. 16)

$$C^+(x) = \frac{1}{K} \frac{\sqrt{Q_0}}{\sqrt{a\,x} - \sqrt{Q_0}} = \frac{1}{K} \frac{\sqrt{L_f}}{1 - \sqrt{L_f}}, \quad \text{with } L_f = \frac{Q_0}{a\,x} = \frac{K\,Q_0}{k_o'\,x} \tag{28}$$

L_f, the loading factor, is easily seen to be the dimensionless ratio of the amount injected to the monolayer capacity of the column length x.

Eqn. 28 deserves a comment: in view of the denominator, peak height is infinite for $L_f = 1$, that is, for some $Q_0^*(x)$, which implies that, for $Q_0 > Q_0^*(x)$, the peak will have an infinitely short plateau of infinite height, which is represented, mathematically, by a Dirac measure, the mass of which is $Q_0 - Q_0^*(x)$. So, for a sufficiently short length of column, any pulse injection of a finite quantity must be completed with such a Dirac measure -it is no more a function in the usual sense. The physics behind this is that, for very high solute concentration, the sites of the adsorbent are saturated and the pulse moves at the carrier speed without deformation, a fact that does not seem to have been noticed up to now.

From eqns. 27, 28, we can draw, as usual, the chromatograms that would be obtained at different locations in the column (fig. 10a). However we have not yet followed through the complete procedure of separation evolution and we are left to look at several quantities of interest linked with the separation state and the x-evolution problem.

3.2.2 Computation of level sets and zonoid

We now calculate the volume and content associated with the level set τ for a band with an indeterminate maximum C^+. Note that we use the time to measure the 'volume' of level sets instead of u t which would give the quantity of carrier.

Pulse injection

$$|\omega_\tau| = \frac{x}{u} \frac{k_o'\,K\,(C^+ - \tau)\,(2 + K\,C^+ + K\,\tau)}{(1 + K\,C^+)^2\,(1 + K\,\tau)^2} \tag{29}$$

As regards the content of level sets, we must make the distinction according to if the peak must be completed by a Dirac or not:

$$Q_\tau = x \frac{k_o'\,K\,(C^+ - \tau)\,(C^+ + \tau + 2\,K\,C^+\,\tau)}{(1 + K\,C^+)^2\,(1 + K\,\tau)^2}, \quad x \geq \frac{K\,Q_0}{k_o'} \tag{30}$$

$$Q_\tau = Q_0 - x \frac{k_o'\,K\,\tau^2}{(1 + K\,\tau)^2}, \quad x < \frac{K\,Q_0}{k_o'} \tag{30a}$$

From eqns. 29,30 or 30a, we draw the zonoids (fig. 10b) for the x-evolution problem.

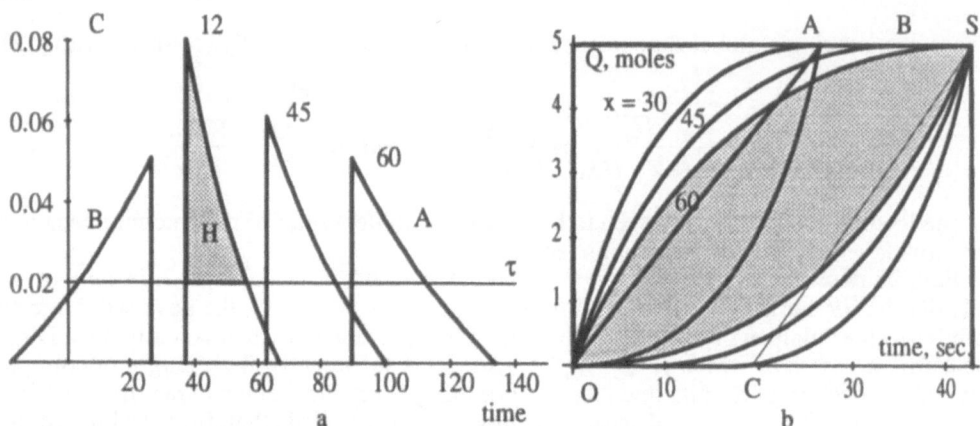

Figure 10 a) Chromatograms at different points in the column
    b) solute-carrier zonoids passing through these points. t_c = 50 sec.

To compare separation states at x and x', x'> x, we must add pure carrier, as indicated by segments AS or BS on fig. 10b, to get zonoids with the same distal point (t_c, Q_o), t_c being the cycle time. Therefore the flow-rate family has a corresponding discrete vector, composed only of pure carrier, whose components are (t_c - $|\omega_d|$, 0). Geometrically, this amounts, e.g., to 'insert' the parallelogram ASCO into the zonoid with distal point A. Let us note that, without this addition, the separation states are not comparable, no inclusion holds and the 2-volume of the smooth family zonoid actually increases. In the case of a long injection band, a plateau remains at the top of the peak, which amounts to add another vector, to the family and, by virtue of Lemma I-9 an edge to the zonoid. We have also figured in light grey, fig. 10a, H, the geometric interpretation of H(τ,x). Clearly this quantity is a decreasing function of x. Peaks A and B will be referred to later on.

We see that separation state is a decreasing function of column length, which means that x-evolution is a carrier-solute mixing process. This result is, of course, in line with the general Theorem I-11. Clearly on fig. 10b, the decrease in the 2-volume loss, indicative of the mixing rate, is a decreasing function of length. Therefore, the loss of separation increases at its maximum rate at the beginning of the column. Note that, for the pulse injection, the initial zonoid, Z(x=0) is the rectangle of total separation, but, for the broad injection plug, it is a parallelogram, thus indicating a lower separation state. The mixing process which induces this lower state is the precolumn dilution device which produces the band at the concentration C_o. Finally, one note that for short values of x, given by the inequality in eqn. 30, we must add a vector $\left(0, Q_o - \dfrac{k_o' x}{K}\right)$ to the smooth 1-d flow family.

Broad injection band
We have $C^+ = C_o$, therefore it is readily calculated

$$|\omega_d| = \left(\Delta t - \frac{x}{v(C_o,0)}\right) + \frac{x}{v(\tau)} = \Delta t - \left(\frac{k_o' K (- C_o + 2 \tau + K \tau^2) x}{(1 + K C_o) (1 + K \tau)^2 u}\right) \qquad (29a)$$

$$Q_\tau = Q_o u \Delta t - x \frac{k_o' K \tau^2}{(1 + K \tau)^2} \qquad (30a)$$

To complete the description of the family, we need, the distal vector of the smooth family, N_{pk} and the vector N_{pl} representing the plateau. Therefore, regrouping these vectors in a matrix, we get

$$[N_{pk}\ N_{pl}] = \begin{bmatrix} |\omega_o| & |\omega C_o| \\ Q_o - C_o |\omega C_o| & C_o |\omega C_o| \end{bmatrix} = \begin{bmatrix} \Delta t + \dfrac{k_o' K C_o x}{(1 + K C_o)\, u} & \Delta t - \dfrac{k_o' K C_o x}{(1 + K C_o)^2\, u} \\[2ex] C_o \dfrac{k_o' K C_o x}{(1 + K C_o)^2\, u} & Q_o - C_o \dfrac{k_o' K C_o x}{(1 + K C_o)^2\, u} \end{bmatrix}$$

These equations hold only for $x \le x_c = \dfrac{u\, \Delta t\, (1 + K C_o)^2}{k_o' K C_o}$. For $x > x_c$, we come back to the pulse case.

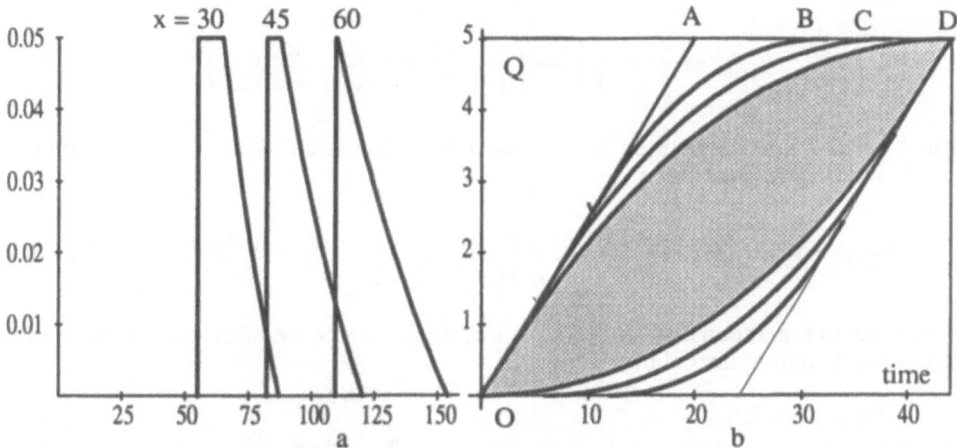

Figure 11 Chromatograms of a broad injection and the corresponding evolution of the solute-carrier separation state. Same conditions as fig. 10, $t_c = 50$, $\Delta t = 20$ sec, $C_o = 0.05$.

On fig. 11a, the zonoid boundary has an edge collinear to OA, which corresponds to the plateau on the chromatogram. An horizontal segment such as BD, at the top, corresponds to the carrier quantity, unused at this stage of the propagation. The zonoids in grey on fig. 10 and 11 are equal, to a good approximation, and so are the peaks at 60 cm, because the initial plateau has almost disappeared. However, the peaks on fig. 11 appear 20 sec later because of the injection time.

2-volume of separation

The smooth subfamily 2-volume, comes from an appropriate rewriting of eqn. I-60, taking the time and the moles of species as the conservative quantities. However, in the family of flow vectors, we must provide for the pure carrier vector, and, if necessary, for the remaining plateau vector. Let us denote these contributions by Ξ_1, Ξ_2, Ξ_3, respectively. Looking at the smooth family first, inserting eqn. 26, 27 into eqn. I-59, we get

$$\Xi_1(x) = \frac{1}{3} \frac{x^2}{u}\, k_o'^2\, \frac{1}{K} \left(\frac{K C^+}{1 + K C^+} \right)^3 > 0 \tag{32}$$

Therefore this rate of 2-volume variation is an increasing function, both of x and C^+.

Broad injection band
First contribution to 2-volume, comes by straight integration of eqn. 31,

$$\Xi_1(x) = \frac{1}{3}\frac{x^2}{u} k_o^{'2} \frac{1}{K}\left(\frac{K\,C^+}{1+K\,C^+}\right)^3 > 0 \tag{33}$$

Second contribution to 2-volume, arises from the unused carrier

$$\Xi_2(x) = Q_o\left(t_c - \Delta t\right) - \frac{k_o^{'}\,K\,C_o\,x}{(1+K\,C_o)\,u}$$

Third contribution to 2-volume, arises from the plateau:

$$\Xi_3(x) = \frac{k_o^{'}\,K\,C_o\,x}{(1+K\,C_o)\,u} - \frac{x^2}{u} k_o^{'2} \frac{1}{K}\left(\frac{K\,C_o}{1+K\,C_o}\right)^3$$

to which we add the contribution Ξ_1, and Ξ_2 to get, after simplifications, the 2-volume, a quadratic decreasing function of x

$$\Xi(x) = Q_o(t_c - \Delta t) - \frac{2}{3}\frac{x^2}{u} k_o^{'2} \frac{1}{K}\left(\frac{K\,C_o}{1+K\,C_o}\right)^3 \quad,\quad x \leq \frac{u\,\Delta t\,(1+K\,C_o)^2}{k_o^{'}\,K\,C_o} \tag{34}$$

The decrease in separation 2-volume for the broad injection is therefore proportionate to the square of the length of the column.

Pulse injection
In the elution case, we obtain Ξ_1 by integrating eqn. 31, after inserting the value of C^+ on x given by eqn. 28,

$$\Xi_1(x) = \frac{4}{3}\sqrt{x}\,\sqrt{K\,k_o^{'}}\,(Q_o)^{\frac{3}{2}} \tag{35}$$

Ξ_2 is given by :

$$\Xi_2 = Q_o\left(t_c - \frac{x}{u}\frac{k_o^{'}\,K\,C^+\,(2+K\,C^+)}{(1+K\,C^+)^2}\right) = Q_o\left(t_c + \frac{KQ_o}{u}\right) - \frac{\sqrt{x}}{u}\,2\sqrt{k_o^{'}\,K}\,Q_o^{\frac{3}{2}}$$

from which we must subtract the contribution of the Dirac mass. Finally, after simplifications, we get, after the Dirac measure has disappeared,

$$\Xi(x) = Q_o\,t_c - \frac{2}{3}\sqrt{x}\sqrt{K\,k_o^{'}}\,Q_{o2}^{\frac{3}{2}} \quad,\quad x \geq \frac{K\,Q_o}{k_o^{'}} \tag{36}$$

whereas, in the initial stage of the propagation we have, from eqn. 34,

$$\Xi(x) = Q_o\,t_c - \frac{2}{3}\frac{x^2\,k_o^{'2}}{u\,K} \quad,\quad x < \frac{K\,Q_o}{k_o^{'}} \tag{36a}$$

The rate of decrease of Ξ with x is only in $\sqrt{x}$, which, compared to the previous case, is much. This is easily understood since $C^+(x)$ is a decreasing function of x. The following figure allows to compare evolution of the separation 2-volume vs x in the two different cases.

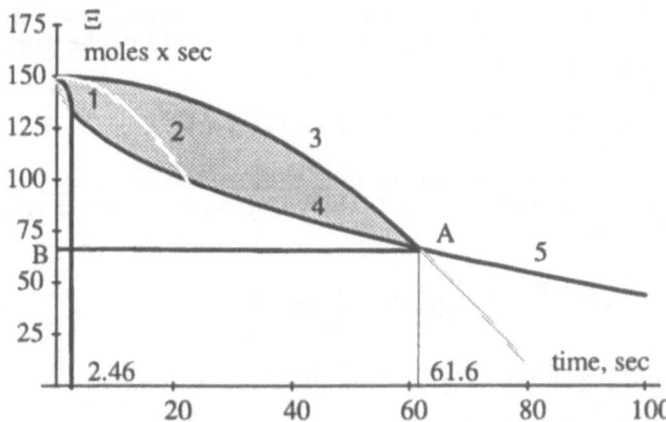

Figure 12 x-Evolution of the separation 2-volume. Curves of variation of 2-volume; 1, 4, 5: pulse injection, 2: broad injection, $C_o = 0.1$, 3: broad injection,$C_o = 0.05$. For 2, 3, $\Delta t = 20$ sec.

All the curves have been translated vertically to get the same initial 2-volume, which means that the loss of 2-volume by premixing has been taken out. Curves 1, 2 or 3 intersect: with curve 4 at the point where the plateau has just disappeared, depending on the plateau concentration. Curve 1 shows the initial decrease of 2-volume for the pulse, when the initial Dirac measure subsists to some extent: although steep, it is not infinitely steep. After intersection point, the 2-volume follows same curve 4 or 5, which has an inverted curvature and nothing more can distinguish the two solutions. We have figured in grey, the domain reached when the initial concentration varies.

A discussion of the separation states themselves is of interest:

The initial separation state is higher for a pulse injection, since the initial state, for the broad injection case, can be obtained, from the former, by mixing. This adds a vertical segment of height $Q_o \Delta t$ to the initial 2-volume on fig. 12. Then, it decreases steadily, whereas the separation state for the band injection decreases slowly, but at an increasing pace. Actually, the two separation states match together. Note that, if Δt is finite, the final regime where no more plateau remains, is always attained, after a finite column length. Therefore, even for the pulse injection, the peak and the separation state follow always two regimes.

3.2.3 Reversible, Irreversible propagation   Let us point out on this example how the irreversibility appears in the propagation course. Reversibility means that, if, after a period of time, we reverse the carrier flow direction, all other things being equal, we shall recover the initial data after same elapsed time period. We can infer the irreversibility of the process directly from the zonoid evolution: since zonoid is decreasing in the course of evolution, starting back from the final state, we get an equal or even smaller zonoid when the given period of time has elapsed. The one case left for reversibility is that zonoid be invariant. This is the case as we have shown if no shock wave arise. In this case also, reversibility is clear since we may follow the characteristics backwards. This is clear on fig. 12: starting back at point A, the reverse propagation is easily shown to be reversible since the front discontinuity now fans out into characteristics. It gives the peak shape B shown on fig. 10a and point B on fig. 12. Note that the discontinuity on peak B is not really a shock: it is a discontinuity produced by the happy circumstance that all characteristics cross together at a given point (x,t).

We could apply, almost without change, the previous development to a parabolic isotherm or to any isotherm which is convex or concave. The case of a sigmoid isotherm, although also tractable by the same method, is complicated by the fact that the shocks cannot be localized a priori at the front or at the rear of the peak.

3.3 ZONOID EVOLUTION FOR THE IDEAL 2-D CHROMATOGRAPHY

3.3.1 Model An example of chromatography in two space dimensions is paper chromatography. We make the assumption of equilibrium and constant carrier speed vector, $\mathbf{u} = (u_1, u_2)$, that is, $\mathbf{J} = \mathbf{u}\,C$:

$$\left(1 + \frac{(1-\varepsilon)}{\varepsilon}\frac{dq}{dC}(C)\right)\frac{\partial C}{\partial t} + \frac{\partial}{\partial x^1}(u_1 C) + \frac{\partial}{\partial x^2}(u_2 C) = 0 \tag{37}$$

$$C(x,t) = C_0, \|x\| \le r \; ; \; C(x,t) = 0, \|x\| > r$$

Initial data mean that a circular solute spot of uniform concentration is placed on a paper, void of solute elsewhere. This model and a short study using characteristics can be found in [13]. We deepen further this subject by showing that the case of shock waves can be treated completely using the condition Z.

This is a t-evolution problem and the Theorem I-11 applies steadily:
the zonoid of accumulated quantities in ideal 2-D chromatography is t-decreasing.

For the sake of simplicity, let us approximate the Langmuir isotherm to the second order; the total accumulated quantity is then a quadratic function of concentration:

$$\tilde{q}(C) = C + a\,K\,C(1 - K\,C) = a\,C + b\,C^2 \,, a > 0, b < 0$$

and the eqn. 36 becomes what is known as the (2-D) Burgers equation (for any constant a and b).

3.3.2 Condition Z and entropic shocks The condition Z can be written, locally,

$$(\mathbf{u},\mathbf{n})\left(\frac{b\,(C^+ - \tau)\,(\tau - C^-)}{a + b\,(C^+ + C^-)}\right) \le 0 \tag{38}$$

which means that, when $b < 0$, only front shocks can exist - in the sense $(\mathbf{u}, \mathbf{n}) \ge 0$, $\mathbf{n}$ the outer normal to the shock curve in the plane (x_1, x_2). Furthermore, with the present initial data, we must have $C^- = 0$.

Clearly, taking a new reference frame, one axe, collinear to $\mathbf{u}$, the other one pointing to an orthogonal direction, we get rid of the 2-D complexity and reduce our problem to a set of independent 1-D problems lying in parallel planes labelled by the transverse coordinate. The only connection between these problems is that the initial conditions for the 1-D problem are the intersection of the initial 2-D data by such a plane. More precisely, taking cylindrical coordinates, θ being the angle with transverse direction, the initial data at transverse distance $r \cos \theta$ is a plug of width $2 r \sin \theta$. Proceeding as in the above subsection, the speed of a characteristic and the Rankine-Hugoniot condition are (u is now a scalar quantity)

$$v(C) = \frac{u}{(a + 2 \, b \, C)} \quad , \quad v(C^+, C^-) \ = \ \frac{u}{a + b \, (C^+ + C^-)}$$

from which a very simple rule, using arithmetic mean value of shock bounds, for shock construction follows.

Let us now look at the level sets of C, and of $\tilde{q}$ as well, at time t : $\partial \omega_\tau^0$, the continuous part, is a circular arc which propagate with speed $v(\tau)$; similarly $\partial \omega_\tau^+$, at least as long as the upper plateau has not disappeared, has a part which is a circular arc propagating with speed $v(C^+, 0) \ = \ \dfrac{u}{a + b \, C^+}$. Remaining part of $\partial \omega_\tau^+$ is a transition part over which the plateau has disappeared and the shock strength is lower. Eventually, the former disappears, leaving only the transition part. At time t, when this event has not yet come out, using the fact that 1-D peak is a triangle of constant area, we obtain a relation giving the maximum concentration, which is easily inverted to give the shock path in the transition case:

$$0 \le \theta = \text{Arcsin} \left(\frac{-t \, u \, b \, C^{+2}}{2 \, C_0 \, r \, a \, (a + 2 \, b \, C^+)} \right) \le \text{Arcsin} \left(\frac{-t \, u \, b \, C_0^2}{2 \, C_0 \, r \, a \, (a + 2 \, b \, C_0)} \right)$$

Note that for $\tau = C^+/2$, the shock and the characteristic speed are equal, hence, for that concentration, the initial circle propagates without changes, except, however, for the transition part.

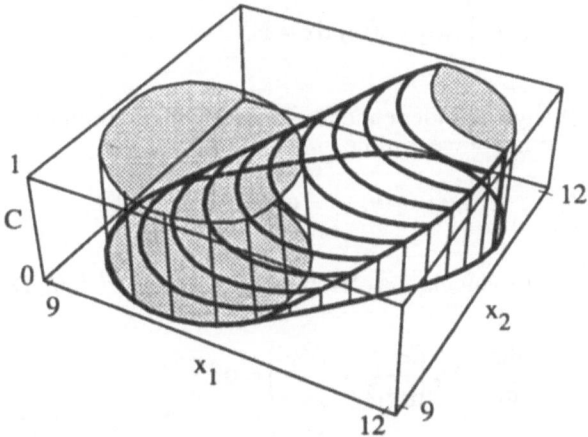

Figure 13 Evolution of the solution of a 2-D nonlinear propagation
$a = 1, b = -0.2, \ t = 10$ sec., $u_1 = u_2 = 1$ cm/s
Initial condition : a circular spot of radius 1, centred at O, $C_0 = 1$

We distinguish on fig. 13, a translate of the initial concentration spot, and the different pieces of the solution at time t: the smooth rear piece, the full developed shock front part, the transition shock side part and, in grey, the upper plateau. One could continue the solution after the plateau has completely disappeared with the same method.

3.4 ZONOID EVOLUTION IN DISPERSIVE CHROMATOGRAPHY

We study now the linear or nonlinear dispersive chromatography of a single solute in several space dimensions and prove that it is a carrier-solute mixing process in the most general case, i.e. 2D, 3D ... chromatography with non linear isotherm and non linear diffusion. Here we have both dispersion and transport terms in the species flux.

3.4.1 Evolution in the general case Consider a t-evolution problem for a model for diffusion/convection of a conservative species in a volume Ω in m-D space (e.g. m = 2, 3) so big that concentration remains negligible on the boundary $\partial\Omega$, which means that this system is isolated (at least for the solute). Furthermore, we suppose that the hydrodynamic flow conserves the volume, i.e. div $\mathbf{u}(x,C) = 0$. Since this is a Problem III type, in application of Theorem I-13, we have the theorem

Theorem The t-evolution problem in nonlinear dispersive chromatography of a single species is a mixing process.

3.4.2 The linear 2-D chromatography The equation of linear 2-D chromatography, with $\tilde{q}(C) = C(1 + k'_o)$, the total accumulated function, assuming an isotropic diffusion, is

$$
\left\{
\begin{array}{l}
(1 + k') \dfrac{\partial C}{\partial t} + \dfrac{\partial}{\partial x_1}(u_1 C) + \dfrac{\partial}{\partial x_2}(u_2 C) - D\left(\dfrac{\partial^2 C}{\partial x_1^2} + \dfrac{\partial^2 C}{\partial x_2^2}\right) = 0 \\[4mm]
C(x,0) = C_o(x), \displaystyle\int_{R^m} \tilde{q}(C_o(x))\, dx = N_o^1 < \infty
\end{array}
\right.
\tag{39}
$$

If we make the following change of independent variable, which amounts to follow the center of the mass of the solute in course of time,

$$
y = x - \frac{u}{(1 + k')} t, \quad C(x,t) = \varphi(y, t) = \varphi(x - \frac{u}{(1 + k')} t, t)
$$

we get the diffusion equation, introducing the effective diffusion coefficient, $D_e = \dfrac{D}{1 + k'}$,

$$
\frac{\partial\varphi}{\partial t} - D_e\left(\frac{\partial^2\varphi}{\partial y_1^2} + \frac{\partial^2\varphi}{\partial y_2^2}\right) = 0
\tag{40}
$$

Let us take the case of a radial symmetric solution in cylindrical coordinates,

$$
\frac{\partial\varphi}{\partial t} - \frac{D_e}{r}\frac{\partial}{\partial r}\left(r\frac{\partial\varphi}{\partial r}\right) = 0, \quad \varphi(r,0) = \varphi_o(r)
$$

the elementary solution for the pulse injection at the initial time of a quantity N_0

$$
\varphi(r,t) = \frac{N_o}{4\pi D_e t} e^{-\frac{r^2}{4 D_e t}}
$$

from which the τ level sets are discs of radius r,

$$r = \sqrt{-4 D_e t \ln\left(\frac{4 \pi D_e t \tau}{N_o}\right)}$$

introducing the dimensionless quantity $\theta = \frac{4 \pi D_e t \tau}{N_o} \leq 1$, which is the ratio of τ to the actual height of the peak, the volume of the level set τ and of its enclosed content are:

$$|\omega_\tau| = -N_o \frac{\theta}{\tau} \ln \theta , \quad Q_\tau = N_o (1 - \theta)$$

from which the peak and the corresponding zonoid are easy to draw (fig. 14)

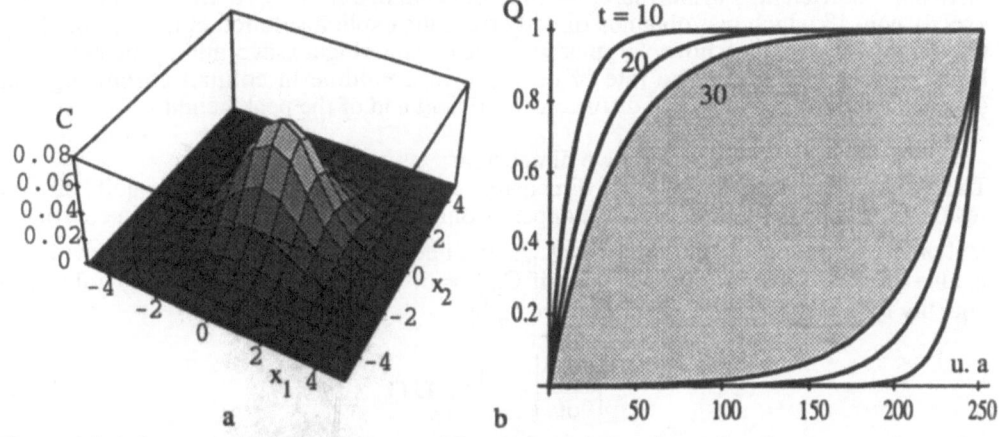

Figure 14 a) the concentration map at t= 10 sec, the origin of axes has been set at the peak symmetry centre. $D_e = 0.1$ cm^2/ sec, $N_o = 1$, b) the zonoids at time 10, 20, 30 sec.

The quantity $H(\tau,t)$, can be evaluated from the above quantities

$$H(\tau,t) = N_o (1 - \theta + \theta \ln \theta) \tag{41}$$

By derivation of eqn. 41, or using eqn. I-57, we find

$$\frac{\partial H}{\partial t}(\tau,t) = 4 D_e \pi \tau \ln \theta < 0 \tag{42}$$

which proves that separation state, i.e. the zonoid of the solute accumulated quantity, decreases in the course of time, in agreement with Theorem I-13.

3.4.3 2-Volume evolution in different space dimensions The precise shape of 2-volume decrease in linear dispersive chromatography is now studied, in one or two dimensions, but generalization to higher dimensions would be straightforward. Of course one will verify that results on separation evolution in the linear case matches the general conclusions gained by the a priori estimates methods.

The Gaussian model in one dimension:

This model is a solution of the diffusion equation in one dimension. More specifically, Let the initial data, $u_o(x)$ be a Dirac distribution. Hence Z_o is a rectangle. The elementary

solution is $C = \dfrac{1}{\sqrt{2\pi}\,\sigma}\,e^{-\frac{x^2}{2\sigma^2}}$, with $\sigma = \sqrt{2Dt}$, the standard deviation. Inserting into eqn.

I.66, $C_{max}(t) = \dfrac{1}{\sqrt{2\pi}\,\sqrt{2Dt}}$, $C_{min}(t) = 0$, we get

$$\frac{d\Xi(t)}{dt} = -8\,D\,(C_{max}(t) - C_{min}(t)) = -8\,\frac{D}{\sqrt{2\pi}\,\sigma(t)} = -4\sqrt{\frac{D}{\pi t}} \tag{43}$$

The rate of decrease of 2-volume depends only on standard deviation at a given time and diffusion coefficient. Furthermore, its time dependance is in $1/\sqrt{t}$. In fact this result is exactly eqn. 13 which was obtained directly from the explicit solution of a problem of type (II), taking into account, however, that we were considering a x-evolution problem.

Physical meaning is that the rate of decrease in 2-volume in column chromatography depends only on the product of diffusion coefficient and of the peak height.

The Gaussian model in two dimensions:

The case of chromatography in two dimensions is similarly treated but gives quite different result: the rate of 2-volume decrease depends only on the product of diffusion coefficient. and on $\mathcal{V}$, the volume enclosed above C_{min} but under the graph of C in space $R^2 \times R$, that is also the total quantity Q_o of species if $C_{min} = 0$. This equation (eqn. I-67) holds as an equality for a Gaussian peak in chromatography:

$$\left|\frac{d\Xi(t)}{dt}\right| = 8\,\pi\,D\,Q_o \tag{44}$$

From eqn. 41, 42, using eqn. I-42, we get after integration,

$$\Xi(t) = \Xi(0) - 8\,D_e\,\pi\,N_o\,t$$

in agreement with eqn. 44.

A consequence of the constant rate of 2-volume decrease if $u_{min} = 0$, that is if $\Omega = R^2$, is that the loss of 2-volume cannot be finite since this would be imply negative values of ζ after a finite time. Conversely, if Ω is bounded, $u_{max}(t) - u_{min}(t)$ must tend to 0 when t tends to the infinite.

It is interesting to note that the factor *eight* in above eqns. 43, 44 results from two very different contributions: in both $8 = 2 \times 4$ with the factor 2 associated with symmetry of the zonoid but in the former, 4 comes from the square of the volume of unit ball in R, in the latter, 4 comes from space dimension through factor m^2.

Therefore the law of 2-volume decrease depends quite strikingly on dimension of space in which chromatography occurs: in 1-D chromatography, the minimum rate of separation decrease depends on peak height, but not on the injected quantity, whereas in 2-D chromatography, the converse occurs.

In closing the present section, we may advance a general conjecture: all effects which create asymmetry on a previously symmetric peak increase the rate of mixing and this, in turn, has a stabilizing effect on the solution. Conversely, if some separation occurs, one may believe that instability will occur conjointly.

The case of several solutes is of course the goal of our investigation. However, in the non linear case, there are specific difficulties: in the presence of shock waves, the principle of decrease of concentrations along the column needs not hold. Consequently, inclusion cannot be hope in this case.The multicomponent chromatography with a Langmuir isotherm appears, for several reasons, to be the next good candidate to evaluate the potential of the theory.

4 Setting and solving problems with zonoids

A good theory should first integrate previous knowledge and then ask new questions the answers to which open new avenues. Many separation problems can be recast geometrically and given an innovative answer. Below four problems will be postulated in order of complexity, but clearly they just scratch the surface of such a vast subject. One may believe that for many of practical problems, the fundamental property of convex set of mixtures will be the critical property. Meanwhile, we shall encounter extent of separation, proposed by Rony [3] and show how well it fits into zonoid theory. Although economy (e.g. marginal costs and linear programming) is supposed to give answers to these problems, this is not so in most cases, due to the overwhelming complexity of calculations and, more fundamentally, to the absence of rules to assess costs of intermediate products in plants. We then need first to answer on a technical basis.

4.1 A PROBLEM IN RECOVERY WITH MINIMAL DILUTION

Given a process delivering a family of mixtures by use of a conservative separation agent, and given a solute mixture, find the sampling (if any) which will make it possible to recover the solute mixture with minimal dilution in the separation agent.

The answer is given below for dim $E = 3$, but can be generalized. Let us remark before that it is not commonly considered that a sampling function λ may oscillate wildly between 0 and 1. There is no real reason why only gentle λ, i.e. involving only a limited number of oscillations should be the best ones.

Consider a family $\mathscr{F}$ and $M = (M^1, M^2)$, a mixture of solutes. Iff $M \in Z_{12}(\mathscr{F})$, M is feasible, the problem amounts to geometrically determine the point on boundary ∂Z^- laying above M. In the case of 3-regular selectivity, in view of eqn. I-28, it is equivalent to find t_1, t_2 such that

$$M = \int_{t_1}^{t_2} F(\tau)d\tau \ , \ 0 \leq t_1 \leq t_2$$

or, a segment equipollent to M can be put as a chord on ∂Z_{12}. Minimum quantity of separation agent will then be

$$M_{min}^3 = \int_{t_1}^{t_2} F^3(\tau)d\tau$$

If selectivity is not 3-regular, a more complicated sampling function, deduced from a variant of eqn I-29, will be necessary, but it remains gentle.

4.2 A PROBLEM IN YIELD OF RECOVERY

Given a family $\mathcal{F}$, characterize all mixtures M with a given content M^i of a given solute i (i.e. all mixtures with a given yield of a solute). Geometrically, this is equivalent to finding the intersection of $Z_{12}(\mathcal{F})$ by a plane, equation of which is $N^i = M^i$ and must be solved by computer to be represented graphically.

4.3 A PROBLEM IN APPROXIMATION

Given a family $\mathcal{F}$, find a criterion for the optimum recovery of a fixed number p of cuts and find corresponding set of optimal p-1 cut-points. Answer is:
      take the p cuts to minimize loss of n-volume.

This is a problem of approximation of a differential family by a discrete family. Clearly we may consider the approximation in the full space in n dimensions (i.e. take in account the separation agent), or a projected subspace defined by r "preferred species", e.g. solutes. We must have, $p \geq n$. or $p \geq r$ depending on the case.
    If n= 3, r = 2, second procedure would give: minimize 2-volume lost in the plane of solutes 1,2. In the following we will take p=2. We have to find one cut-point. Now, recall that for a binary mixture N_0, split into two mixtures N_1, N_2, Rony has defined [3,4] the extent of separation as

$$\xi = \mathrm{Abs}\left(\frac{1}{N_0^1 \, N_0^2} \begin{vmatrix} N_1^1 & N_2^1 \\ N_1^2 & N_2^2 \end{vmatrix} \right) = \frac{1}{N_0^1 \, N_0^2} \left| N_1 \wedge N_2 \right|$$

and this definition has been generalized by Valentin [2] to a **vector** extent of separation, covering the general multicomponent case. We shall take $N_0 = (1, 1)$. In the present theory we should call ξ a yield of 2-volume.
    Indeed zonoid theory gives a physical interpretation which was lacking to date in Rony index

    ξ is the yield of 2-volume of $Z\{N_1, N_2\}$ (a parallelogram).

Now, it is natural, if N_1 and N_2 are obtained from a 1-d family of parameter t by a 1-cut point sampling at τ to ask for the maximum of $\xi(\tau)$, say ξ^*. Clearly

    ξ^* is the 2-volume of the best approximation of Z_{12} by a parallelogram.

In the case of linear chromatography, Rony [6] shows that optimum cut point is at $\tau^* = \frac{\delta}{2}$ and

$$\xi^* = 2\,\mathrm{erf}(\frac{\delta}{2}) - 1 \tag{25}$$

From corresponding eqn. 25 it is clear that loss of 2-volume induced by the two-cut trapping procedure is then independent of δ, as long as the linear expansion is valid and therefore,

$$\frac{\xi^*}{\Xi} = \frac{1}{\sqrt{2}}$$

Comparing eqn. 25, 11 we find a formulation holding in all cases, setting into perspective effect of recovering only two cuts on length of column:

> To get same separation 2-volume between two solutes, chromatography with 1-cut point needs a twice as long a column than differential chromatography.

Thus ξ^*, the **extent of separation** [7] reflects only part of the separation produced by the column.

Furthermore, coming back to the full dimensional approximation, one may take in account the solutes dilution in carrier, i.e. the ternary separation. Now it is clear that a 2-cut procedure cannot be optimal since the 3-volume recovery would then be zero. This strongly suggests that optimizing a 3-cut recovery should be considered in order to recover a better part of the separation produced by the column.

4.4 A PROBLEM IN BASESTOCK MANAGEMENT

Given a stock of basic mixtures, which should be the criterion for optimization of the fabrication of any required mixture if the next mixture to produce is unknown ?

This problem, also called the **paint dealer problem,** has been solved in [9] using the fact that probability of hitting a zonoid by a point is proportionate to its n-volume for a uniform density of probability of demand.

4.5 A PROBLEM OF DESIGN

From all solution u of Problem (I) or (II) which have same initial zonoid, find those which make extremal some criterion, e.g. rate of variation of 2-volume.

$$
\begin{aligned}
&\underset{u_0(x)}{\text{Min}} \quad \phi(u(x,t)) \\
&\left|
\begin{aligned}
&Z_{u_0(x)} = Z_0 \\
&|\Omega| = V_0
\end{aligned}
\right.
\end{aligned}
$$

One may put Ω as fixed or variable, under the constraint $|\Omega| = V_0$.

Solutions to problems of this type would have potentially important practical applications in design of physico-chemical systems by making possible to find the most efficient shape of the system under consideration.

5 Conclusions

Zonoid theory has been applied to a very simple type of chromatographic model, the linear diffusional model. Results apply, of course, to all processes which can be modelled in this way, e.g. liquid-liquid extraction. Method is quite general and should be a safe guide to optimization or design of any process, including those which deliver families diffuse in 3-D space and time. Let us summarize the method :

1. write and solve the equations of balance in conservative form. This defines the accumulated and flow families.

2. Evaluate condition Z and verify the sign of $\frac{\partial H}{\partial t}$.

3. Compute the level sets of the solution for given initial conditions.

4 .Write the equation of evolution for the zonoid boundary and for its 2-volume.

5. Study the evolution of the separation state.

170

Step 2 to 5 supplement the usual method of solving problems in separation engineering.

Although tools may seem too sophisticated and applications remote at this stage, one must point out the strong potential of zonoid theory for the Separation Science.
Mathematical foundations and developments reflect an inherently complex situation and shed a clear light on the specific features of chromatography. To the exception of linear chromatography, we have restrained developments to the single component case: clearly, it is necessary to settle down the case of the interaction between a single solute and the separating agent before attempting to treat the full multicomponent problem.

Chromatography, in the linear multicomponent case, exhibits how the interplay between separation and mixing operations may be a delicate one. Indeed each species mixes with carrier at the same time it separates from other species.

In view of present results, zonoids would also appear to be a promising tool, even for mathematical purposes. They present a unique blending of both local and global character, retaining only salient features of the solution of an evolution problem. Applications of zonoid theory to systems of conservation laws is a challenging problem both in theory of PDE and physical meaning and relevance of these PDE models for study of physico-chemical processes.

Another interesting development would be to apply zonoid theory to classification of separation processes. In this way one should gain a more quantitative look on the work of J.C. Giddings [13].

ACKNOWLEDGMENT

Figure 7 was created using a program for computer manipulation of zonohedra created by N. Szafran (LMC, IMAG, Univ. Joseph Fourier, Grenoble). This work was supported by Société Nationale Elf-Aquitaine.

SYMBOLS

Latin Alphabet

C	concentrations (line or volume)
D, D_e	diffusion, effective diffusion coefficient
E	vector space of quantities
e	base vector of E
F	flow-rate vector
$\mathcal{F}$	a family
H	HETP
$H(\tau)$	excess quantity function over level τ
I	interval of natural numbers, $\{1,..., n\}$
i	chemical species
J	interval of natural numbers
J	flux vector
k'_0	retention factor at infinite dilution
K	Langmuir constant
N	number of theoretical plates
M, N	quantity vector

N_o	overall (or sum) quantity vector, distal point
n	dimension of the space E, number of species
n	outer unit vector at the boundary of a level set
O	origin
p	number of regions or of cuts
q	concentration in the adsorbed phase
$\tilde{q}$	total accumulated solute density
Q	content of species, quantity of solute
Q_o	solute injected quantity
R	vector space of real numbers $(-\infty, +\infty)$, resolution between two peaks
R^3	usual Euclidean space
t	current parameter in a family, retention time
t_c	cycle time
u	carrier speed
v	shock or characteristic speed
W	Wronskian determinant
X	molar fraction vector
Z	zonoid
z	abscissa along column

Greek Alphabet

α	relative volatility
γ	curve of molar fraction in Δ_2
Γ	the cumulated family curve in E
Δ_{n-1}	n-1 dimensional simplex (of molar fractions in E, dim E = n)
Δt	cycle time, injection time interval
∂	operator taking boundary of an open set
δ_{ij}	dimensionless translation parameter between solute i, solute j
ε	bed porosity
Ξ	n-volume of Z, $(Z \subset E, \dim E = n)$ $\quad$ (moles)^n
ξ	extent of separation, dimensionless n-volume
ω	open set in R^m, level set
Σ	system state
σ	variance
τ	reduced time, parameter on a curve, level of a level set
θ	dimensionless height of a 2-D peak
ζ	yield in loss of 2-volume

Indices

i, k	solutes
j	region (lower index)
+,-	upper, lower bound of a discontinuity

Signs

[ ]	matrix
{...}	a set of elements

+	sum or Minkowski sum
∈	belongs to
| |	the volume of a set in R^m
‖ ‖	the Euclidean norm
∧	Exterior product of vectors
n-D	n dimensional (space)
1-d	1-differential (family)

REFERENCES

[1] Valentin, P., Is chromatography a separation process, The zonoid answer, J. Chromatog., **556**, 25-80 (1991)

[2] Valentin, P., Zonoids as a geometric tool for first order quasilinear equations in several space dimensions, Preprints, 13 th. IMACS Congress, Miller, Vichnevski ed. Dublin 22-27 July 1991, p. 400-403.

[3] Valentin, P., The principle of Non Increase of Separation, zonoids as a geometric tool for a parabolic quasi-linear equation, 1st. European Congress on Elliptic and Parabolic Equations, Pont-à-Mousson, 17-22 June 1991

[4] Rony, P.R., The extent of separation, a universal separation index, *Separation Science*, 3, 239, (1968)

[5] Bandle, C., Isoperimetric Inequalities and applications, Pitman, New-York, 1980

[6] Bandle, C., Sperb, R.P., Stakgold, I., Diffusion and Reaction with monotone kinetics, Nonlinear Analysis, Theory, Methods and Applications, **8**, 321-333 (1983)

[7] Girard, D., Valentin P., Zonotopes and Mixtures Management, in *New Methods in Optimization and their Industrial Uses*, Penot (Ed), International Series of Numerical Mathematics, Birkhauser Verlag, Basel, Vol 87, 57-71 (1989)

[8] Lacolle, B., Valentin P., Modélisation Géométrique de la Faisabilité de Plusieurs Mélanges, submitted to M^2 AN: Modélisation et Analyse Numérique.

[9] Szafran, N., Zonoèdres: de la Géométrie Algorithmique à la Théorie de la Séparation, Thèse de Doctorat en Mathématiques Appliquées, Université J. Fourier, Grenoble, Oct. 1991.

[10] Lin, B.C., Ma, Z., Golshan-Shirazi, S., Guiochon G., Theoretical analysis of non linear preparative liquid chromatography, J. Chromatogr. **500**, 185-213; (1990)

[11] Golshan-Shirazi, S., Guiochon G., The Ideal Model of Chromatography, NATO Advance Study Institute Series, 1991, pp

[12] Golshan-Shirazi, S., Guiochon G., The Equilibrium-dispersive Model of Chromatography, NATO Advance Study Institute Series, 1991, pp

[13] Rhee, H. K., Aris R., Amundson N.R, First Order partial Differential Equations, Vol I, Theory and Application of Single equations, Prentice-Hall Ed., 1986

[14] Giddings, J.C., Unified Separation Science, J. Wiley, New-York, 1991.

[15] Rouchon P., Schœnauer M., Valentin P., Guiochon G., Numerical Simulation of Band Propagation in Nonlinear Chromatography, in Preparative-Scale Chromatography, Grushka Ed, Chromatographic Science Series, Vol. 46, Marcel Dekker, 1989, pp 1-41

STOCHASTIC THEORY OF CHROMATOGRAPHY:
The Characteristic Function Method and the Approximation of Chromatographic Peak Shape.

F. DONDI, G. BLO, M. REMELLI, P. RESCHIGLIAN
Department of Chemistry, University of Ferrara
via L. Borsari, 46
I-44100 Ferrara
Italy

ABSTRACT. The Stochastic Theory of Chromatography is reviewed in its fundamental aspects and achievements. Basic elements of the probability theory for linear-finite-time chromatographic model description are presented and the mathematical method of "Characteristic Function" (CF) for solving probabilistic models is described. Solutions are found in terms of CF for general chromatographic models of increasing complexity. These models are the constant mobile phase velocity model with general mechanisms of stationary phase entry and desorption process and non-constant mobile phase processes with Poisson mechanism of the stationary phase entry process. Among the solved models those belonging to the class of stochastic processes with stationary and independent increments are identified for which general description of the peak profile by the Edgeworth-Cramér series holds true. The convergence towards the Gaussian Law through the Berry-Esséen Theorem is discussed and the basic conditions for peak parameter determination by non-linear least squares fitting are discussed.

1. Introduction

The central core of the theoretical development of Chromatography consists of the modeling of differential transport equations and in finding deterministic solutions for differential equations. In these equations the solute concentration c is a function of the coordinate l, the distance from the column inlet and t, the time. Once boundary and initial conditions have been established, solution is obtained by analytical or numerical techniques. This "mass balance approach" is extensively discussed by Guiochon and Golshan-Shirazi in the first three chapters of this book. Such a basic approach requires a modeling step where the existence of rate and equilibrium constants are assumed and defined. An alternative means of modeling a chromatographic process is to start by watching a single solute molecule (= "microscopic approaches") and putting into the model the basic "elements" which describe its behaviour. The purpose is to derive the probability density f of finding a molecule at l, t. The modeling step consists of describing

173

F. Dondi and G. Guiochon (eds.),
Theoretical Advancement in Chromatography and Related Separation Techniques, 173–210.
© 1992 *Kluwer Academic Publishers.*

the molecule behaviour from a "microscopic" point of view. To do so there are two possible "microscopic approaches": the "probabilistic" approach requires definition of the probability for transition of a molecule from one state to another state; the second is the "dynamic" approach which starts, instead, from the equation of motion. The two most famous and foremost examples of these approaches are respectively the celebrated Einstein and Langevin descriptions of the Brownian diffusion process. Both approaches result in the same differential equation (diffusion equation) of $f(l, t)$ [1].

The "mass balance" and the "microscopic" approaches must be equivalent for the same system: final equivalence can be stated through a correspondence between the probability of a molecule being at point l at time t and the concentration c at the same point and time. In chromatographic theory the "microscopic probabilistic" approach is often followed even for introducing the basic concepts of chromatography. This is the so-called "random walk" approach [2,3], able to give a picture of the chromatographic band dynamics even to freshman and to derive in a simple way the terms of the van Deemter equation. The other example of the microscopic probabilistic approach is the classical stochastic theory of chromatography presented by Giddings and Eyring in 1955 [4], right at the time of the great development of gas chromatographic technique and subsequently further developed by McQuarrie [5]. The great appeal of the Giddings-Eyring approach was that the obtained solution -- the Bessel Function -- accounts for the existence of the tailing effect, thus no longer giving the solution only in terms of gaussian peak shape or band width. It is a complete solution of a model of linear finite-time chromatography, i.e. the case of solute molecules which do not interact (=linear) and in a column of finite length (finite-time). The term "infinite-time" refers instead to infinitely long columns where all the approaches give gaussian peak shapes under linearity conditions [6]. Essentially the present paper will develop the original Giddings - Eyring approach (microscopic approach of probabilistic type) by using the powerful mathematical technique provided by the characteristic function (**CF**, see at the end the glossary) method [7], which was of extraordinary importance in the development of the modern Probability Theory [8,11].

2. A simple probabilistic model

The migration of an individual molecule in the chromatographic column is a random process of high complexity. Its most general description consists of a complex chain of random processes of different kinds: ordinary diffusion, flow pattern effects (eddy diffusion) and sorption-desorption kinetics. However, among all these complex processes it is the sorption-desorption which determines the essence of the differential chromatographic migration. It will thus be considered first.

Fig. 1 describes, from a stochastic point of view, the behaviour of a single molecule sorbing from the mobile (gas) phase on a single sorption site. This may be e.g. an adsorption site of a surface or a liquid phase. It is a basic physicochemical fact that the time spent by a single molecule on a single site, (Δts), is a random variable. The symbol "s" will denote a stationary phase quantity. The origin of this random behaviour can be simply explained as follows [12,13]. A molecule flying near a sorption site is "captured" by the sorption site when the sorption energy, Es, overcomes the molecule's kinetic energy, Ek:

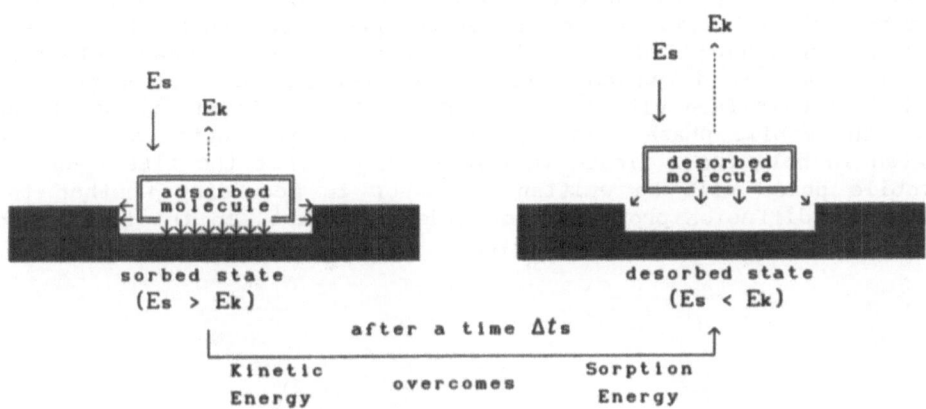

Fig. 1: Kinetics of the sorption-desorption process.

The reverse process -- that is the desorption step -- will appear when the kinetic energy of the sorbed molecule exceeds the sorption energy once more ($Ek > Es$). The probability, per unit of time, of such an event is essentially related to a process of continuous energy exchange between the sorbed molecule and sorption site and to the fluctuations in energy. According to this simple picture the total sorption time is an example of a "waiting time" stochastic process [9]. The simplest reported expression for the frequency function of Δts is:

$$f(\Delta ts) = ks \ exp \ (- \ \Delta ts \ ks) \tag{1a}$$

where ks is the process time constant related to the average time spent by the molecule on the sorption site which is:

$$\overline{\Delta ts} = 1 \ / \ ks \tag{1b}$$

This quantity is not easy to measure experimentally [12]. Its theoretical evaluation can be obtained from the Frenkel equation [13]:

$$\overline{\Delta ts} = \tau_0 \ exp \ (Es \ / \ RT) \tag{2}$$

where τ_0 is approximately 1.7×10^{-14} s and R the gas constant per mole and T the absolute temperature. For example, in physical adsorption processes, Es ranges from 1 to 20 kcal/mol and the mean sorption time varies from 1×10^{-12} up to 100 s [12]. In the case of liquid phase partition, the most appropriate description is be obtained as a diffusion process. The problem now is not to describe in detail the kinetics of the desorption step, but to recall the fact that the basic step of all chromatographic processes is a random process and that no single molecule repeats this process in the same way over equivalent sites, nor do different molecules perform it identically on the same site. Please observe that a random quantity such as the time spent by a molecule on a sorption site, Δt_s, the total time spent, t, the total distance covered , l, the number of steps, n_s, etc. are usually denoted in *italic*. Instead **mean** values of random quantities such as the mean time spent on a sorption site, $\overline{\Delta t_s}$, several types of mean retention times, $\bar{t}_r$, $\bar{t}_R$, $\bar{t}_m$, $\bar{t}_s$ which will be defined in the following, are indicated in **bold** face with the superscript " ¯ ". Constant quantities such as the mobile phase velocity, $\mathbf{v_m}$, the column length, $\mathbf{L}$ etc. are indicated in **bold** face. Please thus note, e.g., that the time spent in the mobile phase will be written as t_m or $\mathbf{t_m}$ according whether the mobile phase diffusion process is considered or not. In the first case the mean value will be written as: $\mathbf{\bar{t}_m}$.

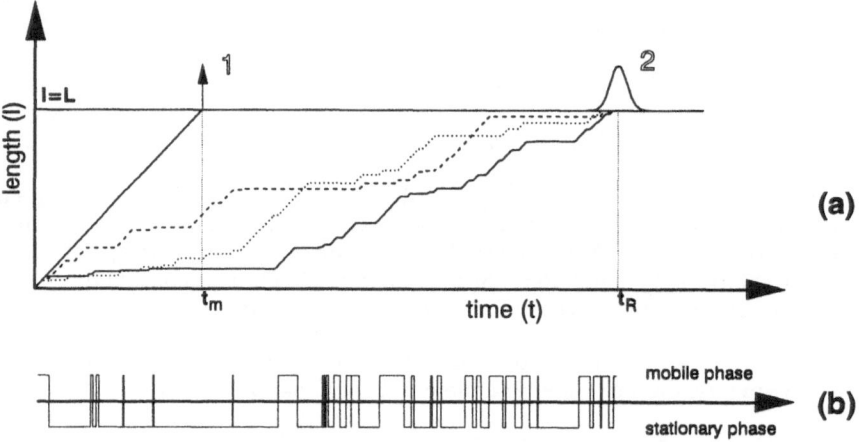

Fig. 2: (a) Example of random trajectories (t, l) of solute molecules moving forward in chromatographic medium. 1, 2: unsorbed, sorbed solute. (b) molecule status during migration inside the column, referred to full line case in (a).

Fig. 2 represents the general features of the simplest stochastic model of chromatography, that in which all the molecules are assumed to migrate in the mobile phase at constant velocity (=constant mobile phase stochastic model). In the lower part of this figure the "state" of a single molecule on the the "chromatographic trip" through the column is described as constantly changing between that of being in the stationary phase and that of being in the mobile phase. The random movement is represented in the upper part of Fig. 2 where, as an example, three trajectories of three different molecules are drawn (obviously, in order to make the point clear, highly improbable trajectories, with respect to the traced band, were reported in Fig. 2). The solute progression through the column is described by these trajectories in the coordinate plane time-column length (t, l), respectively the time elapsed from the beginning of the process (injection) and the column distance covered. When the solute molecule reaches a column distance equal to the column length L, the process ends. The retention time of a single solute molecule is simply the intersection of the trajectories with a horizontal axis located at l = L. The chromatographic elution peak is the probability density function of trajectory cross sections. The cross section with a vertical axis located at t is the band profile inside the chromatographic medium at a given time t. In the first case (elution mode), the random quantity to be considered is time while in the second case (development mode) it is the distance. With reference to Fig. 2, the elution peak of an unretained compound will be a Dirac function located at t = t_m = L / v_m, where v_m is the (constant) mobile phase velocity.

Let us consider the case of elution. A horizontal segment Δts represents the solute sorption time in a sorption site. This quantity is random and independent since the molecule has no memory of his past history. All the slanting segments represent the molecule in the mobile phase. All these segments have a common slope, v_m, the mobile phase velocity. The column distance covered between two successive sorption steps is a random quantity since the time which elapses between one desorption step and the subsequent sorption step is random. Moreover, the total number of entries in the stationary phase (n_m) is a random quantity of integer type. The total time spent in the column is, thus, a sum of random quantities. However the number of terms in this sum can differ. The mathematics of the simplest stochastic model of chromatography reduces to the problem of finding a solution for the sum of a random number of random independent quantities. What characterizes a specific chromatographic system is the distribution of these random quantities: that is the distribution of the number of entry processes (n_m) and the distribution of the sorption time over a single sorption site (Δts).

With reference to the model in Fig. 2, it must be observed that if the process starts with the molecule in the stationary phase, the number of desorption steps n_s is the same as the number of n_m. In this treatment it will be assumed:

$$n_s = n_m = n \qquad (3a)$$

$$\overline{n}_m = \overline{n}_s = \overline{n} \qquad (3b)$$

where $\overline{n}$ is the mean value. The **CF** method is able to find the general solution for this problem, independently of the specific distribution type of n and Δts, provided some very general conditions are respected. In order to describe this approach, some basic definitions and properties in the probability theory are to be recalled [8, 11].

3.   Basic Definitions [1, 8-9, 14]

3.1. DISTRIBUTION AND FREQUENCY FUNCTIONS

Integer or continuous random quantities, like those appearing in the above described stochastic model of chromatography, are completely characterized by the (cumulative) distribution function $F(x)$ defined as:

$$F(x) = Pr(\eta \leq x) \tag{4}$$

where $Pr(\eta \leq x)$ is the probability that the random quantity η will be less or equal to x. The derivative first $F(x)^{(1)}$ is called frequency function (or probability density):

$$f(x) = F(x)^{(1)} \tag{5}$$

For continuous random variables:

$$F(x) = \int_{-\infty}^{x} dF(x) = \int_{-\infty}^{x} f(x) \, dx \tag{6a}$$

while for discontinuous variables F is a step function:

$$F(x_1) = \sum_{j=0}^{j=1} f(x_j) \, , \quad (x_{j-1} < x_j \, , \; all \; j). \tag{6b}$$

The basic property of a distribution function is:

$$\int_{-\infty}^{\infty} dF(x) = 1 \tag{7}$$

In the probabilistic model of chromatography, one example of the continuous random quantity η is the sorption time Δt_s on a single sorption step. In Fig. 2 Δt_s is the length of each horizontal segment which can assume different values. These time values are the variable x of Eqs. 4-7. In the simplest case the Δt_s variable is assumed as exponentially distributed (see Eq. 1). One example of the discontinuous random variable is the number of entries (n_m) for each trajectory in Fig. 2.

If two events A and B are simultaneously considered, ($A \cap B$, i.e. A and B), the probability of simultaneous realization of both $Pr(A \cap B)$ -- called *joint probability* -- is derived from the conditional probability law [1]:

$$Pr(A|B) = Pr(A \cap B) \, / \, Pr(B) \tag{8}$$

where $Pr(A|B)$ is the *conditional probability* and it means the occurrence probability of A given that it is known that B occurred.

If A and B are independent of one another:

$$Pr(A|B) = Pr(A) \qquad (9a)$$

the joint probability from Eq. 8 is simply the product of $Pr(A)$ and $Pr(B)$:

$$Pr(A \cap B) = Pr(A) \, Pr(B) \qquad (9b)$$

One example of general application of this law in linear chromatography is constituted by a two entry process in the stationary phase, referred to as events A and B respectively, giving rise to specific values of $\Delta t_{s,A}$ and $\Delta t_{s,B}$ for each of the two entries. The event determined by the joint realization of both these events ($A \cap B$, i.e. A and B) is the product of the probabilities of the two separate events. This is one of the basic assumptions of the linear chromatography (the other is the independent behaviour of the single molecules). Note that this expression alone does not allow us to determine the probability of a given total sorption time since the combination of sorption times giving rise to the same total value can be different (see below).

If one event consists of the realization of one of several mutually exclusive events, the total probability is the sum of the probability of each individual event:

$$Pr(A_{tot}) = \sum_i Pr(A_i) \qquad (10)$$

Let us consider the application of these basic probability laws to chromatographic events. One example is the sorption behaviour of a molecule performing one sorption step either on the site of type 1, with access probability p_1 and sorption time frequency function $f_1(\Delta t_s)$, or on the site of type 2, with access probability p_2 ($= 1-p_1$) and sorption time frequency function $f_2(\Delta t_s)$. p_1 and p_2 can also express the abundance of these sites in a zone of the column. A_1, A_2 and A_{tot} are the separate sorption processes in the site 1 or 2 and the total process, respectively; moreover both A_1 and A_2 consist in the joint realization of two events: the site access (B) and the site rest (C). Taking into account the described model and by using Eq. 10 $Pr(A_{tot})$ is:

$$Pr(A_{tot}) = Pr(A_1) + Pr(A_2) = Pr(B_1 \cap C_1) + Pr(B_2 \cap C_2) \qquad (11a)$$

The events B and C are assumed to be independent of one another and, by using Eq. 9b, the following relationship is obtained:

$$Pr(A_{tot}) = Pr(B_1) \, Pr(C_1) + Pr(B_2) \, Pr(C_2) \qquad (11b)$$

or, by introducing the frequency functions:

$$f_{tot}(\Delta t_s) = p_1 \, f_1(\Delta t_s) + p_2 \, f_2(\Delta t_s) \qquad (11c)$$

It must be observed that the independent variable of all the frequency functions of Eq. 11c is the same (Δt_s) since all the events A_{tot}, A_1 and A_2 correspond to the same value of the time spent in the stationary phase sorption process. The independent variable can be omitted and the following short form will be used:

$$f_{s,tot} = p_1 f_{s,1} + p_2 f_{s,2} \qquad \qquad (11d)$$

A similar case holds true for a bifunctional molecule (e.g. methanol, with the hydrophobic CH_3 part and the polar, hydrogen bonding OH function) over a two energy sorption site (1 or 2). In this case $f_{s,1}$ and $f_{s,2}$ in Eq. 11d are the frequency functions of the sorption condition with energy E_1 or E_2, respectively. p_1 and p_2 are the statistical or configuration factors of the sorption states.

It can be seen that, in the Eqs. 11c and 11d, the probabilities p_1 play the role of weighting factors. The cases above described are examples of a more general class of processes which are said to be a "mixture" of several processes 1, 2, 3, ...n, with their own weights p_1, p_2, p_3, ...p_n. Another important example is the exit of a single molecule from the column at a given time t_s. With reference to the model described above (Fig. 2), this event can be realized in a many different ways, according to the different number n of the sorption-desorption steps performed by the solute molecule. The *total* event is the sum of many mutually exclusive events since the different solute molecules are assumed to behave independently of one another. With reference to the chromatographic model described in Fig. 2, it can be said that the single trajectories are independent of one another and the following relation will hold:

$$f_{tot}(t_s) = \sum_n p_n f_n (t_s) \qquad \qquad (11e)$$

or:

$$f_{s,tot} = \sum_n p_n f_{s,n} \qquad \qquad (11f)$$

where p_n are the probabilities of performing n sorption desorption steps and $f_n(t_s)$ is the frequency function of the total time spent in n sorption steps.

3.2. EXPECTATIONS

If $g(\eta)$ is a function of η, the expectation or the mean value of g is defined by:

$$\mathbb{E}[g(\eta)] = \int g(x) \, dF(x) \qquad \qquad (12a)$$

If $F(x)$ has the derivative $f(x)$ (= frequency function), Eq. 12a can be more conventionally written as:

$$\mathbb{E}[g(\eta)] = \int g(x) \, f(x) \, dx \qquad \qquad (12b)$$

If $F(x)$ is a step function with steps f_1 at the points x_1, the expectation assumes the form:

$$\mathbb{E}[g(\eta)] = \sum_{i} g(x_i) \, f_i \tag{12c}$$

3.3. MOMENTS

Moment of order j are specific expectations. When

$$g(x) = x^j \tag{13}$$

the jth moment about origin μ_j' is defined, thus:

$$\mu_j' = \int x^j \, f(x) \, dx \tag{14}$$

The mean $\bar{m}$ is:

$$\bar{m} = \mu_1' \tag{15}$$

If :

$$g(x) = (x-\bar{m})^j \tag{16}$$

the jth central moments are defined as follows:

$$\mu_j = \int (x-\bar{m})^j \, f(x) dx \tag{17}$$

If the moments of all orders are assigned, a probability function is completely defined [7]. The second central moment is the variance:

$$\sigma^2 = \mu_2 \tag{18}$$

3.4. CHARACTERISTIC FUNCTION

The characteristic function (CF) $\phi(\gamma)$ is the expectation of a particular function $g(x)$:

$$g(x) = \exp(i \, \gamma \, x) \tag{19}$$

where i is the imaginary unit and γ an additional quantity (auxiliary variable). The characteristic function is thus:

$$\phi(\gamma) = \mathbb{E}[\exp(i \, \gamma \, \eta)] = \int \exp(i \, \gamma \, x) \, f(x) \, dx \tag{20a}$$

and, in the case of the discontinuous random variable:

$$\phi(\gamma) = \sum_{k} \exp(i \, \gamma \, x_k) \, f_k \tag{20b}$$

The name "characteristic" function derives from the fact that if it is known, it fully defines the probability function. $\phi(\gamma)$ is also called the first characteristic function (**1stCF**).

3.5. RELATIONSHIP BETWEEN MOMENTS AND CHARACTERISTIC FUNCTIONS

There is a definite relationship between moments and characteristic function derivatives at $\gamma=0$:

$$\phi^{(j)}(0) = \left[d^j \phi(\gamma)/d\gamma^j \right]_{\gamma=0} = i^j \mu_j' \tag{21}$$

Because of this property, the **1stCF** is also called the moment generating **CF**.

3.6. SECOND CHARACTERISTIC FUNCTION. CUMULANT COEFFICIENTS.

For many applications the logarithm of characteristic function (called second characteristic function, **2ndCF**) is considered:

$$\psi(\gamma) = ln\left[\phi(\gamma)\right] \tag{22}$$

The jth derivative defines the cumulant of order j, K_j:

$$\psi^{(j)}(0) = \left[d^j \psi(\gamma)/ d\gamma^j\right]_{\gamma=0} = i^j K_j \tag{23}$$

The dimension of μ_j', μ_j and K_j is the jth power of the considered random variable. By deriving Eq. 22 the following relationship is obtained:

$$\psi^{(1)}(\gamma) = \frac{\phi^{(1)}(\gamma)}{\phi(\gamma)} \tag{24}$$

since:

$$\phi(0) = 1 \tag{25}$$

by considering Eqs. 21, 22, 24 and 15, one obtains:

$$\psi^{(1)}(0) = \phi^{(1)}(0) = i\,\overline{m} \tag{26}$$

Linear relationships exist between cumulants and central moments [14]. The first three cumulants are:

$$K_2 = \mu_2 \tag{27a}$$

$$K_3 = \mu_3 \tag{27b}$$

$$K_4 = \mu_4 - 3K_2^2 \qquad (27c)$$

Because of the property of Eq. 23, the **2ndCF** is also called the cumulant generating **CF**. If the cumulants are normalized with respect to standard deviation, the cumulant coefficients are defined:

$$\gamma_{j-2} = K_j / (\sigma)^j \qquad (28)$$

It can be shown that the cumulant coefficients are the cumulants of the standardized random variable c:

$$c = (x - \bar{m}) / \sigma \qquad (29)$$

The first two cumulant coefficients are called respectively Skewness **S** and Excess **E**:

$$\gamma_1 = \mathbf{S} \qquad (30)$$

$$\gamma_2 = \mathbf{E} \qquad (31)$$

S and **E** are commonly referred to as the shape properties of the frequency function, respectively the tailing and the extent of flatness with reference to the Gaussian peak shape, for which both **S** and **E** are zero [8]. Expressions of $f(x)$, **1stCF**, **2ndCF** as well as the moments and cumulants for typical distributions such as exponential, rectangular, normal, Poisson, gamma, compound Poisson have been reported [15].

3.7. RELATIONSHIPS BETWEEN CHROMATOGRAPHIC QUANTITIES AND MOMENTS

The three fundamental chromatographic quantities employed in column efficiency studies are the plate height, **H**, specific asymmetry, **Z**, and specific flatness, **F**, [16]:

$$H = (\mu_2 / \bar{t}_R^2) \, L \qquad (32a)$$

$$Z = (\mu_3 / \bar{t}_R^3) \, L^2 \qquad (32b)$$

$$F = (\mu_4 / \bar{t}_R^4) \, L^3 \qquad (32c)$$

where:

$$\bar{t}_R = \bar{t}_s + \bar{t}_m \qquad (33)$$

where $\bar{t}_R$ is the retention time, i.e. the first moment of the frequency function of the time spent by the molecule in the column (i.e. the elution peak, see Fig. 2). $\bar{t}_m$ is the retention time of an unretained component. According to this equation, $\bar{t}_s$ is simply defined as the difference between $\bar{t}_R$ and $\bar{t}_m$. μ_2, μ_3 and μ_4 are the central moments of the elution peak. By introducing the retention ratio **R**

$$R = \bar{t}_m / \bar{t}_R \qquad (34)$$

and by using basic moment-cumulant relationships, Eqs. 27a-c, and the mean mobile phase velocity definition:

$$\bar{v}_m = L / \bar{t}_m \qquad (35)$$

the following alternative equivalent expressions are obtained:

$$H = R (1 - R) (K_2 / \bar{t}_s) \bar{v}_m \qquad (36a)$$

$$Z = R^2 (1 - R) (K_3 / \bar{t}_s) \bar{v}_m^2 \qquad (36b)$$

$$F = R^3 (1 - R) \left[(K_4 + 3K_2^2) / \bar{t}_s\right] \bar{v}_m^3 \qquad (36c)$$

Since moments and cumulants can be obtained by CF derivation (see Eqs. 21, 23 and 26), a chromatographic model is also specified once it has been described in terms of CFs.

4.   Mathematics of elementary chromatographic events. [7-11].

The total time, t, spent by a molecule in the stationary phase (s subscript) when it has performed two entries, is a sum of two random quantities. For a given value of the time t spent, the separate times spent in the two sorption processes (s,1) and (s,2) are related to one another:

$$\Delta t_{s,1} = t - x, \qquad \Delta t_{s,1} > 0 \qquad (37a)$$

$$\Delta t_{s,2} = x, \qquad \Delta t_{s,2} > 0 \qquad (37b)$$

The frequency function of the total time is represented by the convolution integral since all the possibilities are to be summed (= integrated), with the proper weights, thus resulting in a given total time spent equal to t:

$$f_{s,tot}(t) = \int f_{s,1}(t-x) \, f_{s,2}(x) \, dx \qquad (38a)$$

The convolution integral (Eq. 38a) is also symbolically written as:

$$f_{s,tot} = f_{s,1} * f_{s,2} \qquad (38b).$$

Note that the convolution integral, which represents the sum of two random quantities, ($\Delta t_{s,1}$; $\Delta t_{s,2}$), is obtained by making use of both the basic probability laws of Eqs. 9b and 10. If n entries are performed:

$$f_{s,tot} = f_{s,1} * f_{s,2} * \ldots * f_{s,n} \qquad (39a)$$

The main difficulty in solving mathematical models based on the addition of random quantities derives from the difficulty in computing multiple convolution integrals. The use of the characteristic function in characterizing convoluting processes enables one to replace the convolution integral by a product:

$$\phi_{s,tot}(\chi) = \Pi_k \ \phi_{s,k}(\chi) \qquad 1 \leq k \leq n \qquad (39b)$$

This useful property derives from the presence of the exponential function in the **1stCF** and it is common to many other transforms (Fourier, Laplace) [9]. If the **2ndCF** is employed the repeated convolution will result in a sum:

$$\psi_{s,tot}(\chi) = \sum_k \psi_{s,k}(\chi). \qquad 1 \leq k \leq n \qquad (39c)$$

By combining Eq. 23 and Eq. 39c , the following basic relationship is obtained:

$$K_{s,tot,j} = \sum_k K_{s,k,j} \qquad 1 \leq k \leq n \qquad (39d)$$

valid for cumulants of all orders. Since the second cumulant is equal to the variance, the following well-known variance addition theorem for independent random variables [8] is obtained:

$$\sigma_{s,tot}^2 = \sum_k \sigma_{s,k}^2 \qquad 1 \leq k \leq n \qquad (39e)$$

When the sites are equal, Eqs. 39a-e can be simplified as follows:

$$f_{s,tot} = f_s^{n*} \qquad (40a)$$

$$\phi_{s,tot}(\chi) = \phi_s(\chi)^n \qquad (40b)$$

$$\psi_{tot}(\chi) = n \ \psi_s(\chi) \qquad (40c)$$

$$K_{s,tot,j} = n \ K_{s,j} \qquad (40d)$$

$$\sigma_{s,tot}^2 = n \ \sigma_s^2 \qquad (40e)$$

in Eq. 40a n^* means the n-fold convolution of the same frequency function f_s with itself. By combining Eq. 28 with Eqs. 40d and 40e:

$$\gamma_{s,tot,(j-2)} = \frac{K_{s,j}}{(\sigma_s)^j} \frac{1}{n^{(j-2)/2}} = \frac{\gamma_{s,(j-2)}}{n^{(j-2)/2}} \qquad (40f)$$

where $\gamma_{s,(j-2)}$ is the (j-2)th cumulant coefficient referred to the unitary sorption process. From Eq. 40f it can be seen that the repeated process of sorption produces a decrease in the cumulant coefficient of the total sorption time distribution. For **S** and **E** -- respectively the first and the second cumulant coefficient -- the decrease is on the order of $1 / \sqrt{n}$ and $1 / n$.

Since the **1stCF** is a linear transformation, from Eq. 11d the characteristic function of the sorption time frequency function for a molecule performing one alternative stop on a site 1 or 2, is derived:

$$\phi_{s,tot}(\gamma) = p_1 \, \phi_{s,1}(\gamma) + p_2 \, \phi_{s,2}(\gamma) \tag{41}$$

The same relationship holds true for a bifunctional molecule interacting with a sorption site in two different ways -- as above mentioned in connection with Eq. 11a -- or as far as two different trajectories followed by the molecule during its trip along the column. The extension of Eq. 41 to an n-functional molecule case and to n different possible trajectories is obvious. By comparing Eq. 11d and 41 it can be seen that to a "mixed" [9] or weighted frequency function corresponds "mixed" or weighted **1stCF**.

5. Constant mobile phase velocity models [7].

5.1 CHARACTERISTIC FUNCTION OF THE TOTAL SORPTION TIME

According to the chromatographic model described in Fig. 2 the time spent in the mobile phase and the mobile phase velocity are constant quantities:

$$\overline{t}_m = t_m \tag{42a}$$

$$\overline{v}_m = v_m = L \, / \, t_m \tag{42b}$$

The total time spent by a molecule in the stationary phase, when it performs exactly n entries and under the hypothesis that all the sorption sites are equal, is the sum of n equally distributed random variables. The resulting sorption time frequency function is the n-fold convolution of the same frequency function (f_s) with itself (Eq. 40a). Since n sorption step values can be different still resulting in the same value of time spent in the stationary phase, the total frequency function will result in a "mixture" of f^{n*} with weighting factors p_n, the probability of having exactly n entry processes in the chromatographic migration. Thus:

$$f_{s,tot} = \sum_n p_n \, f_s^{n*} \tag{43a}$$

and the corresponding **1stCF**:

$$\phi_{s,tot}(\gamma) = \sum_n p_n \, \phi_s(\gamma)^n \tag{43b}$$

The sum is to be extended to all the possible n entries in the stationary phase. In practice n is considered as a random variable and written as n, since p_n has the property of a frequency function, (see Eq. 7), that is:

$$\sum_n p_n = 1 \qquad\qquad (43c)$$

From Eqs. 43a-b and in the following the symbol "tot" will denote a quantity referred to the whole column length **L**. For example, $\phi_{s,tot}(\gamma)$ is the **1stCF** of the total time spent by a molecule in the stationary phase during its journey over the total column. The characteristic function of the integer variable n, the number of stationary phase entry processes is:

$$\phi_{n,tot}(\gamma) = \sum_n p_n \exp(i\gamma n) \qquad\qquad (44)$$

By using the identity:

$$x^r = \exp\left[r \ln(x)\right] \qquad\qquad (45)$$

Eq. 43b can be written as:

$$\phi_{s,tot}(\gamma) = \sum_n p_n \exp\left[i\, n\, \ln(\phi_s(\gamma))\,/\,i\right] \qquad\qquad (46a)$$

and, by using Eq. 20b, the solution is obtained for the most general chromatographic model under conditions of constant mobile phase velocity:

$$\phi_{s,tot}(\gamma) = \phi_{n,tot}\left[\ln(\phi_s(\gamma))\,/\,i\right] \qquad\qquad (46b)$$

In writing Eq. 46b, we recognized in practice the right hand term of Eq. 46a as a characteristic function of the integer random variable, ($\phi_{n,tot}$), having as its auxiliary variable not a simple variable, but the more complex function $\ln(\phi_s(\gamma))\,/\,i$. The **2ndCF** is:

$$\psi_{s,tot}(\gamma) = \psi_{n,tot}\left[\psi_s(\gamma)/i\right] \qquad\qquad (46c)$$

It can be observed that according to Eqs. 46b or 46c, the total sorption time **CF** is expressed as **CF** of the total number of entry processes in the stationary phase having as argument the **CF** of a single sorption process. This result is absolutely general since no hypothesis has been made either as to the entry or character of the sorption process structure. It is worth mentioning, in this regard, that Eqs. like 46b and 46c hold true for the general class of stochastic compound processes; that is for those stochastic processes for which the addition of the number of random variables fluctuates. Eq. 46a is reported as solution for a problem in the fundamental reference [14].

5.2. CHARACTERISTIC FUNCTION OF THE TOTAL RETENTION TIME

In the present model of constant mobile phase velocity (Fig. 2), the total retention time of an unretained component can be represented by a Dirac function:

$$f(t_m) = \delta(t - t_m) \tag{47}$$

The total retention time frequency function will be shifted on the time axis by a constant term t_m

$$t_R = t_s + t_m \tag{48a}$$

$$\bar{t}_R = \bar{t}_s + t_m \tag{48b}$$

In the following the subscript "R" (the same used to denote the retention time, t_R) will denote a retention quantity, including both the contributions of the mobile phase (m) and the stationary (s) phases. The above mentioned shifting can be formally represented by a convolution product with the Dirac function:

$$f_{R,tot} = f_{s,tot} * \delta(t - t_m) \tag{48c}$$

and by making use of the fundamental properties of Fourier transform [10], the **1stCF** and the **2ndCF** will be respectively:

$$\phi_{R,tot}(\gamma) = \phi_{n,tot}\left[\ln(\phi_s(\gamma)) / i\right] \exp(i \gamma t_m) \tag{48d}$$

$$\psi_{R,tot}(\gamma) = \psi_{n,tot}\left[\psi_s(\gamma) / i\right] + i \gamma t_m \tag{48e}$$

It must be observed that Eqs. 48a-d are equivalent since they express the same basic fact.

5.3. CHARACTERISTIC FUNCTION OF CONSTANT MOBILE PHASE VELOCITY MODELS OF POISSON TYPE

The model described in Fig.2 for which the solution was found is too general to be of real interest. Some hypotheses can forwarded concerning behaviour in the mobile phase. For example, we can assume that the entry process is homogeneous along the chromatographic column. In this case a common mean value $\overline{\Delta t_m}$ of the waiting time Δt_m will exist and the following equations will hold true:

$$\bar{n} = t_m / \overline{\Delta t_m} \tag{49a}$$

or, by using Eqs. 35 and 42a,b:

$$\bar{n} = L\, k_m / v_m \tag{49b}$$

where:

$$k_m = 1 / \overline{\Delta t_m} \tag{50}$$

is the rate constant of the entry process and $\bar{n}$ is the mean number of entry processes and v_m is the mobile phase velocity ($= L/t_m$). If the Δt_m variable is exponentially distributed:

$$f(\Delta t_m) = k_m \exp(- \Delta t_m \, k_m) \qquad (51a)$$

the number of entry processes, n, will be distributed according to the Poisson law [9]:

$$f_n(n=r) = \exp(-\overline{n}) \; \overline{n}^{\;r} / \; r! \quad r=0, \; 1, \; 2 \ldots \infty \qquad (51b)$$

with **1stCF**:

$$\phi_{n,tot}(\gamma) = \exp\left\{\overline{n} \; \left[\exp(i \; \gamma) - 1\right]\right\} \qquad (51c)$$

By introducing this expression into the general solution, Eq. 46c, the **1stCF** for the Poisson entry model is obtained:

$$\phi_{s,tot}(\gamma) = \exp\left\{\overline{n} \; \left[\phi_s(\gamma) - 1\right]\right\} \qquad (52a)$$

or, in terms of the **2ndCF**:

$$\psi_{s,tot}(\gamma) = \overline{n} \; \left[\phi_s(\gamma)-1\right] \qquad (52b)$$

and, by using the Eq. 49:

$$\psi_{s,tot}(L, \; \gamma) = L \; (k_m / v_m) \; \left[\phi_s(\gamma) - 1\right] \qquad (52c)$$

The total retention time **CF** are written making reference to the Eqs. 48c and 48d:

$$\psi_{R,tot}(t_m,\gamma) = t_m\left\{k_m\left[\phi_s(\gamma) - 1\right] + i\gamma\right\} \qquad (53a)$$

$$\psi_{R,tot}(t_m,\gamma) = \frac{L}{v_m} \left\{k_m\left[\phi_s(\gamma) - 1\right] + i\gamma\right\} \qquad (53b)$$

In Eqs. 52c, 53a-b, the linear dependence of the **2ndCF** on the continuous parameter L, the column length, or t_m, the mobile phase time, or $1/v_m$, is emphasized. This mathematical property is fundamental since it allows us to place this model within the important class of stochastic processes with independent and stationary increments (briefly : **s.i.i.** process), for which an interesting series of general mathematical properties are reported [10]: the existence of well defined approximating functions and the rate of convergence towards the Gaussian Law.

5.4. CHROMATOGRAPHIC QUANTITIES OF CONSTANT MOBILE PHASE MODELS

5.4.1. *General Model.* Chromatographic quantities such as **H** and **F** are obtained by introducing expressions of moments, cumulants, t_R or t_s calculated from **CFs** derivatives (see Eqs. 21, 23 and 26) into Eqs. 32a–c or 36a–c. In the present case Eqs. 36a–c containing cumulants and t_s, are to be preferred since solutions in terms **CFs** of the sorption time t_s have been obtained for these models. Only the **H** expressions will be discussed in the following. For a more complete discussion and **Z** and **F** derivation, see ref. [7].

By applying the Eq. 26 to the Eq. 46c **2ndCF** and further developing with the conventional derivation rules:

$$\bar{t}_s = \psi_{s,tot}(0)^{(1)}/\,i = \psi_{n,tot}\big[\psi_s(0)\,/\,i\big]^{(1)}\,\psi_s(0)^{(1)}\,/\,(i)^2 \quad (54)$$

If the **1stCF** basic property of Eq. 25 is applied to Eq. 22, for the **2ndCF** one has:

$$\psi(0)=0 \qquad\qquad\qquad\qquad\qquad (55)$$

and Eq. 54, by correct application of Eq. 26, becomes:

$$\bar{t}_s = -\,\psi_{n,tot}(0)^{(1)}\,\psi_s(0)^{(1)} = K_{1,n,tot}\,K_{1,s} \qquad (56a)$$

$$\bar{t}_s = \bar{n}\,\overline{\Delta t}_s \qquad\qquad\qquad\qquad\qquad (56b)$$

where $\bar{n}$ and $\overline{\Delta t}_s$ are respectively the mean number of sorption–desorption steps and the mean sorption time on a single site:

$$K_{1,n,tot} = \psi_{n,tot}(0)^{(1)}/i = \bar{n} \qquad\qquad (57a)$$

$$K_{1,s} = \psi_s(0)^{(1)}/i = \overline{\Delta t}_s \qquad\qquad (57b)$$

The result of Eq. 56b which states that the total sorption time is the product of two mean quantities is absolutely general for stochastic compound processes and it is known as the Wald relation [9]. The second central moment is, likewise, obtained from Eqs. 54 and 23:

$$K_{s,2,tot} = \left\{\,\psi_{n,tot}(0)^{(2)}\big[\psi_s(0)^{(1)}\big] - \psi_{n,tot}(0)^{(1)}\psi_s(0)^{(2)}/\,i\right\} \quad (58a)$$

and thus, by recalling Eq. 18 and 27a:

$$\sigma_{s,tot}{}^2 = K_{2,n,tot}\,K_{1,s}{}^2 + K_{1,n,tot}\,K_{2,s} \qquad (58b)$$

or

$$\sigma_{s,tot}{}^2 = \underbrace{\sigma_{n,tot}{}^2\,\overline{\Delta t}_s{}^2}_{\substack{\text{entry process} \\ \text{dispersion}}} + \underbrace{\bar{n}\,\sigma_s{}^2}_{\substack{\text{stationary} \\ \text{phase dispersion}}} \qquad (58c)$$

The first term is identified as the "entry process dispersion" term, since it becomes increasingly important the more "disordered" the stationary entry process is. In fact, $\sigma_{n,tot}$ is the standard deviation of the number of sorption processes: the mean value of the time spent in each sorption process ($\overline{\Delta t_s}$) being simply the scaling factor. The second term is related to the stationary phase sorption process, being simply the total sorption time variance of $\overline{n}$ sorption events (see Eq. 40e). An equation similar to Eq. 58e was reported in the renewal theory description of the chromatographic process [17].

It must be observed that the variance or the second cumulant of the total retention time t_R and of the total sorption time t_s are the same since the second derivatives of their 2ndCFs (see Eq. 46c, 48d) are equal for the constant mobile phase model. Thus:

$$K_2 = \sigma_{s,tot}^2 = \sigma_{R,tot}^2 \qquad (59)$$

By combining Eq. 36a with Eqs. 56a, 58c and 59, the H expression is obtained:

$$H = R\ (1 - R)\ \left\{ \frac{\sigma_{n,tot}^2}{\overline{n}} + \left[\frac{\sigma_s}{\overline{\Delta t_s}} \right]^2 \right\}\ \overline{\Delta t_s}\ v_m \qquad (60)$$

In this expression, the general C term of the van Deemter equation can be identified. In fact, it is expressed the linear dependence on both the mobile phase velocity, v_m, and the mean sorption time, $\overline{\Delta t_s}$, in addition to the retention ratio term R (1 - R). The two terms in the brackets have a simple meaning: they are the H proportionality contribution of statistical type coming from the mobile phase and the stationary phase process, respectively . In fact, the first term in the brackets is the ratio of the variance of the total number of the sorption-desorption steps ($\sigma_{n,tot}^2$) with respect to its mean value ($\overline{n}$) . The second term is the squared ratio of the standard deviation of the sorption time in a single sorption site (σ_s) with respect to its mean value ($\overline{\Delta t_s}$). Both these terms act as "amplifying" factors of the mean sorption time value, ($\overline{\Delta t_s}$), in addition to the mobile phase velocity effect. The kinetic peak broadening in chromatography is thus strongly dependent on the degree of "disorder" of these two key steps. The latter point is one important result of the stochastic approach here presented.

5.4.2. *The Giddings-Eyring-McQuarrie Model.* In the case of the Poissonian entry process, the first term in the brackets in Eq. 60 is equal to one. In fact, in this case, the variance is equal to the mean [8]:

$$\sigma_{n,tot}^2 = \overline{n} \quad \text{(Poisson entry process)} \qquad (61)$$

In the case of exponentially distributed sorption time (see Eq. 1), the second term too is equal to one since:

$$\overline{\Delta t_s} = \sigma_s \text{ (Exponentially distributed sorption time)} \qquad (62)$$

When both these conditions hold true, by considering Eq. 1b, the classical C term of the van Deemter equation is derived:

$$H = 2 R (1 - R) v_m / k_s \qquad (63)$$

It must be observed that this last case corresponds to the same model described by Giddings and Eyring [4] and by McQuarrie [5]. The factor 2 which appears in this expression has a clear statistical meaning.

5.4.3 *The Martin-Synge Model.* If, in Eq. 60, the entry process dispersion term $\sigma_{n,tot}$ vanishes and the sorption time is exponentially distributed (Eq. 62), one has a chromatographic process which is the sum of a constant number of exponentially distributed random variables. From a mathematical point of view, this is equivalent to the Martin-Synge classical model of theoretical plates [7, 18].

6. Non-constant mobile phase velocity models.

6.1 CHARACTERISTIC FUNCTION OF NON-CONSTANT MOBILE PHASE VELOCITY MODELS

The model described in Fig. 2 assumes that the solute molecule travels in the mobile phase with constant velocity (= the mobile phase velocity) and thus the total time spent in the mobile phase is constant ($=t_m$) for all the trajectories. The model does not take into consideration longitudinal diffusion and flow pattern effects which induce variations in both the total time spent in the mobile phase and in the mobile phase velocity. Therefore the mobile phase time will be a random variable (= t_m) with frequency function $f_{m,tot}(t)$. With reference to Fig. 2, the trajectory plot will have slanting segment slopes which are no longer constant and equal to v_m, rather they are variable (even with negative values, which means a back-walk due to diffusion processes). Owing to the independence of the single trajectories the problem can be handled in the following way: the total process is partitioned into elementary processes, each of which has the same mobile phase time t_m and is, thus, formally represented by the previously considered model of constant mobile phase velocity. Because of the above-mentioned local mobile phase velocity variations, each elementary process has the same **mean** mobile phase velocity. Keeping this in mind, all the solutions previously derived in the paragraph n.5 will hold true for the elementary processes, by simply putting t_m in the place of t_m and by considering v_m as the mean quantity computed over the total column length (see Eq. 48d, e; 52c; 53a, b).

According to above description, frequency function of the retention time can be written by applying the conditional probability law (Eq. 8):

$$f_{R,tot}(t, t_m) = f_{m,tot}(t_m) f_{R,tot}(t | t_m) \qquad (64a)$$

where $f_{R,tot}(t, t_m)$ is the joint probability of the simultaneous event of a given retention time t *and* of a given mobile phase time t_m.

$f_{R,tot}(t|t_m)$ is the retention time frequency function given t_m, i.e. the case discussed in the paragraph n. 5. Since the same retention time value t can be obtained with different values of t_m, the *unconditional* frequency function of the retention time will be:

$$f_{R,tot}(t) = \int f_{m,tot}(t_m)\, f_{R,tot}(t|t_m)\, dt_m \qquad (64b)$$

The retention time frequency function is thus a "randomized" [9] sum or a "mixture" [14] of the elementary processes frequency functions of common t_m value, the function $f_{m,tot}(t_m)$ being the weighting factor.

It is very difficult to work out a solution in term of **CF** for the general model of chromatographic process represented in Eq. 64b. Nonetheless, a solution can be achieved if Eq. 64b can be represented in a way like Eq. 43a, for which the **CF** solution was found by the log-exp transformation. It can be observed that the simultaneous presence of the same quantity (n) in both the weighting factor (p_n) and in the exponent (n^*) was the condition for the solution found above. Likewise general solution in terms of **CF** can be found for Eq. 64b if the retention time frequency $f_{R,tot}(t|t_m)$ can be expressed as a t_m-fold "continuous" convolution of a unitary process frequency function $f_r(t)$, referred to a unit t_m value:

$$f_{R,tot}(t|t_m) = f_r(t)^{t_m*} \qquad (65)$$

and thus:

$$f_{R,tot}(t) = \int f_{m,tot}(t_m)\, f_r(t)^{t_m*}\, dt_m. \qquad (66)$$

To the formal position of Eq. 65 corresponds a **1stCF** relationship of the following type:

$$\phi_{R,tot}(z|t_m) = \phi_r(z)^{t_m} \qquad (67a)$$

or:

$$\phi_r(\zeta) = \phi_{R,tot}(\zeta|t_m)^{1/t_m} \qquad (67b)$$

or, in terms of **2ndCF**

$$\psi_r(\zeta) = \frac{\psi_{R,tot}(\zeta|t_m)}{t_m} \qquad (67c)$$

Thus the subscript "r" here employed refers to a retention quantity including both the mobile phase and the stationary phase retention processes, but in reference to a unit value of time spent in the mobile phase, as opposed to "R" which, instead, refers to the total time spent in the column. In writing Eqs. 65 and 67a-b, the non conventional concept of "continuous" convolution and the unusual definition of a frequency function of a unitary t_m value through its **1stCF** (see Eq. 67b) was employed. This is substantially based on the equivalence of autoconvolution and of power operations (see Eq. 40a-b) as well as on the extension of this equivalence from integer to real numbers. This procedure is mathematically correct for the so-called "infinitely divisible" distributions, that is for those distributions for which the position of Eq. 67c holds true. The most common distributions such as the the exponential distribution belong to this class and thus this position should hold true for the most common retention time frequency functions. Eq. 66 can be written as **1stCF** because of the **1stCF** is a linear transformation (see Eq. 41):

$$\phi_{R,tot}(\chi) = \int f_{m,tot}(t_m) \, \phi_r(\chi)^{t_m} \, dt_m \qquad (68a)$$

To this expression the same procedure used in deriving the solution for the constant mobile phase velocity model is applied. This procedure consists of applying the exp-log transformation (Eq. 45) and using the basic definition of **1stCF** (see Eqs. 43-46). The only difference here is that since a continuous weighting function $f_{m,tot}(t)$ in Eq. 68a is present, the integral **1stCF** of Eq. 20a for continuous random variable is to be used instead of the discontinuous case. The **2ndCF** of the total retention time for the most general model of non-constant mobile phase time and general structure of both the entry and the sorption process will be:

$$\psi_{R,tot}(\chi) = \psi_{m,tot}\left[\psi_r(\chi) \, / \, i\right] \qquad (68b)$$

where $\psi_{m,tot}(\chi)$ is the **2ndCF** of the time spent in the mobile phase. See the analogy with Eq. 46c.

One important case for which the solution of Eq. 68b holds true is a chromatographic process in which the stationary phase entry process is of Poisson type. In fact, in this case, each "elementary process" at a given t_m value has a retention time **1stCF** which can written by simply replacing t_m with t_m in Eq. 53a and by taking the exponential:

$$\phi_{R,tot}(\chi|t_m) = \exp\left\{t_m \left\{k_m \left[\phi_s(\chi) - 1\right] + i\chi\right\}\right\} \qquad (69)$$

which is of the following type:

$$\phi(\chi, t) = \phi(\chi)^t \qquad (70)$$

The $\psi_r(\chi)$ required for Eq. 68b is obtained by taking the logarithm and by considering Eqs 22 and 67c:

$$\psi_r(\chi) = k_m \left[\phi_s(\chi) - 1\right] + i\chi \qquad (71)$$

As regards the frequency function of retention time spent in the mobile phase, one can consider e.g. the mobile phase diffusion model whose frequency function is the Gaussian Function [1]:

$$f(t_m) = \frac{1}{\sigma_D \sqrt{2\pi} \ \bar{t}_m^{1/2}} \exp\left\{ -\frac{1}{2\bar{t}_m}\left[\frac{t_m - \bar{t}_m}{\sigma_D}\right]^2 \right\} \qquad (72a)$$

whose **1stCF** is [10]:

$$\psi_{m,tot}(\chi) = i\ \chi\ \bar{t}_m - (\sigma_D\ \chi)^2\ \bar{t}_m\ /\ 2 \qquad (72b)$$

where σ_D^2 is the variance (on time basis) of the diffusing band for unit $\bar{t}_m$ value:

$$\sigma_D^2 = 2\ D\ /\ \bar{v}_m^2 \qquad (73a)$$

D is the mobile phase diffusion coefficient. Thus, by introducing Eqs. 71 and 72b into Eq. 68b, a solution is obtained under the hypothesis of longitudinal diffusion and a Poisson entry process in the stationary phase. It must be noted that no hypothesis has been put forward for the sorption process.

6.2. DERIVING CHROMATOGRAPHIC QUANTITIES.

The chromatographic quantities can be obtained by using the same procedure based on the **2ndCF** derivative procedure previously employed for the constant mobile phase velocity model. By observing that Eqs. 46c and 68b are formally similar if the subscript equivalencies "s=R", "n=m" and "s=r" are considered, from Eqs. 56 and 57b one has:

$$\bar{t}_R = K_{1,m,tot}\ K_{1,r} \qquad (74)$$

$$\sigma_{R,tot}^2 = K_{2,tot} = K_{2,m,tot}\ K_{1,r}^2 + K_{1,m,tot}\ K_{2,r} \qquad (75)$$

Readers can note that the dimension homogeneity is verified because the K_r cumulants are referred to unit mobile phase time. Specific expressions are obtained if specific model hypotheses are made regarding the form of $\psi_{m,tot}(\chi)$ and $\psi_r(\chi)$. By assuming, for the former, the mobile phase diffusion model represented by the Eqs 72a-b, the mobile phase quantities are computed:

$$K_{1,m,tot} = \psi_{m,tot}(0)^{(1)}/\ i = \bar{t}_m \qquad (76)$$

$$K_{2,m,tot} = -\psi_{m,tot}(0)^{(2)} = \sigma_D^2\ \bar{t}_m = \sigma_{D,tot}^2 \qquad (77)$$

Assuming as retention process per unit t_m the process described by Eq. 71, (i.e. entry process = Poisson and sorption process of general type), the retention first cumulant is:

$$K_{1,r} = \psi_r(0)^{(1)} / i = k_m \phi_s(0)^{(1)} / i + 1 = \bar{t}_r \qquad (78a)$$

or by considering Eq. 57b and 26:

$$K_{1,r} = k_m \overline{\Delta t_s} + 1 \qquad (78b)$$

The retention second cumulant in Eq. 75 is obtained by deriving twice Eq. 71 and by taking into account Eq. 23:

$$K_{2,r} = - \psi_r(0)^{(2)} = - k_m \phi_s(0)^{(2)} \qquad (79)$$

In this last expression the second order derivative of the **2ndCF**, ψ_r, is expressed as a function of the second order derivative of the **1stCF** ϕ_s. According to Eq. 21 this last derivative gives the moments about origin which are related to the cumulants as follows [14]:

$$\mu_{2,s}' = K_{2,s} + K_{1,s}^2 \qquad (80)$$

Eq. 79 thus becomes:

$$K_{2,r} = k_m (K_{2,s} + K_{1,s}^2) \qquad (81)$$

The retention time expression is obtained by combining Eqs. 74, 76c and 78c:

$$\bar{t}_R = \bar{t}_m \bar{t}_r = \bar{t}_m (k_m \overline{\Delta t_s} + 1) = \bar{t}_s + \bar{t}_m \qquad (82a)$$

or, by using Eq. 50:

$$t_R = t_m \left[\frac{\overline{\Delta t_s}}{\overline{\Delta t_m}} + 1 \right] \qquad (82b)$$

If the classical chromatographic equation is recalled:

$$\bar{t}_R = \bar{t}_m (k' + 1) \qquad (83)$$

the capacity factor k', is expressed as a function of kinetic quantities:

$$k' = \overline{\Delta t_s} / \overline{\Delta t_m} = k_m / k_s \qquad (84)$$

By introducing Eqs. 77, 78a and 82a in the first term of Eq. 75 and Eqs. 76 and 81 in the second term of the same equation and recalling that:

$$\bar{n} = k_m \bar{t}_m \qquad (85)$$

$$\bar{t}_s = K_{1,s} \bar{n} \qquad (86)$$

one obtains:

$$\sigma_{R,tot}{}^2 = K_{2,tot} = \underbrace{\sigma_D{}^2 t_R{}^2 / t_m}_{\substack{\text{mobile phase} \\ \text{diffusion}}} + \underbrace{K_{1,s} t_s (1 + K_{2,s} / K_{1,s}{}^2)}_{\text{stationary phase kinetics}} \qquad (87)$$

where the dimension homogeneity of the first term in the right hand member can be verified recalling that $\sigma_D{}^2$ is the mobile phase diffusion variance per unit mobile phase time (see Eq. 73). In Eq. 87, as it will be shown here below, the origin of the two classical terms of the van Deemter mobile phase diffusion, H_B, and stationary phase desorption kinetic term, H_C, are identified. In fact, from Eq. 34 the following relationship holds true:

$$\overline{t}_s = R (1 - R) \overline{t}_R{}^2 / \overline{t}_m \qquad (88)$$

and by using Eq. 32a and 73:

$$H = \underbrace{\frac{2 D}{\overline{v}_m}}_{H_B} + \underbrace{R (1 - R) \left\{ 1 + \left[\frac{\sigma_s}{\Delta t_s} \right]^2 \right\} \overline{\Delta t}_s \, \overline{v}_m}_{H_C} \qquad (89)$$

It must be underlined that, in this expression, the mobile phase velocity term $\overline{v}_m$ is really a **mean** mobile phase velocity since it is computed by dividing the column length L by the **mean** time spent in the mobile phase, $\overline{t}_m$. Different symbols are not used here so as to avoid introducing additional notations! . The A term of the van Deemter equation can likewise be derived by applying the same procedure to the flow pattern effects.

7. Describing Chromatographic peak shape

7.1. THE ROLE OF THE GAUSSIAN SHAPE

Having obtained the general solutions in terms of characteristic functions, (see e.g Eqs. 46a-c, 52a-c, 53a-c, 68b with 71 and 73), peak shape expressions can be derived by inversion. This route has been followed by McQuarrie [5], but involves complex integrations and only leads to simple solution in a limited number of cases. The stochastic theory of chromatography can avail itself of very general results of the probability theory for peak shape characterization, thus avoiding complex integration or numerical inversions of the solutions expressed as **CFs**. These general results are the most important content of the Central Limit Theorem and of its related asymptotic expansions [7]. Both of these refer to the fact that the limiting solution for all linear chromatographic models is the Gaussian Peak shape $Z(c)$:

$$Z(c) = 1/\sqrt{(2\pi)} \exp(-c^2 / 2) \qquad (90)$$

whose cumulative distribution function is :

$$P(c) = \int_{-\infty}^{c} Z(c) \, dc \qquad (91)$$

The most appropriate and exhaustive explanation of the central role of the Gaussian Function is given in the stochastic theory of chromatography: it is, in fact, a general phenomenon that the sum of a great number n of random independent variables or, its equivalent, the n-fold repeated convolution of distribution functions with finite variances, progressively builds up the Gaussian curve, when n is great enough. This fundamental property (and the other properties which will be discussed later) holds true even for **s.i.i.** process [10] and thus for the above described Poisson chromatographic model.

7.2. THE CONVERGENCE TOWARDS THE GAUSSIAN PEAK SHAPE AND THE IMPROVED APPROXIMATION BY THE EDGEWORTH-CRAMER SERIES

7.2.1. *Building up the Gaussian Function.* The process of building up the Gaussian function can be seen in Fig. 3 where the process of repeated convolution of an exponential function is shown. It can be seen that with an increase in the number of repeated convolutions, the peak shape is even better approximated by a Gaussian function of the same mean and standard deviation. There are two main points linked to the convergence towards the Gaussian function.

The first point is to establish how fast the process of attaining Gaussian peak shape is and its dependence on the properties of the convoluted function. These questions are indeed not purely academic but they are most important for chromatographic practice. In fact, establishing the dependence on n of the rate of convergence towards the Gaussian Law means describing its dependence on column length because of the proportionality between these two quantities. In addition throwing light on the rate of convergence with respect to the convoluting function properties means focusing on the dependence of the degree of tailing on sorption kinetics.

The second point is to find approximating functions with an approximation degree better and faster than the Gaussian function.

7.2.2. *The rate of convergence: the Berry Esséen Theorem.* The question of the rate of convergence towards the Gaussian shape is answered in Probability Theory by the Berry-Esséen Theorem, which states that for a sum of n random and identically distributed variables with finite variance, the following relationship holds true:

$$|F_n(c) - P(c)| \leq T \rho / n^{1/2} \tag{92}$$

where ρ is the reduced third absolute moment of the single random variable:

$$\rho = \int |t|^3 f(t) \, dt / \sigma^3 \tag{93}$$

and $F_n(c)$:

$$F_n(c) = F(c)^{n*} \tag{94}$$

is the resulting cumulative distribution in the standardized variable c. The constant T is equal to 3 [10]. It has been proved that a similar theorem holds true for **s.i.i** processes and thus for all the chromatographic models under linearity conditions, the column being homogeneous in length. In this case the constant T in Eqs. such as Eq. 92 is not well defined, but the role of the n quantity will be played by the mean number of sorption-desorption steps n or by the quantity L k_s / v_m, which is equivalent. Thus, it is shown that in the general case the rate of convergence towards the Gaussian law depends on $1 / L$ and on a "specific" property of the sorption site, i.e. the ρ_s value. If ρ_s is abnormally high, the gaussian peak shape cannot be attained with the column length values normally employed and the peak will exhibit tailing: this is the origin of the so called kinetic tailing in linear chromatography.

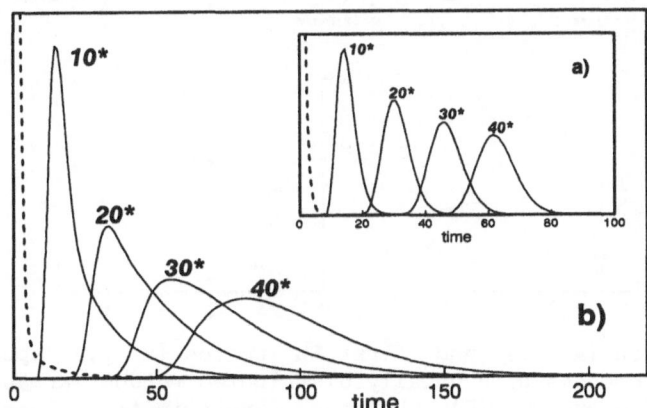

Fig. 3. Convergence to the Gaussian shape. (a) repeated n-convolutions of an exponential function (Eq. 1a, $k_s=1$) with itself. (b) repeated n-convolutions of a "mixture" (Eq. 11c) of two exponential functions (k_1 = 1; p_1 = 0.9; k_2 = 0.1, p_2 = 0.1). The number n of convolutions is reported as : n*. Dotted line: convoluted function.

The point is illustrated in Fig. 3 where three cases of repeated convolutions are reported (n* = 10* -- 40*). The Gaussian shape is rapidly attained only for the simple exponential case (Eq. 1a), whereas for the case of mixed exponential functions (Eq. 1a and 11c), the peak exhibits strong tailing even for the greatest number of repeated convolutions. The case of the mixed exponential can be referred to the sorption of a bifunctional solute on a two energy sorption site [4,5].

7.2.3. *Peak Shape Approximation by Edgeworth-Cramér Series.* The above mentioned second question -- that is the problem of an improved approximation is far from having attained a Gaussian shape -- is answered by the so-called Edgeworth-Cramér series (**EC** series) [10,11,14,15]. This can be seen in Fig. 4 where the approximation by **EC** series is also reported. It can be seen that: (1) the approximation degree is improved by using the **EC** series rather than the Gaussian function; (2) the approximation degree is better with an increase in n. Let us now see how the **EC** series is defined. The **EC** series are fundamental approximating functions of the chromatographic peak shape [18-28].

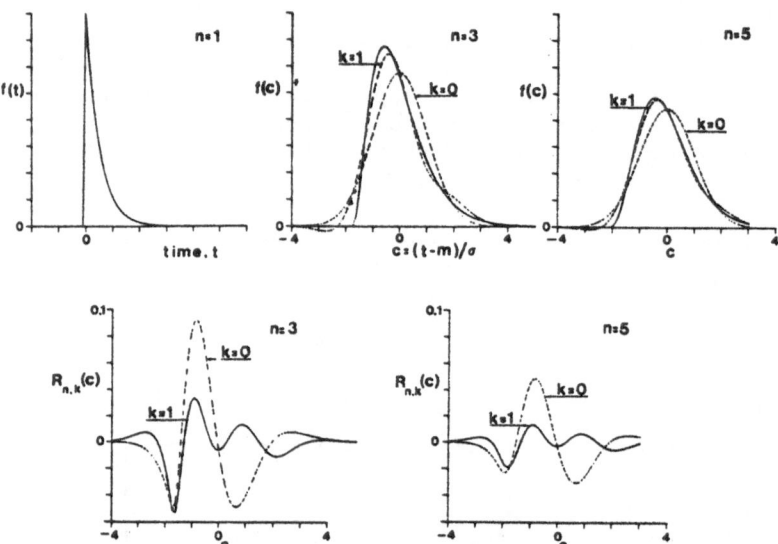

Fig. 4. Gaussian (k = 0) and first **EC** series (k = 1) approximation in the addition process of n equally distributed random variables (upper part) and differences (R) between approximated function and approximating functions (lower part).

The *kth* **EC** series development of a distribution function **F(c)** is formally written [10]:

$$F(c)=P(c)+\sum_{v=1}^{v=k} Q_v(-P)+R_k(c) \qquad (95a)$$

where the first two terms $Q_v(-P)$ are :

$$Q_1(-P) = -(S / 3!) P^{(3)}(c) \qquad (95b)$$

$$Q_2(-P) = (E / 4!) P^{(4)}(c) + (10 S^2 / 6!) P^{(6)}(c) \qquad (95c)$$

As can be inferred from Eqs. 95a-c, the $Q_v(-P)$ terms are linear aggregates of the derivatives of $P(c)$ of maximum order $3v$ containing cumulant coefficients of maximum order v. If we consider the frequency function f instead of the distribution function F, the corresponding **EC** series is obtained from Eq. 95a-c by substituting the normal distribution $P(c)$ with the normal frequency function $Z(c)$ and its derivatives. The simplest form of the **EC** series - the zeroth order - is obtained by taking $k = 0$ in Eq. 95a. In this case the **EC** series is reduced to the Gaussian Distribution $P(c)$ or frequency function $Z(c)$. Detailed expressions of the $Q_v(-P)$ or $Q_v(-Z)$ terms are reported in refs. [14, 15, 22, 28].

7.2.4. *Approximation properties of the* **EC** *series* [10, 21]. If the considered distribution function refers to a case of a sum of independent random variables, that is if it is expressed by the Eq. 40a. and thus its **2ndCF** is expressed according to Eq. 40c with the exponent n integer, the corresponding **EC** development is :

$$F_n(c) = P(c) + \sum_{v=1}^{v=k} Q_{n,v}(-P) + R_{n,k}(c) \qquad (96a)$$

and thus the terms of the development have the following properties:

$$Q_{n,v}(-P) = \frac{q_{1,v}(-P)}{n^{k/2}} \qquad (96b)$$

$$|R_{n,k}(c)| < \frac{M_k}{n^{(k+1)/2}} \qquad (96c)$$

The term $q_{1,v}(-P)$ is formally written as the corresponding term $Q_{n,v}(-P)$, see e.g. Eqs. 95b-c, although they refer to the parent distribution of the individual random variables added, i.e. S, E and the other cumulant coefficients are those of the parent distribution $F(c)$. The quantity M_k is a constant depending on $F(c)$. Given the above reported properties, one can say that the **EC** series expansions are of **asymptotic type** since a given value of the **asymptotic quantity** n can be found for which the remainder term $R_{n,k}(c)$ is less than the last neglected term $Q_{n,v}(c)$, for all values of the independent variable c.

Another way of considering the above asymptotic properties is to take the logarithm of the remainder term and to report it against the logarithm of the asymptotic quantity n:

$$\ln R_{n,k}(c) < \ln M_k - \frac{(k + 1)}{2}\ln n \qquad (96d)$$

Straight lines (in the limiting case) with slopes of -1/2, -1, -3/2 etc. are expected if the normal approximation or if the first or second **EC** expansion is considered. Similar properties also hold true for frequency functions [10] and thus **EC** series expansion can be used for approximating them as well.

7.2.5. *The role of the* **EC** *series in the Stochastic Theory of Chromatography* [7]. Most important in the present context are the approximation properties towards **s.i.i** processes since important chromatographic models were proved to behave as **s.i.i.** processes. In this case the role of the asymptotic quantity n in Eqs. 96a-c is played by the continuous variable of the process. In the sections n.s 5 and 6 it has been shown that if the entry process in the stationary phase is Poissonian in type, both the linear chromatographic models at constant and non-constant mobile phase velocity have the property of being **s.i.i.** processes: L, $1/v_m$ or t_m being the continuous process variable playing the role of n in Eqs. 96a-c. The distribution function of the retention time admits **EC** series asymptotic development and can thus be represented accordingly. In practice in Eqs. 96a-c, the terms $Q_{n,v}(-P)$ and $R_{n,k}(c)$ are written respectively as $Q_{L,v}(-P)$ and $R_{L,k}(c)$ if the linear increase with respect to L is considered. In this last instance the terms $q_{1,v}(-P)$ -- written as in Eqs. 95b-c for k equal to 1 and 2 respectively -- contains the cumulant coefficients **S**, **E** etc. of the parent distribution for unit L quantity. In a similar manner, the frequency function can be approximated by the **EC** series by simply replacing the normal distribution P with the normal frequency function Z. This fact is most important for theoretical chromatography since in the case of homogeneous models like the Poisson models, the column length L plays a definite role in establishing the rate of convergence towards the Gaussian Law and the general **EC** series approximation of chromatographic peak shape holds true.

In the past other general expansions were suggested like the Gram Charlier of type A (**GCA**) [5, 16, 22]. With respect to the **GCA** series, the **EC** series is simply a special case since it can be shown to derive from the former by partial ordering of the **GCA** terms. However the **GCA** series does not have the above described asymptotic properties and thus cannot be employed in a definite way to approximate chromatographic peak shapes [20].

7.2.6 *Approximation Measures and Chromatographic Model Approximation.* Eq. 96d gives only an upper limit to the difference between the actual function and the corresponding **EC** series of order k. In functional analysis different measures are employed for the "distance" between two functions f_1 and f_2. The most common is:

$$d(f_1, f_2) = \sqrt{\int [f_1(t) - f_2(t)]^2 \, dt} \qquad (97)$$

In probability theory a particular distance, the Lévy distance is employed, $L(f_1, f_2)$, to describe the weak convergence between two distributions F and G. This distance is the infimum of all h, i.e. the smallest h value, so that for all c such condition is fulfilled:

$$F(c - h) - h \leq G(c) \leq F(c + h) + h \qquad (98)$$

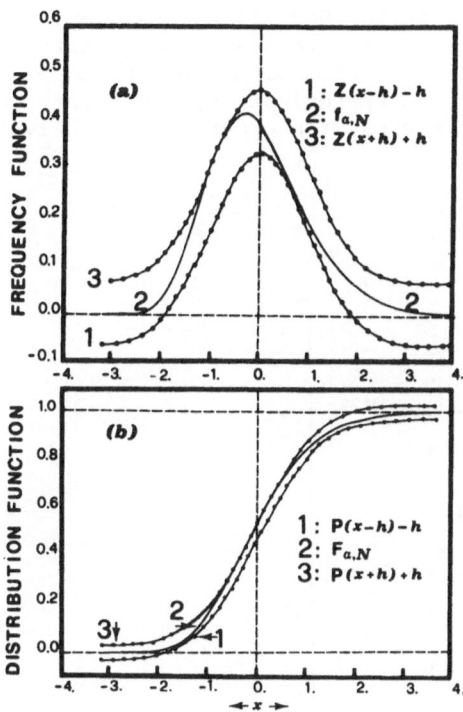

Fig. 5 Lévy distance between a theoretical chromatographic model function and a reference function. (a) Frequency function case (in this case the distance is referred as "improper Lévy distance"). (b) Distribution function case. Particular function: Martin-Synge model ($f_{\alpha, N}$ or $F_{\alpha, N}$, where N are the number of theoretical plates). Reference Function: Normal function (Z or P). Reprinted from [18] with permission. Copyright [1982] American Chemical Society.

An example is reported in Fig. 5, lower part, where the reference function F of Eq. 98 is the Normal distribution function P, and the particular function G is labelled as $F_{\alpha, N}$ and corresponds to the Gamma distribution function of order N. According to this picture the Lévy distance has an intuitive meaning and represents a strip in which the specific distribution G is located near two shifted reference distributions. This shift is expressed in units normalized with respect to the standard deviation. This same idea has been applied to the frequency function in the upper part of the same Fig. 5 (reference function = Z, particular function = $f_{\alpha, N}$, respectively the Normal and the Gamma of order N frequency functions). In this case the Lévy distance was called "improper" Lévy distance, since the Lévy distance was originally defined only for distribution functions [11]. In Fig. 5 one

204

must note that the shift effect of h (= 0.06) on the reference curves is more evident in the y-axis because of the employed scale.

In Fig. 6 the Lévy distance is used to show the approximation power of the EC expansion with respect to two chromatographic models [18]: the Martin-Synge and the Giddings-Eyring-McQuarrie models. In this figure the bilogarithmic relationship between the remainder term (L, the Lévy distance) and the asymptotic quantity n (either the mean number of adsorption-desorption steps $\bar{n}$, for the Giddings-Eyring-McQuarrie model, or the number of plates N in the Martin-Synge model) was reported, according to the expected behaviour of the Eq. 96d. It can be seen that the expected pattern is followed by the two theoretical chromatographic models. In fact straight lines are followed, with slopes b near to -1/2, -1, and -3/2 respectively for the Gaussian approximation, the first and the second EC expansion, for which the k expansion order is respectively 0, 1 and 2.

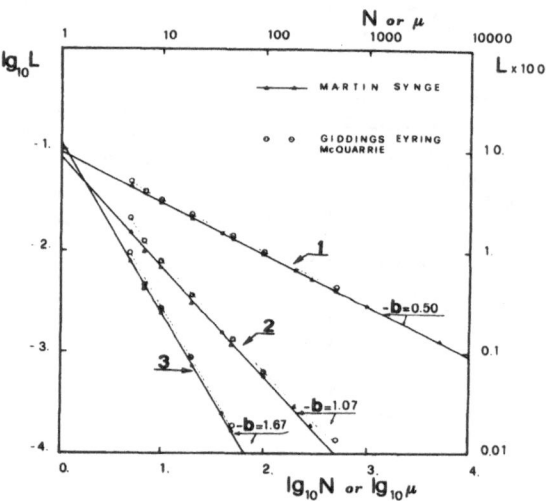

Fig. 6. Lévy distance (L) between theoretical models of chromatography, Giddings-Eyring-McQuarrie with k_s=1 or Martin-Synge model, as a function of the number of theoretical plates (N) or of the mean number of adsorption-desorption steps (μ, reported as $\bar{n}$ in the present paper), respectively. Line 1: the reference function in L computation is P, the Normal distribution function. Line 2 and 3: the reference functions in L computation are the 1st and 2nd EC series expansion, respectively. b is the slope according to Eq. 96d. Reprinted with permission from [18]. Copyright [1982] American Chemical Society.

The real interest of these approximation studies is that it is not really important to invert the solutions expressed in terms of CFs, but, instead, it is only important to derive the cumulant coefficients and with these parameters obtain peak shapes by EC approximation. Another example of study of the shape properties by the EC series has been presented elsewhere [7] where the effect of the complex sorption kinetics can be accounted for by both the EC series expansion and the Berry-Esséen theorem of Eq. 93.

7.2.7. *Approximation of extra-column band broadening factors by EC series.* Extra-column band broadening factors [29] act as convoluting factor and can thus be formally interpreted as additional random variables summing to chromatographic peak shape, the only difference being that they have own proper, different distributions. EC series can represent in a certain extent the process of addition of random variables with different distribution [10], the condition being that this number be great enough and that the difference among the different random variables be not too strong. Under these conditions, EC series can be employed to represent peak shape effects of extra-column band broadening phenomena. The cumulant coefficients, necessary to correctly employ EC series expressions, are computed by using Eqs. 28 and 39d.

8.   Obtaining Peak Parameters by EC Series approximation.

The EC series approximation properties may constitute the basis for peak parameter determination of experimental chromatographic peak shapes. In fact as variable parameters they contain the statistical peak parameters $\bar{m}$, σ, S, E and higher cumulant coefficients γ_i, see Eq. 95a-c which could be determined by non-linear least squares fitting. The higher the expansion order of the EC series, the higher the number of the statistical peak parameters obtained. Obviously, for an acceptable degree of fitting, the greater the peak skewness is, the higher the k expansion order of the EC series will be. This follows, in fact, from the above described approximation properties. However such a procedure would require the EC series to have another property; that is to be **minimum** approximations of the peak shape with respect to the statistical parameters. In other words among the all possible EC series having one different numerical value of the parameters $\bar{m}$, σ, S, E etc, the one containing the true value for these parameters is the one closest to the experimental peak. Furthermore the distance type must be defined well. An additional problem is that of experimental errors.

In a series of papers [7, 18, 20] it has been shown that the solution is not straightforward. In fact, the total squared deviations do not exhibit minimum properties. This property seems to be exhibited by the Lévy distance [18], but the computation of this distance is too time consuming to be applied in practice. However it has been found that the difference between parameters which minimize the sum of the squared errors (Eq. 97) and the true parameters decreases with the EC series expansion order k [17]. Moreover under defined conditions of the skewness value (S < 1) and noise, the peak parameters determined by non-linear least squared fitting by **EC** series are unbiased [23].

The **EC** series non-linear least squares fitting has been applied in column efficiency studies [25], in characterizing extra-column band broadening phenomena [24, 25] and non-linearity phenomena onset in gas-liquid chromatography [24], in Sedimentation Field-Flow Fractionation [26] and in Ion chromatography [27].

9.   Conclusions

The approach to the Stochastic Theory of Chromatography by the **CF** method is able: a) to explain, in terms of models, the most general case of linear finite-time chromatography; b) to represent the dynamics of band broadening and, thus, c) to account for the kinetic tailing in terms of the rate of convergence towards the Gaussian Law of the band profile. Moreover, the asymptotic Edgeworth-Cramér series expansions can be correctly employed as improved peak shape approximation functions. Under given conditions these same peak shape expressions can be employed in peak parameter evaluations making it possible thus to face experimental studies of the kinetics of chromatography under linear conditions.

Glossary

b	slope in the Lévy distance plot of Fig. 6 $[=(k+1)/2]$.
c	concentration.
c	standardized random variable, $(x-\overline{m})/\sigma$.
CF	Characteristic Function.
d	distance; total squared differences.
D	diffusion coefficient.
E_s, E_k, E_1, E_2	sorption Energy, kinetic Energy and Energy of type 1 and 2.
E	Excess.
f, f_j, f_{tot}, f_s, f_R	probability density functions respectively undefined, of type j, of type total, referred to a stationary phase process (s), to a retention process (R).
F	distribution function.
F	specific flatness.
$g(x)$	function of x.
GCA	Gram Charlier series of type A.
G	distribution function.
h	normalized distance.
H, H_B, H_C	plate height and terms of the van Deemter equation..
k	EC series expansion order.
k_s, k_m	rate constant respectively of the desorption process and of the stationary phase entry process.
k'	capacity factor.
i	imaginary unit.
l	distance from the column inlet.
L	Lévy distance.
L	column lenght.

$\overline{m}$	mean.
M	constant in the expression of the **EC** series remainder term.
n_s, n_m	total number respectively of desorption and sorption processes performed by a solute molecule (random quantities).
$\overline{n}_s$, $\overline{n}_m$	average number respectively of n_s and n_m.
$\overline{n}$	common value of $\overline{n}_s$ and $\overline{n}_m$ in the present approximation.
n	number of equally distributed random variables.
N	Number of theoretical plates.
p_j	probability or weight referred to j.
P	Normal distribution function.
Pr( )	Probability of the event contained in parenthesis.
q	corresponding Q term referred to the parent distribution of the added random variable.
Q	term of the **EC** series expansion of a distribution function.
r	random variable of the Poisson Law.
R	gas constant.
R	retention ratio.
R	remainder term in the **EC** series development.
S	Skewness.
t, t_s, t_m, t_R	random time, respectively unspecified, spent in the stationary phase, in the mobile phase, in the column (retention time).
$\overline{t}_m$, $\overline{t}_s$, $\overline{t}_R$	average values of respectively t_m, t_s, t_R.
T	Temperature $^\circ$K.
T	constant in the Berry-Esséen theorem.
$\mathbf{v}_m$	constant mobile phase velocity (L/t_m).
$\overline{\mathbf{v}}_m$	mean mobile phase velocity ($L/\overline{t}_m$)
Z	specific asymmetry.
Z	Normal frequency function.
x	value of a random variable.
1stCF	first characteristic function.
2ndCF	second characteristic function.
γ_j	cumulant coefficient of order j.
δ	Dirac function.
Δt_s, Δt_m	time spent by the solute molecule respectively in the sorption site and in the mobile phase between two subsequent sorption steps.
$\overline{\Delta t_s}$, $\overline{\Delta t_m}$	average values of respectively Δt_s and Δt_m.
η	random variable.
K_j	cumulant of order j.
μ_j, μ_j'	central moment and moment about the origin of order j, respectively.
ρ	reduced third absolute moment.
σ	standard deviation.
ϕ	first characteristic function, **1stCF**
ψ	second characteristic function, **2ndCF**.
$\mathcal{y}$	auxiliary variable.

Superscript:

$^{(j)}$	derivative of order j.
^{j}	j power.
$^{-}$	mean quantity.

Subscript:

D	referred to a diffusion process.
j	order (of the central moment, of the moment about the origin, of the cumulant coefficient) or type.
k	order in the **EC** series.
L	referred to the column lenght.
m	referred to a mobile phase process.
n	referred to the number of stationary phase entry process.
r	referred to a retention quantity per unit t_m value.
R	referred to a retention quantity.
s	referred to a stationary phase process.
tot	referred to a total quantity, in general to the whole column process (mobile phase + stationary phase).

Symbols:

E[ ]	Expectation or mean value of the quantity in parenthesis.
*	symbol for convolution integral.
|	conditioning.
| |	absolute value.
∩	simbol for joint events.

Acknowledgements

This work was made possible by the financial support of the Italian Ministry of the University and the Scientific Research (MURST) and the Italian Research Council (CNR).

8. References

[1] Gardiner, (1990) C.W Handbook of Stochastic Methods for Physics, Chemistry and the Natural Sciences, Second Ed., Springer-Verlag, Berlin.

[2] Giddings, J.C. (1958) 'The random downstream Migration of Molecules in Chromatography', J.Chem.Educ., 35, 588-591.

[3] Giddings, J.C. (1965) Dynamics of Chromatography, M.Dekker, New York.

[4] Giddings, J.C. and Eyring, H. (1955) 'A molecular dynamic theory of Chromatography', J.Phys.Chem.,59,416-421.

[5] McQuarrie, D.A. (1963), 'On Stochastic Theory of Chromatography', J.Chem.Phys., 38, 437-445.

[6]   Giddings, J.C. 'A   critical   Evaluation   of   the   Theory of
      Chromatography', in : Gas Chromatography 1964, Goldup, A. ed.
      Elsevier: Amsterdam, 1964, 3-25.
[7]   Dondi, F. and Remelli, M. (1986) 'The characteristic Function
      Method in the Stochastic Theory of Chromatography',
      J. Phys. Chem., 90, 1885-1891.
[8]   Cramér, H. (1974) Mathematical Method of Statistics, Princeton
      University Press, Princeton, NJ.
[9]   Feller, W. (1968) An Introduction to Probability Theory and its
      Applications, Wiley, New York.
[10] Cramér, H. (1961), Random Variables and Probability Distributions,
      Cambridge University Press, Cambridge, U.K.
[11] Gnedenko B.V. and Kolmogorov (1954) Limit Distributions for sums
      of Independent Random Variables, Addison-Wesley Publ. Comp.,
      Reading, MS.
[12] Steele, W.A. (1974) The Interaction of Gases with Solid Surfaces,
      Pergamon Press, Oxford, U.K., Chap. 6.
[13] De Boer, J.H. (1968) The Dynamical Character of Adsorption,
      Clarendon Press, Oxford, Chap. III.
[14] Kendall, M. and Stuart, A. (1958) The Advanced Theory of Statistics,
      Vol. I., Charles Griffin and Comp. Ltd., London.
[15] Abramowitz, M. and Segun, I., (1965), Handbook of Mathematical
      Functions with Formula, Graphs and Mathematical Tables, Dover,
      New York.
[16] Grubner, O. (1968) 'Statistical Moments Theory of Gas Solid
      Chromatography: Diffusion Controlled Kinetics', in Giddings, J.C.
      and Keller, R.A. Eds. 'Advances in Chromatography', vol.6,
      173-208.
[17] Scott, D.M. and Fritz, J.S. (1984) 'Model for Chromatographic Sepa-
      rations Based on Renewal Theory', Anal.Chem., 56, 1561-1566.
[18] Dondi. F. (1982) 'Approximation Properties of the Edgeworth-Cramér
      Series and Determination of Peak Parameters of Chromatographic
      Peaks', Anal.Chem., 54, 473-477.
[19] Kaminski, V.A. Timashev and S.F. Tunitskii, N.N. (1965), 'The shape
      of chromatographic peaks', Russ.J.Phys.Chem., 39, 1354-1357.
[20] Kelly,   P.C.   and   Harris,   W.E.   (1971),   '   Estimation
      of Chromatographic Peaks with Particular Consideration of Effects
      of Base-Line Noise', Anal.Chem., 43, 1170-1183.
[21] Kelly, P.C. and Harris, W.E. (1971), 'Application of Method
      Posterior Probability to Estimation of GasChromatographic
      Peak Parameters', Anal.Chem., 43, 1184-1195.
[22] Dondi, F Betti, A., Blo, G. and Bighi, C. (1981), 'Statistical
      Analysis of Gas Chromatographic Peaks by the Gram-Charlier
      Series of Type A and the Edgeworth-Cramér Series', Anal.Chem.,
      53, 496-504.
[23] Dondi, F. and Pulidori, F. (1984) 'Applicability of   the
      Edgeworth-Cramér Series in Chromatographic Peak Shape
      Analysis', J.Chromatogr., 284, 293-301.
[24] Dondi, F. and Remelli, M. (1984), 'Characterization of Extracolumn
      and Concentration Dependent Distortion of Chromatographic
      Peaks by Edgeworth-Cramér Series', J.Chromatogr., 315, 67-73.

[25] Remelli, M Blo, G., Dondi, F Vidal-Madjar, C. and Guiochon, G. (1989) 'Fluidic and Syringe Injection by Peak Shape Analysis', Anal.Chem., 61, 1489-1493.
[26] Reschiglian, P Blo, G. and Dondi F. (1991) 'Peak Shape Analysis in Sedimentation Field-Flow Fractionation', Anal.Chem., 63, 120-130
[27] Blo, G., Remelli, M., Pedrielli, F., Balconi, L., Sigon, F. and Dondi, F. (1991) 'Peak Shape Analysis and Noise Evaluation in Suppressed Ion Chromatographic Analysis for Ultra-Trace Ion Analysis', J.Chromatogr., 556, 249-262.
[28] Olivé, J. and Grimalt, J.O. (1991) 'Gram Charlier and Edgeworth-Cramér series in characterization of chromatographic peaks.' Anal.Chim.Acta., 249, 337-348.
[29] Sternberg, J.C. (1956) 'Extracolumn Contributions to Chromatographic Band Broadening', in J.C.Giddings and R.A. Keller (Eds.) Advances in Chromatography, M.Dekker, New York, Vol. 2. 205-270.

APPLICATIONS OF A MECHANISM-INDEPENDENT RETENTION EQUATION - SYSTEM PEAKS

François Riedo and Ervin sz. Kováts
Ecole d'Ingénieurs de Fribourg, CH-1700 Fribourg, Switzerland,
and Laboratoire de Chimie Technique de l'Ecole Polytechnique
Fédérale de Lausanne, CH-1015 Lausanne, Switzerland

ABSTRACT. In the mathematical treatment of chromatography with multi-component eluents by De Vault retention volumes are found to be eigenvalues of a matrix. Elements of this matrix are partial derivatives of partition isotherms which can be given in a simple form by using the thermodynamic function named the isocratic capacity. This function is the material content of the column in equilibrium with an infinite reservoir of a multi-component mixture, the eluent. For binary eluents the expression for the calculation reduces to a simple partial differential equation. Its versatility is demonstrated by applying to two examples : i. gas/liquid chromatography with a binary eluent and ii. liquid/solid chromatography with a binary eluent. Retention volumes of system peaks, labelled components of the eluent and solutes are calculated. For ternary mixtures the solution is given as the eigenvalues of a 2x2 matrix. Calculation of the retention volumes of the system peaks is demonstrated on the simplest possible example : gas/liquid chromatography with a ternary mixture of three ideal gases as eluent.

1. Introduction

In isothermal, isocratic, analytical chromatography the column is first equilibrated with the eluent (mobile phase, μ) of composition, $\tilde{x}_\mu^0$, then an infinitesimal amount of a sample is introduced. This injection perturbes the equilibrium and induces a concentration signal which traverses the column and appears at its outlet after a given volume of eluent has passed i.e., the retention volume. Let us first consider the column in its pre-injection state where it is in equilibrium with an infinite reservoir of a fluid mixture of constant composition. This thermodynamic column model is considered an open system as components of the eluent are free to enter at the inlet. If there is a stationary phase inside the column its material content introduced by the eluent will be the sum of the material content in the mobile phase and that absorbed in and /or adsorbed at the interfaces of the stationary and mobile phase. Let us now suppose that we do not know the mechanism by which material is retained by the column filling but we know from experiment the quantity of each component of the eluent in the column. In fact, this material content can be measured experimentally. After equilibrium is established the fluid is purged from the column and its mass and composition, is determined. The number of moles of a given component, $n_{\kappa,i}$[mol], is a function of the eluent composition

F. Dondi and G. Guiochon (eds.),
Theoretical Advancement in Chromatography and Related Separation Techniques, 211–226.
© 1992 Kluwer Academic Publishers.

$$n_{\kappa,i} = n_{\kappa,i}(\vec{x}_\mu) \tag{1}$$

The total molar isocratic capacity is given by

$$n_{\kappa,tot} = \sum_{i=1}^{m} n_{\kappa,i} \tag{2}$$

Also the total isocratic capacity together with the composition of the column content, $\vec{x}_\kappa$, equally and unequivocally characterizes the material content.

An adequate model of the chromatographic column is necessary for establishing mathematical relationships. This idealized column has a uniform cross-section at any distance, z, from the inlet, z = 0, to the outlet, z = L. It is filled with a quasi-continuum of a column packing having no flow resistance. A fluid m-component mixture is advancing in the column with a plug flow profile at constant temperature. In the case of a perturbation of the column equilibrium by injection at z = 0 a new equilibrium state is arrived at instantaneously in each cross-section; however, inspite of existing concentration gradients in the z-direction there is no axial diffusion. For such a column the column capacity referred to unit column length, $n_{\kappa,i}/L$, is an intensive function.

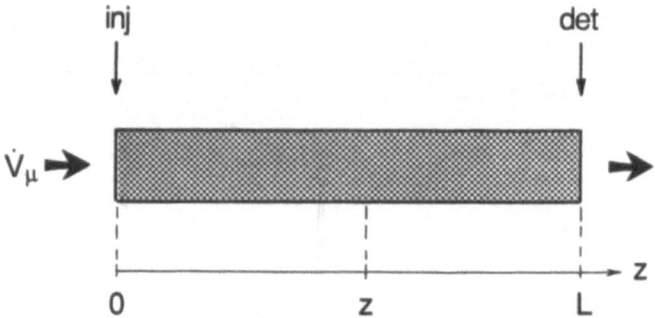

Figure 1. *The idealized chromatographic column of uniform cross-section from column inlet, z = 0, to column outlet, z = L, filled with a quasi-continuum of a stationary phase having no flow resistance. Mobile phase advances in the column with a plug flow profile.*

We are now able to address the problem of retention volume of a concentration signal introduced at the beginning of the column. (For details of the mathematical treatment see ref. 1). First, for compound, i, the material balance in a differential section, dz, can be written as a function of the time, t, and the position in the column, z. This treatment results in m - 1 partial differential equations for an m-component mixture. By considering infinitesimal perturbations, the retention volumes will appear as eigenvalues of a matrix with elements given by this set of partial differential equations as was originally shown by de Vault[2]. For the case of a binary eluent a solution of general validity can be given in terms of column capacities as shown in ref.1. The retention volume of a concentration perturbation by component, i, is given in eq 3 as

$$V_{R,i} = v_\mu^o \left[\left(\frac{\partial n_{\kappa,i}}{\partial x_{\mu,i}} \right)_{\bar{x}_\mu^o} - x_{\mu,i}^o \left(\frac{\partial n_{\kappa,tot}}{\partial x_{\mu,i}} \right)_{\bar{x}_\mu^o} \right] \tag{3}$$

where, $V_{R,i}$, is the retention volume of the concentration perturbation, v_μ^o, is the molar volume of the mobile phase and all other symbols are as explained before. The superscript, o, refers to time $t = 0$, i.e. to the state of the system before injection of the sample. If the sample, i, is a component of the binary mixture, $x_{\mu,i}^o \neq 0$ and the peak will be called "concentration peak"[1] or "system peak"[3] or "eigenpeak"[4]. If, i, is not contained in the mobile phase at $t = 0$, i.e. $x_{\mu,su}^o = 0$, the component is called a solute, su. Under the pretense of infinite dilution solute molecules do not interact with one another and their retention volumes are independent; therefore, only the case of one solute needs to be considered. Let us insist that eq 3 is a mechanism-independent retention equation of general validity for isothermal, isocratic, analytical chromatography if the eluent is a binary mixture[1]. For the retention volume of solutes ($x_{\mu,su}^o = 0$) it is also valid for any multicomponent eluent.

Before applying eq 3 to two specific examples, let us consider an interesting case of a specific solute, the injection of a labelled component, i*, in a multicomponent eluent. For such a component, $x_{\mu,i}^o \neq 0$ but $x_{\mu,i*}^o = 0$. Let us suppose that the physical properties of, i*, are in every respect identical with those of, i, with the exception of one permitting its identification. For such a component eq 4 holds

$$\frac{n_{\kappa,i*}}{n_{\kappa,i}} = \frac{x_{\mu,i*}}{x_{\mu,i}} \tag{4}$$

Substituting eq 4 into eq 3 while considering that, $x_{\mu,i*}^o = 0$ and $v_{\mu,i}/x_{\mu,i} = 1/c_{\mu,i}$ [l mol^{-1}], the retention volume of a labelled component of the eluent is related to the isocratic capacity of the unlabelled component by

$$V_{R,i*} = v_\mu^o n_{\kappa,i}^o / x_{\mu,i}^o = n_{\kappa,i}^o / c_{\mu,i}^o \tag{5}$$

Eq 5 is the necessary relationship for the determination of the isocratic capacity of the column by a chromatographic experiment.

In order to demonstrate the versatility of eq 3 it will now be applied to two relatively simple cases of gas/liquid and liquid/solid chromatography. From these examples it will be seen that the retention mechanism is introduced by the model which yields the explicit expression for the isocratic capacity of the column.

2. Gas/Liquid Chromatography with Binary Eluent

In the model of the idealized column there is a mobile phase occupying a volume, W_μ, and a stationary solvent occupying volume, W_ϑ. The column is now equilibrated with a binary mixture of two ideal gases slightly soluble in the stationary phase containing an infinitesimal amount of a third ideal gas, su. Neither components of the eluent, 1 and 2, nor the solute, su, are adsorbed at the gas/liquid (or liquid/solid support) interface. In this case the isocratic capacity of 1,2 and su and the total isocratic capacity of the column is given by

$$n_{\kappa,1} = \frac{W_\mu}{v_\mu^o} x_{\mu,1} + W_\vartheta c_{\vartheta,1} \tag{6}$$

$$n_{\kappa,2} = \frac{W_\mu}{v_\mu^o} x_{\mu,2} + W_\vartheta c_{\vartheta,2} \tag{7}$$

$$n_{\kappa,su} = \frac{W_\mu}{v_\mu^o} x_{\mu,su} + W_\vartheta c_{\vartheta,su} \quad ; \quad x_{\mu,su} \to 0 \tag{8}$$

$$n_{\kappa,tot} = \frac{W_\mu}{v_\mu^o} + W_\vartheta (c_{\vartheta,1} + c_{\vartheta,2} + c_{\vartheta,su}) \tag{9}$$

where, W_ϑ, is for the volume of the stationary liquid. The retention volume of labelled components, 1^* and 2^*, are calculated by substituting eqs 6 and 7 into eq 5 while employing $v_\mu^o/x_{\mu,i}^o = 1/c_{\mu,i}^o$.

$$V_{R,i*} = W_\mu + W_\vartheta \left(\frac{c_{\vartheta,i}}{c_{\mu,i}} \right)_{x_{\mu,2}^o} = W_\mu + W_\vartheta K_{D,i}(x_{\mu,2}^o) \tag{10}$$

where, $K_{D,i}(x_{\mu,2}^o)$, is the distribution coefficient of component, i, between the gas and the liquid phase at the composition, $x_{\mu,2}^o (= 1-x_{\mu,1}^o)$. The retention volume of a concentration perturbation can be calculated by supposing that the volume of the stationary phase, W_ϑ, is independent of the dissolved quantity of 1 and 2. Let us suppose further that the distribution coefficients of 1, 2 and any solute, su, are independent of each other. For this case use of eq 6 or 7 together with eq 9 in eq 3 and use of the relationship $x_{\mu,1} + x_{\mu,2} = 1$ gives, after rearrangement the same result for the retention volume of the concentration peak for the injection of either 1 or 2.

$$V_{R,cc} = V_{R,1} = V_{R,2} = W_\mu + W_\vartheta \left[x_{\mu,1}^o \left(\frac{dc_{\vartheta,2}}{dc_{\mu,2}} \right)_{x_{\mu,2}^o} + x_{\mu,2}^o \left(\frac{dc_{\vartheta,1}}{dc_{\mu,1}} \right)_{x_{\mu,2}^o} \right] =$$

$$= W_\mu + W_\vartheta \left[x_{\mu,1}^o \widetilde{K}_{D,2}(x_{\mu,2}^o) + x_{\mu,2}^o \widetilde{K}_{D,1}(x_{\mu,2}^o) \right] \tag{11}$$

where $\widetilde{K}_{D,i}(x_{\mu,2}^o)$ is the derivative of the concentration of the ith component in the stationary liquid, $c_{\vartheta,i}$, with respect of the concentration of the same component in the mobile phase at the composition, $x_{\mu,2}^o$, of the mobile phase. Finally, use of eq 8 in eq 3 results in the prediction of the retention volume of a solute.

$$V_{R,su} = W_\mu + W_\vartheta \left(\frac{dc_{\vartheta,su}}{dc_{\mu,su}}\right)_{\bar{x}_\mu^o} = W_\mu + W_\vartheta K_{D,su}^o \tag{12}$$

where, $K_{D,su}^o$, is the distribution coefficient of the solute, su, at infinite dilution; which was assumed not to depend on the quantity of 1 and 2 dissolved in the stationary liquid.

As a general rule, injection of any solute will also introduce a concentration perturbation in the eluent, therefore two peaks will appear, the system peak and the peak of the solute. In fact, by injection of a solute the binary system becomes ternary. In a ternary system there are two independent variables for the composition, therefore, there will be two concentration signals.

By comparing eqs 10, 11 and 12 it is seen that the retention volume of a sample has the general form of

$$V_{R,i} = W_\mu + W_\vartheta K_{D,i}^* \tag{13}$$

where the dimensionless quantity $K_{D,i}^*$ is a function of the composition of the mobile and the stationary phase and is related to the distribution properties of the individual components. In fact, $K_{D,i}^*$ is the net retention volume measured on a column containing 1 ml stationary phase. The experimental determination of the holdup volume, W_μ, is possible with the aid of a sample having $K_D = 0$.

In Fig 2 application of eqs 10 and 11 is illustrated on a simplest possible example. The iso-therm of distribution of the soluble components of the binary eluent, 1 and 2, is linear, therefore,

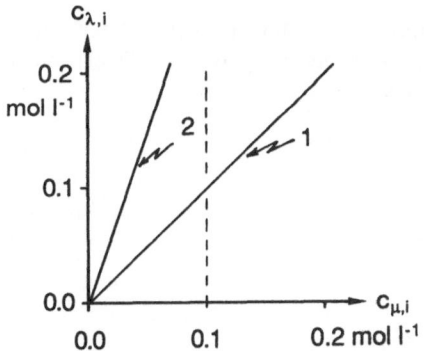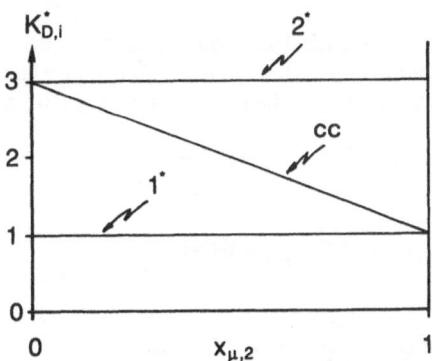

Figure 2. Illustration of the numerical value of the retention of labelled components, 1* and 2*, of the binary eluent and that of the system peak, cc, as a function of the composition of the eluent $x_{\mu,2}$ (= 1-$x_{\mu,1}$). Absorption isotherms of the components, $c_{\vartheta,i} \equiv c_{\lambda,i}$ versus $c_{\mu,i}$ (λ is for liquid), are linear, therefore, $K_{D,i} = \tilde{K}_{D,i}$, independently of the concentration, $c_{\mu,i}$; $K_{D,1} = 1$ and $K_{D,2} = 3$. The dashed line shows that the concentration, $c_{\mu,i}$, is limited to a value of approximately 0.1 mol l^{-1}, corresponding to a column pressure of about 2 bar.

216

the distribution coefficients, $K_{D,1}$ and $K_{D,2}$, are independent of the concentration. In Fig 3 a chromatogram is presented for a given composition of the binary eluent where for simplicity all peaks are marked as positive concentration perturbations which is not necessarily true.

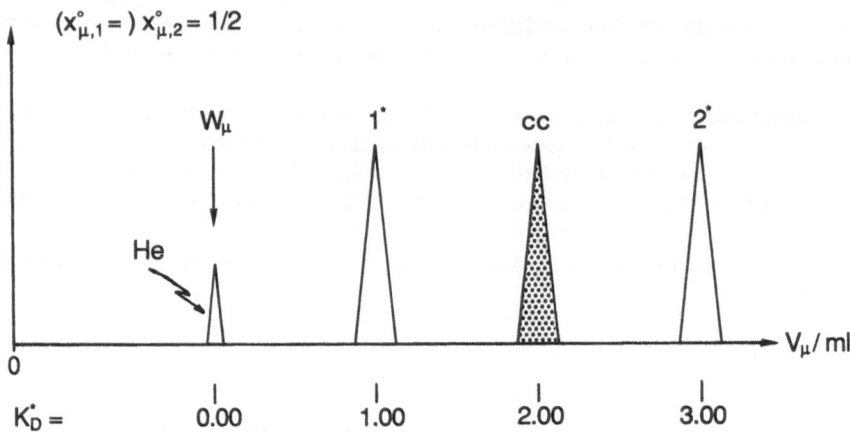

Figure 3. *Retention volume in gas/liquid chromatography of the labelled components of the binary eluent, 1* and 2*, and that of the system peak, cc, depicted as a chromatogram for an eluent composition of $x_{\mu,2} = 1/2$. The absorption isotherms of the individual components are as in Fig 2. The holdup volume of the column, W_μ, can be determined by the retention volume of helium, considered to be insoluble in the stationary liquid.*

3. Liquid/Solid Chromatography with Binary Eluent

In the idealized column there is a liquid mobile phase and a solid adsorbent. Material in the column is contained in the mobile phase and at the solid/liquid interface. The isocratic capacity of, 1, 2 and su, and the total isocratic capacity of the column is given by

$$n_{\kappa,1} = \frac{W_{\mu/CX}}{v_\mu^0} x_{\mu,1} + S\Gamma_{1/CX} \tag{14}$$

$$n_{\kappa,2} = \frac{W_{\mu/CX}}{v_\mu^0} x_{\mu,2} + S\Gamma_{2/CX} \tag{15}$$

$$n_{\kappa,su} = \frac{W_{\mu/CX}}{v_\mu^0} x_{\mu,su} + S\Gamma_{su/CX} \qquad ; \qquad x_{\mu,su} \to 0 \tag{16}$$

$$n_{\kappa,tot} = \frac{W_{\mu/CX}}{v_\mu^0} + S\left(\Gamma_{1/CX} + \Gamma_{2/CX} + \Gamma_{su/CX}\right) \tag{17}$$

where, $S[m^2]$ is the surface area of the adsorbent in the column and, $\Gamma_i[\mu mol\ m^{-2}]$, is the molar quantity of substance, i, adsorbed per unit surface area. The subscript, CX, is for "Convention-X". In fact, in the case of reversible adsorption it is not possible to determine the absolute adsorbed quantity and/or the volume of the non-adsorbed part of the bulk liquid. In other terms, in the material balance summarized in the two eqs 14 and 15 there are 3 unknowns, thus the adsorption problem remains undefined. Therefore, it will be necessary to introduce a convention permitting to write an additional equation. With this reserve in mind the retention volume of component i (= 1, 2 or su) is found by using the material balances (eqs 14 - 17) in eq 3 giving

$$V_{R,i} = W_{\mu/CX} + S\, v_{\mu}^{o} \left[\left(\frac{\partial \Gamma_{i/CX}}{\partial x_{\mu,i}} \right)_{\bar{x}_{\mu}^{o}} - x_{\mu,i}^{o} \left(\frac{\partial \Gamma_{tot/CX}}{\partial x_{\mu,i}} \right)_{\bar{x}_{\mu}^{o}} \right] \tag{18}$$

At this point we will introduce the convention that the sum of adsorbed number of moles is zero,

$$\Gamma_{tot/nNA} = \Gamma_{1/nNA} + \Gamma_{2/nNA} = 0 \tag{19}$$

i.e. in terms of number of moles Nothing is Adsorbed at the surface, the convention nNA. In this case the last term in the brackets in eq 18 is zero. For labelled components of the eluent the relationship in eq 20, analogous to eq 4 ,holds

$$\frac{\Gamma_{i*/CX}}{\Gamma_{i/CX}} = \frac{x_{\mu,i*}}{x_{\mu,i}} \tag{20}$$

By employing eq 20 in conjunction with eq 18, the retention volumes of the labelled components of the binary eluent can be shown to be

$$V_{R,1*} = W_{\mu/nNA} + S\, v_{\mu}^{o} \frac{\Gamma_{1/nNA}^{o}}{x_{\mu,1}^{o}} \tag{21}$$

$$V_{R,2*} = W_{\mu/nNA} + S\, v_{\mu}^{o} \frac{\Gamma_{2/nNA}^{o}}{x_{\mu,2}^{o}} \tag{22}$$

Multiplying eq 21 by, $x_{\mu,1}^{o}$, and eq 22 by, $x_{\mu,2}^{o}$ and summing up the results while applying the nNA-convention the following relationship results

$$x_{\mu,1}^{o} V_{R,1*} + x_{\mu,2}^{o} V_{R,2*} = W_{\mu/nNA} \tag{23}$$

Eq 23 is the necessary relationship for the experimental determination of the holdup volume, $W_{\mu/nNA}$. Let us insist that the holdup volume depends on the specific convention chosen. In fact, it fixes the zero point in a chromatogram but in particular the holdup volume, $W_{\mu/nNA}$, cannot be identified with any physical existing volume inside the column. Nevertheless, the retention volume of the concentration peak (system peak) and that of a solute can be given as

$$V_{R,cc} = V_{R,1} = V_{R,2} = W_{\mu/nNA} + S\, v_{\mu}^{o} \left(\frac{\partial \Gamma_{2/nNA}}{\partial x_{\mu,2}} \right)_{x_{\mu,2}^{o}} \tag{24}$$

$$V_{R,su} = W_{\mu/nNA} + S\, v_{\mu}^{o} \left(\frac{\partial \Gamma_{su/nNA}}{\partial x_{\mu,su}} \right)_{\bar{x}_{\mu}^{o}} \tag{25}$$

In order to find an expression for the holdup volume which can be identified with a physical volume inside the column let us convert eqs 18 - 25 to equations where adsorption is given in terms of adsorbed volume per unit surface area, $\Psi_{i/CX}$. It will be supposed that the liquid eluent is an ideal mixture of its components. For such a mixture the following relations hold true :

$$\Psi_{i/CX} = v_i \, \Gamma_{i/CX} \; ; \; \; d\Psi_{i/CX} = v_i \, d\Gamma_{i/CX} \tag{26}$$

$$\varphi^o_{\mu,2} \, (= 1 - \varphi^o_{\mu,1}) \; = \; x^o_{\mu,2} \, v_2 / \, v^o_\mu \tag{27}$$

$$\frac{d\varphi^o_{\mu,2}}{dx^o_{\mu,2}} = v_1 v_2 / (v^o_\mu)^2 \tag{28}$$

$$\varphi_{\mu,su} = x_{\mu,su} \, v_{su} / v^o_\mu \tag{29}$$

$$\left(\frac{\partial \varphi_{\mu,su}}{\partial x_{\mu,su}} \right)_{\vec{x}^o_\mu} = v_{su} / v^o_\mu \tag{30}$$

where, v_1 and v_2, are molar volumes of the components, 1 and 2, of the eluent independent of the eluent composition, $x^o_{\mu,2}$, whereas, the molar volume of the solute, v_{su}, must be constant in the proximity of, $x^o_{\mu,2}$.

Let us first calculate the position of the system peak. Using eqs 26, 27 and 28 in eq 18 yields upon rearrangement

$$V_{R,cc} = V_{R,1} = V_{R,2} = \; W_{\mu/CX} + S \left[\left(\frac{d\Psi_{2/CX}}{d\varphi_{\mu,2}} \right)_{\varphi^o_{\mu,2}} - \varphi^o_{\mu,2} \left(\frac{d\Psi_{tot/CX}}{d\varphi_{\mu,2}} \right)_{\varphi^o_{\mu,2}} \right] \tag{31}$$

It is now seen that it is particularly attractive to use the convention "vNA"

$$\Psi_{tot/vNA} = \Psi_{1/vNA} + \Psi_{2/vNA} = 0 \tag{32}$$

i.e. in terms of volumes, Nothing is Adsorbed at the surface. This convention states that, in terms of volumes, the positively adsorbed excess of component, 1, is counterbalanced by the missing volume of component, 2. Therefore, if the molar volumes of 1 and 2, v_1 and v_2, are the same in the surface phase and the adsorbent does not change its volume with adsorbate adsorbed at its surface, the holdup volume, $W_{\mu/vNA}$ can be identified as the interstitial volume in the column. With this imposed convention eq 32 simplifies to

$$V_{R,cc} = V_{R,1} = V_{R,2} = \; W_{\mu/vNA} + S \left(\frac{d\Psi_{2/vNA}}{d\varphi_{\mu,2}} \right)_{\varphi^o_{\mu,2}} \tag{33}$$

Substitution of eqs 26, 29 and 30 into eq 18 results

$$V_{R,su} = \; W_{\mu/vNA} + S \left(\frac{\partial \Psi_{su/vNA}}{\partial \varphi_{\mu,su}} \right)_{\vec{\varphi}^o_\mu} \tag{34}$$

For the labelled components of the eluent a relationship analogous to that given in eq 4 holds

$$\frac{\Psi_{i^*/CX}}{\Psi_{i/CX}} = \frac{\varphi_{\mu,i^*}}{\varphi_{\mu,i}} \tag{35}$$

The use of eq 35 along with 34 gives

$$V_{R,1^*} = W_{\mu/\nu NA} + \frac{\Psi^o_{1/\nu NA}}{\varphi^o_{\mu,1}} \tag{36}$$

$$V_{R,2^*} = W_{\mu/\nu NA} + \frac{\Psi^o_{2/\nu NA}}{\varphi^o_{\mu,2}} \tag{37}$$

Combination of eqs 36 and 37 gives the necessary relationship for the experimental determination of the holdup volume, $W_{\mu/\nu NA}$,[5]

$$\varphi^o_1 V_{R,1^*} + \varphi^o_2 V_{R,2^*} = W_{\mu/\nu NA} \tag{38}$$

Finally, let us note, that the general form of the retention equation is given by

$$V_{R,i} = W_{\mu/CX} + S\chi^*_{CX,i} \tag{39}$$

where $\chi^*_{CX,i}$ [μl m^{-2} = nm] is a Henry coefficient or a function of Henry coefficients. In fact, it is the areal retention volume of the concentration signal after perturbation by component, i.

Equations for real mixtures are derived and given in ref 1 which result in complicated expressions. However, for non-electrolyte mixtures deviations from ideality result in deviations of only a few percent in the mean molar volume, v_μ. Experimental determination in acetonitrile/water mixtures with non-polar type adsorbents showed that, in the limits of experimental error, the holdup-volume, $W_{\mu/\nu NA}$, was independent of composition[5-7].

In Fig 4 two excess isotherms, $\Psi_{2/\nu NA}$, are given as a function of the composition, $\varphi_{\mu,2}$. In Fig 4A and B examples of U- and S-type excess isotherms are plotted in units of μl m^{-2} which is equivalent to a distance in units of nm, the mean thickness of the adsorbed layer. The isotherm, $\Psi_{1/\nu NA}$, is also presented in the same plot as a dashed line. The symmetry seen here is due to the choice of the vNA-convention, i.e., $\Psi_{1/\nu NA} = -\Psi_{2/\nu NA}$. Fig 5A and B show the value of, $\chi^*_{\nu NA}$, in units of nm $\equiv$ μl m^{-2} which is the areal retention volume. In Fig 6 is shown the chromatogram of a sample composed of a mixture of labelled components of the eluent, 1* and 2*, which will also perturbs the adsorption equilibrium and induces a system peak. In this example the composition of the eluent is, $\varphi_{\mu,2} = 1/2$, and the value of the individual Henry coefficients, $\chi_{i/\nu NA}$, is based on the S-shaped isotherm of Fig 5B.

Let us insist that there is no substance which would elute at, $\chi^*_{\nu NA} = 0$, indepedently of the composition of the eluent. Of further interest is the fact that the retention volume of labelled 1* and that of the system peak is negative. There are several examples which follow the general feature of this example. S-shaped isotherms are observed on silica surfaces covered with a dense layer of non-polar subsitutents where a small amount of silanols are accessible for water adsorption from water/acetonitrile, water/methanol or water/tetrahydrofurane mixtures[5-8]. For such surfaces at

220

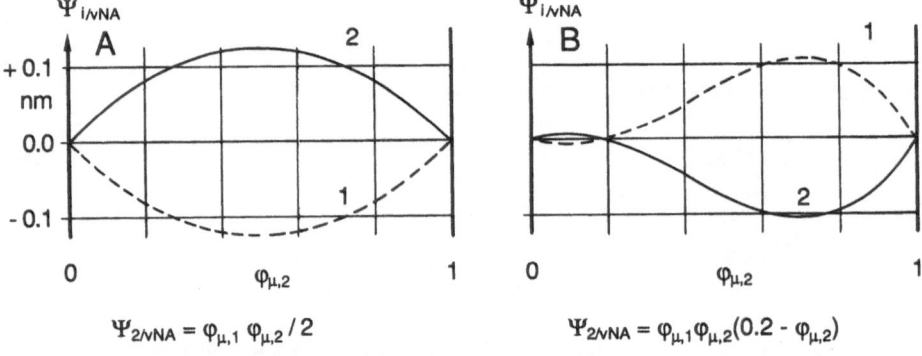

$$\Psi_{2/\nu NA} = \phi_{\mu,1} \, \phi_{\mu,2} / 2 \qquad\qquad \Psi_{2/\nu NA} = \phi_{\mu,1}\phi_{\mu,2}(0.2 - \phi_{\mu,2})$$

Figure 4. *In Fig 4A is shown an example of a U-type excess isotherm of component 2, $\Psi_{2/\nu NA}$ [$\mu l\ m^{-2} = nm$], as a function of the volume fraction of component 2 in the binary mobile phase, $\phi_{\mu,2}$. In Fig 4B an S-type isotherm is shown. The dashed line is trace of the isotherm, $\Psi_{1/\nu NA}$. With the vNA-convention $\Psi_{1/\nu NA} = -\Psi_{2/\nu NA}$.*

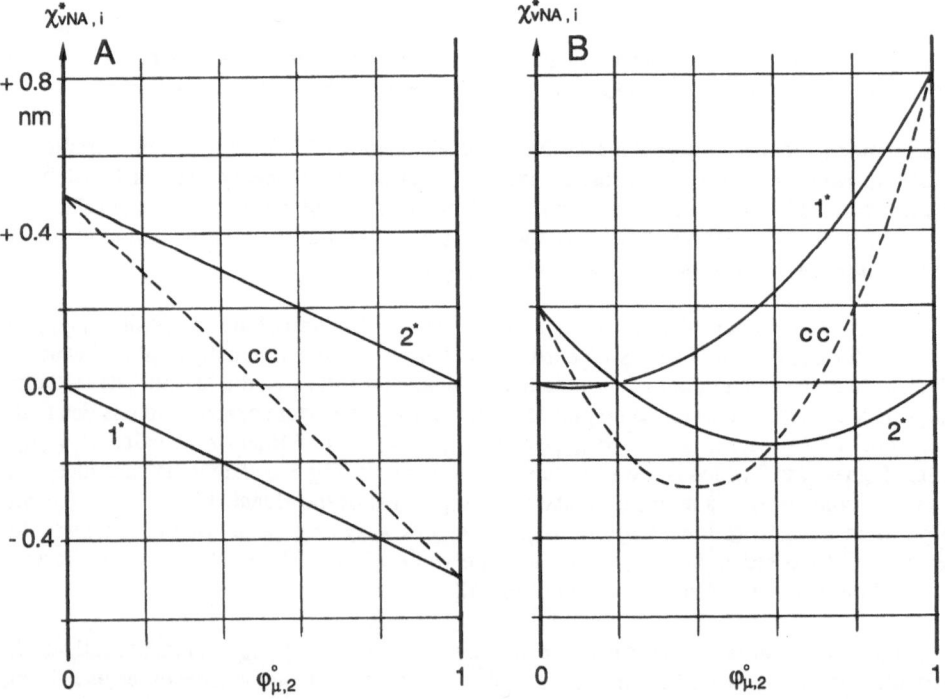

Figure 5. *Areal retention volumes of labelled components of the eluent, 1* and 2*and that of the system peak (concentration peak), cc, as a function of the eluent composition, $\phi_{\mu,2}$. In Fig 5A eluent components are supposed to be adsorbed by the adsorption law given in Fig 4A; in Fig 5B by that of Fig 4B.*

very low water contents water is positively adsorbed whereas at higher water concentrations the organic component is preferred by the surface.

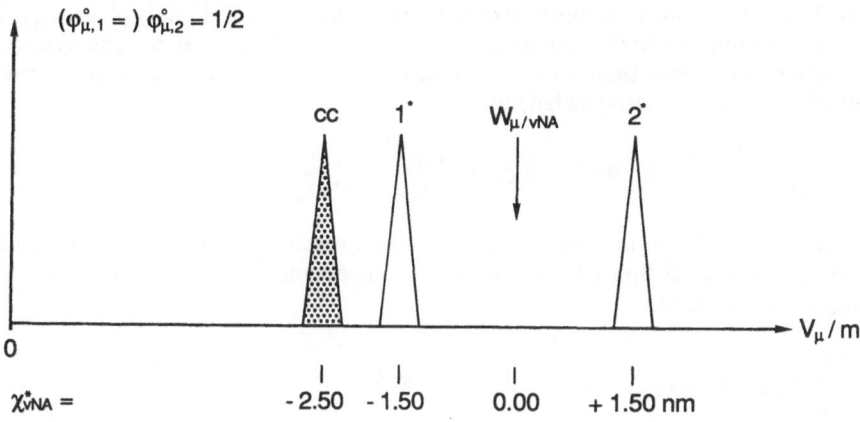

Figure 6. Retention volumes in liquid/solid chromatography of the labelled components of the binary eluents, 1* and 2*, and that of the system peak, cc, depicted as a chromatogram for an eluent composition, $\varphi_{\mu,2}$ (= $\varphi_{\mu,1}$) = 1/2. The underlying excess isotherm of the binary eluent is that given in Fig 4B. The holdup volume cannot be determined by injection of a solute and it depends on the convention chosen. The holdup volume, $W_{\mu/vNA}$, was calculated using eq 38. Concentration perturbations are always depicted as positive deviations which is not necessarily always true.

4. System Peaks in a Ternary Eluent

When evaluating system peaks in a ternary eluent eq 3 is no longer valid. For the calculation of the propagation of the concentration signals in the column it is now necessary to refer to the 2x2 matrix, V, with elements, V_{ij}, given in eq 40[1,2]

$$V_{ij} = v_{\mu}^{o} \left[\left(\frac{\partial n_{\kappa,i}}{\partial x_{\mu,j}} \right)_{\tilde{x}_{\mu}^{o}} - x_{\mu,i}^{o} \left(\frac{\partial n_{\kappa,tot}}{\partial x_{\mu,j}} \right)_{\tilde{x}_{\mu}^{o}} \right] \qquad ; \qquad i,j = 2,3 \tag{40}$$

where the mole fractions of components 2 and 3, $x_{\mu,2}$ & $x_{\mu,3}$, are considered to be independent variables and $x_{\mu,1} = 1 - x_{\mu,2} - x_{\mu,3}$. The retention volumes of the system peaks are the eigenvalues of the matrix, V, i.e., they are solutions of eq 41

$$\text{Det} \left(V - V_R I \right) = 0 \tag{41}$$

where, I, is the 2x2 identity matrix. Eq 41 is of general validity for systems with ternary eluents but solutions will be more complex than the binary case described by eq 3. As a general rule, an infinitesimal perturbation of the composition in the ternary eluent at the column inlet shall produce *two* propagating signals with retention volumes, $V_{R,cc1st}$ & $V_{R,cc2nd}$, given as solutions of the quadratic eq 41.

A simple illustration of the use of eq 41 shall be presented, based on the following premisses. The eluent is composed of three ideal gases slightly soluble in the stationary phase so that the volume of the saturated stationary phase is equal to that in the absence of the eluent mixture. Furthermore, the concentration of each component in the stationary phase, $c_{\vartheta,i}$, depends exclusively and linearly on the concentration of the same component in the gas phase, $c_{\mu,i}$. The final assumption is that there is no adsorption at the interfaces. For such a system, the relationships summarized in eq 42 hold true.

$$K_{D,i} \equiv \frac{c_{\vartheta,i}}{c_{\mu,i}} = v_\mu^o \frac{c_{\vartheta,i}}{x_{\mu,i}} = const = \widetilde{K}_{D,i} \equiv \frac{dc_{\vartheta,i}}{dc_{\mu,i}} = v_\mu^o \frac{dc_{\vartheta,i}}{dx_{\mu,i}} \tag{42}$$

The isocratic capacities of the column are given by expressions similar to those given in eqs 6-9 without the restriction that one of the three components is not present in the eluent at $t = 0$ ($x_{\mu,su} \to 0$ in eq 8).

$$n_{\kappa,i} = \frac{W_\mu}{v_\mu^o} x_{\mu,i} + W_\vartheta c_{\vartheta,i} \qquad ; \qquad i = 1, 2 \text{ and } 3 \tag{43}$$

$$n_{\kappa,tot} = \frac{W_\mu}{v_\mu^o} + W_\vartheta (c_{\vartheta,1} + c_{\vartheta,2} + c_{\vartheta,3}) \tag{44}$$

By choosing the mole fractions of components 2 and 3, $x_{\mu,2}$ and $x_{\mu,3}$, as independent variables ($x_{\mu,1} = 1 - x_{\mu,2} - x_{\mu,3}$) the derivatives of eqs 43 and 44 give upon rearrangement

$$\left(\frac{\partial n_{\kappa,i}}{\partial x_{\mu,j}}\right)_{\tilde{x}_\mu^o} = \frac{1}{v_\mu^o} (W_\mu + W_\vartheta \widetilde{K}_{D,i}) \delta_{ij} \qquad ; \qquad i,j = 2 \text{ and } 3 \tag{45}$$

$$\left(\frac{\partial n_{\kappa,tot}}{\partial x_{\mu,j}}\right)_{\tilde{x}_\mu^o} = \frac{W_\vartheta}{v_\mu^o} (\widetilde{K}_{D,j} - \widetilde{K}_{D,1}) \qquad ; \qquad j = 2 \text{ and } 3 \tag{46}$$

where the Kronecker symbol, δ_{ij}, has the usual meaning, i.e., $\delta_{ij} = 0$ if $i \neq j$ and $\delta_{ij} = 1$ if $i = j$. The elements of the matrix, V, are calculated by substituting eqs 45 and 46 in eq 40 to give

$$V_{ij} = W_\mu \delta_{ij} + W_\vartheta \left[\widetilde{K}_{D,i} \delta_{ij} - x_{\mu,i}^o (\widetilde{K}_{D,j} - \widetilde{K}_{D,1})\right] \tag{47}$$

which can be also written as

$$V_{ij} = W_\mu \delta_{ij} + W_\vartheta K_{ij} \tag{48}$$

where the dimensionless quantity, K_{ij}, represents the expression in brackets in eq 47. An element of matrix, V, is the sum of two terms given in eq 48, therefore, matrix, V, can be written as

$$V = W_\mu I + W_\vartheta K \tag{49}$$

where, K, is a 2x2 matrix with elements, K_{ij}, defined in eq 48. Combination of eqs 49 and 41 gives the solution of the eigenvalue problem as follows

$$\text{Det} \left(\mathbf{K} - \frac{V_R - W_\mu}{W_\vartheta} \mathbf{I} \right) = 0 \tag{50}$$

Comparison of eq 50 with eq 13 reveals that the eigenvalues of matrix, $\mathbf{K}$, give the apparent partition coefficients, $K_{D,i}^*$, in the present case, $K_{D,cc}^*$, as a function of the partition coefficients of the individual components, 1,2 and 3. This leads to the following determinant :

$$\begin{vmatrix} K_{11} - K_{D,cc}^* & K_{12} \\ K_{21} & K_{22} - K_{D,cc}^* \end{vmatrix} = 0 \tag{51}$$

yielding

$$\left(K_{D,cc}^* \right)^2 + A K_{D,cc}^* + B = 0 \tag{52}$$

where the coefficients A and B are given in eqs 53 and 54 respectively

$$- A = K_{11} + K_{22} = (1 - x_{\mu,1}^o) \, \widetilde{K}_{D,1} + (1 - x_{\mu,2}^o) \, \widetilde{K}_{D,2} + (1 - x_{\mu,3}^o) \, \widetilde{K}_{D,3} \tag{53}$$

$$B = K_{11} K_{22} - K_{12} K_{21} = x_{\mu,1}^o \, \widetilde{K}_{D,2} \, \widetilde{K}_{D,3} + x_{\mu,2}^o \, \widetilde{K}_{D,3} \, \widetilde{K}_{D,1} + x_{\mu,3}^o \, \widetilde{K}_{D,1} \, \widetilde{K}_{D,2} \tag{54}$$

At this point let us make the following remarks

i.   An asymmetry is introduced at the beginning of the derivation by choosing one of the mole fractions, $x_{\mu,1}^o, x_{\mu,2}^o$ or $x_{\mu,3}^o$, as the dependent variable. The final solution is independent of this choice.

ii.   If any one of the components has zero initial concentration the solution to eq 52 is the same as that given in eqs 11 and 12 for the binary eluent (e.g. if $x_{\mu,3}^o = 0$ the solutions are : $\widetilde{K}_{D,3}$ and $x_{\mu,1}^o \widetilde{K}_{D,2} + x_{\mu,2}^o \widetilde{K}_{D,2}$).

iii.   If the differential partition coefficient, $\widetilde{K}_{D,i}$, of any two components are equal, eq 52 also reduces to the case of binary eluent.

The solution to eq 52 gives the apparent partition coefficients of the two system peaks

$$K_{D,cc1^{st}}^* = \left(- A + \sqrt{A^2 - 4B} \right)/2 \tag{55}$$

$$K_{D,cc2^{nd}}^* = \left(- A - \sqrt{A^2 - 4B} \right)/2 \tag{56}$$

As an illustration, the apparent partition coefficients of the two concentration peaks were calculated for the following example. In addition of assuming the independence of the individual partition coefficients, it was supposed that Henry's law was valid for every component up to the column pressure (1-3 bar corresponding to about $c_{\mu,i} \approx 0.1$ mol l^{-1}). In this case $K_{D,i} = \widetilde{K}_{D,i}$. The value of the apparent partition coefficients of the two system peaks are shown in Fig 7; where the following numerical values were used for the partition coefficients of the components : $K_{D,1} = 1, K_{D,2} = 2$ and $K_{D,3} = 4$. The two surfaces, corresponding to $K_{D,cc1^{st}}^*$ and $K_{D,cc2^{nd}}^*$, touch at only one point in the plane where $x_{\mu,2} = 0$ i.e., the component with the middle partition

224

coefficient. It is also of interest, that $K_{D,1} \le K_{D,cc1}^* \le K_{D,2}$ and $K_{D,2} \le K_{D,cc2}^* \le K_{D,3}$ for any composition.

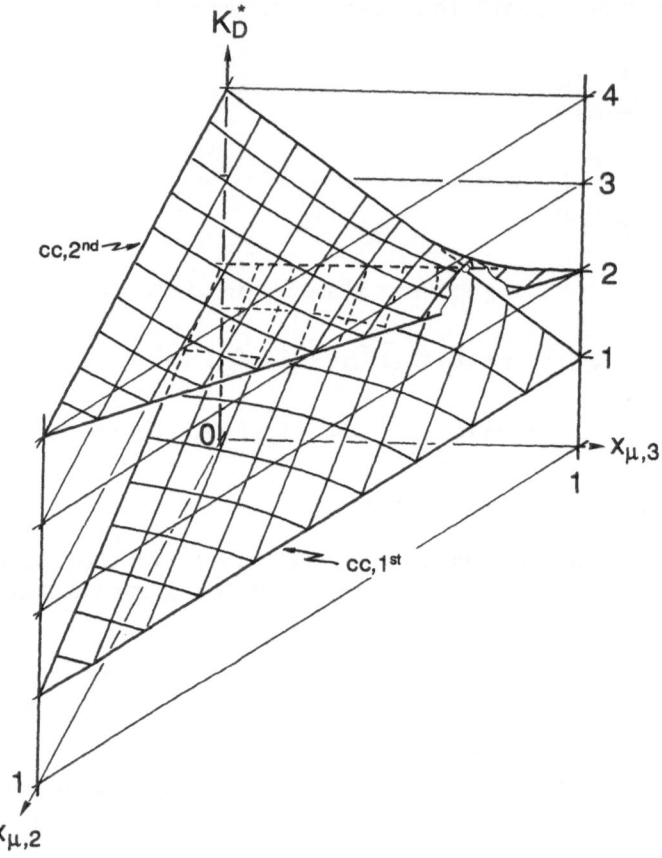

Figure 7. *Apparent partition coefficient of the system peaks, $K_{D,cc1}^*$ and $K_{D,cc2}^*$ in gas chromatography with a ternary mixture of ideal gases as eluent. On injection of either components of the eluent, 1, 2 or 3, both peaks will appear. Partition coefficients of the individual components : $K_{D,1} = 1$, $K_{D,2} = 2$ and $K_{D,3} = 4$.*

Based on this example, a hypothetical chromatogram is shown in Fig 8 for the particular composition, $(x_{\mu,1}^o =) x_{\mu,2}^o = x_{\mu,3}^o = 1/3$. The holdup volume can be determined by injection of a fourth component which is insoluble in the stationary liquid such as helium (i.e., $K_{D,He} \approx 0$; see eq 12). Retention volumes of the labelled components of the eluent, 1^*, 2^* and 3^* are calculated by using the isocratic capacities in eq 5. Their position in the chromatogram is independent of the composition.

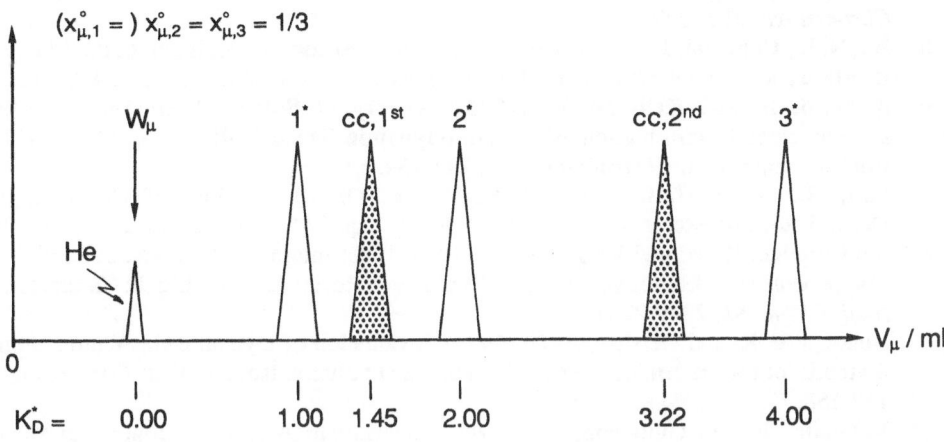

Figure 8. *Retention volumes of labelled components of the ternary eluent in gas chromatography and those of the system peaks for the eluent composition $(x^o_{\mu,1} =) x^o_{\mu,2} = x^o_{\mu,3} = 1/3$. Absorption properties of the individual components are as in Fig 7.*

5. Final Remarks

The calculation of retention volumes was illustrated for two simple cases for elution with binary eluents. Experimental evidence has been presented for both cases. Eq 11 has been used for the determination of equilibrium absorption isotherms (the "method of minor disturbance"[9]). Its validity has been demonstrated by Valentin and Guiochon[10] for the determination of equilibrium isotherms by using helium ($K_{D,He} = 0$) as a diluent for a soluble gas in gas chromatographic experiments. Several papers describe the determination of binary adsorption isotherms in liquid chromatography[5,6,7,11]. The last example, gas chromatography with a ternary eluent, has never been proven experimentally.

6. Acknowledgement

Work reported in this paper was financed by the *"Fonds National Suisse de la Recherche Scientifique"*.

7. References

1 Riedo , F. and sz. Kováts, E. (1982) "Adsorption from Liquid Mixtures and Liquid Chromatography", *J. Chromatogr.*, **239**, 1-28.
2 De Vault, D. (1943) "The Theory of Chromatography", *J. Amer. Chem. Soc.*, **65**, 532-540.
3 Levin, S. and Grushka, E. (1986) "System Peaks in Liquid Chromatography : Their Origin, Formation and Importance", *Anal. Chem.*, **58**, 1602-1607.

4   Melander, W. R., Erard, J.-F. and Horváth, Cs. (1983) "Movement of Components in Reversed-Phase Chromatography; II. Eigenpeaks in Reversed-Phase Chromatography with Silica-Bound Hydrocarbonaceous Stationary Phases : Effect of the Eluite Structure" *J. Chromatogr.*, **282**, 229-248.

5   Ha, N. L., Ungvárai, J. and sz. Kováts, E. (1982) "Adsorption Isotherm at the Liquid-Solid Interface and the Interpretation of Chromatographic Data" *Anal. Chem.*, **54**, 2410-2421.

6   Knox, J. H. and Kaliszan, R. (1985) "Theory of Solvent Disturbence Peaks and Experimental Determination of Thermodynamic Dead-Volume in Column Liquid Chromatography" *J. Chromatogr.*, **349**, 211-234.

7   Fóti, G., de Reyff, Ch. and sz. Kováts, E. (1990) "Method of Chromatographic Determination of Excess Adsorption from Binary Liquid Mixtures" *Langmuir*, **6**, 759-766.

8   Mc Cormick, R. M. and Karger, B. L. (1980) "Distribution Phenomena of Mobile Phase Components and Determination of Dead Volume in Reversed-Phase Liquid Chromatography" *Anal. Chem.*, **52**, 2249-2257.

9   Huber, J. F. K. and Gerritse, R. G. (1971) "Evaluation of Dynamic Gas-Chromatographic Methods for the Determination of Adsorption and Solution Isotherms" *J. Chromatogr.*, **58**, 137-158.

10   Valentin , P. and Guiochon, G. (1976) "Determination of Gas-Liquid and Gas-Solid Equilibrium Isotherms by Chromatography : I. Theory of the Step-and-Pulse Method, and, II. Apparatus, Specifications, and Results" *J. Chromatogr. Sci.*, **14**, 56-63 and 132-139.

11   Findenegg, G. H. and Köster, F. (1986) "A New Equation for the Retention of Solutes in Liquid-Solid Adsorption Chromatography with Mixed Mobile Phases", *J. Chem. Soc., Faraday Trans. I*, **82**, 2691-2705.

PRINCIPLES OF ADSORPTION AT SOLID SURFACES AND THEIR SIGNIFICANCE IN GAS/SOLID AND LIQUID/SOLID CHROMATOGRAPHY

G.H. FINDENEGG
I.N. Stranski-Institute of Physical and Theoretical Chemistry
Technical University Berlin
Straße des 17. Juni 112
D 1000 Berlin 12, Germany

ABSTRACT. The basic principles of gas adsorption and adsorption from liquid mixtures onto solid surfaces are outlined, and the analogies and differences between the two phenomena are explained. In both cases there exists a gradual transition from the low-temperature regime in which the adsorbate forms a well-defined adsorbed layer, to the situation at high temperatures and high bulk densities (or high bulk concentrations) when the adsorbed amount can be expressed only as a surface excess. Monolayer models for single and mixed gas adsorption (based on the Scaled-Particle theory) and for the adsorption from binary liquid mixtures (Parallel-Layer model) are presented. The significance of such models for gas/solid and liquid/solid adsorption chromatography is discussed. In particular, the retention of a small pulse of a given component at finite concentrations of this or other adsorbable components in the eluent (elution on a plateau) and the elution of solutes by a binary liquid eluent is considered in some detail.

1. Introduction

The physical adsorption of gases and the preferential adsorption from the liquid phase onto solid surfaces represent important retention mechanisms in chromatographic separation processes. The equilibrium theory of multicomponent isothermal adsorption columns was developed by Glueckauf [1] and generalized by Helfferich and Klein [2] and by Valentin and Guiochon [3]. When the chromatographic column is packed with an adsorbent and a small pulse of an adsorbable component is injected into the mobile phase, the retention of this pulse is closely related to the *slope* of the *adsorption isotherm* of that component at given concentrations of this and any other adsorbable components in the mobile phase. Therefore, it is of great importance to understand the factors which affect the shape of physisorption isotherms of single gases and gas mixtures, as well as the composite adsorption isotherms of adsorption from the liquid phase. In this article we will consider mainly adsorption on energetically homogeneous surfaces, although surface heterogeneity is known to have a pronounced influence on gas adsorption isotherms and, to a lesser extent, on adsorption from the liquid phase [4,5].

F. Dondi and G. Guiochon (eds.),
Theoretical Advancement in Chromatography and Related Separation Techniques, 227–260.
© 1992 Kluwer Academic Publishers.

An important question in adsorption chromatography concerns the nature of the so-called *adsorbed phase*. As we shall see, the structure of the boundary layer of the gas or liquid phase against a solid surface depends strongly on temperature and on the concentration of the adsorbable component. At low temperature and low concentrations in the bulk phase the *density profile* (or concentration profile) perpendicular to the surface generally exhibits a rather sharp step from the layer next to the surface to the bulk phase. In such a case the concept of an adsorbed phase is well justified. On the other hand, at high temperatures and high bulk concentration, the density profiles exhibit a more gradual decay from the surface into the bulk solution, hence the concept of an adsorbed phase becomes questionable. In such cases one has to go back to the most general definition of adsorption as a *surface excess amount*. Furthermore, in adsorption from liquid mixtures one is primarily interested in the *composition*, rather than the volume concentrations of the individual components, in the surface layer. The relevant quantity to account for such composition changes is the *reduced surface excess amount*. There has been some confusion in the literature about the significance of this quantity.

In Section 2 we first deal with some basic features of gas adsorption, and explain how the observed behaviour can be understood in terms of gas-solid and intermolecular interactions. We then consider in some detail adsorption isotherm equations for monolayer adsorbed films on homogeneous solid surfaces. On the basis of such an equation it is then shown how the chromatographic retention is related to the adsorption isotherms of single gases and two-component gas mixtures.

In Section 3 we turn to adsorption from the liquid phase and explain the definition and meaning of various surface excess quantities. Next we introduce the concept of a surface phase and show how the observed surface excess isotherms can be understood in terms of of this concept. Finally we present a quantitative model for the elution of solutes in liquid/solid chromatography with binary eluents which accounts for differences in the strength of adsorption of the two components and nonideality effects in the mobile phase.

2. Gas Adsorption

2.1 DIFFERENT TYPES OF BEHAVIOUR

2.1.1 *Henry's Law Limit.* Physical adsorption is due to the attractive van der Waals interaction of gas molecules with solids [6]. The gas-solid potential of interaction u_s exhibits a minimum close to the surface and tends to zero for large distances z (see Figure 1). Accordingly, gas molecules will accumulate near the surface; i.e. the local number density $c \equiv \delta N/\delta V$ in a volume element δV at distance z near the surface will be greater than the bulk density $c^g = c(z \to \infty)$. For an ideal gas of non-interacting molecules in contact with the surface, the *density profile* $c(z)$ is given by the Boltzmann law

$$c(z) = c^g \exp\{-u_s(z)/k_B T\} \tag{1}$$

where $c^g = p/k_B T$ is the bulk density of the gas at a pressure p and a temperature T, and

k_B is the Boltzmann constant. Note that the range of the density profile is determined by the range of the gas-solid interaction $u_s(z)$; accordingly, the width of the inhomogeneous boundary layer of the fluid amounts to only a few molecular diameters (except for special cases). Note also that $c(z)$ approaches c^g asymptotically for $z \to \infty$; therefore, it is not possible to distinguish in a strict sense between adsorbed and non-adsorbed molecules.

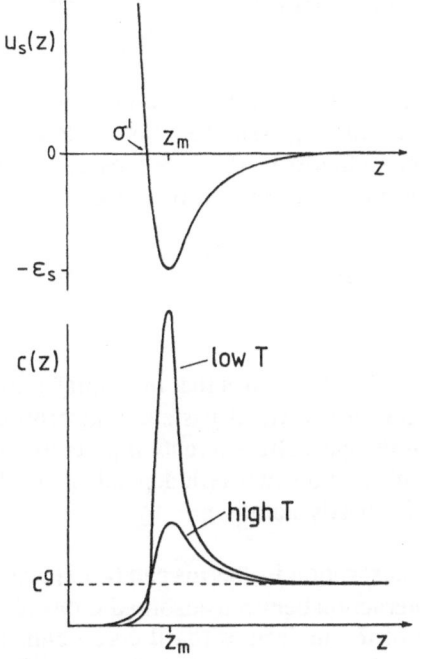

Figure 1. Gas-solid potential $u_s(z)$ with the parameters ϵ_s, σ' and z_m (upper graph); density profiles $c(z)$ of an ideal gas interacting with the surface by the potential $u_s(z)$; c^g is the bulk density of the gas (lower graph).

Instead, the amount adsorbed per unit surface area is defined in a general way as the *surface excess* Γ relative to a reference system in which the density of the gas remains constant and equal to c^g up to the solid surface at $z=0$

$$\Gamma = \int_0^\infty [c(z) - c^g] dz. \qquad (2)$$

The corresponding *operational definition* of the surface excess amount n^σ in a real macroscopic system containing a total amount n of gas in a volume V^g in contact with an adsorbent of surface area A_s is [7]

$$\Gamma \equiv n^\sigma / A_s = (n - c^g V^g)/A_s. \qquad (3)$$

The *areal surface excess* Γ (or *surface excess concentration*) and the corresponding bulk density c^g may be expressed in molecular units (SI unit m^{-2} and m^{-3}, respectively) as in eqn. (2), or in molar units (SI unit· mol m⁻² and mol m⁻³, respectively) as in eqn. (3). For simplicity, we will use the same symbols for these closely related definitions of Γ and c.

From eqs.(1) and (2) it follows for the adsorption of ideal gases that the surface excess Γ is a linear function of the gas density,

$$\Gamma = c^s \int_0^\infty [\exp\{-u_s(z)/k_B T\}-1]dz = c^s F(T) \qquad (4)$$

where the integral F(T) depends on the gas-solid potential and on temperature. For real gases, the ideal gas law p = $c^s k_B T$ holds as a limiting law for low pressures. Accordingly, eqn. (4) applies to any physisorption isotherm in the low-pressure limit (Henry's law limit) and the initial slope of physisorption isotherms at p = o is

$$K_i = \left(\frac{\partial \Gamma_i}{\partial p_i}\right)_{p_i=0} = F_i(T)/k_B T \qquad (5)$$

where the subscript i refers to the given adsorptive. Eqn. (5) implies that the initial slope of physisorption isotherms (and thus the net retention in analytical gas/solid adsorption chromatography; see Section 2.3) depends solely on the gas-solid interaction potential u_{si} and on temperature. In this limit, the adsorption of component i is independent of all other components, i.e. adsorption of gas mixtures is strictly additive.

2.1.2 *Monolayer and Multilayer Adsorption.* Linear adsorption isotherms can be expected only at very low surface concentrations, when the interaction between adsorbed molecules is negligible. Beyond this limit, the discrete nature of the molecules (hard core volume) and the attractive intermolecular energy lead to a diverse adsorption behaviour. The interaction between spherical molecules is given approximately by the Lennard-Jones (LJ) pair potential

$$u(r) = 4\varepsilon \left[\left(\frac{\sigma}{r}\right)^{12} - \left(\frac{\sigma}{r}\right)^6 \right] \qquad (6)$$

where r is the center-to-center distance, ϵ is the depth of the potential well and σ is that distance for which u=o (i.e. a measure of the molecular diameter). The term proportional to r^{-6} represents the attractive interaction, the r^{-12} term is an approximation for the sharply increasing short-range repulsion. The gas-solid interaction is given approximately by

$$u_s(z) = b\varepsilon_s \left[\left(\frac{\sigma'}{z}\right)^9 - \left(\frac{\sigma'}{z}\right)^3 \right] \qquad (7)$$

where b is a numerical factor and ϵ_s is the depth of the gas-solid potential well (see

Fig.1). Typically, $\epsilon_s/\epsilon \sim 10$ for simple systems (noble gas molecules interacting with the graphite surface).

At *low temperatures* (i.e. when $\epsilon_s/k_BT \gg 1$) adsorbed molecules will accumulate strongly near the minimum of the gas-solid potential well. In this case the exponential $\exp(-u_s/k_BT)$ is a sharply peaked function around the potential minimum ϵ_s at z_m, where $c(z_m) = c^g \exp(-\epsilon_s/k_BT) \gg c^g$ (see Figure 1). This fact has two important consequences: First, in the integral on the r.h.s. of eqn.(2) the term c^g can be neglected against $c(z)$ at distances around z_m; this means that the surface excess Γ may be considered as the total number of *adsorbed molecules* N^a (i.e., the number of molecules in a loosely defined *adsorption space* near the surface) per unit surface area A_s

$$\Gamma \sim N^a/A_s \qquad (for\ low\ T). \qquad (8)$$

N^a/A_s (or, in molar units, n^a/A_s) is called *absolute surface concentration* [7].

The second consequence of the sharply peaked nature of $c(z)$ at low T is that adsorption proceeds by a layer-by-layer mode as the pressure increases. At sufficiently low pressures, essentially all adsorbed molecules are confined to a *monolayer*. The behaviour of this adsorbed layer can be described by a suitable *monolayer equation of state* and the corresponding *monolayer adsorption isotherm*. A monolayer equation of state relates the *spreading pressure* Π of the adsorbed film to the surface concentration N^a/A_s or its reciprocal, the mean area per molecule $a = A_s/N^a$. Model equations of state and the corresponding adsorption isotherm equations are presented in Section 2.2.

Due to the long-range nature of the gas-solid attractive interaction (which decays as z^{-3}), physisorption is not limited to a single monolayer. After completion of the first layer molecules can form a second and in turn a third and further layers as the pressure increases. In this way a multilayer adsorbed film may be formed when the vapor pressure approaches the saturation pressure p_o of the liquid or solid adsorptive. The transition from monolayer to multilayer adsorption can be represented by the isotherm equation of Brunauer, Emmett and Teller (*BET equation*). A physically more appropriate model for thick adsorbed films at $p/p_o > 0.5$ is the isotherm equation by Frenkel, Halsey and Hill (*FHH equation*), which treats multilayer adsorption as a layer-by-layer condensation of the vapor in the external field of the adsorbing surface [6]. As the film thickness l increases, the gas-solid interaction $u_s(z)$ at the outer surface of the film (at $z=l$) becomes weaker and thus condensation occurs closer to the bulk saturation pressure p_o. As an example for this layer-by-layer adsorption behaviour, Figure 2 shows low-temperature adsorption isotherms of krypton on graphitized carbon [8]. Because the multilayer adsorbed film is essentially liquid-like, there is a rather sharp step in the density profile $c(z)$ at the outer surface of the film, where $c(z)$ changes from a liquid-like density at $z < l$ to the gas density c^g at $z > l$. Because of this distinct step in the density profile at $z=l$, the adsorbed amount can still be expressed as the absolute surface concentration of molecules per unit area of the adsorption space $V^a = lA_s$. It should be noted, however, that multilayer adsorption is not always observed. For example, at low temperatures the adsorption of benzene on graphitized carbon does not exceed one monolayer even at pressures close to p_o.

232

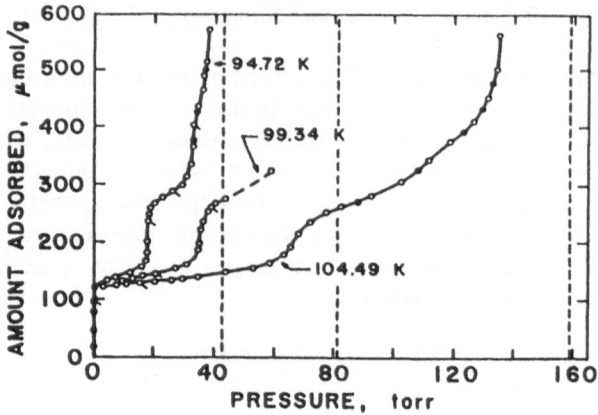

Figure 2. Low-temperature adsorption isotherms of krypton on graphitized thermal carbon black, illustrating the layer-by-layer mode of adsorption on atomically flat surfaces; the dashed vertical lines mark te saturation pressure p_o at the respective temperatures. At the experimental temperatures of this study, the first adsorbed layer is completed at pressures below about 1 mbar. (Fig.1 of ref. [8], reproduced with permission of the American Chemical Society).

2.1.3 *Supercritical Fluid Adsorption.* The layer-by-layer mode of adsorption discussed above applies only at low temperatures, when $\epsilon_s/k_B T$ is large. At high temperatures (say, when $\epsilon_s/k_B T < 5$) the function $\exp(-u_s/k_B T)$ is no longer sharply peaked and thus the adsorption is much weaker than at low temperatures. On the other hand, as the saturation pressure p_o of the adsorptive increases with temperature, the adsorption isotherms extend over a wider pressure range and, accordingly, to significantly higher densities c^s than at low temperatures. In particular, above the critical temperature T_c, the density of the fluid increases in a monotonic manner with the pressure from the dilute gas state up to nearly liquid-like densities.

The density profile $c(z)$ of a supercritical fluid against a flat surface exhibits an oscillatory behaviour indicating a layered structure of the fluid close to the surface but, except for these details, the density profile approaches the bulk density c^s in an asymptotic way [9]. Accordingly, in this supercritical region of the fluid there exists no longer a natural distinction between an "adsorption space" and the bulk fluid, and one has to go back to the most general definition of the adsorbed amount as a surface excess amount Γ, as in eqn. (2). Surface excess isotherms for supercritical fluids ($T > T_c$) on nonporous adsorbents generally exhibit a maximum, at a pressure greater than the critical pressure p_c and a corresponding bulk density somewhat below the critical density of the fluid, as shown in Figure 3 for sulfur hexafluoride (SF_6) on graphitized carbon [10]. The marked difference between the low-temperature and high-temperature adsorption behaviour of gases is seen by comparing Figure 2 and Figure 3.

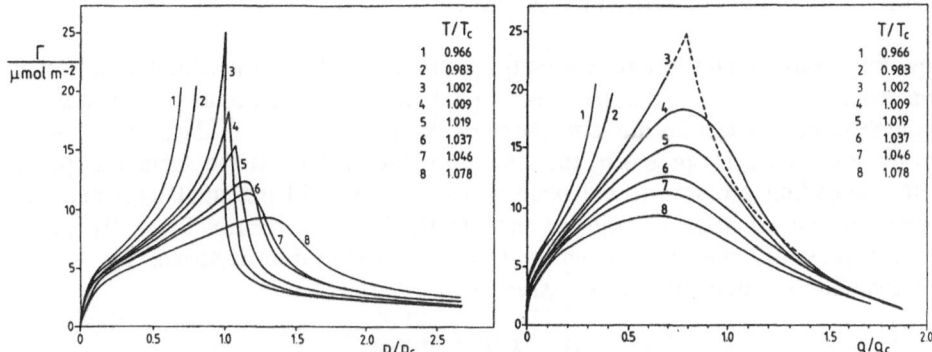

Figure 3. High-temperature adsorption isotherms of SF_6 on graphitized carbon black; the experimental surface excess concentration Γ is plotted as a function of the reduced pressure p/p_c (left) and as a function of the reduced density ρ/ρ_c (right). The figures show the transition in the adsorption behaviour from subcritical temperatures ($T < T_c$; isotherms 1 and 2) to supercritical temperatures ($T > T_c$; isotherms 3-8) in a temperature range from 0.96 T_c to 1.07 T_c. The critical parameters of SF_6 are $T_c = 318.7$ K; $p_c = 37.5$ bar; $\rho_c = 0.74$ kg dm^{-3}. Note that the crossing of the isotherms exhibited in the plot of Γ against pressure disappears when Γ is plotted against the density. The dashed section of the near-critical isotherm 3 ($T/T_c = 1.002$) is subject to large systematic errors on the density axis.

The observed behaviour of surface excess isotherms can be understood in a qualitative way as follows [11]:

(a) At sufficiently low bulk densities (Henry's law limit) the adsorption behaviour of ideal gases is approached, i.e., the surface excess Γ increases proportional to the gas density (see Section 2.1.1).

(b) At higher bulk densities, when the local density near the surface is already high, repulsive interactions between the molecules in the boundary layer become appreciable and will outweigh the attractive gas-surface interaction. Accordingly, the compressibility of the fluid near the surface will become less than the compressibility of the bulk fluid, hence the surface excess reaches a maximum and then decreases as the bulk density is increased further.

Of special interest is the transition from the low-temperature adsorption behaviour with multilayer formation on approaching the saturation pressure (Section 2.1.2), to the high-temperature behaviour outlined above. This transition takes place in a relatively narrow temperature range (typically 10 K) above T_c. The breakdown of multilayer adsorption in this temperature range is evidenced by a pronounced decrease of the maximum of the surface excess isotherms between T_c and $T_c + 10$ K, as shown in Figure 3 [10]. This change in the structure of the fluid boundary layer may be of significance for the retention of solutes in supercritical-fluid chromatography (SFC).

2.2 MONOLAYER ISOTHERM EQUATIONS

At low temperatures and low pressures the adsorbed molecules accumulate in a narrow region around the plane at $z=z_m$ (see Section 2.1.2). In the absence of specific adsorption sites the adsorbed molecules move freely in the xy-plane parallel to the surface and cause a two-dimensional (2D) pressure, the *spreading pressure* Π. If the surface is perfectly flat, the spreading pressure is independent of the gas-solid potential $u_s(z)$ and can be expressed by an *equation of state* of the 2D fluid, i.e. a relation $\Pi = \Pi(\Gamma,T)$. The spreading pressure can be calculated from the adsorption isotherm by a general thermodynamic relation, the *Gibbs equation*

$$\Pi = k_B T \int_0^P \Gamma d\ln p \qquad (9)$$

where Γ is expressed in molecular units. Eqn. (9) applies to ideal gas phase behaviour. In the case of nonideal gases the pressure p has to be replaced by the fugacity.

The Gibbs equation implies a one-to-one relationship between a given 2D equation of state and a corresponding adsorption isotherm equation. For example, the adsorption isotherm of an ideal gas is $\Gamma = Kp$ (see Section 2.1.1); inserting this relation into eqn. (9) yields

$$\Pi = k_B T \Gamma \quad or \quad \Pi A_s = N^a k_B T \qquad (10)$$

which is the equation of state of the *2D Ideal Gas*. Below we will show that useful adsorption isotherm equations can be obtained when one starts from suitable 2D equations of state of nonideal gases.

2.2.1 *Pure Gases.* An important family of equations of state is based on the generalized *van der Waals* model, which assumes that the free energy of a fluid is given by a sum of two terms, one arising from the hard-core repulsive interactions, the other from the attractive interactions between the molecules. The thermodynamic properties of a 2D fluid of hard disks can be obtained on the basis of the *Scaled Particle* (SP) *theory*; this theory yields an expression for the 2D pressure Π in terms of the *packing parameter* $\eta = a\Gamma$, where $\Gamma = N^a/A_s$ is the number of molecules per unit area and $a = \pi\sigma^2/4$ is the area covered by a molecule of diameter σ. Using the *Carnahan-Starling* expression for the pressure of the hard-disk system, and the mean-field expression for the (negative) pressure resulting from the attractive interactions between the molecules, one obtains the *Scaled Particle* equation of state

$$\Pi/\Gamma k_B T = \frac{1}{(1-\eta)^2} - \frac{\alpha}{ak_B T}\eta \qquad (11)$$

where $\alpha = c_o a\epsilon$ is the attractive parameter of the 2D van der Waals equation, ϵ is the depth of the Lennard-Jones pair potential (eqn.6) and c_o is a numerical constant ($c_o =$

2.40). Eqn. (11) reduces to the 2D ideal gas equation of state (eqn. 10) when η (and thus the surface concentration Γ) approaches zero. At low temperatures (below a 2D critical temperature T_{2c}) this model yields a two-dimensional condensation due to the attractive lateral interactions; the critical parameters are $T_{2c} = 0.41\epsilon/k_B$ and $\eta_{2c} = 0.215$. The maximum possible value of the packing parameter is $\eta_{max} = 0.906$, corresponding to a hexagonal close packing of the disks. For an irregular close packing of hard disks, $\eta \approx 0.65$. The adsorption isotherm corresponding to this 2D equation of state is

$$Kp = \frac{\Gamma}{1-\eta} \exp\left(\frac{\eta(3-2\eta)}{(1-\eta)^2} - \frac{2\alpha}{ak_BT}\eta\right). \tag{12}$$

This isotherm equation is explicit in the gas pressure p and implicit in the surface concentration variable $\eta = \Gamma a$. The constant K represents the initial slope of the isotherm at $\eta = 0$ (Henry's Law constant); in addition to K, the isotherm equation contains two parameters characterizing the lateral interactions in the adsorbed layer, viz. the cross-sectional area of the molecules $a = \pi\sigma^2/4$ and the van der Waals constant $\alpha = c_o a\epsilon$, which is proportional to the strength of the lateral interactions.

Figure 4 shows experimental adsorption isotherms for cyclohexane on graphitized carbon black, and a fit of these data by eqn. (12). The differential molar enthalpy of adsorption in the limit of zero coverage, $\Delta_a H°$, or its negative, the limiting isosteric heat of adsorption $q^0_{st} = -\Delta_a H°$, can be obtained from the temperature dependence of the Henry's Law constant

$$\partial lnK/\partial(1/T) = q^0_{st}/R . \tag{13}$$

Table 1 summarizes the LJ pair potential parameters σ and ϵ of three C_6-hydrocarbons as derived from their adsorption isotherms on graphite, using eqn. (12), together with the corresponding values as derived from properties of the bulk liquids [12]. For hexane and cyclohexane the two sets of parameters are in reasonably good agreement; for benzene, however, the value of ϵ/k obtained from the monolayer adsorption data is significantly smaller than that derived from liquid-state properties, indicating that the interaction between the benzene molecules in the adsorbed monolayer is weaker than in the bulk liquid state.

Table 1. Comparison of the pair potential parameters σ and ϵ of three hydrocarbons as derived from adsorption data on g.c.b. (ads) and from bulk liquid data (liq), on the basis of the Scaled-Particle theory.

Substance	σ, A		(ϵ/k), K	
	ads	liq	ads	liq
n-Hexane	6.0	5.95	470	500-520
Cyclohexane	5.3	5.65	650	570-590
Benzene	4.7	5.25	260	500-530

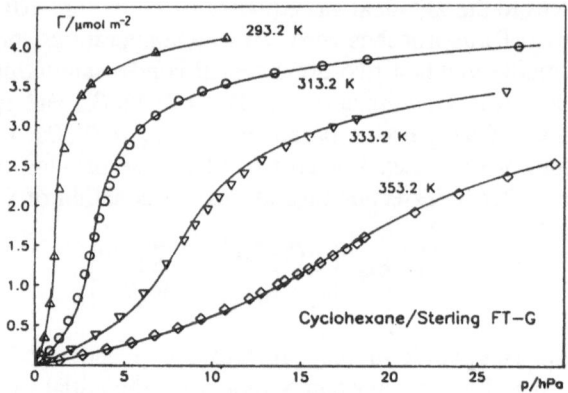

Figure 4. Adsorption isotherms of cyclohexane on graphitized thermal carbon black, and a fit of the experimental data by the Scaled-Particle isotherm equation (12); the resulting best-fit parameters are as follows:

Temperature	20$°$C	40$°$C	60$°$C	80$°$C
K(μmol m^{-2} bar^{-1}	428	171	82	43
σ(nm)	0.537	0.532	0.529	0.523
ϵ/k$_B$ (Kelvin)	668	659	643	631
q$_{st}^{°}$ (kJ mol^{-1})		31.7		

(Fig.1 of ref. [12], reproduced with permission of the Engineering Foundation).

The above example shows that the isotherm equation (12) gives a very satisfactory fit of monolayer adsorption data on homogeneous surfaces like graphite, on which the adsorbed molecules move freely in the xy-plane parallel to the surface. In the other extreme case, when the adsorbed molecules are localized on specific surface sites, the surface concentration is expressed by the *surface coverage* $\theta = \Gamma/\Gamma_m$, where Γ_m represents the number of surface sites per unit area. A well-known isotherm equation for localized monolayer adsorption with lateral interactions between adsorbed molecules is the *Fowler-Guggenheim equation* [6]. It has a formal resemblence with eqn. (12) when η is replaced by θ and the hard-disk term in the exponential of eqn. (12) is abandoned. If the term for ther lateral interactions is also removed (e.g., by setting $\alpha = 0$) the exponential in eqn. (12) disappears and eqn. (12) reduces to the *Langmuir equation*; this isotherm equation is most popular as it can be inverted to an equation which is explicit in the surface concentration Γ or surface coverage θ.

2.2.2 *Gas Mixtures.* The SP equation of state for 2D fluids (eqn. 11) can be adapted for two- and multicomponent mixtures by chosing appropriate combining rules for the parameters a and α. For a binary gas mixture (components 1 and 2) one assumes

$$\alpha = x_1^2 \alpha_1 + x_1 x_2 \alpha_{12} + x_2^2 \alpha_2$$

$$a = x_1 a_1 + x_2 a_2$$

(14)

where $a_i = \pi \sigma_i^2 / 4$ and $\alpha_i = c_o \epsilon_i a_i$ are related to the LJ parameters σ_i and ϵ_i of the pure components (i = 1,2), and α_{12} accounts for the lateral interaction of a pair of unlike molecules; x_i is the mole fraction of component i in the adsorbed layer, i.e., $x_i = \Gamma_i / \Gamma$ where $\Gamma = \Gamma_1 + \Gamma_2$ is the overall surface concentration. The dependence of the surface concentration of component i on the respective partial pressure p_i is given implicitly by the *partial isotherm* or *individual isotherm* of component i

$$K_i p_i = \frac{\Gamma_i}{1-\eta} \exp\left(\frac{\eta(1-\eta)+\eta(2-\eta)a_i/a}{(1-\eta)^2} - \frac{2\eta}{ak_BT}(x_i\alpha_i + x_j\alpha_{ij})\right). \quad (15)$$

Eqn. (15) in combination with eqn. (14) can be used to predict the partial adsorption isotherms of a binary gas mixture if the single-gas adsorption parameters K_i, a_i and α_i of the two constituents and the parameter α_{ij} for the interaction of a pair of unlike molecules are all known. For non-specific interactions between the two components, α_{ij} is expected to be approximately equal to the geometric mean of α_i and α_j by analogy with bulk fluid mixtures (Berthelot rule). Thus

$$\alpha_{ij} = \xi_{ij}(\alpha_i \alpha_j)^{1/2} \quad (16)$$

where ξ_{ij} is expected to be somewhat less than 1 for dispersion interactions between spherical or nearly spherical molecules. A test of Berthelots rule for adsorption systems is an important goal because if $\xi_{ij} = 1$ we can predict adsorption isotherms of gas mixtures from single-gas adsorption data without any additional information. Note that for ideal mixtures $\xi_{ij} = (\alpha_i + \alpha_j)/2(\alpha_i\alpha_j)^{1/2}$, which is greater than 1 if α_i and α_j are different.

2.3   RETENTION IN GAS/SOLID CHROMATOGRAPHY

In this section we discuss briefly how the adsorption behaviour of gases and gas mixtures is related to the retention in gas/solid elution chromatography. Two situations will be considered: first, the retention of a pulse of component i at finite surface concentrations of this component on the adsorbent surface; and second, the retention of a pulse of component i at finite surface concentrations of a second adsorbable component. These situations refer to the chromatographic *Elution on a Plateau* (or *Step and Pulse*) technique, in which a column packed with the adsorbent is equilibrated with a carrier gas stream containing a steady concentration of the adsorbable vapor. A small pulse of this or a different vapor is then injected and its retention volume is measured.

2.3.1 *Single-Component Retention*. If there is only one adsorbable substance (say, component 1) the retention of a pulse of this component is related to the slope of its adsorption isotherm, $d\Gamma_1/dp_1$, at a partial pressure $p_1 = y_1p$ of the vapor by [3,13,14]

$$R(y_1) = \int_1^{P_I} \left(\frac{d\Gamma_1}{dp_1}\right) \frac{3P^2 dP}{P_I^3 - 1} \tag{17}$$

where $P = p(l)/p°$ and $P_1 = p^i/p°$; $p(l)$ represents the total pressure at a position l along the column $(0 < l < L)$, $p^i = p(0)$ is the pressure at column inlet and $p° = p(L)$ the pressure at column outlet. For a given mole fraction y_1 of the adsorbable vapour in the carrier gas, the retention of the pulse is related to its net retention time t_1-t_M and other experimental quantities by

$$R(y_1) = (t_1-t_M) \frac{jv°}{(1-y_1)RTa_s m} . \tag{18}$$

Here a_s and m are the specific surface area and the mass of the adsorbent in the column, $v°$ is the volume flow rate of the gas phase measured at column outlet at the column temperature T, and j is the well-known pressure correction factor

$$j = (3/2)(P_I^2-1)/(P_I^3-1). \tag{19}$$

The implications of eqn. (17) will be discussed together with the situation for two adsorbable components in the next paragraph.

2.3.2 *Two-Component Retention*. If a pulse of a second adsorbable vapor (component 2) is injected on a plateau of component 1, two signals with retention times t' and t'' appear in the chromatogram. For sufficiently small pulses, one of these signals (t') has the same retention time t_1 as a pulse of component 1 and can be attributed to the disturbance of the adsorption equilibrium of this component. Correspondingly, the retention R'(y_1) obtained from eqn. (18) by setting t' = t_1, is related to the slope of the adsorption isotherm of component 1 by eqn. (17), so long as the partial pressure of component 2 in the pulse is negligibly small. The other signal (t'') is also dependent on the concentration of component 1 (plateau height y_1) and approaches the analytical retention time t_2 of pure component 2 as $y_1 \to 0$. Correspondingly, the retention R'' is related to the initial slope ($p_2 \to 0$) of the partial isotherm of adsorption of component 2 at finite partial pressures p_1 [12,14]

$$R''(y_1) = \int_1^{P_I} \left(\frac{\partial \Gamma_2}{\partial p_2}\right)_{p_2 \to 0, p_1} \frac{3P^2 dP}{P_I^3 - 1} \tag{20}$$

where R'' is related to the net retention time t''-t_M of the pulse and other experimental parameters by

$$R''(y_1) = (t''-t_M) \frac{jv^o}{RTa_sm} . \tag{21}$$

For infinite dilution of the preadsorbed component 1 ($y_1 \to 0$) and sufficiently small pulses of component 2, the retention R' approaches K_1, the Henry's Law constant of pure component 1, and R'' approaches K_2, the Henry's Law constant of pure component 2. This holds true so long as the maximum partial pressure $p_i = y_i p$ of vapor i in the pulse at column inlet ($p = p^i$) is still in the initial linear region of the adsorption isotherm, so that in eqn. (20) $\partial\Gamma_i/\partial p_i$ becomes independent of P. In this case one has

$$\lim_{y_1 \to 0} R'' = \left(\frac{\partial\Gamma_2}{\partial p_2}\right)_{p_2 \to 0, p_1 = 0} = K_2 . \tag{22}$$

For finite concentrations of the preadsorbed component ($y_1 > 0$) the pressure gradient along the column has to be taken into account. If the overall pressure drop p^i-p^o is sufficiently small, the slopes $\partial\Gamma_i/\partial p_i$ in eqs. (17) and (20) can be calculated for a mean column pressure P = p/p^o

$$\overline{P} = (3/4)(P_I^4-1)/P_I^3-1) \tag{23}$$

and thus for a mean partial pressure $p_1 = y_1 p$ of the preadsorbed vapor:

$$R'(\overline{p_1}) = (\partial\Gamma_1/\partial p_1)_{\overline{p_1}} ; \quad R''(\overline{p_1}) = (\partial\Gamma_2/\partial p_2)_{\overline{p_1},p_2 \to 0} . \tag{24}$$

On the basis of eqn. (24) experimental retention data R' and R'' obtained for a series of partial pressures $p_1 = y_1 p^o P$ can be correlated by model isotherm equations for single gas adsorption (R' = R_1, given by eqn. 17) or for mixed gas adsorption (R'', given by eqn. 20). From the Scaled Particle isotherm for single gas adsorption, eqn. (12), one finds for pure component i [12]

$$\left(R(\overline{p_i})\right)^{-1} = \frac{1}{K_i}\left(\frac{1+\eta}{(1-\eta)^4} -w_i\frac{\eta}{1-\eta}\right) \exp\left(\frac{\eta(3-2\eta)}{(1-\eta)^2} -w_i\eta\right) . \tag{25}$$

240

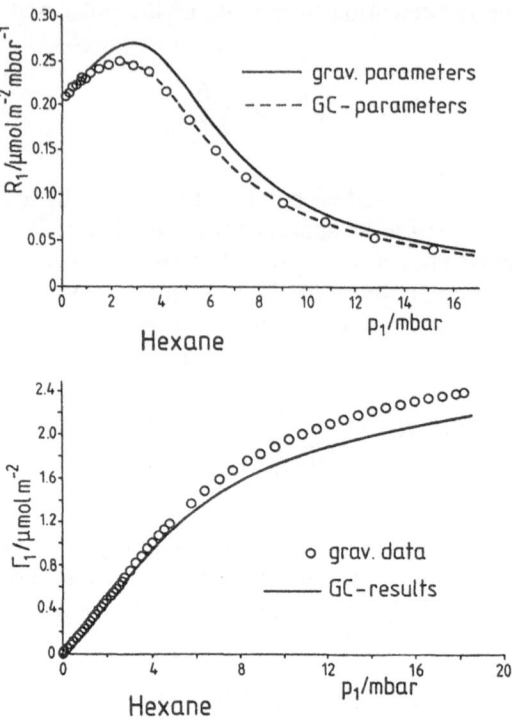

Hexane

Hexane

Figure 5. Determination of adsorption isotherms by the chromatographic elution-on-a plateau technique (hexane on Sterling FT-G graphitized carbon at 70°C). Upper graph: Chromatographic retention $R_1(p_1)$ (circles) and a fit of these data by the Scaled Particle isotherm (eqn.25) (dotted); the full curve shows the function $R_1(p_1)$ as derived from gravimetric adsorption data. Lower graph: Adsorption isotherm resulting from the chromatographic study (full curve), and the corresponding gravimetric adsorption data (circles). (Fig.3 of ref [12], reproduced with permission of the Engineering Foundation).

where $\eta = a_i\Gamma$ and $w_i = c_o\epsilon_i/k_B T$. The three parameters K_i, a_i and w_i (or ϵ_i/k_B) can thus be obtained from the experimental retention data $R_i(p_i)$ by a least-square fitting procedure. Such a precedure yields an excellent representation of experimental retention data for single adsorbable vapors on graphitized carbon, as shown in Figure 5 for hexane. However, the parameters ϵ_i/k_B and σ_i obtained by this chromatographic technique differ from those obtained by direct (gravimetric) determination of the adsorption isotherm. This deviation reflects some systematic error of the dynamic method, most likely due to incomplete equilibration of the adsorbate with the transient concentration pulse in the mobile phase. The *frontal analysis* technique which does not require instantaneous equilibration is therefore preferred for the determination of single gas adsorption isotherms.

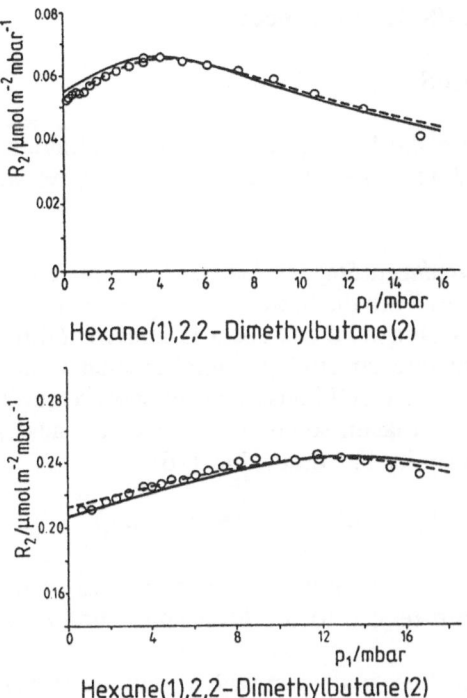

Figure 6. Chromatographic study of mixed gas adsorption: Retention $R''(p_1)$ of a pulse of component 2 plotted vs. the mean partial pressure of component 1 (p_1). The curves represent fits of the retention data by an equation analogous to eqn. (25): parameters K_2, a_1, a_2 all taken from single-gas adsorption data (full curve), and using an adjusted value of K_2 (dashed). Upper graph: pulse of 2,2-dimethylbutane on plateau of n-hexane; lower graph: vice versa. Adsorbent: Sterling FT-G, $70\,^\circ C$. (Fig.4 of ref. [12], reproduced with permission of the Engineering Foundation).

The retention R'' of a small pulse of a second component on preadsorbed component 1 exhibitis a similar dependence on mole fraction y_1 as the retention R' of a pulse of component 1 itself. Typical results for a pair of C_6 hydrocarbons on graphitized carbon, with either of the two alkanes taken as the preadsorbed component 1 and the other injected as a pulse (component 2), are shown in Figure 6. The maximum in $R''(y_1)$ corresponds to a maximum of the initial slope of the partial isotherm $\Gamma_2(p_2)$ at some finite precoverage of the surface by molecules of component 1, as a consequence of attractive lateral interactions between molecules of component 2 and preadsorbed molecules of component 1. The corresponding interaction parameter α_{12} can be obtained from a quantitative analysis of these data on the basis of the Scaled Particle theory for binary adsorbates [12].

3.   Adsorption at Liquid/Solid Interfaces

3.1   GENERAL FEATURES

In this section we introduce the basic phenomena and definitions of adsorption from solution at the liquid/solid interface and stress the analogies and differences to the adsorption of gases [15,16].

3.1.1 Surface Excess and Reduced Surface Excess Amounts. The concept of adsorption from solution by a solid stems from the basic observation that the composition of a liquid mixture after equilibration with a solid adsorbent is somewhat different from its original composition. If a liquid mixture containing a total amount n^o at a mole fraction x_i^o is equilibrated with a mass m of the solid adsorbent of specific surface area a_s, and if the final mole fraction in the supernatant solution is x_i^l, then the adsorption of component i may be expressed by the quantity $\Gamma_i^{(n)}$ defined as [16]

$$\Gamma_i^{(n)} = n_i^{\sigma(n)}/A_s = n^0\Delta x_i^l/(ma_s) \qquad (26)$$

where $\Delta x_i^l = x_i^o - x_i^l$ and $A_s = ma_s$ is the total surface area of the solid. The question arises how $\Gamma_i^{(n)}$ and $n_2^{\sigma(n)}$ are related to the local concentration of component i in the interfacial region.

Consider a binary liquid mixture in contact with a plane, inert surface and let molecules of component 2 have a stronger attractive interaction with the solid than those of component 1. This preference will cause an accumulation of component 2 and a corresponding depletion of component 1 in the boundary layer of the liquid in thermodynamic equilibrium. This situation can be represented by the profiles of the local number density c_i of the two components i as a function of the distance z from the surface. As in the case of gas adsorption the *density profiles* or *concentration profiles* $c_i(z)$ can be expressed in *molecular* units or in *molar* units. Figure 7 shows schematic concentration profiles for the two components of a binary mixture of molecules of similar sizes, from which component 2 is preferentially adsorbed. Due to space filling requirements in dense liquids the concentration profiles of the two components are not independent of each other (as they were in the Henry's law limit of gas adsorption). Therefore, it is convenient to replace the two concentration profiles by two related functions, viz. the *overall concentration profile* c(z) and the *composition* (mole fraction) *profile* $x_2(z)$:

$$c(z) = c_1(z)+c_2(z)$$
$$x_2(z) = c_2(z)/c(z). \qquad (27)$$

These profiles are also shown in Figure 7. As in the case of pure fluids, the density profiles c(z) and $c_i(z)$ exhibit an oscillatory behaviour due to the layering of the molecules near the surface.

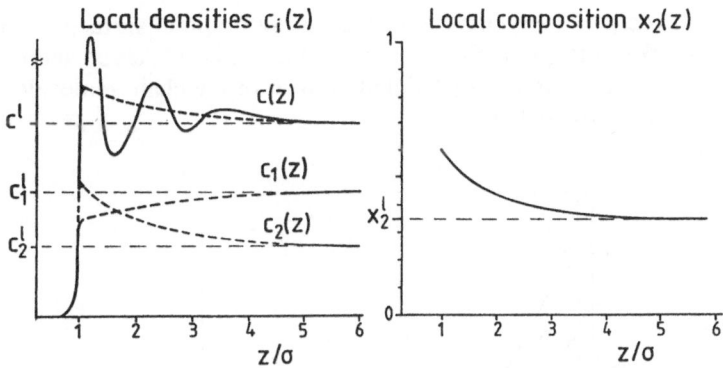

Figure 7. Density profiles $c_i(z)$ of the two components and the overall density profile $c(z)$ for a mixture of molecules of similar size (left), and the corresponding composition profile $x_2(z)$ (right) as a function of the reduced distance z/σ from the surface where σ is the molecular diameter. The density profiles exhibit maxima and minima due to layering effects of the molecules close to the wall; the dashed curves are schematic profiles without these details.

The local concentration and composition profiles introduced in eqn. (27) are related to the following integral *surface excess concentrations*:

$$\Gamma_i = \int_0^\infty [c_i(z) - c_i^l] dz$$

$$\Gamma_i^{(n)} = \int_0^\infty [x_i(z) - x_i^l] c(z) dz \tag{28}$$

where c_i^l and x_i^l represent the respective concentration and mole fraction of component i in the bulk liquid mixture. The surface excess Γ_i defined in eqn.(28) is entirely equivalent to the definition of Γ in the context of gas adsorption (eqn.2), hence its definition in terms of measurable quantities is (cf. eqn. 3)

$$\Gamma_i \equiv n_i^\sigma / A_s = (n_i - c_i^l V^l) / A_s \tag{29}$$

where n_i is the total amount of component i in the volume V^l available to the liquid phase; V^l is taken as the difference between the measurable total volume of liquid + adsorbent and the dead space volume of the solid adsorbent, $V^l = V - V^s$; usually, V^s is determined by helium displacement measurements.

For pure liquids, the surface excess Γ_i at the liquid/solid interface can be measured on the basis of eqn. (29). A positive Γ_i corresponds to a somewhat higher density of the liquid in the boundary layer as compared with the bulk liquid. As an example, Figure 8 shows the surface excess of long-chain n-alkanes at the graphite surface at temperatures near the respective freezing point (T_f) of the alkanes. The observed increase of Γ at temperatures near T_f is attributed to a partial alignment of the chain molecules parallel to the surface (prefreezing effect) [17].

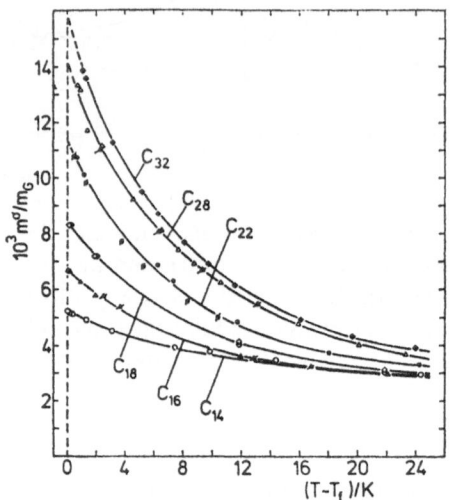

Figure 8. Surface excess amount of pure liquid alkanes at the liquid/graphite interface as a function of the temperature difference $T-T_f$, where T_f is the freezing point of the given alkane; the surface excess amount is expressed as an excess mass (m^σ) per unit mass of adsorbent (m). (Fig.3 of ref.[17], reproduced with permission of Academic Press).

Similar effects will also exist in the boundary layer of liquid mixtures. In this case, however, one is interested mainly in changes of chemical composition (rather than density changes) in the interfacial layer. The relevant quantity accounting for such composition changes is the *reduced surface excess* $\Gamma_i^{(n)}$; it represents the difference between the amount of a given component i actually present in the system (n_i) and that in a reference system containing the same total amount of liquid mixture (n^0) in which the composition of the liquid phase (mole fraction x) is constant throughout the liquid phase up to the surface:

$$\Gamma_i^{(n)} \equiv n_i^{\sigma(n)}/A_s = (n_i - x_i^l n^0)/A_s. \tag{30}$$

The corresponding definition of $\Gamma_i^{(n)}$ in terms of the microscopic profiles of the local composition, $x_i(z)$, and overall density, $c(z)$, is given in eqn. (28); this reduced surface excess can be measured easily by determining changes of mole fraction of the bulk mixture before and after equilibration with the solid, as explained above (see eqn.26).

$\Gamma_i^{(n)}$ is called *reduced* surface excess because it represents a linear combination of the individual surface excess amounts Γ_i of the mixture:

$$\Gamma_i^{(n)} = \Gamma_i - x_i^l \sum_i \Gamma_i \tag{31}$$

where the sum is taken over all components of the liquid mixture. The fundamental difference between Γ_i and the reduced surface excess $\Gamma_i^{(n)}$ becomes evident when we consider pure fluids: Γ_i will generally have some (small) positive or negative value (depending on the strength of the interaction with the surface), but $\Gamma_i^{(n)}$ is zero by definition for pure liquids (see eqn. 31). Accordingly, isotherms of $\Gamma_i^{(n)}$ for binary liquid mixtures necessarily exhibit a maximum or minimum (or both) somewhere along the composition axis x_2^l, as shown in Figure 9. Furthermore, for any bulk composition we have from eqn. (31)

$$\sum_i \Gamma_i^{(n)} = 0 \tag{32}$$

or, for binary mixtures, $\Gamma_2^{(n)} = - \Gamma_1^{(n)}$.

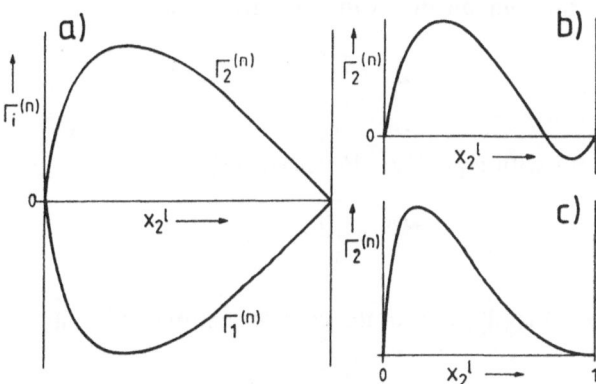

Figure 9. Isotherms of the reduced surface excess concentration $\Gamma_2^{(n)}$ of binary liquid mixtures as a function of the mole fraction x_2^l (schematic): (a) simple behaviour; (b) surface azeotrope; (c) complex behaviour (see also Figure 11).

3.1.2 *Volume additivity*. An important approximation often made in analysing the adsorption at liquid/solid interfaces is the assumption that the volume occupied by a molecule is independent of its surroundings. In this case, the molecular volumes are strictly additive and thus the molar volume of a binary liquid mixture is

$$v_m = \frac{1}{c^l} = x_1 v_1 + x_2 v_2 \tag{33}$$

where $v_i = 1/c_i$ is the molar volume of (pure) component i. For the interfacial region this implies that the local concentration $c(z)$ at any given local composition $x(z)$ is equal to $c^l(x^l)$, the concentration of the bulk mixture at the corresponding composition $x^l = x(z)$:

$$c(z) = c[x(z)] = c^l(x^l) . \tag{34}$$

The equations (33) and (34) are important consequences of the *volume additivity approximation* which is particulary useful when dealing with the adsorption from polymer solutions, i.e., when the molar volumes of the two components are greatly different. In this case it is convenient to introduce the local volume fractions $\phi_i(z) = (v_i/v_m)x_i(z)$, and to express the adsorption in terms of the *volume-reduced surface excess* $\Gamma_i^{(v)}$. By inserting eqs. (33) and (34) in (28) one obtains

$$\Gamma_i^{(v)} = \frac{1}{v_i} \int_0^\infty [\phi_i(z) - \phi_i^l] dz \tag{35}$$

which is often used to compute the surface excess in lattice models of polymer adsorption [18]. Summation over all components of the mixture yields

$$\sum_i v_i \Gamma_i^{(v)} = 0 \tag{36}$$

which is to be compared with eqn. (32). Furthermore,

$$\Gamma_i^{(v)} = \frac{v_i}{v_m} \Gamma_i^{(n)} . \tag{37}$$

Note that eqs. (35) - (37) apply only in the case of volume additivity.

3.2   SURFACE PHASE MODELS

3.2.1 *Concept of a Surface Phase.*

The most commonly employed model for correlating adsorption from solution data approximates the composition profile $x_2(z)$ by a step profile, with a uniform composition x_2^a in the surface layer ($0 < z < L$) and the composition of the bulk mixture (x_2^l) at $z > L$ (see Figure 10). The layer near the surface, which is called the *surface phase* or *adsorbed phase* (superscript a) contains an amount n^a which is proportional to the volume of the surface phase $V^a = LA_s$. In terms of this model, the areal reduced surface excess amount $\Gamma_i^{(n)}$ is given by

$$\Gamma_i^{(n)} = (x_i^a - x_i^l)n^a/A_s = (n_i^a - x_i^l n^a)/A_s . \tag{38}$$

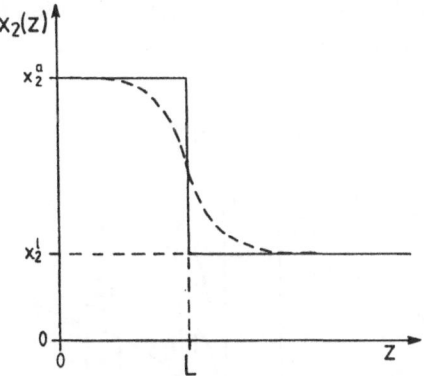

Figure 10. Surface phase model. The real mole fraction profile $x_2(z)$ (dashed curve) is replaced by a step function (full lines). L is the nominal thickness of the surface phase.

In a qualitative way, this simple model accounts for all features of experimental surface excess isotherms in dilute and concentrated solutions:

(a) In dilute solutions of the preferentially adsorbed component 2 ($x_2^l \ll 1$) one has $n_2^a \gg x_2^l n^a$ and thus $\Gamma_2^{(n)}$ is (nearly) equal to n_2^a/A_s, the amount of component 2 in the surface phase per unit surface area. In this regime, adsorption from solution has many similarities with gas adsorption (see Section 3.3.1).

(b) At higher bulk concentration x_2^l, the term $x_2^l n^a$ becomes appreciable and thus $\Gamma_2^{(n)}$ is less n_2^a/A_s. Furthermore, when x_2^a is already high, it becomes increasingly difficult to accommodate more molecules of component 2 in the surface phase so that x_2^a will increase less than the bulk mole fraction x_2^l; hence $\Gamma_2^{(n)}$ passes through a maximum in this region.

(c) At even higher bulk concentrations x_2^l, when the surface phase consists almost entirely of component 2 ($x_2^a \rightarrow 1$), $\Gamma_2^{(n)}$ becomes a nearly linear decreasing function of x_2^l, with a (limiting) slope $-n^a/A_s$ (see Fig.9a and Fig.11a).

(d) In some cases $\Gamma_2^{(n)}$ is positive in the region of low mole fractions x_2^l and negative at high x_2^l, as shown in Fig.9b. At the point at which $\Gamma_2^{(n)}$ passes through zero, the compositions of the surface and bulk phase are equal ($x_2^a = x_2^l$); such a point is called *surface azeotrope*. It is commonly observed when the surface consists of two different types of sites, which preferentially adsorb molecules of component 1 and 2, respectively.

To be more specific, let us consider a mixture of two components with different cross-sectional molar areas, a_1 and a_2, but equal molecular width in the direction perpendicular to the surface. We may think, for example, of a mixture of n-alkanes of different chain length adsorbed in such a way that the major axis of all molecules is oriented parallel to the surface. If the surface phase has a thickness of t such layers of molecules, then the condition of volume additivity requires that

$$n_1^a a_1 + n_2^a a_2 = tA_s . \tag{39}$$

Inserting this relation into eqn. (38) yields the following equation for x_2^a as a function of the mole fraction of the bulk solution, x_2^l:

$$x_2^a = \frac{tx_2^l + a_1 \Gamma_2^{(n)}}{t - (a_2 - a_1)\Gamma_2^{(n)}}.$$

(40)

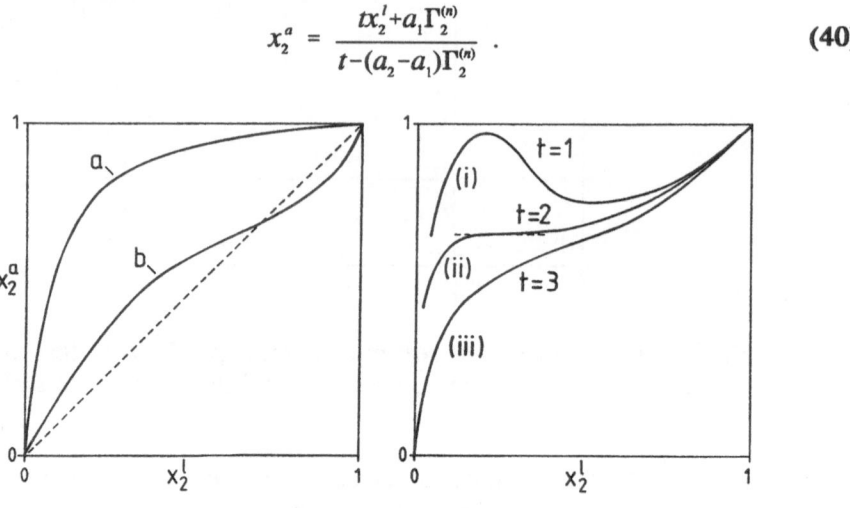

Figure 11. Schematic equilibrium diagrams for adsorption of binary mixtures, giving the composition of the adsorbed phase (x_2^a) as a function of the bulk composition (x_2^l). The diagrams correspond to the surface excess isotherms of Figure 9: (a) simple behaviour; (b) surface azetrope. The diagram on the r.h.s. illustrates the determination of the minimum surface layer thickness t_{min} of a complex system on the basis of eqn. (41).

If the molar areas a_1 and a_2 and the surface layer thickness t are known, eqn. (40) can be used to calculate x_2^a from measured surface excess data ($\Gamma_2^{(n)}$ vs. x_2^l). The dependence of x_2^a on x_2^l for the surface excess isotherms of Figure 9 is shown schematically in Figure 11. For a given binary system, the cross sectional areas of the adsorbed molecules can be estimated from molecular models, but it is difficult to estimate the layer thickness t. If it is assumed that t remains constant over a considerable composition range of the mixture, a minimum value of the layer thickness (t_{min}) can be derived from the general thermodynamic criterion for two-phase stability

$$(\partial x_2^a / \partial x_2^l)_T > 0.$$

(41)

To determine t_{min} by this criterion, one calculates isotherms of x_2^a vs. x_2^l by eqn. (40) for a series of values of t as shown in Figure 11. In this example, curve (i) for $t=1$ exhibits a region in which x_2^a decreases with increasing x_2^l, which is unacceptable on the basis of eqn. (41); curve (ii) for $t=2$ exhibits no such region but a point of inflection with a horizontal tangent, i.e. $\partial x_2^a / \partial x_2^l = 0$; in curve (iii), which is calculated from the same set of data as the curves (i) and (ii), but assuming $t=3$, the mole fraction x_2^a is a monotonic increasing function of x_2^l. Accordingly, $t_{min}=2$ for the example shown in Fig.11.

3.2.2 *Critique of the Surface Phase Model.* Experimental studies of simple liquid mixtures have shown that in many cases the solid surface affects only the composition of the first monolayer (i.e. the molecules which are in direct contact with the surface) [15]. In such cases, eqn.(40) with t = 1 yields physically reasonable values of the surface phase composition; i.e., x_2^s is a monotonic increasing function of x_2' and, in the case of strong preferential adsorption of component 2, x_2^s attains values close to unity already in relatively dilute bulk solutions. In this case the surface phase model applies as a *monolayer model* (see Section 3.3.1).

In other cases, when the analysis of experimental surface excess isotherms yields a minimum surface layer thickness $t_{min} > 1$, the surface phase model becomes questionable. Such a situation often arises with mixtures exhibiting strong positive deviations from ideality, in particular when the mixture is in a state close to demixing. In such cases, when there is an intrinsic tendency for segregation of the two components, one of them (the one which interacts preferentially with the surface) will accumulate near the surface, causing a thick, diffuse surface layer in which the composition profile $x_2(z)$ approaches x_2' slowly, over a range of several layers. Theoretical treatments of this surface segregation [19] predict that the thickness of this layer tends to infinity as the composition of the mixture approaches the solubility limit of the preferentially adsorbed component. This effect is in fact analogous to multilayer adsorption of gases as the gas pressure approaches the saturation pressure (see Section 2.1.2). In both cases the surface layer thickness is strongly varying with composition or pressure, respectively, near the limit of phase stability, and thus the assumption of a constant thickness of the surface layer, on which the determination of the minimum surface layer thickness t_{min} is based, is clearly not justified.

Figure 12 shows surface excess isotherms of the binary system water (1) + acetonitrile (2) at the surface of a graphitic carbon [20]. The water + acetonitrile system exhibits pronounced positive deviations from Raoults law and an upper critical solution temperature T_c at ca. $0\,°C$. Due to the hydrophobic nature of the graphite surface, acetonitrile is preferentially adsorbed from aqueous solutions. In this water-rich region the reduced surface excess $n_2^{\sigma(n)}$ is large and strongly increasing with decreasing temperature: The isotherm at $-5\,°C$ ($T < T_c$) is interrupted by the miscibility gap of the system. The branch in the water-rich region (low x_2') terminates at the solubility limit of acetonitrile in the aqueous phase; the pronounced increase of $n_2^{\sigma(n)}$ in this region is a signature of multilayer adsorption of acetonitrile. The branch in the acetonitrile-rich region ends at the solubility limit of water in this phase; as water is displaced from the surface by acetonitrile, the surface excess isotherm exhibits no pronounced features in this region. At $5\,°C$ ($T > T_c$) the maximum surface excess is still very high; an analysis of the water-rich region of this isotherm yields a minimum surface layer thickness $t_{min} \approx 5$. On the other hand, in the acetonitrile-rich region of this system the surface excess becomes rather small and nearly independent of temperature. In this region the isotherms do not suggest the existence of a thick surface layer.

The acetonitrile + water system is an example of a class of binary mixtures consisting of a polar organic compound and water, which are widely used as mixed eluents in reversed-phase liquid chromatography. Such systems may exhibit a complex behaviour, with phase separation at low temperatures (below an upper critical solution temperature, UCST) and also at high temperatures (above a lower critical solution temperature LCST). As outlined above, the adsorption behaviour of such complex systems cannot be correlated in terms of simple surface phase models. Another class of systems for which surface phase models may become meaningless are polymer solutions.

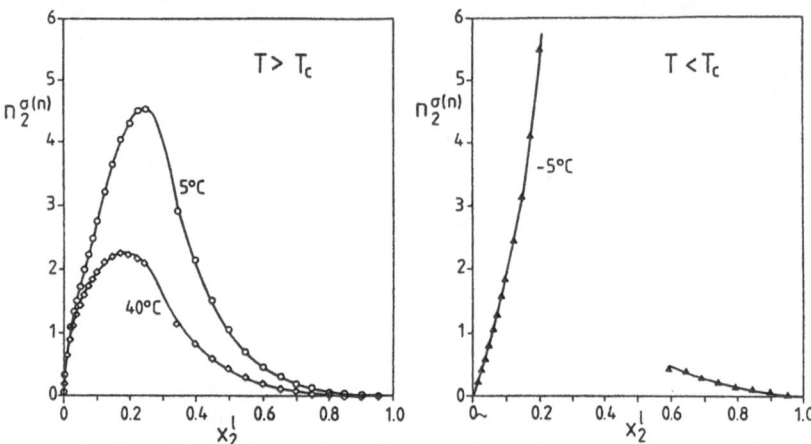

Figure 12. Reduced adsorption $n_2^{\sigma(n)}$ in the system water(1)+acetonitrile(2) on graphitic carbon above and below the upper critical solution temperature ($T_c = 0\,°C$). At temperatures below T_c the two branches of the isotherm end at the coexistence curve of the two phases. Above T_c the surface excess isotherms exhibit a strong temperature dependence (Pesch and Findenegg, to be published).

3.3 MODEL ISOTHERM EQUATIONS

In this section we introduce quantitative models for adsorption at the liquid/solid interface. We start from the surface-phase model outlined in Section 3.2 and consider three important cases: monolayer models for the adsorption from dilute solutions; adsorption of binary liquid mixtures, covering the entire composition range; and the dependence of the Henry's law constant of an adsorbed solute on the composition of a binary solvent mixture. All of these cases are of importance in the context of liquid/solid adsorption chromatography.

On the basis of the two-phase model outlined above, adsorption from a binary mixture can be treated as an exchange reaction of the two components 1 and 2 between the bulk solution (superscript l) and the adsorbed phase (superscript a), viz [18]

$$(2)^1 + r(1)^a = (2)^a + r(1)^1 \ . \tag{42}$$

Here, the stoichiometric coefficient r represents the ratio of the molar areas of the two components, $r = a_2/a_1$. The equilibrium state is then governed by the relation

$$\frac{\phi_2^a \gamma_2^a}{\phi_2^l \gamma_2^l} \left(\frac{\phi_1^l \gamma_1^l}{\phi_1^a \gamma_1^a}\right)^r = K_{21} \tag{43}$$

where ϕ_i represents the volume fraction and γ_i the corresponding (volume-fraction-based) activity coefficient of component i in the phase indicated by the superscript, and K_{21} is the equilibrium constant of the exchange reaction (42).

3.3.1 *Adsorption from Dilute Solutions.* In order to model preparative liquid/solid chromatographic processes one needs physically realistic model isotherm equations for the adsorption from dilute solutions. An important class of monolayer isotherm equations follows from eqn. (43) by assuming ideal dilute solution behaviour for the bulk phase. For the activity coefficients γ_i of the two components (i = 1,2) in the adsorbed phase one adopts the *Flory-Huggins* expression

$$\ln\gamma_i^a = (a_i/a_1)\chi(1-\phi_i^a)^2 \tag{44}$$

where χ is the *Flory-parameter* accounting for the interaction of a solvent molecule with one of the r segments of the solute molecule ($r = a_2/a_1$). By inserting these expressions into eqn. (43), and setting $\phi_2^l = 1$ and $\gamma_1^l = \gamma_2^l = 1$, one obtains the following isotherm equation

$$\phi_2^l = K_{21}\frac{\phi_2^a}{(1-\phi_2^a)^r} \cdot \exp(-2r\chi\phi_2^a) \tag{45}$$

Figure 13 shows an application of this equation to the adsorption of docosane (n-$C_{22}H_{46}$) from heptane onto graphitized carbon black [21]. These isotherms exhibit a striking similarity to the low-pressure region of gas adsorption isotherms on graphite (see Fig.4). This similarity suggests that in cases of strong preferential adsorption of the solute, the solvent acts as a nonspecific "background medium" for the solute. Indeed, by formally setting r = 1 and by replacing the segment interaction parameter χ by the pair potential parameter ϵ/k_BT, the surface-phase concentration ϕ_2^a by the surface coverage θ, and the bulk concentration ϕ_2^l by the gas pressure p, eqn. (45) transforms into the *Fowler-Guggenheim equation* for localized monolayer adsorption of gases (see Section 2.2). In the absence of preferential solute-solute interactions in the adsorbed layer (i.e. when $\chi = 0$), the exponential in eqn. (45) disappears. Further, if solvent and solute molecules are of similar sizes ($a_1 \approx a_2 = a^*$), $r \approx 1$ so that volume fractions become equal to mole fractions. In this special case eqn. (45) reduces to a simple *Langmuir-type equation* which - unlike the previous equations - may be written as an explicit expression for the surface concentration x_2^a. In this special case of monolayer adsorption (t = 1) from

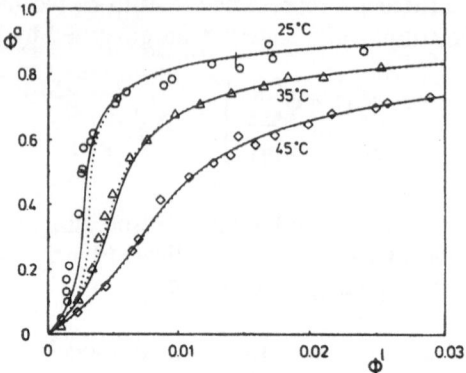

Figure 13. Adsorption of docosane ($C_{22}H_{46}$) from dilute solutions in heptane onto graphitized carbon: surface concentration ϕ^a as a function of bulk concentration (volume fraction) ϕ^l of docosane. The curves show a fit of the data by eqn.(45). (Fig.2 of ref. [21], reproduced with permission of Pergamon Journals Ltd.).

solutions of equal molecular size ($r = 1$), eqn. (40) reduces to

$$x_2^a = x_2^l + a^* \Gamma_2^{(n)} \tag{46}$$

and, for sufficiently strong adsorption from dilute solutions of component 2 ($x_2^a \gg x_2^l$), we obtain

$$\Gamma_2^{(n)} = \frac{1}{a^*} \frac{x_2^l}{K_{21} + x_2^l} \tag{47}$$

where $\Gamma_2^{(n)} = n^0 \Delta x_2^l / A_s$ is the *measured* areal adsorption of the solute and $1/a^* = n^s/A_s$ is the monolayer capacity per unit surface area of the adsorbent. Note that this simple *Langmuir* equations is justified only under the restricted conditions mentioned above. Otherwise one has to go back to eqn. (45), possibly with $\chi = 0$, and use eqn. (40) instead of eqn. (46) to convert the measured adsorption to x_2^a.

3.3.2 *Adsorption from Binary Solvent Mixtures.* In the context of liquid/solid elution chromatography with binary eluents it is of interest to correlate the composition of the adsorbed layer (ϕ_2^a) with the composition of the eluent (ϕ_2^l). A general relation of this kind is eqn. (43). If it is assumed that the adsorbed phase represents a truely autonomous phase the activity coefficients in the bulk and the adsorbed phase can be expressed by the Flory-Huggins expression for $ln\gamma_i$ (see eqn. 44). It has been argued that, due to the influence of the surface on the adsorbed layer, the interaction parameter χ for the adsorbed phase may differ from that for the bulk phase. For example, in some cases it was found that experimental adsorption data can be fitted in a satisfactory way by eqn. (43) by assuming ideality of the adsorbed phase, even if the bulk mixture exhibits pronounced deviations from ideality [15]. In other cases (e.g. for adsorption on graphite or other atomically flat surfaces) the χ-parameter for the surface phase appears to be larger than for the bulk solution [15]. Note, for example, that eqn. (45) is based on the assumption of ideal behaviour of the bulk solution.

In the so-called *Parallel Layer Model* [18] it is explicitly assumed that the adsorbed phase represents a single monolayer in contact with the bulk phase. Accordingly, molecules in the adsorbed layer interact not only with neighbouring molecules within the adsorbed phase but also with neighbours in the adjacent layer of the bulk mixture. Therefore, in this model the activity coefficients of the two components in the adsorbed phase depend not only on the composition of the surface phase but also on the composition of the bulk phase. For this reason the adsorbed layer is said to be a non-autonomous phase.

Most adsorbents have energetically heterogeneous surfaces and in general it is not possible to decide, if the observed deviations from a given model isotherm are due to nonideality effects in the adsorbed phase or to surface heterogeneity effects, or both. If the energetic heterogeneity of the surface can be characterized by a Gaussian distribution

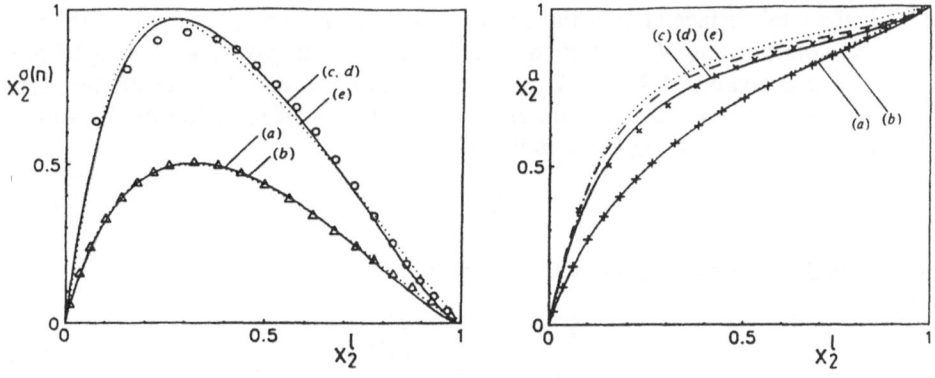

Figure 14. Determination of the surface phase composition from experimental surface excess isotherms of the binary systems coyclohexane+benzene (C+B) and cyclohexane+ 1,2-dichloroethane (C+D) on silica gel at 25 °C. Left: Fit of the experimental surface excess data by eqn.(48) with different sets of parameters (curves a and b refer to the system B+C, curves c-e to the system C+D). Right: Surface phase composition x_2^s vs. bulk phase comsosition x_2^l calculated by eqn. (40) with t = 1, using the same sets of parameters as in the fit of the surface excess isotherms. (Figs. 1 and 2 of ref. [23], reproduced with permission of the Royal Society of Chemistry).

of adsorption energies, with a width c of the distribution function, the so-called condensation approximation yields the following isotherm equation [22]

$$x_2^a = \frac{K_{21}(x_2^1 \gamma_2^1 g_{12})^m}{(x_1^1 \gamma_1^1)^{mr} + K_{21}(x_2^1 \gamma_2^1 g_{12})^m}$$

(48)

where $g_{12} = (\gamma_1^s/\gamma_2^s)^{1/r}$ contains the activity coefficients in the adsorbed phase, and m = RT/c is a heterogeneity parameter. Figure 14 shows a fit of eqn. (48) to the surface excess data for the binary solvent systems cyclohexane+benzene and cyclohexane+1,2-dichloroethane on μ-Porasil silica gel, and the determination of the mole fractions x_2^s for these systems, using eqn. (40).

3.4 RETENTION OF SOLUTES IN LIQUID/SOLID CHROMATOGRAPHY

3.4.1 *Definitions*. In analytical liquid/solid chromatography, the retention volume $V_{R,j}$ of a small pulse of solute j in a column with a dead-space volume V_μ is given by [24]

$$V_{R,j} = V_\mu + A_s v^l (\partial \Gamma_j^{(n)}/\partial x_j^l)_0$$

(49)

where A_s is the total surface area of the column packing, n° the total amount of the liquid eluent in the column, and $V_\mu = n^\circ v^l$ the total volume occupied by an eluent of molar volume v^l ; the derivative $(\partial \Gamma_j^{(n)}/\partial x_j^l)_0 = K_j$ represents the initial slope of the surface excess isotherm of component j in the eluent. The measured retention of the solute is commonly expressed by the *capacity factor* k_j'

$$k_j' \equiv (V_{R,j} - V_\mu)/V_\mu$$

(50)

which is related to the adsorption of the solute by

$$k_j' = \frac{A_s}{n^0} \left(\frac{\partial \Gamma_j^{(n)}}{\partial x_j^l} \right)_0 = \left(\frac{A_s}{n^0} \right) K_j$$

(51)

Thus, in adsorption chromatography k_j' is dependent on the column-specific experimental parameters A_s and n^0; hence it is more appropriate to introduce the *surface specific net retention volume* of the solute [25],

$$V_{S,j} = (V_{R,j} - V_\mu)/A_s = v^l K_j$$

(52)

which is independent of the experimental column parameters.

3.4.2 *Application of the Surface Phase Model.* Introducing the surface-phase concept outlined in Section 3.2, we can express the surface-specific net retention volume $V_{s,j}$ in terms of the distribution coefficient of the solute at infinite dilution, viz.

$$V_{S,j} = d\left[(x_j^a/x_j^l)_{x_j \to 0} - 1\right] \tag{53}$$

where $d = v'n^s/A_s = v'/a^*$ is related to the molar volume (v') and the molar cross-sectional area (a^*) of the solvent forming the eluent. Thus, if the adsorbed phase consists of a single monolayer $(t=1)$, one expects d to be of the order of the molecular diameter of the solvent molecules.

The distribution coefficient of the solute, $(x_j^a/x_j^l)_{x \to 0}$ can be expressed by the corresponding mole fractions of the solvent in the two phases. If the eluent is a pure solvent (say, component 2) we have the displacement reaction

$$(j)^l + r_j(2)^a = (j)^a + r_j(2)^l \tag{54}$$

with an equilibrium constant K_{j2}. For a binary eluent of solvents 1 and 2, we have also to consider the solvent displacement reaction

$$(1)^a + (2)^l = (1)^l + (2)^a . \tag{55}$$

By introducing the *Parallel-Layer Model* one can derive the following expressions for the surface-specific net retention volume in terms of the above adsorption equilibria [23]:
a) Pure eluent (solvent 2):

$$V_{S,j}^{(2)} = d(K_{j2} f_j^{-1} - 1) \tag{56}$$

where K_{j2} is the equilibrium constant of the exchange reaction (54) and $f_j = (\gamma_j^a/\gamma_j^l)_{x_j \to 0}$.
b) Mixed eluent (solvent mixture 1+2): assuming that component 2 is preferentially adsorbed from the solvent mixture one obtains

$$V_{Sj} + d = \left(V_{Sj}^{(2)} + d\right) \left[(x_2^a/x_2^l) F(x_2)\right]^{r_j} \tag{57}$$

with

$$F(x_2) = \exp[-S_{21}(1 + 2x_2^a - 3x_2^l)/4] \tag{58}$$

where S_{21} represent a combination of *Flory*-parameters for the pairs j-2, j-1 and 1-2, viz., $S_{21} = \chi_{2j} - \chi_{1j} + \chi_{12}$. These relations summarize the various factors affecting the surface-reduced net retention volume $V_{s,j}$ and its dependence on the composition of a binary eluent. For a single eluent (eqn.56), $V_{s,j}$ increases with the equilibrium constant K_{j2} of the displacement reaction (54). In the case of strong preferential adsorption of the solute the term -1 on the r.h.s. of eqn. (56) can be neglected; in this case it is justified to argue about retention in terms of *absolute* adsorbed amounts of the solute. Alternatively, in the case of a preferential adsorption of the solvent, $V_{s,j}$ becomes negative.

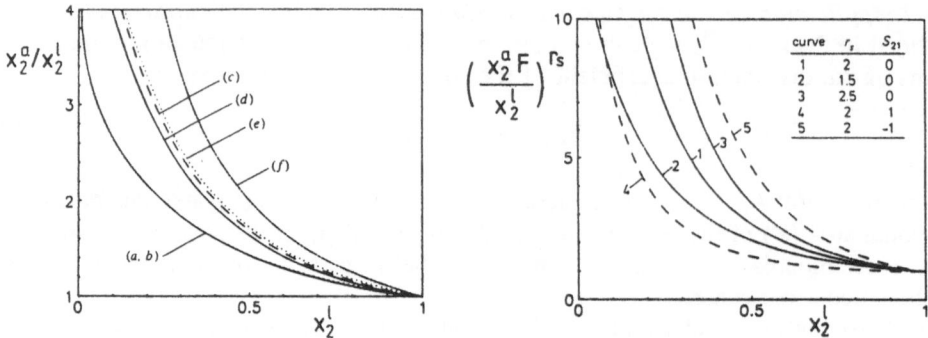

Figure 15. Analysis of eqn. (57) for the dependence of the surface-specific net retention volume of solutes on the composition of a binary eluent. Left: The function x_2^a/x_2^l vs. mole fraction x_2^l. The curves (a)-(e) correspond to the surface concentration isotherms (x_2^a vs. x_2^l) shown in Figure 14; curve (f) shows the function $1/x_2^l$ which corresponds to $x_2^a = 1$ for all liquid phase compositions (complete displacement of component 1 from the surface layer). Right: The function $[(x_2^a/x_2^l)F(x_2)]^r$ for typical combinations of the parameter r and S_{21}. The curves are based on the surface concentration isotherm $x_2^a = f(x_2^l)$ for the system C+D [curve (d) of the graph on the left]. (Figs. 3 and 4 of ref. [23], reproduced with permission of The Royal Society of Chemistry).

Turning now to the dependence of $V_{s,j}$ on the composition of a binary eluent, remember that component 2 is assumed to be more strongly adsorbed than component 1. In the case of strong peferential adsorption of the solute j from solvent 2, $V_{s,j}$ will be much greater than d even in the case of pure solvent 2, so that d can be neglected against $V_{s,j}$ and $V_{s,j}^{(2)}$, respectively, on both sides of eqn. (57).

In this case, the dependence of the net retention volume of the solute on the composition of the eluent is described essentially by the function $(x_2^a/x_2^l)^r$, while the factor $F(x_2)$ accounts for deviations from ideality, being either greater or less than unity depending on the sign of the interaction parameter S_{21}. The function x_2^a/x_2^l resulting from the isotherm equation (48) with the parameters for the solvent systems cyclohexane + benzene (C+B) and cyclohexane + 1,2-dichloroethane (C+D) on μ-Parasil (cf. Figure.14) is shown in Figure 15. The function $1/x_2^l$ (corresponding to $x_2^a = 1$ for all x_2^l, i.e. infinitely strong preferential adsorption of component 2) is shown for comparison. Figure 15 also illustrates the influence of the size parameter r_j and the interaction parameter S_{21} on the function $[x_2^a/x_2^l) F(x)]^r$.

For a quantitative analysis of experimental retention data of organic solutes in the above eluent systems the parameter d in eqn. (57) has to be estimated from the mean molar volume, v, and the corresponding mean molar area, a, of the binary solvent system.

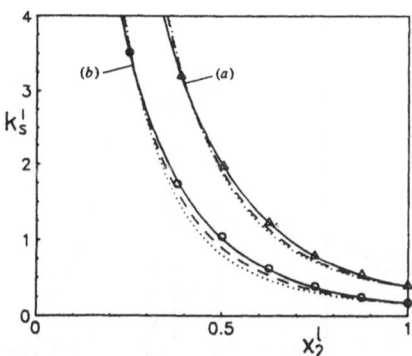

Figure 16. Capacity ratio k_s^j of 1,4-dinitrobenzene vs. mole fraction x_2^l of the binary eluents C+B (a) and C+D (b): Experimental data and fits by an equation equivalent to eqn. (57) for k_s^j instead of $V_{s,j}$. The smaller values of k_s^j in the eluent C+D can be attributed to the stronger adsorption of dichloroethane as compared with benzene from mixtures with cyclohexane (cf. Fig. 14). (Fig. 7 of ref. [23], reproduced with permission of the Royal Society of Chemistry).

Figure 16 illustrates how the retention of a solute (here expressed by the capacity ratio k^j) is affected by the precoverage of the surface by the preferentially adsorbed component of the solvent mixture. At any given composition of the bulk phase the retention of the solute is smaller for the eluent mixture C+D as compared with C+B because D is more strongly adsorbed than B from mixtures with C. Incidentally, if in eqn. (57) the function x_2^a/x_2^l is approximated by $1/x_2^l$ [curve (f) in Figure 15] and if S_{21} and d are both taken as zero, equation (57) reduces to the simple *Snyder-Soczewinski* (SS) *equation*,

$$V_{S,j} = V_{S,j}^{(2)} \left(1/x_2^l\right)^{r_j} \tag{59}$$

which gives a poor fit of the experimental data in Figure 16.

As an example of systems exhibiting a more complex retention behaviour, Figure 17 shows the retention volumes of quaternary ammonium salts in the binary eluent system water+methanol on a column packed with a porous graphitic carbon [26]. For tetramethylammonium bromide (TMAB) in the water-rich domain, where the salt is dissociated, V_R is less than V_μ, i.e. the net retention volume V_R-V_μ is negative, indicating a *negative adsorption* of the salt at the solution/graphite interface. In the methanol-rich region, where the salt forms ion pairs, the net retention volume and hence the adsorption of TMAB is positive. For octyltrimethylammonium bromide (C_8TAB) the net retention volume is positive over the entire composition range. In the methanol-rich region, the dependence of V_R on the composition of the eluent is similar to that of TMAB, but in the water-rich domain V_R becomes very high, indicating that the hydrophobic interaction of

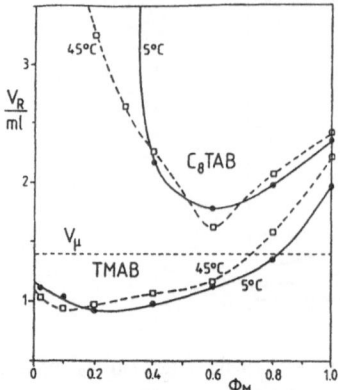 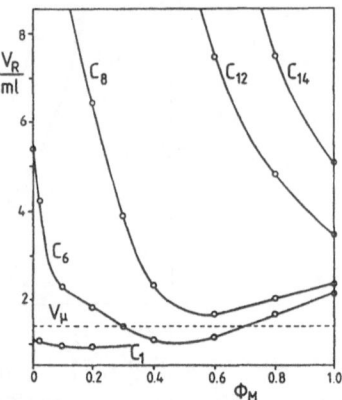

Figure 17. Retention volume V_R of tetramethylammonium bromide (TMAB), and of a series of n-alkyltrimethylammonium bromides (C_nTAB) vs. volume fraction ϕ_M of methanol in the binary eluent water + methanol (W+M) on a column packed with graphitic carbon. The dashed horizontal line marks the column dead-space volume V_μ. Left: TMAB and C_8TAB at $5°C$ and $45°C$. Right: Several C_nTAB compounds (n=1,6,8,12,14) at $25°C$. (Rasch and Findenegg, to be published).

the hydrocarbon chains with the graphite surface leads to a strong adsorption of the amphiphilic molecules. The systematic changes in the retention of n-alkyltri-methylammonium bromides with increasing chain length of the alkyl are also shown in Figure 17. Obviously, such a complex behaviour (which is also manifested by the temperature dependence of V_R) cannot be accounted for by the model outlined above.

3.4.3 *Conclusion.* We have seen that the retention of solutes in liquid/solid chromatography with binary eluent mixtures can be explained in terms of a monolayer model of adsorption from three-component systems. The surface-reduced net retention volume $V_{s,j}$ depends on the adsorption equilibrium of the solvent system (x_2^s/x_2^l); it is strongly affected by the size ratio $r_j = a_j/a_2$, where a_j and a_2 represent the cross-sectional areas of solute j and solvent 2, the preferentially adsorbed component of the solvent mixture. $V_{s,j}$ also depends on the interaction parameter S_{21} which represents a linear combination of the two binary Flory parameters χ_{ji} (i = 1,2) for solvent-solute interactions and the Flory-parameter χ_{12} for solvent-solvent interactions.

When the solute is strongly preferentially adsorbed from both solvents, the surface excess concentration $\Gamma_j^{(n)}$ becomes nearly equal to the total surface concentration of the solute, n_j^s/A_s. In this case the surface-reduced net retention volume $V_{s,j}$ is positive and significantly larger than $d = v^l/a^*$. In this case, the term d in eqn. (57) can be omitted. On the other hand, when the solute is only weakly or even negatively adsorbed, the term d cannot be neglected against $V_{s,j}$. In this case it is essential to obtain a realistic estimate of the parameter d, and thus to acknowledge the fact that the surface excess of the solute can be positive or negative, depending on the strength of competitive adsorption of the solvent.

References

[1] Glueckauf, E. (1947) J. Chem. Soc. (Lond.), 1302-1347; see also (1949) Discuss. Faraday Soc. 7, 12

[2] Helfferich, F. and Klein, G., (1970) Multicomponent Chromatography, Marcel Dekker, New York

[3] Valentin, P. and Guiochon, G. (1976) 'Determination of Gas-Liquid and Gas-Solid Equilibrium Isotherms by Chromatography', J. Chromatogr. Sci. 14, 56-139.

[4] Jaroniec, M. and Madey, R. (1988) Physical Adsorption on Heterogeneous Solids. Elsevier, Amsterdam

[5] Rudzinski, W. and Everett, D.H. (1991) Adsorption of Gases on Heterogeneous Surfaces, Academic Press, London

[6] Steele, W.A. (1974) The Interaction of Gases with Solid Surfaces, Pergamon Press, Oxford

[7] Everett, D.H., Ed. (1972) Manual of Symbols and Terminology for Physicochemical Quantities and Units, Appendix II, Part 1, Definitions, Terminology and Symbols in Colloid and Surface Chemistry. Pure Appl. Chem. 31, 579-638.

[8] Putnam, F.A. and Fort, T. Jr. (1975) 'Physical Adsorption on Patchwise Heterogeneous Surfaces. Heterogeneity, Two-Dimensional Phase Transitions, and Spreading Pressure of the Krypton-Graphitized Carbon Black System near 100 K.' J. Phys. Chem. 79, 459-467.

[9] Wendland, M., Salzmann, S., Heinbuch, U. and Fischer, J. (1989), 'Born-Green-Yvon results for adsorption of a simple fluid on plane walls. Structure,adsorption isotherms, and surface area determination', Molec. Physics 67, 161-172.

[10] Lewandowski, H., Michalski, T. and Findenegg, G.H. (1991) 'Multilayer Adsorption and Pore Condensation of Fluids near their Critical Point in Graphite Substrates! In: Fundamentals of Adsorption, Vol. 3 (A.Mersmann, Ed.), American Institute of Chemical Engineers, New York, p.497-506.

[11] Blümel, S., Köster, F. and Findenegg, G.H. (1982) 'Physical adsorption of krypton on graphite over a wide density range. A comparison of the surface excess of simple fluids on homogeneous surfaces'. J. Chem. Soc. Faraday Trans. 2 78 , 1753-1764.

[12] Findenegg, G.H. (1987) 'Experimental and thermodynamic aspects of mixed gas adsorption'. In: Fundamentals of Adsorption, Vol. 2 (A.I. Liapis, Ed.) American Institute of Chemical Engineers, New York, p. 53-71.

[13]   Dondi, F., Gonnord, M.-F. and Guiochon, G. (1977) 'Chromatographic determination of gas-solid adsorption isotherms by the step-and-pulse method', J. Colloid Interface Sci. 62, 303-328

[14]   Von Rybinski, W., Albrecht, M., and Findenegg, G.H. (1980) 'Study of the Interaction between Adsorbed Hydrocarbon Molecules on Graphitized Carbon Black using the Chromatographtic Step-and-Pulse Method. Faraday Symposia Chem. Soc. 15, 25-37.

[15]   Reviews of the literature on Adsorption at the Solid/Liquid Interface (Nonelectrolyte Systems) are published in Colloid Science, A Specialist Periodical Report (D.H. Everett, Ed.), The Royal Society of Chemistry, London: Everett, D.H. (1973), Vol.1. 49-102; Brown, C.E. and Everett, D.H. (1975), Vol. 2, 52-100; Everett, D.H. and Podoll, R.T. (1978), Vol. 3, 63-149; Davis, J. and Everett, D.H. (1983), Vol. 4, 84-149.

[16]   Everett, D.H. (1986) 'Reporting data on adsorption from solution at the solid/solution interface', Pure Appl. Chem. 58, 967-984.

[17]   Kern, H., von Rybinski, W. and Findenegg, G.H. (1977), 'Prefreezing of liquid n-alkanes near graphite surfaces', J. Colloid Interface Sci. 59, 301-307

[18]   Ash, S.G., Everett, D.H. and Findenegg, G.H. (1968) 'Thermodynamics of Adsorption from Solution: The Parallel Layer Model', Trans. Faraday Soc. 64, 2639-2644.

[19]   Sullivan, D.E. and Telo da Gama, M.M. (1986) 'Wetting Transitions and Multilayer Adsorption at Fluid Interfaces', In: Fluid Interfacial Phenomena (C.A.Croxton, Ed.), Wiley, London, p.45-134.

[20]   Pesch, W. and Findenegg, G.H., to be published.

[21]   Findenegg, G.H. and Liphard, M. (1987), 'Adsorption from solution of large alkane and related molecules onto graphitized carbon', Carbon 25, 119-128.

[22]   Rudzinski, W., Narkewicz-Michalek, J., Suprynowicz, Z. and Pilorz, K. (1985) 'Effects of Surface Heterogeneity on Liquid Adsorption Chromatography with Mixed Mobile Phases', J. Chem. Soc. Faraday Trans. 1, 81, 553-563.

[23]   Findenegg, G.H. and Köster, F. (1986), 'A new equation for the retention of solutes in liquid/solid adsorption chromatography with mixed molebile phases', J. Chem. Soc. Faraday Trans. 1, 82, 2691-2705.

[24]   Riedo, F. and Kováts, E. (1982) 'Adsorption from liquid mixtures and liquid chromatography', J. Chromatogr. 239, 1-28

[25]   Kováts, E. (1985) 'Retention in liquid/solid chromatography', In: The Science of Chromatography (F. Bruner, Ed.), J. Chromatogr. Library 32, 205-217

[26]   Rasch, R. and Findenegg, G.H., to be published.

UNIFIED APPROACH TO THE THEORY OF CHROMATOGRAPHY.
I. "INCOMPRESSIBLE.", BINARY MOBILE PHASE (LIQUID CHROMATOGRAPHY)

DANIEL E. MARTIRE
Department of Chemistry
Georgetown University
Washington, D. C. 20057-2222
U. S. A.

ABSTRACT. A lattice model is used to derive equations for the solute (a) distribution coefficient in a mixed-solvent mobile phase (b+c), $K_{a(b+c)}$, and the solvent sorption isotherms in liquid chromatography (LC). Entropic and interaction energy effects are included. These equations, which are applicable to liquid-like and solid (adsorbent) stationary phases, LLC and LSC, respectively, are found to depend on molecular size and interaction energy parameters, and on the mobile- and stationary-phase volume fractions of one of the solvent components, $\theta_{b(m)}$ and $\theta_{b(s)}$, respectively. It is seen that the equations pertaining to LLC are formally similar to those for LSC. The significance and utility of these theoretical results are discussed and illustrated. The widely used linear expression relating $\ln K_{a(b+c)}$ and $\ln \theta_{b(m)}$ is examined in the light of the present theory and recent experimental findings.

The equations derived herein, which provide the basis for a unified theory of chromatography, are employed in part II to obtain a companion theory of supercritical fluid chromatography (SFC).

1. Introduction

In 1987 we presented a unified molecular theory of solute retention in fluid-liquid (absorption) chromatography [1]. A year later this theory was extended to fluid-solid (adsorption) chromatography [2]. Both studies focused on the equilibrium properties of the respective chromatographic systems. Here we present a refined and more compact version of this unified approach, applicable to both absorptive and adsorptive stationary phases, or any combination thereof. In part I, the basic equations are derived for a model system which regards the binary mobile phase as being incompressible. These equations are used to describe liquid chromatography (LC), for which compressibility effects on solute retention are relatively unimportant. In part II, use is made of the results from part I, and of isomorphic elements of binary-liquid and single-fluid critical behavior, to derive, in a simple manner, the corresponding equations for a system operating with a neat mobile phase which may be both dense and compressible, i.e., supercritical fluid chromatography (SFC). In the limit of low fluid density, these equations are seen also to apply to gas chromatography (GC). The utility of this unified approach is discussed and illustrated.

The initial impetus for this work was the need for a simple and useful molecular theory of solute retention in SFC - one that was properly based on statistical thermodynamics and wherein retention behavior could be explicitly related to the reduced variables of the mobile phase and other accessible pure-component physical properties and/or molecular parameters. In the process of developing and testing a theory based on a lattice-fluid model came the recognition of certain interrelationships and formal similarities among the equations describing SFC and previously derived equations (based on kindred models) for GC and LC.

261

F. Dondi and G. Guiochon (eds.),
Theoretical Advancement in Chromatography and Related Separation Techniques, 261–273.
© 1992 *Kluwer Academic Publishers.*

The end result of this realization is the present unified theory of chromatography, which encompasses all three types of mobile phases, and absorptive and adsorptive stationary phases, and which describes the equilibrium distribution of solute and solvent molecules between the two phases. Despite their conceptual and practical advantages, such unified theoretical approaches have been lacking in the area of chromatography. The notable exceptions, all addressing the dynamics of chromatography, are a classic study by Giddings [3] and two very recent works [4,5].

2. Theoretical Model

In common with our earlier treatments of liquid-solid (adsorption) and liquid-bonded phase chromatography [6-9], statistical thermodynamics and a mean-field lattice model are utilized to derive the relevant equations to describe the equilibrium distribution of solute (a) between a binary liquid mobile phase (b+c) and a principal stationary phase consisting of a solid surface (t) which may also support a bulk liquid or polymer (d). Both phases are assumed to be incompressible. As before, this treatment examines the competitive equilibrium at the molecular level among solute and solvent molecules distributed between generally nonideal mobile and stationary phases, all components being nonelectrolytes. Entropy and interaction energy effects are rigorously incorporated using the Bragg-Williams (random-pairing) approximation [10].

In general, the entire stationary phase is assumed to consist of a bulk liquid or polymer (d) residing on a planar support surface (t) which is energetically and structurally homogeneous, plus sorbed solute and solvent components (a+b+c). For a sufficiently thick film of component d and a relatively inert surface, the system becomes a liquid-liquid distribution system (LLC). If component d is absent from the system, the (now active) solid surface becomes the adsorbent in a liquid-solid chromatographic (LSC) system, where the sorbed solute and solvent molecules form, in general, a multilayer of uniform thickness and composition. (Presumably, a chemically bonded phase or any other thin-film, surface-modified system would fall somewhere in between these two extremes.)

Although a variety of molecular structures may be considered within the general framework of the model [1,6-8,11], for present purposes, the molecules are assumed to be completely flexible chains, each having r_i singly connected (terminal) or doubly connected (internal) cubic segments, each of volume v_o. (Therefore, the hard-core volume of a molecule of type i is $r_i v_o$.) Nonetheless, it will be seen that one form of the equation describing the solute distribution coefficient is apparently independent of molecular structure.

The mobile phase (subscript m) is modelled as a three-dimensional lattice having a nearest-neighbor coordination number z_m (with a simple-cubic lattice, e.g., $z_m = 6$) and M_m cubic cells or sites, each of volume v_o. Occupying these sites in a restricted random walk [1,10] are $N_{a(m)}$ solute molecules, each occupying r_a sites, $N_{b(m)}$ and $N_{c(m)}$ mobile-phase solvent molecules, each occupying r_b and r_c sites, respectively, where $M_m = r_a N_{a(m)} + r_b N_{b(m)} + r_c N_{c(m)}$ and where it is assumed that component d is not present in the mobile phase [1].

The stationary phase (subscript s) is modelled as a two-to-three dimensional lattice having, in general, a nearest-neighbor coordination number of z_s. In one extreme (absence of component d and parallel monolayer adsorption [2]), we would have a two-dimensional lattice (with a square-planar lattice, e.g., $z_s = 4$) containing M_t equivalent surface sites, each of area $(v_o)^{2/3}$. In the other extreme (sufficiently thick film of component d), we would have a three-dimensional lattice. Accordingly, $z_s \leq z_m$ and, in general, $M_s = \tau_s M_t = r_a N_{a(s)} + r_b N_{b(s)} + r_c N_{c(s)} + r_d N_{d(s)}$, where the number of stationary-phase sites, M_s, is equal to the number of surface sites, M_t, times the total number of stationary-phase liquid layers, τ_s, each of which has a thickness $(v_o)^{1/3}$. Note that M_m and M_s are proportional to the total volume of the mobile and stationary phases, respectively.

Let us now formulate, for each phase in this model system, the total Helmholtz free energy, which turns out to be the sum of three contributions [1,2].

The dimensionless configurational entropy of the stationary phase (x=s) or the mobile-phase (x=m) mixture, S_x/k_B, is given by [1,2,10]

$$-\frac{S_x}{k_b} = \sum_{i=a}^{d} (N_{i(x)}\ln N_{i(x)}-N_{i(x)})-M_x\ln M_x+M_x$$

$$-\sum_{i=a}^{d} N_{i(x)}(r_i-1)\ln \{(z_x-1)/M_x\}$$

(1)

where

$$M_x = \sum_{i=a}^{d} r_i N_{i(x)}$$

and k_B is the Boltzmann constant.

If the attractive interaction energy between nearest-neighbor segments on molecules i and j is denoted by ε_{ij}, and between a molecular segment on i and a surface site by ε_{it}, the dimensionless total interaction energy in phase x, E_x/k_BT, is [1,2,10]

$$\frac{E_x}{k_BT} = (\frac{z_e f_x}{k_BT})\sum_{i=a}^{d} r_i N_{i(x)}e_{it}+[\frac{z_e(1-f_x)}{2k_BT}]\sum_{i,j=a}^{d} r_i N_{i(x)}r_j N_{j(x)}e_{ij}$$

(2)

where T is the absolute temperature, ε_{ij} and ε_{it} are negative (attractive), and z_e denotes the number of nearest-neighbor, external contacts of a molecular segment [1,2]. Also, f_x is the fraction of the molecular surface which is in contact with the solid surface, where $0 \le f_x < 1$. For the mobile phase (x=m), $f_m = 0$ and only the second term on the right hand side (r.h.s.) of eqn. 2 survives. For the stationary phase (x=s), f_s depends on τ_s, where $f_s \to 0$ as $\tau_s \to \infty$. The first term on the r.h.s. of eqn. 2 represents the total (dimensionless) molecular energy of adhesion to the solid surface. Therefore, the remaining (intermolecular) interactions in the stationary phase are reflected in the second term, where, in contrast to our previous treatment [2], it is now assumed that that portion of a molecule not in contact with the surface $(1-f_x)$ is bathing in other stationary phase molecules and is not in contact with any mobile phase molecules. This assumption is quite reasonable for sufficiently large τ_s (LLC and multilayer LSC), but becomes marginal for LSC with monolayer adsorption.

Finally, if we denote the cell partition function per segment of the molecule i in phase x by $q_{i(x)}$ and include the contribution [1]

$$\frac{A_x'}{k_BT} = -\sum_{i=a}^{d} r_i N_{i(x)}q_{i(x)}$$

(3)

the entire dimensionless Helmholtz free energy of the mixture, A_x/k_BT, consistent with the Bragg-Williams approximation, is simply given by the sum of eqs. 1, 2 and 3:

$$\frac{A_x}{k_BT} = (\frac{A_x'}{k_BT})+(\frac{E_x}{k_BT})-(\frac{S_x}{k_B})$$

(4)

The key relationship, that for the dimensionless chemical potential of the hth component in phase x,

$\mu_{h(x)}/k_B T$, is determined from

$$\frac{\mu_{h(x)}}{k_B T} = [\partial(A_x/k_B T)/\partial N_{h(x)}]_{T,N_{g(x)}} \tag{5}$$

where $N_{g(x)}$ denotes the number of molecules of components other than the hth one in phase x. From eqs. 1-5, one obtains

$$
\begin{aligned}
\frac{\mu_{h(x)}}{k_B T} = {}& \ln(\frac{\theta_{h(x)}}{r_h}) + r_h \sum_{i=a}^{d} \theta_{i(x)}(1 - r_i^{-1}) \\
& - (r_h - 1)\ln(z_x - 1) - r_h \ln q_{h(x)} + (z f_x r_h/k_B T)e_{hi} \\
& + [z_e(1 - f_x)r_h/k_B T] \sum_{i=a}^{d} \theta_{i(x)} e_{hi} \\
& - [z_e(1 - f_x)r_h/2k_B T] \sum_{i,j=a}^{d} \theta_{i(x)} \theta_{j(x)} e_{ij}
\end{aligned}
\tag{6}
$$

where h = a, b, c or d, x = m or s, and where $\theta_{i(x)} = r_i N_{i(x)}/M_x$ is the volume fraction of component i in phase x. At equilibrium, components a, b, c each must satisfy the following condition at the operational temperature:

$$\frac{\mu_{h(s)}}{k_B T} = \frac{\mu_{h(m)}}{k_B T} \tag{7}$$

For component a, eqs. 6 and 7 yield, in the limit of infinite dilution of component a, the solute distribution coefficient, K. For components b and c, they lead to the respective sorption isotherms. These results are presented in the next section.

Finally, it is mentioned that the chromatographic retention quantities normally measured and reported are the net retention volume, V_N, or the capacity factor, k'. These extensive quantities are related to the intensive quantity, K by

$$K = \frac{V_N}{V_s} = k'(\frac{V_m}{V_s}) \tag{8}$$

where V_m and V_s are the total volumes of the mobile and stationary phases, respectively. In general, $V_s = V_{b(s)} + V_{c(s)} + V_{d(s)}$.

3. Retention and Sorption Isotherm Equations

The chromatographic distribution coefficient, K, is defined as the ratio of the equilibrium concentration of solute in the stationary phase, $c_{a(s)}$, to that in the mobile phase, $c_{a(m)}$, in the limit of infinite dilution of the solute ($N_a \to 0$ or $\theta_a \to 0$). Clearly, in the model system, this ratio is also equal to the ratio of the respective θ_a's, i.e.,

$$K = \lim_{N_a \to 0}(c_{a(s)}/c_{a(m)}) = \lim_{N_a \to 0}(\theta_{a(s)}/\theta_{a(m)})$$ (9)

Recalling that $f_m = 0$ and $\Theta_{d(m)} = 0$, and denoting the distribution coefficient of solute (a) in the binary mobile-phase system (b+c) by $K_{a(b+c)}$, one obtains from eqs. 6, 7 and 9

$$
\begin{aligned}
\ln K_{a(b+c)} = & -(r_a-1)\ln[(z_m-1)/(z_s-1)]-(z_s f_s r_a/k_B T)e'_{at} \\
& +r_a\sum_{i=b}^{c}\theta_{i(m)}(1-r_i^{-1})-r_a\sum_{i=b}^{d}\theta_{i(s)}(1-r_i^{-1}) \\
& +(z_s r_a/k_B T)\sum_{i=b}^{c}\theta_{i(m)}e_{ai}-(z_s r_a/2k_B T)\sum_{i,j=b}^{c}\theta_{i(m)}\theta_{j(m)}e_{ij} \\
& -[z_s(1-f_s)r_a/k_B T]\sum_{i=b}^{d}\theta_{i(s)}e_{ai}+[z_s(1-f_s)r_a/2k_B T]\sum_{i,j=b}^{d}\theta_{i(s)}\theta_{j(s)}e_{ij}
\end{aligned}
$$ (10)

where, in eqn. (10), the cell partition function term, $\ln(q_{a(m)}/q_{a(s)})$, has been incorporated into the surface interaction term, as follows [7]

$$(z_s f_s/k_B T)e_{ht}+\ln(q_{h(m)}/q_{h(s)}) = (z_s f_s/k_B T)e'_{ht}$$ (11)

thus making ε'_{ht} an adhesion <u>free energy</u> per molecular segment in eqs. 10 (h = a) and 11.

3.1 PRELIMINARY EQUATIONS

We start by rewriting eqn. 10 in terms of interaction parameters, χ_{ij} [8,9]:

$$
\begin{aligned}
\ln K_{a(b+c)} = & -(r_a-1)\ln[(z_m-1)/(z_s-1)]-(z_s f_s r_a/k_B T)[e'_{at}-(e_{aa}/2)] \\
& +r_a\sum_{i=b}^{c}\theta_{i(m)}(1-r_i^{-1})-r_a\sum_{i=b}^{d}\theta_{i(s)}(1-r_i^{-1}) \\
& +r_a\sum_{i=b}^{c}\chi_{ai}\theta_{i(m)}-r_a\chi_{bc}\theta_{b(m)}\theta_{c(m)} \\
& -r_a(1-f_s)\sum_{i=b}^{d}\chi_{ai}\theta_{i(s)}+r_a(1-f_s)\sum_{i,j=b,i\neq j}^{d}\chi_{ij}\theta_{i(s)}\theta_{j(s)}
\end{aligned}
$$ (12)

where

$$\chi_{ij} = (z_s/2k_B T)(2e_{ij}-e_{ii}-e_{jj})$$ (13)

In general, eqn. 12, by itself, is incomplete since explicit account has not yet been taken of the sorption of solvent components b and c in the stationary phase. (Strictly, the $\Theta_{i(s)}$ values are required to evaluate eqn. 12.) Nonetheless, to begin to develop the final equations and formal similarities, and to obtain equations which will be used in part II, we shall consider certain special cases of eqn. 12.

3.1.1. *Liquid-Liquid Chromatography (LLC)*. As discussed earlier, to obtain the salient equation for a bulk-liquid stationary phase we let $f_s \to 0$, $z_m \approx z_s$, and, as a further, but reasonable simplification (say, if component d is a hydrocarbon and component c is water, as in a reversed-phase system), we let $\Theta_{c(s)}$

= 0. Accordingly, eqn. 12 simplifies to give [1]

$$\ln K_{a(b+c)} = r_a[\theta_{b(m)}(1-r_b^{-1})+\theta_{c(m)}(1-r_c^{-1})]$$
$$+r_a[\chi_{ab}\theta_{b(m)}+\chi_{ac}\theta_{c(m)}-\chi_{bc}\theta_{b(m)}\theta_{c(m)}]$$
$$-r_a[\theta_{b(s)}(1-r_b^{-1})+\theta_{d(s)}(1-r_d^{-1})]$$
$$-r_a[\chi_{ab}\theta_{b(s)}+\chi_{ad}\theta_{d(s)}-\chi_{bd}\theta_{b(s)}\theta_{d(s)}] \tag{14}$$

For pure c as a mobile phase ($\theta_{b(m)} = \theta_{b(s)} = 0$)

$$\ln K_{a(c)}^* = r_a(r_d^{-1}-r_c^{-1})+r_a(\chi_{ac}-\chi_{ad}) \tag{15}$$

and for pure b as a mobile phase ($\theta_{c(m)} = 0$)

$$\ln K_{a(b)} = \ln K_{a(b)}^*+r_a(r_b^{-1}-r_d^{-1})\theta_{b(s)}$$
$$+r_a[(\chi_{ad}-\chi_{ab}+\chi_{bd})\theta_{b(s)}-\chi_{bd}\theta_{b(s)}^2] \tag{16}$$

$$\ln K_{a(b)}^* = r_a(r_d^{-1}-r_b^{-1})+r_a(\chi_{ab}-\chi_{ad}) \tag{17}$$

where $K_{a(b)}^*$ would be the solute partition coefficient if component b were also excluded from the stationary phase ($\theta_{b(s)} = 0$). Equation 15 (or 17) expresses, in terms of molecular parameters, the equilibrium distribution of solute between two immiscible, pure liquid chromatographic phases [1,8,12,13].

A particularly useful form of $\ln K_{a(b+c)}$ can be obtained by combining eqs. 14 and 15:

$$\ln K_{a(b+c)} = \ln K_{a(c)}^*+r_a(r_b^{-1}-r_d^{-1})\theta_{b(s)}-r_a(r_b^{-1}-r_c^{-1})\theta_{b(m)}$$
$$+r_a[(\chi_{ad}-\chi_{ab}+\chi_{bd})\theta_{b(s)}-\chi_{bd}\theta_{b(s)}^2]$$
$$-r_a[(\chi_{ac}-\chi_{ab}+\chi_{bc})\theta_{b(m)}-\chi_{bc}\theta_{b(m)}^2] \tag{18}$$

Equation 18, which is <u>quadratic</u> in both $\theta_{b(s)}$ and $\theta_{b(m)}$, describes how, relative to pure c (which we now refer to as the "poor" solvent) as the mobile phase, lnK varies with the mobile- and stationary-phase volume fractions of component b (which we now refer to as the "good" solvent; e.g., acetonitrile in a reversed-phase system). The second and third terms on the r.h.s. of eqn. 18 stem from the combinatorial entropy contributions to A_x/k_BT, while the last two terms reflect the various interaction energies in the system. The second and fourth terms couple the mobile and stationary phases, since $\theta_{b(s)}$ depends on $\theta_{b(m)}$ through the sorption isotherm of solvent component b [1] (see later). Therefore, since χ_{ij} is a function of temperature, eqn. 18 may be written in terms of the system variables in the following condensed form

$$\ln K_{a(b+c)} = \ln K_{a(c)}^*+F(T,\theta_{b(m)})+\Delta(T,\theta_{b(s)}) \tag{19}$$

where Δ is the coupling term.

To conclude this subsection, let us consider a further restriction, viz., the virtual absence of both solvent components from the stationary phase ($\theta_{b(s)} = \theta_{c(s)} = 0$; hence, $K_{a(b)} = K_{a(b)}^*$ in equation 16). Accordingly, we retrieve from eqs. 14, 15 and 17 two familiar forms [8] describing the partitioning of solute between two immiscible phases b+c and d, respectively

$$\ln K_{a(b+c)} = \ln K^{*}_{a(c)} + r_{a}(r_{c}^{-1} - r_{b}^{-1})\theta_{b(m)} + r_{a}[(\chi_{ab} - \chi_{ac} - \chi_{bc})\theta_{b(m)} + \chi_{bc}\theta_{b(m)}^{2}] \tag{20}$$

$$\ln K_{a(b+c)} = \theta_{b(m)}\ln K^{*}_{a(b)} + \theta_{c(m)}\ln K^{*}_{a(c)} - r_{a}\chi_{bc}\theta_{b(m)}\theta_{c(m)} \tag{21}$$

both of which would hold with the corresponding capacity factors, k′, substituted for the K's. Also, although chemically bonded phases and bulk liquids differ in their structures and related entropic contributions to ln K [8], eqs. 20 and 21 may still apply to the former stationary phases (provided $\theta_{b(s)}$ and $\theta_{c(s)}$ are negligible, as in the "collapsed chain limit" [8]), since these differences are wholly contained in the respective $K^{*}_{a(b)}$ and $K^{*}_{a(c)}$ terms. Lastly, we note the formal similarity between eqn. 21 (mixed mobile phase, b+c, and single component stationary phase, d) and the lattice-model equation one can derive for a mixed-liquid stationary phase, b+d, and an ideal-gas mobile phase in GLC [1]:

$$\ln K^{o}_{a(b+d)} = \theta_{b(s)}\ln K^{o}_{a(b)} + \theta_{d(s)}\ln K^{o}_{a(d)} + r_{d}\chi_{bd}\theta_{b(s)}\theta_{d(s)} \tag{22}$$

where K° refers to a GLC partition coefficient. In eqs. 21 and 22, the last r.h.s. term represents the effect of mixed-solvent interactions on K and, hence, on solute retention. Unfavorable mixed-solvent interactions (positive χ) reduce K in the LLC system and increase K in the GLC system.

3.1.2. Liquid-Solid Chromatography (LSC).
As discussed earlier, to obtain the salient equation for LSC, we let $\theta_{d(s)} = 0$ and recall that $f_{s} > 0$ and $z_{s} < z_{m}$, and that $f_{s} \to 0$ and $z_{s} \to z_{m}$ as the number of adsorbed solvent layers, τ_{s}, becomes large. Accordingly, eqn. 12 simplifies to give

$$\begin{aligned} \ln K_{a(b+c)} = &-(r_{a}-1)\ln[(z_{m}-1)/(z_{s}-1)] - [z_{s}f_{s}r_{a}/k_{B}T][e'_{at} - (e_{aa}/2)] \\ &+ r_{a}[\theta_{b(m)}(1-r_{b}^{-1}) + \theta_{c(m)}(1-r_{c}^{-1})] \\ &+ r_{a}[\chi_{ab}\theta_{b(m)} + \chi_{ac}\theta_{c(m)} - \chi_{bc}\theta_{b(m)}\theta_{c(m)}] \\ &- r_{a}[\theta_{b(s)}(1-r_{b}^{-1}) + \theta_{c(s)}(1-r_{c}^{-1})] \\ &- r_{a}(1-f_{s})[\chi_{ab}\theta_{b(s)} + \chi_{ac}\theta_{c(s)} - \chi_{bc}\theta_{b(s)}\theta_{c(s)}] \end{aligned} \tag{23}$$

which bears considerable resemblance to eqn. 14. If we let $\theta_{b(m)} = 0$ ($\theta_{c(m)} = 1$) and $\theta_{b(s)} = 0$ ($\theta_{c(s)} = 1$) in eqn. 23 and employ eqn. 13, we obtain

$$\begin{aligned} \ln \kappa_{a(c)} = &-(r_{a}-1)\ln[(z_{m}-1)/(z_{s}-1)] \\ &- [z_{s}f_{s}r_{a}/k_{B}T][e'_{at} - e_{ac} + (e_{cc}/2)] \end{aligned} \tag{24}$$

The constant, $\kappa_{a(c)}$, does not represent the true solute distribution coefficient with pure solvent c as the mobile phase, because the adsorption isotherm of component c and, hence, the competitive equilibrium between components a and c have not yet been taken into account. However, to obtain a preliminary equation which will be used in the next section and then revisited in part II, we combine eqs. 23 and 24 to obtain

$$\ln K_{a(b+c)} = \ln \kappa_{a(c)} + r_a(r_b^{-1} - r_c^{-1})\theta_{b(s)} - r_a(r_b^{-1} - r_c^{-1})\theta_{b(m)}$$
$$+ r_a(1 - f_s)[(\chi_{ac} - \chi_{ab} + \chi_{bc})\theta_{b(s)} - \chi_{bc}\theta_{b(s)}^2] \tag{25}$$
$$- r_a[(\chi_{ac} - \chi_{ab} + \chi_{bc})\theta_{b(m)} - \chi_{bc}\theta_{b(m)}^2]$$

which has a form quite similar to eqn. 18 and, hence, to eqn. 19.

3.2. FINAL EQUATIONS

In this section the sorption isotherms for solvent components b and c are derived. These equations are then combined with the equations derived in the previous section to obtain the final expressions for ln $K_{a(b+c)}$ in LLC and LSC.

3.2.1. *Solvent Sorption Isotherms.* We start by observing that the l.h.s. of eqn. 12 is simply $\ln(\theta_{a(s)}/\theta_{a(m)})$. Therefore, the required expressions for $\ln(\theta_{h(s)}/\theta_{h(m)})$, where h = b or c, may be readily generated by replacing the subscript a, wherever it appears on the r.h.s. of eqn. 12, by the subscript b or c and recalling that $\chi_{bb} = \chi_{cc} = 0$ (eqn. 13). The same can be done by manipulating eqn. 14 (LLC, with $\theta_{c(s)} = 0$) and eqn. 23 (LSC). Also recall that in the former case, $\theta_{b(s)} + \theta_{d(s)} = 1$, and in the latter, $\theta_{b(s)} + \theta_{c(s)} = 1$.

The sorption isotherm for the LLC case (component b only, since $\theta_{c(s)} = 0$) is [1]

$$\ln \theta_{b(s)} + r_b(r_d^{-1} - r_b^{-1})\theta_{b(s)} + r_b\chi_{bd}(1 - \theta_{b(s)})^2 =$$
$$\ln \theta_{b(m)} + r_b(r_c^{-1} - r_b^{-1})\theta_{b(m)} + r_b\chi_{bc}(1 - \theta_{b(m)})^2 + r_b(r_d^{-1} - r_c^{-1}) \tag{26}$$

For component b in LSC, it is

$$\ln \theta_{b(s)} + r_b(r_c^{-1} - r_b^{-1})\theta_{b(s)} + (1 - f_s)r_b\chi_{bc}(1 - \theta_{b(s)})^2 =$$
$$\ln \theta_{b(m)} + r_b(r_c^{-1} - r_b^{-1})\theta_{b(m)} + r_b\chi_{bc}(1 - \theta_{b(m)})^2 \tag{27}$$
$$+ (1 - r_b)\ln[(z_m - 1)/(z_s - 1)] - [z_s f_s r_b/k_B T][e_{bt}' - (e_{bb}/2)]$$

For component c in LSC, it is

$$\ln \theta_{c(s)} + r_c(r_b^{-1} - r_c^{-1})\theta_{c(s)} + (1 - f_s)r_c\chi_{bc}(1 - \theta_{c(s)})^2 =$$
$$\ln \theta_{c(m)} + r_c(r_b^{-1} - r_c^{-1})\theta_{c(m)} + r_c\chi_{bc}(1 - \theta_{c(m)})^2 \tag{28}$$
$$+ (1 - r_c)\ln[(z_m - 1)/(z_s - 1)] - [z_s f_s r_c/k_B T][e_{ct}' - (e_{cc}/2)]$$

Note that eqs. 27 and 28 would yield absolute isotherms. It is possible to relate these model results to experimentally accessible excess isotherms [14].

Another useful LSC expression can be obtained by multiplying eqn. 28 by r_b/r_c and subtracting the result from eqn. 27

$$\ln \theta_{b(s)} - (r_b/r_c)\ln(1 - \theta_{b(s)}) - 2(1 - f_s)r_b\chi_{bc}\theta_{b(s)} =$$
$$\ln \theta_{b(m)} - (r_b/r_c)\ln(1 - \theta_{b(m)}) - 2r_b\chi_{bc}\theta_{b(m)} + \ln K_{b(c)} \tag{29}$$

where

$$\ln K_{b(c)} = [1-(r_b/r_c)]\ln[(z_m-1)/(z_s-1)]$$
$$+(z_s f_s r_b/k_B T)(e'_{ct}-e'_{bt}+e_{bc}-e_{cc}) \tag{30}$$

is the logarithm of the infinite-dilution distribution coefficient of component b with a mobile phase of pure solvent c. Also,

$$\ln K_{b(c)}+(r_b/r_c)\ln K_{c(b)} = 2f_s r_b \chi_{bc} \tag{31}$$

where

$$\ln K_{c(b)} = [1-(r_c/r_b)]\ln[(z_m-1)/z_s-1)]$$
$$+(z_s f_s r_c/k_b T)(e'_{bt}-e'_{ct}+e_{bc}-e_{bb}) \tag{32}$$

is the logarithm of the infinite-dilution distribution coefficient of component c with a mobile phase of pure solvent b.

Equation 26 may now be used in conjunction with eqn. 14 (or 8) to define fully the LLC system. Similarly, eqn. 27 or 28 may be combined with eqn. 23 (or 25) to produce a simplified LSC expression. As will be seen in the next subsection, the final LLC and LSC expressions share the same universal form.

3.2.2. *Combined Equations.* Multiplying eqn. 26 by r_a/r_b and subtracting the result from eqn. 18 gives

$$\ln K_{a(b+c)} = r_a(\chi_{ac}-\chi_{ad}+\chi_{bd}-\chi_{bc})+(r_a/r_b)\ln(\theta_{b(s)}/\theta_{b(m)})$$
$$+r_a(\chi_{ab}-\chi_{ac}+\chi_{bc})\theta_{b(m)}-r_a(\chi_{ab}-\chi_{ad}+\chi_{bd})\theta_{b(s)} \tag{33}$$

Equation 33, which takes into account the competitive equilibrium between components a and b in both phases of the LLC system, has the following form

$$\ln K_{a(b+c)} = (r_a/r_b)[A+\ln(\theta_{b(s)}/\theta_{b(m)})+B\theta_{b(m)}-C\theta_{b(s)}] \tag{34}$$

where $\theta_{b(s)} < 1$ ($\theta_{b(s)} + \theta_{d(s)} = 1$) and

$$A = r_b(\chi_{ac}-\chi_{ad}+\chi_{bd}-\chi_{bc}) = C-B \tag{35a}$$

$$B = r_b(\chi_{ab}-\chi_{ac}+\chi_{bc}) \tag{35b}$$

$$C = r_b(\chi_{ab}-\chi_{ad}+\chi_{bd}) \tag{35c}$$

and where the size ratio, r_a/r_b, may be estimated from the ratio of van der Waals volumes, V_a^*/V_b^* [1,2,8]. Equation 34, which is given in terms of molecular interaction parameters and composition variables, and contains only two adjustable parameters (B and C), shows no explicit dependence on molecular structure (the combinatorial entropy terms have canceled out). Presumably, then, it should be applicable to solvent and solute molecules which are not necessarily chainlike, and, perhaps, even to an immobilized and/or cross-linked, polymeric stationary phase of sufficient film thickness.

The following LSC equations also take into account the competitive equilibrium in the chromatographic system, are in terms of molecular interaction parameters and composition variables, and should be essentially neutral with respect to solvent and solute molecular structure.

Multiplying eqn. 27 by r_a/r_b and subtracting the result from eqn. 25 gives

$$\ln K_{a(b+c)} = \ln K_{a(b)} + (r_a/r_b)\ln(\theta_{b(s)}/\theta_{b(m)})$$
$$+ r_a[\theta_{b(m)} - (1-f_s)\theta_{b(s)} - f_s](\chi_{ab} - \chi_{ac} + \chi_{bc}) \tag{36}$$

where

$$\ln K_{a(b)} = [1 - (r_a/r_b)]\ln[(z_m - 1)/(z_s - 1)]$$
$$+ (z_s f_s r_d/k_B T)(e'_{bt} - e'_{at} + e_{ab} - e_{bb}) \tag{37}$$

is the logarithm of the solute distribution coefficient with a mobile phase consisting of pure solvent b. If we assume that $z_m \approx z_s$, then the first term on the r.h.s. of eqn. 37 may be neglected with respect to the second term. Accordingly, from eqs. 13, 36 and 37, we obtain the same form as eqn. 34:

$$\ln K_{a(b+c)} \approx (r_a/r_b)[A + \ln(\theta_{b(s)}/\theta_{b(m)}) + B\theta_{b(m)} - C\theta_{b(s)}] \tag{38}$$

where

$$A \approx (z_s f_s r_b/k_B T)(e'_{bt} - e'_{at} + e_{ac} - e_{bc}) \tag{39a}$$

$$B = B(eqn.35b) = r_b(\chi_{ab} - \chi_{ac} + \chi_{bc}) \tag{39b}$$

$$C = (1-f_s)B = (1-f_s)r_b(\chi_{ab} - \chi_{ac} + \chi_{bc}) \tag{39c}$$

and where eqn. 38 contains three adjustable parameters (A, B and $(1-f_s)$). Therefore, the logarithmic form of $\ln K_{a(b+c)}$ appears both in the LLC case (eqn. 34) and in LSC (eqn. 38). It is reasonable to expect that it should also provide a decent description of the composition dependence of ln K in such systems as LC with chemically bonded phases (see next section).

Finally, multiplying eqn. 28 by r_a/r_c and subtracting the result from eqn. 25 gives, in terms of the composition of the "good" solvent (b), another useful form for LSC [2]:

$$\ln K_{a(b+c)} = \ln K_{a(c)} + (r_a/r_c)\ln[(1-\theta_{b(s)})/(1-\theta_{b(m)})]$$
$$+ r_a[(1-f_s)\theta_{b(s)} - \theta_{b(m)}](\chi_{ac} - \chi_{ab} + \chi_{bc}) \tag{40}$$

where

$$\ln K_{a(c)} = [1 - (r_a/r_c)]\ln[(z_m - 1)/(z_s - 1)]$$
$$+ (z_s f_s r_d/k_B T)(e'_{ct} - e'_{at} + e_{ac} - e_{cc}) \tag{41}$$

is the logarithm of the (true) solute distribution coefficient with a mobile phase consisting of pure solvent c.

4. Discussion

In the preceding section, "preliminary" equations describing the dependence of the logarithm of the solute (a) distribution coefficient with a binary liquid (b+c) mobile phase, ln $K_{a(b+c)}$, were derived using a lattice model. These equations were found to have the same general (quadratic) dependence on the composition variables ($\theta_{b(m)}$ and $\theta_{b(s)}$) of the system in the LLC case and in LSC. These preliminary equations were then combined with the respective solvent sorption isotherms (which, in fact, are also formally similar - compare, e.g., eqs. 26 and 27) to generate "final" equations for the LLC and LSC systems, which were again found to have the same general form.

Moreover, all of these equations are open to physical interpretation. They involve rigorous relationships containing molecular size parameters (r_i and r_i/r_j), intermolecular interaction parameters (χ_{ij}) and, in LSC, surface interaction terms (ε'_{ik}). In eqn. 40, for example, we first learn that with a mobile phase consisting solely of the "poor" solvent (c), for which $\theta_{b(m)} = \theta_{b(s)} = 0$, ln $K_{a(b+c)} = $ ln $K_{a(c)}$. From eqn. 41 (again with $z_m \simeq z_s$) we see, e.g., that the solute distribution coefficient (and net retention volume) should increase as each of the following molecular quantities increases: the size of the solute molecule (r_a), the extent of its contact with the adsorbent surface (f_a) and the magnitude of the free energy of interaction between a solute segment and the surface (ε'_{al}) relative to the corresponding solvent quantity (ε'_{cl}) (recall that ε_{ij} and ε'_{ik} are negative).

The second term on the r.h.s. of eqn. 40 represents the logarithm of the probability that a solute molecule can displace r_a/r_c solvent molecules in the stationary phase, relative to that probability in the mobile phase. The last term on the r.h.s. of eqn. 40 reflects the effect of replacing "poor" solvent (c) by "good" solvent (b) through increasing $\theta_{b(m)}$ and, hence, $\theta_{b(s)}$. Solute-solvent and solvent-solvent interaction energies in both phases are involved in this process. Since the solute-poor solvent interaction parameter, χ_{ac}, should be more positive (less favorable) than the solute-good solvent one, χ_{ab}, this interaction energy effect is seen to enhance $K_{a(b+c)}$ through the stationary-phase contribution and reduce it through the mobile phase contribution, more so as the solute molecular size (r_a) increases.

Unfortunately, but as is to be expected for such complex, condensed-phase systems, the ln K equations for LLC and LSC are not particularly simple ones, since knowledge of the stationary phase composition is, in general, required for their evaluation, and, with the exception of r_i/r_j, the molecular parameters cannot be readily predicted from first principles. Rather, the strengths of the present theory are: a) it is rigorous and comprehensive, b) its parameters have physical significance, c) it takes into account the solvent sorption isotherms, d) it unifies LLC and LSC theory, and e) it provides universal forms for ln K and the sorption isotherm that can be used to fit and interpret experimental results.

Although the present theory is just beginning to be applied, it would be useful to relate it to a simple expression that has been widely used in LC and, then, to examine in the light of the present theory, some of the caveats in applying and interpreting this simple expression. Let us start with eqn. 34 (LLC) or 38 (LSC) and make either one of the following sets of assumptions: a) $\theta_{b(m)}$ is sufficiently high so that $\theta_{b(s)}$ has reached a limiting, "saturation" value (not necessarily unity; recall that $\theta_{b(s)}$ in the LLC case must always be less than unity) and that B $\simeq 0$ (through cancellation of the interaction parameter terms constituting B), or b) the stationary phase terms (ln $\theta_{b(s)} - C\theta_{b(s)}$) are approximately offset by the $B\theta_{b(m)}$ term. Accordingly, eqn. 34 or 38 becomes, in either case,

$$\ln K_{a(b+c)} = (r_a/r_b)(D-\ln \theta_{b(m)}) \qquad (42)$$

where D = (A + ln $\theta_{b(s)}$ - $C\theta_{b(s)}$ + $B\theta_{b(m)}$) is now presumably independent of $\theta_{b(m)}$. Equation 42 has the familiar Snyder-Soczewinski form.

To examine eqn. 42 (and the assumptions leading to it) more closely, let us summarize and analyze LC data recently obtained in our laboratory. A monomeric C_{18} column (n-octadecyl chemically bonded phase) was used to obtain net retention volumes (solute retention volume minus column holdup volume [14]) of n-alkylbenzenes (primarily, ethylbenzene through hexylbenzene) at four temperatures (25,35,45 and 55° C) in the range 1.00 $\geq \theta_{b(m)} \geq$ 0.40 (intervals of 0.05 in $\theta_{b(m)}$) for acetonitrile (ACN) + water and methanol (MeOH) + water mixtures, where component b is the organic ("good") solvent component. Rather than analyzing absolute retention volumes (V_N) let us consider the methylene group contribution to $\ln V_N$, since the latter does not require specification of the total stationary-phase volume (V_s) to relate V_N and K, i.e., from eqs. 8 and 42.

$$(\partial \ln V_N/\partial n_a)=(\partial \ln K_{a(b+c)}/\partial n_a)=[\partial(r_a/r_b)/\partial n_a][D-\ln\theta_{b(m)}] \tag{43}$$

where n_a is the number of methylene groups in the alkylbenzene solute and, clearly $\partial(r_a/r_b)/\partial n_a = r_n/r_b = V_n^*/V_b^*$, where the subscript n refers to a methylene group and V_n^*/V_b^* is a ratio of van der Waals volumes [15]. For ACN, this ratio is 0.361 and for MeOH, it is 0.496. Rewriting eqn. 43, we have

$$G=(\partial \ln V_N/\partial n_a)=D'-(V_n^*/V_b^*)\ln\theta_{b(m)} \tag{44}$$

where $D' = (V_n^*/V_b^*) D$ and G is the methylene group contribution to $\ln V_N$ at a given T and $\theta_{b(m)}$.

Therefore, if the assumptions leading to eqn. 44 are valid, one should observe that G is a linear function of $-\ln \theta_{b(m)}$, with a slope which is virtually temperature independent and close to the theoretical value of V_n^*/V_b^*. To test this, $\ln V_N$ was first regressed as a linear function of n_a at each temperature and mobile-phase composition for both solvent mixtures. In every case, the correlation coefficient (R^2) is found to be in excess of 0.999. The results (G values) were then regressed, at each temperature, as a linear function of $-\ln \theta_{b(m)}$. The final results for the ACN + H_2O mobile phase system are summarized in Table 1.

TABLE 1. Results of linear regression of methylene group contribution to $\ln V_N$ as a function of - $\ln \Theta_{b(m)}$ for acetonitrile + water as the mobile phase

t(°C)	Intercept	Slope[a]	R^2
25	0.250	0.363	0.9973
35	0.237	0.367	0.9991
45	0.221	0.370	0.9992
55	0.209	0.372	0.9993

a. expected value: $V_n^*/V_b^* = 0.361$

Although the results in Table 1 are in excellent accord with eqn. 44, the basis of this agreement warrants closer examination (see later). For the MeOH + H_2O mobile phase system, good linear correlations are also found (R^2 values in excess of 0.99). However, the slopes, 0.909 (25°), 0.848 (35°), 0.789 (45°) and 0.753 (55°), are highly temperature dependent and much larger that the expected value of 0.496. This indicates that eqn. 44 does not, in fact, properly describe the MeOH + H_2O system and that one must return to the complete form (eqn. 34 or 38) for a meaningful analysis of the data.

Examining the ACN + H_2O system more closely, first, it seems unlikely that B= 0. This would require that segmental interaction energies for ACN are comparable to those for a methylene group (i.e., $\chi_{ab} \approx 0$ and $\chi_{ac} \approx \chi_{bc}$). Rather, χ_{ab} and B are expected to be positive. Also, employing the sorption isotherm equation for component b (eqn. 26 or 27), it can be shown that there is no realistic solution consistent with a nearly constant $\theta_{b(s)}$. Therefore, the validity of the first set of assumptions which could be invoked to obtain eqn. 44 is suspect.

More likely, there is some compensation occurring in the ACN + H_2O system, as invoked by the second assumption, i. e., $D = A + \ln \theta_{b(s)} - C\theta_{b(s)} + B\theta_{b(m)}$ is virtually independent of $\theta_{b(m)}$. As already stated, $B > 0$ and since $C = (1-f_s)B$ in LSC, then $C > 0$. Also, in the LLC case $C = r_b(\chi_{ab}-\chi_{ad}+\chi_{bd})$ where $\underline{a}$ refers to the methylene group and $\underline{d}$ to the alkyl chain in the C_{18} bonded phase; these are roughly equivalent energetically. Therefore, in the LLC case, $C \simeq 2r_b\chi_{ab} > 0$ also. Finally, since $\ln\theta_{b(s)}$ would become more negative, $-C\theta_{b(s)}$ less negative and $B\theta_{b(m)}$ more positive, as $\theta_{b(m)}$ decreases, then it is conceivable that such a compensation could occur. A more detailed description awaits incorporation of the experimental solvent sorption isotherms [14] in the data analysis.

In concluding this chapter, it should be mentioned that the present theoretical treatment is an adaptable one. For example, it could be extended to: a) a three-component mobile phase, b) a heterogeneous solid surface [16], c) other solute and solvent molecular structures [1,6-8,11]. d) differing f_s values for the solute and solvent molecules , and e) finite solute concentrations. Also, although it has not been pursued here, it would be straightforward to revisit eqn. 6, and allow for both a stationary liquid film $\underline{and}$ an active solid surface, to derive expressions for combined LLC/LSC.

References

1. D. E. Martire and R. E. Boehm, J. Phys. Chem., 91 (1987) 2433, and references cited therein.
2. D. E. Martire, J. Chromatogr., 452 (1988) 17.
3. J. C. Giddings, in A. Goldup (ed.), Gas Chromatography 1964, Elsevier, Amsterdam, 1965, p.3.
4. D. P. Poe and D. E. Martire, J. Chromatogr., 517 (1990) 3.
5. J. C. Giddings, Unified Separation Science, Wiley, New York, 1991.
6. D. E. Martire and R. E. Boehm, J. Liq. Chromatogr., 3 (1980) 753.
7. R. E. Boehm and D. E. Martire, J. Phys. Chem., 84 (1980) 3620.
8. D. E. Martire and R. E. Boehm, J. Phys. Chem., 87 (1983) 1045.
9. M. Jaroniec and D. E. Martire, J. Chromatogr., 351 (1986) 1.
10. T. L. Hill, An Introduction to Statistical Thermodynamics, Addison-Wesley, Reading, MA, 1960, Ch. 20 and 21.
11. D. E. Martire, J. Chromatogr., 406 (1987) 27.
12. D. C. Locke and D. E. Martire, Anal. Chem., 39 (1967) 921.
13. D. E. Martire and D. C. Locke, Anal. Chem., 43 (1971) 68.
14. A. Alvarez-Zepeda and D. E. Martire, J. Chromatogr., 550 (1991) 285.
15. A. Bondi, J. Phys. Chem., 68 (1964) 441.
16. D. E. Martire, J. Liq. Chromatogr., 11 (1988) 1779.

Acknowledgment

This material is based upon work supported by the National Science Foundation under Grant CHE-8902735. Aurelio Alvarez-Zepeda, Richard E. Boehm and Xin Zhang of Georgetown University are also thanked for many helpful discussions.

UNIFIED APPROACH TO THE THEORY OF CHROMATOGRAPHY. II. COMPRESSIBLE, NEAT MOBILE PHASE (SUPERCRITICAL FLUID CHROMATOGRAPHY)

DANIEL E. MARTIRE
REBECCA L. RIESTER
XIN ZHANG
Department of Chemistry
Georgetown University
Washington, D. C. 20057-2222
U. S. A.

ABSTRACT. The theoretical results derived in part I for liquid chromatography (LC) are transformed into equations describing the dependence of the solute distribution coefficient, K, on the reduced density, ρ_R, and reduced temperature, T_R, in supercritical fluid chromatography (SFC). This is readily achieved by exploiting isomorphic elements of critical-point behavior in a binary liquid system and a single-component fluid system. It is seen that the lnK expressions for SFLC and SFSC share a common form, viz., the sum of a zero-density contribution (lnK for ideal GLC or GSC), a mobile-phase contribution which is a quadratic function of ρ_R and a linear function of T_R^{-1}, and a coupling contribution which reflects the respective sorption isotherms. Each of these terms involves molecular size and interaction energy parameters. The significance and implications of the theoretical SFC results are discussed. A quantitative application of the SFLC theory - the analysis of recently obtained capillary-column data - is presented and discussed in detail.

1. Introduction

As was noted in part I, this unified approach stemmed from an effort to develop a rigorous and useful molecular theory of supercritical fluid chromatography (SFC), with the goal of explicitly relating solute retention (through the distribution coefficient, K) to the reduced variables of the mobile phase and other pure-component physical properties and/or molecular parameters. In this chapter we make use of "preliminary" equations, sorption-isotherm equations and "final" equations derived for liquid chromatography (LC) in part I to attain, in a surprisingly simple manner, this goal. The key to this straightforward extension lies in exploiting the isomorphism between the upper critical solution temperature (UCST) in a binary liquid system and the liquid-gas critical point in a single component fluid system [1].

It will be seen that the significant operational variables of the mobile phase are its reduced density and temperature (ρ_R and T_R, respectively) [2]. It will also be seen that, not unexpectedly, the lnK expression for a liquid-like (L) or absorptive stationary phase, i.e., SFLC, and that for a solid (S) or adsorptive stationary phase, i.e., SFSC, share the same common form - the sum of a zero-density contribution to lnK (ideal GLC or GSC), a mobile phase term which depends on ρ_R and T_R, and a coupling term which reflects the respective sorption isotherms. The implications of the theory and its application to the analysis of experimental SFC data are discussed and illustrated.

2. Theoretical Model

We first scale the system by letting $r_c = 1$ (see later) and then note that the critical solution condition

F. Dondi and G. Guiochon (eds.),
Theoretical Advancement in Chromatography and Related Separation Techniques, 275–288.
© 1992 *Kluwer Academic Publishers*.

applicable to a binary mixture of solvent components b+c (the "good" and "poor" solvents, respectively) in the lattice model of part I is [3]

$$T^* = 2T\chi_{bc}r_b/(1+\sqrt{r_b})^2 \tag{1}$$

$$\theta_b^* = (1+\sqrt{r_b})^{-1} \tag{2}$$

where

$$\chi_{ij} = (z_c/2k_BT)(2e_{ij}-e_{ii}-e_{jj}) \tag{3}$$

In eqs. 1 and 2, T^* refers to the UCST (above which the solvent mixture is homogeneous over the entire composition region) and θ_b^* refers to the critical volume fraction of "good" solvent, i.e., the composition corresponding to T^* in the $T - \theta_{b(m)}$ phase diagram [3]. Introducing reduced variables (subscript R), we have

$$T_R = T/T^* = (1+\sqrt{r_b})^2/2r_b\chi_{bc} \tag{4}$$

$$\theta_{b(m),R} = \theta_{b(m)}/\theta_b^* = \theta_{b(m)}(1+\sqrt{r_b}) \tag{5}$$

where $\theta_{b(m)}$ is the volume fraction of "good" solvent in the binary liquid mixture.

By invoking the isomorphism between the critical behavior in a binary liquid system and that in a single component fluid system [1], one may utilize the direct correspondence between the volume fraction of "good" solvent (b) in the former and the volume fraction of physical space occupied by the molecules in the latter, and similarly with the "poor" solvent (c) and unoccupied physical space ("holes" or vacancies occupying single sites ($r_c = 1$) in the lattice model). It follows from eqn. 5 that

$$\theta_{b(m),R} = \rho_{b(m),R} = \rho_{b(m)}/\rho_b^* = \theta_{b(m)}(1+\sqrt{r_b}) \tag{6}$$

where $\theta_{b(m)}$ is now the volume fraction of space occupied by the hard cores of the molecules in the mobile phase (i.e., the fraction relative to what it would be in a hypothetical close-packed molecular arrangement of these cores, for which $\theta_{b(m)} = 1$), $1-\theta_{b(m)}$ is the volume fraction of unoccupied space, $\rho_{b(m)}$ is the actual density of the mobile-phase fluid, ρ_b^* is its critical density and $\rho_{b(m),R}$ is its reduced density. Deleting the subscripts b and b(m) in the density terms, we obtain the first of our transforming equations

$$\rho_R = \rho/\rho^* = \theta_{b(m)}(1+\sqrt{r_b}) \tag{7}$$

which relates the experimental reduced density to the occupied volume fraction and a molecular-size parameter of the model. Therefore, the single-component mobile phase is now a compressible one, where replacing void space by occupied space through increasing the density is comparable to replacing "poor" solvent by "good" solvent in a mixed-liquid mobile phase through increasing the volume fraction of the latter.

Turning to eqn. 4, $T_R = T/T^*$ becomes the usual reduced temperature of a single-component fluid, where T^* is its critical temperature. It also follows that all interactions involving component c (now representing void space in the model) should be set equal to zero, i.e., $\varepsilon'_{ci} = 0$ and $\varepsilon_{ic} = 0$ (i = a,b,c). Accordingly, with the aid of eqn. 3, eqn. 4 becomes

$$T_R = T/T^* = (1+\sqrt{r_b})^2/2r_b\chi_{bc} = -(1+\sqrt{r_b})^2 k_B T/r_b z_c e_{bb} \qquad (8)$$

where eqn. 8, the second transforming equation, relates the critical temperature to molecular parameters of the lattice-fluid model.

One can now transform the LC equations in part I into SFC equations by using eqs. 3, 7 and 8 and by setting equal to zero all segmental interactions involving component c. The important results of this straightforward, but tedious, procedure are summarized in the following sections. First, eqs. 15 and 18 (quadratic form of lnK), eqn. 26 (sorption isotherm of component b) and eqn. 33 (logarithmic form of lnK) from part I are employed to generate equations for an absorptive stationary phase (SFLC), which continues to be incompressible in the model system (through the absence of component c, which is now void space). Secondly, eqs. 24 and 25 (quadratic form of lnK), eqs. 27 and 28 (adsorption isotherms) and eqs. 40 and 41 (logarithmic form of lnK) from part I are used to obtain equations for an adsorptive stationary phase (SFSC), where the adsorbed surface layers of component b are now compressible (through the allowed presence of free space). It is shown that, in both cases, equations describing ideal and moderately nonideal gas chromatography (GLC and GSC) stem from the SFC expressions.

3. Supercritical Fluid-Liquid (Absorption) Chromatography, SFLC

Let us start with eqn. 15 in part I. Anticipating its application to capillary GLC and SFLC with cross-linked and/or immobilized polymeric stationary phases, we let $r_d \to \infty$ (recall that $r_c = 1$) and multiply r_a in the first r.h.s. term by a constant factor u to correct the combinatorial entropy [2], yielding

$$\ln K_o = -r_a u + r_a(\chi_{ac} - \chi_{ad}) \qquad (9)$$

where K_o is now the solute distribution coefficient at zero mobile-phase density, corresponding to ideal GLC. Using eqs. 3 and 8, with $\varepsilon_{ic} = 0$ for i = a,b,c, eqn. 9 becomes

$$\ln K_o = (r_a/r_b)[-r_b u + \{(1+\sqrt{r_b})^2/T_R\}\{(e_{ad}/e_{bb}) - (e_{dd}/2e_{bb})\}] \qquad (10)$$

where K_o is a function of the reduced temperature of the mobile-phase fluid (T_R) and, as will be the case in all of the SFC equations, absolute segmental interactions (ε_{ij}) are replaced by values relative to segmental interactions between the mobile-phase fluid molecules (ε_{bb}). Equation 10 reasonably indicates that $\ln K_o$ increases as the size of the solute molecule increases, as the combinatorial entropy contribution decreases, as the strength of solute-stationary liquid segmental interactions (ε_{ad}) increases and as the strength of stationary liquid segmental interactions (ε_{dd}) decreases.

The other terms in the quadratic form of lnK for LLC, given by eqs. 18 and 19 in part I, may be similarly manipulated using eqs. 3, 7 and 8 (again, with $r_d \to \infty$, $r_c = 1$ and $\varepsilon_{ic} = 0$ to give the corresponding SFLC form [2]

$$\ln K = \ln K_o(T_R) + F(T_R, \rho_R) + \Delta(T_R, \theta_{b(s)}) \qquad (11)$$

where

$$F(T_R, \rho_R) = (r_a/r_b)[(\sqrt{r_b}-1)\rho_R - (1+\sqrt{r_b})(e_{ab}/e_{bb})(\rho_R/T_R) + (\rho_R^2/2T_R)] \tag{12}$$

$$\Delta(T_R, \theta_{b(s)}) = (r_a/r_b)[\theta_{b(s)} + r_b(\chi_{ad} - \chi_{ab} + \chi_{bd})\theta_{b(s)} - r_b\chi_{bd}\theta_{b(s)}^2] \tag{13}$$

and where $\ln K_o$ is given by eqn. 10. The term $F(T_R, \rho_R)$, which depends only on the mobile phase, is a quadratic function of ρ_R and a linear function of T_R^{-1} [2]. It decreases as the strength of the solute-mobile phase segmental interactions (ε_{ab}) increases, more so as the size of the solute molecules increases. The coupling term, $\Delta(T_R, \theta_{b(s)})$, which is also proportional to the size of the solute molecules represents the effect of the swelling of the stationary phase due to sorption of the carrier fluid (component b). Clearly, $\theta_{b(s)} = 0$ and $\Delta = 0$ when $\rho_R = 0$. The evaluation of $\theta_{b(s)}$ and Δ requires the sorption isotherm relating $\theta_{b(s)}$ and ρ_R (see below). Also, for future reference note that the χ_{ij}'s in eqn. 13 may, with the use of eqs. 3 and 8, be written as follows

$$\chi_{ij} = [(1+\sqrt{r_b})^2/2r_bT_R][(e_{ii}+e_{jj}-2e_{ij})/e_{bb}] \tag{14}$$

Therefore, both $F(T_R, \rho_R)$ and $\Delta(T_R, \theta_{b(s)})$ depend on molecular size (r_i) and interaction energy (ε_{ij}) parameters.

Before considering the sorption isotherms, let us examine eqs. 10-13 at very low mobile phase densities where we may neglect terms involving ρ_R^2 and the coupling or swelling term, Δ. Reemploying eqn. 7, eqs. 10-13 give

$$\ln K = \ln K_o + r_a[\{1-r_b^{-1}\} - \{(1+\sqrt{r_b})^2/r_bT_R\}(e_{ab}/e_{bb})]\theta_{b(m)} \tag{15}$$

where $\theta_{b(m)} = V_b^\square/V_b^\circ$ is the occupied volume fraction and where $V_b^\square$ and V_b° are, respectively, the hard-core and actual molar volumes, the latter pertaining to the low-density (and pressure) mobile phase. At sufficiently low density, $V_b^\circ \simeq RT/<P>$, where R is the molar gas constant and $<P>$ is the average mobile-phase pressure in the column; therefore, $\theta_{b(m)} \simeq <P>V_b^\square/RT$. Also, since $r_a/r_b = V_a^\square/V_b^\square$ and the mixed (solute-carrier) or interaction second virial coefficient, B_{ab}, is given by [2]

$$2B_{ab} = r_a[1 - \{(1+\sqrt{r_b})^2/r_bT_R\}(e_{ab}/e_{bb})]V_b^\square \tag{16}$$

then eqn. 15 becomes

$$\ln K = \ln K_o + [(2B_{ab} - V_a^\square)/RT]<P> \tag{17}$$

Once again we see that the general model equations yield a familiar result - in this case, the first order correction for mobile-phase nonideality in GLC [2].

Considering now the sorption isotherm of component b, eqn. 26 in part I can be transformed by employing eqs. 3 (or 14), 7 and 8 (with $r_d \to \infty$, $r_c = 1$ and $\varepsilon_{ic} = 0$), and by multiplying r_a in the last r.h.s. term of eqn. 26 by u to correct the combinatorial entropy. The result for the corresponding sorption isotherm of the carrier fluid in SFLC is

$$\ln \theta_{b(s)} - \theta_{b(s)} + r_b \chi_{bd}(1-\theta_{b(s)})^2 = \ln[\rho_R/(1+\sqrt{r_b})] + (\sqrt{r_b}-1)\rho_R$$
$$+[(1+\sqrt{r_b})^2/2T_R][1-\{\rho_R/(1+\sqrt{r_b})\}]^2 - r_b u \tag{18}$$

where χ_{bd} is related to T_R and molecular parameters through eqn. 14.

The logarithmic form of lnK for LLC, given by eqn. 33 in part I, may be similarly manipulated to obtain the companion SFLC equation:

$$\ln K = (r_a/r_b)[\{(1+\sqrt{r_b})^2/T_R\}\{(e_{ad}/e_{bb})-(e_{bd}/e_{bb})\}+\ln \theta_{b(s)}$$
$$-\ln\{\rho_R/(1+\sqrt{r_b})\}+\{(1+\sqrt{r_b})/T_R\}\{1-(e_{ab}/e_{bb})\}\rho_R$$
$$-\{(1+\sqrt{r_b})^2/T_R\}\{1-(e_{ab}/e_{bb})+(e_{ad}/e_{bb})-(e_{bd}/e_{bb})\}\theta_{b(s)}] \tag{19}$$

which, recalling that $\theta_{b(m)} = \rho_R/(1+\sqrt{r_b})$, retains the same form as eqn. 34 in part I, i.e.,

$$\ln K = (r_a/r_b)[A+\ln(\theta_{b(s)}/\theta_{b(m)})+B\theta_{b(m)}-C\theta_{b(s)}] \tag{20}$$

and where the combinatorial entropy terms have canceled out.

To conclude this section, we relate K to the solute capacity factor, k', where $k' = K(V_s/V_m)$ and where V_s and V_m are the total volumes of the stationary and mobile phases, respectively. From eqn. 11, for example, we obtain

$$\ln k' = \ln k'_o + F(T_R,\rho_R) + \Delta(T_R,\theta_{b(s)}) - \ln(1-\theta_{b(s)}) \tag{21}$$

where $k'_o = K_o(V_s/V_m)_o$, and where $(V_s/V_m)_o$ is the phase-volume ratio at zero density. The last term on the r.h.s. of eqn. 21 corrects the phase-volume ratio for the swelling of the stationary phase (which increases V_s), based on the assumption that $V_m/V_s \gg 1$ (the normal case in capillary SFLC). Alternatively,

$$\ln k' = \ln K + \ln(V_s/V_m)_o - \ln(1-\theta_{b(s)}) \tag{22}$$

Equations 10-14 and 18-21 are the primary SFLC equations derived in this section. With all the molecular parameters involved, they appear to be formidable. However, the treatment simplifies considerably in actual applications, as will be illustrated.

4. Supercritical Fluid-Solid (Adsorption) Chromatography, SFSC

Starting with the quadratic form of lnK for LSC, given by eqs. 24 and 25 in part I, and following the same procedure as in the previous section, one obtains the following SFSC counterparts:

$$\ln K = \ln K_o(T_R) + F(T_R,\rho_R) + \Delta(T,\theta_{b(s)}) \tag{23}$$

where

$$\ln K_o = (1-r_a)\ln[(z_m-1)/(z_s-1)]+(r_a/r_s)[\{f_s(1+\sqrt{r_b})^2/T_R\}(e'_{ad}/e_{bb})]$$ (24)

and where $F(T_R, \rho_R)$ is still given by eqn. 12. Therefore, the equations for SFSC and SFLC (see eqn. 11) have exactly the same form.

$$\Delta(T,\theta_{b(s)}) =$$
$$(r_a/r_b)[(1-r_b)\theta_{b(s)}+\{(1-f_s)(1+\sqrt{r_b})^2/T_R\}\{(e_{ab}/e_{bb})\theta_{b(s)}-(\theta_{b(s)}^2/2)\}]$$ (25)

In eqn. 24, K_o is the solute distribution coefficient for ideal GSC. The first term on the r.h.s. arises from the change in the configurational entropy of the solute when it is transferred from an ideal-gas mobile phase to a bare adsorbent. The second term reflects the interaction free energy of adsorption of an isolated solute molecule on the surface. The SFSC coupling term (Δ) represents the contribution to lnK arising from entropic and energetic effects in the surface phase due to adsorption of the carrier fluid. Note that $\theta_{b(s)}$ depends on ρ_R through the adsorption isotherm.

By again applying eqs. 3 (or 14), 7 and 8 (with $r_c = 1$, $\varepsilon_{ic} = 0$ and $\varepsilon'_{ci} = 0$), the adsorption isotherm of solvent component b in LSC (eqn. 27 in part I) can be transformed into the corresponding expression for the carrier fluid in SFSC:

$$\ln\theta_{b(s)}+(r_b-1)\theta_{b(s)}+[(1-f_s)(1+\sqrt{r_b})^2/2T_R][1-\theta_{b(s)}]^2 =$$
$$\ln[\rho_R(1+\sqrt{r_b})]+(\sqrt{r_b}-1)\rho_R+[(1+\sqrt{r_b})^2/2T_R][1-\{\rho_R(1+\sqrt{r_b})\}]^2$$
$$+(1-r_b)\ln[(z_m-1)/(z_s-1)]+[f_s(1+\sqrt{r_b})^2/T_R][e'_{bd}/e_{bb}]$$ (26)

where the last two terms on the r.h.s. of eqn. 26 constitute the logarithm of the limiting ($\rho_R \to 0$) adsorption isotherm, $(\theta_{b(s)}/\theta_{b(m)})_o$, or the infinite-dilution distribution coefficient of the carrier fluid, $K_{o(b)}$:

$$\ln K_{o(b)} = (1-r_b)\ln[(z_m-1)/(z_s-1)]+[f_s(1+\sqrt{r_b})^2/T_R][e'_{bd}/e_{bb}]$$ (27)

The adsorption isotherm of solvent component c in LSC, given by eqn. 28 in part I, also has a counterpart in SFSC. Following the now familiar procedure, we obtain

$$\ln(1-\theta_{b(s)})+[(r_b-1)/r_b]\theta_{b(s)}+[(1-f_s)(1+\sqrt{r_b})^2/2r_bT_R]\theta_{b(s)}^2 =$$
$$\ln[1-\{\rho_R(1+\sqrt{r_b})\}]+[(\sqrt{r_b}-1)/r_b]\rho_R+(\rho_R^2/2r_bT_R)$$ (28)

The r.h.s. of eqn. 28 is simply the negative of a dimensionless pressure-to-temperature ratio, Φ, of the pure mobile-phase fluid [2]

$$\Phi = Pv_o/k_BT =$$
$$-[\ln[1-\{\rho_R(1+\sqrt{r_b})\}]+[(\sqrt{r_b}-1)/r_b]\rho_R+(\rho_R^2/2r_bT_R)]$$ (29)

where eqn. 29 is essentially the equation of state of the pure fluid and v_o is defined in part I. Therefore, the l.h.s. of eqn. 28 is equal to the negative of Φ for the adsorbed surface layer(s) and eqn. 28 is a statement that the two phases are in equilibrium at the same temperature and pressure [2].

The most useful logarithmic form of lnK to transform from LSC into SFSC is the one given by eqs. 40 and 41 in part I [4]. Following the same procedure, we obtain

$$\ln K = \ln K_o + r_a \ln[(1-\theta_{b(s)})/(1-\theta_{b(m)})] \tag{30}$$
$$+ (r_a/r_b)[\{(1+\sqrt{r_b})^2/T_R\}(e_{ab}/e_{bb})\{(1-f_s)\theta_{b(s)}-\theta_{b(m)}\}]$$

where $\theta_{b(m)} = \rho_R/(1+\sqrt{r_b})$ (eqn. 7), $\ln K_o$ is given by eqn. 24 and where eqn. 30 is analogous to eqn. 20. In eqn. 30, which links ideal GSC to nonideal GSC (see below), SFSC and LSC, the second term on the r.h.s. is associated with the statistics of the displacement process (the relative availability to the solute of free space in the two phases), while the third term reflects the exchange interaction energy associated with the competitive equilibrium [4].

Let us examine eqn. 30 at very low mobile-phase densities. Expanding $\ln(1-\theta_{b(s)})$ and $\ln(1-\theta_{b(m)})$ through first order in ρ_b and collecting terms, we have

$$\ln K = \ln K_o + r_a[1-\{(1+\sqrt{r_b})^2/r_b T_R\}(e_{ab}/e_{bb})][\theta_{b(m)}-\theta_{b(s)}] \tag{31}$$
$$- r_a[f_s(1+\sqrt{r_b})^2/r_b T_R][e_{ab}/e_{bb}]\theta_{b(s)}$$

Making use of $\theta_{b(s)} \simeq \theta_{b(m)} K_{o(b)}$, which holds at very low densities (see eqs. 26 and 27), recalling that $\theta_{b(m)} \simeq V_b^{\square} <P>/RT$ and $r_a/r_b = V_a^{\square}/V_b^{\square}$, and applying eqn. 16, we obtain the following result from eqn. 31:

$$\ln K \approx \ln K_o + [2B_{ab}/RT][1-(1-f_s)K_{o(b)}]<P> - [f_s r_b V_a^{\square}/RT]<P> \tag{32}$$

If, for example, $(1-f_s)K_{o(b)} > 1$ (<1) and $B_{ab} < 0$, the first correction term on the r.h.s. of eqn. 32 would be positive (negative), leading to an increase (decrease) in lnK with increasing $<P>$. The second correction term, which also involves the extent of direct molecular contact with the adsorbent (f_s), is always negative, leading to a decrease in lnK with increasing $<P>$. Equation 32, which describes moderately nonideal GSC, is more complex than its GLC counterpart (eqn. 17). It warrants testing and subsequent interpretation.

In concluding this section, note that eqn. 12 and eqs. 23-30 are the primary SFSC equations generated in this section. Equation 29, the model equation of state for the neat mobile-phase fluid, applies to both SFLC and SFSC. Also, given that the equations derived for SFLC and SFSC share a common form (compare, for example, eqs. 10-13 (SFLC) to eqs. 12, 23-25 (SFSC)), it is reasonable to expect that intermediate stationary phases, such as chemically modified (bonded) phases, should also have this general form.

5. Discussion

The SFC equations for a compressible, neat mobile phase, presented in the previous two sections, were derived simply and directly from corresponding LC equations for a binary liquid mobile phase (part I) by exploiting the isomorphism between the critical behavior in a binary liquid system and that in a single-

282

component fluid system. Although the equations were, for good reason, based on a versatile lattice model, similar transformations could also be carried out on results from any statistical thermodynamic model of LC.

Equations similar to those derived for SFSC have already been successfully applied to the theoretical analysis of the GSC retention behavior of n-butane on graphitized carbon black appreciably modified by (monolayer) adsorption of propane from the carrier-gas stream [4]. Now we consider two preliminary applications of the present theory to the analysis and interpretation of SFC data recently obtained in our laboratory. The first, in the area of packed-column SFC, involves a qualitative comparison and interpretation. The second, in the area of capillary-column SFC, is more detailed.

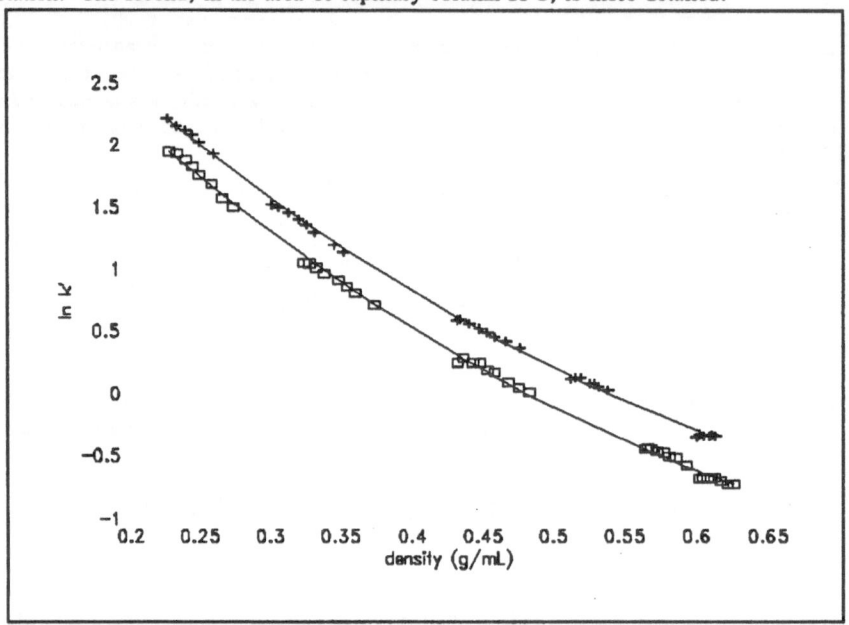

Figure 1. Dependence of the logarithm of the capacity factor, k' , of ethylbenzene at 80°C on the temporal average density, $<\rho>_t$ of CO_2 in packed-column SFC; C_{18}-modified silica (+), bare silica (□).

Shown in Fig. 1 is the SFC retention behavior (lnk′) of ethylbenzene at 80°C, as a function of the average mobile phase density of CO_2, determined using two different columns packed with 10-μm particles: bare silica and C_{18}- modified silica. Each cluster of points for a given column represents one of five different outlet pressure settings. Inlet pressures were varied giving rise to different average column pressures (and densities) for a given outlet pressure setting. The mobile-phase densities plotted in Fig. 1 are the temporal average densities of CO_2, $<\rho>_t$ [5,6].

It is apparent that the sets of data on the two columns run <u>nearly</u> parallel to each other. This behavior is also observed for other solutes and temperatures, and other workers [7,8] have reported it.

Reminiscent of a line of reasoning used in part I, the observed behavior may be rationalized as follows. Consider the universal quadratic form of lnK (eqs. 10-13 in SFLC; eqs. 12 and 23-25 in SFSC) and keep in mind that different intercepts, lnK_o, are to be expected for different columns, but that the dependence of the mobile-phase term, $F(T_R,\rho_R)$, on density at fixed temperature must be the same for the two columns. That in itself would explain the near parallelism seen in Fig. 1, were it not for the coupling term, $\Delta(T_R, \theta_{b(s)})$, which also contributes to lnK. This suggests one of the following possibilities with respect to the coupling term: a) it is negligible compared to $F(T_R, \rho_R)$, b) it has roughly the same dependence on density at fixed temperature for each of the two columns, or c) it is essentially constant for both columns over the operational density range (suggesting virtual saturation of the respective stationary phases by sorbed CO_2 at $\rho \simeq 0.2$ g/mL). The last possibility seems to be the least likely one, for the same reason that it did in the LC case discussed in part I. In any event, by analyzing the entire array of SFC data on the two columns, in conjunction with recently published excess isotherm data for supercritical CO_2 on similar column materials [9], a definitive and quantitative interpretation, based on the theory, is anticipated.

The second example, which involves a quantitative and more detailed application of the unified theory to SFC, concerns retention data (k′ values, based on methane as the unretained marker compound) for n-alkane solutes in a capillary-column SFC system employing supercritical CO_2 as the mobile phase. The stationary phase is SB-OCTYL 50 (cross-linked, 50% n-octyl, 50% methyl polysiloxane) from Lee Scientific, Inc. (Salt Lake City, UT), which has predominantly alkyl character. Measurements were made at seven temperatures (320 to 380 K, in intervals of 10°) and over as wide a mobile-phase density range as was feasible at each temperature.

We again consider the methylene group contribution to lnk′, since it is independent of the total stationary phase volume which is expected to increase with increasing mobile-phase density as sorbed CO_2 swells the stationary phase [9,10]. From eqn. 22

$$(\partial \ln k'/\partial n_a)_{T,\rho} = (\partial \ln K/\partial n_a)_{T,\rho} = G \qquad (33)$$

where n_a is the number of methylene groups in the alkane solute and G is the methylene group contribution to lnk′.

Before presenting, analyzing and discussing the results for G, let us recall the model equation of state (eqn. 29) and the primary equations derived for SFLC, viz, eqs. 10-14, 18 and 19. First, the assignment of r_b, the number of segments in a CO_2 molecule, was previously done by fitting limited SFC data - an indirect method which gave a value of $r_b = 4.61$ [2]. This assignment is now revised by directly fitting the extensive PVT data for CO_2 [6] to the model equation of state, giving a best-fit value of $r_b = 5.04$ which we round to $r_b = 5$. Next, given the nonpolar character of the molecules in the system the geometric mean combining rule is used to assess unlike segmental interaction energies, ε_{ij}, i.e.,

$$\varepsilon_{ij}/\varepsilon_{bb} = (\varepsilon_{ii}\varepsilon_{jj})^{1/2}/\varepsilon_{bb} \qquad (34)$$

This reduces the number of interaction energy parameters in the SFLC equations to a mere two, $x = (\varepsilon_{aa}/\varepsilon_{bb})^{1/2}$ and $y = (\varepsilon_{dd}/\varepsilon_{bb})^{1/2}$

Preliminary analysis of the G results, employing eqs. 10-14 (for the quadratic form of lnK) or eqn. 19 (the logarithmic form of lnK) and eqn. 18 (the sorption isotherm of CO_2), reveals that the best-fit value of y should be close to unity and that the fits are relatively insensitive to y. Accordingly, a value of $y = 1$ is assigned, suggesting that mutual CO_2 segmental interaction energies (ε_{bb}) are the same as mutual stationary-phase ones (ε_{dd}) and that, since the stationary phase is primarily alkyl in nature, x should be close to unity as well (see later). Also, setting $(\varepsilon_{dd}/\varepsilon_{bb})^{1/2} = 1$, from eqs. 14 and 34 one finds that $\chi_{bd} = 0$ and $\chi_{ad} = \chi_{ab}$. Therefore, from eqn. 13, $\Delta(T_R, \theta_{b(s)}) = (r_a/r_b)\theta_{b(s)}$ provides a direct relationship between the coupling term in lnK and the extent of swelling of the stationary phase, as determined by the volume fraction of sorbed CO_2 ($\theta_{b(s)}$).

With $(\varepsilon_{dd}/\varepsilon_{bb})^{1/2} = y = 1$, $(\varepsilon_{aa}/\varepsilon_{bb})^{1/2} = x$ and eqn. 34, the quadratic form of lnK for SFLC becomes considerably simplified:

$$\ln K = \ln K_o(T_R) + F(T_R, \rho_R) + \Delta(T_R, \theta_{b(s)}) \tag{35}$$

where

$$\ln K_o(T_R) = (r_s/r_b)[-r_b u + \{(1 + \sqrt{r_b})^2/T_R\}\{x - (1/2)\}] \tag{36}$$

$$F(T_R, \rho_R) = (r_s/r_b)[(\sqrt{r_b} - 1)\rho_R - (1 + \sqrt{r_b})(x)(\rho_R/T_R) + (\rho_R^2/2T_R)] \tag{37}$$

$$\Delta(T_R, \theta_{b(s)}) = (r_s/r_b)\theta_{b(s)} \tag{38}$$

where x and u are the unknown molecular parameters. From eqs. 33 and 35-38,. the methylene group contribution to lnK is

$$G = G_o + F_n(T_R, \rho_R) + \Delta_n(T_R, \theta_{b(s)}) \tag{39}$$

where G_o, F_n and Δ_n, the constituent parts of the methylene group contribution, are given by eqs. 36,37 and 38, respectively, with $r_s/r_b = r_n/r_b = V_n^{\square}/V_b^{\square}$ (ratio of van der Waals volumes), where the subscript n refers to a methylene group and $r_n/r_b = 0.5193$ [2].

The sorption isotherm becomes

$$\ln \theta_{b(s)} - \theta_{b(s)} = \ln[\rho_R/(1 + \sqrt{r_b})] + (\sqrt{r_b} - 1)\rho_R \\ + [(1 + \sqrt{r_b})^2/2T_R][1 - \{\rho_R/(1 + \sqrt{r_b})\}]^2 - r_b u \tag{40}$$

where u is the only adjustable molecular parameter appearing.

Finally, the logarithmic form of lnK simplifies to

$$\ln K = (r_s/r_b)[\ln \theta_{b(s)} - \ln\{\rho_R/(1 + \sqrt{r_b})\} \\ + \{(1 + \sqrt{r_b})^2/T_R\}(x - 1)(1 - \{\rho_R/(1 + \sqrt{r_b})\})] \tag{41}$$

where x is the only adjustable molecular parameter appearing. The methylene group contribution to lnK can be determined from eqn. 41 by again setting $r_s/r_b = r_n/r_b = 0.5193$. Note that eqs. 35-39 can be generated by combining eqs. 40 and 41.

In eqs. 35-41, $r_b = 5$, and $\rho_R = \rho(g/mL)/0.468$ and $T_R = T(K)/304.2$ (for CO_2) can be calculated from the experimental conditions. However, neither volumetric (swelling) nor composition (sorption) data are available for SB-OCTYL 50, to permit the calculation or reliable estimation of $\theta_{b(s)}$, which appears in the lnK and sorption isotherm equations. Nonetheless, the reasonableness of the model equations for SFLC can be judged by the goodness of the fits (see below), by whether or not the determined molecular parameters (x and u) are physically sensible, and by comparing the calculated $\theta_{b(s)}$ values with available swelling and sorption data on related CO_2/polymer systems [9-12].

To continue, the measured methylene group contributions, G, in conjunction with eqs. 40 and 41, were used to determine the best-fit values of x and u, and consistent $\theta_{b(s)}$ values. This was done at each temperature, resulting in slightly temperature dependent x values [range: 0.974 (320 K) to 1.044 {380

K)] and u values [range: 0.713 (320 K) to 0.702 (380 K)]. At 350 K, for example, x = 1.017 and u = 0.704, and the generated G values fall within ± 0.005 (average difference) of the experimental ones. (This average difference is roughly twice the experimental uncertainty in G.) Note that $x = (\varepsilon_{aa}/\varepsilon_{bb})^{1/2}$ is indeed close to unity and that $u < 1$. Others have also found that the uncorrected formula (u = 1) overestimates the stationary-phase combinatorial entropy [13].

The experimental results at three of the temperatures and the theoretical results at 350 K are summarized in Figures 2 to 5.

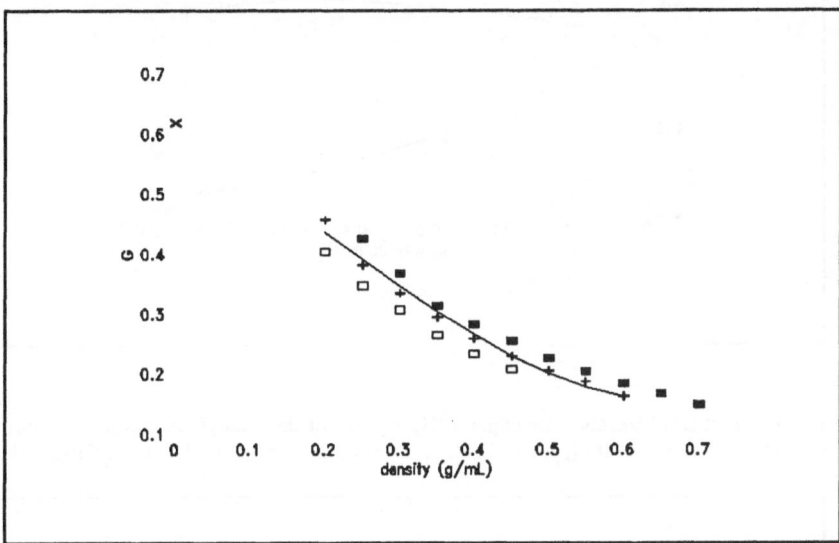

Figure 2. Dependence of the methylene group contribution to lnk' (G) on CO_2 density (ρ) at 330 K (■), 350 K (+) and 370 K (□), for n-alkane solutes and SB-OCTYL 50 stationary phase; solid line: calculated at 350 K using eqs. 35-39, with x = 1.017, u = 0.704 and $\theta_{b(s)}$ values from consistent fit based on eqs. 40 and 41; (x) calculated G_o value at 350 K (0.616).

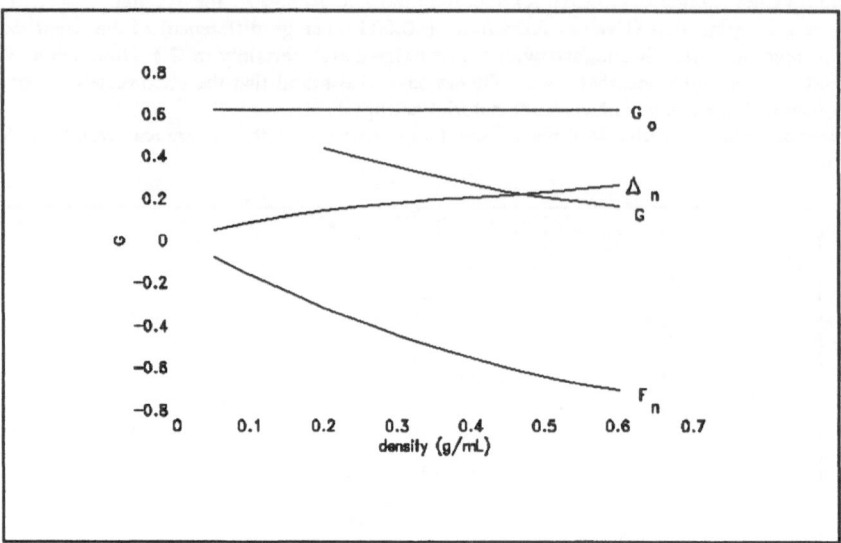

Figure 3. Contribution of the constituent parts (G_o, F_n, Δ_n) of the methylene group contribution to lnk´ (G) as a function of CO_2 density (ρ), calculated at 350 K (for additional details, see Fig. 2 legend).

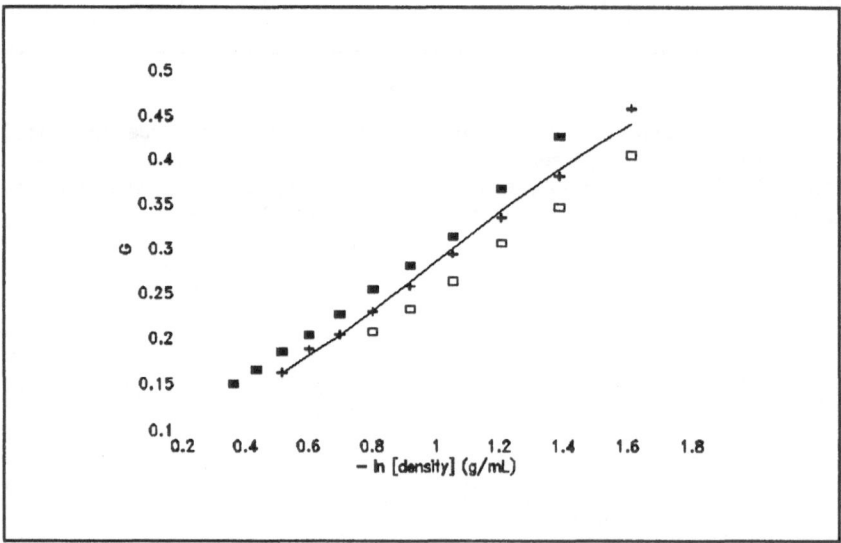

Figure 4. Dependence of the methylene group contribution to lnk´ (G) on the logarithm of CO_2 density (- ln ρ); solid line: calculated at 350 K using eqn. 41 (for additional details, see Fig. 2 legend).

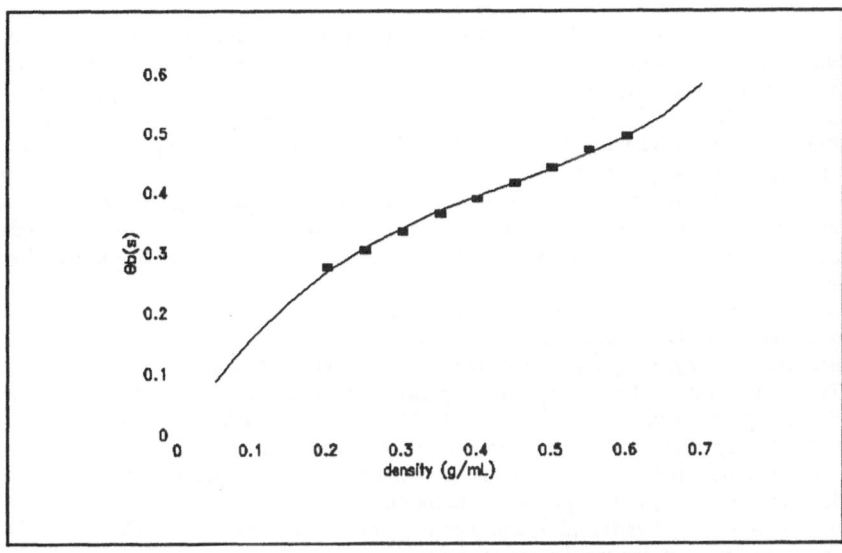

Figure 5. Dependence of the volume fraction of CO_2 in the stationary phase ($\theta_{b(s)}$) on CO_2 density (ρ) at 350 K; solid curve: calculated from eqn. 40, with u = 0.704; (■) $\theta_{b(s)}$ values from consistent fit based on eqs. 40 and 41 (for additional details, see Fig. 2 legend).

Considering Figs. 2 and 3, we observe, as expected [2], that the methylene group contribution to lnk´ decreases with increasing temperature, but otherwise defer discussion of temperature effects. As was mentioned earlier and is evident in Fig. 2, the theoretical equations provide a good fit to the experimental G values. According to eqs. 35-39, the density dependence of G is governed by $F_n(T_R,\rho_R)$ and $\Delta_n(T_R,\theta_{b(s)})$ - the coupling term, which reflects the sorption isotherm. That F_n decreases rapidly with increasing ρ indicates that the dominant mobile-phase term is the central "solubility" term in eqn. 37. This decrease is partly offset by the increase in $\theta_{b(s)}$, and, hence Δ_n with increasing ρ (see Fig. 3). At 350 K and ρ = 0.6 g/mL, e.g., F_n = -0.71 and Δ_n = +0.25. However, the predicted increase in $\theta_{b(s)}$ with increasing ρ is more pronounced than expected (see below), having the consequence
that, at a higher density (fortunately, above any truly practical operating density), a minimum in the G vs. ρ plot is still predicted [2]. Such minima have not been observed in our laboratory or others', for capillary-column SFLC systems. Parenthetically, it warrants mentioning that a previous assignment of x = 0.884 was based on fitting limited SFC data and neglecting the Δ_n term [2]. Obviously, inclusion of the Δ_n term in our present treatment leads to a larger and apparently more realistic x value.

Plots of solute retention vs. mobile-phase density in a log-log format, such as shown in Fig. 4, have been displayed and analyzed before in SFC [14]. Given the small uncertainty (± 0.002) in the experimental G results, it is clear from Fig. 4 that such plots are, in general, not linear over an extended density range. Moreover, as was discussed in part I (LC representation of log retention vs. log volume fraction of "good" solvent), one can be
misled into attaching undue physical significance to such plots. For example, at 350 K a linear regression of G vs. -ln ρ gives a slope of 0.262 (with R^2 = 0.998), compared to r_n/r_b = 0.519, which is the theoretical coefficient of the - ln ρ term in eqn. 41. Even setting x $\simeq$ 1, a further requirement for a truly linear relationship between G and ln ρ over the experimental density range is that either $\theta_{b(s)}$ $\simeq$ constant or $\theta_{b(s)}$ $\propto$ ρ^n, where n is some positive exponent. Neither possibility seems likely (see

below).

Finally, we comment on Fig. 5. At the highest density studied at 350 K, $\rho = 0.6$ g/mL, a value of $\theta_{b(s)} = 0.492$ is calculated using the present model, which takes the stationary phase to be incompressible. This indicates that the total stationary phase volume at this density, $V_{b(s)} + V_{d(s)} = V_{\rho(s)}$, is about twice the volume at zero density, $V_{d(s)} = V_{o(s)}$, i.e., according to the model, $V_{\rho(s)}/V_{o(s)} = (1-\theta_{b(s)})^{-1} \simeq 2.0$. Although this value is in the vicinity of indirectly determined swelling factors on polysiloxane stationary phases [10], it is much higher than directly determined ones on silicone rubber [11,12]. At $\rho \simeq 0.6$ g/mL and in the temperature range 323 to 373 K, e.g., the values determined by Shim and Johnston [12] cluster around $V_{\rho(s)}/V_{o(s)} \simeq 1.5$, which would convert to $\theta_{b(s)} \simeq 1/3$ in our model. Also, although a plot of $V_{\rho(s)}/V_{o(s)}$ vs. ρ constructed from our predicted $\theta_{b(s)}$ values strongly resembles theirs, our slopes are much higher.

In summary, the model equations derived for SFLC from the unified approach do take into account the sorption isotherm of CO_2 and swelling of the stationary phase, and provide a good fit to and physical interpretation of the methylene group contributions to $\ln k'$. The model is very good, but clearly not perfect. The predicted dependence of $\theta_{b(s)}$ on ρ appears to be too pronounced. This, combined with evidence that the partial molar volume of sorbed CO_2 in silicone rubber eventually increases with increasing CO_2 density, suggests that we return to equations derived for a compressible stationary liquid phase [2]. In the present treatment this could be achieved by allowing component c ("poor" solvent in LLC; "holes" or void space in SFLC) to penetrate into the stationary phase.

A further extension of our earlier [2,4] and present theoretical treatments would be to introduce still another component into both the mobile and stationary phases, enabling one to consider ternary liquid mobile phases in LC and modified mobile phases in SFC.

Acknowledgment

This material is based upon work supported by the National Science Foundation under Grant CHE-8902735. Richard E. Boehm of Georgetown University is also thanked for many helpful discussions.

References

1.   R. B. Griffiths and J. C. Wheeler, Phys. Rev. A, 2 (1970) 1047.
2.   D. E. Martire and R. E. Boehm, J. Phys. Chem., 91 (1987) 2043.
3.   T. L. Hill, An Introduction to Statistical Thermodynamics, Addison-Wesley, Reading, MA, 1960, Ch. 20 and 21.
4.   D. E. Martire, J. Chromatogr., 452 (1988) 17.
5.   D. E. Martire, J. Chromatogr., 461 (1989) 165.
6.   D. E. Martire, R. L. Riester, T. J. Bruno, A. Hussam and D. P. Poe, J. Chromatogr., 545 (1991) 135.
7.   H. Engelhardt, A. Gross, R. Mertens and M. Peterson, J. Chromatogr., 477 (1989) 169.
8.   J. J. Shim and K. P. Johnston, J. Phys. Chem., 95 (1991) 353.
9.   J. R. Strubinger, H. Song and J. F. Parcher, Anal. Chem., 63 (1991) 98.
10.   S. R. Springston, P.David, J. Steger and M. Novotny, Anal. Chem., 58 (1986) 997.
11.   G. K. Fleming and W. J. Koros, Macromolecules, 19 (1986) 2285.
12.   J. J. Shim and K. P. Johnston, AIChE J., 35 (1989) 1097; 37 (1991) 607.
13.   M. Roth, J. Phys. Chem., 94 (1990) 4390, and references cited therein.
14.   D. R. Luffer, W. Ecknig and M. Novotny, J. Chromatogr., 505 (1990) 79.

THE POSITION OF SUPERCRITICAL-FLUID CHROMATOGRAPHY BETWEEN GAS- AND LIQUID CHROMATOGRAPHY FROM A KINETIC POINT OF VIEW

Carel A. Cramers[1], Peter J. Schoenmakers[2] and Hans-Gerd Janssen[1]
[1] Eindhoven University of Technology, Laboratory of Instrumental Analysis, P.O. Box 513, 5600 MB Eindhoven, The Netherlands.
[2] Philips Research Laboratories, P.O. Box 80000, 5600 JA Eindhoven, The Netherlands.

ABSTRACT.   In the first part of this work the chromatographic performance of packed and open tubular columns in GC, SFC and LC is compared. Equations are given for the speed of analysis, pressure drop and plates per bar pressure drop. These equations are derived under the simplifying assumption of non-compressibility of the mobile phases.
In part II the effect of the compressibility of the mobile phase is included. The effect of high pressure drop in GC is briefly discussed. The emphasis in the second part is on SFC, where the pressure drop over the column affects the chromatographic process, because a supercritical fluid is both highly compressible and highly non-ideal. With such fluids the solvating and transport properties and thus in principle the chromatographic retention and efficiency are a function of pressure and thus may vary along the column length.
   No exact analytical expressions exist to determine the influence of the pressure drop across the column on retention, efficiency and speed of analysis in SFC. Fluid flow through packed and open-tubular columns can, however, be described accurately by numerical expressions derived from the Darcy equation for laminar flow. Due to density gradients significant variations of the capacity factor along the column can occur. In open columns, the effect of the increasing linear velocity on the plate height is effectively compensated by an increase in the diffusion coefficients. In packed columns it seems that both the increase in the linear velocity and in the capacity factor cause the plate height to increase along the column.

1. Introduction

By definition all (column-) chromatographic methods have in common the use of a mobile phase that carries the analytes through the column. The differences in partition coefficient K of the sample components between the mobile phase and the stationary phase affect the separation. As mobile phase (fluid) respectively a gas (gas chromatography, GC), a supercritical fluid (supercritical fluid chromatography, SFC), or a liquid (liquid chromatography, LC) may be used.
   The differences in chromatographic behaviour between GC, SFC and LC are largely determined by pure fluid properties like density, "polarity", diffusivity and viscosity of the selected mobile phase. Fig. 1 illustrates the conventional definition of fluid states of a pure substance (in this case for carbon dioxide).
   The gas (g) and liquid regions (l) are separated by the vapour pressure curve between the triple point (Tr) and the critical point (CP). In the region sf both the pressure and the temperature exceed the critical value, this is the region of supercritical fluids.

F. Dondi and G. Guiochon (eds.),
Theoretical Advancement in Chromatography and Related Separation Techniques, 289–314.
© 1992 *Kluwer Academic Publishers.*

It should be noted, however, that in many applications of SFC the selected conditions (for P- or T-programming) are such that during part of a run either the pressure (line 2) or the temperature (line 1) is subcritical.

The dashed lines bordering the sf-region indicate no phase changes. The changes in density, diffusivity and viscosity in passing these lines are very smooth. Important is that in any programmed separation never the vapour pressure curve between Tr and CP should be passed, the then occurring phase change will spoil any separation.

Fig. 1 clearly demonstrates that depending on the temperatures and pressures selected, CO_2 (and many other substances) can be used for either GC, SFC or LC, with a proper design even in one instrument.

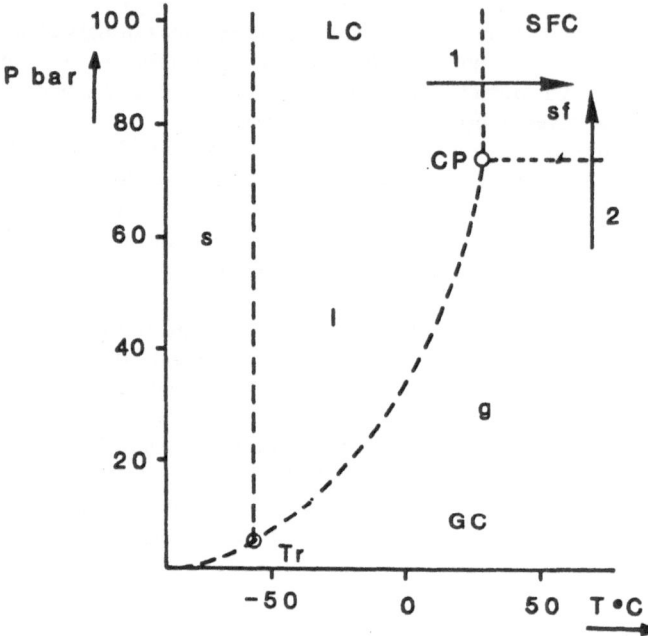

Fig. 1.   Phase diagram for carbon dioxide.

With respect to physical properties, the supercritical fluids generally take an intermediate position between those of a gas and a liquid, as is shown in Table 1.

Table 1.   Some properties of supercritical CO_2 in comparison to those of typical gases and liquids.

	Gases	Low density supercritical fluid	High density supercritical fluid	Liquids
Density [kg/m^3]	1	300	800	1000
Viscosity [kg/ms]	10^{-5}	10^{-5}-10^{-4}	10^{-4}	10^{-2}
Diffusion coefficients [m^2/s]	10^{-5}	10^{-7}	10^{-8}	10^{-9}

It should be noted that in order to exploit the dissolving power of supercritical fluids for low volatility compounds, generally a high mobile phase density is required. This is caused by the observation that retention decreases roughly exponentially with density, therefore, for this type of applications of SFC generally the high density data given in Table 1 are applicable.

2. Non-compressible Fluids. Comparison of Chromatographic performance columns in fluid chromatografie

2.1. CAPILLARY COLUMNS

In a first approach the fluids are considered to be incompressible. In order to compare chromatographic conditions in different forms of fluid chromatography, it is very convenient to use dimensionless parameters. This concept will be used in the following sections.

2.1.1. *Reduced plate height equation for capillary columns.* For open tubular columns the relation between the plate height, H (m), and the linear velocity of the mobile phase, u (m.s^{-1}), is given by the Golay equation [1]:

$$H = \frac{2D_m}{u} + f(k)\frac{d_c^2 u}{D_m} + g(k)\frac{d_f^2 u}{D_s} \tag{1}$$

where D_m, D_s are respectively the diffusion coefficient of the solute in mobile and stationary phase (m^2.s^{-1}); d_f is the thickness of the stationary phase film (m); k is the capacity factor (dimensionless retention time).

$$f(k) = \frac{1+6k+11k^2}{96(1+k)^2} \tag{2}$$

$$g(k) = \frac{2k}{3(1+k)^2} \tag{3}$$

By using the dimensionless parameters as defined in Table 2, the Golay equation can be rewritten in a simple reduced form valid for all forms of "fluid" chromatography [2,3].
The dimensionless plate height equation is given by:

$$h = \frac{2}{v} + f(k)\,v + g(k)\delta_f^2 v \tag{4}$$

where

$$\delta_f^2 = \frac{d_f^2}{d_c^2}\frac{D_m}{D_s} = \frac{\tau_s}{\tau_m} \tag{5}$$

δ_f^2 is the ratio of the diffusion times in stationary and mobile phase.

Table 2.   Definitions of reduced (dimensionless) parameters.

Parameter	Open columns[a]	Packed columns[a]
Reduced plate height	$h = \dfrac{H}{d_c}$	$h = \dfrac{H}{d_p}$
Reduced velocity	$v = \dfrac{u \cdot d_c}{D_m}$	$v = \dfrac{u \cdot d_p}{D_m}$
Dimensionless film thickness	$\delta_f = \dfrac{d_f}{d_c} \sqrt{\dfrac{D_m}{D_s}}$	-----
Capacity factor	$k = K \cdot \dfrac{V_s}{V_m}$	$k = K \cdot \dfrac{V_s}{V_m}$

[a] d_c, column diameter [m]; d_p, particle size [m]; K, thermodynamic distribution coefficient; V_s, V_m, volumes of stationary and mobile phase, respectively.

At this point, it should be emphasized that both Eqn. 1 and 4 are valid only for non-compressible fluids. The applicability is limited to liquids (in all situations) and low pressure drop columns in SFC and GC. This is not a severe restriction in the daily practice of SFC and GC with open tubular columns, unless extreme column lengths and/or very small column diameters are used.

The optimum (reduced) linear velocity can be found by differentiation of Eqn. (4). Differentiation yields for the optimum velocity, v_{opt},

$$v_{opt} = \sqrt{\frac{2}{f(k) + g(k)\delta_f^2}} \qquad (6)$$

and for the minimum (reduced) plate height, h_{min},

$$h_{min} = 2 \sqrt{2 \cdot (f(k) + g(k)\delta_f^2)} \qquad (7)$$

Owing to the use of dimensionless parameters, the optimum (reduced) velocity and the minimum plate height are independent of either the column diameter or the diffusion coefficients of the solutes in the mobile phase.

If the stationary phase contribution to band broadening can be neglected Eqs. 6 and 7 simplify to:

$$v_{opt} = \sqrt{\frac{2}{f(k)}}$$

(8)

and

$$h_{min} = 2\sqrt{2 \cdot f(k)}$$

(9)

2.1.2. *Speed of analysis*. Based on reduced parameters (Table 2), an equation for the speed of analysis in all forms of column chromatography can be derived. Starting from the retention-time equation:

$$t_R = t_0 \cdot (1+k) = \frac{L}{u} \cdot (1+k)$$

(10)

where t_R, t_0 are the retention times of a solute and an unretained component, respectively. Substituting L = N.H and the reduced parameters from Table 2 it follows that:

$$t_R = N \cdot \frac{h}{v} \cdot \frac{d_c^2}{D_m} \cdot (1+k)$$

(11)

In this expression N is the plate number. Under identical conditions (N, h, v and k), the analysis time is proportional to d_c^2/D_m (or d_p^2/D_m for packed columns). The order of magnitude of D_m in SFC is about 10^{-8} m^2s^{-1} compared with 10^{-5} and 10^{-9} for gaseous and liquid mobile phases, respectively. From Eqn. 11 and the respective diffusion coefficients, it can be concluded that in order to obtain equal speeds of analysis in GC, SFC and LC, the diameters of the columns must be chosen in the ratio 1.0: 0.03: 0.01. Table 3 gives a comparison of d_c^2/D_m and thus the relationship between the analysis time for the three forms of open-tubular chromatography. For the purpose of these comparisons a dimensionless film thickness of 0.3 was used for all columns. The diffusion coefficient in the stationary phase was assumed to be 10^{-10} m^2/s for all three forms of chromatography.

Table 3. Comparison of retention times in open-tubular chromatographic methods ($\delta_f = 0.3$).

Method	d_c (μm)	d_f (μm)	D_m (m^{2}s^{-1})	$\dfrac{d_c^2}{D_m}$ (s)	Relative time
GC	250	0.24	10^{-5}	0.006	1
SFC	50	1.5	10^{-8}	0.250	40
	8	0.24	10^{-8}	0.006	1
LC	10	0.95	10^{-9}	0.100	16
	5	0.47	10^{-9}	0.025	4

The very narrow columns essential in open-tubular LC possess a very limited sample capacity. This, together with the requirement for extremely small detector volumes poses enormous technical problems. Therefore, it can be concluded that open-tubular SFC, because of the larger column diameters involved, is more within the scope of current technology than is open-tubular LC. An additional advantage of SFC is the fact that if the correct mobile phase is selected, sensitive GC detectors with low effective dead volumes may be used (e.g. flame ionization, mass spectrometric or nitrogen-phosphorous specific detectors).

For the exact calculation of retention times by means of Eqn. 11 (at a given value of N), values for h and v have to be included (see Table 4). For fairly high k values, v = 45 and h = 4.5 are reasonable values for open-tubular columns with average film thicknesses ($\delta_f \approx 0.3$), as used in SFC. A further increase in v in order to increase the speed of analysis does not make sense, because at high velocities h is proportional to v and hence h/v remains constant. Fig. 2 shows a plot of h/v vs. v. As can be seen in this figure, h/v approaches asymptotically to a minimum value. An appreciable reduction of the analysis time can be obtained by increasing v from v_{opt} to about 5 - 10 times v_{opt}. Any further increase does not give a further reduction of the analysis time.

Table 4. Comparison of open-tubular and packed columns (dimensionless parameters) under optimal and daily practical conditions.

Type of column	ϕ_0^*	h	h_{min}	v	v_{opt}	E*	E_{min}
Open-tubular	32	4.5	0.8	45	5	650	20
Drawn (packed) capillary	150	3	2	10	5	1350	600
Packed capillary	500	3	2	10	2.5	4500	2000
Packed	1000	3	2	10	2.5	9000	4000

* ϕ_0 is the column resistance factor. E the separation impedance.

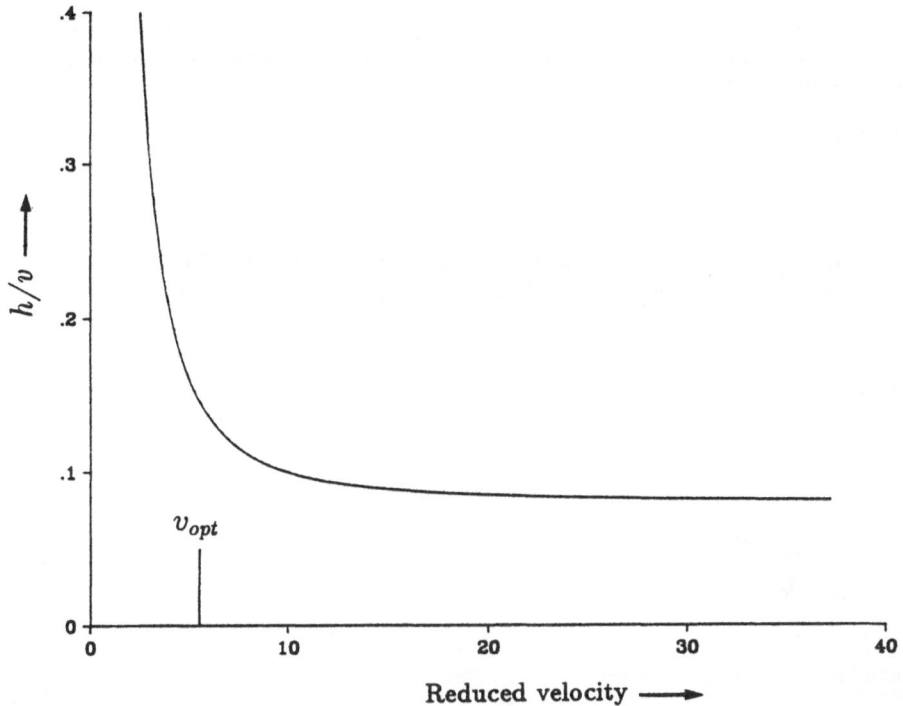

Fig. 2.   Influence of the reduced velocity on the analysis time ($\propto$ h/v). Conditions: k = 2, $\delta_f = 0.3$.).

2.1.3. *Speed and pressure drop.* In the previous calculations the pressure drop over the chromatographic column was neglected. Fluid flow through a chromatographic column, however, requires a pressure difference over the column as the driving force for the flow. In LC this pressure drop has no influence on the chromatographic separation process in the column. The liquid flowing through the column is virtually non-compressible, i.e. the chromatographically relevant properties of the mobile phase (e.g. the viscosity, the density and the diffusion coefficients of the solutes in the mobile phase) are not affected by the pressure difference across the column. In GC with an ideal gas as the carrier, the density of the gaseous mobile phase is directly proportional to the pressure. Thus, the density of the mobile phase decreases along the column length, thereby altering the diffusivity and the linear velocity. The ratio of these two parameters, however, remains constant. At the relatively low pressures commonly employed in GC, the carrier gases show (almost) ideal gas behaviour, which means that retention is neither a function of the nature of the carrier gas nor its pressure. In SFC the situation is more complex. The viscosity, the density and the diffusivity of the mobile phase all are complex functions of the pressure and the temperature of the mobile phase. Furthermore, retention in SFC is a strong function of the mobile-phase density (and, hence, pressure).

Below, a brief comparison of the speed of analysis and the pressure drop in packed and open columns for SFC is given. As all properties of supercritical fluids strongly depend on the pressure, the nature of the calculations is only approximative. The effects of the column pressure drop on retention, efficiency and speed of analysis in SFC will be discussed in more detail in the second part of this article.

For non-compressible fluids (liquids) and for the conditions of relatively low pressure drops in GC and SFC, the pressure drop is described by the Darcy equation for pressure drop in systems with laminar flow:

$$\Delta P = B_0 \cdot \eta \cdot L \cdot u \tag{12}$$

The permeability B_0 for open-tubular columns equals $32/d_c^2$, or expressed in general terms ϕ_0/d_c^2, where ϕ_0 is the column resistance factor. Rewriting with the dimensionless parameters h and v:

$$\Delta P = \frac{\phi_0}{d_c^2} \cdot \eta \cdot N \cdot h \cdot v \cdot D_m \tag{13}$$

or with

$$v = \frac{u \cdot d}{D_m} = \frac{L}{t_0} \cdot \frac{d}{D_m} = \frac{N \cdot h \cdot d^2}{t_0 \cdot D_m} \tag{14}$$

Eqn. 14 can be rewritten as

$$\Delta P = \frac{\phi_0 \cdot N^2 \cdot h^2 \cdot \eta}{t_0} = \frac{E \cdot N^2 \cdot \eta}{t_0} \tag{15}$$

E is defined as the separation impedance given by $E = h^2 \cdot \phi_0$. From Eqn. 15, it follows that

$$\Delta P \cdot t_0 = constant \tag{16}$$

For a given fluid state, the analysis time (proportional to t_0) can be reduced proportionally to d_c^2 (Eqn. 11) at the price of a proportionally increasing pressure drop. For a given plate number N, the proportionality constant from Eqn. 16 depends on the fluid state (η). For high pressure drops (packed columns) with high plate numbers in SFC, η decreases along the column length. In contemporary open-tubular columns this effect can generally be neglected and Eqn. 11 can be applied.

A simple expression for the number of plates per bar pressure drop (N/ΔP) can be derived from Eqn. 13. From this equation it follows that:

$$\frac{N}{\Delta P} = \frac{d_c^2}{32\eta \cdot h \cdot v \cdot D_m} \tag{17}$$

Introduction of the plate height equation (Eqn. 4) yields:

$$\frac{N}{\Delta P} = \frac{d_c^2}{32\eta \cdot D_m (2 + f(k)v^2 + g(k)\delta_f^2 v^2)} \tag{18}$$

From this equation it is evident that working at high reduced velocities leads to a relatively low number of plates per bar pressure drop. A plot of the number of plates obtained per bar pressure drop as a function of the reduced velocity is given in Fig. 3.

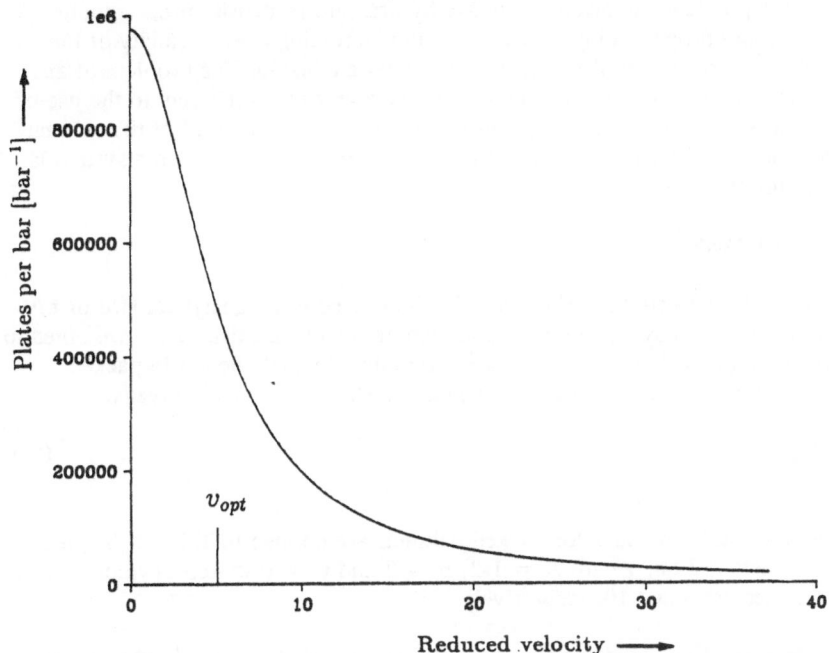

Fig. 3.   Plates per bar pressure drop as a function of the reduced velocity in a 25 μm open column. Conditions: $D_m = 10^{-8}$ m^2.s^{-1}, $\eta = 10^{-4}$ kg.m^{-1}.s^{-1}, $k = 2$, $\delta_f = 0.3$.

Table 5.   Plates per bar pressure drop in open-tubular columns (h = 4.5; v = 45; $\phi_0 = 32$) and packed columns (h = 3; v = 10; $\phi_0 = 1000$). $\eta = 10^{-4}$ kg.m^{-1}.s^{-1}; $D_m = 10^{-8}$ m^2.s^{-1}.

Conditions	Open-tubular columns		Packed columns	
	d_c (μm)	Plates per bar pressure drop	d_p (μm)	Plates per bar pressure drop
Supercritical CO$_2$, 40°C, 200 bar	50 8	38,600 1000	10 5	333 83

Table 5 gives values for N/ΔP (plates per bar pressure drop) for two column diameters in open-tubular SFC (after Eqn. 13). For the calculation of the data in this table, values of h = 4.5 and v = 45 were used. From Table 5 it can be concluded that the pressure drop is relatively small in open-tubular SFC; even with 8 μm columns about 10,000 plates can be achieved with a pressure drop of only 10 bar (column inlet 210 bar, outlet 200 bar). In SFC, in contrast to GC and LC, retention is very dependent on the pressure (density) of the mobile phase. A large pressure drop will result in a density gradient along the column. This effect is similar to a negative temperature gradient along the axis of a GC column. Or, a negative organic modifier gradient along an RP-HPLC column, which is more difficult to imagine. In all instances the migration of the components will slow down and (in extreme cases) will eventually stop. This problem can be solved partly by pressure or density programming. A disadvantage of pressure programming, however, is the increasing linear velocity of the mobile phase and thus a reduction of the number of plates attainable. The problem of an increasing linear velocity in pressure-programmed analysis in SFC is inherent to the use of fixed outlet restrictors. A variable-restrictor system for independent control of the column pressure and the linear velocity is necessary to circumvent this problem. Such a system is described and evaluated in ref. [4].

2.2. PACKED COLUMNS

2.2.1. *Reduced plate-height equation.* For packed columns, no exact analytical H/u or h/v equation analogous to the Golay equation for open-tubular columns exists. Here we opted to use the Knox equation [5]. It is the general observation that the plate height in packed columns in LC is adequately described by this equation. The Knox equation reads:

$$h = A\,v^{1/3} + \frac{B}{v} + C\,v \qquad\qquad (19)$$

The dimensionless quantities h and v for packed columns are defined in Table 2. Typical values for the constants in Eqn. 19 are A = 1-2, B = 2 and C = 0.05-0.5. Typical values for h and v in practice are 3 and 10, respectively.

2.2.2. *Speed of analysis.* Eqn. 11 describes the speed of analysis for packed columns, if d_c is replaced by the particle size, d_p, and the appropriate values for h and v as given above are used. Modern packed columns contain particles of 10, 5 or even 3 μm. From Eqn. 11 it can be concluded that open-tubular fluid chromatography cannot compete with packed columns with respect to the speed of analysis unless very small inner diameters are used.

2.2.3. *Speed and pressure drop.* Eqs. 11-16 are applicable to packed columns by inserting d_p instead of d_c and using the values for h, v and E from Table 4. It should be borne in mind, however, that these equations are valid only for incompressible fluids. For GC and SFC this limits the validity to situations of low pressure drop and, hence, low plate numbers. In SFC the linear velocity (u), the diffusivity (D_m) and the viscosity (η) are dependent on the pressure drop. In high-pressure GC, pressure-drop correction factors have to be included in the plate-height equations. One of the results is that the speed of analysis is no longer proportional to d_c^2 or d_p^2 and eventually, at high-pressure drops, it becomes proportional to d_c or d_p [6,7]. Table 5 gives values for the number of plates per bar pressure drop for packed columns (ϕ_0 = 1000) and open-tubular columns (Eqn. 13). For the open columns ϕ_0 = 32 was used. For drawn (packed) capillary columns, the numbers given in the table have to be multiplied by 1000/150, i.e. the ratio of the column resistance factors of the respective column types (Table 4). For packed capillary columns, the ratio is close to 1000/500.

From Table 5, it is clear that for a given allowable pressure drop, ΔP, open-tubular columns always allow much higher plate numbers to be achieved. The maximum number of plates that can be obtained from any type of column in the three forms of fluid chromatography depends on the maximum allowable pressure drop. In GC and LC this is dependent on the instrument design. In SFC retention and efficiency depend heavily on pressure (or density). Here, the allowable pressure drop is determined by fundamental aspects.

A more detailed study on the effects of the pressure drop across the column on retention and plate height in SFC is described in ref. [8].

3. Compressible Fluids. Effects of the column pressure drop on retention and efficiency in packed and open-tubular GC and SFC

3.1. HIGH PRESSURE-DROP GAS CHROMATOGRAPHY

The effects of the column pressure drop on efficiency in GC were first described by Giddings. He presented a theoretical study in which it was shown that even severe pressure drops lead to little loss in resolution in GC. The column pressure drop in GC may also influence the capacity factors. A small, but measurable influence of the carrier-gas pressure on the capacity factors in GC has been observed by several authors [9,10]. For routine GC, however, this effect is not a major source of concern.

Under the conditions of non-zero pressure drop the average carrier gas velocity, $\bar{u}$, has to be used in Eqn. 10 for the analysis time. $\bar{u}$ can be calculated from:

$$\bar{u} = u_0 \cdot f_2 \tag{20}$$

u_0 being the carrier gas velocity at the column outlet. f_2 is the Martin-James correction factor (discussed in more detail in the High Pressure-Drop SFC section).

For very high values of P_R, f_2 reduces to:

$$f_2 = \frac{3}{2P_R} \tag{21}$$

where P_R is the ratio of the inlet pressure over the outlet pressure.

Together with the equation for pressure drop for ideal gases, the following equation can be derived [11]:

$$t_R = \left[\frac{8}{9} \cdot N^{3/2} \cdot \frac{h^{3/2}}{v^{1/2}} \cdot d \cdot \phi_0^{1/2} \left[\frac{\eta}{D_{m,a} P_{atm}} \right]^{1/2} \cdot (1+k) \right. \tag{22}$$

Where:

d $=$ the characteristic diffusion distance (being d_p for packed and d_c for capillary columns).

$D_{m,a}$ $=$ the diffusion coefficient in the mobile phase at atmospheric pressure.

ϕ_0 $=$ the column resistance factor.

η $=$ the dynamic viscosity of the carrier gas

P_{atm} $=$ the atmospheric pressure.

When comparing Eqn. 22 with Eqn. 11 (incompressible fluids) it can be seen that t_R is no longer proportional to d_c^2 or d_p^2, but simply to d_c or d_p. Also the column resistance factor and the carrier gas velocity are included in the equation. The influence of N, h, v and $D_{m,a}$ are different from the non-compressible case.

3.2. HIGH PRESSURE-DROP SFC

For incompressible, ideal fluids, the time required to generate a certain number of plates is given by Eqn. 11:

$$t_R = N \cdot \frac{h}{v} \frac{d_c^2}{D_m} (1+k) \tag{11}$$

In SFC, under the conditions of a significant pressure drop, this equation can not be used as the parameters h, v, D_m and k may vary along the column. The influence of the pressure drop on retention and efficiency in SFC has been the subject of a number of recent publications [e.g. 12,13]. We have recently described the influence of the pressure drop across the column on these four parameters in SFC with pure CO_2 as the mobile phase [8]. Theoretical models were described that enable the calculation of the pressure gradient and the variation of the capacity factor of a solute along the column. Furthermore, the influence of the pressure gradient on the plate height and efficiency in open-tubular SFC was discussed in that article. Here, we will present a short summary of the theory from this previous work and extend the plate height and efficiency theory, which at that time was only applicable to open columns, to packed column SFC.

3.2.1. *Pressure gradients in packed and open columns.* Fluid flow through packed and open-tubular columns can be described starting from the well-known Darcy equation (Eqn. 12).

$$\Delta P = B_0 \cdot \eta \cdot u \cdot L \tag{12}$$

Here u is the superficial linear velocity calculated from the diameter of the empty column:

$$u = \frac{4F_m}{\pi \cdot \rho \cdot d_c^2} \tag{23}$$

where F_m is the mass flow rate through the chromatographic column and d_c is the diameter of the empty column. Substitution of Eqn. 23 in Eqn. 12 yields:

$$\Delta P = \frac{4B_0 \cdot \eta \cdot F_m \cdot L}{\pi \cdot \rho \cdot d_c^2} \tag{24}$$

For supercritical fluids, Eqn. 24 can only be applied over short segments of the column because both the density and the viscosity are functions of the pressure. If the column is divided in n segments of length ΔL, the pressure drop over the ith segment is given by:

$$\Delta P_i = \frac{4B_0 \cdot \eta_i \cdot F_m \cdot \Delta L}{\pi \cdot \rho_i \cdot d_c^2} \tag{25}$$

The subscript i indicates that the values of these parameters pertain to the conditions in segment i. The pressure in segment j is given by:

$$P_j = P_{in} - \sum_{i=1}^{j} \frac{4B_0 \cdot \eta_i \cdot F_m \cdot \Delta L}{\pi \cdot \rho_i \cdot d_c^2} \tag{26}$$

From this equation the pressure gradient along the column can be calculated. Because accurate expressions describing the density and the viscosity of CO_2 are extremely complex, Eqn. 26 can only be solved numerically. Outlet pressures calculated using this method have been shown to be in very good agreement with experimental data [8].

The approach described above allows the calculation of the pressure and density in every segment of the column. From the mass-flow rate through the column in combination with the density in a segment the residence time in that segment can be calculated. The residence time $t_{0,i}$ of an unretained component in segment i is given by:

$$t_{0,i} = \frac{\pi \cdot \rho_i \cdot d_c^2 \cdot \epsilon \cdot \Delta L}{4F_m} \tag{27}$$

Here ϵ is the void fraction (porosity) of the column. For open-tubular columns ϵ equals unity. The linear velocity in segment i is given by:

$$u_i = \frac{4F_m}{\pi \cdot \rho_i \cdot d_c^2 \cdot \epsilon} \tag{28}$$

From Eqn. 27 the total hold-up time of the fluid in the column can be calculated. Comparisons of experimental and calculated data again showed an excellent agreement between experiments and theory [8].

3.2.2. *Capacity factors.* From the pressure (or density) gradient in the column the variation of the capacity factor along the column can be calculated. This requires knowledge of the capacity factor of the solute as a function of the mobile phase density. Such a function can be established by measuring capacity factors at very low flow rates where the effect of the pressure drop on the capacity factor is negligible. At constant temperature the capacity factors thus obtained can be described accurately by an equation of the form:

$$\ln(k) = a + b\rho + c\rho^2 \tag{29}$$

Martire gave a theoretical basis for this expression in adsorption chromatography [12]. The residence time of a retained solute in the ith segment, $t_{R,i}$, is given by:

$$t_{R,i} = t_{0,i} \cdot (1 + k_i) \tag{30}$$

Here k_i is the local capacity factor of the solute in segment i, which can be obtained from Eqn. 29. Summation of the $t_{R,i}$ values in the n segments eventually gives the elution time of the solute.

3.2.3. *Plate-height and efficiency.* If the length of the segments is chosen sufficiently small, a local plate height can be defined that can be considered constant within a segment. Plate-height expressions derived for non-compressible fluids can then be used to calculate the band broadening in every segment. The chromatographic band broadening (in time units) in segment i, σ_i is given by:

$$\sigma_i^2 = \frac{H_i \cdot t_{R,i}^2}{\Delta L} \tag{31}$$

Here, H_i is the local plate height. The total observed bandwidth can be obtained by applying the rule of the additivity of variances:

$$\sigma_t^2 = \sum_{i=1}^{n} \sigma_i^2 \qquad (32)$$

where σ_t^2 is the total variance of the peak observed at the column outlet (in time units). Summation of the individual contributions according to Eqn. 32 yields the total band width upon elution, which is thus:

$$\sigma_t^2 = \sum_{i=1}^{n} \frac{H_i \cdot t_{R,i}^2}{\Delta L} \qquad (33)$$

The overall observed plate height, H_{obs}, is

$$H_{obs} = \frac{L \cdot \sum_{i=1}^{n} (H_i \cdot t_{R,i})}{\Delta L \cdot \left[\sum_{i=1}^{n} t_{R,i} \right]^2} \qquad (34)$$

Evaluation of Eqs. 33 and 34 requires knowledge of the individual plate heights and the residence times in each of the segments. Appropriate plate-height equations are required to calculate the local plate height in every segment.

3.2.3.1. Open-tubular columns. For open-tubular columns the well-known Golay equation can be used to calculate the local plate height in every segment:

$$H_i = \frac{2D_{m,i}}{u_i} + \frac{f(k_i) \cdot d_c^2 \cdot u_i}{D_{m,i}} \qquad (35)$$

Here, $D_{m,i}$ and u_i are the binary diffusion coefficient of the solute in the mobile phase and the linear velocity in the ith segment, respectively. In this expression the stationary phase contribution to band broadening is neglected. The function $f(k_i)$ is given by:

$$f(k_i) = \frac{1 + 6k_i + 11k_i^2}{96 \cdot (1 + k_i)^2} \qquad (36)$$

The value for u_i can be calculated from Eqn. 28. The diffusion coefficient can be estimated from the Wilke and Chang equation [14]:

304

$$D_m = 7.4 \cdot 10^{-8} \cdot \frac{\sqrt{M_b} \cdot T}{\eta \cdot V_a^{0.6}} \tag{37}$$

where M_b is the molecular mass of the solvent, T the absolute temperature, η the dynamic viscosity (cp) and V_a the molar volume of the solute at its boiling point (cm^3/mole). Substitution of η_i in the Wilke and Chang equation yields the diffusion coefficient in segment i. Applying Eqn. 35 with the input data calculated from Eqs. 28, 36 and 37 yields the plate heights in the individual segments.

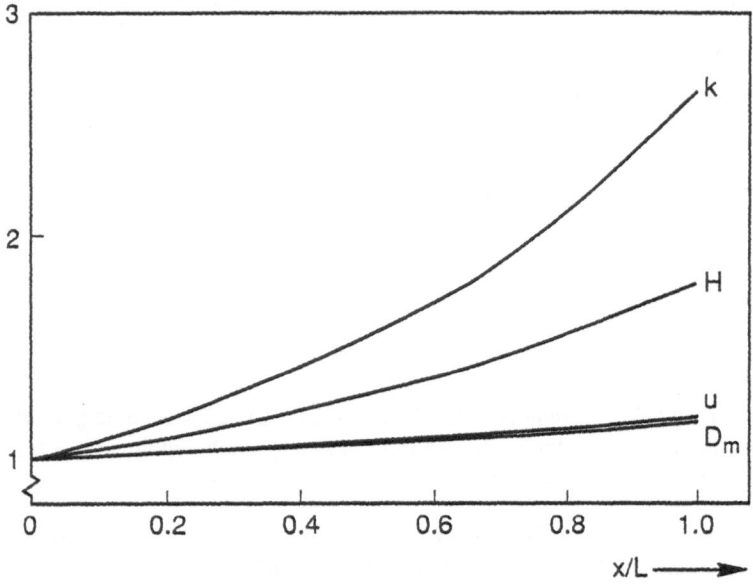

Fig. 4.   Variation of the local parameters u, D_m, k and H as a function of the dimensionless distance along the column. Conditions: Open column, T = 70°C, P_{in} = 165 bar, Mass flow rate = 6.5 * 10^{-7} kg/s. Calculated outlet pressure = 145 bar, Calculated (average) linear velocity = 51 cm/s. Solute: C_{28}.

In Fig. 4 the calculated changes in the linear velocity, the diffusivity and the capacity factor along an open-tubular column are plotted in the case of a considerable density gradient. In this figure the local plate height is also indicated. All parameters were normalized to one at the column inlet. As can be seen in the figure, all parameters increase across the column. It is noteworthy that the changes in the velocity and the diffusion coefficient are almost identical, which is similar to the situation in GC. The linear velocity at constant mass flow is inversely proportional to the fluid density, which is apparently also (almost) the case for the diffusion coefficient. This implies that the ratio of the velocity to the diffusivity is almost constant along the column. As can be seen in the plate height-equation for open-tubular columns (Eqn. 35), only the ratio of u over D_m (or D_m over u) is important. This means that the variation of the local plate height along an open column is mainly due to variations in the capacity factor. The effects of the increase in the linear velocity and the diffusion coefficient on the local plate height effectively cancel.

3.2.3.2. Packed columns. For packed columns no exact analytical plate-height equation comparable to the Golay equation for open-tubular columns exists. For liquid chromatography several plate height expressions for packed columns have been reported. Schwartz et al. [15] selected the Horvath and Lin equation [16] for use in SFC. Schoenmakers [2] used the Knox equation in a comparative study of the speed of analysis in packed and open-tubular SFC. More recently, Poe & Martire [13] discussed the influence of density on the various terms of the Knox, and the Horvath and Lin plate-height equations. In any case, some empirical coefficients appear in the plate-height equation, which need to be established experimentally.

Here, it was decided to use the Horvath and Lin equation. The column constants that appear in this equation were taken from the original article by Horvath and Lin [16]. During subsequent derivations we used the empirical constants from the original article by Horvath and Lin without experimental verification. It should be emphasized that it is not the intention of the discussion below to give a complete, quantitative description of the plate height in packed-column SFC with non-zero pressure drops. This can only be achieved when accurate plate-height equations are available. As this is not yet the case for packed columns (in any form of chromatography), only preliminary, qualitative conclusions can be drawn here.

The complete Horvath and Lin plate-height equation under zero-pressure-drop conditions reads

$$H = H_{disp.} + H_{e.diff.} + H_{i.diff.} + H_{kin.}$$ (38)

The term $H_{disp.}$ expresses the plate height increment due to axial dispersion of the solute in the interstitial space. This term is believed to be unaffected by the retention of the solute. The plate-height increment $H_{e.diff.}$ is related to the "film" resistance at the boundary layer between the mobile phase and the stagnant mobile phase. The third term, the intraparticular diffusion resistance to mass transfer, is represented by the term $H_{i.diff.}$. The last term, $H_{kin.}$, represents the kinetic resistance for solute binding. The four terms are given by the following four expressions:

$$H_{disp.} = \frac{2\gamma \cdot D_m}{u_e} + \frac{2\lambda \cdot d_p \cdot u_e^{1/3}}{u_e^{1/3} + \omega \cdot (D_m/d_p)^{1/3}}$$ (39)

Here γ, ω and λ are structural parameters of the column packing. u_e is the interstitial mobile phase velocity. The first term on the right-hand side of Eqn. 39 is due to longitudinal diffusion, whereas the second term expresses the combined effect of the velocity profile and diffusion in the interstitial spaces.

$$H_{e.diff.} = \frac{\kappa \cdot (\psi + k + \psi \cdot k)^2 \cdot d_p^{5/3} u_e^{2/3}}{(1+\psi)^2 \cdot (1+k)^2 \cdot D_m^{2/3}}$$ (40)

In this equation k is the capacity factor of the solute and κ is an additional structural parameter of the packing material. The structural parameters γ, λ, ω and κ can significantly vary from column to column as their values are determined by the properties of the column material, the tube dimensions and materials, as well as by the packing procedure. ψ is the ratio of the intraparticulate void volume over the interstitial void space in the column.

The value of ψ is given by $\epsilon_i(1-\epsilon_e)/\epsilon_e$, where ϵ_i and ϵ_e are the appropriate intraparticular and interstitial porosities, respectively.

$$H_{i.diff.} = \frac{\theta \cdot (\psi + k + \psi \cdot k)^2 \cdot d_p^2 \cdot u_e}{30 D_m \cdot \psi \cdot (1+\psi)^2 \cdot (1+k)^2} \qquad (41)$$

Here θ is the tortuosity factor. The fourth term is given by

$$H_{kin.} = \frac{2k^2 \cdot u_e}{(1+\psi) \cdot (1+k)^2 \cdot \beta \cdot k_a} \qquad (42)$$

Here β is the phase ratio of the column defined by Horvath and Lin as the concentration of surface sites per unit volume of the mobile phase in the column. k_a is the rate constant for the adsorption step. The mechanism and kinetics of solute adsorption onto or desorption from the stationary phase are largely unexplored. Hence, it is difficult to estimate the magnitude of the plate-height contribution arising from slow adsorption and desorption kinetics. On the other hand, if the other three terms are known, the importance of the slow kinetics term can be established from experimental data.

In the same article in which they derived their general plate-height equation for (packed-column) LC, Horvath and Lin also gave an extensive discussion of the relative importance of the four individual contributions to band broadening. The most striking conclusion was that for LC systems with particle sizes below about 3 - 5 μm, the term describing the adsorption and desorption kinetics may easily outweigh all the other terms. On close examination of the above equations it can be seen that if this conclusion is valid in LC, it is even more so in SFC, because there the diffusion coefficients in the mobile phase are about one order of magnitude higher than in LC.

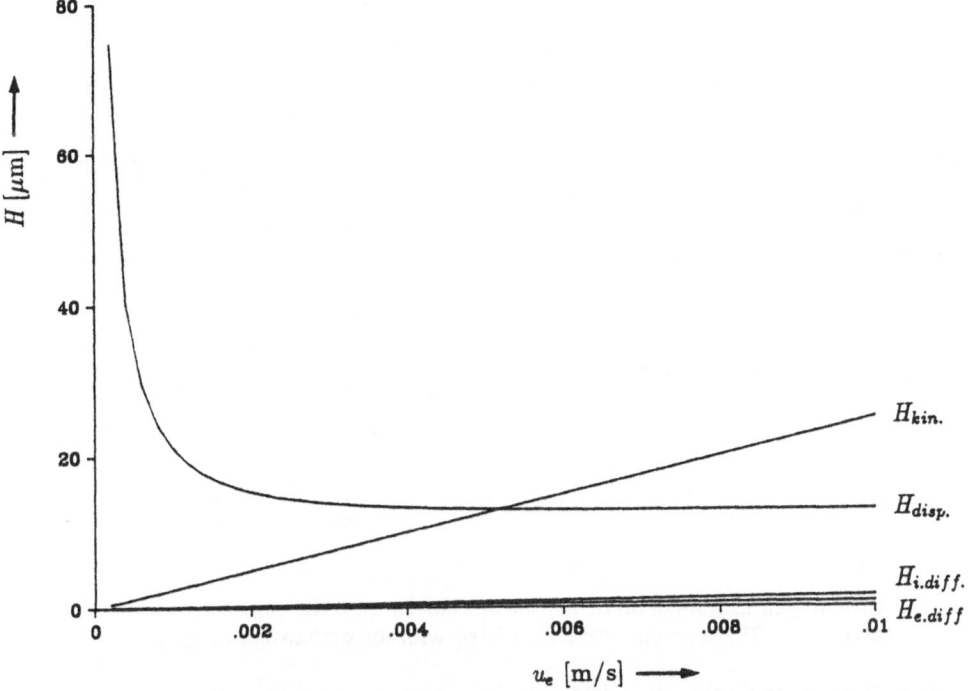

Fig. 5.   Graph illustrating the individual plate-height increments as a function of the
linear velocity. The data used in the calculations (Eqs. 39-42) are as follows:
$\gamma = 0.7$, $\lambda = 2.5$, $\omega = 2$, $\kappa = 1/15$, $\psi = 0.8$, $\theta = 2$, $\beta k_a = 2.5 * 10^2$ sec^{-1},
$k = 5$, $d_p = 5$ μm, $D_m = 10^{-8}$ m^{2}sec^{-1}.

Fig. 5 illustrates the relative contribution of the four different terms in the Horvath and Lin
equation to the plate height under typical SFC conditions. All constants required in the
calculations were taken from the original Horvath and Lin article. Only the value of the
diffusion coefficient in the mobile phase was adapted to the situation in SFC. The values for
the respective parameters are given in the caption to the figure. All parameters were assumed
to be unaffected by the pressure drop over the column. It can be seen in the figure that the
kinetic resistance term is indeed much larger than the diffusion term. In fact, only the kinetic
term and the dispersion term are important. It is also important to note that the dispersion
term remains almost constant at velocities above ca. 0.002 m.sec^{-1}. Based upon these
observations the Horvath and Lin plate-height equation can be simplified to:

$$H = \frac{2\gamma \cdot D_m}{u_e} + \frac{2\lambda \cdot d_p \cdot u_e^{1/3}}{u_e^{1/3} + \omega \cdot (D_m/d_p)^{1/3}} + \frac{2k^2 \cdot u_e}{(1+\psi) \cdot (1+k)^2 \cdot \beta \cdot k_a} \tag{43}$$

In Fig. 6 a number of plate-height curves calculated according to Eqn. 43 are plotted with
the capacity factor as the parameter. As can be seen in this figure, the capacity factor has a
marked influence on the total plate height. It is clear from Eqn. 43 that the capacity factor
dependence must arise from the kinetic resistance term.

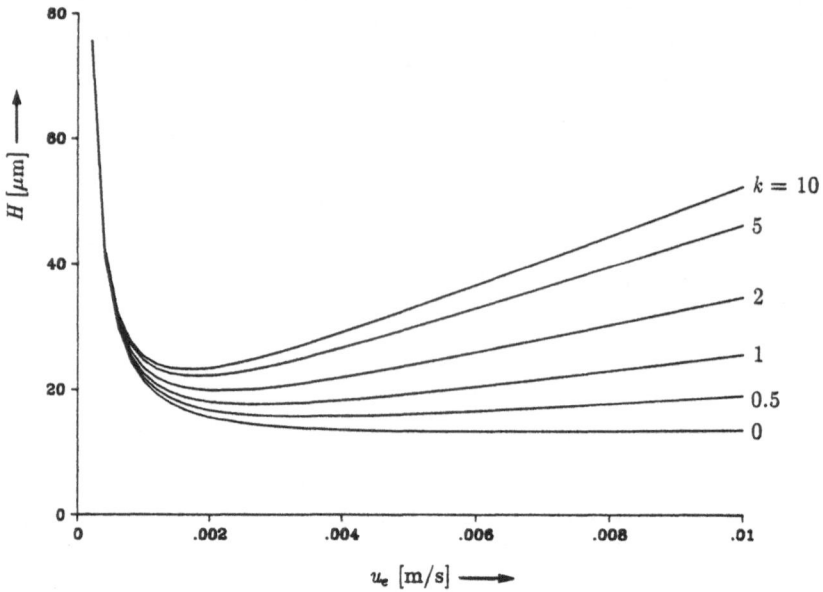

Fig. 6.   Plate height as a function of the linear velocity with the capacity factor as the parameter. These results were calculated with the data used in Fig. 5.

It should be borne in mind that all conclusions drawn above are based on the assumption that the Horvath and Lin equation is valid. Not only is it assumed that the Horvath and Lin equation algebraically fits the (H-u)-data set, it is also assumed that the equation provides the correct physical interpretation of the dispersion processes that take place. Whether this really is the case is still questionable. Different opinions on the validity of the Horvath and Lin equation and other plate-height equations for packed-column LC have been published in literature. Arnold et al. [17] for example, discussed the validity of the second term on the right-hand side of Eqn. 43. According to these authors the power (1/3) for u_e is not correct. They presented a slightly modified, alternative plate-height equation, which contained the same adsorption kinetics term as the Horvath and Lin equation. Katz et al. [18] tested a number of plate-height equations against 25 data sets of experimental values of H and the linear velocity obtained for columns packed with silica gel. In this study, the Van Deemter equation [19] was observed to yield the best fits. To illustrate the complexity of the matter, Giddings [20] argued that this equation is fundamentally incorrect.

If we assume the Horvath and Lin equation to be valid, it can be concluded that variations in the local plate height along a packed column can arise from variations in the diffusion coefficient, the linear velocity and the capacity factor of the solute, which is similar to the situation in open-tubular columns. In the first term of the simplified Horvath and Lin equation (Eqn. 43), the increase of D_m is effectively cancelled by an increase in u_e. Also the second term remains (almost) constant along the column as variations in u_e are again outweighed by changes in D_m. The first two terms do not contain the capacity factor, so they are unaffected by the increase of k along the column. For the last term the situation is more complicated. Here the increase in u_e is no longer compensated by an increase in D_m, which causes this term to increase along the column due to the increase in the linear velocity. In addition, the third term of the simplified Horvath and Lin equation is capacity-factor dependent. Hence, if there is a significant pressure drop across the column the increase in the capacity factor will cause an increase of the plate height along the column. Especially for capacity factors below 5 this increase can be significant.

In conclusion, the local plate height along a packed column in SFC will increase due to the increasing value of the kinetic resistance term. Two parameters affect the increase of this term, i.e. the increase in the linear velocity and the increase of the capacity factor. Under most circumstances, the second effect will be the predominant one.

3.2.4. *Speed of analysis in open-tubular SFC.* In the first part of this article the performance of packed and open-tubular columns in GC, LC and SFC was compared with respect to the speed of analysis. At that time the mobile phase was treated as an incompressible fluid and the effect of the pressure drop on the plate height was neglected. Using the numerical methods described above, the efficiency of a column and the speed of analysis can be calculated without the need to treat the supercritical mobile phase as incompressible. Currently, accurate expressions for the plate height are only available for open columns. Hence, only this type of column is considered. In this section a summary of the results from the plate-height and analysis-time calculations for open-tubular columns in SFC with a significant pressure drop is presented. A more comprehensive discussion will be published elsewhere [21].

The efficiency observed in open-tubular columns is affected by the diffusion coefficient of the solute in the mobile phase and by its capacity factor. Moreover, the linear velocity has a strong influence on the efficiency. In case of a significant pressure drop across the column these three parameters will vary. For open-tubular columns the influence of variations in D_m, u and k on the plate height can be described quantitatively. For unretained components, the effect of the pressure drop on the analysis time is very similar to the effects of the expansion of the carrier gas along the column in GC. These effects can be accounted for by the introduction of the appropriate compressibility correction factors in the plate height equation. The extended Golay equation for gas chromatography including the f_1 and f_2 correction factors reads:

$$H = \left[\frac{2D_{m,o}}{u_o} + \frac{f(k) \cdot d_c^2 \cdot u_o}{D_{m,o}} \right] \cdot f_1 + \left[\frac{g(k) \cdot d_f^2 \cdot u_o}{D_s} \right] \cdot f_2 \qquad (44)$$

The subscript o refers to column outlet conditions.
In gas chromatography f_1 is given by:

$$f_1 = \frac{9}{8} \frac{(P_R^4 - 1)(P_R^2 - 1)}{(P_R^3 - 1)^2} \qquad (45)$$

and f_2 by:

$$f_2 = \frac{3}{2} \cdot \frac{P_R^2 - 1}{P_R^3 - 1} \qquad (46)$$

P_R is the ratio of the column in- and outlet pressure. The limiting values for f_1 in GC are 1 for $P_R \downarrow 1$ and 9/8 for $P_R \rightarrow \infty$. This means that even large pressure gradients only marginally affect the column efficiency.

In SFC no exact analytical expressions describing the f_1 and the f_2 factor as a function of the pressure drop or the ratio of the inlet to the outlet pressure can be derived. The values of these factors can, however, be calculated numerically. Numerical expressions for calculating the f_1 and the f_2 factor have been derived elsewhere [21]. The value of the f_1 factor can be calculated from:

$$f_1 = \frac{n \cdot \sum\limits_{i=1}^{n} \left[\frac{1}{u_i} \right]^2}{\left[\sum\limits_{i=1}^{n} \frac{1}{u_i} \right]^2} \tag{47}$$

In deriving this expression it was assumed that the ratio of D_m over u remains constant along the column. It has been shown in previous work that this is a reasonable assumption [8]. Numerical evaluation of Eqn. 47 requires knowledge of the entire velocity gradient in the column. To get an impression of the order of magnitude of the f_1 factor, a simple relationship for the linear velocity profile can be introduced in Eqn. 47. It has been shown by several authors that density gradients along SFC columns are almost linear [8,22]. This leads to the following expression for the linear velocity, u_l, as a function of the position along the column:

$$u_l = \frac{\rho_0 \cdot u_0 \cdot L}{\rho_0 \cdot L - (\rho_0 - \rho_L) \cdot l} \tag{48}$$

In this equation the subscripts refer to the position along the column (0 = inlet, L = outlet). l varies between 0 at the column inlet to L at the outlet. Substitution of Eqn. 48 in Eqn. 47 and subsequent integration yields:

$$f_1 = \frac{4 \cdot (1 + v_l + v_l^2)}{3 \cdot (1 + 2v_l + v_l^2)} \tag{49}$$

Here v_l is the fractional density loss defined as:

$$v_l = \frac{\rho_0}{\rho_L} \tag{50}$$

Note that ν_l is very similar to $P_R = P_{in}/P_{out}$ in GC. Both parameters reflect the ratio of the inlet- to the outlet density.

Under the assumption of a linear density gradient along the column, the f_2 correction factor can be shown to be given by [21]:

$$f_2 = \frac{2}{1+\nu_l} \tag{51}$$

where ν_l again reflects the ratio of the inlet- to the outlet density. In Fig. 7 a plot of the f_1 and the f_2 factor vs. ν_l is given. It is difficult to define a limit for the maximum allowable density drop in SFC. Even with pressure drops of almost 200 bar good separations can be obtained [23]. In that case, however, very high inlet pressures, far beyond the limits of current instrumentation, are required to elute high-molecular-weight solutes with capacity factors in the optimum range. A density drop of 20% ($\nu_l = 1.2$) appears to be a reasonable estimate for the maximum permissable density drop under average practical conditions. Hence, it can be concluded from Fig. 7 that both the f_1 and the f_2 factor are always very close to unity. This implies that the expansion of the fluid along the column length does not cause a significant increase of the observed plate height.

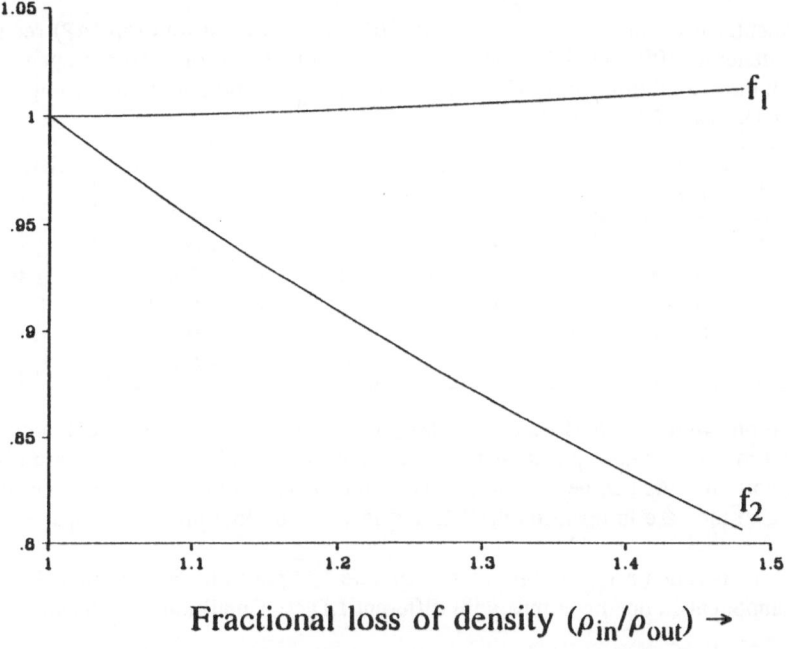

Fig. 7. f_1 and f_2 compressibility correction factor in open-tubular columns for SFC with a linear density gradient as a function of the fractional loss of density, ν_l.

It is interesting to compare the value of ν_l with typical values for P_R in GC. The decompression effect in SFC is extremely low when compared to the situation in GC. Routine GC columns (25 m, 320 μm I.D.) are typically operated at a column head pressure of about 1.5 bar which means that P_R already equals 1.5. For narrow-bore GC columns of for example, a length of 10 meter and 50 μm internal diameter this increases to about 25.

For vacuum-outlet GC, P_{in}/P_{out} can easily be as large as 10^5. In conclusion, some compressibility effects occur in supercritical media. In general, however, the behaviour of supercritical media seems to closely resemble that of non-compressible media. Under practical conditions, only small errors are introduced by neglecting the compressibility of the fluid.

Using the numerical procedures described above the efficiencies and the analysis times in SFC can be calculated accurately. Here we will first consider an <u>unretained</u> component. For such a component the pressure drop across the column affects the efficiency only via the f_1 and f_2 correction factor. Table 6 gives the column lengths, analysis times and the pressure drops required to generate 100,000 plates for an unretained component on a 10 μm I.D. open column operated at the optimum velocity and at 2, 3 ,5 and 10 times the optimum flow. In the table also the calculated values for the f_1 and f_2 factor are given. As can be seen in the table the effects of the pressure drop on the plate height in SFC are very small. For this very narrow column, even if operated far above optimum, both the f_1 and f_2 factor are very close to 1. This means that the plate height for this unretained component is virtually unaffected by compressibility effects. f_1 and f_2 factors significantly different from unity were only observed in the direct vicinity of the critical point. As this region is of no practical interest for SFC it is left out of consideration here.

Table 6.   Calculated column length (L), analysis time (t) and pressure drop (ΔP) required to generate 100,000 plates for an unretained component on a 10 μm open-tubular column. Conditions: pure CO_2, T = 70°C, P_{in} = 160 bar. Molar volume used for D_m calculations: 644 cm^3/mole.

Velocity	L (m)	t (s)	ΔP (bar)	f_1	f_2
u_{opt}	0.289	30.3	0.45	1.00000	1.000
$2u_{opt}$	0.364	19.3	1.14	1.00000	1.000
$3u_{opt}$	0.486	17.1	2.26	1.00000	0.999
$5u_{opt}$	0.756	15.9	5.87	1.00003	0.991
$10u_{opt}$	1.45	14.8	22.3	1.00058	0.958

Similar calculations were also performed for 100 μm, 50 μm and 25 μm I.D. columns [20]. In Table 7 the times required to generate 100,000 plates at the optimum velocity are given for these four columns. As can be seen in the table, the analysis time reduces proportional to d_c^2, which is also the case in open-tubular LC and in GC with low pressure drops.

Table 7.   Analysis time (at u_{opt}) required to generate 10^5 plates for an unretained component in open columns with different I.D.'s. Conditions: see Table 6.

Column diameter (μm)	Analysis time (s)	Relative time (-)	$1/d_c^2$ (normalized)
100	3111	1.00	1.00
50	789	0.25	0.25
25	195	0.063	0.0625
10	30.3	0.0097	0.010

For <u>retained</u> species not only the compressibility effect but also the increase of the capacity factor and the concomitant increase of the plate height along the column may affect the speed of analysis in SFC columns with a significant pressure drop. To asses the influence of the capacity factor increase along the column on the speed of analysis, two series of calculations were performed. In the first series, the capacity factor of the test solute (n-C_{28}) was fixed at the value under column inlet conditions (0.708). In doing so, the effect of the fluid compressibility on the efficiency and speed of analysis is taken into account, whereas the effect of the capacity factor increase along the column is not considered. Table 8 lists the retention times and the observed plates numbers for these calculations. The column lengths from Table 6 were used.

Table 8. Influence of the pressure drop on the chromatographic parameters of retained components in open-tubular SFC. Conditions: pure CO_2, T = 70°C, P_{in} = 160 bar, ρ_{in} = 0.549 g/mL. Column diameter: 10 μm. Solute: C_{28}, molar volume for D_m calculations: 644 cm^3/mole. k_{obs} = observed capacity factor.

Conditions		k fixed at 0.708		k variation included			Relative time
Velocity	L (m)	t_r^{fix} (s)	N^{fix} (10^4)	k_{obs} (-)	t_r^{incl} (s)	N^{incl} (10^4)	$\dfrac{t_r^{incl.} N^{fix}}{t_r^{fix} N^{incl}}$
u_{opt}	0.289	42.3	4.28	0.715	42.4	4.26	1.01
$2u_{opt}$	0.364	26.5	3.19	0.727	26.8	3.14	1.03
$3u_{opt}$	0.468	23.4	2.96	0.748	23.9	2.86	1.06
$5u_{opt}$	0.756	21.6	2.81	0.824	23.1	2.57	1.17
$10u_{opt}$	1.45	19.7	2.73	1.55	29.4	1.60	2.55

In the second series of calculations the variation of the capacity factor along the column is also taken into account. The results from these calculations are also shown in Table 8. The test component was n-C_{28}. The coefficients for the relationship between the capacity factor and the density (Eqn. 29) were taken from [8]. The last column in Table 8 represents the ratio of the analysis times for the situation in which the capacity factor increase is included relative to the case where it is not included.

Table 8 nicely illustrates the increase of the observed capacity factor with increasing linear velocity. It is interesting to note that the effects of the pressure drop on the retention time and the plate height for this 10 μm column already start to become visible at the optimum velocity, although only for higher velocities the effects really become significant. For example at 10 u_{opt}, the increase of the capacity factor along the column causes the analysis time to increase by a factor 2.55 relative to the hypothetical case in which the pressure drop would not lead to increased capacity factors. Similar calculations were also performed for columns with larger inner diameters. For these columns the effects of the pressure drop on the performance are marginal, if velocities not to far above optimum are used [20]. From this it can be concluded that the use of open-tubular columns in SFC is limited to columns with inner diameters above approximately about 10 μm. For lower I.D. columns the optimum performance may not be obtainable due to the adverse effects of the pressure drop.

314

REFERENCES

1. M.J.E. Golay, in "Gas Chromatography 1958", D.H. Desty (Ed.), Butterworths, London, 1958, 35.
2. P.J. Schoenmakers, J. High Resolut. Chromatogr. & Chromatogr. Commun., 11 (1988) 278.
3. H.-G. Janssen and C.A. Cramers, J. Chromatogr., 505 (1990) 19.
4. H.-G. Janssen, J.A. Rijks and C.A. Cramers, J. Microcolumn Sep., 2 (1990) 26.
5. G.J. Kennedy and J.H. Knox, J. Chromatogr. Sci., 10 (1972) 549.
6. C.P.M. Schutjes. E.A. Vermeer, J.A. Rijks and C.A. Cramers. J. Chromatogr., 253 (1982) 1.
7. C.P.M. Schutjes, Ph.D. Thesis, Eindhoven University of Technology, The Netherlands, 1983.
8. H.-G. Janssen, P.J. Schoenmakers, H.M.J. Snijders, J.A. Rijks and C.A. Cramers, J. High Resolut. Chromatogr., 14 (1991) 438.
9. R.J. Laub, Anal. Chem., 56 (1984) 2115.
10. J.A. Rijks, Ph.D. Thesis, Eindhoven University of Technology, the Netherlands, 1973.
11. A.J.J. van Es, Ph.D. Thesis, Eindhoven University of Technology, The Netherlands, 1990.
12. D.E. Martire, J. Chromatogr., 461 (1989) 165.
13. D.P. Poe and D.E. Martire, J. Chromatogr., 517 (1990) 3.
14. R.C. Reid, J.M. Prausnitz and T.K. Sherwood, The properties of Gases and Liquids, 3rd Ed., McGraw Hill, New York, 1977, p. 544.
15. H.E. Schwartz, P.J. Barthel, S.E. Moring and H.H. Lauer, LC/GC, 5 (1987) 490.
16. C. Horvath and H.-J. Lin, J. Chromatogr., 149 (1978) 43.
17. F.H. Arnold, H.W. Blanch and C.R. Wilke, J. Chromatogr., 330 (1985) 159.
18. E. Katz, K.L. Ogan and R.W.P. Scott, J. Chromatogr., 270 (1983) 51.
19. J.J. van Deemter, F.J. Zuiderweg and A. Klinkenberg, Chem. Eng. Sci., 5 (1956) 271.
20. J.C. Giddings, Dynamics of Chromatography, Part I, Marcel Dekker, New York, 1965.
21. H.-G. Janssen, P.J. Schoenmakers, H.M.J. Snijders and C.A. Cramers, in preparation.
22. P.J. Schoenmakers, P.E. Rothfusz and F.C.C.J.G. Verhoeven, J. Chromatogr., 395 (1987) 91.
23. D.E. Gere, R.D. Board and D. McManigill, Anal. Chem., 54 (1982) 736.

PHYSICOCHEMICAL MEASUREMENT BY CHROMATOGRAPHY:
Overview and Solution Thermodynamics.

J. R. CONDER
Chemical Engineering Dept
University of Wales
Swansea,   SA2 8PP
U.K.

ABSTRACT. Chromatography is used for physicochemical measurement in a
great variety of fields of study. Characteristically, it is fast, can
cope with extreme experimental conditions, and requires only a small
amount of material, which need not be pure and can be reactive. On the
other hand, the limitations of the technique need to be clearly
understood for reliable measurements. The types of physicochemical study
which can be conducted by chromatography are classified and experimental
requirements for successful measurements are outlined. A major area of
study is solution thermodynamics. Some of the practical problems that
need to be overcome to obtain reliable solution data are examined.
Examples are given of applications to determination of activity
coefficients and partial molar entropies of solution, and to study of
complexing behaviour. Studies can be carried out with precision at both
infinite dilution and finite concentration of the solute component.
Both gas and liquid chromatographic techniques are discussed.

1.   Introduction

This year is the fiftieth anniversary of the Nobel Prize-winning paper
of Martin and Synge (1941). The paper was remarkable in two aspects. It
introduced not only liquid-liquid chromatography but also the first
application of chromatography to physicochemical measurement - the
determination of liquid-liquid distribution coefficients. Since then the
scope of the technique has expanded enormously. There is now a great
variety of fields of physico-chemical study which can be explored
chromatographically, as set out in Fig. 1 and Table 1.

Both gas and liquid chromatography are used for physico-chemical
measurement. All the applications listed in Table 1 can be carried out
by an appropriate gas chromatographic (GC) method (the only exception
being studies of adsorption at liquid-liquid interfaces), but only a
limited number have been conducted by a liquid chromatographic (LC)
method.

The   principal   LC   applications   concern   aspects   of   solution

315

F. Dondi and G. Guiochon (eds.),
Theoretical Advancement in Chromatography and Related Separation Techniques, 315–337.
© 1992 *Kluwer Academic Publishers.*

316

interactions and kinetic and transport properties. To these may be added adsorption at liquid-liquid and liquid-solid interface, while pore size distributions can be determined by the size-exclusion form of LC with a series of polymers of known molecular weight. Except in this last case, the GC and LC methods of measuring a particular property are usually analogous in principle, though the LC method may present greater problems in selecting compatible materials and interpreting the results (Section 8.1). In cases where a choice is available between GC and LC forms of the technique, GC is often to be preferred, but LC is better suited to volatile liquid phases.

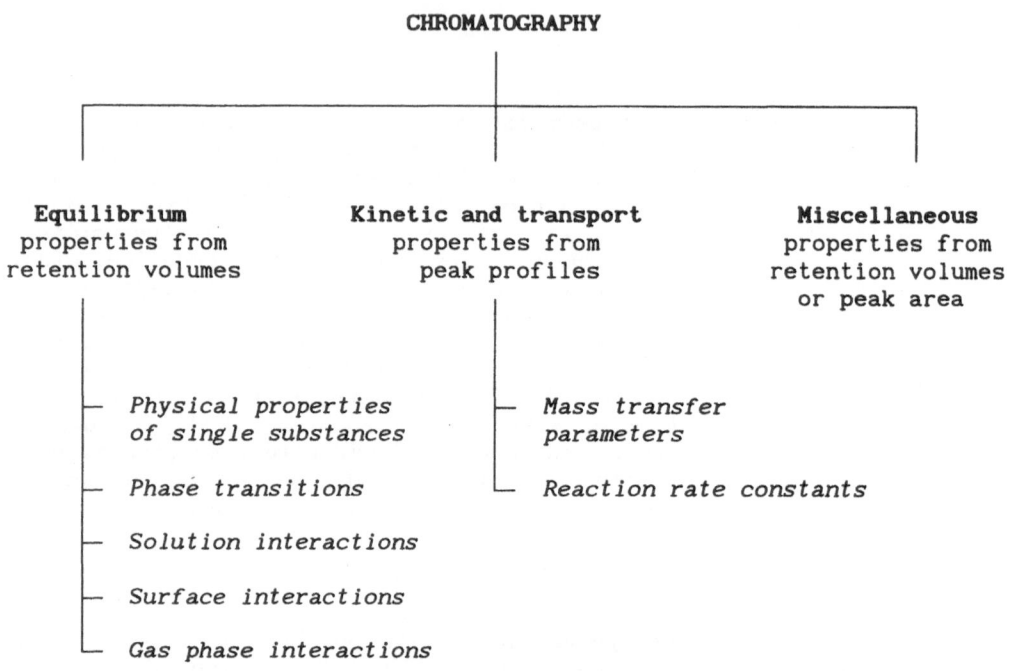

Fig. 1 Physicochemical applications of chromatography.

This article falls into two parts. In the first, we shall consider the general characteristics of chromatographic methods and outline the principal practical requirements for obtaining reliable physicochemical data. In the second, we shall concentrate on the largest single area of application, the study of solution thermodynamics. Only methods which depend fundamentally on the dynamics of the chromatographic process will be discussed. We shall not consider measurements in which chromatography is used purely as a means of chemical analysis, even though this approach is sometimes well established, e.g., in the study of solution thermodynamics by headspace analysis.

TABLE 1   Physicochemical parameters measured by chromatography.

```
Physical properties of single substances
    latent heat
    boiling point
    vapour pressure
Phase transitions
    liquid crystals
    melting transitions
    glass transition in polymers
    adsorbed liquid films
    solid-solid transitions
Solution interactions
    partition coefficient
    activity coefficient
    enthalpy and entropy changes
    complexing constants
    hydrogen bond strength
    polymer solution interactions
    solubility of gases in liquids
    liquid-liquid distribution coefficient
Surface interactions
    gas-solid adsorption coefficient
    liquid surface adsorption coefficient
    liquid-liquid interface adsorption coefficient
Gas phase interactions
    second virial coefficient of gas mixtures

Mass transfer parameters
    diffusivity of gases
    diffusivity of liquids
    diffusivity in micropores
    interfacial resistances
    adsorption and desorption rate constants
    obstruction factor
    extra particle voidage
Reaction rate constants
    in liquids
    at surfaces

Miscellaneous properties
    molecular weight
    surface area
    pore size distribution
    liquid film thickness
    polymer crystallinity
    structural assignment
```

It is estimated that a few thousand papers have been published since 1941 on physicochemical measurement by chromatography. A brief survey such as this must inevitably be highly selective.

2. Characteristics of the Method.

As an alternative to other means of studying physicochemical behaviour, chromatography has its own characteristic virtues and vices. The virtues are listed in Table 2.

TABLE 2 Advantages of chromatographic methods of physicochemical measurement.

Speed. Due to very short path lengths for mass transfer.

Small amounts of material required. E.g., 10^{-5} μg of solute; 1 mg of stationary phase with an open tubular column.

Suitable for impure and mixed solutes. Are separated on the column, so do not interfere.

Suitable for reactive solutes. Separation of reactants and products on the column allows forward reaction to be studied alone without interference from the reverse reaction, autocatalysis or product inhibition.

Wide temperature range. Is easily controlled from below 100^{0} K to above 1200^{0} K.

Wide pressure range. Up to at least 2000 bar.

Commercial equipment readily available. Analytical chromatographs and components are suitable for all except the most precise physico-chemical work.

Particularly significant is the speed with which large volumes of data can be determined, e.g. in the comparison of liquid solutions differing systematically in composition, concentration or chemical structure. This has led to the discovery of a number of valuable generalizations about the behaviour of chemical systems. In many cases, however, the value of chromatography lies not so much in offering an alternative or better method, but in the scientific progress which results from trying to understand why different methods of studying apparently the same property sometimes give different results.

Despite the many virtues of the method, its limitations also need to be appreciated for reliable measurements. The remarkable versatility of the method evident in Fig. 1 and Table 1 is partly due to the large number of components in a chromatographic column. Thus a GLC or LLC column contains at least three bulk and two interfacial phases. Herein lies a potential difficulty: the study of one interaction may be complicated by the presence of another. An example of this all too common problem is adsorption at phase interfaces, encountered as interfering processes in the study of solution thermodynamics. Another is lack of inertness in the carrier gas in GC. The presence, or potential presence, of an interfering process is the principal disadvantage of chromatographic methods. In most cases these problems can be overcome, but usually at the cost of greater complexity of experimental procedure or data analysis.

The precision of chromatographic measurements varies with the application. Measurements based on relative retention can be made with a precision of about 0.2%, or 0.04% with special provision. When measurements depend on the specific retention the limiting factor is the determination of the volume or area of the stationary phase (Laub et al., 1978) and the precision worsens to 1-2%. When the primary data source is peak height or area, the precision is lower than for retention. In comparing chromatography with other methods, much therefore depends on the field of study. Thus partial molar enthalpies of solution are usually obtained with better precision by calorimetry than chromatography, whereas the precision of chromatographically determined second virial coefficients of gas mixtures (Cruickshank et al., 1966) is superior to that of classical volumetric methods.

3.   Experimental Requirements for Chromatography.

Although physicochemical measurements can often be conducted with commercial chromatographs designed for chemical analysis, some modification or addition to the equipment is usually necessary (Conder and Young, 1979). For example, pressure and flow control arrangements in analytical gas chromatographs are usually rudimentary and need to be replaced. In addition, a number of precautions and procedures need to be observed for accurate measurements. Table 3 summarises the principal experimental requirements. I would like to elaborate briefly on two of these requirements.

The need to deal with asymmetrical peaks often causes difficulty. Peak shape can be described mathematically in various ways (Jonsson, 1987). Asymmetry and tailing peaks are very common and can arise from several causes. It is not safe merely to assume that the cause is nonlinear (concentration-dependent) chromatographic behaviour in the column. One must first diagnose the causes and then, as far as possible, eliminate them (Conder, 1982). If the residual cause does prove to be nonlinear behaviour, various measures can be adopted to achieve linearity (infinite dilution). The simplest of these, where feasible, is to reduce the solute concentration. In a typical 4 mm bore packed column, the maximum sample sizes for infinite dilution are of the order of:

$10^2 - 10^3$ µg   in GLC, LLC, bonded-phase LC, and size exclusion chromatography (when free from adsorption effects)

$10^1 - 10^3$ µg in liquid-solid chromatography and ion-exchange chromatography

$10^{-3} - 10^1$ µg in gas-solid chromatography

In open tubular GLC, the maximum size is reduced by about four orders of magnitude.

TABLE 3 Experimental requirements for good physicochemical measurements.

Precision pressure and flow control
Precision flow rate measurement and Pressure measurement (GC)
Good temperature control in space and time (GC)
Packed rather than open-tube columns where feasible
Appropriate choice of support and (in GLC) liquid loading
Careful determination of the amount of stationary phase present (preferably by more than one method)
Avoidance of packing voids (in LC)
Minimal dead spaces in the apparatus
Use of appropriately sized samples (very small for infinite dilution)
Control of peak asymmetry and tailing; use of peak mass centre rather than peak maximum if these differ in retention
Comprehensive reporting of experimental conditions.

The second point concerns the measurement of retention time. If infinite dilution can be achieved, retention should be measured to the mass centre of the peak; this position is more accurately determined as the mean of the initial and final retention times than from the peak maximum. If linear behaviour cannot be achieved, empirical procedures for correcting the measured retention time are available (Conder et al., 1986).

4.   Solution Thermodynamics at Infinite Dilution.

Studying solution thermodynamics is probably the most common of all the chromatographic applications to physical measurement. The chief interest is usually to provide information on molecular interactions (Phillips, 1973) for it is in this field that the ability of chromatography to generate large amounts of data rapidly confers the greatest advantage; a great deal of information can be obtained by systematic, and sometimes subtle, variations in structure of solute and stationary phase. Chemical engineers, however, also need vapour liquid equilibrium data for process design and the potential of the method for this purpose is still much less well appreciated than its capacity to shed light on molecular interactions.

The chromatograpic method is particularly advantageous when applied at infinite dilution of the solute component in the liquid stationary phase. Other methods, which depend on measuring vapour pressure, suffer from diminishing precision of measurement at infinite dilution. In chromatography, the precision of measurement of retention is not inherently concentration-dependent. The method provides a unique way of studying interactions with precision at infinite dilution, which is the most important concentration region for studying molecular interactions.

The technique in its GC form is simplest to operate if
(i)   the solute component is at infinite dilution,
(ii)   the solvent component (stationary phase) is involatile,
(iii) only one component (the solute) is distributed between the gas and stationary phases, and
(iv)   the solute is not absorbed at the liquid surface (Gibbs adsorption).
All these restrictions, however, can be removed with some modification of method. The LC technique will be described in Section 8.

4.1   TYPES OF GC MEASUREMENTS.

Several thermodynamic parameters for simple binary solutions can be calculated from the measured retention volume of a solute. The partition coefficient defined by the equation

$$K_L = q/c \tag{1}$$

can be obtained from the net retention volume V_N and the volume V_L of liquid stationary phase in the column by the equation

$$K_L = V_N / V_L \tag{2}$$

Thermodynamically, the activity coefficient $\gamma(0)$ of the solute in the stationary phase at zero total system pressure is of more interest. It can be calculated from V_N by the equation.

$$\ln \gamma(0) = \ln \frac{RT \, w_L}{V_N p_1^0 M_L} - \frac{(B_{11} - v_1^0) p_1^0}{RT} + (2B_{12} - v_1^\infty) P_o J_3^4 / RT \tag{3}$$

(Notation is defined in section 9).

The partial molar heat and entropy of mixing can be obtained from the temperature dependence of the activity coefficient. The heat quantity is usually obtained with less precision than from calorimetry, but with care ± 40 J/mol can be achieved (Meyer and Baiocchi, 1978). The activity coefficient, however, can be determined to 1-2% or better (section 2). Equation (3) also provides the basis of an accurate method of determining mixed second virial coefficients (B_{12}) of gas mixtures (Cruickshank et al., 1966). An alternative to the conventional retention-based technique is to obtain the thermodynamic parameters from measurements of peak height using reversed flow GC (Katsanos et al., 1986). Two other types of solution parameter can also be determined: the solubility of gases in liquids, i.e. Henry's law constants (Parcher et al., 1984), and the distribution coefficient of a solute between two immiscible solvents, a parameter which is important in liquid-liquid extraction (Sheehan and Langer, 1968).

4.2 PITFALLS IN MEASUREMENT

There are a number of pitfalls to be avoided, or at least taken account of, in investigating solution behaviour by GC.

4.2.1. *Gas imperfection correction*. This is the contribution made by the second and third terms on the right-hand side of equation (3) to the activity coefficient calculated from the net retention. Without this correction the value of the activity coefficient is typically 1-5% too low. The literature contains many examples of the neglect of this correction, of use of incorrect versions of the equation or of adoption of a false assumption that the choice of helium as a carrier gas allows the correction to be neglected. Neglection may be justified in cases, such as complexing systems (Section 5) where the chief interest lies in relative retentions, but should always be made when accurate values of thermodynamic parameters are required.

4.2.2. *Adsorption on the solid support*. If the solute undergoes any adsorption by the solid support, there is an extra contribution to retention to add to that caused by the bulk liquid. The result is to lower the apparent activity coefficient. The problem has been reviewed (Conder and Young, 1979). The main conclusion is that the effect can be rendered virtually negligible for many solution systems by choosing an appropriate support for the solute and solvent system under study.

4.2.3. *Liquid surface adsorption*. A further possibility with some systems is Gibbs adsorption of the solute at the gas-liquid interface. The existence of this source of interference is less under the experimenter's control than adsorption on the support, since its presence or absence depends on the nature of the solute and solvent alone. It is therefore necessary to use procedures for diagnosing and evaluating adsorption and solution contributions to retention (Section 7).

4.2.4. *Ageing of liquid phases*. This occurs in many liquids with time. The process can be monitored by repeatedly checking the retention of standard solutes. The problem can be reduced by using only the highest grade of oxygen-free carrier gas and avoiding excessive conditioning and operating temperatures for the liquid phase in question.

TABLE 4   Some examples of binary solution systems studied by GC.

n-alkane + n-alkane mixtures (Cruickshank, Windsor
and Young, 1966).
Aromatic solute systems (Janini and Martire, 1974).
Chloroalkane systems (Alessi et al., 1982).
Polar solutes (Defayes et al., 1990).
Isomeric systems (Karger, Stern and Zannucci, 1968).
Polymer solutions (Gray, 1977).
Metal chelates (Wolf et al., 1972).
Volatile solvents, incl. water (Thomas et al., 1982).

4.3   EXAMPLES OF APPLICATION

Many types of solute and stationary phase system have been investigated
over the last 35 years and activity continues to expand.   Table 4
provides a small selection from a large literature to give an idea of
the scope of the method.

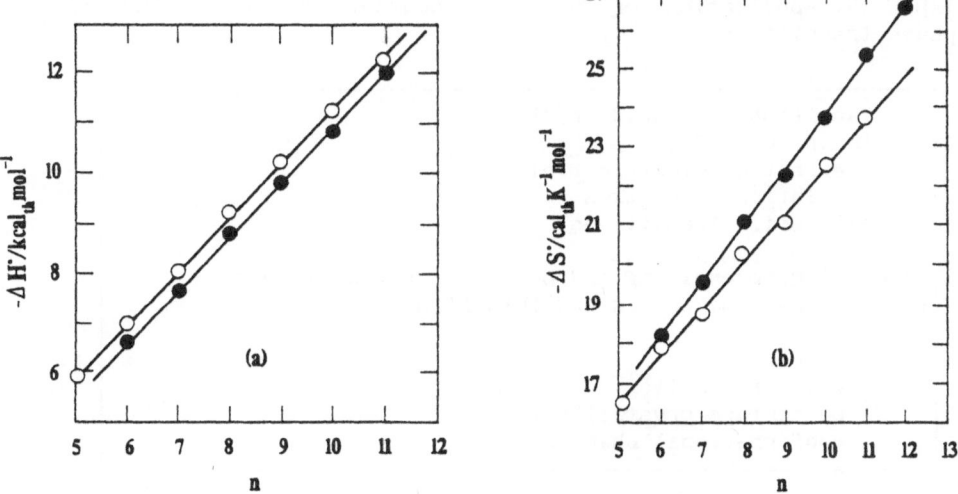

Fig. 2 (a) enthalpy and (b) entropy of solution at 317.15^{0} K in $n-C_{36}H_{74}$
function of the number of carbon atoms in the solute.   ● n-alkanes;
○ cyclic alkanes. (Reprinted with permission from: Meyer and Weiss,
1977, Copyright [1977] Academic Press Ltd.)

A good example of what can be achieved by chromatographic methods is
Meyer and Weiss's (1977) work on cyclic alkanes in a long straight chain
paraffin. Fig. 2 shows the interesting result that the partial molar

entropy of solution behaves differently from the entropy. Whereas the partial molar enthalpy of solution of the cyclic alkanes is less than that of the corresponding normal alkanes by a constant amount, independent of the carbon number of the alkane, the partial molar entropy diverges progressively from that of the normal alkanes. The implications of these results for the molecular interactions involved were extensively discussed by Meyer and Weiss.

4.4   POLYMER SOLUTIONS AND PHASE TRANSITIONS: INVERSE CHROMATOGRAPHY.

Polymer solutions have been investigated at both infinite dilution and finite concentration.   The solvent component in which polymer would normally be dissolved to form the solution is here employed as the solute, since it is injected into the mobile phase with the polymer as the stationary phase. This is an example of the general technique of "inverse chromatography" in which the solute is used as a probe to investigate the behaviour of the stationary phase.   Inverse chromatography at infinite solute dilution has also been extensively adapted to explore various types of phase transition (Table 5) and solid surfaces; see, e.g., Lloyd et al (1989), Rays and Waksmundski (1984), Gray (1977) and Williams (1991).

TABLE 5   Applications of Inverse Chromatography in the study of phase transitions.

```
Transition in liquid crystals:
    surface effects
    effect of electric fields
    solution properties
Solid-solid transitions

Melting, premelting and postmelting transitions
    in bulk materials and thin films

Polymers:
    glass transition
    percentage crystallinity
    conformational changes
```

5. Complexes.

An area of solution thermodynamics where infinite dilution chromatographic methods have proved particularly fruitful is the study of complexes in solution.   The method is essentially simple. Consider the original example of the complexing reaction

$$\text{olefin} + Ag^+ \longrightarrow Ag^+.\text{olefin} \tag{4}$$

studied in ethylene glycol solution by Gil-AV and Herling (1962). The stability constant of the complex is

$$K_1 = \frac{[Ag^+.olefin]}{[Ag^+][olefin]} \tag{5}$$

This is related to two partition coefficients K_L and K_L^* at infinite solute dilution by the equation

$$K_L = K_L^* (1 + K_1 [Ag^+]) \tag{6}$$

The partition coefficient K_L is determined from the retention volume of the olefin on a column of silver nitrate and ethylene glycol, and K_L^* from the retention in ethylene glycol alone. K_1 is then calculated from the last equation.

Several classes of complex formation can be studied, as shown in Table 6 (Purnell, 1966).

TABLE 6 Classes of complexing system studied by gas chromatography.

Class A. Solute (X) reacts with a stationary-phase additive (A) to give complexes.
Class B. Solute (X) reacts with the stationary-phase solvent (S) to give complexes.
Class C. Solute (X) reacts with itself.
Class D. Additive (A) reacts with the solvent (S).

TABLE 7 Types of complexing interaction that have been studied chromatographically.

Cation-liquid complexing
Charge-transfer interactions
Hydrogen bonding
Complexing between inorganic compounds and fused metal salts
Complexes with surface modifiers on solid surfaces

The nature of the complexing interaction can be various; Table 7 shows the main categories to which most investigations to date belong.

Purnell and co-workers (Harbison et al., 1979) have found that the behaviour of a wide variety of systems of the Class A (ternary) type can be described by a linear equation,

$$K = \phi_A K^0_{L(A)} + \phi_S K^0_{L(S)} \qquad (7)$$

where K_L is the solute partition coefficient between the gas phase and a binary stationary phase of components A and S, $K^0_{L(A)}$ is the same parameter in pure component A, and $K^0_{L(S)}$ in pure component S; ϕ_A and ϕ_S are the volume fractions of A and S. The validity of this equation has been the subject of much debate since it does not agree with accepted models of molecular interactions in solution (Martire, 1974) and some systems have been found showing substantial deviations from the equation, as shown in Fig. 3.

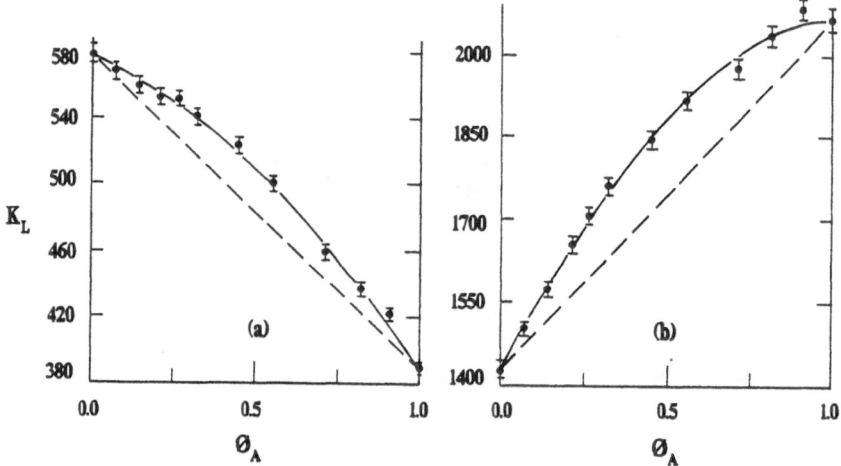

Fig. 3. Plots of K_L against ϕ_A for (a) cyclohexane and (b) toluene in mixtures of di-nonyl phthalate (A) and squalane (S) at 30.0°C. Broken line, eqn. (7). Solid line, additive contribution model (Martire, 1974) of interactions in solution. (Reprinted with permission from Harbison et al. 1979, Copyright [1979] American Chemical Society).

The current consensus is that equation (7) is most reliable when used to predict relative, rather than absolute retentions (Harbison et al., 1979). The equation then provides a particularly simple basis for calculating the composition of mixed stationary phases needed for optimum resolution in analytical GLC.

6. Solution Thermodynamics at Finite Concentration.

6.1 METHODS OF MEASUREMENT.

Whereas only one or two chromatographic methods are available for studying infinite dilution of the solute component, there are many methods for finite solute concentrations. They are listed in Table 8. The same techniques are used to determine gas-solid adsorption isotherms and are described by Conder and Young (1979).

TABLE 8   Chromatographic methods for finite solute concentrations.

Frontal analysis by characteristic point (FACP)
Frontal analysis (FA)
Elution by characteristic point (ECP)
Elution on a plateau (EP) (step and pulse method)
Elution of an isolope on a plateau (EIP)
Limiting transition point method
Variation of peak maximum with sample size
Sorption effect chromatography
Reversed-flow gas chromatography

The most often used methods in Table 8 are the first five. ECP has the advantage that it requires no purpose-built apparatus to produce sharp step changes in mobile phase composition or maintain steady concentrations in the mobile phase. Modified commercial chromatographs serve ECP as well as they do infinite dilution. The ECP technique is best suited to relatively low solute concentrations. The remaining techniques allow a greater concentration range. Of these, FACP is the quickest, since it is the only one, along with ECP, which allows a complete concentration range to be studied in a single run. FACP does, however, require a reasonably large number of theoretical plates in the column for accuracy. EP is used more often than EIP for binary systems, but EIP is better suited to multicomponent systems and can also be used without an inert carrier gas.

6.2 APPLICATIONS.

The greater precision offered by chromatography at infinite dilution in comparison with alternative static gravimetric or volumetric techniques has already been pointed out (Section 4). At finite concentrations static methods can be made, with care, to give greater precision than chromatography, though they cannot match the latter's speed. When studying gas-solid adsorption, considerations of precision are, generally speaking, a little less pressing than in solution studies. Moreover, GSC, unlike static methods, can not only permit adsorption on catalysts to be followed close to the normal catalytic operating temperature, but also distinguish adsorption from bulk absorption into polymers. For these reasons chromatography at finite concentrations has been employed more for gas–solid adsorption than for solution studies. A number of solutions have however, been studied, particularly multicomponent systems (Section 6.3). Adsorption on liquid surfaces has also been investigated (e.g., Dorris and Gray, 1981).

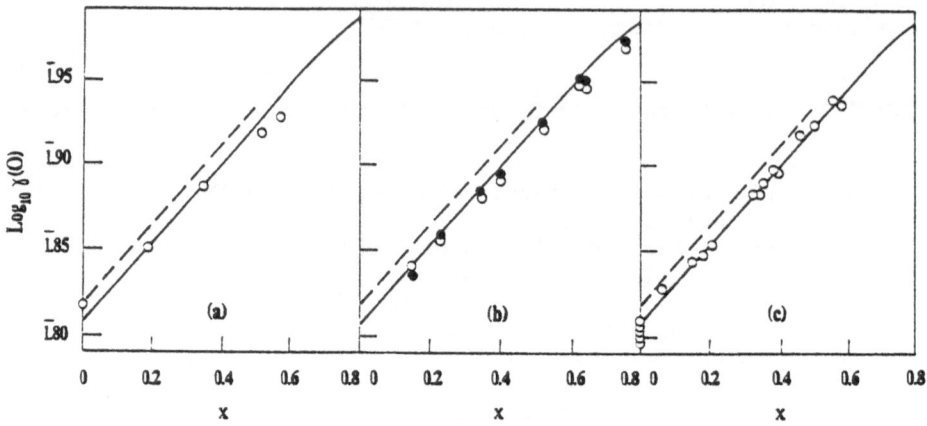

Fig. 4. Logarithm of activity coefficient of hexane in squalane against mole fraction x of hexane, at 30°C. (a) ○ FACP; (b)○FA: diffuse boundary, ● self sharpening boundary; (c) ○ EP: continous curve represents best line, extrapolated to x=o, through experimental static data of Ashworth and Everett. Broken curve represents static data of Martire, Pecksok and Purnell. The experimental points for EP were compiled from runs with widely varying flow rates (72–110 ml/min) and column lengths (0.25–2.5 m). (Conder and Purnell, 1969, reprinted with permission, Copyright [1969], The Royal Society of Chemistry).

Fig. 4 shows a comparison of data for the system n-hexane and squalane. The fact that the FACP points lie on a curve of slightly lower slope than the FA, EP and static data indicates that FACP results can be significantly affected by an inadequate number of plates in the column. Otherwise the agreement between the three chromatographic results and between the chromatographic and static results is excellent and is within the 2.5 % difference between the two static plots. Similar comparisons have been made between FACP and static data by Parcher (e.g. Hussey and Parcher, 1973) and between EP and static data by Valentin and Guiochon (1976).

Finite concentration techniques are useful in lifting the restriction, implied up to this point, that only one component, the solute, is distributed between both phases. If the gas stream consists of two components, 1 and 2, which are both sorbed by an involatile stationary phase, 3, the natural approach is to use a small concentration pertubation and use the EP method. At first sight the difficulty is that this produces only one peak and hence only one retention value with which to determine two isotherm slopes, for components 1 and 2, using the equation

$$V = V_s \left[(1 - y_1) \frac{dq_1}{dc_1} + y_1 \frac{dq_2}{dc_2} \right] \tag{8}$$

Several ways have now been devised to overcome this information deficiency. One can use an isotopic tracer method (EIP), or replace one of the two distributed components by a non-sorbed carrier, to obtain one of the two partition coefficients. Alternatively, the retention of the flow rate transient due to the sorption effect can be determined as well as the retention of the concentration peak (Buffham et al., 1985).

7. Multiple Retention Mechanisms

We have already referred to a potential drawback of chromatographic methods, the presence of processes in the column which interfere with the measurements in question. The commonest problem of this type is the presence of retention contributions from absorption processes when the behaviour of the bulk solution is under investigation. The existence of the problem is not surprising. In a gas-liquid column the liquid is spread over a large area of support surface with a very small depth, typically $0.05 - 1$ μm on average on a wetted support. Of course, this means that GLC is also well suited to study of adsorption at the liquid surface (Gibbs adsoption at the gas-liquid interface). Its extensive use for this purpose has recently been reviewed by Jonsson and Mathiasson (1987).

Whichever process is of interest, bulk solution or liquid surface adsorption, techniques are needed to separate the two. These methods start with the basic equation which states that, if the various processes causing retention are independent, their contributions to the measured retention are additive:

$$V_N = K_L V_L + K_I A_I + K_{SL} A_{SL} + K_{SG} A_{SG} \tag{9}$$

The subscripts L, I and SL refer to the bulk liquid, gas-liquid interface and liquid support interface. The term subscripted with SG allows for the possibility that the liquid may not wet the support (e.g. a silanised support), so that the solute can also be adsorbed on the uncovered part of the support surface. Means of detecting adsorption contributions, identifying their type, and separating the solution and adsorption terms are described elsewhere (Conder and Young, 1979). The basis of methods of separating the terms is to vary the liquid loading on the packing and plot V_N/V_L against $1/V_L$. K_L is then obtained from the intercept and the sum of the adsorption terms from the slope (Fig. 5).

Plots of retention against loading for systems of polar solutes in hydrocarbon stationary phases on a silanised support show an unusual peak in the plot, such as seen in Fig. 6. The peak is believed to be due

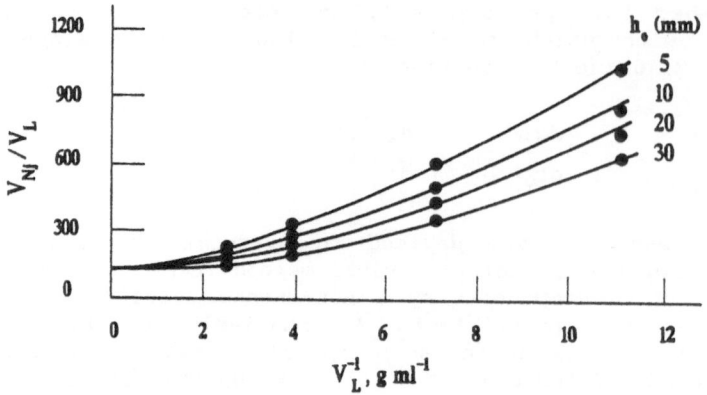

Fig. 5. Plots of V_N/V_L against V_L^{-1} for various values of solute concentration (determined by the height on the peak, h_o). Data for elution of 1-butanol at $60^{\circ}C$ from a column of a mixed stationary phase consisting of 9.4% w/w didecyl sebacate in squalane coated on silanized Sil-O-Cel firebrick. (Reprinted with permission from Cadogan and Purnell, 1969, Copyright [1969] American Chemical Society).

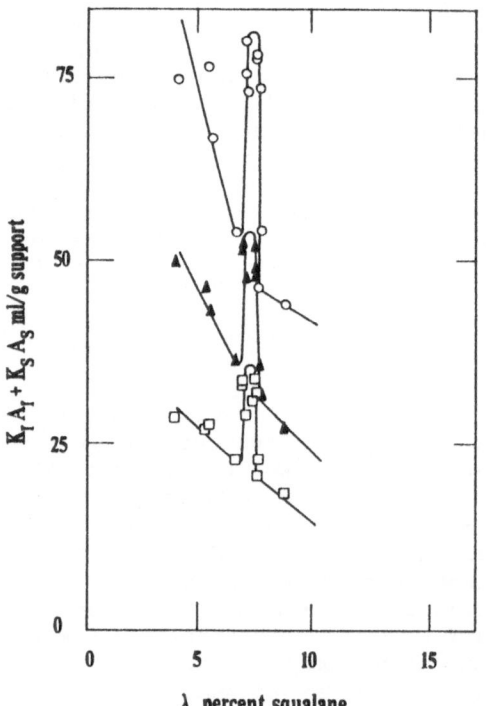

Fig. 6 Graph of total adsorption contribution to retention ($V_N - K_L V_L$) against liquid loading for ethyl acetate in squalane on silanized Chromosorb P. $\circ$ $60^{\circ}C$, $\blacktriangle$ $70^{\circ}C$, $\square$ $80^{\circ}C$. (Reprinted from Conder et al. 1985, Copyright [1985], American Chemical Society).

to a transition between wetting and nonwetting of the support by the liquid phase, evidenced in a change in A_I in equation (9). (Conder et al., 1985; Jonsson and Mathiasson, 1987).

8. Liquid and Supercritical Fluid Chromatography.

8.1 INTRODUCTION

Using LC for physicochemical measurement presents more complex problems of interpretation than GC, with supercritical fluid chromatography (SFC) lying somewhere in between.

There are two reasons. First, solvent effects are present in the mobile phase as well as in the stationary phase. Second, there now arises the question of mutual solubility of the two phases which, as Phillips (1973) has pointed out, causes complications both ways. If the solubilities are large, one is effectively dealing with mixed phases. If they are small, the solute is likely to show concentration effects at the large interface necessitated by normal chromatographic operation. One can try to circumvent this problem by using bonded phases, instead of coated liquid stationary phases, but there is good evidence that the bonded phase environment for the solute can still be much influenced by the mobile phase (Schantz et al., 1988; Cheong and Carr, 1990). In any case, an environment consisting of a thin, more-or-less ordered film cannot be equated with bulk liquid. The best reason for using bonded phases is the practical one that they do not bleed. Moreover, since they are, for this reason, ubiquitous in modern analytical HPLC, physicochemical data obtained with bonded phases have a direct analytical application.

8.2 APPLICATIONS TO SOLUTION THERMODYNAMICS.

Martin and Synge's (1941) original use of the method was the simplest: to obtain the distribution coefficient of the solute between the stationary and mobile phases from the measured retention time. This has its analogue in the later GLC method already mentioned (Section 4.1) and uses the same equation (2) above. This method has frequently been used in more recent times to obtain distribution coefficients by HPLC.

The determination of activity coefficients by LC depends on the following equation (Locke and Martire, 1967) for the specific retention Volume Vg

$$V_g = \frac{\gamma_m^\infty M_m}{\gamma_s^\infty M_s \rho_m} \tag{10}$$

where γ_m^∞ and γ_s^∞ are the activity coefficients of the solute at infinite dilution in the mobile and stationary phases, respectively. To obtain an activity coefficient in one phase therefore requires an independent knowledge of the activity coefficient in the other phase.

This approach to determining activity coefficients is less straightforward than GC. It is also restricted in application by the need to select a second phase which is compatible with the first in degree of immiscibility, yet which permits the solute to have substantial miscibility in both phases.

The attraction of the method is that it is well suited to volatile solvents, which cause practical difficulties in GC. Locke (1968) has used it, for example, to determine activity coefficients in acetonitrile as mobile phase, using the involatile liquid squalane, which is well characterised by GC, as the LC stationary phase. Some of the results are shown in Table 9. The increasing discrepancy between LC and static values for activity coefficients above 15 is thought to be due to adsorption at the liquid-liquid surface, illustrating the problem discussed in Section 8.1 above.

TABLE 9   Comparison of limiting activity coefficients in acetonitrile by LLC and static techniques (Locke, 1968)[c]

Compound	Temp. (^{0}C)	γ (Lit.)[a]	γ (LLC)[b]
Benzene	25	2.7	3.08
	45	3.08	2.83
	60	2.6	2.66
Toluene	45	3.93	3.90
Pentene-1	25	9.45	9.47
	35	8.68	8.58
n-Butane	65	12.3	12.4
n-Pentane	25	20.40	21.4
	25	17.29	18.8
	45	15.42	16.6
	50	15.5	15.8
n-Hexane	25	25.5	30.7
	60	15.8	18.9
n-Heptane	50	19.5	27.5
CCl$_4$	45	7.54	5.90

[a] Literature values, static techniques.
[b] LLC values.

[c] Reprinted with permission. Copyright [1968] Elsevier Sci.Publ..

Differences in partial molar excess enthalpies and entropies of mixing can also be obtained by this approach, by varying the column temperature. A few applications to determination of complexing constants have been reported (e.g. Horvath et al., 1979), as have finite concentration studies using LC in the modes already discussed for GC (e.g. Dondi et al., 1979; Ha, Ungvarai and Kovats, 1982).

8.3 SUPERCRITICAL FLUID CHROMATOGRAPHY

Interest is as yet only beginning to develop in the physicochemical implications of SFC, but the necessary basis of chromatographic theory is already being laid (e.g. Poe and Martire, 1990). Following previous patterns of development in the physicochemical applications of GC and LC, the initial applications are likely to be driven by the desire to obtain data with which to design analytical separations. We may expect appreciation of the capabilities of the method for specifically physicochemical purposes to follow.

9. Notation.

A_I Area of gas-liquid interface in column

A_{SG} Area of support-gas interface in column

A_{SL} Area of support-liquid interface in column

B_{11} Second virial coefficient of pure solute vapour

B_{12} Mixed second virial coefficient of solute (1) and carrier gas (2)

c   Concentration of solute in mol (or mass) per unit volume of gas phase

$$J_3^4 \qquad \left[\left(\frac{P_i}{P_o}\right)^4 - 1\right] \qquad \left[\left(\frac{P_i}{P_o}\right)^3 - 1\right]$$

K_I Distribution coefficient for solute between gas-liquid interface and gas phase

K_L Partition coefficient at column pressure for solute between liquid and gas phase (equation 1)

K_L^* Partition coefficient of uncomplexed complexing solute

K_{SG} Distribution coefficient for solute between solid support and gas phase

K_{SL} Distribution coefficient for solute between solid support and liquid phase

M_L Molecular weight of liquid phase

M_m Molecular weight of mobile phase

M_s	Molecular weight of stationary phase
P_i	Column inlet pressure
P_o	Column outlet pressure
p_1^o	Saturation vapour pressure of solute (1)
q	Concentration of solute in stationary phase in mol (or mass) per unit volume
R	Gas Constant
T	Column absolute temperature
V_g	Specific retention volume
V_L	Volume of liquid phase in column
V_N	Net retention volume
V_s	Volume, area or mass of stationary phase in column (defined consistently with q)
v_1^o	Molar volume of pure liquid solute (1)
v_1^∞	Partial molar volume of solute (1) at infinite dilution in liquid phase
W_L	Mass of liquid in column
y_1	Mole fraction of solute (1) in gas phase
γ_m^∞	Activity coefficient of solute at infinite dilution in mobile phase
γ_s^∞	Activity coefficient of solute at infinite dilution in stationary phase
ρ_m	Density of mobile phase (eluent)

10. References.

Alessi, P. Kikic, I. Alessandrini, A. and Fermelia, M. (1982) "Activity coefficients at infinite dilution by gas-liquid chromatography. 1. Hydrocarbons and n-chloroparaffins in organic solvents," J. Chem. Eng. Data 27, 448-450.

Buffham, B.A. Mason, G. and Yadav, G.D. (1985) "Retention volumes and retention times in binary chromatography," J. Chem. Soc. Faraday Trans. I 81, 161-173.

Cheong, W.J. and Carr, P.W. (1990) "Study of partition models in reversed-phase liquid chromatography based on measured mobile phase solute activity coefficients," J. Chromatog. 499, 373-393.

Cadogan, D.F. and Purnell, J.H. (1969) "Concurrent solution and adsorption phenomena in chromatography. III. The measurement of formation constants of H-bonded complexes in solution," J. Phys. Chem. 73, 3849-3854.

Conder, J.R. (1982) "Peak distortion in chromatography," J. High Res. Chromatog & C.C. 5, 341-348, 397-404.

Conder, J.R. Ibrahim, N.K. Rees, G.J. and Oweimreen, G.A. (1985) "Wetting transition in squalane on a silanised diatomaceous gas chromatographic support," J. Phys. Chem. 89, 2571-2576.

Conder, J.R. McHale, S. and Jones, M.A. (1986) "Evaluation of methods for measuring gas-solid chromatographic retention on skewed peaks," Anal. Chem. 58, 2663-2668.

Conder, J.R. and Purnell, J.H. (1969) "Gas chromatography at finite concentrations. Part 4 - Experimental evaluation of methods for thermodynamic study of solutions," Trans. Faraday. Soc. 65, 839-848.

Conder, J.R. and Young, C.L. (1979) "Physicochemical Measurement by Gas Chromatography, John Wiley and sons, Chichester.

Cruickshank, A.J.B. Windsor, M.L. and Young, C.L. (1966) "The use of GLC to determine activity coefficients and second virial coefficients of mixtures. II. Experimental studies on hydrocarbon solutes," Proc. Roy. Soc. A 295, 271-287.

Defayes, G. Fritz, D.F. Tatiana, G. Huber, G. de Reyff, C. and Kovats, E. (1990) "Organic solutes in paraffin solvents; influence of the size of the solvent molecule on solution data," J. Chromatog. 500, 139-184.

Dondi, F. Betti, A. Blo, G. Bighi, C. Versino, B. (1979) "Comparison of liquid-chromatographic methods for isotherm determination on active carbons," Ann. Chim. (Rome) 68, 293.

Dorris, G.M. and Gray, D.G. (1981) "Adsorption of hydrocarbons on silica-supported water surfaces," J. Phys. Chem 85, 3628-3635.

Gil-Av, E. and Herling, J. (1962) "Determination of the stability constants of complexes by gas chromatography," J. Phys. Chem. 66, 1208-1209

Gray, D.G. (1977), "Gas chromatographic measurements of polymer structure and interactions," Prog. Polym. Sci. 5, 1-60.

Ha, N.L. Ungvarai, J. and Kovats, E. (1982) "Adsorption isotherm at the liquid-solid interface and the interpretation of chromatographic data," Anal. Chem. 54, 2410-2421.

Harbison, M.W.P. Laub, R.J. Martire, D.E. Purnell, J.H. and Williams, P.S. (1979) "Solute infinite-dilution partition coefficients with mixtures of squalane and dinonyl phthalate solvents at 30.0^{0}C," J. Phys. Chem. 83, 1262-1268.

Horvath, C. Melander, W. and Nahum, A. (1979) "Measurement of association constants for complexes by reversed-phase HPLC", J. Chromatog. 186, 371-403.

Hussey, C.L. and Parcher, J.F. (1974) "A modified gas chromatograph for thermodynamic measurements by frontal chromatography," J. Chromatog. 92, 47-54

Janini, G.M. and Martire, D.E. (1974) "Measurement and interpretation of activity coefficients for aromatic solutes at infinite dilution in n-octadecane and n-hexadecyl halide solvents," J. Phys. Chem. 78, 1644-1648.

Jonsson, J.A. (1987) "Dispersion and peak shapes in

Chromatography," in J.A. Jonsson (ed.), Chromatographic Theory and Basic Principles, Marcel Dekker, New York, 27-102.

Karger, B.L. Stern, R.L. and Zannucci (1968) Anal. Chem. 40, 727.

Katsanos, N.A. Karaiskakis G. and Agathonos P. (1986) "Measurement of activity coefficients by reversed-flow gas chromatography," J. Chromatog 349, 369-376.

Laub, R.J. Purnell, J.H. Williams, P.S. Harbison, M.W.P. and Martire, D.E. (1978) "Meaningful error analysis of thermodynamic measurements by gas-liquid chromatography," J. Chromatog. 155, 233-240.

Lloyd, D.R. Ward, T.C. and Schreiber, H.P. (1989) (eds.) Inverse Gas Chromatography, ACS Symps. Ser. 391, A.C.S., Washington, D.C.

Locke, D.C. (1968) "Chromatographic study of solutions of hydrocarbons in acetonitrile," J. Chromatog. 35, 24-36.

Locke, D.C. and Martire, D.E. (1967) "Theory of solute retention in liquid-liquid chromatography," Anal. Chem. 39, 921-925.

Martin, A.J.P. and Synge, R.L.M. (1941) "A new form of chromatography employing two liquid phases. 1. A theory of chromatography. 2. Application to the micro-determination of the higher monoamino-acids in proteins," Biochem. J. 35 1358-1368.

Martire, D.E. (1974) "Determination and comparison of association constants for weak organic complexes by thermodynamic, resonance and optical methods," Anal. Chem. 46, 1712-1719.

Meyer, E.F. and Baiocchi (1978) "A comparison of gas-liquid chromatographic enthalpies of solution with calorimetric values for four alkane systems," J.Chem. Thermod. 10, 823-828.

Meyer, E.F. and Weiss, R.H. (1977) "Thermodynamics of solution of cyclic alkanes from C_5 through C_{11} in n-$C_{36}H_{74}$ at 373.15^0 K using GLC," J.Chem. Thermod. 9, 431-438.

Parcher, J.F. Bell, M.L. and Lin, P.J. (1984) "Determination of the solubility of gases in liquids by GLC" in J.C. Giddings, E. Grushka and P.R. Brown (eds.), Advances in Chromatography, Vol 24, Marcel Dekker, New York.

Phillips, C.S.G. (1973) "Chromatography and intermolecular forces," Ber. Bunsen-gas. 77, 171-177.

Poe, D.P. and Martire, D.E. (1990) "Plate height theory for compressible mobile phase fluids and its application to gas, liquid and supercritical fluid chromatography," J. Chromatog. 517, 3-29.

Purnell, J.H. (1966) "Gas chromatographic study of chemical equilibria" in A.B. Littlewood (ed.), Gas Chromatography 1966, Institute of Petroleum, London, 3-18.

Rayss, J. and Waksmundski, A (1984) "Comparison of the properties of liquid crystal films on silica gel and on a graphite surface," J. Chromatog. 292, 207-216.

Schantz, M. Barman, B.N. and Martire, D.E. (1988), J. Res. Natl. Bur. Stand. U.S. 93, 161.

Sheehan, R.J. and Langer, S.H. (1971) "Determination of liquid-liquid distribution coefficients by GLC," Ind. Eng. Chem. Process Des. Develop. 10, 44-46.

Thomas, E.R. Newman, B.A. Long, T.C. Wood, D.A. and Eckert, C.A. (1982) "Limiting activity coefficients of nonpolar and polar solutes in both volatile and nonvolatile solvents by gas chromatography," J. Chem. Eng. Data 27, 399–405.

Valentin, P. and Guiochon, G. (1976) "Determination of gas–liquid and gas–solid equilibrium isotherms by chromatography: II. Apparatus, specifications and results," J. Chromatog. Sci. 14, 132–139.

Williams, D.R. (1991) "Inverse gas chromatography of solid surfaces," Chromatog and Analys., Feb. 1991, 9–11.

Wolf, W.R. Sievers, R.E. and Brown, G.H. (1972) "Vapor Pressure Measurements and Gas Chromatographic Studies of the Solution Thermodynamics of Metal β-Diketonates," Inorg Chem 11, 1995.

PHYSICOCHEMICAL MEASUREMENTS BY CHROMATOGRAPHY: DIFFUSION AND RATE PROCESSES

N.A.KATSANOS
Physical Chemistry Laboratory
University of Patras
P.O. Box 1045
261 10 Patras
Greece

ABSTRACT. After the description of the basic equations of gas chromatography, the two main broadening factors are described and the van Deemter equation is given. Physicochemical Measurements by conventional gas chromatography are reviewed, and the carrier gas flow-rate perturbation is introduced in two forms, the stopped-flow and the reversed-flow techniques. The general principles of the latter are presented, and then the latest theoretical developments and applications concernig measurement of rate coefficients and distribution constants are given. These include gas diffusion coefficients, obstructive factors and external porosities in solid beds, diffusion coefficients of gases in liquids together with their partition coefficients, interfacial gas-liquid resistances to mass transfer, and finally the simultaneous determination of adsorption, desorption and reaction rate constants of gases on solids, together with mass transfer coefficients and distribution constants.

1. INTRODUCTION

1.1. The Basic Equations of Gas Chromatography

The quantitative theory of gas chromatography can be understood by reference to Fig.1. It is based on a simple first-order partial differential equation, the so-called first-order conservation equation:

$$\frac{\partial c_G}{\partial t} + r\,\frac{\partial c_s}{\partial t} = -v\,\frac{\partial c_G}{\partial x} \tag{1}$$

where

c_G = concentration of the solute vapour in the gas phase of the column (mol cm^{-3})

c_s = concentration of the solute in the stationary phase (mol cm^{-3})

F. Dondi and G. Guiochon (eds.),
Theoretical Advancement in Chromatography and Related Separation Techniques, 339–367.
© 1992 *Kluwer Academic Publishers*.

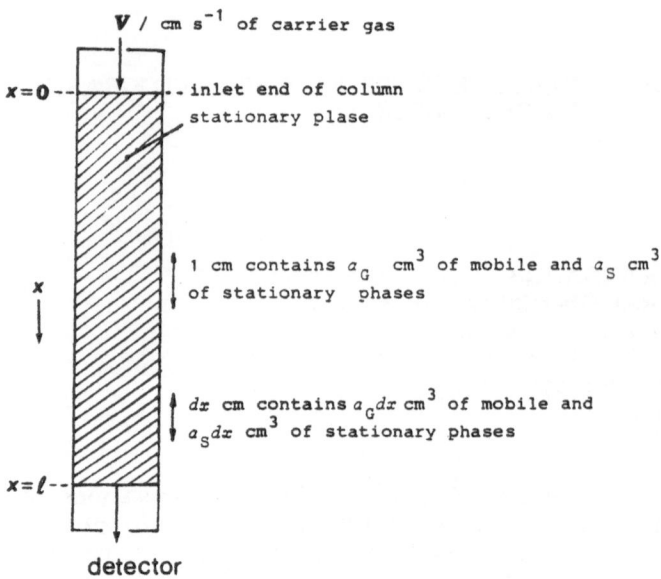

Figure 1. Schematic representation of the chromatographic column.

$\hbar = a_s/a_G$, a_s and a_G representing the volume of stationary and gas phase (both moving and stagnant), respectively, per unit length of column (cm²)

v = linear velocity of the carrier gas averaged over the interparticle and intraparticle carrier-filled space of the column (cm s⁻¹)

t = time variable (s)

x = distance coordinate along the column.

The derivation of eqn(1) was done under two assumptions: (1) Axial diffusion of the solute component in the gas phase is negligible, which is not unrealistic for high enough linear velocities, v, and (2) equilibration of the solute between the two phases is instantaneous.

The solution of eqn (1) depends on: (1) The relation between c_G and c_s, and (2) the initial and boundary conditions. In *linear* chromatography, the relation between c_G and c_s is a linear isotherm:

$$c_s = Kc_G \qquad (2)$$

where K is the partition coefficient or distribution constant (here dimensionless). Substitution of Kc_G for c_s in eqn (1) gives, after some rearrangement

$$(1 + k)\ \frac{\partial c_G}{\partial t} = -v\ \frac{\partial c_G}{\partial x} \qquad (3)$$

where k is given by the relation

$$k = Kr \tag{4}$$

and is termed *the partition ratio*.

The initial condition is

$$c_G(x, 0) = c_s(x, 0) = 0 \tag{5}$$

since at $t = 0$ the column is empty of any solute component along its entire length. As regards the boundary condition at $x = 0$, this depends on the mode of solute introduction onto the chromatographic column. The most common input distribution is an instantaneous injection of a pulse of solute, described mathematically by Dirac's delta function $\delta(t - a)$ with $a \geqslant 0$. This has non-zero value only for $t = a$, and thus the boundary condition at $x = 0$ is

$$c_G(0, t) = \frac{m}{a_G v} \delta(t) \tag{6}$$

where m is the amount (mol) of the component solute.

The solution of eqn (3) under the initial conditions (5) and subject to the boundary condition (6) is

$$c_G = \frac{m}{a_G v} \delta\left(t - \frac{1+k}{v} x\right) \tag{7}$$

According to the properties of the delta function, the right-hand side of this equation has a non-zero value only for

$$t = \frac{1+k}{v} x \tag{8}$$

This relation forms the basis for chromatographic analysis because it predicts that each solute component moves along the column with a certain linear velocity $u = x/t$, which depends on the value of the partition ratio k:

$$u = \frac{x}{t} = \frac{v}{1+k} \tag{9}$$

The time t_R required by a component to travel the entire column length $x = \ell$ (*retention time*) is therefore $t_R = \ell/u$, or using eqn (9):

$$t_R = \frac{\ell}{v} (1+k) = t_M(1+k) \tag{10}$$

where $t_M = \ell/v$ is the so-called *hold-up time* or *dead time*. The difference $t_R - t_M = k t_M$ is the *adjusted retention time* t_R', and if eqn (10) is multiplied by the volume flowrate $\dot{V} = a_G v$ of the mobile phase, e.g., the carrier gas, through the column, one obtains the *corrected retention volume* V_R^O for each solute component:

$$V_R^O = \dot{V}t_R = \dot{V}t_M(1+k) = V_M^O (1+k) \tag{11}$$

where V_M^O is the corrected *hold-up volume* or *dead volume*. The *net retention volume* V_N is the difference $V_R^O - V_M^O$, which from eqn (11) is seen to be equal to $V_M^O k$. Taking into account eqn (4), we obtain

$$V_N = V_M^O k = \dot{V}t_M K \frac{a_s}{a_G} = a_G v \frac{\ell}{v} K \frac{a_s}{a_G} = \ell a_s K$$

But ℓa_s is the total volume V_s of the stationary phase (of the whole phase if this is a solid adsorbent, or only of the liquid if this is the stationary phase held on a solid support). Thus, we arrive at the important relation

$$V_N = V_R^O - V_M^O = V_s K \tag{12}$$

If the solute concentration in the stationary phase, c_s, is expressed per unit mass of that phase, the partition coefficient in the isotherm (2) would have the units cm^3g^{-1}, and eqn (12) would become

$$V_N = W\beta \tag{13}$$

where W is the total weight in g of the stationary phase, and β the partition coefficient in cm^3g^{-1}.

The direct experimental data in gas chromatography are the volume flowrate $\dot{V}_f$, measured at the temperature of the flowmeter T_f, and the chromatogram. The flowrate requires: (1) Temperature and vapour pressure corrections, to bring the flowrate at the column temperature T_c, and take into account the vapour pressure of the soap solution p_w (equal to that of pure water at the flowmeter temperature):

$$\dot{V}_c = \dot{V}_f \frac{T_c}{T_f} (1 - \frac{p_w}{p_o}) \tag{14}$$

where p_o is the pressure at the flowmeter; (2) compressibility correction $\dot{V} = j\dot{V}_c$, where the correction factor j for the pressure drop along the column is given by the relation

$$j = \frac{3}{2} \cdot \frac{(p_i/p_o)^2 - 1}{(p_i/p_o)^3 - 1} \tag{15}$$

p_i and p_o being the pressures at the inlet end $(x = 0)$ and the outlet end $(x = \ell)$ of the column, respectively. Then, $\dot{V}$ is used in eqn (11) to find the corrected retention volume V_R^O, by multiplication with the retention time t_R.

A typical chromatogram is shown in Fig.2. The retention time t_R of an injected solute and the dead time t_M are usually measured

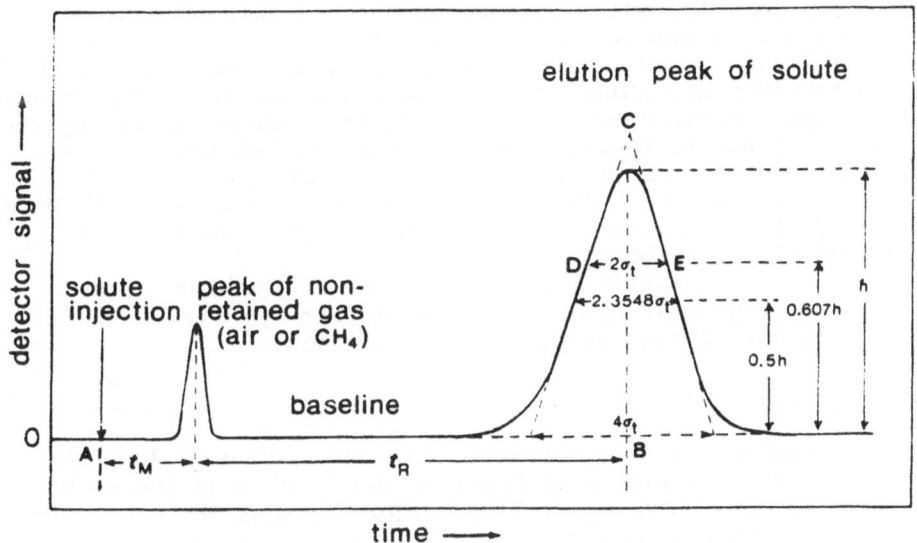

Figure 2. Typical appearance of an elution chromatogram, after the injection of a single solute.

from the respective distances z_R and z_M on the recorder chart and division by the chart speed v_c:

$$t_R = \frac{z_R}{v_c} \quad , \quad t_M = \frac{z_M}{v_c} \tag{16}$$

These are then used in eqn (11) to find v_R^o and v_M^o, and from these the v_N value [eqn (12)], from which the partition coefficient K or β is easily calculated using eqns (12) or (13), respectively.

1.2. Broadening Factors in Gas Chromatography

The elution curve of a solute component (cf. Fig.2) is a plot of the concentration c_G of this component at the end of the chromatographic column $(x = \ell)$ versus time t. Theoretically the function $c_G = c_G(t)$ is given by eqn (7) with $x = \ell$, which predicts a very sharp and narrow peak at $t = (1+k)\ell/v = t_R$, according to eqn (10). The peak is supposed to be so sharp that for t a little different from t_R the concentration c_G is zero. In spite of that, the elution curve is broadened compared to that predicted by the theory, and this is due to various so-called *broadening factors*. The most important of these are related to non-fulfilment of the assumptions under which eqn (7) was derived, namely:

(1) Non-linearity of the isotherm.

(2) Non-negligible axial diffusion of the gas in the chromatographic column.

(3) Unsharp input distribution of the vapour.

(4) Non-instantaneous equilibration of the solute components between the mobile and the stationary phases.

If the elution curve $c_G = c_G(t)$ is taken as a continuous statistical distribution in t, the broadening of the curve is proportional to the standard deviation σ_t of the distribution, as measured by the width of the peak at various heights from the baseline (cf. Fig.2). If the shape of the peak is Gaussian, the width at the inflection points D, E, located at 0.607 h, equals $2\sigma_t$ whereas at the peak base the width is $4\sigma_t$. At half of the maximum height the width is equal to $2\sqrt{2\ln 2}\sigma_t = 2.3548\ \sigma_t$.

Each of the broadening factors produces a variance σ_i^2 on an instantaneously introduced zone of sample. The total variance is equal to the sum of the variances due to all factors:

$$\sigma_{total}^2 = \sum_i \sigma_i^2 \tag{17}$$

The retention time t_R represents the mean value of the distribution of t, and therefore it is given by the abscissa of the center of gravity of the elution peak. This is approximately the same with the abscissa of point C in Fig.2.

Let us now examine in more detail some of the broadening factors.

1.2.1. Longitudinal Diffusion in the Gas Phase. The term longitudinal diffusion is used to denote axial diffusion of the gas in the chromatographic column. This includes true longitudinal molecular diffusion and apparent or eddy diffusion. The first occurs because of concentration gradients within the carrier gas along the column, but eddy diffusion results from uneven velocity profiles because of zigzag paths of unequal lengths and widths. Here we assume only true longitudinal diffusion obeying the one dimensional diffusion equation (Fick's second law):

$$\frac{\partial c_G}{\partial t} = \mathcal{D}_G\ \frac{\partial^2 c_G}{\partial y^2} \tag{18}$$

where c_G is the gas phase solute concentration, $\mathcal{D}_G$ its diffusion coefficient in that phase (carrier gas), and y the length coordinate along the column axis similar to x of Fig.1, but with its zero at the peak maximum in the centre of the vapour zone. We assume that initially the zone is a very narrow one as that described by a delta function, and diffusion of this narrow band occurs whether the zone is stationary or being eluted through the column. Ignoring the movement of the band along the column, i.e., the chromatographic process itself, and assuming that the elution of the vapour does not affect its diffusion, we come to the solution of eqn (18) as

$$c_G = \frac{m/a_G}{2(\pi D_G t)^{\frac{1}{2}}} \exp(-y^2/4D_G t) \qquad (19)$$

The mathematical details of the solution can be found elsewhere [1]. Comparison with the probability density function of the normal or Gauss distribution:

$$\varphi(y) = \frac{1}{\sigma(2\pi)^{\frac{1}{2}}} \exp[-\frac{(y-\bar{y})^2}{2\sigma^2}] \qquad (20)$$

shows that c_G is a Gaussian function of y with mean $\bar{y} = 0$ and variance

$$\sigma_x^2 = 2D_G t \qquad (21)$$

This relation, known as Einstein's law of diffusion, together with eqn (19), explains why the chromatographic column is a Gaussian operator and the solute vapour zone, introduced as a delta function, will come out at the other end of the column as a Gaussian function, had longitudinal diffusion been the only broadening factor.

The variance in eqn (21) and the respective standard deviation σ_x are expressed in length units along the column coordinate. To transform it in time units, it is necessary to multiply it by $(1+k)/v$ as eqn (8) shows:

$$\sigma_t = \sigma_x \frac{1+k}{v} = \sigma_x \frac{(1+k)a_G}{\dot{v}} \qquad (22)$$

The normal distribution curve, described by eqn (20), is *symmetrical* about the line $y = 0$ through the mean of the distribution. This also applies to the chromatographic peaks coming out of very long columns. Further, the ordinates decrease rapidly as $|y|$ increases. Lastly, by equating to zero the second derivative of c_G in eqn(19), one can easily verify that the points of inflection on the curve are given by $y = \pm \sigma_x$. This is why the width at the inflection points D, E in Fig.2 is equal to $2\sigma_t$.

1.2.2. Non-Instantaneous Equilibration of Solute between the two Phases. If the rate of exchange of solute between the mobile and the stationary phase is not infinite, equilibrium between phases will not be established instantaneously. This effect is often called resistance to mass transfer. The mass balance equation is again eqn (1), but the isotherm (2) cannot be substituted into it, as was done there, since equilibration between phases is slow. If equilibrium is reached according to a first-order law, the rate of change of the solute concentration in the stationary phase is

$$\frac{\partial c_s}{\partial t} = k_1(c_s^* - c_s) = k_1(Kc_G - c_s) \qquad (23)$$

where k_1 is a rate constant and K the partition coefficient of the solute between the two phases. Thus, Kc_G is the concentration c_S^* in the stationary phase which would be in equilibrium with the actual concentration in the gas phase.

Substituting the right-hand side of (23) for $\partial c_S / \partial t$ in eqn (1):

$$\frac{\partial c_G}{\partial t} + \pi k_1 (Kc_G - c_S) + v \frac{\partial c_G}{\partial x} = 0 \tag{24}$$

and solving the system of partial differential equations (23) and (24), under the same initial and boundary conditions as before, namely, eqns (5) and (6), one obtains c_G as a function of x and t. After various rearrangements and approximations (see reference 1, pp.23-26), the following equation is obtained:

$$c_G = \frac{m}{v} \left[\frac{k_1}{4\pi(t-t_M)} \right]^{\frac{1}{2}} \exp\left[- \frac{k_1(t-t_R)^2}{4(t-t_M)} \right] \tag{25}$$

Comparing eqn (25) with eqn (20), we conclude that it is a Gaussian function of t with mean $\bar{t} = t_R$ and variance

$$\sigma_t^2 = \frac{2(t-t_M)}{k_1} \tag{26}$$

To transform the variance in length units we make use of eqn(8):

$$\sigma_x^2 = \frac{2kvx}{k_1(1+k)^2} \tag{27}$$

In conclusion, eqn (25) shows that a finite rate and non-instantaneous equilibration of a solute component between the mobile and the stationary phases broadens the delta function input distribution into an approximately normal or Gaussian distribution, having the same mean with that in the absence of the broadening factor.

There remains to specify the nature of the rate constant k_1 for the transfer of solute between the stationary and the mobile phase. There are several mechanisms to explain the non-instantaneous equilibration of solute, the most common of which seems to be the slow diffusion of the solute vapour in the stationary phase. If the diffusion coefficient in this phase is $\mathcal{D}_S$ and the depth of the phase d (thickness of the liquid layer on solid support in GL chromatography or depth of pores of uniform bore in GS chromatography), the mass transfer coefficient in this phase is $\mathcal{D}_S/d$. The rate constant k_1 is proportional to this and to the effective area per unit volume of the phase, $A/V = A/Ad = 1/d$:

$$k_1 = \frac{\pi^2}{4} \cdot \frac{\mathcal{D}_S}{d} \cdot \frac{1}{d} = \frac{\pi^2 \mathcal{D}_S}{4d^2} \tag{28}$$

The constant of proportionality $\pi^2/4$ has been derived by van Deemter et al. [2]. Substituting this into eqn (27) we obtain

$$\sigma_\chi^2 = \frac{8}{\pi^2} \cdot \frac{d^2 v}{\mathcal{D}_S} \cdot \frac{kx}{(1+k)^2}$$ (29)

1.3. Theoretical Plates and van Deemter Equation

The analogy between fractional distillation and chromatography leads to the definition of theoretical plates N for the chromatographic column by the relation

$$N = \frac{t_R^2}{\sigma_t^2} = \frac{\ell^2}{\sigma_\chi^2}$$ (30)

The apparent HETP $\hat{H}$ is given by

$$\hat{H} = \frac{\ell}{N} = \frac{\sigma_\chi^2}{\ell}$$ (31)

Apparently the separation efficiency of the column depends on the total σ^2 as given by eqn (17) and this in turn on the various broadening factors. For packed columns with relatively large volumes of stationary phase, the major contributions to peak broadening are those previously described, i.e., longitudinal diffusion in the gas phase [eqn (21)] and non-instantaneous equilibration of solute between the two phases due to slow diffusion in the stationary phase [eqn (29)]. The total broadening is measured by the total variance, which is obtained by adding eqns (21) and (29), after some modifications:

$$\sigma_{total}^2 = \frac{2\mathcal{D}_G t_R \gamma}{1+k} + \frac{8}{\pi^2} \cdot \frac{d^2 v}{\mathcal{D}_S} \cdot \frac{k\ell}{(1+k)^2}$$ (32)

The first term was multiplied by the *obstructive factor* γ to account for the fact that diffusion is not exactly along the column axis, but along the tortuous paths between the particles of the column packing. It was also divided by $1+k$ because the fraction of the solute in the vapour phase is $1/(1+k)$. Finally, both terms are specified for the total column time, i.e., t_R, and the total column length ℓ. If now $t_R/(1+k)$ is replaced by ℓ/v according to eqn (10) and the right-hand side of eqn (32) is substituted for σ_χ^2 in eqn (31), the plate height is obtained as a function of the linear velocity of the carrier gas:

$$\hat{H} = A + \frac{2\gamma \mathcal{D}_G}{v} + \frac{8}{\pi^2} \cdot \frac{d^2}{\mathcal{D}_S} \cdot \frac{k}{(1+k)^2} v$$ (33)

The term A is added to account for flow-independent contributions to $\hat{H}$. Equation (33), called the van Deemter equation [2], is classically written

$$\hat{H} = A + \frac{\bar{B}}{\bar{v}} + C\bar{v} \tag{34}$$

where $\bar{B} = 2\gamma \mathcal{D}_G$ and C is given by the coefficient of v in the third term of the eqn (33), and $\bar{v}$ is the average linear velocity of the carrier gas. Giddings [2] has developed a generalized non-equilibrium theory and tabulated a great number of C terms due to various mechanisms.

1.4. Physicochemical Measurements by Gas Chromatography

In addition to chemical analysis, gas chromatography offers many possibilities for physicochemical measurements. Some of these methods lead to very precise and accurate results with relatively cheap instrumentation and a very simple experimental setup. They are widely used today, a fact emphasized by the edition of books [3,4] dealing only with such physicochemical measurements. Until a few years ago these measurements were exclusively based on the traditional techniques of elution development, frontal analysis, and displacement development, under constant gas flow rate. Studies on diffusion and rate processes were based on the broadening factors described in section 1.2, and embraced by van Deemter equation (34). It was through the coefficients $\bar{B}$ and C of this equation, determined from $\hat{H}$ measurements as a function of $\bar{v}$, that most diffusion coefficients in gases and liquids, as well as the coefficients of other rate processes, were calculated. Chapter 4 of reference 2 and chapter 12 of reference 3 describe all the above in detail. An extensive table of diffusion coefficients measured by the chromatographic broadening techniques, together with the underlying theory, was published by Maynard and Grushka [5].

Another approach to extract physicochemical parameters from the elution peaks is based on the analysis of the statistical moments of the peaks [6].

Other physicochemical measurements are adsorption studies relating to determination of adsorption isotherms and thermodynamic parameters for adsorption. Details of this application can be found elsewhere [7]. Here it is worthmentioning that the simple observation of a chromatographic elution peak gives qualitative information about the shape of the adsorption isotherm. The peaks are symmetrical only when the isotherm is linear, whereas a convex or a concave isotherm produces peak asymmetry either in the back or in the front of the peak profile, respectively. Some advances on determination of gas-solid adsorption isotherms by the so-called step and pulse method have been made by Guiochon and his co-workers [8,9]. Also, Jaulmes et al. [10-12] made a thorough study of peak profiles in non-linear gas chromatography. They proposed a theoretical model for the profile of elution peaks accounting both for the influence of the isotherm curvature at zero concentration, and for the perturbation of the flow rate due to solute exchange between the mobile and the stationary phase. The physical reliability of the theoretical model was demostrated by the fact that the parameters of the adsorption isotherm determined from their experimental data are in good agreement with the results obtained by independent techniques of isotherm determination.

1.5. The Carrier Gas Flow-Rate Perturbation

Although there would be no gas chromatography without a mobile gas phase, i.e. a carrier gas, its linear velocity v or volume flow-rate $\dot{V}$ remains constant throughout a single experiment in most gas chromatographic studies, or analytical applications. Thus, this magnitude is usually treated as an adjustable parameter of gas chromatographic equations. Following, however, the widespread use of temperature programming in gas chromatographic analysis, the programming of the carrier gas inlet pressure, and hence its flow-rate, had also been reported and reviewed [13,14]. In spite of the development of various programming modes (e.g.,step programming, continuous linear and non-linear programming), and the existence of commercial units permitting the general use of the technique, flow (pressure) programming has not been used to extract information of a physicochemical nature in gas chromatography. Its uses have been limited to analytical applications.
    Except flow programming, there are two other kinds of flow-rate perturbations imposed on the carrier gas. These are *the stopped-flow* and *the reversed-flow* technique. Both are very simple to apply and consist in either *stopping* the carrier gas flow for short time intervals, or *reversing* the direction of its flow from time to time. Experimentally, this is most easily done by using shut-off valves in the first technique and a four- or six-port gas sampling valve in the second. Thus, sophisticated mechanical, pneumatic or other special systems are not required as in flow programming gas chromatography.
    To the best of our knowledge, the first who used the stopping of the carrier gas flow for varying time periods were Knox and McLaren [15], with the purpose of producing extra broadening of the chromatographic peaks for measuring gas diffusion coefficients. However, the stopped-flow method was substantially introduced in 1967 by Phillips and his co-workers (see reference 4, pp.551-555) to study chemical reactions on the chromatographic column. The reversed-flow technique was introduced in its preliminary form in 1980 [16]. Both these techniques have solely been used for physicochemical measurements and constitute the object of reference 1. The rest of this paper will be devoted to the latest theoretical developments and applications concerning measurement of rate coefficients and distribution constants for diffusion and other rate processes by Reversed-Flow Gas Chromatography (RF-GC).

2. THE REVERSED-FLOW TECHNIQUE

2.1. General Principles

Instead of basing physicochemical measurements on retention volumes

of elution peaks, their broadening and their shape distortion, due to
the physicochemical processes under study, one can perform such measurements accurately and easily if the chromatographic column, being
at a steady-state condition, is perturbed so that it deviates from equilibrium for a short time interval and then is left to return to the original state. This is analogous to relaxation techniques. The perturbation chosen was the change in the direction of flow of the carrier gas,
and it was done by using a four- or six-port valve connected as shown
in Fig.3. The carrier gas is turned to flow in the opposite direction
either for a long time, or only for a short time interval t' (10-60 s),
smaller than the gas hold-up time of a solute in the sections ℓ' and
ℓ of the sampling column. Then, it is restored to its original direction
of flow. The question arising is what would we observe on the chromatographic elution curve? If pure carrier gas were passing through the
sampling column, nothing would happen on reversing the flow. But if a
solute comes out of the diffusion column L_1 as the result of its diffusion into the carrier gas, the flow reversal records its concentration
in the junction $x = \ell'$. This concentration recording has the form of
extra chromatographic peaks (*sample peaks*), superimposed on the
otherwise continuous detector signal. An example is given in Fig.4.
The peaks can be made as narrow as we want, since the width at their
half-height is equal to the duration t' of the backward flow of the carrier gas through the empty sampling column.

The sample peaks are predicted theoretically by the so-called
chromatographic sampling equation:

$$c = c_1(\ell', t_0 + t' + \tau) . u(\tau)$$
$$+ c_2(\ell', t_0 + t' - \tau) . [1 - u(\tau - t')] . [u(\tau) - u(\tau - t'_R)]$$
$$+ c_3(\ell', t_0 - t' + \tau) . u(t_0 + \tau - t')$$
$$. \{u(t - t')[1 - u(\tau - t'_R)] - u(\tau - t')[u(\tau) - u(\tau - t'_R)]\} \qquad (35)$$

where c = concentration of solute at the detector

t_0 = time from injection of a solute at $x = \ell'$

t' = time interval of backward flow

t = time from last restoration to original direction of flow

$\tau = t - t_R$

t_R, t'_R = retention times on column sections ℓ and ℓ', respectively.

Equation (35) describes the concentration-time curve of the sample
peaks created by the flow reversals, and has been derived [1,18]
using mass balances, rates of change etc., and integrating the resulting partial differential equations under given initial and boundary
conditions. It gives the concentration of the solute at the detector as
the sum of three terms, denoted by c_1, c_2 and c_3. They all refer
to the junction of the sampling and the diffusion column $x = \ell'$ of
Fig.3, but to different values of the time variable. Each of the concentration terms is multiplied by a combination of unit step functions,
so that it appears in a certain time interval and vanishes in all others.
Equation (35) predicts the sample peaks theoretically and its predictions

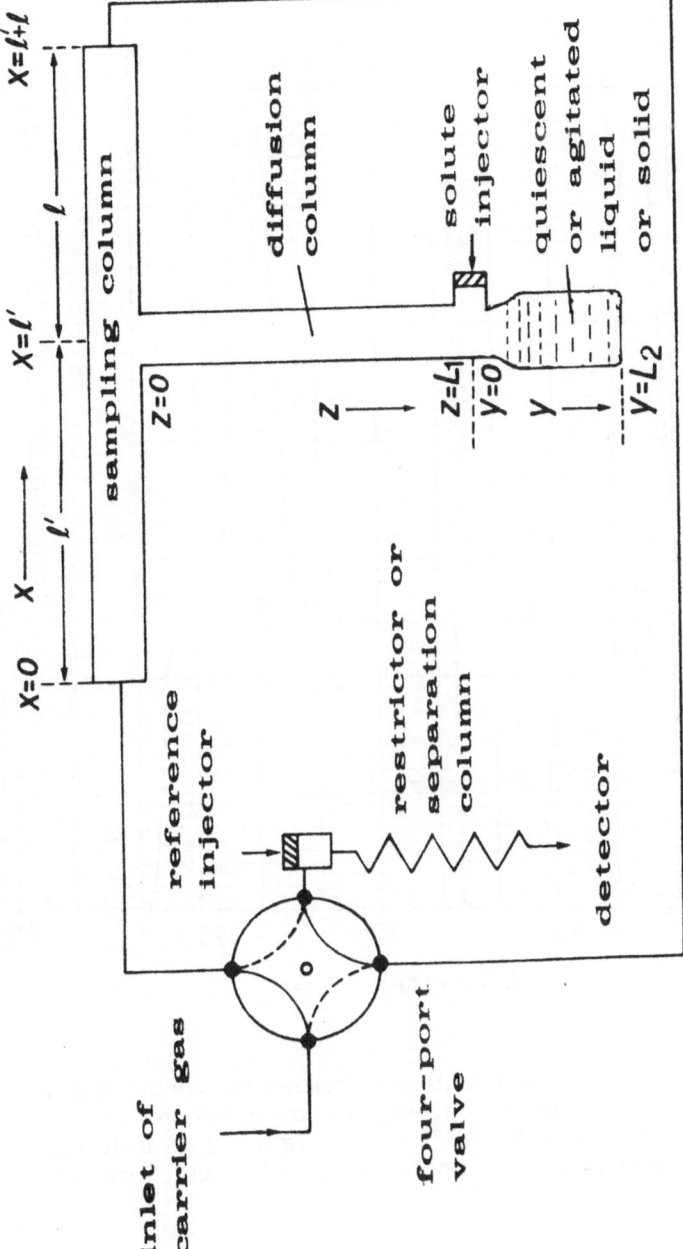

Figure 3. Schematic representation of columns and gas connections for studying diffusion and other rate processes by the Reversed-Flow technique.

352

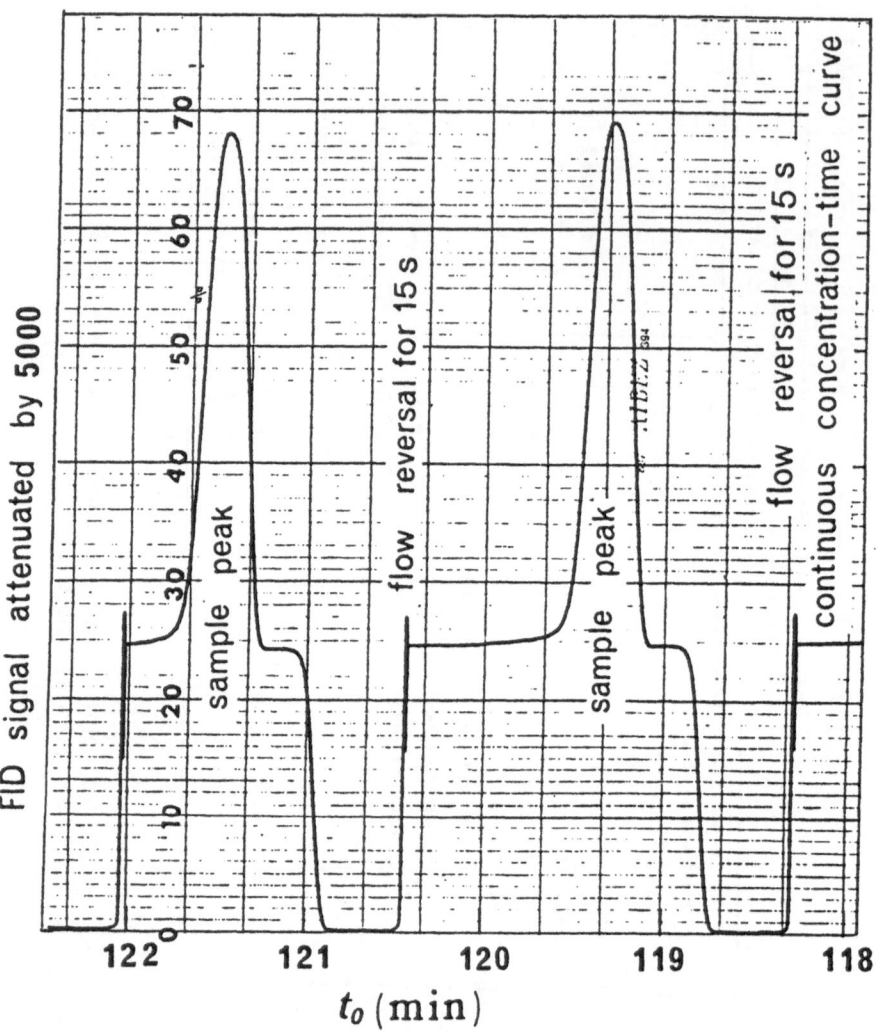

Figure 4. A reversed-flow chromatogram with two sample peaks due to the diffusion of propene into nitrogen (corrected volume flow-rate $\dot{V} = 0.36$ cm^3 s^{-1}), at 324.7 K and 1 atm (Figure 2 of Reference 17, reproduced with permission of the American Chemical Society).

coincide with the experimental sample peaks shown in Fig.4, the only difference being that the peaks predicted are square, whereas those actually found are not square owing obviously to non-ideality.

The area under the curve or the height h from the continuous signal of the sample peaks, measured as a function of time, is proportional to the concentration of the substance under study at the junction $x = \ell'$ of the sampling cell. Therefore, it can be used for the determination of the rate or equilibrium coefficient of the phenomenon responsible for this concentration. The relation between h and $c(\ell', t_0)$ is given by the equation

$$h \simeq [2c(\ell', t_0)]^m \tag{36}$$

where m is the response factor for the detector, being unity for a flame ionization detector. There remains $c(\ell', t_0)$, and therefore h, to be found theoretically as an analytic function of time for every phenomenon under study. Then, measuring h experimentally as a function of time, one can construct what is termed a *diffusion band*. Examples are shown in Fig.5. It is the shape and the distortion of these bands, rather than of the old elution peaks, mentioned in the INTRODUCTION, that lead to the calculation of various physicochemical parameters. In most cases, this is done in a simple, though accurate way, using cheap conventional gas chromatography instrumentation, without the need of doing any kind of proper gas chromatography operations and measurements.

2.2. Gas Diffusion Coefficients and Related Parameters

The first example of the application of RF-GC method is provided by the accurate measurement of gaseous diffusion coefficients [18-20], without bothering about the main difficulties associated with the traditional gas chromatographic methods, e.g., the disadvantages inherent in operation at low flow-rates, the difficulty of correctly allowing for the instrumental spreading of the chromatographic band outside the column, to mention only a few difficulties. The same and other problems are met in the determination of the obstructive factor γ, since it is the product $\gamma \mathcal{D}_G$ which is usually determined from plate height measurements in packed columns, employing a range of carrier gas velocities around the optimum value. Then, γ is usually found by assuming a theoretical value for the diffusion coefficient $\mathcal{D}_G$. An additional disadvantage of this method is that the experimental data are fitted to the van Deemter equation (34), assumed correct.

The arrested elution method of Knox and McLaren [15] bypasses some of the experimental and theoretical difficulties of the standard continuous elution method, but it still lies heavily on the time of passage along the column and the accurate measurement of the outlet elution velocity.

The RF-GC technique does not have any of the disadvantages connected with the carrier gas flow and the instrumental spreading of the chromatographic bands, because the phenomena being studied are taking place inside the diffusion column L_1 and the vessel L_2

354

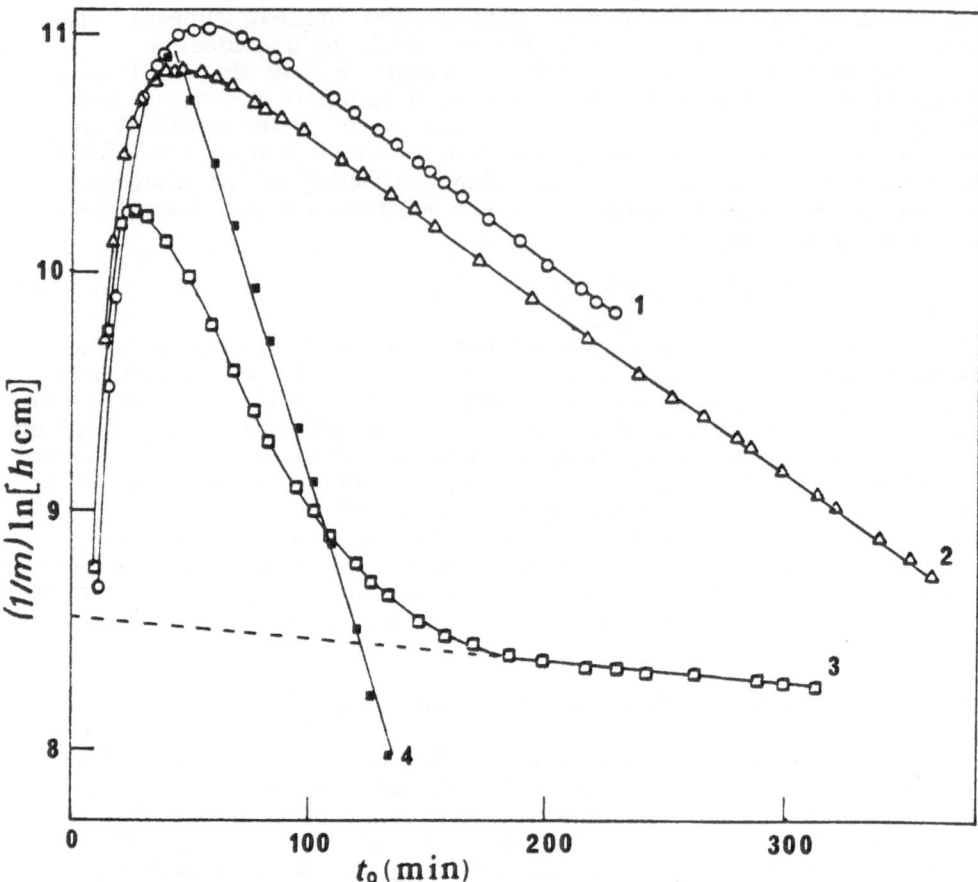

Figure 5. Diffusion bands obtained with 1 cm³ of butane injected into the diffusion column L_1 at 327 K. Curve 1 was obtained with an empty L_2 vessel, curve 2 with L_2 containing 10.4 cm³ water, and curve 3 (plotted as $1 + \ln h$) with L_2 containing 10.4 cm³ hexadecane. Line 4 (plotted as $2 + \ln h$) was obtained by subtracting from the experimental points of curve 3 after the maximum the points corresponding to the extrapolated (dashed line) last linear part of it (Figure 3 of Reference 17, reproduced with permission of the American Chemical Society).

and no carrier gas flows through these vessels. The gas flows only through the column $\ell' + \ell$, and is merely used as a means for repeated sampling of the concentrations at the point $x = \ell'$, i.e., at the exit of the column L_1. This is done with the help of the narrow and symmetrical sample peaks mentioned before (cf. Fig.4), without measuring their elution velocity, or the carrier gas flow-rate, provided it is steady. The experimental data recorded are the height h of the sample peaks in arbitrary units (say in cm), and the time t_o elapsing between the solute injection and the respective flow reversal, the duration of the latter being always the same (say 30 s). If one plots $(\ln h)/m$ against t_o, a diffusion band is obtained, like curves 1 and 2 of Fig.5. An obvious difference between the old elution gas chromatography and the RF-GC is that in the former longitudinal gaseous diffusion currents are parallel with the chromatographic current and the diffusion coefficients $\mathcal{D}_G$ or $\gamma \mathcal{D}_G$ are extracted from this mixed current by mathematical analysis. In the second method, the diffusion current is from the outset physically separated from the chromatographic current, and this is done by placing the diffusion process perpendicular to the chromatographic process. A diffusion band, rather than an elution band, is now mathematically analyzed to yield diffusion coefficients, or other physicochemical parameters from its distortion, in the same way that a distorted elution chromatographic band permits similar calculations. It must be pointed out that instrumental or other spreading of the sample peaks does not influence the results, as it is the same in all peaks of the same run. If the duration of the flow reversals is changed, the above spreading changes, but the physicochemical quantity extracted from the diffusion band comes out the same, provided that the same duration is maintained in all flow reversals in the same experiment.

2.2.1. The Function for the Diffusion Band. The mathematical function describing the diffusion band can be derived for two general situations:(1) When the gaseous solute injected at the lower end of the diffusion column L_1 (cf. Fig.3) does not interact in anyway with the solid or liquid materials filling column L_1 and/or vessel L_2; (2) when the injected solute interacts physically or chemically with the filling materials. For gas diffusion coefficients and related parameters measurements, situation (1) applies and the diffusion band has been derived [21] for three subcases: (a) When both, column L_1 and vessel L_2, are empty of any solid or liquid material; (b) when both, column L_1 and vessel L_2, are packed with a solid material that does not interact with the injected solute; (c) when L_1 is empty and only L_2 is packed with the above solid.

Case (c) gives a general solution comprising (a) and (b) as special cases. This solution is

$$c(\ell', t_o) = N_1 [\frac{(A r_1 - 1)(r_1 - 1)}{(r_1 - r_2)(r_1 - r_3)} \exp(-r_1 \beta t_o)$$

$$+ \frac{(A r_2 - 1)(r_2 - 1)}{(r_2 - r_1)(r_2 - r_3)} \exp(-r_2 \beta t_o) + \frac{(A r_3 - 1)(r_3 - 1)}{(r_3 - r_1)(r_3 - r_2)} \exp(-r_3 \beta t_o)]$$

(37)

where

$$N_1 = \frac{m\beta}{\dot{V}A(1.29 + 1.87R)} \tag{38}$$

m is the amount of solute injected at $z = L_1$, A is given by the relation

$$A = 4L_2^2/L_1^2\gamma \tag{39}$$

R is the ratio of the gaseous volumes of vessel L_2 (V_G') and column $L_1(V_G)$:

$$R = V_G'/V_G \tag{40}$$

β is the diffusion parameter:

$$\beta = \pi^2 \mathcal{D}_G/L_1^2 \tag{41}$$

and $-\lambda_1$, $-\lambda_2$ and $-\lambda_3$ are the roots of the equation

$$A(1.29 + 1.87R)\lambda^3 + (1.29 + 4.29A + \pi^2R + 1.87AR)\lambda^2$$

$$+ (4.29 + A + \pi^2R)\lambda + 1 = 0 \tag{42}$$

when solved for λ.

If the right-hand side of eqn (37) is substituted for $c(\ell', t_0)$ in eqn (36), the height h of the sample peaks as a function of time is obtained, i.e., the function describing the diffusion bands (cf. Fig.5), when the column L_1 is empty and the vessel L_2 is packed with a solid not interacting with the injected solute.

Equation (37) also describes the diffusion band when column L_1 and vessel L_2 are both empty of any solid material or both packed with non-reacting material. In these two cases, however, $A = 4L_2^2/L_1^2$ and the parameter R will have the same value irrespective whether L_1 and L_2 are both empty or packed with solid. Therefore, the value of A and R are in these cases characteristic of the cell dimensions, and with the help of them the roots $-\lambda_1$, $-\lambda_2$ and $-\lambda_3$ of eqn (42) can be found with any desired precision. These roots differ considerably from one another, making the exponential coefficients $-\lambda_1\beta$, $-\lambda_2\beta$ and $-\lambda_3\beta$ of the three functions in eqn (37) very different, and therefore easily determinable from the experimental diffusion band. For example, the absolutely smaller root, say $-\lambda_3$, describes the diffusion band at long enough times, i.e., after its maximum (cf. Fig.5), when the other two exponential functions have already decayed to negligibly low values. It corresponds to the last linear part of the band, as the latter is a semilogarithmic plot. The slope of this part gives $-\lambda_3\beta$ and, using eqn (41), the diffusion coefficient $\mathcal{D}_G$ of the solute in the carrier gas is easily calculated.

2.2.2. Two Limiting Cases of Eqn (37). The first arrises when $L_2 = 0$ and $V_G' = 0$, i.e., when vessel L_2 is absent. Then, $A = 0$ and $R = 0$, eqn (42) reduces to $1.29\lambda^2 + 4.29\lambda + 1 = 0$, with roots $-\hbar_1 = -3.073$ and $-\hbar_2 = -0.2522$, and eqn (37) becomes

$$c(\ell', t_0) = \frac{m\beta}{\dot{V}} [0.2055 \exp(-0.2522\beta t_0) + 0.5697 \exp(-3.073\beta t_0)] \quad (43)$$

Therefore, the slope of the last linear part is $-0.2522\beta = -0.2522\pi^2 D_G / L_1^2 = -\pi^2 D_G / 3.97 L_1^2$, which coincides with that predicted by eqn (4-39) of reference 1. The second limiting case of eqn (37) is obtained when $L_2 \ll L_1$ and thus A can be set equal to zero. In that case only the volume ratio $R = V_G' / V_G$ determines the roots of eqn (42) which reduces to

$$(1.29 + \pi^2 R)\lambda^2 + (4.29 + \pi^2 R)\lambda + 1 = 0 \quad (44)$$

If these roots are $-\hbar_1$ and $-\hbar_2$, instead of eqn (37), the expression is obtained

$$c(\ell', t_0) = N_1' [\frac{\hbar_1 - 1}{\hbar_1 - \hbar_2} \exp(-\hbar_1 \beta t_0) + \frac{\hbar_2 - 1}{\hbar_2 - \hbar_1} \exp(-\hbar_2 \beta t_0)] \quad (45)$$

where

$$N_1' = \frac{m\beta}{\dot{V}(1.29 + \pi^2 R)} \quad (46)$$

2.2.3. Measurement of Gas Diffusion Coefficients. Injecting a small gaseous volume (0.5 to 1 cm³ at atmospheric pressure) of a solute at the lower end of the diffusion column (cf. Fig.3), and then performing repeated flow reversals for 10-60 s, we obtain a series of sample peaks (cf. Fig.4). The height h of these peaks from the ending baseline are plotted as $(1/m)\ln h$ vs. the time of each reversal t_0 measured from the injection moment, when a diffusion band is obtained (cf. Fig.5). The slope of the last linear part of this plot gives the diffusion coefficient D_G of the injected solute into the carrier gas:

$$\text{Slope} = -\hbar_2 \beta = -\hbar_2 \pi^2 D_G / L_1^2 \quad (47)$$

where $-\hbar_2$ is the absolutely smallest root of eqn (42) if both L_1 and L_2 are empty of solids, or $-\hbar_2 = -0.2522$ if vessel L_2 is absent, or $-\hbar_2$ is the absolutely smallest root of eqn (44) if $L_2 \ll L_1$.

Results of diffusion coefficient measurements, using the method described above, in binary and multicomponent gas mixtures, are collected in reference 1 (pp. 132-140). The precision of the method, defined either as the relative standard deviation (%) or as the relative standard error (%) associated with each value, is better than 1% in most cases.

358

Comparison of the diffusion coefficients found with those calculated theoretically, either using the Hirschfelder-Bird-Spotz equation [22] or the Fuller-Schettler-Giddings equation [2, p.240] permits the calculation of the method's accuracy defined as

$$\text{Accuracy (\%)} = \frac{|D_{found} - D_{calcd}|}{D_{found}} \cdot 100 \tag{48}$$

With the exception of two pairs containing methane as solute and three values for the pair C_2H_4/He, the accuracy is better than 7.5% in 55 cases. Finally, a comparison with the values determined by broadening techniques [5] leads to the conclusion that, with the exception of C_2H_4/N_2, the values of diffusion coefficients determined by the reversed - flow method are closer to the theoretically calculated values than are the experimental values found by broadening techniques, under similar conditions of temperature and pressure.

By plotting $\ln D_G$ versus $\ln T$, one can calculate the exponent n in the relationship

$$D_G = AT^n \tag{49}$$

to which all theoretical and semiempirical equations lead for the dependence of D_G on T. For carrier gases helium and nitrogen, a mean value of 1.61 ± 0.01 and 1.74 ± 0.02, respectively, was found [18]. The mean values of n found from similar plots of calculated diffusion coefficients [22] are 1.679 ± 0.001 and 1.808 ± 0.004 for helium and nitrogen, respectively. The reversed-flow method for measuring gas diffusion coefficients can be extended to simultaneous determination of diffusion coefficients in multicomponent gas mixtures [20], an experimental problem which has practical, as well as theoretical importance. This extension of the method was done by filling part or all of the sampling column (cf. Fig. 3) with a chromatographic material, e.g. silica gel, which can effect the separation of some or all components of the gas mixture. When the chromatographic sampling is then performed, two or more sample peaks appear in the chromatogram. These correspond to two or more different components of the mixture. For each of these components eqns (36) and (37) hold true, and therefore the maximum height h of each peak, measured from the ending baseline, can be plotted in the form $(1/m) \ln h$ versus t_0 to yield from the slope its *effective* binary diffusion coefficient in the mixture. A comparison between the experimental and the calculated values shows differences of about the same magnitude as the accuracies for the diffusion of the same solutes in pure carrier gases.

2.2.4. Determination of the Obstructive Factor, γ. This factor is defined by Giddings [2] and also stressed by Knox and McLaren [15] as arising from two effects, namely, the tortuosity of the paths through the medium and the alternating constriction and widening of the paths. Two experimental plots at the same temperature, using the same cell, are required for this determination: (a) A diffusion band with both the diffusion column L_1 and the vessel L_2 empty of any solid and (b) a diffusion band with both L_1 and L_2 packed with the solid material under study, provided that a gaseous solute, that is not sorbed by the solid or does not interact with it in any way, is used in the diffusion experiment. If the slopes of the last linear parts after the maximum of the above bands are b (empty) and b (packed), their ratio gives directly the value of the obstructive factor, without any other measurement or correction:

$$\frac{b(\text{packed})}{b(\text{empty})} = \frac{-\varkappa_2\beta(\text{packed})}{-\varkappa_2\beta(\text{empty})} = \frac{\pi^2\gamma\mathcal{D}_G L_1^2}{\pi^2\mathcal{D}_G/L_1^2} = \gamma \tag{50}$$

This relation is based on the fact that the parameters A and R have the same value, given by eqns (39) and (40), respectively, in both experiments (a) and (b) described above. Therefore, all roots $-\varkappa_1$, $-\varkappa_2$ and $-\varkappa_3$ of eqn (37) are the same whether the cell is empty or packed. The root $-\varkappa_2$ is taken as having the smaller value, and thus describing the last linear part of the diffusion band. The value of the diffusion parameter β is given by eqn (41), with the diffusion coefficient being $\mathcal{D}_G$ when the cell is empty, and $\gamma\mathcal{D}_G$ when it is packed with solid.

It must be noted that the simpler functions described by eqns (43) and (45) lead to exactly the same eqn (50). This means that the experiments could be conducted only with column L_1, without the presence of vessel L_2, although in this case only the obstructive factor, but not the porosity of the solid bed, could be determined, as is shown below.

2.2.5. Determination of the External Porosity ε. One more diffusion band, in addition to those described under (a) and (b) above, is required for this determination: (c) a band obtained with the diffusion column L_1 empty and vessel L_2 packed with the solid under study. This is the case leading to the most general solution, i.e., eqn (37). The slope of the last linear part of the band, b (semi-packed), is again required. The steps to be taken for calculating the porosity are to find the experimental slopes b (empty) and b (semi-packed), the lenghts L_1 and L_2, the volumes V'_G (empty) and V_G, and the value of γ from the previous determination. Using the values of A (empty) = $4L_2^2/L_1^2$ and R (empty) = V'_G (empty)$/V_G$ in eqn (42), its roots are found, the absolutely smaller one $-\varkappa_3$ being retained. Then, using the relation

$$\frac{b(\text{semi-packed})}{b(\text{empty})} = \frac{-\varkappa_{3S}\beta}{-\varkappa_3\beta} = \frac{-\varkappa_{3S}}{-\varkappa_3} \tag{51}$$

$-\varkappa_{3S}$ is computed. This is used, together with the new value of A (semi-packed) = $4L_2^2/L_1^2\gamma$, to find R (semi-packed) by the relation

$$R(\text{semi-packed}) = \frac{1.29A\varkappa_{3S}^3 - (1.29 + 4.29A)\varkappa_{3S}^2 + (4.29+A)\varkappa_{3S} - 1}{-1.87A\varkappa_{3S}^3 + (\pi^2 + 1.87A)\varkappa_{3S}^2 - \pi^2\varkappa_{3S}} \tag{52}$$

Finally, the porosity ε is calculated from the ratio

360

$$\varepsilon = \frac{R(\text{semi-packed})}{R\,(\text{empty})} = \frac{V'_G(\text{packed})}{V'_G\,(\text{empty})} \tag{53}$$

More details and results obtained can be found in the original paper [21].

2.2.6. Other Related Parameters. In another work [23] it was shown that, in the same experiment, one can determine, together with the diffusion coefficient, relative molar responses of the solutes to the thermal conductivity detector, collision diameters and critical volumes. These are especially important to theoreticians, and are experimentally obtained only by elaborate apparatuses and with great difficulty, because no reliable estimation methods have been developed.

The major advantage of the RF-GC method is the estimation of the above four fundamental parameters, for the chromatographic solute under study, in one single experiment, with simplicity, short duration and fairly good accuracy. Following the determination of diffusion coefficients simply and accurately, a treatment of them was undertaken [24] to successfully calculate Lennard-Jones potential parameters and parachor values.

2.3. Resistance to Mass Transfer and Other Rate Processes

So far the RF-GC technique was applied to measure gas diffusion coefficients, obstructive factors, external porosities and some other parameters, all related to longitudinal diffusion in the gas phase. The old method to measure these parameters is based on the $\bar{B}$ term of van Deemter equation (34) which embraces such parameters, as it equals $2\gamma D_G$, according to eqn (33). This term constitutes the first broadeding factor of elution peaks in gas chromatography adding to HETP (cf. Section 1.2.1). What about the second main broadening factor, namely, non-instantaneous equilibration of solute between the mobile and the stationary phases (cf. Section 1.2.2)? This gives rise to the C term in van Deemter equation, upon which term one bases the determination of liquid diffusion coefficients, the interfacial resistance to mass transfer, gas-solid adsorption/desorption kinetics and other rate processes. Can these measurements be made by using the RF-GC method? They can, and this objective constitutes the situation (2) mentioned in Section 2.2.1, i.e., when the injected solute interacts physically or chemically with the liquid or solid materials filling column L_1 and/or vessel L_2 (cf. Fig. 3). Such interactions distort, not the sample peaks (cf. Fig. 4) which remain narrow and symmetrical, but the diffusion band, changing its shape and/or its slope. Compare, for instance, in Fig. 5 curves 1, 2 and 3 obtained with 1 cm³ of butane injected into: an empty cell, a cell containing in L_2 10.4 cm³ liquid water, or containing in L_2 10.4 cm³ hexadecane, respectively. In the second case, where butane dissolves very little in water, the distortion of the diffusion band (curve 2) is negligible, as regards shape and slope, compared to curve 1. In the third case, however, where the solubility of butane in hexadecane is big, the distortion of the diffusion band (curve 3) is very big. It is this distortion which is used to measure liquid diffusion coefficients, interfacial mass transfer coefficients, adsorption/desorption and reaction rate constants, and other rate coefficients, combined with the simultaneous determination of distribution constants, e.g., partition coefficients. Naturally, all these can be achieved by first deriving the relevant mathematical function for the diffusion bands, as was done previously in Section 2.2.1.

2.3.1. The Function for the Diffusion Band with a Quiescent Liquid. This has been derived recently [25] and can be followed by referring to Fig. 3. The diffusion equation (18), given in the INTRODUCTION, was first written for the gas phase in region z under the boundary conditions at $z = 0$, where the carrier gas is passing in the x direction, i.e., perpendicularly to the direction of gaseous diffusion. Then, the diffusion equation was written for the liquid phase in region y, and the solutions of the equations in regions z and y were linked using the boundary conditions at $z = L_1$ and $y = 0$. The final equation gives the concentration of the injected solute $c(\ell', t_0)$ at the junction $x = \ell'$ and time t_0. This, combined with eqn (36), gives the height h of the sample peaks as a function of time:

$$h^{1/m} = N_2[(1+\frac{Z}{y})\exp(-\frac{X+y}{2}t_0)+(1-\frac{Z}{y})\exp(-\frac{X-y}{2}t_0)] \qquad (54)$$

where

$$N_2 = \frac{3m\mathcal{D}_G}{\dot{V}L_1^2(1+3V_G'/V_G)} \qquad (55)$$

$$X = \frac{3\beta + 72K\alpha V_L/V_G}{\pi^2(1+3V_G'/V_G)} + 25\alpha \qquad (56)$$

$$\frac{X^2 - y^2}{4} = \frac{75\alpha(\beta + 16K\alpha V_L/V_G)}{\pi^2(1+3V_G'/V_G)} \qquad (57)$$

$$Z = X - 50\alpha \qquad (58)$$

m = amount of solute injected
V_G' = gaseous volume in L_2 above the liquid
V_G = gaseous volume in the diffusion column L_1
V_L = volume of the liquid
K = partition coefficient of the injected solute between the liquid and the carrier gas
α,β = diffusion parameters given by the relations $\alpha = \pi^2 \mathcal{D}_L/4L_1^2$ and $\beta = \pi^2 \mathcal{D}_G/L_1^2$, $\mathcal{D}_L$ and $\mathcal{D}_G$ being the diffusion coefficients of the injected gas into the liquid and into the carrier gas, respectively.

2.3.2. Calculation of $\mathcal{D}_L$ and K. Equation (54) describes the descending branch of the diffusion band as a sum of two exponential functions. The exponential coefficients $(X+Y)/2$ and $(X-Y)/2$, together with the respective preexponential factors, can be determined by means of a non-linear regression analysis computer programme or, if these coefficients are sufficiently different, by finding first the slope $-(X-Y)/2$ and the intercept $\ln[N_2(1-Z/Y)]$ of the last linear part of $(1/m)\ln h$ vs. t_o plot, and then replotting the initial data of the non-linear part after the maximum as $\ln\{h^{1/m} - N_2(1-Z/Y)\exp[-(X-Y)t_o/2]\}$ vs. t_o. Using the new straight line thus obtained, one finds $-(X+Y)/2$ from its slope and $\ln[N_2(1+Z/Y)]$ from the intercept. An example of this kind is given in Fig. 5 (curves 3 and 4).

There are two ways to calculate $\mathcal{D}_L$ and K (or the Henry's Law constant) from the diffusion band: either by using the two exponential coefficients $(X+Y)/2$ and $(X-Y)/2$, and the diffusion parameter β, or by employing the above exponential coefficients and the respective preexponential factors $N_2(1+Z/Y)$ and $N_2(1-Z/Y)$ of eqn (54), found from the intercepts, as mentioned above. In both ways the gaseous volumes V'_G and V_G, the volume of the liquid V_L and the height of the liquid L, are required. Results and a discussion of them can be found in the original paper [25].

After having studied diffusion of gases into liquids, one naturally thinks of extending the study to chemical interactions of the dissolved gas with the liquid itself or with a solution of other substances. An example of this kind of rate processes has been investigated only recently, and the relevant paper is in press [26].

2.3.3. The Function for the Diffusion Band with an Agitated Liquid. If the experiment is conducted under the same conditions as just previously described, but with an agitated liquid, the overall mass transfer coefficients of the injected solute in the gas phase, K_G, and in the liquid phase, K_L, come into play, adopting the two-film theory for the gas-liquid interface. The function for the diffusion band comes out as having exactly the same form with eqn (54), although the physical content and meaning of the parameters N_2, X, Y and Z, is now different. Thus, $N_2 = \pi m \mathcal{D}_G / \dot{V} L_1^2$, and

$$X = \alpha + \frac{2K_G A_L}{V_G} + \frac{K_L A_L}{V_L} \tag{59}$$

$$\frac{X^2 - Y^2}{4} = \frac{\alpha K_L A_L}{V_L} \tag{60}$$

$$Z = X - \frac{2K_L A_L}{V_L} \tag{61}$$

where $\alpha = \pi^2 \mathcal{D}_G / 4 L_1^2$, A_L is the free surface area of the liquid, and V_G and V_L have the same meaning as before.

2.3.4. Calculation of K_G, K_L and K. These three physicochemical parameters are easily found from the diffusion band, like curve 3 of Fig. 5, by first determining the exponential coefficients $(X + Y)/2$ and $(X - Y)/2$ of eqn (54), as described in Section 2.3.2. By adding these coefficients, we obtain the value of X, and by multiplying them, the value of $(X^2 - Y^2)/4$. From these and the diffusion parameter α, determined with a cell empty of liquid as described in Section 2.2.3, we calculate K_G and K_L, using eqns (59) and (60). Finally, the partition coefficient K is found by the ratio

$$K = K_G/K_L \tag{62}$$

The Henry's law constant $H^{\ddagger}$ is related to K by $H^{\ddagger} = RT/K$.

The details of the mathematical derivations and some results can be found in the original publication [17]. It should be pointed out that the determination of the mass transfer coefficients K_G and K_L in an agitated liquid amounts to measuring interfacial resistance to mass transfer. This contrasts with $\mathcal{D}_L$ determined with a quiescent liquid and pertaining to the normal resistance to mass transfer, which arises from diffusion in the bulk liquid phase. The conventional methods of measuring interfacial resistance are very difficult and relatively inaccurate.

The inclusion of a chemical reaction in an agitated solution has also been studied [26].

2.3.5. The Function for the Diffusion Band in the Presence of a Reactive Solid. This is the last case to be examined, and it includes the rate processes due to the presence of a solid material in vessel L_2 of Fig. 3. The small volume of gaseous solute injected into the cell may interact with the solid physically and/or chemically. Thus, adsorption/disorption kinetics, gas-solid chemical reactions and heterogeneous catalysis are important rate processes belonging to this experimental arrangement. Let it be noted that to study adsorption /desorption kinetics by conventional gas chromatography, i.e., using broadening techniques, the elution peak broadening due to the adsorption must be separated from that due to longitudinal diffusion and other dispersion processes. This requires either a negligible dispersion contribution or an increased amount of experimental information, like results with varying gas velocity and particle size. With the RF-GC technique, adsorption is most easily separated from longitudinal diffusion, since there is no carrier gas running inside the diffusion column L_1 to cause mixing of the two processes. In fact, the one process is used as a carrier of the other, giving rise to a diffusion band.

The mathematical function describing this band is given here for the most general case comprising the following four basic steps in series:

(1) Mass transfer of the gaseous solute to the gross exterior surface of the solid material.

(2) Diffusional and flow transfer of the solute in and out of the pore structure of the solid.

(3) Activated adsorption of the gas at the interface.

(4) Surface chemical reaction of the adsorbed solute.

Steps (1), (2) and (3) can be simplified by considering two overall mass transfer coefficients, one in the gas phase, K_G, and one in the solid phase, K_s, their ratio K_G / K_s giving the equilibrium constant K for the distribution of the solute between the bulk solid and gaseous phases, according to eqn (62).

The rate constants connected with the overall mass transfer coefficients are given by the relations

$$k_1 = K_G A_s / V_G' \tag{63}$$

$$k_{-1} = K_s A_s / V_s \tag{64}$$

where

A_s = total external surface area of the solid
V_G' = gaseous volume in the near vicinity of the solid
V_s = volume of the solid.

The parameter k_1 represents the rate constant for the deposition of the solute on the gross exterior surface of the solid, while k_{-1} gives the overall rate constant for the desorption of the gas from the whole solid.

To the above, one must add a first-order rate constant k_2 for the possible reaction of the adsorbed gas with the solid material acting as a catalyst or as a reactant in a gas-solid reaction.

All parameters, k_1, k_{-1}, k_2, K_G, K_s, K can be measured simultaneously in the same experiment, under nonsteady-state conditions, by using the technique of RF-GC. This can offer valuable information on the detailed mechanism of the action of air pollutants on buildings, thus providing the scientific basis for their conservation. Although partial equations pertaining to the above rate processes have been derived in the past [1,27], the most general equation is formulated as follows [28]:

$$h^{1/m} = N_2 (1 + \frac{Z}{Y}) \exp(-\frac{X+Y}{2} \beta t_o) + N_2 (1 - \frac{Z}{Y}) \exp(-\frac{X-Y}{2} \beta t_o) \tag{65}$$

where N_2 is given by eqn (55), $\beta = \pi^2 D_G / L_1^2$, and X, Y, Z are dimensionless functions of the rate constants k_1, k_{-1} and k_2, of β, and of R, the latter being the ratio of the gaseous volume in vessel $L_2 (V_G')$ to that in column $L_1 (V_G)$ (cf. Fig. 3). Instead of writing these functions analytically, we give later on the inverse relations, i.e., k_1, k_2 and k_{-1} as functions of X, Y and Z [cf. eqns (68)-(70)].

If there is no solid in region y of the sampling cell, the slope of the last linear part depends only on the diffusion parameter β, which can be calculated by means of eqn (47) or using the simpler relation

$$\text{Slope} = -\beta/\pi^2 (1/3 + R) \tag{66}$$

When an experiment is conducted under the same conditions as above, the only difference being that vessel L_2 now contains solid material, the diffusion band is distorted in shape and/or its slope, compared to that without solid present, and it is now described (after the maximum) by eqn (65). This has the same form with eqn (54), X, Y and Z having now a different physical content than that given by eqns (56), (57) and (58) or by eqns (59), (60) and (61), respectively.

2.3.6. Calculation of k_1, k_{-1}, k_2, K_G, K_s and K. From the experimental band, the two exponential coefficients $(X+Y)\beta/2$ and $(X-Y)\beta/2$, together with the two preexponential factors $N_2(1+Z/Y)$ and $N_2(1-Z/Y)$ are computed, as previously described in Section 2.3.2. Dividing now the two exponential coefficients by β, we find the values of $(X+Y)/2$ and $(X-Y)/2$. The addition of these gives the value of X, and their subtraction the value of Y. From the ratio ρ of the two preexponential factors, calculated from the intercepts, we find $\rho = (1 - Z/Y)/(1 + Z/Y)$, and from this

$$Z = \frac{1-\rho}{1+\rho} Y \tag{67}$$

Finally, from the values of the dimensionless parameters X, Y and Z, we compute the values of k_1, k_2 and k_{-1} with the help of the relations

$$k_1 = \frac{(X+Z)(1/3+R) - 2\pi^{-2}}{2R} \beta \tag{68}$$

$$k_2 = \frac{(X^2 - Y^2)(1/3+R)-2(X - Z)\pi^{-2}}{2(X + Z)(1/3 + R) - 4\pi^{-2}} \tag{69}$$

$$k_{-1} = [X - \pi^{-2}(1/3 + R)^{-1}]\beta - R(1/3+R)^{-1}k_1 - k_2 \tag{70}$$

The value of R used here is V'_G (packed)$/V_G$, in contrast to R of eqn (66) which is V'_G (empty)$/V_G$.

The values of K_G, K_s and K are calculated from k_1 and k_{-1}, by means of eqns (63), (64) and (62), respectively.

The method described here was applied for the action of SO_2 on marble particles of two different sizes (22-30 and 100-120 mesh), at various temperatures [28]. It was also applied for studying the effects of SO_2 and NO_2 on single pieces of marble having different geometrical forms, like sphere, cylinder, etc. [29].

A limiting case of eqns (68)-(70) arises when the injected solute does not react chemically with the solid. Then eqn (68) is unaffected, eqn (69) becomes $k_2 = 0$, and in eqn (70) k_2 is deleted from the right-hand side. Only the adsorption (k_1) and desorption (k_{-1}) rate constants are left, together with K_G, K_s and K. Such a study was made with propene and butane acting on γ-alumina and on 80-100 mesh chromosorb P covered with hexadecane [30].

3. REFERENCES

1. Katsanos, N.A. (1988) 'Flow Perturbation Gas Chromatography', Dekker, New York-Basel.
2. Giddings, J.C. (1965) 'Dynamics of Chromatography', Dekker,

New York, pp.190-193.

3. Conder, J.R. and Young, C.L. (1979) 'Physicochemical Measurement by Gas Chromatography', Wiley, Chichester.

4. Laub, R.J. and Pecsok, R.L. (1978) 'Physicochemical Applications of Gas Chromatography', Wiley, New York.

5. Maynard, V.R. and Grushka, Eli (1975) 'Measurement of Diffusion Coefficients by Gas Chromatography Broadening Techniques: A Review', Adv. Chromatogr. 12, 99-140.

6. Suzuki, Motoyuki and Smith, J.M. (1975) 'Transport and Kinetic Parameters by Gas Chromatographic Techniques', Adv. Chromatogr. 13, 213-263.

7. Kiselev, A.V. and Yashin, Ya.I. (1969) 'Gas Adsorption Chromatography', Plenum Press, New York.

8. Dondi, Francesco, Gonnord, Marie-France and Guiochon, Georges (1977) 'Chromatographic Determination of Gas-Solid Adsorption Isotherms by the Step and Pulse Method I. Apparatus and Data Processing', J.Colloid Interphace Sci. 62, 303-315.

9. Dondi, Francesco, Gonnord, Marie-France and Guiochon, Georges (1977) 'Chromatographic Determination of Gas-Solid Adsorption Isotherms by the Step and Pulse Method II. Choise of a Model for the Adsorption Isotherm of Benzene and Cyclohexane on Graphitized Carbon Black', J.Colloid Interface Sci. 62, 316-328.

10. Jaulmes, A., Vidal-Madjar, C., Ladurelli, A. and Guiochon G. (1984) 'Study of Peak Profiles in Nonlinear Gas Chromatography 1. Derivation of a Theoretical Model', J.Phys.Chem. 88, 5379-5385.

11. Jaulmes, A., Vidal-Madjar, C., Gaspar, M. and Guiochon, G. (1984) 'Study of Peak Profiles in Nonlinear Gas Chromatography 2. Determination of the Curvature of Isotherms at Zero Surface Coverage on Graphitized Carbon Black', J.Phys.Chem. 88, 5385-5391.

12. Jaulmes, A., Vidal-Madjar, C., Colin, H. and Guiochon, G. (1986) 'Study of Peak Profiles in Nonlinear Liquid Chromatography', J.Phys.Chem. 90, 207-215.

13. Scott, R.P.W. (1968) 'Flow Programming' in J.H.Purnell (ed.), Progress in Gas Chromatography, Interscience, 6, 271-287.

14. Ettre, Leslie S., Májor, László and Takács, József (1969) 'Pressure (Flow) Programming in Gas Chromatography', Adv.Chromatogr. 8, 271-325.

15. Knox, J.H. and McLaren, L. (1964) 'A New Gas Chromatographic Method for Measuring Gaseous Diffusion Coefficients and Obstructive Factors', Anal.Chem. 36, 1477-1482.

16. Katsanos, N.A. and Georgiadou, I. (1980) 'Reversed-flow Gas Chromatography for Studying Heterogeneous Catalysis', J.Chem. Soc., Chem.Commun. 242-243.

17. Katsanos, N.A. and Dalas, E. (1987) 'Mass Transfer Phenomena Studied by Reversed-flow Gas Chromatography 1. Mass Transfer and Partition Coefficients Across Gas-Liquid Boundaries', J.Phys. Chem. 91, 3103-3108.

18. Katsanos, N.A. and Karaiskakis, G. (1983) 'Temperature Variation of Gas Diffusion Coefficients Measured by the Reversed-flow

Sampling Technique', J.Chromatogr. 254, 15-25.

19. Katsanos, N.A. and Karaiskakis, G. (1982) 'Measurement of Diffusion Coefficients by Reversed-flow Gas Chromatography Instrumentation', J.Chromatogr. 237, 1-14.

20. Karaiskakis, G., Katsanos, N.A. and Niotis, A. (1983) 'Measurement of Diffusion Coefficients in Multicomponent Gas Mixtures by the Reversed-flow Technique', Chromatographia 17, 310-312.

21. Katsanos, N.A. and Vassilakos, Ch. (1989), 'A New Method for Measuring Obstructive Factors and Porosity by Gas Chromatography', J.Chromatogr. 471, 123-137.

22. Bird, R.B., Stewart, W.E. and Lightfoot, E.N. (1960)'Transport Phenomena', Wiley, Chichester, p.511.

23. Karaiskakis, G., Niotis, A. and Katsanos, N.A. (1984) 'Characterization of Gases by the Reversed-flow Gas Chromatography Technique', J.Chromatogr. Sci. 22, 554-558.

24. Karaiskakis, G. (1985) 'A Reversed-flow GC Technique: Lennard--Jones Parameters', J.Chromatogr. Sci. 23, 360-363.

25. Katsanos, N.A. and Kapolos, J. (1989) 'Diffusion Coefficients of Gases in Liquids and Partition Coefficients in Gas-Liquid Interphaces by Reversed-flow Gas Chromatography', Anal. Chem. 61, 2231-2237.

26. Stolyarov, B.V., Katsanos, N.A., Agathonos, P. and Kapolos, J. (1991) 'Homogeneous Catalysis Studied by Reversed-Flow Gas Chromatography', J.Chromatogr., in press.

27. Kapolos, J., Katsanos, N.A. and Niotis, A. (1989) 'Physicochemical Quantities in Catalytic Reactions Measured Simultaneously by Gas Chromatography', Chromatographia 27, 333-339.

28. Katsanos, N.A. and Vassilakos, Ch. (1991) 'Theoretical Analysis for Measurement of Buildings Pollution Parameters by Gas Chromatography', J.Chromatogr., in press.

29. Vassilakos, Ch., Katsanos, N.A. and Niotis, A. (1991) 'Physicochemical Damage Parameters for the Action of SO_2 and NO_2 on Single Pieces of Marble', Atmos. Environ., in press.

30. Katsanos, N.A., Agathonos, P. and Niotis, A. (1988) 'Mass Transfer Phenomena Studied by Reversed-Flow Gas Chromatography. 2. Mass Transfer and Partition Coefficients Across Gas-Solid Boundaries', J.Phys.Chem. 92, 1645-1650.

GAS CHROMATOGRAPHIC TECHNIQUES TO ELUCIDATE THE WORKING MECHANISM OF GRAPHITIZED CARBON BLACK-LIQUID MODIFIER-ELUATE INTERACTIONS

F. BRUNER
Istituto di Scienze Chimiche, University of Urbino
Piazza Rinascimento, 6
I- 61029 Urbino,
Italy

ABSTRACT. The modification in adsorption properties caused by very small amounts of organic compounds when preadsorbed on the surface of graphitized carbon blacks are discussed in terms of the heats of adsorption of different compounds on the modified surface. The results obtained by deposition of large molecular weight molecules, such as those used as stationary phases in gas chromatography, are compared with those obtained when the preadsorbed molecules belong to the same compound. The heats of adsorption are measured at different surface coverages. As a consequence of the surface modifications, the graphite surface becomes highly homogeneous. This, in turn, leads to better performance of the graphitized carbon blacks as stationary phases for gas chromatography. The main advantage is that with this technique selective stationary phases can be obtained making fast analysis possible in terms of linear gas velocity while maintaining very low HETP when compared to gas-liquid systems.

1. Introduction

The idea that profound modifications in gas adsorption mechanisms could be obtained by depositing controlled amounts of organic molecules on the surface of adsorbents was first forwarded by Eggertsen, Knight and Groennings in 1956 (1). These authors modified Pelletex, a furnace carbon black, with 1.5% squalane deposited on the adsorbent surface. In this manner important results were obtained in terms of reduced peak tailing in the GC elution of some hydrocarbons and enhanced resolution of GC columns toward isomeric pairs. In a more extensive paper, the same group stated that "...The separations obtainable with liquid-modified solids (up to a few percent liquid) depend upon the amount as well as the type of supported liquid" (2). This technique has been called Gas-Liquid-Solid Chromatography (GLSC) or better liquid-modified adsorption chromatography.

These pioneering studies indicated the practical and theoretical potentional of detailed knowledge as to the effect such techniques have on the three-way adsorbent-stationaryphase-eluate interactions in gas chromatographic processes. Unfortunately, for many years, as is often

369

F. Dondi and G. Guiochon (eds.),
Theoretical Advancement in Chromatography and Related Separation Techniques, 369–395.
© 1992 *Kluwer Academic Publishers.*

370

the case in scientific research, these results have been neglected or
forgotten. For this reason, it is worth reporting (Fig. 1) the data
Eggertsen and Knight obtained in modifying retention parameters on the
basis of the amount of polar and non-polar liquid phases added to the
adsorbent. Years later, our group was deeply involved in attempts to
separate isotopic molecules by gas adsorption chromatography and good
results were obtained using capillary columns made of soda glass with
inner wall egched with a hot solution of concentrated sodium hydroxide
(3,4,5) as per the procedure established by Mohnke and Saffert (6).

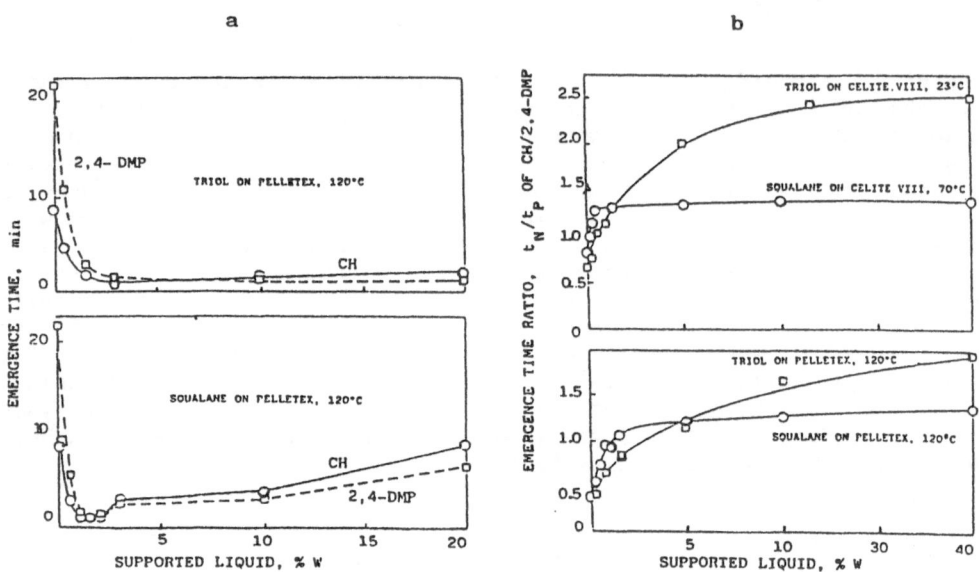

Fig. 1. Effect the amount of stationary phase has on retention for
Pelletex Carbon Black (from ref. 2). a)Retention time of cyclohexane
(CH) and 2,4-Dimethylpentane (2,4-DMP) vs. percentage of squalane.
b)Naphthalene/pentane retention time ratio for the same adsorbent with
the percentage of squalane given. (Reprinted with permission).

In a previous paper (3) the silica gel surface of the inner walls of
a glass capillary column were coated with different amounts of squalane.
This liquid phase coating had an effect on hydrocarbon retention similar
to what is reported in Figure1. A good effect on the separation of
isotopic pairs was also observed. In these early papers, however, a
systematic approach to the effect of different amounts of liquid phase
on the chromatographic behaviour of the adsorbent was not observed.
During a vist to our laboratory, A. V Kiselev suggested that for such
gas adsorption chromatography studies on isotope separations graphitized
carbon black might be a better adsorbent than silica gel, both because
it presents greater homogemeity of active site energy and because

greater separation of deuterated and hydrogenated species is possible (7). Considering this precious advice, as well as the results obtained by Eggertsen and Knight (2), we started a study of the modifications induced on graphitized carbon blacks by surface coating with different amounts and types of organic molecules. Gas chromatographic measurements were chosen to study such modifications, through the evaluation of retention parameters and changes in adsorption heat.

This technique proved the most precise and sensitive experimental method for such research and very detailed information was obtained on the effects of preadsorbed molecules on organic molecule adsorption on the surface of carbon blacks. At the same time, based on information obtained through thermodynamic studies, a number of practical applications have been found, not only for the separation of isotopic molecules (8,9), but also for using liquid modified adsorption chromatography in analysis of very polar compounds such as primary amines and carboxylic acids (19), alcohols (11), highly reactive gases (12) as well as trace concentrations of sulphur compounds (13,14). These applications fully demonstrated the flexibiity and usefulness of liquid modified adsorption chromatography. It was shown that adsorption chromatography can be used for both the separation of hydrocarbons and non polar compounds. Moreover it showed that, with a proper modifier, it can also be used for the analysis of any type of compound, even at extremely low concentrations. In fact, today, two international firms produce GC columns based on liquid-modified adsorption chromatography and this technique is widely used by analytical chemists. The present paper makes a systematic approach to understanding the working mechanism of liquid-modified adsorption chromatography and attempts a comparison with the results obtained by other groups.

2. The unique properties of graphitized carbon blacks.

Physical chemists became interested in graphitized carbon blacks in the early 1950's, most likely in an attempt to find properties of greater industrial application than previously possible with the then widely used ungraphitized blacks. Hundreds of papers were published in the 1950's and 1960's on the properties of graphitized carbon blacks as adsorbents. Mention of even the most important papers on this topic is all but impossible although three fundamental books on physical adsorption will help the reader wishing to learn more to locate the original contribiutions. Such information can be found in references 15, 16 and 17.

GCBs can be defined through their most important physical and geomet rical properties; these are:
1) homogeneity in crystalline structure;
2) non porous or at least macroporous, with a pore diameter exceeding 150 A x ;
3) non specific, which means that the active adsorptive sites act only through London dispersion forces or Van der Waals forces, avoiding directional active sites such as electron donors or acceptors,
   i.e. mainly metal ions or functional groups such as -OH, -COOH, =NH, -NH2, etc..

Should the previous conditions be met, the adsorption energy distribu tion and the GCB surface can be considered "Gaussian". It is worth noting that an ideal adsorbent should strictly meet the above-mentioned pre-requisites, although such an adsorbent has not been invented yet. However, the GCBs approach these requirements much better than do other adsorbents and this has made it possible to satisfactorily measure the thermodynamics of physical adsorption, especially the effect of preadsorbed molecules on adsorption energy.

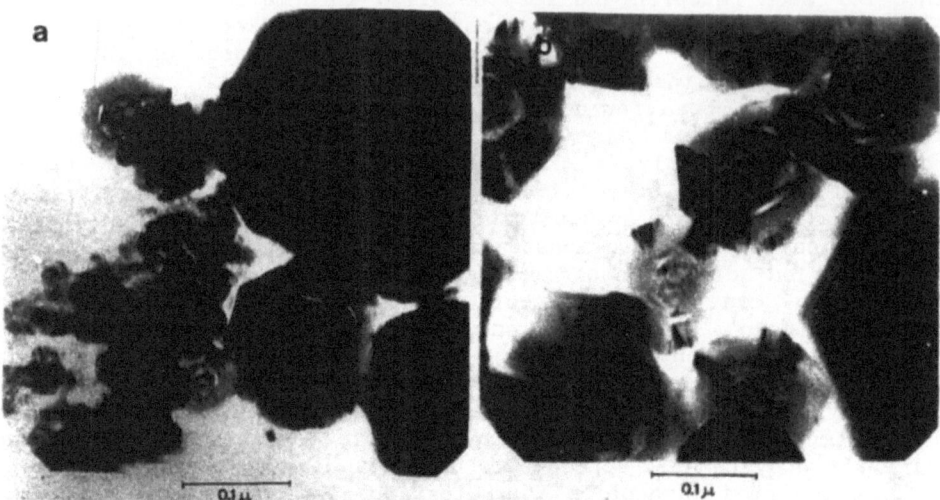

Fig. 2.Electron microscope micrographs showing the microcrystalline structure of graphitized carbon black. a) Sterling FT; b) Carbopack C (P-33).

The main differences observed between an "ideal", non specific adsorbent and the carbon blacks presently available are due to the following:

1) Imperfections in the graphitization process. In fact, to obtain the maximum possible graphitization, a temperature of 3000 C x should be reached by the entire mass of material and this may not be the case if large, non cylindrical ovens are used.
2) During graphitization, and more important during cooling to room temperature, the material should be in contact with an inert, highly pure gas such as helium. If impurities of oxygen or other elements, especially sulphur and metals, are present during these critical phases a portion of them will be retained in the graphitic structure. This yields specific active sites so that condition 3 above, one of the most important for successful use of the adsorbent, can no longer be met.
3) Despite its crystalline graphitic structure, the adsorbent surface of even very well-made graphitic crystals (Fig. 2) contains cracks, crevices and edges. Thus the surface is not homogeneous since it contains sites with dense electron clouds which enhance the

adsorption energy; it no longer shows a perfect gaussian distribution. The most exhaustive method for studying homogeneity of an adsorbent in regard to the above-mentioned characteristics is to measure the change in isosteric heat of adsorption, Qs, as suface coverage of well known molecules of the same substance (i.e. argon) is varied.

In Figure 3 the change in Q s vs. Argon surface coverage is reported for the same carbon black graphitized at different temperatures (18). A detailed discussion of Qs behaviour in terms of surface coverage will be made later. For the moment, it is worth noting that a very good adsorbent homogeneity (achieved by high temperature graphitization) can be observed through the vertical jump in Qs at the monolayer and by the peak in this parametre just before the monolayer is obtained.

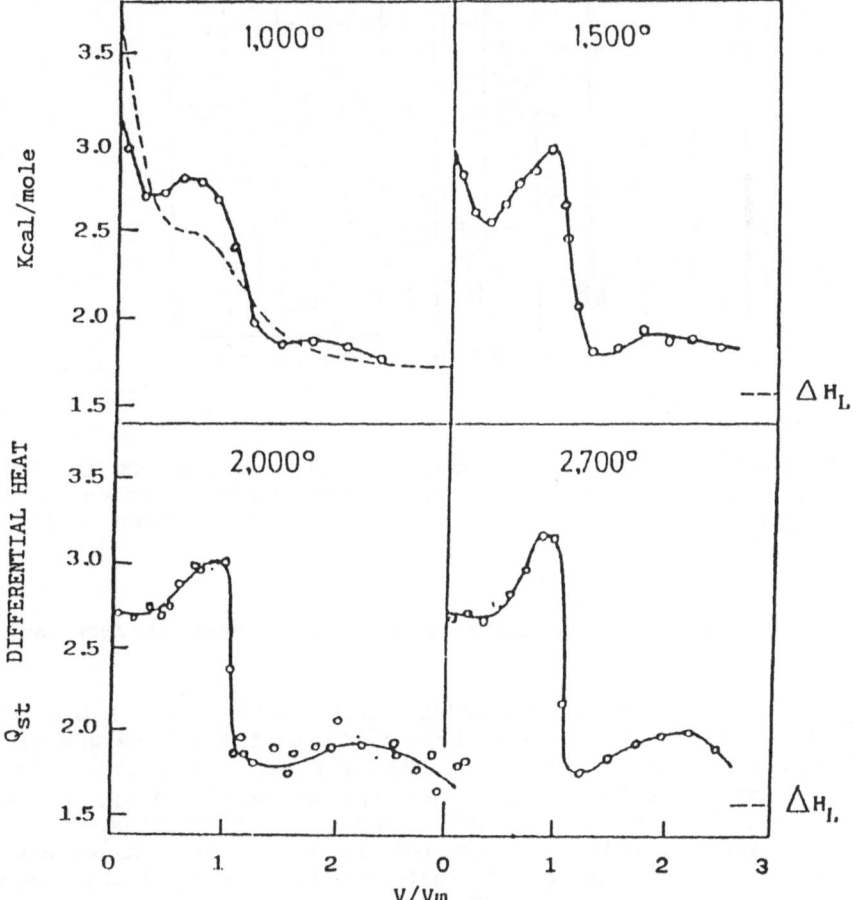

Fig. 3. Calorimetrically measured heats of adsorption of argon on Spheron carbon blacks at -195xC, plotted against surface coverage, for blacks graphitized at progressively higher temperatures. The broken line represents untreated black (from ref. 18).(Reprinted with permission)

374

Changes in the heat of adsorption obtained with different grades of
graphitization yield stepwise isotherms as shown in Figure 4 (19).   It
is, however, the author's opinion that the situation can be better
observed through the measurements reported in Figure 3.   For the sake of
clarity, and to make treatment simpler, this type of experiment will be
dealt with below.

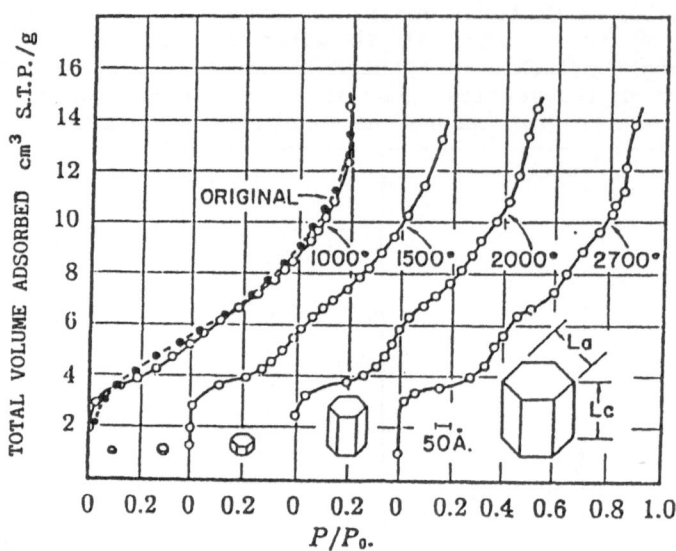

Fig. 4. Adsorption isotherms of argon on P-33 carbon black at -195xC.
The effect of graphitization temperature on the isotherm and on the size
of the crystallites is shown (from ref. 19). (Reprinted with
permission).

3. Gas Chromatographic Measurements of Adsorption Heat at zero surface coverage.

Over the last twenty years, gas-solid chromatography (GSC) has often
replaced the traditional static methods for studying the nature of the
adsorption process.   Thus, adsorbent surface area, adsorption isotherms
and thermodynamic functions such as adsorption heat, entropy and free
energy have been determined with this technique.   Methods based on GSC
(i)are less time consuming, (ii)do not require elaborate apparatus,
(iii)need only small amounts of adsorbate and (iv)yield reliable results
with very low surface coverage.   The latter applies when the adsorbent
is sufficiently homogeneous so that, with respect to the remaining
surface, the effect of heterogeneous sites can be neglected.   This
effect makes it possible to carry out measurements on the linear portion
of the adsorption isotherm, where lateral interactions are negligible
and   the   thermodynamic   parameters   depend   exclusively   on

adsorbate-adsorbent interations.

GBCs have proved to possess a sufficiently homogeneous surface and several of them have been used as standard absorbents, i.e. Graphon, Sterling FTG and Sterling MTG.   In their exhaustive paper, Crescentini et al. (20) report thermodynamic data determined by gas chromatography performed on some commercially available graphitized carbon blacks, such as Carbopack B, C and F.   The data obtained are compared with those previously determined on other carbon blacks by either GSC or conven tional methods. The thermodynamics of the adsorption process can be conveniently described by the well known equation

$$\Delta G^o = \Delta H^o - T \Delta S^o \qquad (1)$$

where ΔG^o, ΔH^o and ΔS^o are the integral changes in standard molar enthalpy, free energy and entropy, respectively. ΔG^o and ΔH^o can be obtained from gas chromatographic data (i.e. the net retention volume) and ΔS^o can easily be obtained from eq. 1.   This is true if the adsorption process takes place within the region governed by the Henry law so that adsorbate-adsorbate interactions are negligible and the net retention volume does not depend on sample size.   Under these conditions, if we consider n mol of a gas at a given pressure p in equilibrium with n_a mol of gas adsorbed per unit volume, the Henry law and the Clausius-Clapeyron equation yield the following relationships:

and

$$K_c = \frac{dn_a}{dn_p} \qquad (2)$$

$$Q_{st} = RT^2 \left(\frac{\delta \ln p}{\delta T} \right)_{n_a} = -R \left(\frac{\delta \ln p}{\delta 1/T} \right)_{n_a} \qquad (3)$$

where K_c is the Henry constant and the isosteric heat of adsorption Q_{st} is a differential entity which, at near-zero surface coverage, is to the integral change in enthalpy by means of the following equation:

$$\Delta H^o = -Q_{diff} = -Q_{st} + RT \qquad (4)$$

From eq. 2, and taking into account the material balance for an elemen tary layer, Kiselev and Yashin showed that

$$K_c = \frac{V'_R}{m} = \frac{V_R - V_0}{m} = V'_m = S V_s \qquad (5)$$

where V'_R is the net retention volume, m the adsorbent mass and V'_m the retention volume per unit of adsorbent mass, S is the total adsorbent surface area contained in the column and V_s the retention volume per unit of adsorbent surface area or specific retention volume. According to Kiselev and Yashin the following relationships are obtained:

$$K_c = V_s \tag{6}$$

$$K_p = \frac{V_s}{RT} \tag{7}$$

On the other hand, the adsorbed layer can be considered as a two-dimensional gas and, if the adsorbate behaves ideally in both the gaseous and adsorbed states, the following equations can be written:

$$\pi S = n_a RT \tag{8}$$

$$P V = n_p RT \tag{9}$$

$$K_c = \frac{\pi}{p} \tag{10}$$

Taking into account eq.5, it can easily be seen that π , the two-dimensional pressure, is related to the three-dimensional pressure, p, and to the specific retention volume, V_s, by the relationship

$$\pi = v_s p \tag{11}$$

Eq. 3 can now be written in the following form

$$Q_{st} = -RT^2 \frac{d\ln K_p}{dT} = R \frac{d\ln K_p}{d(1/T)} = RT^2 \frac{d\ln(V_s/T)}{dT} \quad R \frac{d\ln(V_s/T)}{d(1/T)}$$

and, as S and m are independent of T:

$$Q_{st} = R \frac{d\ln(V_R/T)}{d(1/T)} \tag{12}$$

Hence, the isosteric heat of adsorption can be obtained by plotting $\ln V_r/T$, $\ln V_s/T$ or $\ln V_m/T$ versus $1/T$, where T is the column temperature. Plots of $\ln V_r/T$ versus $1/T$ were used in the present calculations. Heats of adsorptions calculated in this manner are independent of the surface area and density of the various graphitized carbon blacks.

Integral change in standard free energy, ΔG^0, at a fixed temperature, T , of 1 mol of adsorbate on adsorption from a standard gaseous state at a pressure p' to a value p in equilibrium with a standard adsorbed state at a pressure π is given by

$$\Delta G^0 = - RT_1 \ln \frac{p'}{p} \qquad (13)$$

and combining eq. 11 and 13

$$\Delta G^0 = -RT_1 \ln \frac{pV_s}{\pi} \qquad (14)$$

if the standard state values are chosen following De Boer's approach so that p=101 kN/m^2=1.01 10^{6}dyne/cm2 and π=0.338dyne/cm, eq. 13 becomes

$$\Delta G^0 = -RT \ln (2.99 \; 10^6 \; V_s) \qquad (15)$$

where V s is expressed in cm.

The integral change in standard entropy can be calculated from eq. 1

$$\Delta S^0 = \frac{\Delta H^0 - \Delta G^0}{T_1}$$

From eq. 3 and 14 one gets

$$\Delta S^0 = \frac{- Q_{st} + RT_1 \ln(2.99 \; 10^6 \; V_s)}{T_1}$$

$$= \frac{- Q_{st}}{T_1} + R + R \ln (2.99 \; 10^6 + R\ln V_s)$$

$$\Delta S^0 = \frac{- Q_{st}}{T_1} + 39.62 + R \ln V_s \qquad (16)$$

The entropy of adsorption is also independent of the surface area of the adsorbent. As previously mentioned the isosteric heat of adsorption can be calculated from the slopes of the straight lines obtained by plotting lnV$_R$/T versus 1/T. A typical example is shown in Figure5 for propane, butane, pentane, hexane, benzene on Carbopack B at a zero surface coverage. As an example, in Table1, the Q$_s$ values obtained at zero surface coverage by several authors (21,30) are compared with those obtained in ref. 20. Considering the different experimental conditions and the various origins of the material studied, a very good agreement amoung the values obtained can be observed.

The reliability of the gas chromatographic method in determining ad sorption heats is also shown by the values obtained by calorimetric methods which, in the little data data available, coincide to about 10%.

Table I. Isosteric heats of adsorption (Kcal/mol) at zero surface coverage.[1]

ADSORBENT	Surf. area m^2/g	C_3H_8	C_4H_{10}	C_5H_{12}	C_6H_{14}	C_6H_6
Carbopack B	95	7.20	8.84	10.34	12.91	11.23
Carbopack C	9	6.10	7.35	8.45 9.80(25)	10.23	9.52
Carbopack F	6	5.80	7.00	8.44	9.65	8.97
Hydrogen-treated Cabopack B		7.00	9.81	14.40	18.63	17.84
Hydrogen-treated Carbopack C		6.82	9.78 9.9(26)	11.37 8.4(26) 9.25(26)	14.23 9.9(26)	12.85 9.3(26)
Graphon	138	8.4	12.20	11.02	15.08	
Hydrogen-treated Graphon		5.9	7.35	8.41	9.87	
Sterling FTG	14	6.10(24) 5.87(28) 6.7(30)	7.54(24) 7.45(28) 8.1(30)	8.82(24) 9.15(21) 8.88(22) 9.3(30) 9.35(25)	10.50(24) 10.2(30)	9.98(22)
Hydrogen-treated Sterling FTG		6.15(24)	7.45(24)	8.55(24) 9.15(21) 8.84(22) 9.25(21)	10.25(24)	9.83(22)
Sterling MTG	8	6.26(23)	8.10(23)	8.61(23) 8.84(23) 8.90(27) 8.90(29)	9.83(23) 10.10(27) 10.40(29)	9.87(22) 9.40(26)
VG3					8.08	

[1] Numbers in parentheses are literature references, the others are taken from ref. 20.

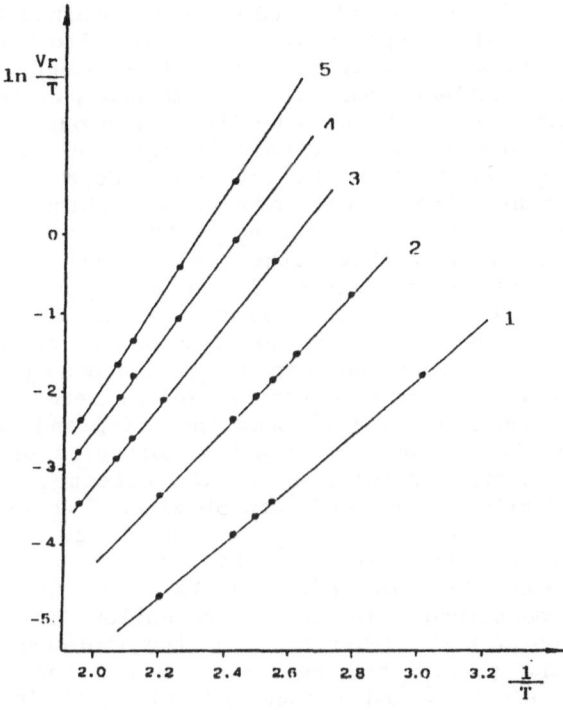

Fig. 5. Variation in ln V r /T with reciprocal temperature on Carbopack B at zero surface coverage. (1) propane; (2) butane; (3) pentane; (4) benzene; (5) hexane.

It may be noted that the values obtained with static (calorimetric) methods are consistently higher than those obtained by chromatography. This may be attributed to the fact that the sensitivity of the static method makes it impossible to reach a situation where coverage can really be zero; therefore, the effect of lateral interactions is still present. However, it is the author's opinion that the difference may be better attributed to the experimental conditions which are completely different in static and chromatographic measurements.

4. The effect of preadsorbed Molecules on Retention Parameters and Heats of Adsorption.

Determination of the adsorption isotherm is the very best way to investigate any modifications in the adsorptive properties of a particular material caused by the presence of preadsorbed molecules. There are essentially two chromatographic methods with which this goal can be achieved. One, exploited by Kiselev and coworkers (32), is based on the shape and area of the chromatographic peak obtained by injecting in creasing amounts of the compound for which the isotherm is to be determined into a GC column containing the adsorbent under study and fed by pure carrier gas. Another, more recent, method was introduced by Valentin and Guiochon (33,34) and used by the same group (35,36). This method, called the "step and pulse" method, is based on the use of a GC column containing the adsorbent to be studied. Here, the gas phase contains a known concentration of the compound for which one wishes to determine the adsorption isotherm. Once equilibrium between the vapour and adsorbed phase has been reached within the column for a given P/P value (step), the retention parameters of the same substance injected into the system (pulse) are measured from the chromatographic peak observed. Retention changes depend on the value of the surface coverage and, through proper calculations, one can obtain the adsorption isotherm together with the Henry constant by extrapolation at zero coverage.

More recently Findenegg et al. (37) used the "step and pulse" method to study the interactions between hdyrocarbon molecules on graphitized carbon black and interesting results as to the effect lateral interac tions have on the retention parameters were obtained. An example of the results obtained is given in Figure 6 showing the change in retention of several hydrocarbons vs. the molar fraction of n-hexane adsorbed on graphitized carbon black Sterling FT. It is worth noting that the maximum effect of preadsorbed molecules on retention is achieved just prior to the formation of a monolayer and that the increase in retention is maximized when the preadsorbed molecule is the same as the one injected. From calorimetric and chromatographic methods it is well known (15-17) that the heats of adsorption are largely modified by the effect of the lateral interactions of preadsorbed molecules and that a sharp decrease is achieved when a monolayer of such molecules is present on the adsorbent. Such behaviour has already been shown in Figure 3 above.

Figure 7 indicates the results obtained by Dondi et al. (35). It is interesting to note that the difference in retention for lateral inter actions found by these authors, and thus the decrease in this parameter after monolayer formation, is much more evident in the work of Dondi et al. than in that of Findenegg et al. Since the experimental techniques used in these works are quite similar and both sets of data fit very well with the theory, the differences observed in the shape of the curves can be ascribed to the greater homogeneity of the adsorbent used by Dondi (Sterling MT) than the one used by Findenegg (Sterling FT). In fact, these adsorbents differ essentially inregard to surface area (6.5m^2/g and 12.5m^2/g, respectively) and carbon purity (Sterling MT is almost pure graphite).

From the data given in Figures 6 and 7 one can also observe that the shape of such curves depends not only on adsorbent purity but also on the structural similarity of the preadsorbed and adsorbed molecules. A demonstration of this fact will be given below.

A totally different approach to the study of the changes preadsorbed molecules induce on vapour adsorption has been developed by the author's research group. The method is described in detail for different adsor adsorbents and molecules preadsorbed (20,30,38). The technique used is called Gas-Liquid-Solid Chromatography (GLSC) or more properly liquid-modified adsorption chromatography. This technique originates in the works of Eggertsen et al. (1,2). The method consists of choosing an adsorbent (usually one of the graphitized carbon blacks available) and a non volatile organic molecule (usually one of the stationary phases available for GC) having a well-known structure and dimensions (30). Then, several GC columns with different, well known, amounts (or percentages) of the stationary phase are prepared. On these columns the retention parameters are measured for different compounds depending on the type of study being carried out. In ref. 30 squalane, the most non polar substance available having a definite structure and glycerol, the most polar, least volatile compound available were used as stationary phases. Isosteric heats of adsorption were then measured with the method described above (20) which is of general use.

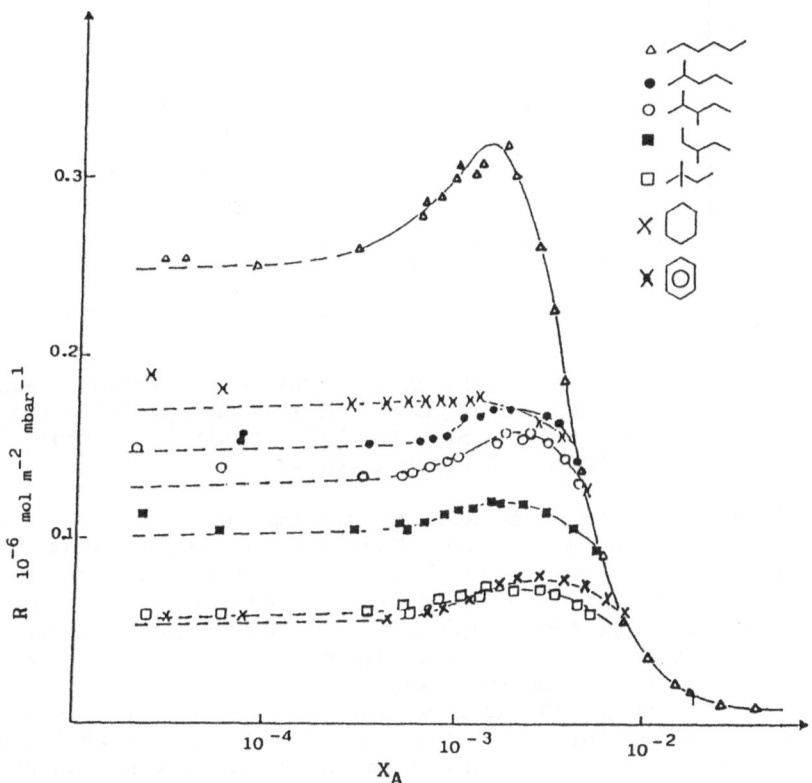

Fig. 6. Calculated (___) and experimental (▲) variation of R(X) versus the molar fraction in the gas phase: Cyclohexane on graphitized carbon black Sterling MTG (T=70°C, P=1.6atm, Po=1 atm) (logarithmic fitting). Best parameter values: Van der Waals (solid line); K_H =9.858, β=34.17Ax2 2α/β=3033cal/mole; Viral (broken line). (Reprinted with permission).

382

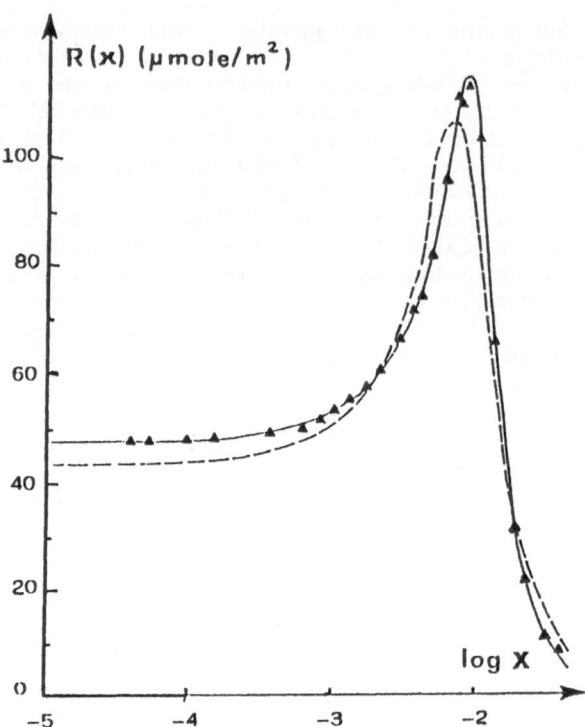

Fig. 7. Retention R(X) of small samples of the isomeric hexanes, cyclohexane and benzene as a function of the n-hexane X mole fraction in mobile phase. The adsorbent is Sterling FTG graphitized carbon black; column temperature 342.2°K, P=2.0. (Reprinted with permission).

With this technique, a great deal of information has been obtained regarding adsorbents and the effect different degrees of coverage of molecules differing in structure has on the behaviour of various molecules. Figure 8 (30) reports the results obtained for the heat of adsorption of some linear aliphatic saturated hydrocarbons with different percentages of squalane and glycerol on Sterling FT, as well as the behaviour of some alcohols on glycerol. From a qualitative point of. view, Beebe and young (18) used calorimetric methods or calculations to obtain curves very similar to those obtained by gas chromatography. These results are reported in Figure 9 and have been taken from ref. 15 although the original paper is by Pace (39).

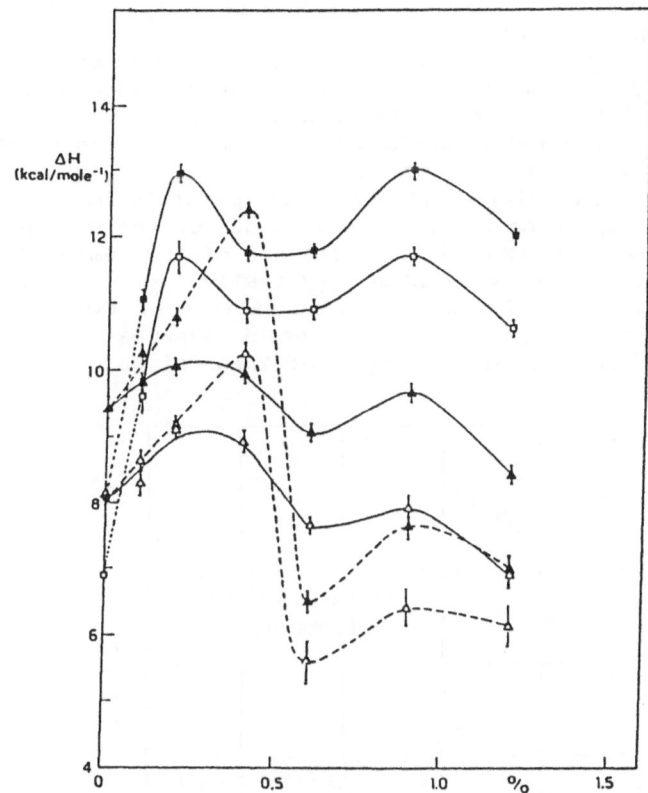

Fig. 8. Isosteric heats of adsorption of n–butane, n–pentane, ethanol and propyl alcohol at various percentages of liquid phase on Sterling FT. (△) butane; (□) ethanol; (___) glycerol; (▲) pentane; (■) propanol; (·····) squalane. (Reprinted with permission).

For the sake of comparison, Figure 10 reports the dependence of the heat of adsorption on the surface concentration of preadsorbed mole cules. The curve obtained in ref. 30 for n–hexane on Sterling FT dif ferently covered with squalane is superimposed on the same abscissa, assuming that the point of inflection for the descending portion of the curve corresponds to the formation of a monolayer of preadsorbed molecules.

Now let us discuss the advantages of the gas chromatographic technique introduced by our group in detail; i.e. coating the surface of homogeneous adsorbent with different amoutns of a non volatile, well knwon substance and studying the progression in retention data and the heats of adsorption of a volatile compound under such conditions.

384

a) The experimental technique is simple and only requires a gas chromatograph for packed columns.
b) The surface coating can be controlled with a high degree of accuracy. In fact, the instrument used is a balance, which, when feasible, is always preferable to pressure or volume measurements.
c) More detailed information on surface characteristics can be obtained, especially at very low surface coverage.
d) The technique is flexible and makes it possible to observe the interactions occurring among the three components of the system
(i.e. carbon surface, preadsorbed molecules and eluted compound).
More over, the last two can be changed very easily.
e) Study of the behaviour of the separation factors of two very similar compounds (i.e. isotopic pairs), at different degrees of preadsorbed molecule coverage, yields a higher degree of information.

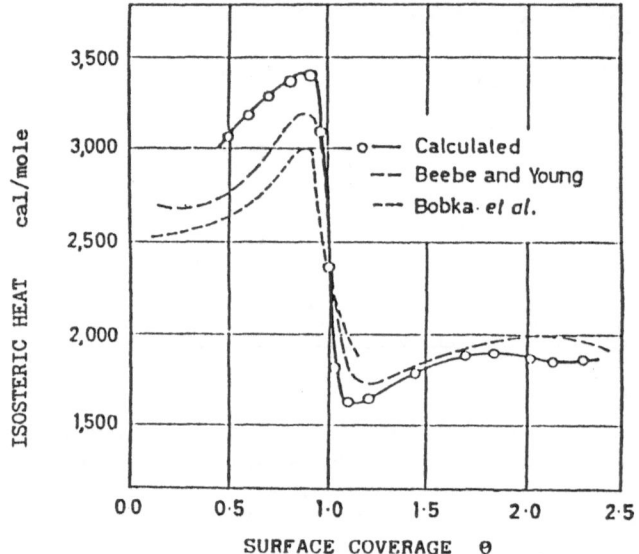

Fig. 9. Theoretical and experimental isosteric heats of adsorption of argon on graphite as a function of coverage (from ref. 39). (Reprinted with permission).

4.1 POLAR AND NON-POLAR MODIFIERS

From the graphs in Figure 8, completed with other measurements on propane and butane and by comparing them with similar measurements on other graphitized blacks, the following information has been obtained:

i) The surface of graphitized Sterling FT is very homogeneous since the decrease in the heat of adsorption at the monolayer is very sharp when the modifier is squalane, a molecule quite similar in structure to the molecules eluted.

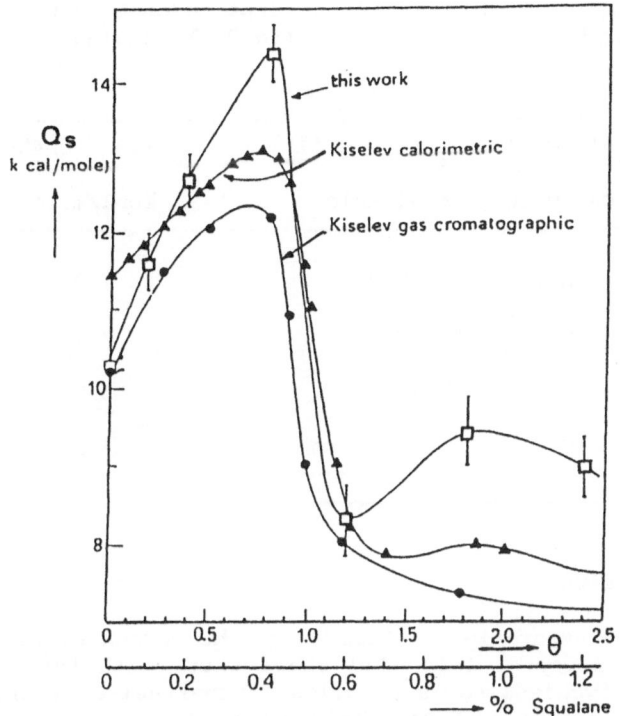

Fig. 10. Isosteric heat of adsorption for n-hexane on graphitized carbon black at various coverages calculated by GLSC and other methods. (Ref. 30, reprinted with permission).

ii) From the difference in the heat of adsorption at zero and at maximum coverage, the effect of lateral interactions can be measured as can be inferred from the results shown in Table II. This difference can be estimated as approximately 1 Kcal/mole per methylene group. In the cases of ethanol and propanol, analogous calculations lead one to the conclusion that the contribution of lateral interactions to the heat of adsorption is simply due to the formation of one hydrogen bond between the alcohol molecule and the preadsorbed glycerol. In fact, for both compounds, the difference between the two values at zero coverage and at the maximum of the curve is 5Kcal/mol (i.e. just the average energy of a hydrogen bond).

iii) The different shape of the curves obtained for hydrocarbons when squalane or glycerol are the preadsorbed molecules shows that glycerol is deposited more irregularly on the carbon surface, giving rise to a less homogeneous surface in the monolayer region. This is proved by the much lower decrease in adsorption heat in that region.

Table II. Gas Chromatographic heats of adsorption of C_3 - C_6 hydro-carbons on Sterling FT pure and treated with 0.4% squalane.[6]

Compound	Q_s (% = 0) Kcal/mole	Q_s (% = 0.4) Kcal/mole	ΔQ_s [(Δ = 0.4%-0.0%)] Kcal/mole
C_3H_8	6.7	7.7	1.0
$n-C_4H_{10}$	8.1	10.2	2.1
$n-C_5H_{12}$	9.3	12.3	3.0
$n-C_6H_{14}$	10.2	14.4	4.2

4.2. ISOMER SEPARATION.

Measurements of the separation of isomers on the carbon surface modified with squalane and glycerol (Fig. 11) shows that, with the latter, the separation of the two isomers is practically unaffected by the presence of polar compounds on the surface. This may be interpreted in two ways. One is that glycerol is not uniformly distributed over the carbon surface as has been proposed above. For this reason, even at high nominal surface coverage the non polar hydrocarbon molecules are still adsorbed on the pure adsorbant surface which is, in turn, responsible for the high separation factors. Another explanation may be that, since in the case of hydrocarbons on preadsorbed glycerol the lateral interactions are hardly signficant (see graphs in Fig. 8) and molecules involved are of different chemical natures, adsorption on the bidimensional glycerol layer takes place after the monolayer has been formed. Of course one explanation does not rule out the other and the truth may well be depicted by a combination of both hypotheses.
  In the case of separation of hydrocarbon isomers on a squalane-modified surface, a tremendous decrease in separation factors is ob served after formation of the monolayer. This may be explained by the fact that, since the modifier and eluate molecules are quite similar in structure, the isomers interact preferentially with the squalane molecule rather than the adsorbent surface. This greatly reduces the differentation between isomeric molecules.

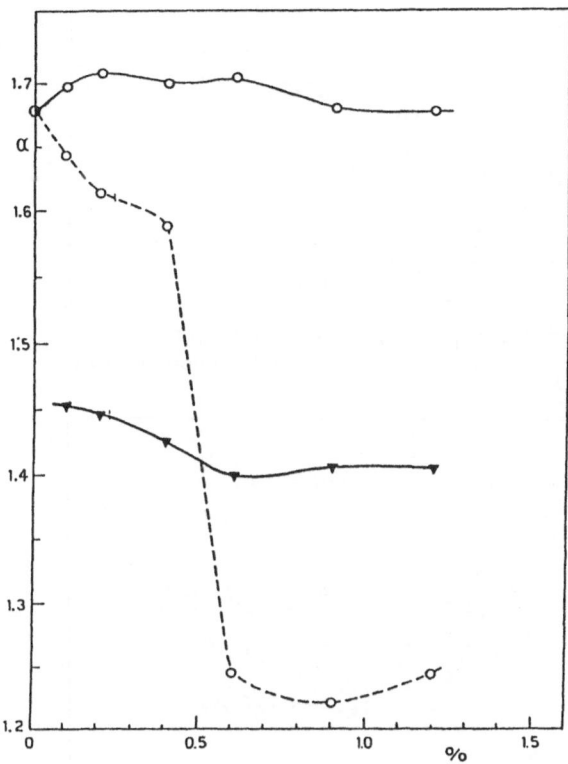

Fig. 11. Separation factors at 46xC vs % of liquid phase on Sterling FT for the isomeric pairs. (o= n-pentane/isopentane; (-) glycerol; (▼) n-propyl alcohol/isopropyl alcohol; (···) squalane (from ref. (30)). (Reprinted with permission).

4.3. THE USE OF ISOTOPIC MOLECULES

Figure 12 shows the results of a study of the modifications induced by different modifiers performed through observation of the behaviour of the separation factors for deuterated and hydrogenated species. As is well known (40), the separation between deuterium- and hydrogen-containing isotopic pairs is due to the differences in molecular polar izability yielding the so-called "abnormal" isotope effect (i.e. the heavier species has a higher vapour pressure and is less adsorbed than the hydrogenated species). Our research group used the separation

388

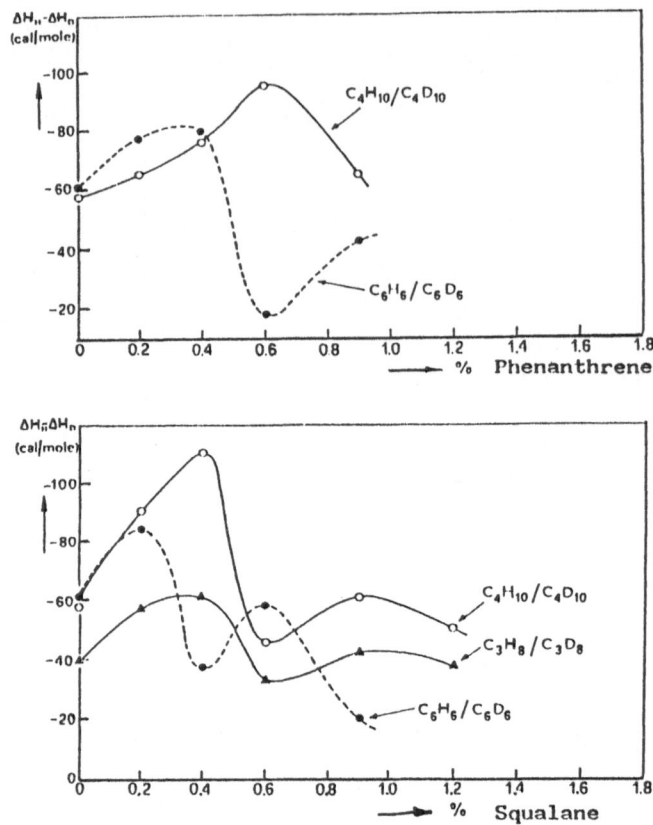

Fig. 12. $\Delta H_H - \Delta H_D$ vs. % of squalane and phenanthrene on Sterling FT for different isotopic pairs. (from Ref. 41). (Reprinted with permission).

factors and the differences in the heats of adsorption of such pairs of molecules (in particular C_6H_6 and C_6D_6) to study the carbon surface coated with different modifiers.

In Figure 12 the separation factors of the two species are reported as a function of the percentage of the two modifiers (squalane and phenanthrene) at different temperatures. First, one can observe that in the case of GLC the maximum value for the separation factor α (using squalane as liquid phase) is 1.04 (41) while with squalane-modified carbon black the α value is 1.08.

An interesting feature concerning the secondary maximum (second step) is that it becomes evident as the temperature rises. This happens with both modifiers but is more evident with phenanthrene which was chosen because its chemical and steric structure is similar to that of graphite. The prsence or absence of a second step in the adsorption isotherm which corresponds to the presence of a secondary maximum in the present type of measurements, is an indication of surface homogeneity. This can also be seen in Figure 1 where the second step is only observed for those carbon blacks with a higher degree of graphitization. The fact that the secondary maximum is more evident with phenanthrene is further confirmation of the statement made by Polley et al. (19).

The fact that the second step is more pronounced at higher temperatures demonstrates this uniforming effect this parameter has on the distribution of preadsorbed molecules (42). Finally, one must note that the assertion that the sharp decrease in the heat of adsorption (or of the difference) yields indication of formation of the monolayer has only relative meaning and interpretation is limited to the particular modifier and to the particular molecule eluted.

In Figure 12 the usual curves have been determined for the two pairs C_6H_6/C_6D_6 and C_4H_{10}/C_4D_{10} (30). A very substantial difference is found in the percentage where the monolayer occurs for the two pairs. Actually, one can say that the monolayer occurs when the molecule coming from the gas phase can no longer reach the adsorbent surface. This explains the drop in heat of adsorption values. In fact, this occurs at a lower percentage of modifier for benzene, which is a "larger" molecule than butane. This has proved true for most of the systems investigated (20), but with the isotopic molecules finer measurements can be obtained.

5. Kinetic considerations.

The availability of very homogeneous adsorptive surfaces makes it possi possible to obtain very efficient chromatographic columns, even from the kinetic point of view. The C term of the Van Deemter equation becomes small because the exchange between gas and condensed phase is faster. Giddings (43) treated resistance to mass transfer in the condensed phase and modified the classical Van Deemter equation for GSC. Taking into account that the adsorption-desorption process is much faster than the solution-evaporation process, he obtained the following expression for a capillary column:

$$H = \frac{B}{u} + C_g \, \bar{u} + C_k \, \bar{u} \qquad (17)$$

where C_k, the kinetic mass transfer term, replaces the liquid-phase mass transfer term, C_1.

The terms related to diffusion in the gas-phase (B) and to the resistance to mass transfer in the gas phase (C_g) refer to the usual expression, namely:

$$B = 2D_g \quad \text{and} \quad C_g = \frac{1 + 6k' + 11k'^2}{3(1 + k')^2} \, \frac{r_0^2}{8D_g} \qquad (18)$$

where r_0 is the radius of the capillary and D_g the diffusion coefficient in the gas phase.

A comparison of the efficiency of the two capillary columns used in GLC and GSC can be made provided the geometry of the colums is the same and the same coumpound, showing analogous k' values, is eluted at the same temperature. Then, both the C and B terms of the Van Deemter equation can be deemed as yielding the same contribution to the magnitude of H. Under these conditions, the differences between the overall C terms can be ascribed to the difference between C_1 and C_k.

According to Giddings,

$$C_k = 2 \frac{k'}{(1+k')^2} \bar{t}_d \tag{19}$$

where $\bar{t}_d$ the mean adsorption time of an equilibrium population of sorbed molecules. If the adsorbent is sufficiently homogeneous in terms of adsorption site energy (as is the case in linear elution from liquid-modified graphitized carbon black) then

$$C_k = 2 \frac{k'}{(1+k')^2} K_d \tag{20}$$

where K_d is the first-order desorption constant.

From further considerations and acceptable approximations, Giddings came to the conclusion that the term C_k can be well expressed by the equation

$$C_k = 10^{-4} \frac{V_g}{\beta A} \tag{21}$$

where V_g is the gas phase volume of the column, a the fraction of the molecules striking a unit area in unit time actually adsorbed and A the overall area of the inner walls coated with the adsorbent. For a highly porous, homogeneous adsorbent Giddings forecast a value of approximately 10^{-3}, while a should range between 0.1 and 1.0. In this instance a theoretical value of 10^{-7} to 10^{-8} for C_k can be assigned if the surface area of the adsorbent is very high ($10^3 m^2/g$). Unfortunately, it is impossible to obtain an adsorbent which is highly porous and, at the same time, homogeneous in terms of adsorption site energy distribution. Therefore, the best compromise can be achieved by using a non porous, homogeneous adsorbent such as graphitized carbon black, which has the advantage of exhibiting a linear adsorption isotherm. Working with this kind of adsorbent, one can assume a C_k value of about 10^{-4}. In chromatographic practice, however, this value, also forecast by Giddings, is hardly reached for C_1.

In Figure 13 the value of the C term of the Van Deemter equation is plotted for geometrically identical GC columns but which contain differ ent percentages of squalane or glycerol (30). It is interesting to note that using squalane as modifier the maximum C value is obtained after the first monolayer of preadsorbed molecules is formed. This means that under such conditions the exchange between gas and condensed phases is slow. In fact, the molecules coming from the gas phase 'feel' the interactions of both the adsorbent and the molecules of squalane, similar in structure to the eluate n-pentane.

The contrary occurs when the modifier is glycerol. In this the C term approaches zero. A simple explanation is that under these conditions two situations occur:

a) The adsorbent is deactivated by the molecules of glycerol adsorbed and becomes gaussian.

b) The polar molecules of glycerol do not interact to any significant extent with the non polar molecule of pentane which, in turn, is adsorbed on the bidimensional layer; therefore, the exchange between the two phases is very fast.

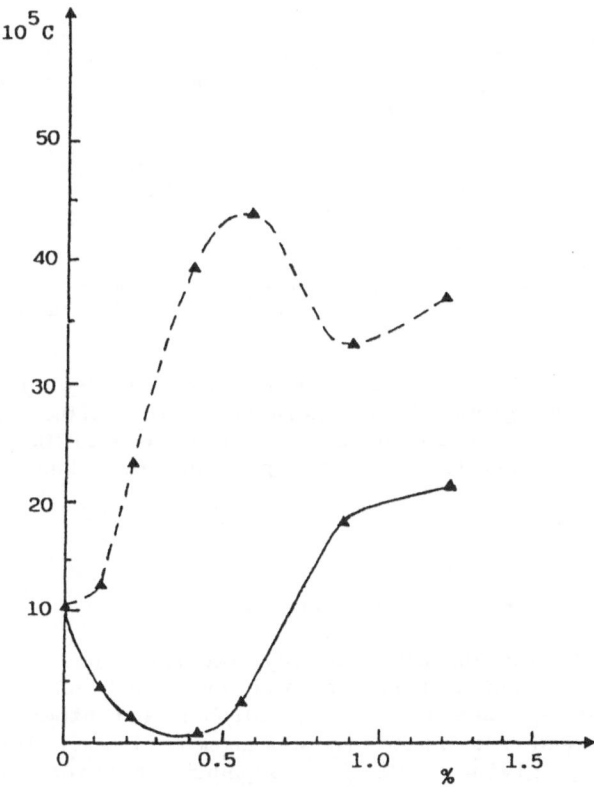

Fig. 13. Plots of C for n-pentane vs. % of liquid phase on Sterling FT coated with squalane.

392

Analogous results have been obtained with capillary columns (44,45). In Figure 14 a comparison of the Van Deemter plots obtained for two capillary columns of the same dimensions but working with GLC and GLSC; the advantage of the latter is evident. The value of 2.3×10^{-4} found for C in the GLSC capillary is in good agreement with the Giddings forecast.

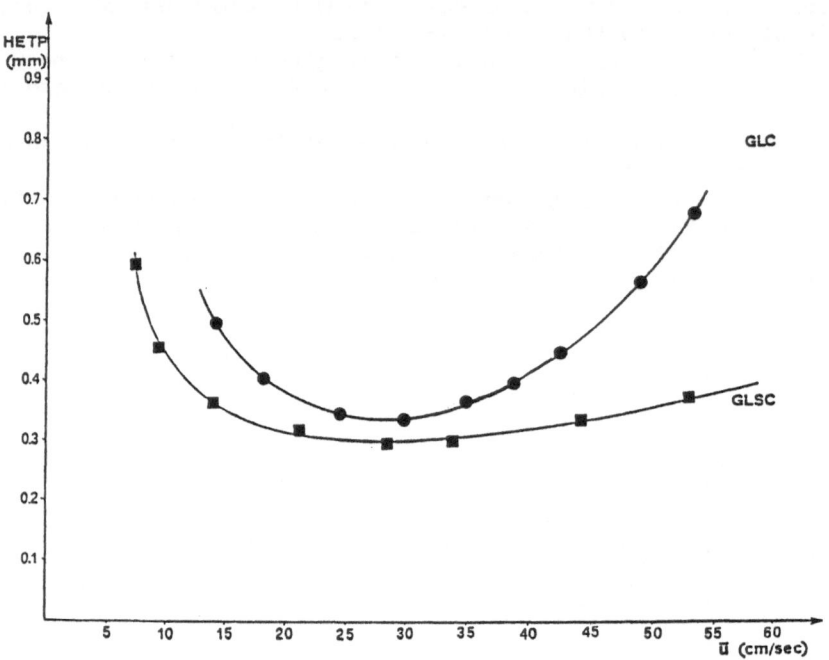

Fig. 14. Van Deemter plots obtained with two fused-silica capillary columns of the same geometrical characteristics (20mx0.25mm I.D.). Coatings: ● = graphitized carbon black, Carbopack F + SP100 (for details see text) ■ = SP1000, film thickness 0.35μm. sample, n-hexadecane.

6. Chemical Impurities on the Carbon Surface.

Very interesting studies on the chemical imperfections of graphitized carbon black have recently been carried out (46) showing that the chemical impurities are due to metals, sulphur and other heteroatoms. In particular, oxygen was found to be chemically combined with the graphit ic carbon, giving rise to compounds bearing chromene-like structures. It has also been hypothesized that the creation of benz-pyran and benz-pirylium salt is likely.

Most of these impurities can be eliminated by hydrogen treatment at about $1000^{\circ}C$ (48). With this treatment the adsorption isotherms of very polar or even acidic or basic organic compounds can be linearized at low coverage. As an example, in Figure 15 (49,24) the modifications of the heat of adsorption and of the adsorption isotherm of methanol on Sterling FT induced by hydrogen treatment at 1000xC is reported. With such modification the surface of graphitized carbon blacks becomes perhaps the most homogeneous surface available. This fact, together with the flexibility obtained by coating the graphitic adsorbents with proper amounts of a non volatile modifier is the reason for the great success these materials have had in gas chromatography and related analytical and preanalytical techniques (50,51,52). A good review of the analytical applications has been made by Di Corcia and Liberti (53). A detailed treatment of this matter appears beyond the scope of the present paper.

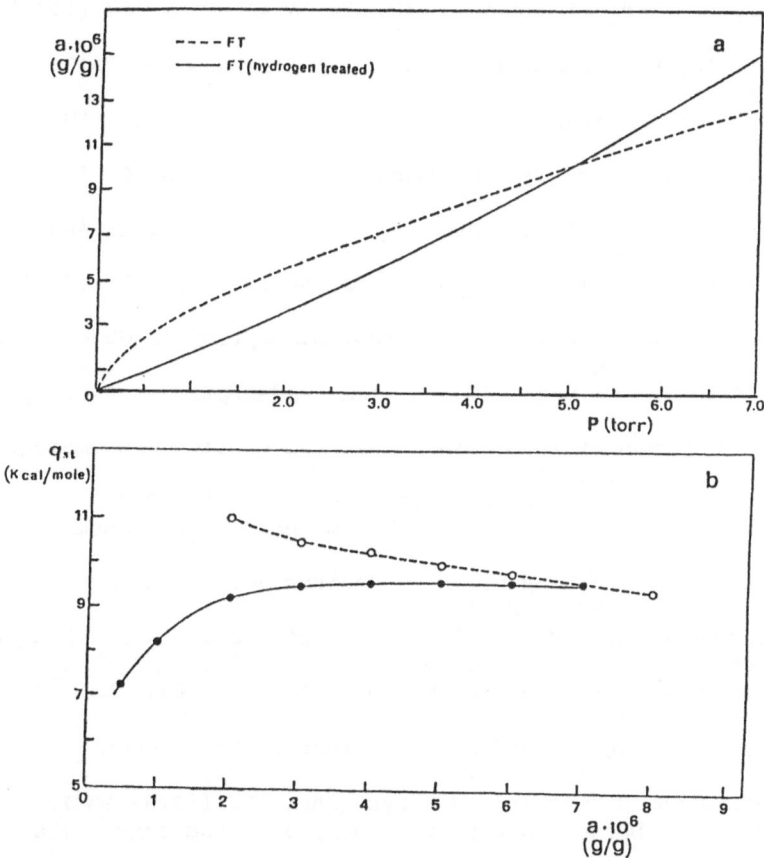

Fig. 15. (a) Isotherms for methanol at 33xC on FT and hydrogen-treated FT. (b) Isosteric heats of adsorption of methanol on FT and on hydrogen-treated FT.

394

7. References.

1. F.T. Eggertsen, H.S. Knight and S. Groennings; Anal. Chem. 28 (1956) 303-306.
2. F.T. Eggertsen, H.S. Knight; Anal. Chem. 30 (1958) 15-20.
3. F. Bruner and G.P. Cartoni; Anal. Chem. 36 (1964) 1522-1526.
4. F. Bruner, G.P. Cartoni and A. Liberti; Anal. Chem. 38 (1966) 298-303.
5. F. Bruner, G.P. Cartoni and M. Possanzini; Anal. Chem. 41 (1969) 1122-1124.
6. M. Mohke and W. Saffert; Gas Chromatography 1962, M. Van Swaay, Ed., p. 216, Butterworths, London 1963.
7. A.V. Kiselev, Private Communication, Rome 1966.
8. F. Bruner, C. Canulli, A. Di Corcia, A. Liberti; Nature 231 (1971) 175-177.
9. F. Bruner, P. Ciccioli and A. Di Corcia; Anal. Chem. 44 (1972) 894-898.
10. A. Di Corcia, D. Fritz and F. Bruner, Anal. Chem. 42 (1972) 1500-1504.
11. A. Di Corcia, A. Liberti, R. Samperi; Anal. Chem. 45 (1973) 1228-1235.
12. A. Di Corcia, P. Ciccioli, F. Bruner; J. Chromatog. 62 (1971) 128-131.
13. F. Bruner, A. Liberti, M. Possanzini and I. Allegrini; Anal. Chem. 44 (1972) 2070-2074.
14. F. Bruner, P. Ciccioli and F. Di Nardo; Anal. Chem. 47 (1975) 141-144.
15. D.M. Young and A.D. Crowell, "Physical Adsorption of Gases" Butterworths, London, 1962.
16. S. Ross and J.P. Olivier, "On Physical Adsorption", Interscience, New York, 1964.
17. A.V. Kiselev and Y.I. Yashin, "Gas Adsorption Chromatography", Plenum Press, London, 1969.
18. R.A. Beebe and D.M. Young; J. Phys. Chem 58 (1954) 330.
19. M.H. Polley, W.D. Shaeffer and W.R. Smith; J. Phys. Chem 57 (1953) 469.
20. G. Crescentini, F. Mangani, A.R. Mastrogiacomo and P. Palma; J. Chromatogr. 392 (1987) 83-94.
21. E.V. Kalaschnikova, A.V. Kiselev, R.S. Petrova and K.D. Shcherbakova, Chromatographia 4 (1978) 495.
22. C. Vidal-Majar, M.F. Gonnord, M. Goedert and G. Guiochon, J. Phys. Chem. 79 (1975) 732.
23. S. Ross, J.K. Saelens and J.P. Olivier; J. Phys. Chem. 66 (1962) 696.
24. A. Di Corcia and R. Samperi; J. Phys. Chem. 77 (1973) 1301.
25. F. Bruner, G. Bertoni and P. Ciccioli; J. Chromatogr. 120 (1976) 307.
26. F. Bruner, G. Bertoni, R. Montali and C. Severini; Ann. Chim. (Rome) 68 (1978) 565.
27. P.A. Elkington and G. Curthois, J. Phys. Chem. 73 (1969) 2321.
28. S.E. Hoory and J. Prausnitz; Trans. Farad. Soc. 63 (1967) 4550.

29. A.V. Kiselev and Y.J. Yashin; "Gas Adsorption Chromatography", Plenum Press, New York, 1969, p. 23.
30. F. Bruner, P. Ciccioli, G. Crescentini and M.T. Pistolesi; Anal. Chem. 45 (1973) 1851.
31. A.V. Kiselev and Y.J. Yashin; "Gas Adsorption Chromatography", Plenum Press, New York, 1969, pp. 125-131.
32. Ibid. pp. 105-141, and references therin.
33. P. Valentin and G. Guiochon; J. Chromatogr. Sci. 14 (1976) 56.
34. P. Valentin and G. Guiochon; J. Chromatogr. Sci. 14 (1976) 132.
35. F. Dondi, M.F. Gonnord and G. Guiochon; J. Colloid Interface Sci., 62 (1977) 303.
36. F. Dondi, M.F. Gonnord and G. Guiochon; J. Colloid Interface Sci., 62 (1977) 316.
37. W. Von Rybinski, M. Albrecht and G.H. Findenegg; Faraday Symposia of the Faraday Society, No. 15, "Chromatography, Equilibria and Kinetics ", p. 25 (1980).
38. A. Di Corcia, A. Liberti and R. Samperi; Anal. Chem. 45 (1974) 1228.
39. E.L. Pace; J. Phys. Chem. 27 (1957) 1341.
40. A. Di Corcia and A. Liberti; Trans. Farad. Soc. 66 (1970) 967.
41. A. Liberti, G.P. Cartoni and F. Bruner; J. Chromatogr. 12 (1963) 8.
42. D.M. Young and A.D. Crowell, "Physical Adsorption of Gases", Butterworths, London, 1962, pp. 172-174 and references therin.
43. J.C. Giddings; Anal. Chem. 36 (1964) 1170.
44. F. Bruner, G.Crescentini, F. Mangani, P. Palma and M. Xiang; J. Chromatogr. 399 (1987) 87.
45. F. Bruner, G. Crescentini, F. Mangani, and L. Lattanzi; J. Cromatogr. 517 (1990) 123.
46. L. Campanella, A. Di Corcia, R. Samperi and A. Gambacorta; Materials Chemistry 7 (1982) 429.
47. F. Andreolini, C. Borra, F. Caccano, A. Di Corcia and R. Samperi; Anal. Chem. 59 (1987) 1720.
48. B. Millard, E.G. Caswell, E.E. Leger and D.R. Mills; J. Phys. Chem. 59 (1955) 876.
49. A. Di Corcia and F. Bruner; Anal. Chem. 43 (1971) 1634.
50. P. Ciccioli, G. Bertoni, E. Brancaleoni, R. Fratarcangeli and F. Bruner; J. Chromatogr. 126 (1976) 757.
51. G. Bertoni, F. Bruner, A. Liberti and C. Perrino; J. Chromatogr. 203 (1981) 263.
52. F. Mangani, G. Crescentini and F. Bruner; Anal. Chem. 53 (1981) 1627.
53. A. Di Corcia and A. Liberti, "Advances in Chromatography", vol. 14, J.C. Giddings, ed. Marcel Dekker, New York and Basel, 1976, pp. 305-363.
54. R.J. Bobka, R.E. Dihimmy, A.R. Siebert and E.L. Pace; J. Phys. Chem. 61 (1957) 1646.

MECHANISMS OF THE SEPARATION AND TRANSPORT OF POLYMER SYSTEMS IN CHROMATOGRAPHIC MEDIA

> "You don't know much, and that's a fact"
> The Duchess, in Alice in Wonderland
> (Lewis Carroll)

Robert Tijssen and Jaap Bos
Koninklijke/Shell-Laboratorium, Amsterdam (Shell Research B.V.)
Department AG/2
Badhuisweg 3
1031 CM Amsterdam
The Netherlands

KEYWORDS/ABSTRACT: particles/size/separation/gel permeation/size exclusion/polymer size/confinement/microcapillary/hydrodynamic/chromatography/diffusion/reptation/chemical potential/lift force/mass balance/concentration profile/tubular pinch effect/TPE/pore/mass balance/SSC/GPC/SEC/HDC/separation-by-flow/SBF

"Chromatographic" techniques which are based on the principle of separation according to differences in the size of materials in solution only, belong to the class of SSC methods (Size Separation Chromatography). The ultimate parameter in SSC is, by definition, the aspect ratio λ = (particle size)/(channel size). Notably the size of polymer molecules near impermeable channel walls is described, including reptation of polymer chains. Next SEC/GPC (Size Exclusion/Gel Permeation Chromatography) and HDC (Hydrodynamic Chromatography (or separation-by-flow)) are treated. For the latter technique the so-called "Tubular Pinch" effect, stemming from "lift" forces from inertia in sheared flow, is also discussed. Finally, it is concluded that HDC and SEC are complementing SSC methods.

1. Beginnings

The ever-expanding use of polymeric materials increasingly demands the development of techniques to characterise these materials. The main question to be answered concerns the determination of molecular "weight" (or better: mass, M) and its distribution (MWD) in the often polydisperse polymer materials, as this determines the physical properties of polymers and hence many product properties and process variables.

From this point of view, chromatographic techniques, performing instrumental analyses with unsurpassed selectivity and efficiency, are potential candidate techniques for rapidly solving these questions. Classical fractionation techniques are extremely time-consuming [1]; phase distribution chromatography is better in this respect [1], having inherent advantages in terms of column dispersion, but it is not universally

397

F. Dondi and G. Guiochon (eds.),
Theoretical Advancement in Chromatography and Related Separation Techniques, 397–441.
© 1992 Kluwer Academic Publishers.

applicable [like SEC (Size Exclusion) or GPC (Gel Permeation Chromato-graphy)]. Because of the inherent properties of polymers (no volatility; solubility in liquid solvents) especially the Liquid Chromatographic (LC) techniques are of interest. Among the HPLC (High Performance LC) techniques, GPC, also called SEC, is a powerful method, and by far the most widely used one for separating polymers.

Basically, GPC separates molecules of a polymer in solution according to their sizes, size being the foremost property which distinguishes the various individual polymer molecules within a polymeric sample. GPC is the accepted acronym, while SEC emphasises the main mechanism of separa-tion, *viz.* exclusion. At present GPC is a very useful, rapid and comfor-table routine method for the determination of MWDs which are not extre-mely narrow. Admittedly, zone—spreading phenomena are not easily correc-ted for in GPC but then, it is the challenge and art of separation sci-ence to maximise separation relative to dispersive transport. In this contribution we will not go into details of the latter but concentrate on the understanding of the primary separative mechanism, *i.e.* we confi-ne ourselves to real SEC and related SSC techniques, such as HDC (Hydro-dynamic Chromatography), also called SBF (Separation—By—Flow).

Although SEC is to be regarded as an LC technique, the nature of poly-meric molecules (limited solubility; limited diffusion; number of speci-es occurring exceedingly large; selectivity by size (entropic) rather than by (enthalpic) interactions; influence of flow properties such as shear) explains why "*the difference between macromolecular chromatogra-phy and small-molecule chromatography, from a fundamental point of view, is considerably greater than the difference between gas and liquid chro-matography*", to quote Giddings [2]. This point of view is further stres-sed by Glöckner [1] in his standard text on polymer chromatography.

In ideal GPC (= SEC), macromolecules with a certain size, are eluted from the separation column always in the same period of time, the larger the size the faster they are eluted. As the "size" of macromolecular species is related to the associated molar mass via more or less compli-cated statistical modelling, the measurement of residence times also gives information on both the molar mass (M) and its distribution (MWD). This indirect calibration procedure (we obtain information on size rath-er than on M) is obviously a main problem in GPC methods.

Full characterisation of polymeric materials (including new and/or in-tractable polymers: *e.g.* polyketones, polyolefins) is possible by combi-ning the GPC separation method with a number of independent detection techniques. By using "absolute" detectors based on principles of visco-metry, light scattering and osmometry, the required physico—chemical/ mechanical information is obtained [3, 4].

Although there is no single other method of analysis more effective in the characterisation of polymeric materials than the combined GPC/(div-ers detectors) method, particulate materials (including polymeric micelles and other colloidal species) and very large macromolecules do not lend themselves to analysis by this technique. For such materials other separation techniques (Hydrodynamic Chromatography, HDC; Field Flow Fractionation, FFF) are being developed. FFF techniques are covered in another contribution in this ASI volume. In the present paper typical properties of GPC and HDC will be discussed.

2. Introduction to Size Separations

The characterisation of particle sizes in industrial samples for both
research and routine purposes is an essential part of quality control
procedures. [In this contribution, a particle is defined as an object in
the submicrometre range, thus ranging from molecules, via macromolecules
and their agglomerates to discrete bodies]. This puts ever larger demands on the ability of i.a. separation methods to deal with higher molecular weights and/or larger sized particles, an increase by more than
eight orders of magnitude being observed over the last decades.

If the average particle in a dispersed system is too small to be dis-
criminated by light (say 1 μm), many regular sizing methods become im-
practicable and it is in these submicrometre dispersed systems, that
size separation methods such as GPC, HDC and FFF become ever more important. It is obvious that these methods are primarily based on the prem-
ise that "particle" size and flow channel size are approaching each other, say the aspect ratio: $\lambda = R_p/R_c$ will be $> 10^{-4}$ (in the practice of
GPC pore sizes range from about 50 Å to 10^6 Å, while polymer sizes range
from 5 Å (monomer) to $\approx$ 0.1 μm). As will be seen, however, in capillary
HDC (and also in FFF) flow phenomena themselves may play an important
role in the differential migration mechanism.

2.1. THE MEANING OF "SIZE"

In the case of discrete particles, the term size is intuitively, taken
to mean the diameter of an average particle, an appropriate concept for
macroscopic systems. When dealing with submicrometre, colloidal or polymeric systems, however, the definitions of both "particle" and "size"
become somewhat imprecise.

For macromolecules with known chemical structure, all kinds of characteristic sizes are possible, even though the conformation of the molecule in its surrounding solvent is "known" from more or less well—established polymer model theories. What is more, these size parameters, even
if known, are susceptible to conditions such as temperature, concentration, shear rate, type of solvent, to name but a few. For reasons of
convenience, statistically equivalent sphere diameters are often defined
as the "size" of non—spherical objects such as random—coil, rodlike and
wormlike macromolecules. Also, discrete "particles" such as oil droplets
in water offer problems of definition, and are subject to conditions.

2.1.1. *Size and Shape of Polymer Molecules in Solution.* Concerning this
subject the science of polymer physics has undergone a tremendous development and we refer to the pertinent literature for reviews [5 - 15],
where especially refs [5, 10, 11, 15] offer recommendable accounts.

The simplest possible model of a macromolecule is the picture of a
linear long sequence of (M/M_0) monomer segments of molar mass M_0 each.
Because of the rotational freedom of the chemical bonds between the
monomer units, the chain can assume a huge number of different spatial
arrangements and the "shape" of the whole chain—like structure is continuously changing. The final shape of the macromolecule is a statistically weighted average of the shapes of these conformations. A picture
often used to represent such a real chain is that of the "freely jointed
equivalent chain" (after Kuhn) consisting of a sequence of identical and

rigid segments, with bond angles that can assume any value. The length of a segment (a) and the number of segments (N) are adjusted so, as to mimic the length and flexibility of the real chain, *i.e.* with increasing flexibility of the real chain, N increases and a decreases. Typically each segment represents two to nine monomer units.

As each segment has a characteristic length (a) and a (bond) direction, the chain is conveniently described as a train of vectors. Free rotation about these bond directions caused by the ever present thermal motion, allows an enormously large number of possible chain conformations. These conformations can be simulated by random walks and the average shape of such a chain is a random coil-like structure with an end-to-end distance that is the sum of all segment vectors of length a: $r = {}_i\Sigma\, a_i$, where the summation ${}_i\Sigma$ is from i = 1 to N. Each conformation has its own end-to-end distance, and averaging this distribution for all possible conformations, the average end-to-end distance itself will be zero. Hence, it is custom to use the square of the end-to-end distance (thus eliminating vectors, forming the scalar product $r^2 = {}_i\Sigma\, a_i \cdot {}_j\Sigma\, a_j$) as the characteristic parameter for such chains [8, 10, 11]. From this the average (mean-square) distance $<r^2>$ is obtained as $<r^2> = Na^2$, *cf.* the Appendix. $<r^2>$ has a prominent status among polymer properties being related to experimental characteristics such as *e.g.* viscosity, diffusion, sedimentation and light scattering. Of course, the average end-to-end distance ℓ is now obtained as the root of the mean square end-to-end distance: $\ell = \sqrt{<r^2>}$. We remark, that this result equals the familiar (random-walk) relation $<r^2> = Na^2$, the very relationship which since the days of Einstein has been used for the description of thermal-motion-driven movement of (ensembles of) discrete particles (Brownian movement and diffusion, see the next Section and the Appendix).

2.1.2. *Random Coils represented by Random-Flight Chains*. One of the simplest idealisations of flexible polymer chains consists in replacing it by a random walk, consisting of N steps of length a, starting in the origin (the first segment of the chain) and reaching out ("diffuse" or randomly fly away) to an arbitrary end point (where the last segment of the chain is supposed to be located). The analogy between the two processes (diffusion and the formation of a random coil), which was later expertly employed by Casassa in his renowned SEC theory (Section 4.2.2) is obvious if we imagine that a diffusing particle leaves a trace of its movements in time: a perfect picture of a random coil chain emerges, with the associated density distribution of segments being Gaussian. For a chain of N segments, basic random-walk statistics readily yields the following expression for the probability density function $P(r)$ which expresses the chance of the end segment being found at a fixed spatial position r, when the first segment is fixed in the origin, for N volumeless elements of length ("steps") a each: $P(r) = (\beta/\sqrt{\pi})^3 \exp\{-(\beta r)^2\}$, where $\beta^2 = (3/2Na^2)$ [9]. As we are actually more interested in the probability to find the end segment in a spherical shell of thickness dr at a distance r from the first segment in the origin, the probability function for that becomes: $W(r) = 4\pi r^2 P(r) = 4\pi r^2 (\beta/\sqrt{\pi})^3 \exp\{-(\beta r)^2\}$. This function reaches its maximum value at $r = 1/\beta$, which is obviously the most probable end-to-end distance r_{prob}. For comparison, the (linear) mean value of the end-to-end distance $<r> = {}_0\!\int^\infty r\,W(r)\,dr$ equals $<r> = \ell\sqrt{(8/3\pi)} \approx 0.92 \cdot \ell$, where ℓ is the root mean square end-to-end-distance $\ell = \sqrt{<r^2>}$, which follows from the mean square end-to-end distance: $<r^2> = {}_0\int^\infty r^2 W(r) \, dr = Na^2$.

Volkenstein [12] proves mathematically that the mean-square end-to-end distance (Na^2), on average, is the largest square distance between any pair of segments in the chain, and represents as such a good parameter for the characterisation of the "length" of the chain. The fact that $<r^2> = Na^2$ is a factor ($1/N$) smaller than the squared stretched chain length (Na)^2, indicates that for not too small a number of segments N, the chain is strongly coiled. We recall that $\ell = \sqrt{<r^2>}$ and stress the fact that this is not identical to the (linear) mean end-to-end distance $<r>$, see above.

Another useful size parameter, introduced by Volkenstein [12], is the "effective radius", defined as half of the mean maximal cross-sectional dimension (maximal "width", perpendicular to ℓ) of the coil and equal to: $r_{eff} = (1/2)\ell\sqrt{(\pi/6)}$, see also Sections 2.1.6 and 2.1.8.

2.1.3. *The Hookean Spring Model*. For the Gaussian chain the number of different molecular configurations which are consistent with the end-to-end separation ℓ, Ω, is simply proportional to P(r), so $\Omega(r) = AP(r)$. Now, because the entropy S, by virtue of the Second Law of Thermodynamics, equals $k \ln \Omega$, we can find the tension in the chain, *i.e.* the force between the end segments from $F_S = -T(\partial S/\partial r)_T$ (for the relationship between forces and chemical potentials, *cf.* Section 3.3), which yields: $F_S = 2kT\beta^2 \cdot r = H \cdot r$. Thus a freely-jointed chain acts like a Hookean spring with force constant H, and we could write for the resulting mean-square end-to-end distance: $<r^2>_{spring} = 3kT/H$. This picture of a chain represented by a spring is also used in the polymer kinetics literature, see *e.g.* Bird [16], or Williams [17].

The most probable end-to-end separation of this springlike chain can be envisaged to result from the tendency of the chain ends to equilibrate between two opposing actions: *i.e.* either to diffuse away from one another by the three-dimensional (thermal motion) random walk or to contract according to the springforce: $r_{eq} = r_{prob} = 1/\beta$.

2.1.4. *The Excluded Volume Effect*. However valuable, the intrinsic artificiality of this idealised model is very clear: an obvious difference between a diffusing particle and a polymeric chain is that the particle is free to cross its own path, while the chain is not capable of self-intersection: *i.e.* segments are not allowed to occupy the same space at the same time (so-called long-range interactions). Also, when two segments are in close contact, (short-range) interactions (attraction or repulsion) may occur.

For the description of these phenomena of volume exclusion of polymer chains, we will follow the early and comparatively simple approach of Flory [8, 10, 11] which yields very good results. Here the long-range or excluded volume interactions, which tend to disperse the chain segments over a larger volume than that for the so-called unperturbed chain, are described by the expansion factor α, by which the linear dimension of the configuration is altered, so: $<r^2> = \alpha^2 <r_0^2>$, where the subscript o refers to the unperturbed state (where, by definition $\alpha = 1$). This expansion due to excluded volume effects, makes itself apparent through experiments, *e.g.* viscosity measurements.

Flory & Fox assumed polymer coils to be nonpermeable bodies with a

402

so-called effective hydrodynamic radius r_h (*i.e.* the radius of a solid spherical particle with the same hydrodynamic behaviour as the polymer coil), corresponding to a hydrodynamic volume $v_h = (4\pi/3)<r_h^2>^{3/2}$, to find the intrinsic viscosity [also called the Limiting Viscosity Number, LVN, and from its definition (*cf.* the List of Symbols) representing a measure of compactness (*viz.* inverse concentration) of macromolecules] as $[\eta] = (10\pi N_A/3M)<r_h^2>^{3/2}$ (where N_A is Avogadro's number) or simply $[\eta] = \alpha^3 K_0 M^{1/2}$ where K_0 is constant.

This relationship was experimentally obtained long ago by Mark, Houwink and Staudinger in the general form $[\eta] = K \cdot M^a$, and is currently known as the Mark–Houwink–Sakurada (MHS) equation. The power, **a**, empirically ranges from 0.50–0.85 and abundant tabulations are available, *cf.* the Polymer Handbook [18], where **a** is generally found to be 0.6. On this we base the notion of the *Flory radius* $R_F = aN^{3/5}$, also used by de Gennes *et al.* in their successful Scaling Theory [7, 19–22], to represent a realistic polymer size parameter. For quantitative purposes it should of course be replaced by experimentally obtained (Mark–Houwink, MH) data.

The concept of Mark–Houwink plays an ever more important role in the universal calibration in GPC, which is based on the Flory–Fox notion of the hydrodynamic volume, and which was first introduced by Benoit *et al.* [23]. The basic assumption here is that SEC separation occurs on the basis of size differences only, such that all molecules with the same r_h or v_h, are eluted in equal retention volumes or times. If so, from the above, v_h can be expressed as being proportional to $[\eta]M$, which implies that a universal calibration of SEC data is possible by plotting v_h or $[\eta]M$ *vs* time. This is largely supported by many experiments [1, 3, 4, 5, 23, 24], provided that the chemical structures of the various polymers used are not too widely different.

The deviation of **a** from the value of 0.50 is a clear effect of the excluded volume, and is an indication not only of the expansion of polymer coils but also of the shape and the extent of permeability of the coil. Concerning the shape, **a** = 0.50 corresponds to a spherically symmetric body (such as the unperturbed polymer chain (θ-conditions) for which it has been derived), while at higher **a**-values, the shape becomes ever more wormlike and even rodlike (*e.g.* cellulose derivatives and other polymers with a very stiff backbone: **a** exhibits values up to 1).

2.1.5. *Permeability of Polymer Coils.* Concerning the flow permeability of real polymer chains, the above result of the experimentally confirmed Flory–Fox equation already indicates that the assumption of impermeability is obviously well chosen [5]. Ever since the 1950's, however, there has been a protracted discussion as to what extent polymer coils are hydrodynamically permeable.

Consider the limiting situation of a fully permeable Gaussian coil, which has been treated by Debye and by Debye & Bueche [25, 26]. Because of complete permeability, the flow field of the solvent is not disturbed by the presence of the chain segments. Debye allowed for a simple constant shear field (G = du/dx), *i.e.* a linear velocity profile: ($u = u_0 + Gx$), which causes the coil to rotate about the (hydrodynamic) centre. At low flow rates, forces that attempt to deform the coil are small and the integrity of the coil makes it rotate as a rigid solid body at a constant angular velocity $\omega = G/2$ in the stationary situation.

Calculation of the excess amount of dissipated power needed for this rotation, which is obviously connected to viscosity, shows that the intrinsic viscosity $[\eta]$ is related to molar mass M as: $[\eta] = K_D \cdot M$, i.e. a linear dependence on M. This, as shown by the general validity of the MHS relation, is not in accordance with most of the experiments (except for cellulose derivatives and other "stiff" polymers, as mentioned). Thus it appears that to all intents and purposes permeability of coils (so-called "free draining") can be ruled out, i.e. coils behave as "non-free draining", to use the cautious phrasing in Yamakawa's monograph [9]. Yamakawa presents evidence for the absence of draining effects, irrespective of the nature of the solvent.

Yet, prudence in this complicated matter is called for, as has been shown recently in a theoretical study by Mulderije & Jalink [27]. These authors prove that within the framework of the Kirkwood—Riseman (bead-string) model, the average fluid velocity relative to the stationary molecule and at the centre of mass is only 3.6% of the unperturbed velocity, so practical impermeability is shown. Even at a radial position in the coil beyond which half of the segments are present, the relative velocity is only 36% of the unperturbed fluid velocity. Although this proves that for practical purposes interior segments of long flexible chains are hydrodynamically shielded, it also shows that on going from the centre of mass to the outside the relative permeability is small but definitely increasing by at least one order of magnitude. This is in concordance with the viscometric observations. Yet for purposes of further discussion we rely on the near—impermeable character of flexible coils. Before leaving the problem of size and shape of polymer coils, we need to address several more aspects.

2.1.6. *Size Parameters for Chains*. The concept of the end—to—end distance, having the meaning of a chain length, is of course only meaningful for strictly linear chain molecules, which are hardly ever encountered. Moreover, it loses this meaning for certain configurations (Volkenstein [12]) and most importantly, the end—to—end distances cannot be measured directly by experimental techniques. Another measure of size for chain molecules that obviates both drawbacks mentioned is the radius of gyration, r_G, defined as the root of the square of the average distance of a mass element m_i from the centre of mass of the whole particle, i.e. $r_G = \sqrt{\langle r_G^2 \rangle}$, where $\langle r_G^2 \rangle = {}_i\Sigma(m_i r_i^2)/{}_i\Sigma\, m_i$.

In the case of Gaussian chains it is easily shown that the two size parameters are related by $\langle r_G^2 \rangle = \langle r^2 \rangle/6$. For impermeable coils, where the solvent inside the coil moves together with the macromolecule, the system (molecule + internal solvent) is one integral body from the hydrodynamic point of view. Thus, these bodies are similar to spheres and are to be characterised by a hydrodynamic radius r_H, i.a. through its translational friction by the Stokes Law: $f = 6\pi\eta r_H$. Values for r_H as measured from this relationship are in very good accordance with those obtained from the Kirkwood—Riseman necklace model for Gaussian coils, provided that $r_H \approx (2/3)r_G$ is used. On the other hand, from the famous Flory—Fox viscosity equation, it follows that $r_h \approx (7/8)r_G$ should be taken for the equivalent sphere radius.

The difference between the two values reminds us of the limitations of such an equivalent sphere model. The same applies to values of the equivalent sphere that can (obviously) be obtained from diffusion coefficient experiments, where $D = kT/f$, $i.e.$ $D = K_D M^{-(a_D)}$. Again in practice $a_D \approx 0.6$ ($viz.$ $0.55 - 0.60$) is found. For example, in very accurate light scattering experiments in the polystyrene/THF system, Schulz & Baumann [28] obtain $a_D = 0.588$. Other experiments with light scattering [29] and viscometry [23] yield for the same system $a_D = 0.564$ and 0.567 respectively, close, but not equal to the theoretical value 0.588, refs [6, 7].

So far we have dealt with the (hydrodynamic) behaviour of averaged polymer coils. The time-averaged coil has a spherical symmetry, while, of course, at fixed times individual coils are asymmetric. Volkenstein [12] finds that next to $<r^2> = Na^2$, representing the mean-square length of one chain, the mean square width of the chain is $(1/6)<r^2>$ (so equal to r_G^2), while the mean square thickness reads $(1/24)<r^2>$, in the case of a freely jointed chain, without excluded volume effects. So even for Gaussian coils, the instantaneous shape of one particular coil is not spherically symmetrical but rather bean-like. The time average, however, of the chain, which is fluctuating by thermal motion, is spherical.

In SSC the most important question is not so much what size the polymer coils as such have, but what their size is in the presence of an excluding wall. Volkenstein [12] and others ($cf.$ refs [9, 10, 13, 14]) show how statistical analysis of random coils near such an impermeable wall changes with respect to the unperturbed random coil statistics used so far. Especially Volkenstein stresses the influence of chain dynamics and the role of fast rotation of the coil. It then appears that the largest dimension of a coil near the surface, $i.e.$ the mean distance between the two segments that are farthest apart, is given as: $<H> = \ell\sqrt{(2\pi/3)} = r_G(2\sqrt{\pi}) = <r>(\pi/2)$ while the mean maximal projection $<X>$ of the chain in a given direction reads: $<X> = (2/\pi)H = \ell\sqrt{(8/3\pi)} = <r>$. Time averaging the distance $Q \perp H$ over all the different points of farthest removal from each other in all possible conformations yields the mean of this maximum cross-sectional dimension $<Q>$ as precisely half of the mean maximal distance $<H>$: $<Q> = <H>/2 = \ell\sqrt{(\pi/6)} = r_G\sqrt{\pi}$. Half of this mean maximal cross-sectional dimension was already introduced as the effective radius $r_{eff} = (1/2)\ell\sqrt{(\pi/6)}$ and while the radius of gyration equals $r_G = \ell\sqrt{(1/6)}$ we find that $r_{eff} = (\sqrt{\pi}/2)r_G$.

2.1.7. *Elongation in Shear Fields*. As indicated, the envelope of a polymer chain, so the shape, is not spherically symmetric but such that the width is about half the largest elongation. As a result, a good model of a polymer coil is represented by a prolate ellipsoid with an eccentricity (axial ratio) of 2. The hydrodynamics of such ellipsoids is a complicated matter ($cf.$ $e.g.$ ref [16]) and it has been shown that in a sheared flow these bodies tend to orient themselves, the more strongly with increasing flow rate and in a direction more or less parallel to the direction of flow.

For flexible bodies, the average deformation by the radial forces is not zero and the bodies tend to be stretched as well, $cf.$ $e.g.$ refs [30, 31]. As a result of orientation with increasing velocity gradient, intrinsic viscosities decrease to 50-80% of their initial values, and then level off (non-Newtonian viscosity behaviour). In Section 4.3.3 we will see that this elongational effect will be of no importance to SEC or GPC, as equilibrium is maintained only through the number density of coils outside pores, and not through deformation of shapes [32]. For other SSC techniques, like HDC, elongational effects are of influence, however.

2.1.8. *The Effective Coil Size near a Solid Wall*. The intriguing question, which of the above introduced size parameters is the governing one in SSC, has received a lot of attention in the early GPC literature.

Casassa's theory [33–37], (See Section 4.2.2), quite naturally leads to a relationship between partition coefficient and the radius of gyration of the excluded chains. This is a logical consequence of the mathematical procedure followed by him, which is based on a solution of the diffusion equation. To obtain insight into the actual size of the chain near the walls, Casassa calculated the critical thickness x_c of the layer adjacent to an infinitely flat wall (so the one-dimensional case) in which no chains are allowed, because of restricted conformational freedom, thus possessing a reduced concentration. At the same time a uniform chain concentration is placed at $x > x_c$. This is the same as having an effective exclusion layer near the wall in which no mass centres of polymer molecules are present, intuitively taken to be equal to half of the effective size of those polymers. Indeed, as pointed out by van Kreveld [38, 39], the critical exclusion layer thickness in the one-dimensional case, for random-flight chains, x_c after Casassa (1969), equals just half of the mean maximal projection onto a line ($\langle X \rangle$ as in Section 2.1.6 above), so $x_c = \langle X \rangle/2 = (2/\sqrt{\pi})r_G = (4/\pi)r_{eff}$.

In this context it is important to remark that Giddings *et al*. [40] also interpreted the exclusion layer as half the mean molecular projection (their so-called "mean external diameter"), which after all is trivial for spherical particles, but apparently not obvious for chain-like molecules.

As a general conclusion from this section, it can be stated that the pertinent polymeric size for SSC is to be regarded as the mean maximal projection $\langle X \rangle$, at least for the one-dimensional case. In a real three-dimensional situation, where a polymer molecule passes through numerous (10–100) consecutive spatial configurations in relevant (diffusion) times of about $5 \cdot 10^{-10}$ s, the mean maximal cross-sectional dimension $\langle Q \rangle$ rather than the projection $\langle X \rangle$ of the chain could be used, which was obtained in Section 2.1.6 as: $\langle Q \rangle = \ell/(\pi/6) = 2r_{eff} = r_G\sqrt{\pi}$.

Quite in contrast, the derivation of DiMarzio & Guttman (DG; *cf*. Section 4.3.2) leads to the result that a polymer chain is larger near a solid (flat) wall than in free solution: DG obtain the surprising result that the size near a wall is exactly twice as large as in the non-hindered one-dimensional situation. As the effective radius in the three-dimensional case they find on this basis: $a_{DG}^2 = (3/2\pi)\langle r^2 \rangle = (9/\pi)r_G^2$. Thus, the DG effective polymer size near a surface differs by almost a factor of two from our preferred measures: $a_{DG} = (6/\pi)r_{eff} = (3/2)x_c$.

As also pointed out by Silberberg [41], this case of single chain expansion is not expected to arise in practice. It would require that the contact with the surface produce no energetic effect whatsoever, while yet the chain would be attached at one side, and the other end being *held* at another.

Upon relaxing the confinement to non-attached chains, however, Monte Carlo calculations [42, 43] definitively prove that, provided that the chains are not pressed against the wall (*e.g.* by a second wall), the free space dimension is retained, as expected. The problem with the DG calculation is thus connected to the artificiality of their counting procedure which requires an attached chain. This must be wrong, also from the viewpoint of the Hookean spring model in Section 2.1.3. From that model, a chain that approaches an impenetrable surface and hits that wall by at least one segment, will be hindered in its diffusional action to expand. The associated number of configurations Ω_{trans} becomes smaller, and the entropic force will take over to *shrink*, rather than to expand the molecule to a new near—equilibrium condition to overcome the loss in entropy. All this being said, we leave it to experiments to indicate which effective size parameter is the correct one, *cf. seq.*

2.2. THE BASIC SEPARATION MECHANISM

Although the mechanism behind size exclusion based separations finds its origin in the interaction of "particle" size and channel size, characterised by the aspect ratio $\lambda = R_p/R_c$, the suggestion of "gel filtration" or "molecular sieving", as coined in the older literature, is far too simple. A better description can be given, as exemplified in the classical book by Yau, Kirkland and Bly [24], by interpreting the size exclusion mechanism as an equilibrium, entropy—controlled process. To this end a number of so—called equilibrium theories have been proposed (for reviews see [1] and [24], which agree well with experiments, also in the sense that temperature is of minor influence, as should be the case in entropy driven systems.

As the very concept of separation "...*flies in the face of the second law of thermodynamics.*", to cite another saying by Giddings [44], while in addition the chromatographic separation process is dynamic rather than static in nature, it is not surprising that mass transport which is governed by (relatively slow) diffusion, may become limiting in reaching the required equilibrium conditions. Indeed, there is a close interrelation between transport processes and partitioning, clearly visible in *e.g.* the analogous expressions for diffusion and partitioning in porous media, *cf.* Section 4.2.2. Also, in the case of HDC, the situation will be met, in which flow phenomena interfere with equilibration, and take over as the ruling differential migration principle (see Section 4.3.4 on Tubular Pinch). Also in the case of GPC it is still disputed whether possible flow dependences occur, see later (Section 4.3.3).

As such the best starting position for the discussion on the basic separation mechanism is the dynamic point of view, still allowing for obedience of the thermodynamic requirements. In the case of particulate solutes this approach has been advocated by Brenner *et al.* [45—47], while in the case of molecules the approach fits into the framework as described by the successful non—equilibrium theory, originally put forward by Giddings [44, 48], which stresses the close relationship between equilibrium and transport phenomena. Purely geometrical models, such as those proposed by Cheng [49] and Squire [50] *e.g.*, are far too simple except for those conditions where transport phenomena are always ensuring equilibrium. Approaches like these will never cast light on the intricate phenomena which appear when uncommon conditions are applied, *e.g.* reptation and radial migration.

2.3. THE SEPARATING MEDIUM

As stated above, ideally, size separation is based solely on the interaction of channel and particle sizes. However, according to Murphy's
Law: *"Life is not like that!"*.

Over the years many porous materials, including sulphonated polystyrene resins, starch gels, cross–linked gels such as dextrans and polystyrene gels with controlled pore sizes, acrylamide gels, agar, (rigid) porous glass and silica beads, and even (elastic) rubber have been used as
the separating media. These materials differ largely in "sieving" action
because of differences not only in internal structure (size and shape of
the pores and interstices), but also in polymer–sorbent interactions
(leading to adsorptive chromatographic retention contributions).

Because of this indeterminate nature of the structure of the active
"channels", it is hardly possible to generalise the treatment except for
simple systems such as (inert!) cylindrical channels/pores and more or
less regular (*e.g.* randomly overlapping spheres) internal structures/
networks. As it is difficult, if not impossible, to include the effect
of complicated geometrical shapes of real pores/channels, this modelling
of the separating medium puts a strong limitation on our ability to describe real life systems. Quite generally, the description of these
porous media is so complicated that instead of theoretical modelling it
is more attractive to either introduce a fictive random network representing the medium quite effectively (*cf.* *e.g.* Giddings *et al.* [40] and
Doi [51]) or characterise the media *vice versa* from SEC with known–sized
polymer molecules, see *e.g.* Gorbunov *et al.* [52] and Knox *et al.* [53].

Thus, we take the common stand that pure SEC will be modelled with
idealised pore shapes, such as cylinders and cones, which are at best
randomly distributed (for a comprehensive recent review of the effects
of pore shape on partitioning, *cf.* Deen [54]). As we are more interested
in the particles than in the pore matrices, this is no serious deficiency, especially when comparing the behaviour of two different particles
in a particular porous surrounding.

3. Size Separations in the Framework of Chromatography

Although differential migration methods in general are dynamic in nature
there is abundant evidence that they operate under near–equilibrium conditions [24, 44, 48]. The same applies to SEC as in its normal working
range elution volumes are independent of flowrate of the mobile phase
solvent [1, 3, 24, 37], which indicates that the concentration distributions, accompanying the moving sample zones, reach equilibrium rather
rapidly. Thus, as with all other forms of chromatography it is tempting
to try and describe the migration of samples in SSC systems with a dynamic equilibrium theory.

3.1. THERMODYNAMIC EQUILIBRIUM PARTITIONING.

Under normal (liquid) chromatographic conditions the sample is injected
into the mobile solvent phase and upon entering the separation column
which contains both the mobile solvent and a stationary retentive phase,
the molecules are distributed in a dynamic way, transferring rapidly
back and forth between both phases. The rapid transfer is a very impor-

tant requirement, which ensures that the residence times of all molecules of one kind within each phase are, on average, the same. This allows
the rapid attainment of equilibrium concentration distributions over
both phases and consequently the integrity of sample zones during their
transport.

Thermodynamic equilibrium of the sample zone concentration distribution is generally defined such that the chemical potential (μ) of the sample molecules is the same throughout their environment, i.e. dμ = 0. For
a two-phase (stationary/mobile) system: $\mu_s = \mu_m$, and so with the standard two-factor definition of $\mu = \mu^0 + kT \ln c$ (k = Boltzmann constant,
used here for later convenience; thus μ is defined on a molecular basis
rather than on a molar basis):

$$-\Delta\mu^0 = kT \ln K \tag{1}$$

where $\Delta\mu^0$ is the difference between the standard chemical potentials
(under conveniently chosen conditions) of the sample in the two phases,
while K is the partition coefficient of the sample substance, defined as
$K = c_s/c_m$. Now, while $\Delta\mu^0$ consists of enthalpic (molecular interactions)
and entropic (molecular conformations) parts: $\Delta\mu^0 = \Delta h^0 - T\Delta s^0$, it is
recognised that in normal LC separations of small molecules, entropic
changes are relatively small and so: $K_{LC} = \exp[-\Delta h^0/kT]$. Commonly, en-
thalpy-driven separations are exothermal, i.e. $\Delta h^0 < 0$ and often $K_{LC} > 1$
which leads to increased retention as compared to inert solutes (and
solvents for that matter). In the case of SSC separations, involving
large molecules with (ideally) no enthalpic interactions, obviously the
partition coefficient is entropy-controlled: $K_{SSC} = \exp[\Delta s^0/k]$.

Indeed, in the case of SEC in a porous medium, the permeation of solute into the pores of the packing leads to a decrease in allowed conformations (positions and orientations), i.e. a decrease in entropy and accordingly $K_{SEC} < 1$. While enthalpic interactions promote partitioning
into a stationary medium, entropic factors are clearly unfavourable to
the partitioning of solutes into a porous medium: K_{SEC} being less than 1
implies that concentrations are lower inside the pores than outside,
i.e. the solute is rejected from the pores. This is a well-known observation in, for instance, membrane technology as well and confirms the
near temperature independence of residence times (or volumes) in SEC.

3.1.1. *Conceptual Problems*. Thus, the close relationship between equilibrium driving forces and residence behaviour in SEC, as originally expressed by Giddings et al. [40] and Casassa [33–37] having been established, SEC is to be treated as a normal member of the chromatographic
family.

A conceptual problem, however, is again posed by the separating medium
as it is not immediately clear how the phase boundary between the mobile
(solvent) phase and the stationary phase should be defined. In the case
of SEC the stationary phase consists of the pore space within the porous
medium filled with (stagnant) solvent. There is some obvious ambiguity
in defining this intraparticle part of the solvent, V_i, as being the
volume of "stationary" phase. Especially when the porous medium is partially permeable to flow (see e.g. Section 4.1.1 and further), this will
lead to difficult interpretations.

Equally ambiguous is the definition of the interstitial void (interparticle space) volume V_0 as the mobile phase volume. Yet it is common practice to write: $V_e = V_0 + K_{SEC}V_i$ for the elution volume in SEC. This formula is used to define both V_0 and V_i from experimental elution volumes of very large (totally excluded, so $K = 0$) and very small (totally permeable, so $K = 1$) calibrating samples. This procedure avoids the difficult volumetric definition of the phase volumes, but suffers from experimental errors and interpretational difficulties, *cf. seq.*

3.2. SIMPLE SIZE EXCLUSION MODELS

As to the basic SEC mechanism of porous systems, expressions for illustrative cases of pore models can be easily derived, as has been shown in several good reviews [1, 3, 24, 44, 55]. The simple exclusion model of hard sphere bodies in a cylindrical pore space is of particular interest, because of its later use. In this case, the hard sphere centres are excluded from the impenetrable cylindric pore wall in an annular volume extending from the wall at radial position R_c towards $(R_c - R_p)$. This excluded volume leaves an effective or accessible volume for the spheres to occupy, which volume is less than the true pore volume. Consequently, the partition coefficient in this idealised case would read:

$$K = (c_{pore}/c_{bulk}) = [(\text{accessible/true) volume})] = (1 - \lambda)^2 \quad (2)$$

for $0 < \lambda \leq 1$ (as K is supposedly zero for $R_p > R_c$; Section 4.1. describes an exception to this supposition).

Although we do not treat those cases where electrostatic or other colloidal interactions with the pore walls take place, we note at this stage that eq.(2) in this case is modified such that the apparent aspect ratio is increased by a factor $\delta(U_\epsilon)$, which is a function of the electrostatic interaction energy U_ϵ: $\lambda_{app} = (\lambda + \delta)$ see *e.g.* Skowland [56].

As true pores are no cylinders it is better to replace the pore diameter by the effective mean pore size, which follows from the experimentally measurable and practical characterisation of porous materials (with any pore geometry) $D = 4/s_h$, where s_h is the so-called hydraulic radius (= surface area per unit volume of pore space). Then:
$K = [1 - (s_h/2)R_p]^2$ is a very useful approximation for more complex pore spaces than cylinders [40]. Eq.(2) is a special case of the general relation $K = (1 - \lambda)^n$, derived by Giddings *et al.* [40], for slit shaped ($n = 1$), cylindrical ($n = 2$) and spherical/conical ($n = 3$) pores.

The case of partitioning objects more complex than hard spheres (nonspherical rodlike or flexible chain molecules) is much more complicated as is obvious: orientations and conformations that interact with the excluding walls should then also be taken into account. This case will be introduced in the next chapters, with emphasis on flexible chain molecules, but our main discussion will be restricted to spherically symmetrical objects. Here we refer to the surprisingly simple results found by Giddings *et al.* [40], which were more recently extended (and corrected!) by Limbach *et al.* [57]. For the general case of rigid objects of any shape and size, where an isotropic pore space is modelled as randomly spaced and oriented intersecting planes the partition coefficient reads:

$$K = \exp[-(s_h/2)\mathcal{L}] \quad (3)$$

where the mean projection (or external) length l, takes up the role of object size. l is defined as the average length of projection of the body along an arbitrary axis, averaged over all possible orientations of the body in space. For a sphere, this equals the diameter $2R_p$, and the result is then: $K = \exp(-s_h R_p) = 1 - s_h R_p + (1/2)(s_h R_p)^2 - ..$, the same limiting result as in eq.(2), for small $s_h R_p$ values, i.e. small λ.

3.3. THE CENTRAL ROLE OF (CHEMICAL) POTENTIALS

The above reasoning identifies SSC related separations as being chromatographic in nature, with the recognition that the distribution of partitioning bodies is taking place within the same solvent (rather than over two separate liquids), but at different spatial positions. Once having proved that SSC separations have a common basis in the generalised framework of chromatography, it is perfectly valid to describe partitioning by the universal equilibrium relationship in eq. (1), which is alternatively written as: $K = \exp(-\Delta\mu^0/kT)$, the basic form of the above eqns.(2, 3). From these equations it is clear that for simple exclusion effects in small pores, the chemical potential differences are completely determined by entropic changes. We have also seen how dynamic aspects such as diffusion come into play, ensuring the rapid attainment of equilibrium.

3.3.1. *Transport and Potentials*.
At this point it is useful to point out that transport phenomena such as diffusion are also to be described as being driven by potentials, in the case of diffusion the one originating from concentration gradients. The application of other external fields, such as sedimentation and centrifugal forces in the Ultra Centrifuge and in secondary flow in curved chromatographic flow channels are also well--known external fields, not naturally occurring but applied on purpose.

Less common but growing in importance are external field applications based on e.g. electrical and magnetic forces, temperature or concentration/composition gradients, and even radiation (for a recent review see [44]). All of these external fields are used to produce selective changes in equilibrium distributions, the basic requirement for separation to occur. This implies that they indeed influence the potential energy of a particle or molecule, which is identical to stating that they bring about an additive contribution to the overall chemical potential: $\mu^* = \mu + \mu^{ext}$, cf. e.g. Guggenheim [58]. Here μ represents the classical chemical potential $\mu = \mu^0 + kT \ln c$, which was used in the above to represent the intermolecular interactions (including entropy and attractive forces such as dispersion, orientation, induction and hydrogen bonding). The external field applied adds a special form of (potential) energy, μ^{ext}, to the system, which mathematically speaking can be treated as being identical to the internal μ function.

Consequently the equilibrium partition relation for K should read:

$$K = \exp(-\Delta\mu^*/kT) \tag{4}$$

which replaces eq.(1) and where $\Delta\mu^* = \Delta\mu^0 + \Delta\mu^{ext}$. In the latter relation the internal potential term may change discontinuously (e.g. at a phase boundary), while the external term changes continuously. The accompanying gradients represent a force-like action (just as in the case of a mechanical (conservative) driving force: $F = -dU/dx$, along the x-axis, the classical expression of forces in potential theory). For example, the force-like action of the entropic contribution to μ is caused by the molecular collisions from thermal motion which result in the diffusional Brownian motion.

3.3.2. *Mass Balance*. The use of these *virtual forces* based on a *pseudo-potential* dates back to the treatment of Brownian motion by Einstein [59]. Molecules and colloidal bodies in general are subject to the same forces that are used to describe the motion of macroscopic bodies by Newton's laws. Viscous drag forces are well-known examples. These forces are at low velocities proportional to velocity, *i.e.* $F = f \cdot v$ (where in the one-dimensional case v = linear velocity = dx/dt and f is the friction factor).

An example of such a factor is the Stokes factor $6\pi\eta R_p$ for the friction of spherical solid particles. It was Einstein's proposition to express the frictional force $F = f \cdot v$ alternatively as the gradient of the (chemical) potential μ, where $\mu = \mu^0 + kT \ln c$ (for diffusion alone no external field is present). This leads to the force $F = -d\mu/dx = f \cdot v$ from which the mass flux $J = c \cdot v$ for the diffusional process follows as:

$$J_D = - (kT/f)(dc/dx) = -D(dc/dx) \qquad (5)$$

The latter expression is the well-known Fickian First Law of diffusion, with the important result that the diffusion coefficient is related to the friction factor as $D = kT/f$, the Nernst-Einstein equation.

In the later discussion of particles in relatively wide capillary flow channels, the lateral or radial dimension becomes also of interest, for which purpose the gradient d/dx in eq. (5) should be replaced by the three dimensional operator nabla: ∇ (see the List of Symbols).

Eq.(5) is of special interest when external fields are applied which contribute an additional virtual force $F^{ext} = -\nabla\mu^{ext} = (f \cdot v)^{ext}$, thus modifying eq.(5) into:

$$J^* = J_D + J^{ext} = -D[\nabla c + (c/kT)\nabla\mu^{ext}] \qquad (6)$$

With this formulation of particle mass flux, the basic goal of separations, *i.e.* selective transport and redistribution of component concentration pulses in a separating medium can be described by the general equation of continuity (*cf.* the Appendix): $\partial c/\partial t = -\nabla J$, which serves as the mass balance equation in many fields. The mass flux density J is then the total flux including diffusion, convection and external transport: $J = c \cdot u_p + J^*$, so that from continuity in a circular cylindrical tube we have:

$$\partial c/\partial t = -\nabla J = -\nabla(cu_p) - \nabla J^* =$$

$$= -u_p(\partial c/\partial z) + D_z(\partial^2 c/\partial z^2) +$$

$$+ (1/r)(\partial/\partial r)[r\{D_r(\partial c/\partial r) + (cD_r/kT)(\partial\mu^{ext}/\partial r)\}] \qquad (7)$$

where $u_p = u_p(r)$ is the particle velocity field in the longitudinal (z) direction. This equation can be used to calculate the concentration

profile of individual particle zones during differential migration
through a separating medium, provided that appropriate initial and boundary conditions can be formulated as well as expressions for the particle velocity and the external potential(s) can be found. Eq.(7) is an
extension of the well-known Taylor–Aris–Golay convective dispersion
equation [see e.g. refs 44, 46, 48], allowing for different axial (in z-
direction) and radial dispersion (in r-direction) with transport coefficients D_z and D_r as well as for external potential fields.

The first additional feature concerning D_r is of practical importance
e.g. in radial induced flows, such as in curved flow channels, but also
in packed beds and in cross–flow (FFF) situations. The second additional
feature allows for inclusion of applied potentials or force fields, as
used in many FFF techniques, but also in tube flow with inertial terms
non–negligible, with colloidal wall interactions and with particle shear
deformation, as will be discussed. Eq.(7) proves how (pseudo–)chemical
potentials rule the dynamics of chromatographic separations.

Because they also govern the equilibrium properties, as discussed above, these potential forces are recognised to play the central role in
separative processes, recently formulated explicitly by Giddings [44].
Modifications of the above equations have been published (see e.g. refs
[44, 46, 60–72]). At present we will make use of this mass balance equation in order to calculate particle migration through cylindrical pores,
the case which is highly illustrative for all SSC separations.

4. SSC in Divers Channels

Now that the governing mass balance equation has been formulated, we are
in a position to solve the one for the required concentration profiles
in the various separating channels of interest, starting with a single
very narrow pore where $\lambda \gg 1$, followed by a larger pore where $\lambda \approx 1$ (in
both cases no convection is allowed), and subsequently followed by still
wider pores (channels) in which some convection is allowed. Finally, we
turn to the case of a channel in which prominent flow phenomena occur.

4.1. THE VERY NARROW PORE.

In this elementary case of a pore with a diameter less than the characteristic diameter of the partitioning bodies, it is obvious that the exclusion effect is quite dramatic. Hard spherical particles are expelled
entirely, but fibres could eventually be trapped, provided that they
have the correct orientation. Likewise, chain–like molecules could be
trapped, provided that they can be made to lose a number of possible
conformations (and so entropy) with respect to the free solution.

4.1.1. *Reptation of Confined Polymer Chains*. Especially in de Gennes'
work [7, 19–22], it is shown that, in contrast to the classical Debye–
Bueche approximation [8, 25, 26] (where the chain is modelled as a spon-
ge–like porous object), statistical fluctuations in the local monomer
concentration lead to relatively easy flow of solvent (and thus easy
flow of the chain as a whole) through a confining pore. Of course, in a
capillary of diameter $2R_c < 2R_p$ (where polymer size $2R_p$ is being defined
in terms of $\langle X \rangle$ or $\langle Q \rangle$, see Section 2.1.6) the chain–like polymer molecule in a confining capillary is squeezed into an extended cigar–like

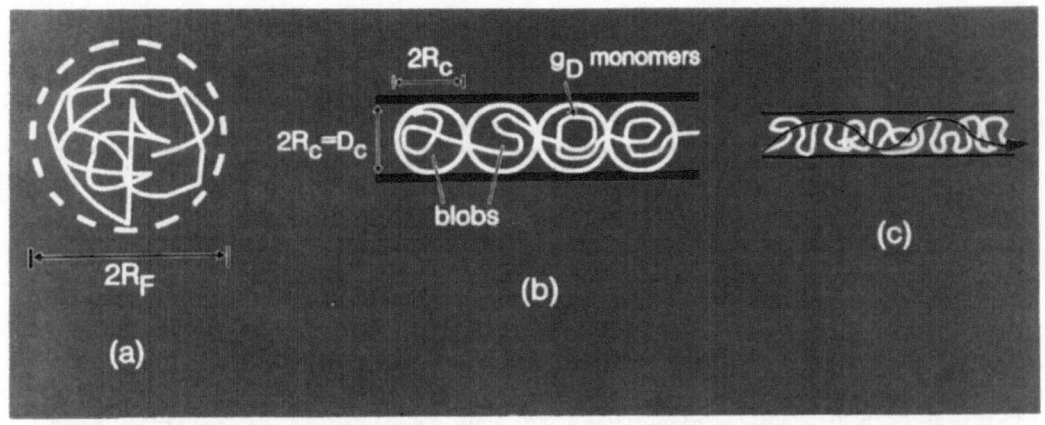

Figure 1. Schematic picture of the size and conformation of a chain
molecule. (a) in free solution, (b) confined cigar-like in a
pore with blob formation, each behaving after Flory, and (c)
confined with complete segregation (solvent flow possible).

configuration, with a length $\mathcal{L}_c > 2R_c$. It is tempting to picture the
trapped and wriggling chain as a train of contacting segments or "blobs"
forming a pearl necklace. Figure 1b illustrates this. Successive blobs
act as hard spheres, which on the scale of the pore size are each allowed to behave as in the unrestricted free solution, i.e. the blobs show a
knotted and partially swollen structure as described qualitatively but
rather well by the Flory radius of gyration $R_F = aN^{3/5}$ (cf. Section
2.1.4), shown in free solution for comparison in Figure 1a.

Thus the chain is represented by a succession of N_b blobs of g segments each and size $D_b = 2R_c = 2R_p$. Particle (= blob) size equals the
channel size as the chain is being squeezed into the pore confinement.
In view of the supposed analogy with the Flory case, we have that each
blob shows the same relationship between size and number of segments as
in free solution. Hence, D_b takes over the role from R_F (omitting numerical factors like 2, as is custom in de Gennes' Scaling Theory), and g
that from N. So: $g = (D_b/a)^{5/3}$ in analogy with $N = (R_F/a)^{5/3}$ and as $D_b =
2R_p = ag^{3/5}$, the total chain length of the cigar is $\mathcal{L}_c = (N/g)D_b =
aN(a/D_c)^{2/3}$.

The associated free energy as measured from the free solution state
($D_b = 2R_p = 2R_F$) present outside the pore is then: $\Delta\mu_N = (N/g)kT$ [20-22]
which with eq.(4) yields the partition ratio for the separation process
potentially to be based on this mechanism: $K = \exp(-N(a/D_c)^{5/3}) =
\exp(-\mathcal{L}_c/D_c)$, which is again eq.(3). In this particular case of strongly
elongated bodies, however, the absolute value of the argument $\mathcal{L}_c/D_c$ is
larger than that for the more common case of more or less spherical bodies. This implies that partitioning of cigar-like bodies in narrow pores
has a smaller K-value than the partition ratio of near-spherical bodies
in the same pores, which is not unfavourable from the point of view of
separations.

In ref.[22] even smaller pore sizes are treated where $D_c < a$ and the

segregation of the chain becomes complete. The stretched chain of N seg-
ments each possessing a volume a^3, now fills a length $\mathcal{L}_c - Na3/D_c^2 \approx Na$.
The confinement free energy change for complete segragation of N seg-
ments is NkT [22], leading to $K - \exp(-N) - \exp(-\mathcal{L}_c D_c^2/a^3)$. It is questionable, however, that the eventual separation process based on this
chain trapping is sufficiently fast and efficient, in view of the potentially limited speed at which the chain is propagated through the pore.
To investigate this, one should study the dynamics of such trapped
chains. The actual mobility of the chain through the tube can be estimated and expressed in terms of a diffusional transport number from the
Nernst–Einstein relation as: $D_{rept} - kT/f_{rept}$. As in free solution
$D_m - kT/f_0$ is valid for the molecular diffusion coefficient, see Section
3.3.2, we find that $D_{rept}/D_m - f_0/f_{rept}$, where after Stokes the friction
factors are given by $6\pi\eta R_{part}$. Because particle sizes in free solution
are much larger than those in confined pores, this indicates that potentially $D_{rept} > D_m$, which gives hope for the requirement of relative fast
transport in the separation process to be designed.

Qualitatively we can estimate the reptational transport as follows. If
the polymer chain is represented by N spherical segments of size a, the
friction factor in free solution reads: $6\pi\eta aN$. Following the same reasoning in the pore we encounter (N/g) spherical blobs of size D_c, so the
friction factor reads: $6\pi\eta(N/g)D_c - 6\pi\eta\mathcal{L}_c - 6\pi\eta Na(a/D_c)^{2/3}$. As a result:
$D_{rept}/D_m - (D_c/a)^{2/3}$ and so $D_{rept} > D_m$, i.e. larger than in free solution. Even in the case of stretched segregated chains where $f - f_0 \approx 6\pi\eta aN$
the moving chain has the same diffusion coefficient as in free solution.

This relatively fast transport process has been termed *reptation*, because in concept it arises from the snake–like wriggling motion of the
trapped chain, which is flexible and acts with elastic energies as in a
spring. Alternatively, it may be visualised as the periodic compression
and dilation of the blobs. In actual fact it can be proven that each
coil inside a blob drags its own volume of solvent with it upon moving
[7, 22].

From this paragraph we conclude that reptational transport in very
narrow pores may proceed with appreciable velocity, in accordance with
experimental results [73–75] (ref.[73] reports an effective diffusion
coefficient that is two orders of magnitude larger than in dilute solution!). Thus the separation process to be based on this (with the above
K–value), may be well feasible and efficient in practice.

4.1.2. *Potentials of Reptational SSC*. Problems, however, arise in the
actual realisation of this obviously feasible SSC technique. For this we
have two options, *viz.*

(a) the single cylindrical channel, and
(b) the porous structure of a microporous medium packed into a column.

Re (a): *the single capillary*. Here we should think of capillary sizes
such that the size of macromolecules ($R_F - aN^{3/5}$, with $a \approx 5–10$ Å) is
larger than the capillary diameter. Thus for a molar mass of about 10^6,
we have $N \approx 10^4$ and $R_F \approx 1250$ Å, i.e. capillaries of less than 0.1 μm
(internal diameter) are required. So far, these are not available in
practice, where the smallest reported capillary has a diameter of 1.2 μm
[76]. It will be shown later, see Section 4.3.4, that capillaries of sizes larger than that are perfect media for size separations (so-called

HDC), but the present reptation mechanism cannot be used then.

For the latter mechanism, typical capillaries of say $D_c < 200$ Å should be made available: then, for $N \approx 10^4$, the number of segments in the reptating blobs ($g \approx 500$) is large and Flory's statistics would be valid. Even if that could be realised one day, we would have the conceptual problem of how to "inject" chains essentially larger than the opening of the capillary entrance into the channel. The only way this can be visualised is by flow, because spontaneous diffusion will fail.

In this respect the preliminary analysis by Brochard [20] is illustrative. Brochard treats the aspiration of a chain into a (very slightly) conical pore, where solvent enters the (slightly) wider pore mouth with a flow or flux $J_s \approx uD_c^2$ (again omitting for simplicity numerical constants such as $\pi/4$). By making a balance of forces (driving force from Stokes vs elastic force due to the free energy upon confinement), the required pressure drop for positive movement of the chain into the pore opening and along the pore length is found. For too small pressure differences, the chain will simply be stopped at position $z = R_F/\psi$ because of steric hindrance of the wall (ψ is the slightly tapering angle D_c/z). For sufficiently large pressure drops, however, the chain will be confined because the confinement free energy $\Delta\mu = T(\partial\Delta s/\partial z) = T \cdot f(D_c/R_F)$ [20] is available, and the chain progresses into the capillary. The required critical pressure to obtain that situation is rather low. For example, for typical D_c values of say 100 Å, the required pressure drop is only 0.1 bar. It is thus realistic to envisage this process as a potential candidate for sampling these tiny pores with macromolecular chains. An interesting feature of the result for the critical conditions is that these conditions are independent of both the size of the chain (N) and the diameter D_c of the pore, i.e. all molecules are transported equally into the pore medium (neglecting chain end effects).

Also of fundamental interest is the case of strictly circular cylindrical pores [21]. Although flexible macromolecules with $R_F > D_c$ will not spontaneously penetrate into the cylindrical pore mouth, with large enough flows the chains are subject to an elongational converging flow and will be sufficiently stretched to enter the pore entrance after all. The chains can be envisaged to show the same behaviour as a fluid element (droplet) in a shearing flow field (cf. e.g. Bird et al. [16]). Upon approach of the pore opening, driven by the elongational flow, at distances $< R_F$, the chain will be stretched to fit the size D_c provided that another (larger) critical value for the flow through the pore is reached, the value again being independent of both molecular size and the pore diameter, i.e. all molecules are trapped alike.

Re (b): microporous media. Here it is easier to obtain small pore sizes of controlled diameter, statistically the pores being on the average conically shaped, but sometimes almost circular cylindrical [73, 74, 77--80]. It is not always clear, however, whether and if the pores are permeable to flow. Yet, as already predicted by the Doi [51] and Casassa [33-37] theories, in the common GPC (SEC) procedure with available porous materials with a range of pore sizes, this process of confined entrapment of chains may be observed. In actual fact this phenomenon extends the selective working area of these techniques by offering separation power for molecules larger than the smallest pore sizes present. At the same time it obscures the precise determination of this working area and the so-called "exclusion limit", *i.e.* the limiting characteristic value where the largest particles are eluted, supposedly all together. It seems from the above, that this supposition should be re-examined.

4.2. THE NARROW PORE

The description of selective partitioning and transport of bodies inside and outside the pore with dimensions larger than the bodies, is less complicated than that above, because especially macromolecular conformations are less restricted. A complication, however, arises as a result of possible flow phenomena in or near these relatively wide pores/channels. Again we can make a distinction between molecular and rigid hard bodies, and we will start with pores only slightly larger than the bodies. This is the typical condition in packed bed GPC/SEC with porous separation media.

4.2.1. Polymer Molecules in Porous Media. For pores with sizes not exceeding those of the partitioning molecules too much, a vast amount of experience has been obtained over the last decades in the SSC methods GPC and SEC (*cf.* refs [1, 3, 4, 24] for critical reviews). Ever since its conception, the SEC separation mechanism has been debated, which is not surprising considering how difficult it is to define both molecular sizes and the structure of the porous media used.

In addition the hydrodynamics in such systems is a very complicating factor: flow is not parallel but typically elongational and flexible polymer coils tend to deform [31]. Neglecting this for the moment, the next section treats the partitioning of random coil chains in pores along the lines of the successful random-flight Casassa theory.

4.2.2. Partitioning of Random-Flight Polymers in (Cylindrical) Pores. Casassa [33-37] and later Doi [51] analysed the partitioning of random-coil molecules by the earlier mentioned analogy between chain conformations and diffusion, *cf.* Sections 2.1.1, 2.1.2 and the Appendix.

As known, the diffusion equation in the non-convective case and in the absence of external fields describes the concentration of point-like particles at a particular location after a given number of random walks. Eq.(7) describes this as: $\partial c/\partial t = \nabla(D\nabla c)$ and for a constant diffusion coefficient: $\partial c/\partial t = D\nabla^2 c$, the well-known Fickian Second Law.

Because each conformation of a freely jointed chain is equivalent to the path trace in such a diffusion-like random walk process, where segments are represented by the random walk steps of length a, the description of the mass balance can be done equally in terms of concentration (c) or in the probability of finding a segment at the same time and spatial position. Thus, solutions of the diffusion problem for appropriate initial and boundary conditions can also be used to yield the probability (P) distribution. Here probability $P(r,j)$ can be defined as the number of ways the j-th segment of a random flight chain can be found at position r. The diffusion equation above, written in terms of probabilities, then reads: $\partial P(r,j)/\partial j = (a^2/6)\nabla^2 P(r,j)$, where the diffusion coefficient D is replaced by the mean square displacement $(a^2/6)$ [9, 51] for each random walk step of length a, see the Appendix. The formal derivation of this equation can be found in refs [7] and [9], while the most complete extension, including external fields, has recently been given by Davidson & Deen [81].

The partition coefficient can now be evaluated from the general eq.(4) which can also be written as: $K = \int \exp\{-\mu(q)/kT\}dq \,/\, \int dq$, where the potential energy $\mu(q)$ is clearly allowed to be a spatial function of the coordinates q (x,y,z,t). For one- to three-dimensional pores, characterised by n = 1 (slit), n = 2 (cylinder) and n = 3 (spherical cavity), the radial coordinate $\rho = r/R$ is the only one of importance and this expression reads: $K = (1/V) \int_V \exp\{-\mu(q)/kT\}dq$. By assuming a hard-core steric exclusion (∞) potential function for the region $1 \geq \rho \geq (1 - \lambda)$ (Heaviside step function), the integral result in this region becomes zero and we are left with the integration range 0 to $(1 - \lambda)$: $K = n \int_0^{(1-\lambda)} \exp\{-\mu(\rho)/kT\}\rho^{(n-1)}d\rho$. The result reported by Giddings et al. [40] for these cases, viz. $K = (1 - \lambda)^n$, is readily obtained from this as should be.

For flexible chain molecules, replacing concentrations by the probabilities P(ρ,j) as defined above, the partition coefficient was found by Casassa from: $K = \{_0\int^1 P(\rho,1)\rho \; d\rho\}/(_0\int^1 \rho \; d\rho)$. P($\rho$,1) measures the number of ways that the last segment (N) can be found at position ρ, given that the first segment (j = 1 at the origin) is free to sample all positions in the pore. This is to be evaluated [33-37, 81-83], while necessary boundary conditions to exclude chain configurations which would penetrate the pore walls, together with suitable initial conditions were used to evaluate the probabilities as solution to the above "diffusion" equation. Following Casassa, the total probability of all chain configurations starting in the plane (x_0,y,z), and excluded from x = 0, is then the integral of P: $P_{tot} = erf(\beta x_0)$. As $(1 - P_{tot})$ is the fraction of conformations not allowed by the presence of the excluding plane, the critical thickness below which depletion of polymer chains takes place is then (see Section 2.1.8): $x_c = {_0}\int^\infty (1 - P_{tot})dx_0 = 1/\beta\sqrt\pi = (2/\sqrt\pi)r_G = (1/2)\langle X \rangle$. K was found along the same lines and analogously to the well-established heat conduction/mass transfer solutions listed in e.g. Carslaw & Jaeger [84] (summation Σ ranges from m = 1 to ∞): $K = \Sigma \; (2n/\beta_m^2(n))\exp\{-[\beta_m(n)\lambda_G]^2\}$, where $\lambda_G = r_G/R_{pore}$, is the aspect ratio based on the radius of gyration of the macromolecule, which appears here naturally. In the cylindrical case (n = 2), β_m is the m-th root of the Bessel function of the first kind and order zero: $J_0(\beta) = 0$, which values are tabulated, cf. e.g. Abramowitz et al. [85].

By plotting K vs λ_G, two limiting cases are visible:

(i) the first limiting case for wide pores (of any dimension), say $\lambda_G < 0.5$: here the K-value reads: $K \approx 1 - (2n/\sqrt\pi)\lambda_G$, i.e. with the hydraulic radius s_h being $s_h = n/R_{pore}$, see Section 3.2, $K \approx 1 - s_h(2/\sqrt\pi)r_G \approx 1 - s_h R_p$, as expected from the Giddings equations [40]. Obviously, if we take for R_p: $(2/\sqrt\pi)r_G = x_c$ as the effective polymer radius, the Casassa theory predicts the same result as the Giddings theory, and is universal (for all types of pores, and also for all types of polymers including star and branched shapes) provided that $\lambda_G < 0.5$.

(ii) the second limiting case arises for $\lambda_G \to 1$ and even > 1: K approaches the value of zero exponentially: $K = (2n/\beta_1^2(n))\exp\{-(\beta_1(n)\lambda_G)^2\} \approx \exp[-(4/\sqrt\pi)\lambda_G]$, emphasising the strength of the theory developed by Casassa, who was the first to predict partitioning also in the range of very small pores with oversized chain molecules ($\lambda_G \geq 1$). The use of these universal equations has been advocated throughout GPC literature, up to these days, see *e.g.* Gorbunov [52]. In that work the inverse relationship $R_{pore,h} - n/s_h$ is used to characterise the equivalent pore radii from SEC experiments with known polymer sizes.

It should be noted that the above universal calibration relationships, have never been accurately confirmed by experiments. Often, deviations from the simple linear K-vs-λ_G prediction are quite strong, say by a factor of about two, see *e.g.* Fig.2.9 in ref.[24], Fig.2 in ref.[36] and Fig.4 in ref.[79]. Nonetheless, on account of the large body of experiments, which show qualitative agreement with Casassa's theory, this be regarded as one of the best available theories for SEC separations in porous media. For quantitative tests, independent methods should be used to measure both macromolecule and pore sizes, which is rather difficult. We will encounter such a situation in the following, when discussing the results which van Kreveld *et al.* [38, 39] and Doi [51] obtained in a well-characterised porous medium.

A further problem to be solved in the Casassa theory concerns the role of the aspect ratio λ, including the influence of the porous medium. While the analysis is valid for a chain with an arbitrary distribution of segment lengths, the results are valid only for an infinite number of (vanishingly small) chain segments. Aubert & Tirrell [86] introduced a linear elastic dumbbell model of the polymer rather than a random chain and found partition coefficients in divers pore geometries to differ appreciably from the Casassa case. For instance, for a cylindrical pore and $\lambda_G - 1$, they obtained a K-value of about 0.2, which is some two orders of magnitude larger than the Casassa-based value, being 0.002. Here the Giddings [40] result for the rigid rod in the same tight fitting pore is much better. The same applies to the result by Priest [87], who considered a confined freely jointed chain in cylindrical pores with $\lambda_s - a/R_{pore} \geq 2$, where λ_s is the aspect ratio based on the statistical segment, rather than on the chain as a whole. At *e.g.* $\lambda_G - 1.5$, Priest predicts a K-value (for three segments) that is larger by four orders of magnitude than the (∞-segment) case of Casassa. This illustrates the importance of the aspect ratio λ_s, which in practical situations ranges from say $5 \cdot 10^{-5}$ to 0.4.

The only systematic study in this practically important range, including finite segment numbers has been carried out recently by Davidson *et al.* [83], using a Monte Carlo numerical technique, and the above outlined probability (P) determination. The results demonstrate that K-values of finite segment chains (say up to even 400 segments) can be substantially higher than the K-values of the ∞-segment chains with the same radius of gyration, and the more so for chain sizes larger than the pore radius.

This observation shows that the all important λ_s-influence already starts to manifest itself at a value below 0.1: *e.g.* for $\lambda_s - 0.09$, the K-value for $\lambda_G - 1$ exceeds the Casassa-value by 40%! Thus, in practical systems using small pore sizes and not extremely large segment numbers, K-values (*i.e.* the extent of pore penetration) are grossly underestimated; this typically applies to relatively inflexible polymers, where a segment contains many monomer units. One may also conclude that the SSC technique termed Reptational Chromatography, see Section 4.1.2, is of

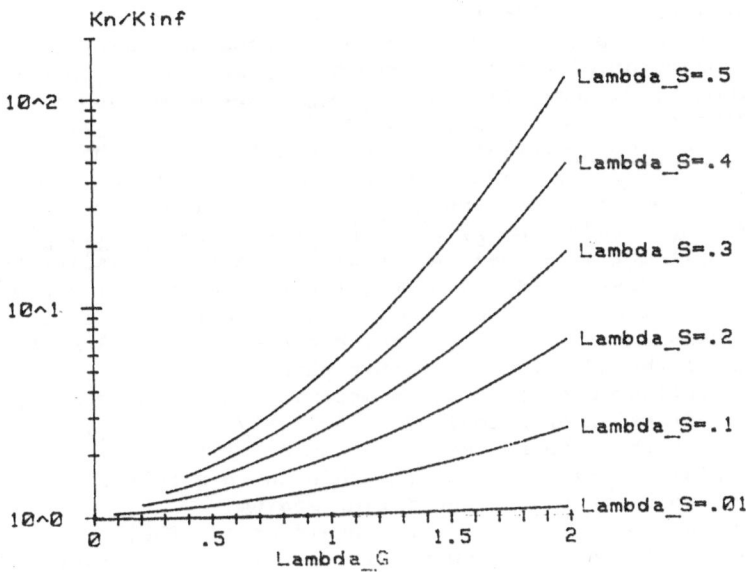

Figure 2. Deviations from Casassa's theory (K(∞) or Kinf) for finite
chain numbers (N) with λ_s (Lambda_S) the aspect ratio based on
segment size as the parameter, drawn after eq.(6) in ref [83]
(remark: λ_s is always less than λ_G, which is why the lines do
not reach the origin)

fundamental practical interest, because it enables the separation of
structurally different polymers of the same radius of gyration.

For $\lambda_s \leq 0.15$, Casassa's predictions are well confirmed, while at constant λ_G it appeared that ln K varies linearly with λ_s. From this the
authors produce an approximate fit relation to their calculated predict-
ions for K: ln $[K(N)/K(\infty)] = \lambda_s(0.49 + 1.09\lambda_G + 1.79\lambda_G^2)$. Here K(∞) is
the Casassa-based K-value, for cylindrical pores equal to:
$K(\infty) = \exp\{-(4/\sqrt{\pi})\lambda_G\} = \exp(-2.26\lambda_G)$. Figure 2 is the resulting graphical representation of this function for K(N). The approximation is re-
ported to be quite good, being < 6% for $\lambda_G \leq 1$ and 25% for $\lambda_G > 1$. Thus,
Casassa's theory is much improved by the above semi-empirical equation,
or by the analytical expressions recently derived for a large number of
segments by the same authors [81]. Fig. 2 shows distinctly that predicted K-values are larger than in the Casassa theory, and especially the
limiting cases are remarkable:

(i) for λ_G going to zero, it shows that partition ratios larger than 1
are possible, particularly for relatively large segments or narrow
pores. So actual "retardation" takes place, as is common in other forms
of chromatography, and this effect is of great importance for the determination of the "total permeation limit" in the practice of GPC.

(ii) the same applies to the other limiting case $\lambda_G > 0.5$ and even
> 1. Here, too, appreciable deviations from "total exclusion" behaviour
are observed, again noticeable for large segments or small pores, and
K-values up and above 1 are seen to be predicted. Again, practical GPC
is largely disturbed by this effect, as it obscures the second charac-

teristic "total exclusion limit".

The extension that Doi [51] gave to the Casassa treatment is essentially the introduction of a polydisperse porous medium, which was pictured as a volume V_p randomly filled with n_{RS} impermeable spherical elements of radius a_{RS}, hence at a concentration of $c_{RS} = n_{RS}/V_p$, while for this so-called Randomly overlapping Spheres Model, RSM, the hydraulic radius $s_h = 4\pi c_{RS} a^2_{RS}$. By solving the analogous "diffusion equation" as used by Casassa, Doi obtained the extended expression:
$K = \exp\{-(4/\sqrt{\pi})\lambda_G - (1/\pi c_{RS} a^3_{RS})\lambda_G^2\} = K_{Cas} \exp\{-(1/\pi c_{RS} a^3_{RS})\lambda_G^2\}$ (written here for cylindrical pores), which always leads to smaller K-values than Casassa's expression (K_{Cas}) alone, and so does not give a better fit of experimental observations. (Doi [51] presents his Fig.4, which gives the impression that $K > K_{Cas}$, which cannot be true.) Anyhow, this effect is due to the additional variable parameter $c_{RS} a^3_{RS}$, which characterises the influence of the porous medium, and which is simply related to the void fraction or porosity $\epsilon = \exp\{-(4/3)\pi c_{RS} a^3_{RS}\} = \exp(-s_h a_{RS}/3)$, so that the Doi equation can be written as:
$K = K_{Cas} \exp\{4/(3\ln\epsilon)\lambda_G^2\}$. This relationship shows a relatively large influence of the porosity of the separating medium.

Two years before Doi, in 1973, van Kreveld *et al.* [38] employed the same RSM model to the description of silica gel beads (Porasils C and D) in use as SEC medium, inspired by SEM-microscopic pictures of that gel. These authors showed that in this hard-gel case the RSM description of the porous medium leads to a quantitative prediction of the SEC residence time behaviour of polystyrenes in THF, whose sizes were supposed to be characterised by the effective radius discussed in Section 2.1.6. As the free void fraction in the RSM is given by $\epsilon = \exp\{-(4/3)\pi c_{RS} a^3_{RS}\}$ the accessible pore volume can be expressed as:
$\epsilon_{acc} = \exp\{-(4/3)\pi c_{RS}(a_{RS} + r_{eff})^3\}$, where it is assumed that each microsphere surface in the RSM is surrounded by an excluded layer of thickness r_{eff} such that polymer coils cannot approach the surface closer than that layer. From both volume fractions the partition coefficient K based on this purely steric exclusion effect reads: $K = \epsilon_{acc}/\epsilon = \exp\{-(4/3)\pi c_{RS}[(a_{RS} + r_{eff})^3 - a^3_{RS}]\} = \epsilon \cdot \exp\{[1 + (r_{eff}/a_{RS})]^3 - 1\}$
which is not easily simplified because size of the microspheres a_{RS} and the effective radii of the polymers are of the same order (*e.g.* for Porasil C, a_{RS} was determined as 160 Å, while for Porasil D 323 Å was found; the test PS polymers had effective radii ($r_{eff} = 0.886 r_G$) ranging from 8.1 to 584.9 Å for the M-range $1250 - 1.80 \cdot 10^6$).

The close match between experimental partitioning data and predicted K-values based on this RSM model is a strong case for the use of the effective radius as the decisive size parameter, although it may be fortuitous because of apparent shortcomings of the RSM model (see the discussion on Doi's result, above). The RSM model was later improved by Knox *et al.* [53], Squire [50] and Cheng [49]. Whatever the situation, this correction to the classical Casassa equation by introducing an RSM-type of model is quite important as it proves to have a large influence. The process of the partitioning of polymer coils in not too wide pores, which is the basic requirement for selective partitioning, is clearly influenced by the curvature of pore space.

4.3. HARD BODIES IN WIDER CHANNELS INCLUDING FLOW PHENOMENA.

In the foregoing several pertinent references for partitioning of hard bodies in non-convective cases were mentioned, see refs [36, 40, 51, 69, 73, 74, 88]. The flexible chain random coil model used above is particularly suited to synthetic polymers, but for most biopolymer studies, the rigid rod, sphere, and wormlike rather stiff chain models are of importance. Also, as already seen, it is very attractive to explain SEC separation of random-coil polymers with the equivalent-sphere approach. Thus, upon allowing for flow in pore space, we treat hard bodies first.

4.3.1 *Pore Entering.*

The problem of pore entering, already discussed for chain molecules in Section 4.1.2, remains a complicated matter even for hard bodies. For tight fitting bodies, Dagan *et al.* [89], have treated this problem, and it was shown that:

(1) the particle trajectory near the pore opening departs significantly from the undisturbed fluid streamlines, and

(2) particles tend to aggregate near the pore wall, thus leaving a diminished concentration in the pore centre.

These interesting predictions are well-known from observations in the flow of blood through relatively wide capillary tubes (i.d. about 500 μm), the so-called Fahraeus phenomenon. An extension of the theoretical framework has recently been given by Wang *et al.* [90].

4.3.2 *Simple Flow Phenomena Inside (Cylindrical) Pores/Channels: Introduction to Hydrodynamic Chromatography.*

Once inside and upstream the tube, in the sheared flow regime of the (Poiseuillean) parabolic flow profile, quite different phenomena may occur, because of particle rotations, apart from mutual hydrodynamic particle-particle (colloidal) interactions, which also lead to concentration gradients in the cross-section (see *e.g.* Davis *et al.* [91]), but only at higher concentrations, which are outside the scope of the present discussion.

The basic phenomena pertaining to the transport of particles through flow channels in the range $\lambda < 1$ have been thoroughly studied by Brenner *et al.*; for an overview *cf.* the standard treatise by Happel & Brenner [45]. We are especially concerned with the situation where particles move in the presence of a wall, which, being an obstacle, will influence the flow field around the body. Even for flow past a planar surface, where in simple shear flow the particle experiences a non-uniform (linear) flow field, a force imbalance on the particle surface perpendicular to the wall arises, which will cause the particle to drift away from the wall by a so-called lift force. Before going into this, see Section 4.3.4, which requires the effects of inertia (acceleration) to be taken into account in the equations of motion, we will have to discuss even more fundamental processes.

A naive approach of transporting small particles through long flow channels results in the leading order correct expression for the residence times of the particles in the system, simply based on the above notion of steric exclusion from a wall region of thickness R_p, the effective particle radius. Based on this exclusion alone, the partition ratio of particles inside and outside the tube (in the fluid reservoir) is $(1 - \lambda)^2$ and so we expect larger molecules to leave the column in a correspondingly faster fashion than the point-like fluid molecules:

$\langle u_p \rangle = \langle u \rangle / K = \langle u \rangle / (1 - \lambda)^2 = \langle u \rangle \{1 + 2\lambda - O(\lambda^2)\}$. However naive, this expression, alternatively expressed in residence times (t_p is the particle residence time, t_m that of the infinitely small molecules of the flowing fluid): $\tau = t_p / t_m = K = (1 - \lambda)^2 = 1 - 2\lambda + O(\lambda^2)$ has proven to be supported by more realistic models, and is only slightly wrong for $\lambda < 0.1$ [92].

A lesser naive approach incorporates the main hydrodynamic factor, the existence of the nonlinear (parabolic) flow profile. Thus, allowing particles to sample all radial positions in the tube cross-section very rapidly (by Brownian motion), except for those in the excluded layer, the particle velocity is then found from the simple averaging over accessible fluid velocities, *provided that the particles follow the fluid flow exactly*: $\langle u_p \rangle = \{_0\int^{(R_c - R_p)} u(r) r \, dr\} / \{_0\int^{(R_c - R_p)} r \, dr\}$. This is readily integrated because the local Poiseuille velocity is known: $u(\rho) = 2\langle u \rangle (1 - \rho^2)$ where $\rho = r/R_c$ and $\langle u \rangle = (2/R_c^2)_0\int^{(R_c)} u(r) r \, dr$ is the average velocity. Hence: $\langle u_p \rangle = \langle u \rangle (1 + 2\lambda - \lambda^2)$. This implies that the term in λ, stemming from purely steric exclusion from the wall, apparently "accelerates" the particle, while due to the presence of the parabolic flow profile, the particle apparently decelerates behind the fluid flow by an negative contribution $-\lambda^2 \langle u \rangle$. These effects are apparent only, because it was presupposed that the *actual local particle velocity is identical with the local fluid velocity*.

In the next better approximation to the problem under discussion this assumption no longer holds. As a result of the velocity gradient, the different sides of the particle surface undergo different velocities, which will make the particle rotate at an angular velocity $\omega = G/2$, just as in the Debye treatment for free-draining chains, *cf.* Section 2.1.5. In contrast to the simple constant shear situation in the Debye case, in a Poiseuille profile the angular velocity $\omega(r) = G(r)/2$ is a function of the radial coordinate r: $\omega(r) = -(2\langle u \rangle / R_c)(r/R_c)$. For later convenience we write: $\omega^2(r) = br^2$, where $b = (2\langle u \rangle / R_c^2)^2$. It shows that the particle as a result of the shear rate, rotates about its centre with a velocity which is zero in the tube centre and maximal ($-2\langle u \rangle / R_c$) at the wall. It is remarked that the surface velocity of the rotating particle, $v_s(r) = \omega(r) \cdot R_p = -2\langle u \rangle \lambda \cdot (r/R_c)$ is appreciable especially near the wall ($-2\langle u \rangle \lambda$) which will influence the flow properties of the particle transport via so-called lift forces, leading to radial migration, *cf.* Section 4.3.4.

Omitting for the moment, these potential influences of lift phenomena, we can find the particle's movement in the neutrally buoyant case (*i.e.* under zero friction force) by arguing that the particle's centre moves with the *surface averaged* Poiseuille flow velocity of all surrounding fluid elements contacting the particle's surface, $u_p = \langle v_p \rangle$. More precisely, using the notion of the Laplacian operator ∇^2 (see the Appendix), we have the "diffusion" equation: $\langle v_p \rangle - u = (R_p^2/6)\nabla^2 u$ for describing the extent to which, one particle radius away from the particle's central element, the average velocity $\langle v_p \rangle$ of the surrounding elements around the the particle, exceeds the velocity of the centre. Here, u is the local Poiseuille velocity and thus: $\nabla^2 u = -8\langle v_p \rangle / R_c^2$. Substitution of this into the "diffusion" equation, gives the local particle migration velocity: $u_p = \langle v_p \rangle = u - (4/3)\lambda^2 \langle u \rangle$. As a result of this elementary calculation we now find that the *particle actually slips behind the local fluid velocities* with the slip velocity: $u_{slip} = -(4/3)\lambda^2 \langle u \rangle$ at all radial positions in the tube cross-section. We also remark that this slip velocity is far smaller than the surface rotational velocity above, indicating that rather strong influences of the latter are still to be expected, see Section 4.3.4. The present description is a far better approximation of the particle migration velocity than the naive ones above and agrees nicely with the Bungay & Brenner [93] theoretical result, and the experimental observations by Happel & Byrne [94], and Goldsmith & Mason [95] for the centre line, where: $(u - u_p)_{r=0} = (4/3)(R_p/R_c)^2\langle u \rangle$.

The resulting cross-sectional average particle migration velocity is calculated by integration over the accessible core region, finding:

$$\langle u_p \rangle = \langle u \rangle \{1 + 2\lambda - (7/3)\lambda^2\} \tag{8}$$

We note here, for later use in Section 4.3.4, that Brownian "diffusion" in this system can also be found from the accompanying friction factor: $f \approx f_{Stokes}(1 + 2\lambda - (7/3)\lambda^2)$. f is seen to be modified by the factor in parentheses, thus approaching the Renkin equation (cf. e.g. [57]). Better approximations are found in refs [46] and [47]. Especially Cox & Brenner [96] and Ho & Leal [97] critically examine wall effects obtaining $f = f_{Stokes}(1 + \lambda f(r))$ where $f(r) = (9/8)(1 - r/R_c)^{-1}$.

The present problem of axial particle migration has been studied more fundamentally by Brenner & Gaydos [46] and DiMarzio & Guttman [98-101]. The latter authors were the first to recognise the concept as a means for separation, which they called "separation-by-flow". In the cited work by DiMarzio & Guttman, transport of single particles in flow through a circular cylindrical tube is considered, with *allowance for particle rotations*. Their main result is that the particle does not follow the fluid flow velocity, and indeed "slips" behind, while the particle velocity is given by the previous equation (8). This being the case it becomes apparent that particle rotation does not influence the axial transport behaviour, unless radial migration is allowed for, cf. Section 4.3.4. It should be remarked that the factor (7/3) in the last term of eq.(8) consists of two contributions, viz. $\{1 + (4/3)\} = 1 + \Gamma$, where the first (i.e. 1) stems from the integration over the quadratic velocity profile, and the second, Γ, represents the action of the slip velocity. More precisely, Brenner [102] and Goldman et al. [103] developed the following expressions for the slip factor Γ:

$\Gamma = (4/3)\{1 + 3(r/R_c)\}$, in the core region, and
$\Gamma = (5/4)\lambda\{1 - (r/R_c)\}^{-2}$ in the wall region; an even more involved expression is given for the region very close to the wall:
$\Gamma = (2/\lambda^2)\{1 - [0.7431/(0.6376 - 0.2\ln(\lambda^{-1}-(r/R_p) - 1))] + \{1 - (r/R_c)^2\}$ (note the misprints in Silebi & McHugh [68]).

For polymer coils, DiMarzio & Guttman (DG) considered the chain as rigid and permeable, in the spirit of Debye, cf. Section 2.1.5. The extension they provided to the simple Debye rotating/translating model is the inclusion of the quadratic velocity field, which was assumed to be undisturbed by the presence of the polymer segments. Their final result can thus be written as:

$$\langle u_p \rangle = \langle u \rangle \{1 + 2\lambda - C\lambda^2\} \tag{9}$$

where C is a constant equal to $C = 1 + 2\gamma$, γ being the slip factor based on the maximum centre line fluid velocity [$u(0) = 2<u>$] rather than on $<u>$ (as above for Γ). $\gamma = (1/2)\Gamma$ equals $(2/3)(r_G/a_{DG})^2$, where $a_{DG} = (3/\sqrt{\pi})r_G$ is the polymer radius after the statistical method of these authors, as discussed in Section 2.1.8. Their final result is so: $\gamma_{DG} = 2\pi/27 = 0.233$, while $C_{DG} = 1.465$. Upon using other size parameters, we arrive at modified slip factors, generally to be expressed as $C_m = 1 + (4/3)(r_G/R_{part})^2$.

Equation (9) has a validity beyond the original DG model, as becomes clear from the more involved analysis of Brenner & Gaydos (BG) [46], recently extended further by Mavrovouniotis & Brenner [47]. As BG point out, the DG analysis fails to account properly for hydrodynamic wall effects, except partially for the slip velocity in the core region. Indeed, the BG approach, being conceptually very simple in that it focuses on the statistical trajectory of single Brownian particles, takes into account the translational-rotational coupling which occurs for particles especially near the wall and which leads to additional retardation. Upon coupling the solutions to an Taylor-Aris type continuity equation such as eq.(7) in Section 3.3.2, in both the core region and the wall region, they obtain a modified eq.(9) with a slip factor: $C_{BG} = [1 + (4/3) + 8C_3]/(1 - \lambda)^2 \approx \{(7/3) + 8C_3(1 + 2\lambda)\}$ or numerically, with $C_3 = 0.32$: $C_{BG} = 4.89 + 2.56\lambda$. This implies that the average particle velocity reaches a maximum value ($1.155<u>$ for $\lambda = 0.14$), increasing with increasing particle size up to that maximum, and subsequently decreasing for still larger particles.

The BG solution breaks down for λ-values above say 0.20; it has only recently been extended to include closely fitting spheres in ref. [47]. These improved results are, however, not readily expressed in simple equations, and so the narrow pore region is best described by the older expressions of Brenner (1966) [102] and Goldman et al. [103], as reported above, where $\Gamma = 2\gamma$ and $C = 1 + 2\gamma = 1 + \Gamma$. DosRamos & Silebi [67], citing the Ph.D. thesis (1988) of the first author, provide the most explicit slip factor expression available. We prefer the use of simpler expressions, especially since the quantitative validity of these involved theories is less certain, vide infra.

In earlier work, Tijssen et al. [76, 104] summarised the above and showed for the first time that the basic idea of DG, that of separation-by-flow, is experimentally feasible in microcapillary tubes with polystyrenes in several solvents as the polymeric test particles. With column radii down to 0.6 μm, the λ-range applied was < 0.2, well within the BG limit. It appeared that the DG equation, eq.(9) with the modified slip factor $C_{m,p} = 1 + (4/3)(r_G/R_p)^2$ fitted the experiments very well, provided that for R_p the effective radius $r_{eff} = (\sqrt{\pi}/2)r_G = 0.866r_G$ is used, rather than $a_{DG} = (6/\pi)r_G$, at least for good solvents. This experimental result supports the theoretical discussion in Section 2.1.8., where a_{DG} is shown to be oversized with respect to the other size parameters. Required data on the size parameters were obtained from precise light scattering experiments [28, 29] which, in THF as the solvent at room temperature, yields the common general Flory-type relationship: $r_F = pM_w{}^q$, e.g. $r_G = (1.39 \cdot 10^{-5})M_w{}^{0.588}$ (μm). Although a_{DG} seems to be of no further interest, the choice between the other size parameters, r_{eff} and x_c, is by no means clear at present. To complicate the matter, in θ-solvents for example, using $C_{m,\theta} = 7/3$ and r_G rather than r_{eff} gives a better match between theory and experiment, when both are plotted in the fashion of molecular weight *vs* dimensionless residence times $\tau = t_p/t_m = \langle u \rangle / \langle u_p \rangle$, which are directly measurable quantities in the experiments [104]. According to eq.(9), it is more general to write:

$$\tau = K = 1/\{1 + 2\lambda_m - C_m\lambda_m^2\} \qquad (10)$$

where the index m refers to the various possible slip velocity theories (m = DG, BG, DGm, BGm etc., *cf.* models I–VIII in ref. [104]), allowing for the use of different size parameters as well.

Inspired by the findings for the θ-solvents, where seemingly the radius of gyration is a better size parameter than the effective radius, we performed more accurate measurements than the ones mentioned in ref. [104]. The experimental results are presented in Fig. 3 as M_w *vs* τ for a tube with radius 1.700 μm. The drawn curves represent eq.(10), the upper one based on the effective radius, the lower one on Casassa's x_c; the intermediate one is based on the radius of gyration. It appears, just as in the case of θ-solvents, that the intermediate curve gives the best fit. In fact $\lambda_m = \lambda_G$ with $C_m = (7/3)$ should have been used here. For this τ-range, however, the exact C_m-value is of lesser importance, and for all curves the same C_{DGm}-value of 2.698 has been used.

As seen in Section 2, there is no theoretical support for the use of λ_G instead of λ. Hence it is more interesting to compare the situations of Casassa's x_c and Volkenstein's r_{eff}. It then appears from Fig.3 that the former, x_c, yields slightly better matches with the theory than r_{eff} especially when higher C_{BGm}-values (up to 5.26) are used, which shift the curves somewhat upwards, resulting in an even better fit for x_c, and a worse one for r_{eff}.

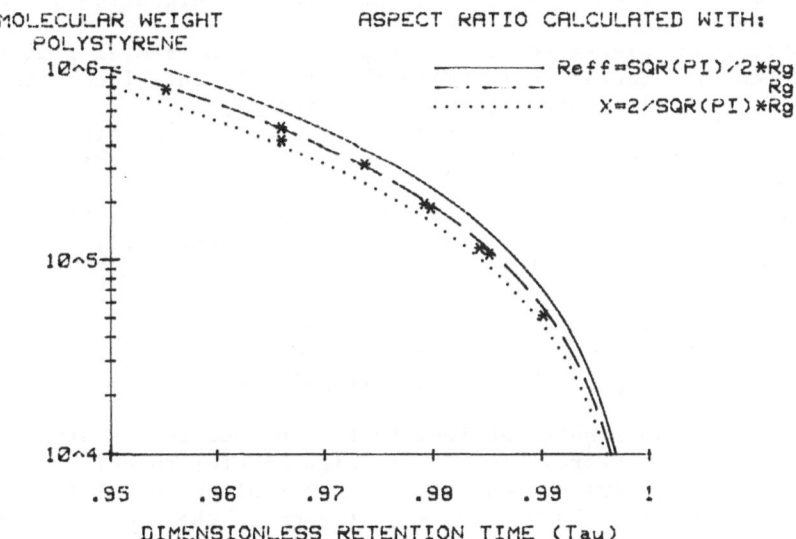

Figure 3. Microcapillary HDC experiments with standard polystyrenes in THF, column radius 1.700 μm. The theoretical lines are based on the modified DG model, with $C_{DGm} = 2.698$ [104]

As shown in ref.[104], it is not easy to discriminate between the DGm and BGm theories, because in the actual working range for practical systems, $\tau \geq 0.83$, *i.e.* $\lambda \leq 0.12$, in which region the two modified theories are only marginally different, because the leading term $\tau \approx 1/(1 - 2\lambda)$ prevails. More accurate experiments are needed to decide in the matter.

In our efforts to test the applicability of this essentially universal calibration relationship λ *vs* τ for practical applications, it appeared that either for very large polymers ($M \gg 10^6$) or for very high fluid velocities ($<u> > 1$ cm/s), deviations occur from the general $\lambda(\tau)$ curve. It is found that under these conditions, τ-values shift to higher values, *i.e.* the particles behave as if they are smaller. A plausible explanation is to be found in the action of fluid inertia, which becomes pronounced in the working ranges mentioned, *cf.* Section 4.3.4.

4.3.3. *Particle Flow in Cylindrical Pores including External Forces.*

DG as well as BG explicitly excluded radial forces and external forces from their solutions. Notably BG, however, in their *derivation* of the framework for the particle velocity function allow in general for external forces (and put these to zero in the final solution) acting in a radial sense, *e.g.* a wall potential. This external field influences the particles' sampling of radial positions, which is no longer random.

This becomes apparent from the discussion in Section 3.3.2 on the mass balance equation, eq.(7). For a steady state situation, *i.e.* a long tube and a sufficiently long residence time, such that the Fourier number (a dimensionless measure of time: process/diffusion time) $Fo = D \cdot t/R_c^2 \gg 1$ the resulting concentration profile is fully developed; the time required for convection along the tube axis is very long then as compared to the restoring mass flux from diffusion in the radial sense. The resulting radial concentration profile does not change by migration in the direction of axial flow.

In this situation, always to be met in practice, the radial flux (eq. (6)) is zero and so: $(\partial c/\partial r) + (c/kT)(\partial \mu^{ext}/\partial r) = 0$. This lends itself to integration to yield general solutions of the type: $c(r) = C \cdot \exp\{-\mu^{ext}/kT\}$, which expresses the probability of finding a particle at a certain radial position, obviously following the expected Boltzmann-type of distribution.

By weighting this concentration distribution with the velocity distribution: $<u_p> = [_0\int^{(R_c)} 2\pi r \cdot u_p(r) \cdot c(r)\, dr]/[_0\int^{(R_c)} 2\pi r \cdot c(r)\, dr]$, we arrive at the average particle velocity:

$$<u_p> = \frac{_0\int^{(R_c-R_p)} u_p(r) \cdot \exp\{-\mu^{ext}(r)/kT\} r\, dr}{_0\int^{(R_c-R_p)} \exp\{-\mu^{ext}(r)/kT\} r\, dr} \tag{11}$$

which is the principle result obtained by BG. In case the external potential also contains external forces, we might prefer to write: $-\mu^{ext}(r) = -U(r) + _0\int^r f(r)\, dr$, where $U(r)$ is a real potential (*e.g.* electrostatic in nature) and $U_f(r) = -_0\int^r f(r)\, dr$ is the pseudo-potential connected with the force field $f(r)$, *cf.* Section 3.3.1. In view of the above, the local particle velocity $u_p(r)$ is known within the limitations of the Brenner-Goldman [102, 103] treatment: $u_p(r) = u(r) - <u>\Gamma\lambda^2$.

A typical example of the application of the above can be found in the problem of the flow of polymer coils in sheared flow fields, as in the present Poiseuille flow through a circular cylindrical capillary. The presence of a stress field, in principle perturbs the chain dimensions and leads to elongation and orientation as discussed in Section 2.1.7. By picturing the coil as a Hookean spring, *cf.* Section 2.1.3, there is a driving force that tends to restore the unperturbed dimensions of the coil. Qualitatively, one may speculate that this force will cause diffusion of the coils, away from the high stress regions, near the wall, towards low stress regions near the tube centre, *i.e.* a radial concentration profile is formed and the residence time behaviour is altered.

Tirrell & Malone [62], followed by Aubert *et al.* [86, 105–108] and Brunn [109], were able to cast this in a framework based on the above pseudo–potentials, in this case entropic in nature. They termed the phenomenon: stress– (or hydrodynamically) induced diffusion, for which experimental confirmations exist, see refs [63–66, 105–107, 110].

In essence, this phenomenon of stress–induced deformation of coils, leading to elongation and subsequent radial migration perpendicular to the flow direction, is based on the existence of the entropic potential associated with the free energy of extension of a Gaussian chain, after Flory [8]. Once the pseudo–potential $U_{spring}(r)$ has been formulated, application of eq.(6) leads to solutions for the steady state concentration profile c(r), which is, again, Boltzmann–like:

$c(r) = C \cdot \exp\{-U_{spring}(r)/kT\}$. Tirrell *et al.* [62] illustrate this with a typical case of macromolecules modelled by a dumbbell (two beads connected with a spring) in dilute solutions in 0.1 cm radius tubes at velocities of 5–10 cm/s. It is shown that the radial concentration profile is largely enriched in the centre of the tube, exponentially decaying towards the wall, where sometimes strong depletion occurs. The radial profile becomes more pronounced at higher velocities, with narrower tubes and for higher molecular weights. Decreasing the channel size is particularly effective in reaching high deformation rates. Reported reductions in apparent viscosities which are observed in capillary flow systems [63–66], may be partly explained by this depletion of the wall layer.

Potentially this effect is, of course, present in SSC techniques such as HDC, SEC, GPC, and FFF, where a pronounced radial concentration profile manifests itself through a change in residence times, and particularly by a flow rate dependence of these times. Flow rate dependence of the average elution volume (*i.e.* K–value) in reliable GPC/SEC experiments is indeed reported, see *e.g.* refs [63–66, 105–107], where K–values readily amount up to a 10% increase for the largest molecular weights (>10^6). Also in studies on (micro)capillary HDC, *cf. seq.*, and packed bed HDC, see Kraak *et al.* [111, 112] and Hoagland *et al.* [31], reference is made of such observations. Quite probably, however, the inertial effect produced by lift forces based on the rotation–translation coupling mentioned earlier, will also contribute, albeit in the opposite sense in that it decreases K, see below.

4.3.4. *Particle Flow in Cylindrical Pores, including Radial Inertial Migration.* Ever since the 1950's, studies of the transport of particles through tubes or other flow channels have shown that the trajectory of a test particle flowing with the fluid is dependent on, among other things, the intensity of the flow. At particle Reynolds numbers larger than 10^{-4}, the common lateral migration of small particles which is controlled by Brownian motion or diffusion, is replaced by a process based on fluid inertia. These inertial effects, systematically studied experimentally first by Segré & Silberberg [113], and later confirmed by a large body of data (for a review, also of early theories, cf. Leal [114]), make particles migrate across the streamlines, irrespective of their initial radial position, such that they tend to assume a preferred equilibrium position at about 60% of the tube radius away from the tube axis. In Poiseuille flow through a circular cylindrical tube, this leads to a concentrated annular region around that preferred radial position; henceforth the phenomenon is called "Tubular Pinch Effect" (TPE).

To date, the most involved theoretical description of spherical particle migration in cylindrical Poiseuille flow including inertial effects in the Navier-Stokes equation, after Cox & Brenner [96], is given by Ishii & Hasimoto (IH) [115–117]. Recently, Ploehn [69] and DosRamos et al. [67] used the IH theory to explain residence time phenomena.

Mullins et al. [118, 119], Brough et al. [120], and McHugh [121] were the first to speculate that these lateral migration mechanisms were active in the SSC technique emerging in those years, Hydrodynamic Chromatography (HDC), either in packed beds, following the pioneering work of Small [122], or in capillary tubes, naturally following the concept of separation-by-flow from DiMarzio & Guttman [98–101]. The only work in microcapillaries with sizes down to about 1 μm, is that of Tijssen et al. [76, 104, 123], as mentioned above. In that work, because of the low velocity conditions, the occurrence of inertial effects is precluded.

Recently, many experiments have been conducted to investigate the limits of this technique, thus allowing for high velocities at which the radial migration effects become readily visible through their influence on the residence time. If radial migration effects are active, the average particle samples more high velocity streamlines in the core region than at low velocities, where they sample the core more or less at random. This, as will be seen in the following, is experimentally observed with the ever shorter residence times at increasing velocities, especially for larger particles. Also, the microcapillary HDC version has been extended to include application onto polymer aggregates, such as diblock copolymer micellar structures [123].

In describing a complete flow-induced separation mechanism, we make use of the formalism discussed in Section 3.3.2, with the final resulting eq.(11) for the particle velocity, as first obtained by BG. The inertial pseudo-potential required, U_{F_L}, just as the spring potential in the former section, is to be found from the lateral (or lift) force F_L as $-(dU_{F_L}/dr)$, see Section 3.3.1. Here the force is defined as the product of velocity and friction factor, the latter to be simply described by the Stokes friction factor. To improve on this friction factor in a later stage, one could take the earlier mentioned Ho & Leal result, see Section 4.3.2, corrected for wall effects, as actually done by Ploehn [69] and DosRamos [67]. Here, for illustrative purposes and in view of the approximations in the velocity profiles to be used, we prefer the simple Stokes friction factor $f = 6\pi\eta R_p$. Thus: $-(dU_{F_L}/dr) = F_L = v_L \cdot f$ can be evaluated, provided the radial lift velocity is known.

For this purpose, one of the best available, but approximate expressions is that by Ishii & Hasimoto [115–117], which predicts, a priori, the existence of a preferred stable equilibrium position, r_{eq}, which is dependent upon solvent flow rate and particle size, as indicated earlier in the critical analysis by Walz & Grün [124] of compiled tubular pinch experiments with neutrally buoyant particles. Again, in view of all other uncertainties, we prefer the simplest approach i.e. a fixed numerical value for r_{eq}, to be established somewhere around $(0.56 - 0.71)R_c$, the range reported in the literature. In doing so, we can write a simple analytical expression for the lift velocity after Ishii & Hasimoto as:

$$v_L(r) = \alpha\lambda^2(<u>^2R_p/\nu)(r/R_c)\{(r_{eq}/R_c) - (r/R_c)\} \qquad (12)$$

where $\alpha - 0(1 - 10^2)$, and ν is the kinematic viscosity of the fluid. Eq. (12), with $\alpha = 3.56$ and $r_{eq} = 0.71R_c$, yields an accurate fit to the numerical results of Ishii & Hasimoto (their Fig. 4a). In accepting for the moment eq.(12), we are able to find the lift force f_L as:
$f_L(r) = 6\pi\eta R_p v_L(r) = 6\pi\alpha\rho_m\lambda^4<u>^2r(r_{eq} - r)$, where ρ_m is the density of the fluid medium. From this the final average particle velocity is obtained via the BG expression eq.(11), in which $u_p(r) - u(r) - \Gamma\lambda^2<u>$, see Section 4.3.2. In view of all approximations associated with our radial velocity profile, eq.(12), it seems unnecessary to use all detailed expressions for Γ, which differ from core region towards wall region and even very close to the wall region, cf. Section 4.3.2. Both DosRamos [67] and Ploehn [69] include these details, [67] even along with the more involved lift velocity expression. Here we proceed using the core relation: $\Gamma(r) - (4/3)[1 + 3(r/R_c)]$ only, to obtain:

$$1/\tau - <u_p>/<u> = \{2 - (4/3)\lambda^2\} - (I_1^{-1})\{(4\lambda^2/R_c)I_2 + (2/R_c^2)I_3\} \qquad (13)$$

where $I_n = {}_0\int^{(R_c-R_p)}\{exp[-(A/kT)r^2(2r - 3r_{eq})]\}r^n$ dr, for $n - 1, 2, 3$, while $A = -\pi\alpha\rho_m\lambda^4<u>^2$. Numerical evaluation of these integrals is required and possible by standard PC programs (like MathCAD).

In order to compare the predicted τ values with experimental ones, we performed microcapillary HDC experiments, such as described in Section 4.3.2, this time as a function of the fluid flow. In experiments with narrow molar mass distributions in the range of $3\cdot10^6 - 8.5\cdot10^6$ Dalton in five fused silica microcapillaries ranging from 7.5 to 38.5 μm in i.d., 219 precise residence time measurements were obtained. The columns were carefully straightened and positioned horizontally in order to eliminate inertial effects based on centrifugal forces, which are easily generated to produce radial (secondary) flow effects, cf. refs [125 - 128], especially with macromolecular particles.

A simple dimensional analysis predicts that the dimensionless group $v - Re\cdot Sc\cdot\lambda^2$ represents the velocity in radial migration conditions {here Re is the (dimensionless) Reynolds number, $Re - 2<u>R_c/\nu$, Sc is the Schmidt number ν/D_m (a measure for the ratio boundary layer thickness in hydrodynamic/diffusive mass transfer; the combined Re·Sc number is also called a Péclet number, $Pe - 2<u>R_c/D_m$, representing convective/ diffusive mass transfer)}. When $v > 1$, this indicates the predominant existence of inertial effects. Consequently, we plotted the deviation from τ-values with respect to the expected value (τ_{HDC}, Section 4.3.2, eq.(10)) as $\tau_{red} - \tau/\tau_{HDC}$ vs this dimensionless velocity group. It is found, see Fig. 4, that the expected deviation occurs for $v > 1$, while, as should be the case in the dimensionless representation, all experiments group together around one common trend. The solid line in Fig.4 is the theoretical trend as predicted from the above reasoning when using realistic values for the numerical factors in the IH velocity function,

430

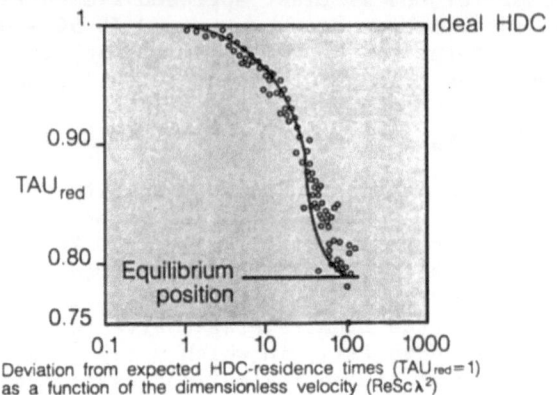

Deviation from expected HDC-residence times (TAU$_{red}$=1)
as a function of the dimensionless velocity (ReScλ²)

Figure 4. Residence time behaviour of high polymers in capillary
transport as a function of (dimensionless) velocity v;
solid line: IH with $\alpha - 10^2$, $r_{eq} - 0.6R_c$, $D - 1.5 \cdot D_{SE}$.

with $\alpha - 10^2$ and $r_{eq} - 0.6 \cdot R_c$. The match between the approximate theory
and the experiments is very reasonable, considering the uncertainties
discussed. Furthermore, phenomena like elongational diffusion (Section
4.3.3) are probably of importance here in view of the very high velocities applied (up to several hundred cm/s), but they are not accounted
for in the theoretical line. Yet, Fig.4 shows that the proposed radial
migration due to the tubular pinch effect is certainly present, and the
realistic values for the parameters α and r_{eq} indicates the remarkably
good agreement of theory and experiment. This implies that the onset of
inertial forces controls the residence time to a large extent.

These lateral migration mechanisms are very strong indeed, because
rapid radial mixing (Fourier numbers > 100), even when supported by secondary flow radial transport, is unable to redistribute the particle
concentration throughout the cross-section. We found that coiling the
tubes strongly, thus generating secondary flow radial transport, did not
influence the first moment of the peaks (the measure of residence time)
but does significantly influence the peak profile, such that peak widths
diminish, as expected [125–128] (to be published elsewhere).

Fig.5 is a plot of such strongly peaked concentration profiles around
the equilibrium position, as a function of the dimensionless velocity v.
This results from the solution of eq.(6) for the steady state concentration profile, which in this case of inertial lift potential U_{F_L} reads:
$(dc/dr) - (c/kT) \cdot 6\pi\alpha\rho_m \lambda^4 <u>^2 r(r_{eq} - r)$ and so after integration:
$c(r)/c(r_{eq}) - \exp[-(\pi\alpha\rho_m \lambda^4 <u>^2 r^2 (2r - 3r_{eq})]$. Fig.5 represents several
two-dimensional cross-sections of this profile at different values of v.

Concerning the understanding of the nature of the lift forces, even
before the famous Segré and Silberberg experiments a lively discussion
had been taken place (for a review see Leal [114]), and the subject is
still actively studied, both theoretically, see refs [129–131], and ex-

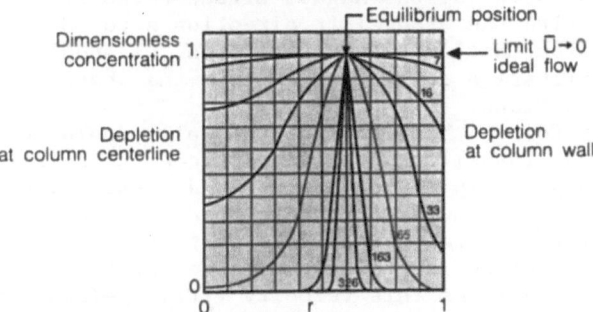

TUBULAR PINCH THEORY

Radial concentration profiles as a function of velocity (ReScλ^2) as the parameter

Equilibrium position

Dimensionless concentration

Limit $\bar{u} \to 0$ ideal flow

Depletion at column centerline

Depletion at column wall

Adapted Ishii/Hasimoto liftvelocity with R= 10 μm and λ= 0.01

Figure 5. Radial concentration profiles as a function of the dimensionless velocity v, using the IH velocity profile, $\alpha = 10^2$,
$r_{eq} = 0.6R_c$

perimentally, refs [132–135]. Ho and Leal [97, 114] describe the two-
sided nature of the lift force cryptically. In more mundane terms we
might interpret this such that moving of a particle in the presence of
an obstacle like the wall, will influence the flow field around the particle: the translating particle displaces fluid also laterally and inertia causes this displacement to be irreversible some distance away from
the particle. The wall resists the fluid displacement and "pushes" the
particle away. This repelling action of the wall may then be partly compensated by the action stemming from the non—uniformity of the sheared
flow (curved velocity profile), which causes the particle to rotate. The
resulting force imbalance makes the particle either drift away from, or
migrate towards the wall, with the particle finally settling in the equilibrium position.

To visualise the problem, we refer to the well—known smoke photographs
of streamlines around (rotating) particles, which clearly show the formation of "eddies" in the wake, stemming from the flow disturbance [136].
The wake is oriented towards the direction of rotation of the particle,
i.e. in the case of Poiseuillean flow, towards the wall, where logically
a repelling action of that wall pushes the rotating particle away.

It is tempting to explain this part of the lift in terms of the pressure differences caused by the rotation of the particle in the flow
field. Based on Bernouilli's theorem then, the pressure near the low—velocity wall—faced side of the spinning particle is high and the particle
is lifted up and away from the wall by this Magnus—effect. How fallacious this kind of reasoning is, has been recognised for a long time and
Birkhoff [137] even speaks of the Magnus Effect Paradox. This paradox is
simply illustrated with the wrong lift direction a top—spinning tennis-
ball would obtain from that, *viz.* upwards! Thus, we should stick to more
realistic, less qualitative reasonings, rather than base ourselves on

field. Based on Bernouilli's theorem then, the pressure near the low-velocity wall-faced side of the spinning particle is high and the particle is lifted up and away from the wall by this Magnus-effect. How fallacious this kind of reasoning is, has been recognised for a long time and Birkhoff [137] even speaks of the Magnus Effect Paradox. This paradox is simply illustrated with the wrong lift direction a top-spinning tennis-ball would obtain from that, *viz.* upwards! Thus, we should stick to more realistic, less qualitative reasonings, rather than base ourselves on tempting easy explanations.

The opposite lift force, then, towards the wall, could be visualised as resulting from the inability of solid particles to follow the deformation that a fluid particle would undergo in a sheared flow field. The necessary displacement of fluid to restore the situation takes place in a thin boundary layer next to the particle surface, with a velocity that depends strongly on the angular velocity of the particle's rotational motion, *cf.* Section 4.3.2. As this velocity $v_s(r) = -(2<u>\lambda(r/R_c)$ is high in the high shear rate region near the wall, particles tend to position themselves there rather than in the non-rotating regions near the centre line.

5. Conclusions

We have covered many of the SSC (Size Separation Chromatography) techniques, which have all their own virtues and shortcomings. As a main conclusion from this work it can be said that the simple open tubular systems (*e.g.* HDC), including complicated hydrodynamic mechanisms, are well understood. This holds especially for hard spherical particles, but to a large extent also for macromolecular particles, provided that these behave as impenetrable particles.

The practical application of these HDC techniques as particle characterisation methods is still far underrated. For example, we proved experimentally that tubular pinch in itself can be used as a separate selective separation mechanism, being able to separate *e.g.* two high molecular weight polystyrenes (not shown here).

In principle, secondary flow phenomena could be treated along the same lines as the above discussed phenomena of elongation and tubular pinch. This subject will be dealt with elsewhere.

The working area of SEC techniques, not applicable to polymer aggregates or solid particles, could well be extended to include reptational phenomena, which offer special selectivities to finite segment number polymers.

6. References

1. Glöckner, G., (1987) Polymer Characterisation by Liquid Chromatography, Journal of Chromatography Library Vol. 34, Elsevier, Amsterdam.
2. Giddings, J.C. (1967), J. Gas Chromatogr., 14, 413-419.
3. Hunt, B.J. and Holding, S.R. (1989) Size Exclusion Chromatography, Blackie, Glasgow.
4. Proceedings of the International GPC Symposium '89, Newton (1989), Waters Millipore, Milford.
5. Munk, P. (1989) Introduction to Macromolecular Science, Wiley-Interscience, New York.
6. Doi, M. and Edwards, S.F. (1986) The Theory of Polymer Dynamics, Int. Series Monographs Physics, Vol. 73, Clarendon Press, Oxford.
7. De Gennes, P. (1985) Scaling Concepts in Polymer Physics, (2nd ed.), Cornell University Press, Ithaca, N.Y.
8. Flory, P.J. (1971) Principles of Polymer Chemistry, Cornell University Press, Ithaca, N.Y.
9. Yamakawa, H. (1971) Modern Theory of Polymer Solutions, Harper & Row, New York.
10. Flory, P.J. (1969) Statistical Mechanics of Chain Molecules, Wiley-Interscience, New York.
11. Flory, P.J. (1974), Brit. Polym. J., March 1976, 1-10.
12. Volkenstein, M.V. (1963) Configurational Statistics of Polymeric Chains, High Polymers Series Vol. XVII, Wiley-Interscience, New York.
13. Birshtein, T.M. and Ptitsyn, O.B. (1966) Conformations of Macromolecules, High Polymers Series Vol. XXII, Wiley-Interscience, New York.
14. Morawetz, H. (1965) Macromolecules in Solution, High Polymers Series Vol. XXI, Wiley-Interscience, New York.
15. Forsman, W.C. (Ed.) (1986) Polymers in Solution, Theoretical Considerations and Newer Methods of Characterisation, Plenum, New York.
16. Bird, R.B., Curtiss, C.F., Armstrong, R.C. and Hassager, O. (1987) Dynamics of Polymer Liquids (2nd ed.), Wiley-Interscience, New York.
17. Williams, M.C. (1975) A.I.Ch.E. J., 21, 1.
18. Brandrup, J. and Immergut, E.H. (Eds.) (1989) Polymer Handbook (Third Edition), Wiley-Interscience, New York.
19. Daoud, M. and de Gennes, P.G. (1977), J. Phys.,(Paris), 38, 85-93.
20. Brochard, F. and de Gennes, P.G. (1977), J. Chem. Phys., 67, 52-56.
21. Daoudi, S. and Brochard, F. (1978), Macromol., 11, 751-758.
22. Brochard-Wyart, F. and Raphael, E. (1990), Macromol., 23, 2276-2280.
23. Benoit, H., Grubisic, Z., Rempp, P., Dekker, D. and Zilliox, J.G. (1966) J. Chim. Phys., 63, 1507.
24. Yau, W.W., Kirkland, J.J. and Bly, D.D. (1979) Modern Size-Exclusion Liquid Chromatography, Wiley-Interscience, New York.
25. Debye, P. (1947) Phys. Rev., 71, 486.
26. Debye, P. and Bueche, A.M. (1948) J. Chem. Phys., 16, 573.
27. Mulderije, J.J.H. and Jalink, H.L. (1987), Macromol., 20, 1152-1158.
28. Schulz, G.V. and Baumann, H. (1968) Makromol. Chem., 114, 122-138.
29. Mandema, W. and Zeldenrust, H. (1977) Polymer, 18, 835.
30. Champion, J.V. and Davis, I.D. (1970) J. Chem. Phys., 52, 381.
31. Hoagland, D.A. and Prud'homme, R.K. (1989), Macromol., 22, 775-781.
32. Aubert, J.H., Prager, S. and Tirrell, M. (1980), J. Chem. Phys., 73, 4103
33. Casassa, E.F. (1967), J. Pol. Sci., Part B, 5, 773.
34. Casassa, E.F. and Tagami, Y. (1969), Macromol., 2, 14.
35. Casassa, E.F. (1976), Macromol., 9, 182-185.
36. Casassa, E.F. (1971), J. Phys. Chem., 75, 3929.
37. Casassa, E.F. (1971), Separ. Sci., 6, 305-319.
38. Kreveld, M.E. van, and van den Hoed, N. (1973), J. Chromatogr., 83, 111.
39. Kreveld, M.E. van (1975), J. Polym. Sci., Part B, Polym. Phys. Ed., 13, 2253-2257.
40. Giddings, J.C., Kucera, E., Russell, C.P. and Myers, M.N. (1968), J. Phys. Chem., 72, 4397-4408.

434

41. Silberberg, A. (1982) J. Colloid Interf. Sci., 90, 86-91.
42. Lax, M. (1976), Chem. Phys., 18, 353-361.
43. Barr, R., Brender, C. and Lax, M. (1980), J. Chem. Phys., 72, 2702.
44. Giddings, J.C. (1991) Unified Separation Science, Wiley, New York.
45. Happel, J. and Brenner, H. (1973) Low Reynolds Number Hydrodynamics with Special Applications to Particulate Media, 2nd revised ed., Noordhoff Int. Publ., Leyden.
46. Brenner, H. and Gaydos, L.J. (1977), J. Colloid Interface Sci., 58, 312.
47. Mavrovouniotis, G.M, and Brenner, H. (1988), J. Colloid Interface Sci., 124, 269-217.
48. Giddings, J.C. (1965) Dynamics of Chromatography, Dekker, New York.
49. Cheng Wei (1986), J. Chromatogr., 362, 309-324.
50. Squire, P.G., Magnus, A. and Himmel, M.E. (1982), J. Chromatogr., 242, 255-266.
51. Doi, M. (1975), J. Chem. Soc. Faraday Trans. II, 71, 1720-1729.
52. Gorbunov, A.A., Solovyova, L.Ya. and Pasechnik, V.A. (1988), J. Chromatogr., 448, 307-332.
53. Knox, J.H. and Scott, H.P. (1984), J. Chromatogr., 316, 311-332.
54. Deen, W.M. (1987), A.I.Ch.E. J., 33, 1409.
55. Cooper, A.R., (Ed.) (1989) Determination of Molecular Weight, Monographs on Analytical Chemistry and its Applications, Vol. 103, Wiley-Interscience, New York.
56. Skowland, C.T. and Kirmse, D.W. (1988), J. Colloid Interface Sci., 126, 547-551.
57. Limbach, K.W., Nitsche, J.M. and Wei, J. (1989), A.I.Ch.E. J., 35, 42.
58. Guggenheim, E.A. (1977) Thermodynamics (6th ed.), North-Holland, Amsterdam.
59. Einstein, A. (1926, 1956) in R. Fürth (Ed.) Investigations on the Theory of the Brownian Movement, Dover, New York.
60. Anderson, J.L. and Reed, C.C. (1976), J. Chem. Phys., 64, 3240-3250.
61. Malone, D.M. and Anderson, J.L. (1978), Chem. Eng. Sci., 33, 1429-1440.
62. Tirrell, M. and Malone, M.F. (1977), J. Polym. Sci., Polym. Phys. Ed., 15, 1569-1583.
63. Cohen, Y., and Metzner, A.B. (1982), AIChE Symposium Series No. 212, Vol. 78.
64. Metzner, A.B., Cohen, Y and Rangel-Nafaile, C. (1979), J. Non-Newt. Fluid Mech., 5, 449-462.
65. Chauveteau, G., Tirrell, M. and Omari, A. (1984), J. Colloid Interface Sci., 100, 41-54.
66. Sorbie, K.S. (1990), J. Colloid Interface Sci., 139, 299-314;315-323
67. DosRamos, J.G. and Silebi, C.A. (1989), J. Colloid Interface Sci., 133, 302-320.
68. Silebi, C.A. and McHugh, A.J. (1978), A.I.Ch.E. J., 24, 204-212.
69. Ploehn, H.J. (1987), Int. J. Multiphase Flow, 13, 773-784.
70. Soodak, H. and Iberall, A. (1978), Am. J. Physiol. 235(1), R3-R17.
71. Buffham, B.A. (1978), J. Colloid Interface Sci., 67, 154-165.
72. Lin, S.P. (1989), J. Colloid Interface Sci., 131, 211-217.
73. Guillot, G., Leger, L. and Rondelez, F. (1985), Macromol., 18, 2531.
74. Colton, C.K., Satterfield, C.N. and Lai, C.-J. (1975), A.I.Ch.E.J., 21, 289-298.
75. Martin, J.E. (1986), Macromol., 19, 922-925.
76. Tijssen, R., Bleumer, J.P.A. and van Kreveld, M.E. (1983), J. Chromatogr., 260, 297-304.
77. Fleisher, R., Price, P. and Walker, R. (1963), Rev. Sci. Instr., 34, 510.
78. Grubisic-Gallot, Z. and Benoit, H. (1971), J. Chromatogr. Sci., 9, 262.
79. Waldmann-Meyer, H. (1985), J. Chromatogr., 350, 1-13.
80. Styring, M.G., Honig, J.A.J. and Hamielec, A.E. (1986), J. Liq. Chromatogr., 9, 3505-3541.
81. Davidson, M.G. and Deen, M.D. (1990), J. Polym. Sci., Part B: Polym. Phys., 28, 2555-2563.
82. Lin, N.P. and Deen, M.D. (1990), Macromol., 23, 2947-2955.
83. Davidson, M.G., Suter, U.W., and Deen, W.M. (1987), Macromol., 20, 1141.
84. Carslaw, H.S. and Jaeger, J.C. (1959) Conduction of Heat in Solids (2nd ed.), Oxford University Press, London.
85. Abramowitz, M. and Segun, I.A. (1965) Handbook of Mathematical Functions, Dover Publ., New York.
86. Aubert, J.H. and Tirrell, M. (1982), J. Chem. Phys., 77, 553.
87. Priest, R.G. (1981), J. Appl. Phys., 52, 5930.
88. Post, A.J. and Glandt, E.O. (1985), J. Colloid Interface Sci., 108, 31.

435

89. Dagan, Z., Weinbaum, S. and Pfeffer, R. (1983), Chem. Eng. Sci., 38, 583.

90. Wang, Y., Kao, J., Weinbaum, S. and Pfeffer, R. (1986), Chem. Eng. Sci., 41, 2845-2864.

91. Davis, R.H. and Leighton, D.T. (1987), Chem. Eng. Sci., 42, 275-281.

92. Prieve, D.C. and Hoysan, P.M. (1978), J. Colloid Interface Sci., 64, 201.

93. Bungay, P.M. and Brenner, H. (1973), J. Fluid Mech., 60, 81.

94. Happel, J. and Byrne, B.J. (1954), Ind. Eng. Chem., 46, 1181.

95. Goldsmith, H.L. and Mason ,S.G. (1966), J. Colloid Interface Sci., 17, 448.

96. Cox, R.G. and Brenner, H. (1968), Chem. Eng. Sci., 23, 147-173.

97. Ho, B.P. and Leal, L.G. (1974), J. Fluid Mech., 65, 365-400.

98. DiMarzio, E.A. and Guttman, C.M. (1969), J. Polym. Sci., Part B, Polym. Letters, 7, 267-272.

99. DiMarzio, E.A, and Guttman, C.M. (1970), Macromol., 3, 131-146.

100. Guttman, C.M. and DiMarzio, E.A (1970), Macromol., 3, 681-691.

101. DiMarzio, E.A. and Guttman, C.M. (1971), J. Chromatogr., 55, 83-97.

102. Brenner, H. (1966), Adv. Chem. Eng., 6, 287-438.

103. Goldman, A.J., Cox, R.G. and Brenner, H. (1967), Chem. Eng. Sci., 22, 653

104. Tijssen, R., Bos, J. and van Kreveld, M.E. (1986), Anal. Chem., 58, 3036.

105. Aubert, J.H. and Tirrell, M. (1980), Rheol. Acta, 19, 252-461.

106. Aubert, J.H. and Tirrell, M. (1983), J. Liq. Chromatogr., 6(S-2), 219-249

107. Aubert, J.H. and Tirrell, M. (1980), Sep. Sci. Technol., 15, 123-130

108. Aubert, J.H. and Tirrell, M. (1980), J. Chem. Phys., 72, 2694-2701.

109. Brunn, P.O. (1983), Int. J. Multiphase Flow, 9, 187-202.

110. Cheng W. and Hollis, D (1987), J. Chromatogr., 408, 9-19.

111. Stegeman, G., Oostervink, R., Kraak, J.C. and Poppe, H. (1990), J. Chromatogr., 506, 547-561

112. Kraak, J.C., Oostervink, R., Poppe, H., Esser, U. and Unger, K.K. (1989), Chromatographia, 27, 585

113. Segré, G. and Silberberg, A. (1962), J. Fluid Mech., 14, 115, 136.

114. Leal, L.G. (1980), Ann. Rev. Fluid Mech., 12, 435-476.

115. Ishii, K. and Hasimoto, H. (1980), J.Phys. Soc. Japan, 48, 2144.

116. Hasimoto, H. (1976), J. Phys. Soc. Japan, 41, 2143-2144; erratum (1977), ibid. 42, 1047.

117. Shinohara, M. and Hasimoto, H. (1979), J. Phys. Soc. Japan, 46, 320.

118. Noel, R.J., Gooding, K.M., Regnier, F.E., Ball, D.M., Orr, C. and Mullins, M.E. (1978), J. Chromatogr., 166, 373-382.

119. Mullins, M.E. and Orr, C. (1979), Int. J. Multiphase Flow, 5, 79-85.

120. Brough, A.W.J., Hillman, D.E. and Perry, R.W. (1981), J. Chromatogr., 208, 175-182.

121. McHugh, A.J. (1984), CRC Crit. Rev. Anal. Chem., 15, 63-117.

122. Small, H. (1974), J. Colloid Interface Sci., 48, 147-161.

123. Bos, J., Tijssen, R. and van Kreveld, M.E. (1989), Anal. Chem., 61, 1318-1321.

124. Walz, D. and Grün, F. (1973), J. Colloid Interface Sci., 45, 467-477.

125. Tijssen, R. (1977), Can.J. Chem. Eng., 55, 225.

126. Tijssen, R. (1978), Sep.Sci. Techn., 13, 681.

127. Tijssen, R. (1980), Anal. Chim. Acta, 114, 71.

128. Sumpter, R. and Lee, M.L. (1991), J. Microcol. Sep., 3, 91-113.

129. Schonberg, J.A. and Hinch, E.J. (1989), J. Fluid Mech., 203, 517.

130. Dandy, D.S. and Dwyer, H.A. (1990), J. Fluid Mech., 216, 381-410.

131. McLaughlin, J.B. (1991) , J. Fluid Mech., 224, 261-274.

132. Berge, L.I. (1990a) , J. Colloid Interface Sci., 135, 283-293.

133. Berge, L.I. (1990b) , J. Colloid Interface Sci., 139, 295-297.

134. Silebi, C.A. and DosRamos, J.G. (1989) , J. Colloid Interface Sci., 130, 14-24.

135. DosRamos, J.G. and Silebi, C.A. (1990) , J. Colloid Interface Sci., 135, 165-177.

136. Merzkirch, W. (1982) Flow Visualisation, II, Symposium Bochum 1980, Hemisphere Publ. (McGraw Hill), New York, p 256.

137. Birkhoff, G. (1950) Hydrodynamics, a Study in Logic, Fact and Similitude, Princeton Univ. Press, Cincinnaty.

436

LIST OF SYMBOLS and ACRONYMS

a_{RS}	radius of spherical elements in the RSM model	(m)
a	segment size in polymer chain model	(m)
a	Mark–Houwink constant in intrinsic viscosity relation	
a_D	MH–type constant in diffusion relation	
a_{DG}	polymer chain size near excluding wall after DG, $(3/\sqrt{\pi})r_G$	(m)
A	constant in IH velocity profile	
A	constant in probability expression for chain configurations	
b	constant in expression for angular rotation velocity	
BG	indicating Brenner–Gaydos [45, 46, 47]	
c	molar concentration	(mole/m³)
c_m	molar concentration of solute in mobile phase	(mole/m³)
c_p	particle concentration	(m⁻³)
c_s	molar concentration of solute in stationary phase	(mole/m³)
C	slip coefficient for λ^2 term in r and $\langle u_p \rangle$	
C_m	modified C coefficient	
C_3	BG correction term in slip coefficient C; $C_3 = 0.32$	
D	diameter (particle, polymer, pore)	(m)
D_b	blob diameter for confined polymer chain	(m)
D_c	channel diameter	(m)
D_p	particle diameter	(m)
DG, DG	indicating DiMarzio–Guttman [98–101]	
D	diffusion/dispersion coefficient	(m²/s)
D_m	molecular diffusion coefficient in free solution	(m²/s)
D_r	radial dispersion coefficient in r–direction	(m²/s)
D_{rept}	reptational diffusion coefficient after kT/f_{rept}	(m²/s)
D_{SE}	diffusion coefficient after SE; $D_{SE} = kT/f$	(m²/s)
D_z	axial dispersion coefficient in z–direction	(m²/s)
erf(y)	error function with argument y: $(2/\sqrt{\pi})_0\int^y \exp(-\zeta^2)\,d\zeta$ [85]	
$f = F/v$	friction factor, kT/D	(N.s/m)
f_0	friction factor in free solution	(N.s/m)
f_{Stokes}	Stokes' friction factor $6\pi\eta R_{part}$	(N.s/m)
f(r)	correction function (in r) in friction factor	
F	mobile phase (solvent) volumetric flow rate	(m³/s)
FFF	Field Flow Fractionation	
Fo	dimensionless time relative to diffusion time, Fourier number $D \cdot t/R_c^2$	
F	force, $-\nabla U$	(N)
$F(r)$	force field in radial direction	(N)
F_L	lift force	(N)
F_S	entropic spring force	(N)
g	number of segments in confined blob	
G	shear rate du/dx	(1/s)
GPC	Gel Permeation Chromatography	
h^0	standard state partial molecular enthalpy	(J/molecule)

H	Hookean spring constant $2kT\beta^2 r$	(J/m^2)
HDC	Hydrodynamic Chromatography	
HPLC	High Performance LC	
$\langle H\rangle$	mean maximal distance of chain segments $(2/\sqrt{\pi})r_G$	(m)
I_n	integral factors in IH velocity profile expression	
IH	indicates Ishii–Hasimoto [115–117]	
J	mass flux density, $cu_p + J^*$	$(kg/m^2 s)$
J^*	mass flux from diffusion and external transport	$(kg/m^2 s)$
J_D	mass flux from diffusion, $-D\nabla c$ (Fick's first law)	$(kg/m^2 s)$
J^{ext}	mass flux from external transport, $-(c/f)\nabla\mu^{ext}$	$(kg/m^2 s)$
$J_0(\beta)$	Bessel function of first kind and zero order [85]	
k	Boltzmann constant	(J/K)
K	partition coefficient, c_s/c_m	
K_{Cas}	partition coefficient after Casassa [33–37], $\equiv K(\infty)$	
K_D	constant after Debye in viscosity relationship	
K_D	MH-type constant in diffusion relation	
K	Mark–Houwink constant in intrinsic viscosity relation	
$K(N)$	partition coefficient for a chain with N segments	
$K(\infty)$	partition coefficient for a chain with ∞ segments, $\equiv K_{Cas}$	
K_0	constant in Flory–Fox viscosity relation	
ℓ	root mean square end-to-end distance in chain, $\sqrt{\langle r^2\rangle}$	(m)
LC	Liquid Chromatography	
LVN	Limiting Viscosity Number $\equiv$ intrinsic viscosity $[\eta]$	(dl/g)
$\mathcal{L}$	characteristic length of elongated particle	(m)
$\mathcal{L}_c$	characteristic length of cigar-like confined chain	(m)
M	molar mass	(kg/mole)
M_i	molar mass of the fraction of the MWD with i monomers	(kg/mole)
M_w	weight average molecular weight, $_i\Sigma\,(N_i M_i{}^2)/_i\Sigma\,(N_i M_i)$	(kg/mole)
M_0	molar mass of monomer unit	(kg/mole)
MWD	Molecular Weight Distribution	
MH, MHS	indicating Mark–Houwink or Mark–Houwink–Sakurada	
n	index	
n_{RS}	number of spheres in RSM model	
N	number of segment units in polymeric chain	
N_A	Avogadro's number	
N_b	number of blobs formed by confined chain	
N_p	number of particles	
N_i	number of molecules in MWD with i monomers	
$O(\lambda^2)$	terms of the order of magnitude λ^2	
$P(r)$	probability density function	(m^{-3})
$P(r,j)$	probability to find segment j at radial position r	
Pe	dimensionless Péclet number, $Re\cdot Sc = 2\langle u\rangle R_c/D_m$, measure of convective/diffusive mass transfer	
PS	polystyrenes	

438

$\langle Q \rangle$	mean maximum cross-sectional chain dimension, $r_G \sqrt{\pi}$ [12]	(m)

r radial coordinate   (m)

$\langle r \rangle$ average linear distance between end segments in chain

                $\langle X \rangle = \ell \sqrt{(8/3\pi)}$ (m)

$\langle r^2 \rangle$ mean square end-to-end distance in chain, Na^2 (m^2)

$\langle r_G^2 \rangle$ mean square radius of gyration, $\langle r^2 \rangle / 6$ (m^2)

$\langle r_0^2 \rangle$ mean square end-to-end distance in unperturbed chain (m^2)

r_{eff} effective polymer chain radius, $(\sqrt{\pi}/2)r_G$ (m)

r_{eq} radial equilibrium position (spring and TPE)   (m)

r_F Flory-type property in $r_F = pM_w^q$

r_G radius of gyration of chain $= \sqrt{\langle r_G^2 \rangle}$ (m)

r_h hydrodynamic radius after Flory-Fox $\approx (7/8)r_G$ (m)

r_H hydrodynamic radius from Stokes friction factor $\approx (2/3)r_G$ (m)

r_{prob} most probable end-to-end distance (spring, chain) $1/\beta$ (m)

$\langle r^2 \rangle_{spring}$ mean square end-to-end distance in spring $3kT/H$ (m^2)

R_c radius of cylindrical channel   (m)

R_F Flory radius , $aN^{3/5}$ (m)

R_p radius of spherical/polymer particle   (m)

R_{part} radius of particles in general   (m)

R_{pore} radius of pores in general   (m)

Re   dimensionless velocity, Reynolds number $2\langle u \rangle R_c / \nu$

RSM   Randomly overlapping Spheres Model

s^0 standard state partial molecular entropy   (J/K.molecule)

s_h hydraulic radius (surface area/pore volume), $4/D$ (m^{-1})

S entropy   (J/K)

Sc   dimensionless Schmidt number ν/D_m, measure of hydrodynamic/

          diffusive boundary layer thickness in flowing mass transfer

SBF   Separation-By-Flow

SEC   Size Exclusion Chromatography

SEM   Scanning Electron Microscopy

SSC   Size Separation Chromatography

t time   (s)

t_m residence time of mobile phase and small molecules   (s)

t_p residence time of finite sized particle   (s)

t_R residence (retention) time   (s)

T temperature   (K)

THF   tetrahydrofuran

TPE   Tubular Pinch Effect

u linear velocity   (m/s)

$\langle u \rangle$ average velocity   (m/s)

$u(0)$ fluid linear velocity in centre line ($r = 0$)   (m/s)

u_0 maximum velocity in velocity profiles [often $= u(0)$]   (m/s)

u_p particle velocity   (m/s)

$\langle u_p \rangle$ average particle velocity

u_{slip} slip velocity of particle behind local fluid velocity   (m/s)

$U(r)$ potential energy (field)   (J)

$U_F(r)$ pseudo-potential connected with force field   (J)

$U_{F_L}(r)$ pseudo-potential connected with lift forces   (J)

U_{spring} spring potential energy   (J)

U_ϵ	electrostatic potential	(J)
v	symbol for velocities	
v	dimensionless velocity $Re.Sc.\lambda^2$ for TPE	
v_h	hydrodynamic volume, $(4/3)\pi r_h^3$	(m³)
v_L	lift velocity for TPE	(m/s)
$<v_p>$	averaged fluid flow velocity over particle surface	(m/s)
v_s	surface velocity due to particle rotation, ωR_{part}	(m/s)
V_e	elution volume for SEC/GPC, $V_e = V_0 + K_{SEC}V_i$	(m³)
V_i	intraparticle void volume	(m³)
V_m	mobile phase volume	(m³)
V_R	retention volume, $F.t_R$	(m³)
V_0	interstitial (interparticle) void volume	(m³)
$W(r)$	probability function for end–to–end distance	(m⁻¹)
x_c	critical exclusion layer thickness (equal to effective chain radius) after Casassa [33–37], $<X>/2 = (2/\sqrt\pi)r_G$	(m)
$<X>$	mean maximal projection of chain [12], $<r>$	(m)
z	longitudinal axis coordinate	(m)
$\propto$	expansion factor for chains with respect to the unperturbed state, $<r^2>/<r_0^2>$	
α	constant in IH velocity profile $O(1-10^2)$	
β	characteristic width parameter of $P(r)$, $\beta = \sqrt{(3/2Na^2)} = 1/r_{prob}$	
β_m	m–th root of Besselfunction $J_0(\beta) = 0$ [85]	
γ	slip factor based on the maximum centre line fluid velocity $u(0)$, $\Gamma/2$	
γ_{DG}	slip factor after DG, $(2/3)(r_G/a_{DG}) = 2\pi/27$	
γ_m	modified slip factor, $(2/3)(r_G/R_p)$	
δ	increment for aspect ratio from electrostatic potentials	
ρ	dimensionless radial coordinate, r/R_c	
κ	constant in exponential testfunction (Appendix)	(m or m.mole)
λ	aspect ratio, R_p/R_c	
λ_m	modified aspect ratio, R_{part}/R_c	
λ_G	modified aspect ratio, R_G/R_c	
λ_s	modified aspect ratio, a/R_c	
$\Omega(r)$	number of configurations for end–to–end distance r	
μ	chemical potential or pseudo potential	(J/molecule)
μ^{ext}	external chemical potential or pseudo potential	(J/molecule)
μ^0	standard state chemical potential	(J/molecule)
μ^*	total chemical potential including internal and external (pseudo–) potentials	(J/molecule)
ν	kinematic viscosity	(m²/s)
η	dynamic viscosity	(N.s/m²)
η_0	dynamic viscosity of pure solvent	(N.s/m²)
$[\eta]$	intrinsic viscosity $=$ LVN, $\lim_{c\to 0}\{(\eta - \eta_0)/\eta_0 c\}$	(dl/g)
Γ	slip factor based on average fluid velocity $= 2\gamma$ [102, 103]	
θ	indicating theta condition in polymer solution	
ϵ	porosity $= V_{pores}/V_{total}$	
ϵ_{acc}	accessible porosity $= (V_{pores} - V_{excluded})/V_{total}$	
τ	dimensionless residence time $= t_p/t_m$	

τ_a	step time in random walk	(s)
τ_{red}	relative dimensionless residence time $= \tau/\tau_{HDC}$	
τ_{HDC}	dimensionless residence time $= t_p/t_m$ after eq.(10)	
ω	angular velocity	(1/s)
Δ	difference between two situations	
∇	gradient, for scalar c: $(\partial c/\partial x + \partial c/\partial y + \partial c/\partial z)$ or $[(\partial c/\partial r) + (\partial c/\partial z)]$ in the cylindrical symmetric case; for vector J: $[(1/r)\{\partial(rJ_r)/\partial r\} + (\partial J_z/\partial z)]$	
∇^2	Laplacian operator $(\partial^2/\partial x^2 + \partial^2/\partial y^2 + \partial^2/\partial z^2)$ or for scalar in cylindrical symmetric coordinates: $[\partial^2/\partial z^2 + (1/r)\{(\partial/\partial r)(r(\partial/\partial r))\}]$	
ψ	tapering angle of conical pore, D_c/z	
Λ	area	(m²)

Appendix. *Basic Mechanisms of Random Walk Diffusion*. The diffusion equation in the non-convective case and in the absence of external fields describes the concentration of point-like particles at a particular location after a given number of random walks. Eq.(7) describes this as: $\partial c/\partial t = \nabla(D\nabla c)$ and for a constant diffusion coefficient: $\partial c/\partial t = D\nabla^2 c$, the well-known Fickian Second Law.

The validity of this important relationship can be illustrated in terms of the random walk model. We consider the simplest one-dimensional case, in which a particle moves to the right or to the left with equal probabilities and equal fixed diffusion steps of length a along a line in the x-direction. The total net displacement x in time t is a statistical quantity equal to $\sum_i x_i$ with $x_i = \pm a$, $+a$ and $-a$ with equal probability. Thus, each x_i has a mean value of $<x_i> = 0$ and a variance of $<x_i^2> = a^2$. The square of x reads: $x^2 = \sum_i x_i^2$ + cross-product terms in $x_i \cdot x_j$. Consequently, as i and j are not correlated: $<x_i \cdot x_j> = <x_i> \cdot <x_j> = 0$, and so: $<x^2> = Na^2$ is the mean square displacement (*cf.* also Section 2.1.1), where $N = t/\tau_a$ is the number of steps and so: $<x^2> = 2Dt$, where the diffusion coefficient is *defined* as $D = a^2/2\tau_a$.

In a three-dimensional random walk of steps a and step time τ_a, diffusion takes place in 6 directions ($\pm x$, $\pm y$ and $\pm z$) with equal probabilities (1/6) and thus: $<x_i> = <y_i> = <z_i> = 0$ and $<x_i^2> = <y_i^2> = <z_i^2> = a^2/3$ because displacement along each coordinate direction in one diffusion step is one of the 6 equal probabilities a, $-a$, 0, 0, 0, 0. The mean square displacement in time t, after $N = t/\tau_a$ steps is then equal for all directions: $<x^2> = <y^2> = <z^2> = Na^2/3 = (a^2/3)(t/\tau_a)$, which, again, can be written in terms of the above definition of the diffusion coefficient $<x^2> = 2Dt = (a^2/3)(t/\tau_a)$. The mean square spatial distance is thus: $<r^2> = <x^2> + <y^2> + <z^2>$ and so: $<r^2> = Na^2 = a^2(t/\tau_a) = 6Dt$, where now $D = <x^2>/2t = <y^2>/2t = <z^2>/2t = <r^2>/6t = a^2/6\tau_a$.

We now show that a collection of such particles obeys Fick's laws of diffusion with the derived transport coefficient $D = a^2/6\tau_a$. Consider in each direction volume elements of thickness equal to the step length a and area Λ. For example in the x-direction the particle count N_p in each segment is $N_p = c_p\Lambda a$, where c_p represents the particle concentration. When we focus on the contact area between segments n and n+1, in the time τ_a needed for one step in the random walk, going from the centre of segment n to that of segment n+1, the net particle current density J_x [*i.e.* the number of passing particles per unit time (divide by τ_a) and unit area (divide by Λ)] is then: $J_x = \{(c_n\Lambda a/2 - (c_{n+1}\Lambda a/2)\}/\tau_a\Lambda$. In terms of the concentration gradient (and omitting index $_p$ for simplicity) $dc/dx = (c_{n+1} - c_n)/a$, J_x is found to be: $J_x = -(a^2/2\tau_a)(dc/dx)$, which is identical with $J_x = -D(dc/dx)$, the one-dimensional form of Fick's first law, eq.(5).

The conservation of mass for the diffusive process requires that the time rate of change $\partial N/\partial t$ equals the net (inward - outward) particle current density, being the gradient of J. As a result, in one dimension (say x): $(\partial N/\partial t) = -dJ_x$ and consequently: $(\partial c/\partial t) = D(d^2c/dx^2)$. The analogous reasoning in three dimensions requires the replacement of the gradient dc/dx by the three-dimensional gradient (nabla): ∇c. The generalised particle flux density then reads: $J = -(a^2/6\tau_a)\nabla c = -D\nabla c$, while the conservation equation becomes: $(\partial c/\partial t) = \nabla(D\nabla c)$ or for constant D: $(\partial c/\partial t) = D\nabla^2 c$, the generalised form of Fick's second law.

Although the Laplacian operator ∇^2 is mathematically well defined, we may well examine the physical meaning of this formalism when applied to a physical property (like concentration). In this context it should be realised that, quite obviously, if particles are uniformly distributed throughout space, their stepwise movements in and out of neighbouring elements do not change the concentration in an element of interest somewhere in that space. Concentration changes with time in the observed element would occur only if the concentration of the surrounding elements differs from its own. The importance of the Laplacian ∇^2 is that it appears to be the appropriate measure of the extent to which the average concentration of the surrounding nearest neighbouring elements ($<c>$) exceeds the concentration of the central element of interest. This is illustrated by the common case where concentration varies exponentially with distance, the well-known Boltzmann distribution: $c(r) = c_0 \cdot \exp\{(\kappa/c_0)r^2\}$. As we are interested in the small distances connected with surrounding neighbour segments, we may approach the exponential function by a quadratic function of radial position: $c(r) = c_0 + \kappa r^2$. For this case, the value of the concentration in the element of interest at $r = 0$ is: $c(0) = c_0$, while the (average) value at a nearby distance one step length a away around $r = 0$ is $<c(a)> = c_0 + \kappa a^2$. Thus: $<c(a)> - c(0) = \kappa a^2$. Calculation of the Laplacian operation $\nabla^2 c$ in this spherically symmetrical case proceeds conveniently in spherical coordinates where: $\nabla^2 c = (1/r^2)\{(\partial/\partial r)[r^2\{\partial c/\partial r\}]\} = 6\kappa$. From this $\kappa = (\nabla^2 c)/6$ is obtained and we observe the fact that: $<c(a)> - c(0) = (a^2/6)\nabla^2 c$. This, of course, is completely analogous to the mass balance equation (Fick's second law), taking into account that for small enough step distances (a) and associated step times τ_a the difference $<c(a)> - c(0)$ can be written in terms of the differential $(\partial c/\partial t) = \{<c(a)> - c(0)\}/\tau_a$. Hence, Fick's second law is obeyed: $(\partial c/\partial t) = D\nabla^2 c$ with $D = a^2/6\tau_a$ as required. In conclusion of this Appendix we restate the important theorem that: *the Laplacian ∇^2 is the appropriate measure of the extent to which the average concentration of the surrounding nearest neighbouring elements ($<c>$) exceeds the concentration of the central element of interest*: $<c(a)> - c = (a^2/6)\nabla^2 c$. This valuable result is also applied to other situations, concerning velocity fields rather than concentration distributions (*cf*. Section 4.3.2).

PHYSICOCHEMICAL BASIS OF HYDROPHOBIC INTERACTION CHROMATOGRAPHY

DIETMAR HAIDACHER and CSABA HORVÁTH
Department of Chemical Engineering
Yale University
New Haven, Connecticut, USA

ABSTRACT. In hydrophobic interaction chromatography (HIC), which is widely used for protein separation, a mildly hydrophobic stationary phase is employed with an aqueous salt solution as the mobile phase, and the magnitude of retention is modulated by the concentration of the salt. Retention is governed by the hydrophobic effect, which is attributed to the strongly ordered structure and high cohesive energy of water. Various theoretical approaches are used to treat chromatographic retention in HIC. The most comprehensive is the solvophobic theory that was first adapted to provide a theoretical framework for reversed-phase chromatography, a technique fundamentally similar to HIC. According to this approach the role of salt as the primary retention modulator in HIC is primarily due to the increase in the surface tension of the mobile phase with the salt concentration. However, when specific salt binding occurs the appropriate preferential interaction parameters have to be considered. The effect of temperature on the retention in HIC is yet to be elucidated. Recent advances in exploring the temperature dependence of the hydrophobic effect are expected to provide a framework for the interpretation of experimental data in HIC and to facilitate further understanding of the role of temperature in hydrophobic interactions.

1. Hydrophobic Interaction Chromatography

1.1. HISTORY

Hydrophobic interaction chromatography (HIC) is a technique of growing importance for the separation and purification of biological macromolecules. It is carried out with mildly hydrophobic stationary phases and with an aqueous mobile phase by gradient elution with decreasing salt concentration [1, 2].

The concept of protein chromatography that is based on hydrophobic interactions for retention and discrimination was first put forward in 1949 by Tiselius [3, 4]. He separated dyes by paper chromatography using salt solutions as the eluent and coined the term "salting-out chromatography" to embrace chromatographic techniques, in which the retention is enhanced upon increasing salt concentration. Along these lines the technique of HIC had been systematically developed in the early seventies for the separation of proteins [5-14]. In these works agarose beads with covalently bound n-alkyl functions of relatively low concentration were employed as the stationary phase. Due to the use of amino functions for bridging groups of the hydrocarbonaceous ligates to the

F. Dondi and G. Guiochon (eds.),
Theoretical Advancement in Chromatography and Related Separation Techniques, 443–480.
© 1992 *Kluwer Academic Publishers.*

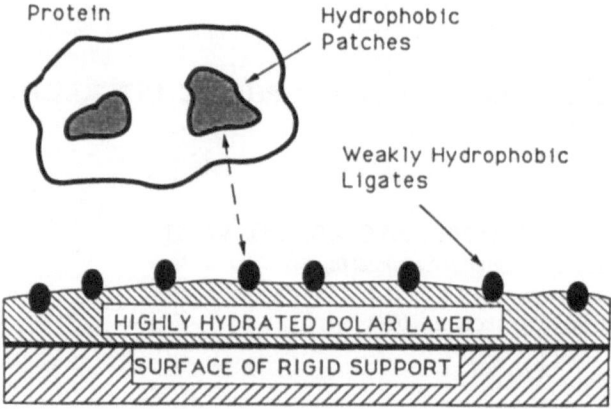

Figure 1. Schematic illustration of the surface of the stationary phases for high-performance hydrophobic interaction chromatography of proteins. The mechanically stable, rigid microparticulate support is made of a crosslinked polymer or silica and can be macroporous, gigaporous or nonporous. The most commonly employed ligates are short chain n-alkyl moieties, aromatic functions or of the polyoxyethylene type.

hydrophilic support, the first sorbents used for HIC contained not only hydrophobic but also positively charged functions and therefore exhibited adsorption characteristics, which involved also electrostatic interactions. In order to practice HIC without the complexity caused by coulombic interactions, the presently used agarose based hydrophobic sorbents are prepared without fixed charges [11, 13]. The low mechanical stability of agarose gels requires the use of relatively low flow rates and consequently the time of separation is rather long in HIC by traditional column chromatography. In order to carry out HIC under conditions typically employed in HPLC, microparticulate and macroporous stationary phases based on siliceous [15-21] and polymeric [22-25] supports of sufficient rigidity were introduced for HIC in the 80's. These "composite" sorbents are prepared with a hydrophilic surface layer to mimic the inertness and highly hydrated nature of polysaccharide gels as well as with mildly hydrophobic ligates, usually widely spaced short n-alkyl or aryl functions or polyether moieties as shown in Fig. 1. The various ligates of sorbents employed in HIC have been reviewed in the literature [1]. Very recently nonporous stationary phases with similar surface architecture have also been developed for rapid separation of biopolymers by HIC [26-28].

1.2. CHROMATOGRAPHIC SYSTEM

In a typical HIC system proteins are first bound to the stationary phase from an aqueous mobile phase at relatively high concentration of a common neutral salt such as ammonium sulfate. Elution and separation are carried out by using a decreasing salt gradient to diminish hydrophobic interactions. This is illustrated in Fig. 2 by the separation of standard proteins on a silica based micropellicular polyether stationary phase [28], which has been synthesized in our laboratory for rapid analysis.

Under conditions typically employed in HIC the interactions between the hydrophobic ligates at

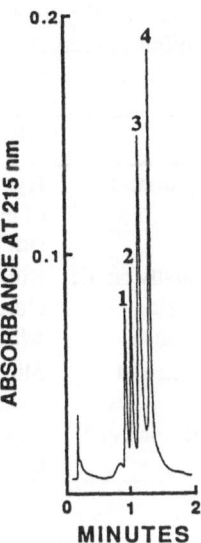

Figure 2. Rapid hydrophobic interaction chromatography of proteins. Column: 30 x 4.6 mm, 2-μm micropellicular polyether; 3 minute gradient from 3 M ammonium sulfate in 100 mM phosphate buffer, pH 7.0, to neat buffer; flow rate, 3ml/min; temp. 50°C. Sample components, (1) cytochrome c, (2) ribonuclease A, (3) lysozyme and (4) α-chymotrypsinogen. Fig. 7 of Ref. [28], reproduced with permission of the publisher.

the mildly hydrophobic chromatographic surface and the "greasy patches" on the surface of the protein molecule are relatively weak. Furthermore, the salt in the eluent has a stabilizing effect so that the structural integrity of proteins is largely preserved. As a result HIC is a chromatographic technique for separating and purifying proteins without loss of their biological activity.

It should be noted that reversed-phase chromatography (RPC), a technique most widely used in HPLC, is also based on hydrophobic interactions. However, RPC employs not only more strongly hydrophobic stationary phases than those customary in HIC but also hydro-organic eluents instead of aqueous salt solutions. For the separation of proteins, the eluent pH is usually low and gradient elution with increasing organic solvent concentration is utilized almost exclusively. Under such harsh conditions unfolding of the protein molecule usually occurs [29, 30]. Thus, in the chromatography of a given protein mixture by one or the other of these techniques, the elution order of the components can be distinctly different: in HIC largely intact proteins whereas in RPC partially unfolded protein molecules are separated [16]. Some of the distinctive features of HIC and RPC that arise from the use of different types of stationary and mobile phases are summarized in Table I.

1.3. PROTEIN STRUCTURE AND ADSORPTION ONTO HYDROPHOBIC SURFACES

According to our present knowledge of the three-dimensional molecular structure of proteins the polypeptide chain of water-soluble globular proteins is folded in such a way that most hydrophobic amino acid residues (e.g., alanine, valine, leucine, isoleucine, phenylalanine, tryptophane, and

Table 1. Distinguishing features of reversed phase and hydrophobic interaction chromatography in HPLC. The support in both cases can be microparticulate silica gel or a highly crosslinked polymer.

Features	RPC	HIC
Stationary phase surface	Dense layer of strongly hydrophobic ligates: "hard" surface	Hydrophilic layer with a low density of mildly hydrophobic ligates: "soft" surface
Mobile phase	Hydro-organic eluents, usually acidic; gradient elution with increasing organic solvent concentration	Buffered aqueous salt solutions of close to neutral pH; gradient elution with decreasing salt concentration
Effect on protein	Major conformation changes and denaturation	Slight, usually reversible conformation changes
Application	Analysis or purification of small and large molecules	Separation and purification of biopolymers, mostly proteins

tyrosine) are buried in the interior. On the other hand most hydrophilic amino acids (*e.g.*, serine, threonine, asparagine, glutamine, and those with ionogenic side chain) are at the surface and thereby exposed to the aqueous solvent. Yet, nonpolar amino acids are also found at the surface of the protein molecule and they constitute the "greasy (hydrophobic) patches" that interact with the hydrophobic ligates of the stationary phase. The concept that the direct contact between the accessible hydrophobic domains on the protein surface and the hydrophobic moieties at the chromatographic surface represent the driving force for protein adsorption in HIC is supported by elegant experimental studies. For instance, the calcium-binding protein calmodulin was found to adsorb to phenyl-Sepharose in the presence of Ca^{2+} [31]. Removal of Ca^{2+} from the bound protein upon the addition of EDTA to the mobile phase, however, induced a conformational change of the protein and as a result it readily eluted from the HIC column [31]. This explanation for the changing retention behavior is strongly supported by findings from extrachromatographic studies that have shown that upon addition of Ca^{2+} certain hydrophobic residues of the calmodulin molecule become exposed to the solvent.

HIC is based on hydrophobic adsorption of proteins on the stationary phase and, like other branches of interactive biopolymer chromatography, it is a surface mediated separation technique. Therefore, the hydrophobic features of the protein surface play a key role in determining the energetics of the retention process. The protein surface has been conveniently defined by employing the concept of the water-accessible surface area [32]. It represents the area that is traced out by the center of an imaginary water molecule of radius 1.4 Å as it rolls around the van der Waals radii of the external atoms of the protein molecule. The nonpolar part of the protein surface area is then defined as the area that embraces carbon and sulfur atoms of the side chains as well as main chain carbon atoms, and the rest is considered to be the polar part of the surface. As a rule of thumb the nonpolar part of the protein surface area represents the same fraction of the total molecular surface area for most polar globular proteins which amounts to approximately 50% [33]. However, the size and/or the proportion of the nonpolar surface area of a protein molecule are only partially responsible for the strength of protein binding to HIC sorbents. Lysozyme and myoglobin, for instance, have similar molecular weights so that the number of stationary phase ligates that come into contact with the protein in the chromatographic retention process are expected to be similar. In view of the nonpolar water-accessible surface areas of the two proteins [32], lysozyme would be expected to elute before myoglobin, yet myoglobin is generally less retained than lysozyme on typical HIC sorbents. Indeed the distribution, size, nature and accessibility of the protein's hydrophobic patches as well as the interplay of nonpolar and polar amino acids in the contact region for binding to the stationary phase surface have strong influence on the hydrophobic adsorption characteristics of proteins. The retention in HIC can also be sensitive to minor changes in the protein structure such as oxidation of a methionine residue or substitution of a single amino acid residue in the polypeptide chain. It has been demonstrated with lysozymes from different sources that when the substitution occurs in the chromatographic contact region of the protein surface the effect on retention can be quite substantial [34]. The three-dimensional macromolecular nature of proteins permits simultaneous interaction with several separate binding sites at the chromatographic surface and thus elicits multipoint attachment. For instance, equilibrium binding studies and desorption kinetic measurements suggest that at least 4-5 residues participate in the binding of phosphorylase b to typical HIC sorbents [35]. The interaction of proteins via multiple binding sites in HIC is also supported by the observation that small peptides containing hydrophobic amino acids do not bind to octyl-Sepharose under conditions where proteins are strongly adsorbed [13].

1.4. APPLICATIONS

Recent developments in molecular biology and the concomitant industrial production of therapeutic proteins require high purity proteins without loss of biological activity. For this reason HPLC is increasingly employed for the purification and analysis of such proteins. Since the introduction of rigid microparticulate stationary phases, high-performance hydrophobic interaction chromatography (HPHIC) has gained wide recognition as a major branch of HPLC in the rapidly growing field of protein separations. The mild operating conditions and the short separation times make HPHIC the method of choice for separating complex biological macromolecules according to the hydrophobic features of their surface with high recovery and retention of their biological activity. As mentioned above the method promises to take advantage of subtle conformational differences for the separation of closely related protein molecules [34]. A few selected applications of HPHIC to both analytical and preparative protein separations are described below to illustrate the practical use of the technique.

HIC has been extensively employed for the analysis of recombinant DNA-derived proteins. As such it has been effectively utilized for the characterization of recombinant human growth hormone (rhGH) variants [36]. Fig. 3 depicts the separation of hGH and met-hGH on an ether-type HIC column. The two recombinant proteins differ only in an extra methionine residue at the N-terminus of met-hGH. The native form and a glycosilated variant of human insulin-like growth factor-I (IGF-I) were separated by HPHIC on a phenyl-type HIC column using a linear gradient from 0.7 M sodium sulfate in 30 mM Tris-HCl of pH 8.0 to 30 mM Tris-HCl of pH 8.0 containing 5% acetonitrile and 0.075% Brij 35 [37]. Among many similar applications HIC has also been incorporated into the purification protocol for recombinant tumor necrosis factor (TNF) [38].

Detergents and/or organic solvents are widely used in the mobile phase in order to enhance the efficiency or the selectivity of the separation in HIC. It is believed that such additives mask "hot spots" at the chromatographic surface and thereby reduce its energetic heterogeneity with the result of enhanced column efficiency and peak symmetry. Our limited knowledge of the properties of stationary phases and proteins, however, seriously hampers a quantitative treatment of these phenomena at this time.

448

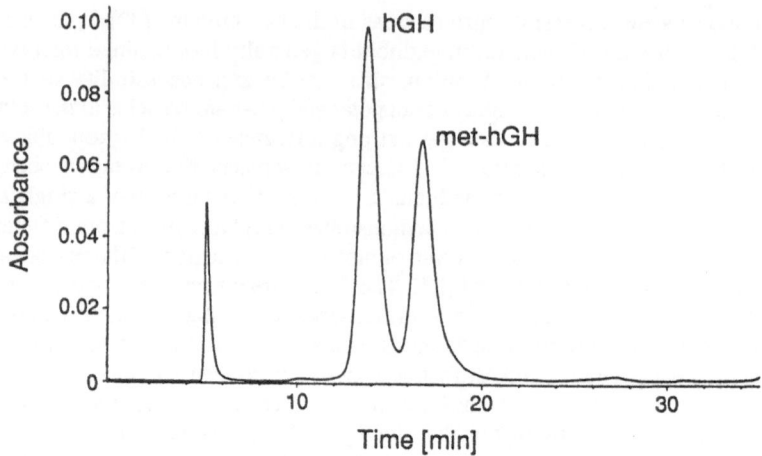

Figure 3. Separation of hGH and met-hGH by hydrophobic interaction chromatography. Column: 75 x 7.5 mm, 10 μm TSK-phenyl-5PW; starting eluent A, 0.5 M sodium sulfate and 0.5 % acetonitrile in 0.03 M Tris-HCl, pH 8.0; gradient former B, 5 % acetonitrile and 0.07 % Brij in 0.03 M Tris-HCl, pH 8.0; 20 min linear gradient from A to B; flow rate, 0.5 ml/min; temp. 25°C. Fig. 5 of Ref. [36], reproduced with permission of the publisher.

Due to the mild operating conditions, HPHIC is frequently employed for rapid purification of very labile proteins. For instance, the multimeric enzymes carbamoyl phosphate synthetase I and ornithine transcarbamoylase from crude mitochondrial extracts of rat liver were separated on a weakly hydrophobic polyether column with a descending salt gradient [39]. The enzymes were recovered with a 3- to 10-fold increase in specific activity in a chromatographic run of 20 min.

For the purification of crude lipoxidase, phosphoglucose isomerase and lactate dehydrogenase a phenyl-type HIC column was used with better than 80% recovery of enzyme activity [40].

HIC has been found to be a particularly useful technique for the separation and isolation of the various isoforms of steroid hormone receptors with high retention of their biological activity [41, 42]. Radioactively labelled estrogen receptor (ER) from human breast tumors was separated by HIC into two isoforms, both having estrogen-binding activity, on a polyether-type stationary phase with a descending salt gradient [42]. The elution profiles suggested significant variation between the breast tumors of different individuals. In addition to its utility in the purification of ER isoforms, HPHIC also may be effective in the elucidation of their structure-function relationships. HPLC analysis of ER variants in the presence of certain protein reagents which modify the hydrophobic moieties of receptors led to the establishment of those receptor domains that interact with the chromatographic surface [43]. The results of this HPHIC study suggest that one isoform interacts through the DNA-binding site whereas the other isoform binds via its steroid-binding domain.

HIC has widely been used for the purification of monoclonal antibodies from mouse ascites fluid after ammonium sulfate precipitation [44, 45]. Monoclonal antibodies were separated from polyclonal immunoglobulin G (IgG) by HIC on a siliceous stationary phase [45]. In IgG purification HIC is often preferred to affinity chromatography, e.g., with a protein A column, because the separation can be carried out under relatively mild conditions and the immunoglobulin thus obtained is not contaminated with traces of the binding protein leaching out from the stationary phase. Another disadvantage of affinity chromatography is that some subclasses of mouse immunoglobulin G are retained only weakly or not at all and, in such cases, HIC offers an excellent alternative.

Although HIC is used almost exclusively for the chromatography of proteins it has been shown to offer a suitable option also for peptide separation in cases where RPC, the most widely used technique for the chromatography of peptides, is not applicable. The mildly hydrophobic sorbents in HIC have been successfully employed for the purification of strongly hydrophobic polypeptides that exhibit poor recovery from a reversed-phase column. The separation of a synthetic analogue of a lipoprotein fragment from its failure sequences and from the incompletely deprotected product by HIC exemplifies such a case [46].

2. The Hydrophobic Effect

2.1. SIGNIFICANCE IN BIOLOGICAL SYSTEMS

The term "Hydrophobic Effect" is often used to refer to the thermodynamic consequences arising from the unfavorable interaction between nonpolar molecular surfaces and water [47, 48]. It has also been used to represent the driving force for the association of nonpolar molecules or binding of hydrophobic moieties in aqueous solutions. The first strong case was made in 1959 for the importance of hydrophobic interactions in determining the architecture and dynamics of biological systems [49]. The hydrophobic effect has also be considered "a unique organizing force, based on repulsion by the solvent instead of attractive forces at the site of organization" [50]. The last description, however, may be misleading, because there is no repulsion between water and hydrophobic molecules. In fact, water and nonpolar molecules do attract each other by ubiquitous van der Waals forces, *vide* Section 2.2.2. It is just that the attraction between water molecules is much stronger than the attraction between water and nonpolar molecules. Thus the total free energy of the system is lower when water molecules stay hydrogen-bonded in the system and the hydrophobic molecules are aggregated. In any case, the hydrophobic effect is a major driving force for the folding of proteins [49, 51-53], for the association of protein subunits, as well as for many aspects of enzyme catalysis and various regulatory and transport processes. Furthermore hydrophobic interactions have been shown to play a prime role in determining the stability of site-specific protein-DNA complexes [48]. They are also responsible for the self association of phospholipids and other lipids to form membranes that comprise the boundaries of cells and intracellular compartments [47, 50]. The low specifity and absence of strong attractive forces associated with hydrophobic interactions make the membranes flexible enough to assure their function in living sytems. In the process of spontaneous folding of proteins hydrophobic interactions act together with strong attractive forces to engender more rigid structures.

2.2. MOLECULAR AND THERMODYNAMIC INTERPRETATION OF HYDROPHOBIC INTERACTIONS

2.2.1. *Effect of Water Structure.* One of the most common treatments of the hydrophobic effect relates this phenomenon to the peculiarities of liquid water that is ubiquitous in nature and shows unique properties. Its structure consists of a distorted network of hydrogen-bonds that gives rise to an anomalously high surface tension. The strong attractive forces between water molecules are believed to cause self-association of nonpolar molecules in water by hydrophobic interactions rather than by mutual attraction between the solutes. The low affinity of nonpolar substances for water is primarily attributed to a large negative entropy change for the process of dissolution associated with a highly ordered water structure around the nonpolar solute molecules [54]. Upon introduction of a solute into water, hydrogen bonds temporarily have to be disrupted due to the formation of a cavity to accomodate the nonpolar solute molecule. It cannot form hydrogen bonds with the surrounding water molecules; thus the water molecules in the immediate vicinity of the nonpolar solute reorient themselves and form as many as possible new hydrogen bonds with themselves. The preferred orientations are illustrated in Fig. 4. The spatial and orientational rearrangement of water molecules surrounding nonpolar solutes thus results in a higher degree of local order than exists in pure water and, therefore, in a decrease in entropy [54]. Although the disruption of hydrogen bonds in the process of creating a cavity in water for a nonpolar solute requires energy, it is partially or fully compensated by the energy gained in the formation of new hydrogen bonds between water molecules at the wall of the cavity. Accordingly, the corresponding net enthalpy change at room temperature has a small positive or even negative value and the solubility of nonpolar substances in water at room temperature is chiefly limited by entropic factors. The transfer of nonpolar solutes into water is also accompanied by a large positive heat capacity change [55] implying that both the enthalpy and entropy changes are strongly temperature dependent. The molecular basis for the observed large heat capacity change associated with the dissolution process of nonpolar molecules can be explained also by extending the above molecular picture. At room temperature the water molecules that form a hydration layer at the hydrophobic surface of the solute molecule have only a few configurations to remain fully hydrogen bonded and therefore occupy a low-energy and low-entropy state. Upon increasing the temperature the water molecules in contact with the nonpolar solute can also adopt other configurations as illustrated in Fig. 4. This occurs at the expense of weakened and deformed hydrogen bonds (melting of the "icelayer"), *i.e.*, more water molecules populate a higher-energy and higher-entropy state. The resulting increase in the system energy with temperature is responsible for the observed large heat capacity change.

Various theoretical approaches to the treatment of the hydrophobic effect have been put forward on the basis of the dissolution process for nonpolar molecules in water [47, 49, 54, 56]. They all postulate that the main driving force for the interaction of hydrophobic moieties in water stems from an increase in disorder due to transfer of water molecules from a highly ordered structure in the surrounding of the nonpolar groups to the less ordered bulk water. It follows then, that contrary to the above described hydrophobic hydration, the reverse process of removal of nonpolar surfaces from water to a nonaqueous (hydrophobic) environment is associated with a large increase in entropy that usually outweighs the accompanying enthalpy change. Accordingly, hydrophobic interactions at room temperature are considered to be entropy-driven and are accompanied by a negative heat capacity change.

2.2.2. *Role of van der Waals Forces.* In the above treatment hydrophobic interactions are attributed to entropic forces that arise from the increased ordering of water molecules in the surroundings of nonpolar surfaces. Another school of thought considers hydrophobic interactions as the result of van der Waals forces, which comprise dispersion, orientation and induction effects [57-60]. Dispersion forces are ubiquitous attractive forces between atoms due to electrodynamic attractions between fluctuating dipols present in all atoms. In contrast to dispersion forces, interactions due to orientation and induction forces occur between permanent dipols and between permanent dipols and induced dipols, respectively. The overall van der Waals forces between any two entities are always attractive. Nevertheless, the van der Waals forces between two objects of different composition

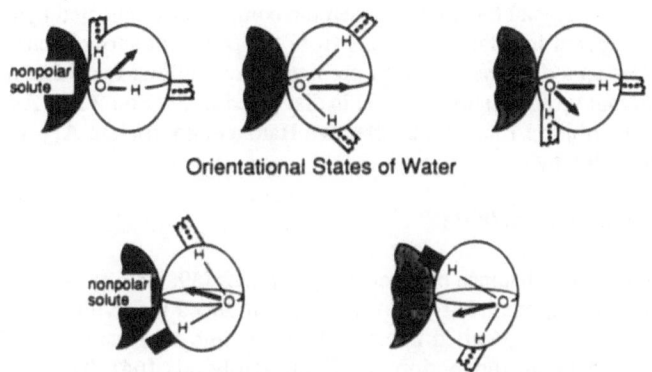

Orientational States of Water

Additional Orientational States (at High Temperature)

Figure 4. Hydrophobic hydration model for the large negative entropy and positive heat capacity effects observed for the transfer of nonpolar solutes into water. Reprinted with permission from [53]. Copyright 1990 American Chemical Society.

(particles, cells or macromolecules), that are immersed or dissolved in a liquid medium, can be attractive or repulsive due to the complex interplay of the corresponding forces between the three components of the system. Repulsion will occur when the van der Waals interactions between the two objects together with those between the molecules of the liquid medium are weaker than the sum of the van der Waals interactions between each object and the liquid medium [58].

In the case of water, which is a highly cohesive, structured and polar liquid, strong orientation and induction effects account for about 80% of the van der Waals interactions between the molecules that are strongly associated by hydrogen-bonds. Water molecules can interact only via dispersion forces with nonpolar molecules because the strong orientation and induction forces, which characterize the interaction among water molecules, are absent between water and nonpolar solutes. Consequently, in a high energy liquid such as water the overall van der Waals interactions between nonpolar objects having low surface energy tend to be strongly attractive [57]. Within this theoretical framework the hydrophobic effect is accounted for by the overall effect of a complex interplay of van der Waals interactions between the nonpolar species themselves, between them and the water and between the water molecules.

A simplifying approach to such a complex interplay of the van der Waals forces has been given in terms of a geometrical factor and the effective Hamaker coefficient [61]. The latter represents the contribution of the molecular properties of the components of the system under investigation such as ionization potentials and polarizabilities to the free energy of interaction. The van der Waals interactions between two objects 1 and 2 which are immersed in a liquid medium 3 will be attractive or repulsive depending on the sign of the corresponding effective Hamaker coefficient A_{132}. A positive value predicts attractive interactions whereas a negative value predicts repulsion between the two objects. The value of A_{132} can be independently derived via a surface thermodynamic approach from surface tension data which can be - at least in principle - experimentally measured [59, 62, 63]. According to this approach the free energy of interaction per unit surface area ΔG_{132} between two objects 1 and 2 in a liquid 3 is described by

$$\Delta G_{132} = \gamma_{12} - \gamma_{13} - \gamma_{23} \tag{1}$$

where γ denotes the interfacial tensions between the components indicated by the subscripts. The evaluation of the relevant interfacial tensions from the pertinent surface tensions and the use of surface tension data as a diagnostic tool to predict the conditions for either retaining or eluting proteins from chromatographic surfaces [59] is discussed in Section 2.3.1. Assuming a flat plate geometry for the two objects 1 and 2, the effective Hamaker coefficient A_{132} can be related to the free energy of interaction by

$$A_{132} = -12\pi d_0^2 \, \Delta G_{132} \qquad (2)$$

where d_0 is the equilibrium distance between the two plates [59, 62, 63].

Another simplifying approach has recently been introduced by dividing the forces involved in hydrophobic interactions, also called interfacial forces, into long range Lifshitz-van der Waals interactions and short range interactions by hydrogen-bonds [64]. The long and short range contributions to the surface tensions of the components have been determined separately by measuring the appropriate contact angles with two different liquids having known long range and short range surface tension components [65]. The results have confirmed that the interfacial attraction between hydrophobic molecules in water arises mainly from the high cohesive energy of water due to hydrogen-bonding and it can be only to a very minor extent attributed to Lifshitz-van der Waals interactions [66].

2.3. HYDROPHOBIC EFFECT IN CHROMATOGRAPHY

As mentioned above the hydrophobic effect dominates the separation process in two major branches of liquid chromatography: reversed-phase chromatography (RPC) and hydrophobic interaction chromatography (HIC). Both techniques employ nonpolar stationary phases of strongly different hydrophobic character. The thermodynamic basis of the chromatographic retention process underlying the separation by these techniques is just as complex a phenomenon as the hydrophobic effect itself. Despite the extensive theoretical and experimental studies, since the role of hydrophobic interactions in maintaining the three-dimensional structure of proteins has been recognized [49], our understanding of the hydrophobic effect is rather meager. Both the concept and the quantitative treatment of hydrophobic interactions are more difficult than those of other non-covalent molecular interactions such as coulombic or hydrogen-bonding. As a consequence the theoretical basis of HIC is also poorly understood in comparison to other chromatographic techniques. Statistical thermodynamics has been applied to the treatment of hydrophobic interactions in general [56, 67] and to hydrophobic interactions associated with retention in RPC [68-70] and protein folding [71, 72]. Other theoretical frameworks of the hydrophobic effect are based on a thermodynamic analysis of the transfer of nonpolar solutes between water and a nonpolar liquid [47]. While informative and illuminating in certain respects, the results of these studies are not readily applicable to the treatment of the retention process in RPC or HIC. The statistical thermodynamic analysis of aqueous solutions usually requires detailed molecular information about the components of the system. Furthermore the results of such analysis focus on the properties of aqueous solutions although in reversed-phase chromatography the eluent is rarely neat water but a hydro-organic mixture or a polar organic solvent. So far these theoretical approaches have not been used successfully to explicate the physicochemical phenomena underlying reversed-phase and hydrophobic interaction chromatography.

An alternative treatment of solute retention is based on the solvophobic theory [73, 74], a thermodynamic approach that has been adapted to both RPC and HIC [75, 76]. Although hydrophobic interactions are solely manifestations of a solvent effect within the framework of the solvophobic theory, it still offers the most rigorous treatment of the physicochemical underpinning of the retention process in both RPC and HIC. Another interesting and somewhat related theoretical approach is based on a surface thermodynamic analysis of "interfacial interactions" between proteins and low energy surfaces in aqueous media [59, 64]. The solvophobic theory and the treatment of interfacial interactions, both of which relate the free energy of binding in chromatography to surface tension or interfacial tension, will be discussed below in some detail.

2.3.1. *Surface Thermodynamic Treatment.* According to this approach [59, 64, 77] the free energy of binding of a protein to a hydrophobic sorbent per unit surface area in the presence of the mobile phase ΔG_{smp} is related to the pertinent interfacial tensions by the following expression

$$\Delta G_{smp} = \gamma_{sp} - \gamma_{sm} - \gamma_{mp} \tag{1a}$$

where γ denotes the interfacial tension and the subscripts s, m and p refer to the sorbent, mobile phase and protein, respectively. The interfacial tensions may be evaluated from the individual surface tensions using an equation-of-state approach [78, 79]. The surface tension of liquids is measured by simple conventional techniques whereas contact angle and adsorption measurements are used to determine protein surface tensions [79, 80]. The surface tension of certain common proteins has been found to range from 65 to 70 mJ/m^2 [60, 80]. The sedimentation volume technique has been utilized to characterize the surface tension of HPLC column packings [81]. The corresponding surface tension values for sorbents used in reversed phase chromatography are in the 32-39 mJ/m^2 range, whereas for typical, more polar, HIC stationary phases they range between 47 and 53 mJ/m^2.

A simple surface thermodynamic model has been used to describe the qualitative features of protein adsorption to hydrophobic surfaces [77] on the basis of the pertinent surface tensions of the protein, mobile and stationary phase and to predict whether attraction or repulsion is the result of interfacial interactions between the protein and sorbent under a given set of conditions [59]. When the surface tension of the mobile phase is higher than that of both the protein and stationary phase, the value of ΔG_{smp} is negative and protein adsorption to the hydrophobic sorbent is expected to occur. Lowering of the surface tension of the mobile phase to a value less than that of the protein elicits the desorption of the protein. When the surface tensions of the mobile phase and of the protein are the same, the value of ΔG_{smp} is zero regardless of the value of the surface tension of the stationary phase, and in this case the extent of protein adsorption is independent of the surface tension of the stationary phase [77, 80]. To stationary phases with relatively high surface tensions, such as those commonly used for HIC, the binding affinity of the protein from water is low. The surface tension of the aqueous mobile phase is conveniently increased by the addition of common inorganic salts whereas upon addition of organic modifiers such as ethylene glycol the surface tension decreases. Thus the experimental conditions required for either retention or elution of proteins in HIC and RPC can be qualitatively predicted on the basis of the pertinent surface and interfacial tensions [59, 60, 82]. The results of such studies indicate that protein retention in HIC and RPC may be modelled qualitatively by a surface thermodynamic approach. The treatment, however, does not provide an explicit relationship between chromatographic retention factor k' and the physicochemical parameters required to express free energy changes associated with the interfacial interactions according to equ. 1a.

2.3.2. *Extended Solvophobic Theory.* To examine the common thermodynamic foundation of both RPC and HIC as far as the energetics of retention is concerned, an adaptation of the solvophobic theory to chromatographic systems [73-76, 83] appears to offer the most useful approach. According to this theory the energetics of interaction between species in solution can be divided into two parts: the interaction in a hypothetical gas phase and the solvent effect. The latter entails the process of solvation of the eluite, the stationary phase ligates, and the complex formed upon binding of the eluite to the stationary phase. Each of them is treated as the sum of two distinct steps: the creation of a cavity of appropriate size and shape for the species in the mobile phase; and their interaction with the surrounding mobile phase after being placed into the cavity. The overall free energy change for the chromatographic retention process [74] is given by the following relationship

$$\Delta G_R^o = \Delta G_{assoc}^o + \Delta G_{cav}^o + \Delta G_{es}^o + \Delta G_{vdw}^o + \Delta G_{red}^o - RT \ln \left(\frac{RT}{PV}\right) \tag{3}$$

where ΔG_{assoc}^o denotes the free energy change for the association process in the hypothetical gas phase, ΔG_{cav}^o, ΔG_{es}^o, ΔG_{vdw}^o are the net free energy changes associated with cavity formation, as well as with electrostatic and van der Waals interactions between solute and solvent molecules, respectively. The term ΔG_{red}^o expresses the reduction of the free energy due to solvent-ligate and solvent-eluite interactions not included in the preceding three terms. The last term accounts for the entropy change arising from the change in free volume, where P and V are the operating pressure and the mean molar volume of the solvent, respectively. The advantage of this eclectic theory is that, in essence, it yields explicit expressions for the retention factor that contain measurable physicochemical quantities only. A detailed discussion of the individual terms in equ. 3 can be found in the literature [75, 76, 84]. According to this theory, hydrophobic interactions are only the manifestation of a solvent effect which is particularly strong in the case of water, a solvent of unusually high surface tension. The adjective "solvophobic" implies that this theory encompasses a treatment of solvent effects in general and is not restricted to water. Nevertheless the application of the theory to chromatography is most appropriate when water rich mobile phases are used in RPC because at high organic solvent concentrations the morphology of the nonpolar chromatographic surface and solvation of the hydrocarbonaceous ligates have also to be considered [85, 86].

Within the hermeneutics of the solvophobic approach the net free energy change related to the reduction of the cavity size as a result of eluite binding, ΔG_{cav}^o, is mainly responsible, together with the van der Waals term, for the hydrophobic effect, *i. e.*, for chromatographic retention. It is usually expressed as

$$\Delta G_{cav}^o = - [N \Delta A_s + 4.8 \, N^{1/3} \, (\kappa-1) \, V^{2/3}] \, \gamma \tag{4}$$

where ΔA_s is the reduction in molecular surface area exposed to mobile phase upon formation of an eluite-ligate complex, which is equivalent to the molecular contact area upon binding, N is Avogadro's number, and κ corrects the macroscopic surface tension of the mobile phase, γ, to molecular dimensions. According to equ. 4 the cavity term ΔG_{cav}^o, which is believed to be the main driving force for eluite binding to the stationary phase, increases with the contact area between the eluite and the stationary phase ligates and with the microthermodynamic surface tension of the mobile phase. Thus the manipulation of the surface tension offers a powerful means to modulate retention in both RPC and HIC. Indeed there is ample experimental evidence that retention in these chromatographic techniques decreases with the surface tension of the eluent. The dependence of the surface tension for salt solutions and hydro-organic mixtures on the concentration of salt and

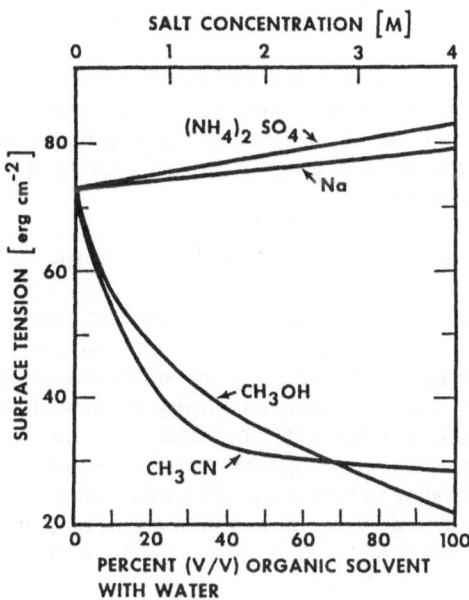

SALT CONCENTRATION [M]

SURFACE TENSION [erg cm^{-2}]

(NH$_4$)$_2$ SO$_4$

Na

CH$_3$OH

CH$_3$CN

PERCENT (V/V) ORGANIC SOLVENT
WITH WATER

Figure 5. Surface tension of salt solutions and hydro-organic solvent mixtures as a function of salt concentration and organic solvent content, respectively. Fig. 10 of Ref. [75], reprinted with permission of the publisher.

organic solvent component, respectively, is depicted in Fig. 5. It is seen that addition of common inorganic salts and organic solvents increases and decreases the surface tension of aqueous solutions, respectively. In chromatographic practice the retention free energy has to be adjusted so that the retention factors fall in the practical range of 0.1 to 20. The stationary phases used in RPC and HIC have widely different hydrophobic character. For this reason the two chromatographic techniques employ mobile phases having widely different surface tensions. In HIC, where the hydrophobic character of the stationary phase is much weaker than in RPC, neat water is usually a strong eluent and retention is brought about by increasing the salt concentration with a concomitant increase in the surface tension of the mobile phase. On the other hand, neat water is a very weak eluent in RPC and a lowering of the surface tension of the eluent by addition of an organic modifier is necessary to diminish retention to a practical value. According to these considerations and as suggested by Fig. 5, HIC and RPC are chromatographic techniques both based on the solvophobic effect but are performed at the two ends of the surface tension spectrum as far as the mobile phase is concerned. Despite the widely different domains of practical application both are governed by the same fundamental physicochemical principles underlying the solvophobic effect.

The solvophobic theory offers a comprehensive framework to analyze various aspects of RPC and HIC. Here our attention is focussed on a quantitative description of the effect of salt on retention in HIC and this subject is discussed in the subsequent section.

3. Role of Salt in HIC

The effect of salt on hydrophobic interactions is one of the most intriguing phenomena, not only because salts are the primary means to regulate retention in HIC and, therefore, the effect is of great chromatographic significance, but also because salting-out of proteins is important from the technological point of view. Furthermore the effect of salts on processes in biological systems has wide ranging physiological consequences.

3.1. SIMPLE SOLVOPHOBIC TREATMENT OF SALTING-OUT AND HIC OF PROTEINS

In order to treat the effect of salt on salting-out of proteins and on the chromatographic retention in HIC, the solvophobic theory has been extended to aqueous salt solutions [76, 83, 84]. Within this framework the retention free energy in HIC is related to the surface tension of aqueous solutions.

According to the hermeneutics of the solvophobic theory the magnitude of retention due to hydrophobic interactions in a given chromatographic system is largely determined by the cavity term and as a result increases with the hydrophobic contact area between the eluite and the chromatographic surface upon binding as well as with the surface tension of the mobile phase. As illustrated in Fig. 5 the surface tension, γ, of aqueous solutions of most neutral salts, which are considered antichaotropic or cosmotropic ("structure making"), increases linearly with the salt concentration, so that

$$\gamma = \gamma_0 + \sigma m \tag{5}$$

where γ_0 is the surface tension of neat water, m is the molal concentration and σ is the molal surface tension increment of the salt. The latter is a convenient measure of the propensity of a salt to increase the surface tension of aqueous solutions. It is noted that the σ values parallel Hofmeister's lyotropic series [87] which has been a solely empirical ranking of salts according to their efficacy for salting-out of proteins. Values of σ for a variety of common neutral salts are given in Table 2. These salts have positive σ, thus, they raise the surface tension of water and enhance hydrophobic adsorption in view of equ. 4. In HIC retention increases with the concentration of such salts in the eluent. This property of these cosmotropic salts is believed to arise from their proclivity to order water molecules at interfaces. Here we consider only the effect of such salts on hydrophobic interactions.

The dependence of the logarithmic retention factor on the salt concentration in HIC has been expressed [76, 83] as

$$\ln k' = \ln k'_0 - \frac{1}{RT} \frac{Bm^{1/2}}{1+Cm^{1/2}} - \frac{1}{RT} D\mu m + \frac{1}{RT} vm + \frac{1}{RT} \Delta A_s \sigma m \tag{6}$$

where k'_0 is a hypothetical retention factor for the eluite that would be obtained with the given chromatographic system in the absence of the salt, and ΔA_s is the molecular contact area upon binding of the eluite to the stationary phase. B and C are constants for a given protein in the Debye term, μ is the dipole moment of the protein, D is a function of the dielectric constant of the medium and v is related to the net free energy change due to van der Waals interactions. According to equ. 6 and as illustrated in Fig. 6, the retention factor in HIC first decreases with increasing salt concentration. This occurs because the binding of the protein is attenuated by the effect of the salt on electrostatic interactions that is expressed by the second and third terms on the RHS of equ. 6,

Table 2. Molal surface tension increments of various salts[a]

$\left(\times 10^3 \dfrac{\text{dyne g}}{\text{cm mol}}\right)$	Salt	Reference	$\left(\times 10^3 \dfrac{\text{dyne g}}{\text{cm mol}}\right)$	Salt	Reference
0.45^b	KSCN	[121]	1.82	$CuSO_4$	[122]
0.55	$NaClO_3$	[122]	1.96	K_2-tartrate	[123]
0.74	NH_4I	[122]	2.0	$Ba(NO_3)_2$	[122]
0.79	LiI	[122]	2.0	LiF	[121]
0.84	KI	[122]	2.02	Na_2HPO_4	[124]
0.85	NH_4NO_3	[122]	2.10	$NiSO_4$	[122]
0.86	$KClO_3$	[121]	2.10	$MgSO_4$	[124]
1.02	NaI	[122]	2.10	$MnSO_4$	[122]
1.06	$NaNO_3$	[122]	2.16	$(NH_4)_2SO_4$	[122]
1.14	NH_4Br	[122]	2.27^d	$ZnSO_4$	[122]
1.16	$LiNO_3$	[122]	2.35	Na_2-tartrate	[123]
1.26	LiBr	[122]	2.58	K_2SO_4	[122]
1.31	KBr	[122]	2.66	Na_3PO_4	[122]
1.32	NaBr	[122]	2.73	Na_2SO_4	[122]
1.39	CsI	[122]	2.78	Li_2SO_4	[122]
1.39	NH_4Cl	[122]	2.78	$FeCl_3$	[122]
1.4	$KClO_4$	[125]	2.93	$BaCl_2$	[122]
1.55	$FeSO_4$	[122]	3.12	K_3-citrate	[123]
1.63^c	LiCl	[122]	3.16^e	$MgCl_2$	[122]
1.64	NaCl	[122]	3.66^f	$CaCl_2$	[122]
1.57	$CsNO_3$	[121, 126]	4.34	$K_3Fe(CN)_6$	[121]

[a]The deviation from equ. 5 is less than 1.5% except in the cases indicated.
[b]0 - 0.5 m only.
[c]0 - 2.0 m only.
[d]Maximum deviation, 8%.
[e]Maximum deviation, 18%.
[f]Maximum deviation, 28%.
[g]Average value of two sets of data.

which are both negative. Above a certain salt concentration hydrophobic interactions become predominant over electrostatic effects and as a result the retention factor increases with further increase in salt concentration. At sufficiently high salt concentrations the second term on the RHS of equ. 6 becomes constant and the dependence of ln k' on salt molality is linear. Under such conditions the logarithmic retention factor can be expressed as a function of the salt concentration [83] according to the simple relationship

$$\ln k' = \text{const.} + \lambda m \qquad (7)$$

where

$$\lambda = \frac{1}{RT}\left(-D\mu + \nu + \Delta A_s \sigma\right) \qquad (8)$$

458

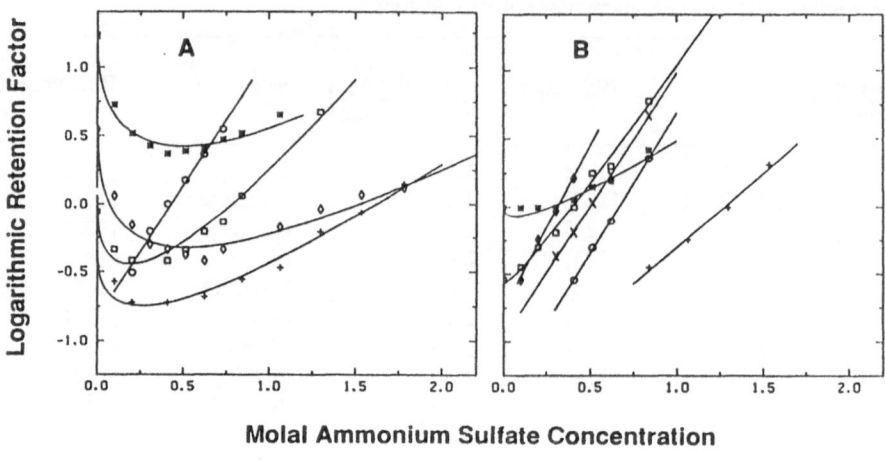

Molal Ammonium Sulfate Concentration

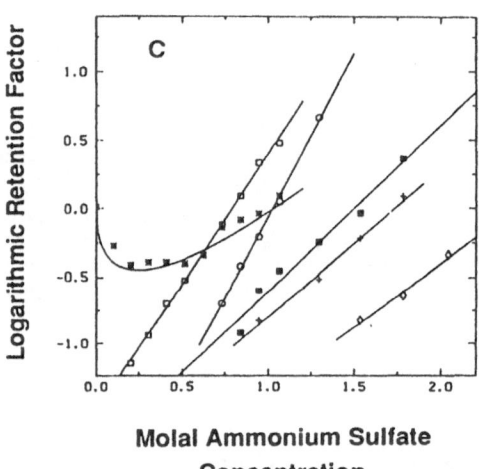

**Molal Ammonium Sulfate
Concentration**

Figure 6. Plots of the logarithmic retention factor of proteins against the molality of ammonium sulfate in 0.1 M phosphate buffer, pH 7.0, used as the eluent. The results were obtained on three different columns packed with homemade HIC sorbents having the following ligates: A, polyvinyl alcohol; B, oligoaminoalcohol; C, polyether. Proteins: (+) ribonuclease A; (◊) cytochrome c; (■) myoglobin; (◖) α-chymotrypsinogen A; (○) ovalbumin; (♦) γ-globulin; (×) bovine serum albumin; (✳) lysozyme. Reprinted from Ref. [88], p. 3264 by courtesy of Marcel Dekker Inc.

The limiting slope λ of the plot ln k' against salt molality has been termed hydrophobic interaction parameter. It is a linear function of both the contact surface area between the protein and the stationary phase ligates upon binding and the molal surface tension increment of the salt. Results of numerous studies concerning the salt effect on the retention of proteins in HIC lend support to key

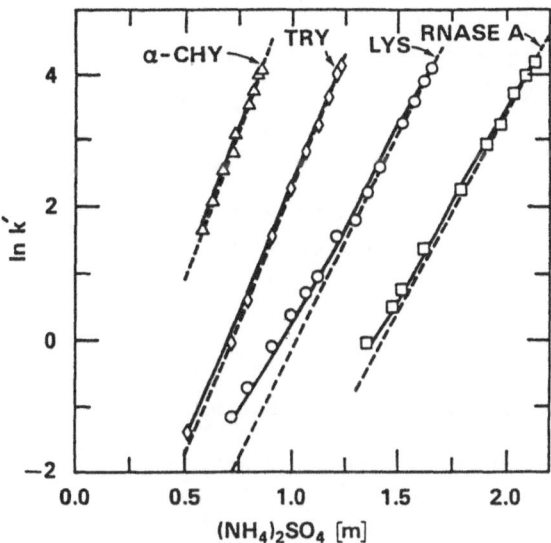

Figure 7. Plots of the logarithmic retention factor against the salt molality in HIC of various proteins. Column: 75 x 7.5 mm, 10 μm TSK-Phenyl-5PW; mobile phase, 50 mM phosphate buffer, pH 7.0, containing ammonium sulfate at different concentrations. Sample components: (△) α-chymotrypsinogen, (◊) trypsinogen, (□) ribonuclease A, (○) lysozyme. The dashed lines represent the limiting slopes of the plots at high salt concentrations, *i. e.*, those of the straight lines obtained by least square fit of the five highest data points. Fig. 1 of Ref. [89], reprinted with permission of the publisher.

features of the theory [34, 76, 88-90].

Plots of ln k' versus salt molality for the retention of lysozymes from related bird species that differ between one and 21 amino acid residues yielded straight lines at sufficiently high salt concentrations [34]. The slopes of the plots obtained with data measured with various lysozymes are nearly alike indicating similar contact surface areas in the light of equ. 8. This finding is consistent with the observation that in this particular case the amino acid substitutions do not drastically alter the three-dimensional protein structure. For any given protein the hydrophobic interaction parameter increases with the surface tension increment of the salt as expected in the absence of specific salt binding effects. In another study the relationship between protein surface area and retention behavior in HIC has also been examined [89]. The molecular surface areas accessible to a 4 Å diameter spherical probe were calculated from crystallographic data for a few proteins of similar molecular size and isoelectric points such as ribonuclease, lysozyme, trypsinogen and α-chymotrypsinogen by using a well established technique [32]. The hydrophobic interaction parameters were evaluated from the limiting slopes of plots of the logarithmic retention factor against the salt molality, as illustrated in Fig. 7, and found to correlate with the protein surface area. In conformity with the theory, the dependence of the hydrophobic interaction parameter on the molal surface tension increment of the salt was also linear as shown in Fig. 8 [89]. The theory predicts that the values of the intercepts in Fig. 8 represent the free energy change due to the net electrostatic and van der Waals interactions whereas the slope yields the contact surface area

460

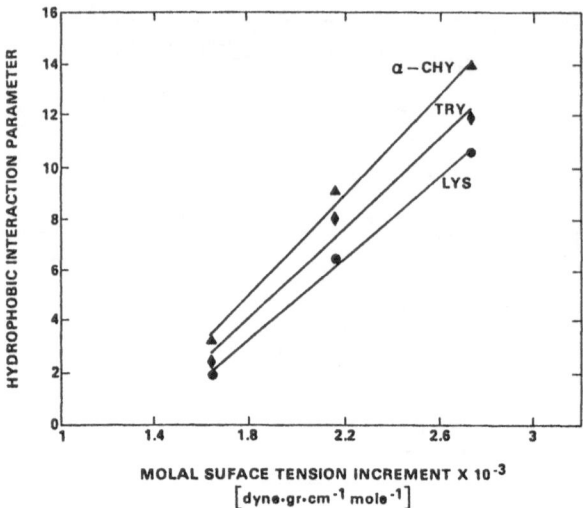

Figure 8. Linear dependence of the hydrophobic interaction parameter of proteins on the molal surface tension increment of the salts in the eluent. Column, 75 x 7.5 mm, 10 μm TSK-Phenyl-5PW; mobile phase, 50 mM phosphate, pH 7.0 containing different concentrations of ammonium sulfate, sodium chloride or sodium sulfate. Proteins: (▲) α-chymotrypsinogen, (♦) trypsinogen, (●) lysozyme. Fig. 4 of Ref. [89], reprinted with permission of the publisher.

upon binding of the eluite to the chromatographic surface. In the case investigated the area calculated from the slope amounts to about 20% of the accessible surface area of the proteins. The predictions of the theory for the dependence of the hydrophobic interaction parameter on the molecular surface area is further supported by the experimental results obtained from the studies of the retention of small oligoriboadenylic acid homologues in HIC [88]. First, from the linear plots of the logarithmic retention factor versus the salt concentration the hydrophobic interaction parameters were calculated for the individual oligonucleotides on two different HIC columns. Then these values were plotted against the number of adenylyl phosphate residues and straight lines were obtained with the data for each column as shown in Fig. 9. According to equ. 8, the slopes of the lines in Fig. 9 are expected to be proportional to the contact area for any single adenylyl phosphate residue which binds to the stationary phase ligates. The linearity of the plots in Fig. 9 indicates that with such relatively small eluite molecules the molecular contact area is indeed the sum of the contact areas of the adenylyl phosphate residues present in the eluite molecule, *i. e.*, each of the residues binds to the chromatographic surface in an equiaccessible fashion. The respective slopes of the lines in Fig. 9 for the two columns which contained polyoxyethylene ligates of different average molecular weight are similar indicating that the contact area for a residue is only slightly dependent on the length of the polyoxyethylene ligates.

The simple surface tension argument breaks down in the case of certain salts such as $MgCl_2$ [91, 92] and this is explained by the proclivity of these salts to bind specifically to the protein surface at neutral pH. Indeed, as has been shown independently [93], *vide* Section 3.3., the simple surface tension argument is not expected to be applicable to such complex phenomena as protein interactions with interfaces when specific salt binding by the protein also occurs. Despite its

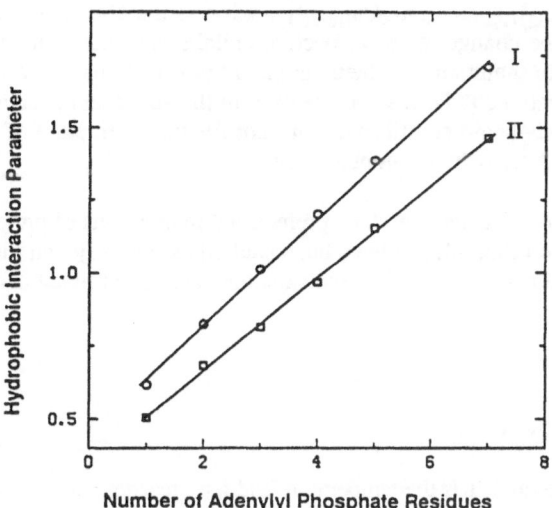

Figure 9. Plot of the hydrophobic interaction parameter measured on two polyether columns against the number of adenylyl phosphate residues in oligoriboadenylic acid homologues used as the eluites. Columns I and II contained polyoxyethylene ligates of the average molecular weight of 400 and 1000, respectively. The hydrophobic interaction parameters have been evaluated from retention data obtained with eluents containing different concentrations of phosphate buffer, pH 6.3. Reprinted from Ref. [88], p. 3256 by courtesy of Marcel Dekker Inc.

simplicity, however, it offers in many cases a very convenient approach to explicate the effect of salt on hydrophobic interactions. The use of preferential interaction parameters for the protein and salt under investigation, if available (cf. Section 3.2.2.), might offer an alternative approach to treat salt effects in HIC [93, 94].

3.2. APPLICATION OF THE THERMODYNAMIC THEORY OF LINKED FUNCTIONS TO HIC

3.2.1. *Wyman's Thermodynamic Theory of Linked Functions.* The effect of salt on protein adsorption in HIC has also been treated by applying Wyman's thermodynamic theory of linked functions for equilibrium reactions [95-97]. The linked functions approach [98] proffers a general theory for ligand binding equilibria and was used successfully in the study of ligand binding by proteins. According to Wyman, "it provides a powerful touchstone for exploring the implications of any proposed mechanism as well as for establishing the consistency of underlying observations at a phenomenological level" [98].

According to this notion, for a given equilibrium reaction, when the reactants and products have sites capable of interacting with an external agent x, the change in the equilibrium constant K of the reaction with changing activity of the external agent a_x is given by

$$\frac{\partial \ln K}{\partial \ln a_x} = \left(\frac{\partial m_x}{\partial m_{prod}}\right)_{T,\mu} - \left(\frac{\partial m_x}{\partial m_{react}}\right)_{T,\mu} \tag{9}$$

where the terms $(\partial m_x/\partial m_{prod})_{T, \mu}$ and $(\partial m_x/\partial m_{react})_{T, \mu}$ express the change in the molality of the external agent, m_x, with the change in the respective molalities of the product, m_{prod}, and the reactant, m_{react}, required to maintain the chemical potential μ of both water and external agent constant at a given temperature, T. Equ. 9 will be used in the subsequent section to express the dependence of the thermodynamic equilibrium constant for the interaction of proteins with the stationary phase on the salt activity in the mobile phase.

3.2.2. *Preferential Interaction Parameter.* The preferential interactions of proteins with salt and water in an aqueous solution [93, 94, 99] have been studied extensively and the values of the so called "preferential interaction parameter", ξ, were experimentally determined in certain cases. This parameter is defined as

$$\xi = \left(\frac{\partial g_s}{\partial g_p}\right)_{T, \mu_w, \mu_s} \tag{10}$$

where μ is the chemical potential, T is the temperature and g represents mass concentration whereas the subscripts w, p, and s refer to water, protein and salt, respectively. In molal units the preferential interaction parameter becomes

$$\left(\frac{\partial m_s}{\partial m_p}\right)_{T, \mu_w, \mu_s} = \left(\frac{\xi M_p}{M_s}\right) \tag{11}$$

where m is the molality and M denotes the molecular weight.

According to equ. 10, ξ is the infinitesimal amount of salt, ∂g_s, required to maintain the chemical potential of both the water and the salt at a constant level upon addition of ∂g_p grams of protein to the bulk solution at constant temperature. Preferential interaction parameters for a limited number of proteins and salts have been directly measured using densimetric techniques [100]. The results suggest a strong correlation between the preferential interaction parameter and the effect of salt on protein solubility [94]. The preferential interaction parameters with antichaotropic salts are negative and therefore indicate that these salts are excluded from the immediate vicinity of the protein, resulting in preferential hydration. This can be viewed as a manifestation of the increasing surface free energy (surface tension) of water with the concentration of an antichaotropic salt. According to IUPAC, the molar reduced excess of salt per unit surface area at the protein/liquid interface, Γ_s, is defined as the excess of the amount of salt in the actual system, over the amount of salt in a reference system containing the same total amount of liquid as the actual system, and in which a constant composition (equal to that of the bulk liquid in the actual system) is maintained throughout the liquid phase. It is related to the surface tension according to the Gibbs adsorption isotherm [101] by the expression

$$\Gamma_s = -\frac{m_s}{RT}\frac{d\gamma}{dm_s} \tag{12}$$

where m_s denotes the salt molality and $d\gamma/dm_s$ is the molal surface tension increment of the salt. When the surface tension increment is positive, the salt is excluded from the protein/water interface, that is, according to equ. 12 the molar reduced excess of salt is negative. It is convenient [94, 99, 101] to relate the molar reduced excess of salt to the preferential interaction parameter by

$$A_p \Gamma_s = \left(\frac{\xi M_p}{M_s}\right)^*$$

(13)

where A_p is the accessible surface area for salt binding per mol of protein and the asterisk on the RHS denotes that the quantity in brackets is assumed to originate from the sole effect of the surface free energy. By combining equs. 11 - 13, we can evaluate that portion of the preferential interaction parameter, ξ^*, that arises solely from the effect of the surface tension by the following relationship

$$\xi^* = \left(\frac{\partial g_s}{\partial g_p}\right)^*_{T,\mu_w,\mu_s} \frac{A_p m_s M_s}{RT M_p} \left(\frac{\partial \gamma}{\partial m_s}\right)_{T, m_p}$$

(14)

Indeed the experimentally obtained preferential interaction parameters could be correlated with the surface tension increments of NaCl and Na_2SO_4, which are not bound by the protein specifically, according to equ. 14 [94, 99].

The preferential interaction parameter is a concentration dependent property. For antichaotropic salts ξ becomes a greater negative number as the salt concentration is increased, and the trend appears to be nonlinear.

The increase of free energy due to preferential hydration of the protein is proportional to its surface area [99]. The binding of a protein to hydrophobic sorbents is accompanied by the removal of a given fraction of its molecular surface area as well as that of the sorbent surface from exposure to solvent. Therefore, the unfavorable free energy change experienced by the protein and the stationary phase upon addition of salt to the solution is opposed by the favorable free energy change associated with the adsorption process. This explains that in HIC protein retention is favored at high concentrations of an antichaotropic salt in the eluent [102].

For proteins the preferential interaction parameters with salts having divalent cations and univalent anions, e.g. $MgCl_2$, have small negative or even positive values [93]. It is believed that these salts preferentially bind to proteins and the specific salt binding is responsible for their irregular behavior that manifests itself in their propensity to salt-in rather than salt-out proteins.

The change in the thermodynamic equilibrium constant K for the interaction of proteins with the stationary phase upon changing the salt activity in the mobile phase can be related to the pertinent preferential interaction parameters using Wyman's thermodynamic theory of linked functions [97] according to the following relationship

$$\frac{\partial \ln K}{\partial \ln a_s} = \xi_{s,c} \frac{M_c}{M_s} - \left(\xi_{s,p} \frac{M_p}{M_s} + n\xi_{s,l} \frac{M_l}{M_s}\right)$$

(15)

where a_s is the activity of the salt in the eluent, $\xi_{s,c}$, $\xi_{s,p}$ and $\xi_{s,l}$ are the respective preferential interaction parameters of the salt with the protein-stationary phase complex, the protein eluite and the protein binding ligates of the stationary phase, and n represents the number of ligates involved in the binding of a protein molecule. M denotes the molecular weight and the subscripts s, p, l and c refer to salt, protein, interacting stationary phase moiety and protein-stationary phase complex, respectively. When the preferential interaction parameters of a given salt with both the protein and

464

the sorbent are available, the trend for the change in capacity factor with salt concentration in HIC can be predicted from Wyman's linkage theory [97]. Salts with negative $\xi_{s,p}$ and $\xi_{s,l}$ for a given protein promote the adsorption of that protein in HIC whereas desorption is favored by salts that preferentially bind to both the protein and stationary phase, i.e., have positive preferential interaction parameters. The advantage of the thermodynamic theory of linked functions is that it also takes into account the role of the stationary phase in determining the retention in HIC. The usefulness of the theory remains limited, however, until a sufficient amount of experimental data becomes available and particularly until the dependence of the preferential interaction parameters on the salt concentration is established.

3.3. SPECIFIC SALT BINDING BY PROTEINS

As mentioned before, when certain salts such as $MgCl_2$ and $CaCl_2$, which are believed to be bound by the protein in a specific way, are used in the eluent, the observed retention in HIC deviates from that predicted by the simple surface tension argument, according to equ. 7. They do not enhance protein binding to the chromatographic surface as much as expected from the increase in surface tension and can even weaken protein binding [91, 92]. Furthermore, despite their effect in increasing the surface tension of aqueous solutions, these salts increase protein solubility, i. e., have salting-in properties. For this reason the case of specific salt binding requires special attention. Fig. 10 illustrates the irregular effect of magnesium salts on the retention behavior of various proteins in HIC [92]. In spite of the nearly identical surface tension increments of ammonium sulfate and magnesium sulfate the slopes of the ln k' versus salt molality plots for a variety of proteins are significantly lower for magnesium sulfate. As shown in Fig. 10C, in the case of magnesium chloride the retention factor of α-chymotrypsin and α-chymotrypsinogen increases, although the enhancement of the retention is much smaller than expected from the surface tension increment of $MgCl_2$. More importantly, the retention factor of lysozyme remains unchanged and that of trypsin decreases slightly with increasing salt concentration. These experimental results corroborate other observations regarding the peculiar effect of certain salts with divalent cations on the solubility and stability of various proteins. In spite of their high surface tension increments, such salts particularly with monovalent anions can have a destabilizing effect on proteins and may even act as salting-in rather than salting-out agents due to their specific binding to proteins. Consequently their effectiveness in salting-in and destabilizing proteins has been interpreted in terms of preferential interactions [93]. Whereas for a series of sodium salts the observed preferential hydration has been shown to be a consequence primarily of the surface tension effect [102], no such correlations between the preferential interaction parameters and surface tension increments have been found for $MgSO_4$ and a fortiori for $MgCl_2$ [93]. In these cases the preferential exclusion of salt due to the increase in surface tension according to equ. 14 is offset by specific salt binding to the protein. This results in a lesser degree of preferential hydration than predicted by the solvophobic theory and concomitantly deviations from the expected salting-out effect occur.

The anomalous retention behavior of several milk proteins in HIC, with ammonium sulfate in the mobile phase with and without urea as an additional mobile phase modifier, was assumed to be due to specific salt binding by the protein. The observed dependence of the capacity factor on the salt concentration in the eluent was fitted to a model based on the sequential binding of salts to proteins according to Wyman's theory of linked functions [103].

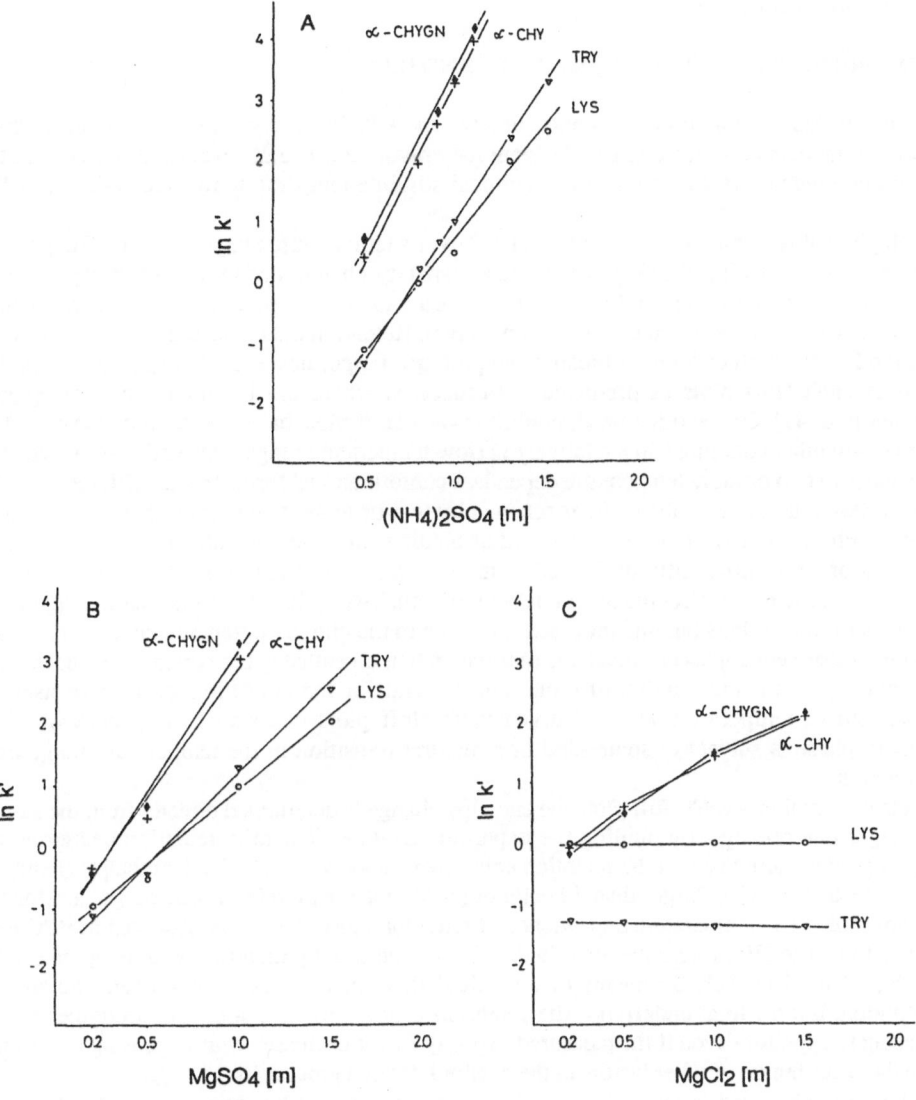

Figure 10. Plots of the logarithmic retention factors of proteins in HIC against the molality of various salts: A, ammonium sulfate; B, magnesium sulfate; C, magnesium chloride. Column: 75 x 7.5 mm, 10 μm TSK-Phenyl-5PW; eluent: 20 mM Tris-HCl, pH 7.0, containing different concentrations of the salt; flow rate, 1.0 ml/min; temp. 25°C. Proteins: (α-CHYGN) α-chymotrypsinogen, (α-CHY) α-chymotrypsin, (LYS) lysozyme, (TRY) trypsin. Figs. 5, 6 and 7 of Ref. [92], reproduced with permission of the publisher.

4. Effect of Temperature

4.1. EXPERIMENTAL OBSERVATIONS IN RPC AND HIC

The effect of temperature on the retention in RPC and HIC is largely determined by the enthalpy change for the transfer of the eluite to the stationary phase. It is usually evaluated from plots of the logarithmic retention factor against the reciprocal absolute temperature that are called van't Hoff plots.

Such plots of retention data measured in RPC over a temperature range of about 30 degrees are often linear [104, 105] indicating that no heat capacity change occurs upon binding. In many practical cases, therefore, the enthalpy change appears to be essentially temperature independent over the narrow temperature range of the experiment. Retention data obtained in RPC with water-rich mobile phases over a much broader temperature range, however, is expected to result in nonlinear van't Hoff plots as predicted for processes where the hydrophobic effect plays a dominant role [47]. On the other hand, nonlinear van't Hoff plots have also been observed in RPC under certain other conditions in a relatively narrow temperature range. One such case is when the eluite exists in two or more temperature dependent conformational forms having different retention factors [106]. Alternatively the eluite is retained by two or more different mechanisms when the stationary phase contains more than one kind of binding sites and the enthalpies of binding at the two kinds of sites sufficiently differ [107]. In such cases the van't Hoff plots can be strongly nonlinear because they reflect the overall retention enthalpy, which is a weighted average of the enthalpies for the various binding processes involved in the chromatographic retention. Nonlinear coupling of these enthalpies can make the observed retention enthalpy temperature dependent, even if the enthalpies for the binding of either species and for the conformation change itself are independent of temperature [105]. Curved van't Hoff plots can also be expected when the stationary phase is subject to some kind of structural transition in the temperature range under investigation.

In the case of linear van't Hoff plots the enthalpy change is determined directly from the slope of the straight line obtained by plotting the experimental data. Generally retention decreases with increasing temperature so that the retention enthalpy is negative. In RPC the enthalpy change for small molecules is rarely larger than 5 kcal/mol [105], but for proteins it can be greater than 20 kcal/mol [108]. The temperature dependence of retention most commonly observed in RPC is the opposite to that in HIC where the magnitude of retention usually increases with temperature [19, 96, 109, 110]. However, by means of statistical thermodynamics, it has been shown, that hydrophobic interactions underlying the retention process in RPC are also strengthened with increasing temperature even if the measured capacity factor decreases with increasing temperature due to the contribution of other factors to the overall retention process [68].

In HIC the observed increase in retention with temperature has been ascribed either to an enhancement of hydrophobic interactions with increasing temperature [47] or to temperature induced conformational changes of proteins and concomitant increase in hydrophobic contact area upon binding to the chromatographic surface or both. The effect of temperature on the conformational behavior of proteins in HIC was studied [96] using gradient elution. Retention data was measured as a function of temperature and on-line UV spectroscopic analysis was employed to monitor conformational changes. The results indicate that certain proteins indeed experience conformational changes under conditions of the HIC experiments. However, even if the integrity of the protein molecules were preserved the use of gradient elution precludes a thermodynamic analysis of the effect of temperature on the retention.

Temperature is an operational variable not yet widely exploited in the HIC of proteins. For this reason its effect has been studied less extensively than that of salt concentration, which is now relatively well documented and understood [76, 83, 102, 111]. The only thorough thermodynamic study was carried out by examining the retention of homologous aliphatic alcohols and carboxylic acids on octyl-agarose [112, 113]. However, no salt was used in the eluent to modulate retention. The standard thermodynamic functions (ΔC_p^o, ΔH^o, ΔS^o) for the interaction of these small amphiphilic substances with the stationary phase were evaluated from isocratic retention data measured at different temperatures. With neat water as the eluent the retention of the eluites usually increased with temperature and the corresponding van't Hoff plots were nonlinear. Most of the ΔC_p^o values calculated from the experimental data were negative and virtually all the ΔS^o values were positive as expected for processes in which hydrophobic interactions play an important role [47].

Historically, the dissolution of small liquid hydrocarbons in water has provided a model system for evaluating the thermodynamics of hydrophobic interactions *per se*. In Section 4.2., therefore, the temperature dependence of the solubility of benzene in water and of the thermodynamic functions associated with the dissolution process are discussed in order to examine the temperature dependence of the retention in HIC and RPC in the light of this simple model system.

Recent studies in biophysical chemistry have addressed the effect of temperature on hydrophobic interactions on the basis of calorimetric data on protein folding. Analysis of the thermodynamic data and their comparison to those obtained for the dissolution of nonpolar compounds revealed the existence of certain extrathermodynamic relationships [114]. We expect the results of these studies to provide a framework also for the treatment of the temperature effect on the retention in HIC. Therefore, the results of these studies are summarized in Section 4.3., and subsequently the possible adaptation of these theoretical approaches to HIC is discussed.

4.2. SOLUBILITY OF NONPOLAR SUBSTANCES IN WATER

A simple model system for investigating the thermodynamics of the hydrophobic effect *per se* is the transfer of small nonpolar molecules (*e.g.*, hydrocarbons) from their aqueous solution to their pure liquid state [48, 51, 52]. The temperature dependence of the standard thermodynamic functions for the transfer of benzene from aqueous solution to pure liquid benzene is depicted in Fig. 11. The data reveals that the standard enthalpy of transfer, ΔH_{tr}^o, decreases linearly with increasing temperature in the range from 15 to 35°C due to the large negative standard heat capacity change: $\Delta C_{p,tr}^o = -53.8$ cal mol^{-1} K^{-1} [48]. This and the observed large positive standard entropy change represent key features of the process of removing the nonpolar benzene molecules from water [47, 54, 55].

The negative logarithm of the solubility of benzene in water, ln $X_{b,w}$, reaches a maximum (corresponding to the minimum solubility) at 15.8°C and this temperature, where ΔH_{tr}^o is zero, is denoted by T_H. The free energy of transfer, ΔG_{tr}^o, decreases with increasing temperature until a minimum is reached around 110°C. This temperature, at which the entropy of transfer, ΔS_{tr}^o, approaches zero, is denoted by T_S. Consequently, hydrophobic interactions, as measured by the value of ΔG_{tr}^o, come to be stronger with increasing temperature.

From the chromatographic point of view the negative logarithmic solubility of benzene, which corresponds to the logarithmic retention factor, is of interest, because both parameters are related to the transfer of molecules from an aqueous to a nonpolar phase. For this reason, the temperature dependence of the negative logarithm of the solubility of benzene is expected to parallel that of the logarithmic retention factor in HIC and in RPC when neat aqueous mobile phases are used.

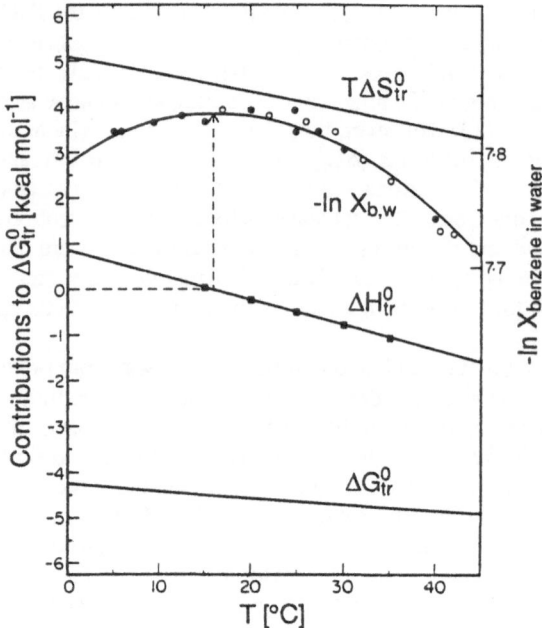

Figure 11. Thermodynamic parameters associated with the transfer of benzene from water to pure liquid benzene. The temperature dependence of the standard thermodynamic parameters was computed from solubility data assuming a constant ΔC_p^o of -53.8 cal mol^{-1} K^{-1}. Fig. 5 of Ref. [48], reproduced with permission of the publisher.

4.3. THERMAL UNFOLDING OF PROTEINS

4.3.1. *Extrathermodynamic Relationships*.

Protein unfolding in aqueous solutions is associated with a large increase in heat capacity [115, 116], and the process is believed to share common features with the exposure of small nonpolar molecules to water [51, 114].

For several globular proteins plots of the specific enthalpy change associated with unfolding against the temperature appear to converge at 110°C, when the experimental values are extrapolated with the assumption of invariant heat capacity change over the whole temperature range under consideration [116]. Similarly, plots of the specific entropy change for protein unfolding against the temperature also show an approximate intersection point near 110°C [116]. These "convergence" temperatures are denoted by T_H^* and T_S^* and the corresponding enthalpy and entropy changes by ΔH^* and ΔS^*, respectively. The temperature dependence of ΔS^o and ΔH^o with reference to the respective T_S^* and T_H^* values can then be expressed as

$$\Delta S^o = \Delta S^* + \Delta C_p^o \ln(T/T_S^*) \tag{16}$$

and

$$\Delta H^o = \Delta H^* + \Delta C_p^o (T - T_H^*) \tag{17}$$

Table 3. The isoentropic temperature, T_S^*, and the corresponding entropy change, ΔS^*, evaluated for certain processes based on the hydrophobic effect from the linear dependence of the entropy change on the heat capacity change at 25°C [114]. Also shown are the isoenthalpic temperature, T_H^*, and the corresponding enthalpy change, ΔH^*, taken from Ref. [52] or calculated from data in Ref. [127].

Process	T_S^* [°C]	ΔS^* (cal mol^{-1} K^{-1}) for T_S^*=112°C	T_H^* [°C]	ΔH^* (kcal mol^{-1})	Substances
Protein denaturation	110	4.3	110	1.5	11 proteins
Liquid hydrocarbon dissolution in water	106	~0	20	~0	6 alkanes and aromatics
Gas dissolution in water	121	-18.8	--	--	rare gases and lower aliphatics
Solid dissolution in water	102	3.8	70	2.7	hydrophobic cyclic dipeptides

For substances that have a common convergence temperature T_S^*, the change in entropy is a linear function of ΔC_p^o according to equ. 16. Quite remarkably, the $\ln(T/T_S^*)$ values have been found to be virtually identical not only for the dissolution in water of substances as different as liquid hydrocarbons, saturated hydrocarbon vapors, rare gases and crystalline cyclic dipeptides with hydrophobic side chains but also for thermal unfolding of proteins, as illustrated in Fig. 12 and Table 3. These apparently widely different processes thus have a very similar common convergence temperature T_S^* in the range from 102 to 126°C [114]. This observation is taken as an extrathermodynamic manifestation of the dominant role of water in determining the thermodynamics of all these processes that involve the interaction of hydrophobic molecules or moieties in aqueous media.

For the dissolution of various liquid aliphatic and aromatic hydrocarbons in water the convergence temperature T_S^* happens to be nearly identical to T_S which has a value of 112.8 ± 2.2°C [51]. We recall that the standard free energy of transfer reaches a maximum at T_S so that it is solely determined by enthalpic contributions and the standard entropy of transfer is zero. This finding corresponds to an empirical relation between ΔS^o and ΔC_p^o observed earlier [115]: the ratio $\Delta S^o/\Delta C_p^o$ measured at 25°C has a constant value of -0.263 ± 0.046 for the transfer of a variety of nonpolar substances from nonaqueous media to water. If T_S^* is equated with T_S, then equ. 16 becomes

$$- \Delta S^o/\Delta C_p^o = \ln(T_S/T) \tag{18}$$

and for a given system based on hydrophobic interactions a constant $\Delta S^o/\Delta C_p^o$ ratio is tantamount to having a constant T_S value.

For protein unfolding the change in the specific enthalpy is also a linear function of the specific heat capacity change and the convergence temperature T_H^* is 110°C. In contrast to the convergence temperature T_S^*, which is very similar for all the processes summarized in Table 3, the value of T_H^* varies with the system comprising the process and the class of compounds under investigation.

The existence of T_H^* and T_S^*, which are referred to as isoenthalpic and isoentropic temperatures [117], appears to be the manifestation of certain extrathermodynamic relationships that facilitate the

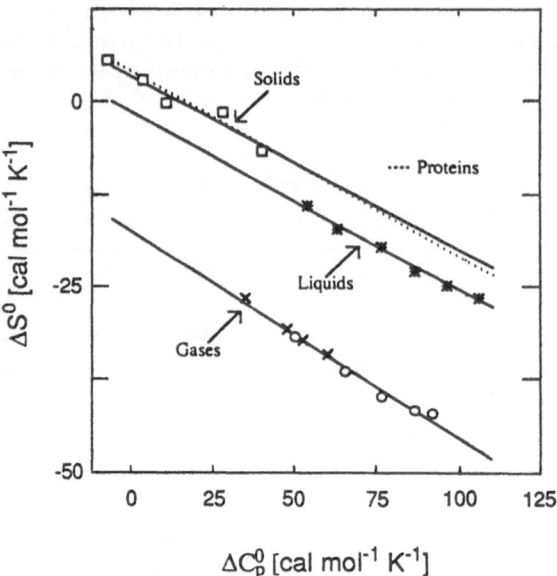

Figure 12. Linear correlation between entropy change and heat capacity change for processes based on the exposure of hydrophobic groups to water at 25°C. The process of protein unfolding is shown by the dotted line, which represents the least-squares fit of thermal denaturation data on 11 proteins. The solid lines are the least-squares fits of data on the dissolution processes of hydrophobic compounds including rare (×) as well as saturated hydrocarbon (○) gases, liquid hydrocarbons (✳) and solid cyclic dipeptides (□). For the process of protein unfolding the entropy and heat capacity changes per mol of amino acid residue rather than per mol of protein are plotted. Fig. 1 of Ref. [114], reproduced with permission of the publisher.

organization and interpretation of experimental data concerning various phenomena dominated by hydrophobic interactions. It is believed that such extrathermodynamic relationships are also useful in the treatment of the retention process in HIC.

4.3.2. *Changes in Water-accessible Nonpolar Surface Area and* ΔC_p^o. The standard heat capacity change for the transfer of a variety of nonpolar aromatic and aliphatic hydrocarbons from water to the pure liquid state has been found to be proportional to the solvent-accessible nonpolar surface area, ΔA_{np}, that is removed from water as a result of hydrophobic interactions [52, 118, 119]. An empirical relationship for the dependence of the heat capacity change on ΔA_{np} [119] is given by

$$\Delta C_p^o = -0.28 \ (\pm 0.04) \ \Delta A_{np} \tag{19}$$

where ΔA_{np} is in $Å^2$ and ΔC_p^o is in cal mol^{-1} K^{-1}.

An identical correlation between ΔC_p^o and the change in water-accessible nonpolar surface area was obtained for the folding of small globular proteins [119] and the results are illustrated in Fig. 13. These findings also support the notion that the hydrophobic effect is the dominant contributor to ΔC_p^o in protein folding. Therefore, the heat capacity change associated with the removal of some

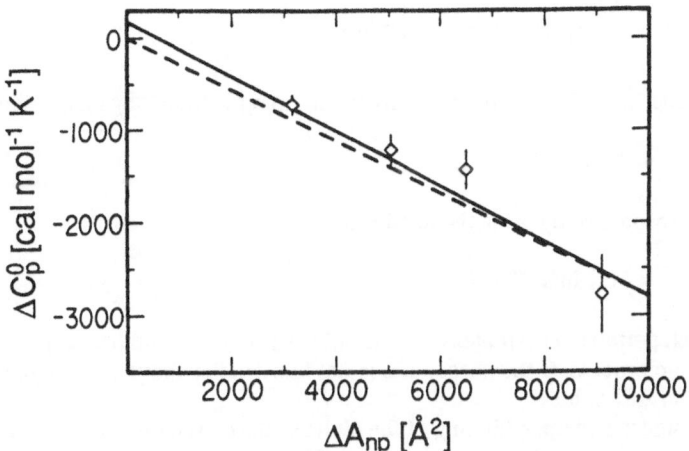

Figure 13. Illustration of the linear relationship between ΔC_p^o and changes in the water-accessible nonpolar surface area ΔA_{np}. The solid line represents data on the folding of four globular proteins. The dashed line represents the relationship for the transfer of hydrocarbons from water to pure liquid hydrocarbon phase. Fig 2 of Ref. [119], reproduced with permission of the publisher.

nonpolar surface moiety of a protein from water in other processes, such as site specific binding to nucleic acids [48], can also be expected to be proportional to the nonpolar surface area removed from exposure to water. In most cases experimental data on the change in nonpolar surface area upon binding is not available. However, it can be estimated from the heat capacity change [48] by the following relationship obtained upon rearranging equ. 19 as

$$\Delta A_{np} = -3.6 \ (\pm 0.6) \ \Delta C_p^o \tag{20}$$

It is believed that equ. 20 can also be useful in estimating the nonpolar contact area in HIC and it provides an important link between the treatment by the solvophobic theory, which is centered around ΔA_{np}, and ΔC_p^o, which is closely related to the corresponding enthalpy and entropy changes. Heat capacity data could be readily obtained from appropriate chromatographic or calorimetric experiments and preliminary results indicate that the change in the heat capacity is temperature independent in the temperature range of chromatographic interest.

4.4. EXTENSION OF THE EXTRATHERMODYNAMIC ANALYSIS TO HIC

4.4.1. *Evaluation of Thermodynamic Parameters from Chromatographic Data.* Chromatographic experiments by using linear elution and isocratic conditions over a sufficiently wide temperature range allow a convenient determination of the retention factors as a function of temperature for a large number of eluites. When the van't Hoff plots are curved the thermodynamic functions for the chromatographic retention process in HIC can be evaluated provided the phase ratio ϕ of the column is known or can be reliably estimated. By using a 3 parameter nonlinear least-square fitting procedure [48], first ΔC_p^o, T_H and T_S are evaluated from the following equation

$$\ln k' = \frac{\Delta C_p^o}{R}\left[\frac{T_H}{T} - \ln\left(\frac{T_S}{T}\right) - 1\right] + \ln \phi \qquad (21)$$

Then the value of ΔH^o is calculated for any temperature from the following expression

$$\Delta H^o = \Delta C_p^o \, (T\text{-}T_H) \qquad (22)$$

Finally, ΔS^o can be estimated by using the relationship

$$\Delta S^o = \Delta C_p^o \, \ln(T/T_S) - R \ln\phi \qquad (23)$$

Thus, this approach permits the estimation of the enthalpy, entropy and heat capacity changes for the chromatographic process. If the phase ratio is not known, the above procedure still yields the appropriate ΔH^o and ΔC_p^o values.

The nonpolar contact area upon binding of the eluite to the chromatographic surface in HIC can be estimated from the value of ΔC_p^o according to equ. 20.

4.4.2. *Extrathermodynamic Relationships for Use in HIC.* As discussed before retention in HIC involves binding of the eluite to the stationary phase by hydrophobic interactions and thus the removal of some hydrophobic surface area from contact with the aqueous environment. Consequently, the chromatographic process ought to share common features with such apparently different phenomena as the transfer of hydrocarbons from water to their pure liquid state and the folding of proteins. By analogy, we also expect the following relationships which have strong experimental support from measurements with widely different systems based on hydrophobic interactions to hold in HIC. For the enthalpy change associated with the retention process the following expression, which is identical to equ. 17, may be used

$$\Delta H^o = \Delta H^* + \Delta C_p^o \, (T\text{-}T_H^*) \qquad (24)$$

For the evaluation of the entropy change germane to retention in HIC the following relationship is expected to apply

$$\Delta S^o = \Delta S^* + \Delta C_p^o \ln(T/T_S^*) - R \ln\phi \qquad (25)$$

These extrathermodynamic relationships based on recent findings described above can provide an eminently suitable framework for the organization and interpretation of data obtained by HIC experiments under controlled conditions. In the light of equs. 24 and 25 the logarithmic retention factor can be expressed as

$$\ln k' = \frac{\Delta C_p^o}{R}\left[\frac{T_H^*}{T} - \ln\left(\frac{T_S^*}{T}\right) - 1\right] - \frac{\Delta H^*}{RT} + \frac{\Delta S^*}{R} + \ln\phi \qquad (26)$$

Here, T_S^* is expected to be about 110°C and T_H^* to be a value characteristic for a given homologuous series and given stationary phase. For this reason, by using the same set of eluites, the T_H^* value may provide a means to characterize stationary phases. The value of ΔS^* may be used to evaluate the phase ratio.

We believe that these extrathermodynamic relationships will facilitate a more profound understanding of the role of temperature in the hydrophobic binding phenomena underlying the separation process. Concomitantly the new theoretical insight emerging in the context of protein folding should spur further advances in the analysis of hydrophobic interactions of chromatographic relevance. The investigation of the effect of temperature in HIC along these lines should expand the potential of chromatography for the measurement of physicochemical data. Thus chromatographic measurements are expected to complement calorimetric studies on the hydrophobic effect and further expand our understanding of the physicochemical phenomena underlying retention in HIC and RPC as well as of the hydrophobic effect at large.

Acknowledgements

This work was supported by grant No. GM 20993 from the National Institutes of Health, U. S. Department of Health and Human Resources, and by grant No. BCS - 9014119 from the National Science Foundation.

References

1. Shansky, R.E., S.-L. Wu, A. Figueroa, and B.L. Karger (1990). Hydrophobic interaction chromatography of biopolymers, in *HPLC of Biological Macromolecules - Methods and Applications*, K.M. Gooding and F.E. Regnier, Editors. Chromatographic Science Series, Vol. 51, Marcel Dekker, New York: 95-144.

2. Eriksson, K.-O. (1989). Hydrophobic interaction chromatography, in *Protein Purification: Principles, High Resolution Methods, and Applications*, J.-C. Janson and L. Rydén, Editors. VCH Publishers, New York: 207-226.

3. Shepard, C.C. and A. Tiselius (1949), in *Chromatographic Analysis, Discussions of the Faraday Society*. Hazell, Watson and Winey, Ltd., London: 275.

4. Tiselius, A. (1949), in *Chromatographic Analysis, Discussions of the Faraday Society*. Vol. 7, Hazell, Watson and Winey, Ltd., London: 9.

5. Hofstee, B.H.J. (1973). Hydrophobic affinity chromatography of proteins. *Anal. Biochem.* **52**: 430-448.

6. Hofstee, B.H.J. (1973). Protein binding by agarose carrying hydrophobic groups in conjunction with charges. *Biochem. Biophys. Res. Commun.* **50**: 751-757.

7. Hofstee, B.H.J. (1976). Hydrophobic adsorption chromatography of proteins, in *Methods of Protein Separation*, N. Catsimpoolas, Editor. Vol. 2, Plenum Press, New York: 245-278.

8. Er-el, Z., Y. Zaidenzaig, and S. Shaltiel (1972). Hydrocarbon-coated Sepharoses. Use in the purification of glycogen phosphorylase. *Biochem. Biophys. Res. Commun.* **49**: 383-390.

9. Shaltiel, S. and Z. Er-el (1973). Hydrophobic chromatography: Use for purification of glycogen synthetase. *Proc. Natl. Acad. Sci. U.S.A.* **70**: 778-781.

10. Shaltiel, S. (1974). Hydrophobic chromatography, in *Methods in Enzymology*, W.B. Jakoby and M. Wilchek, Editors. Vol. 34, Academic Press, New York: 126-140.

11. Porath, J., L. Sundberg, N. Fornstedt, and I. Olsson (1973). Salting-out in amphiphilic gels as a new approach to hydrophobic adsorption. *Nature (London)* **245**: 465-466.

12.   Hjertén, S. (1973). Some general aspects of hydrophobic interaction chromatography. *J. Chromatogr.* **87**: 325-331.

13.   Hjertén, S., J. Rosengren, and S. Pahlman (1974). Hydrophobic interaction chromatography. The synthesis and the use of some alkyl and aryl derivatives of agarose. *J. Chromatogr.* **101**: 281-288.

14.   Hjertén, S. (1976). Hydrophobic interaction chromatography of proteins on neutral adsorbents, in *Methods of Protein Separation*, N. Catsimpoolas, Editor. Vol. 2, Plenum Press, New York: 233-243.

15.   Kato, Y., T. Kitamura, and T. Hashimoto (1983). High-performance hydrophobic interaction chromatography of proteins. *J. Chromatogr.* **266**: 49-54.

16.   Fausnaugh, J.L., L.A. Kennedy, and F.E. Regnier (1984). Comparison of hydrophobic-interaction and reversed-phase chromatography of proteins. *J. Chromatogr.* **317**: 141-155.

17.   Fausnaugh, J.L., E. Pfannkoch, S. Gupta, and F.E. Regnier (1984). High-performance hydrophobic interaction chromatography of proteins. *Anal. Biochem.* **137**: 464-472.

18.   Gooding, D.L., M.L. Schmuck, and K.M. Gooding (1984). Analysis of proteins with new, mildly hydrophobic high-performance liquid chromatography packing materials. *J. Chromatogr.* **296**: 107-114.

19.   Miller, N.T., B. Feibush, and B.L. Karger (1985). Wide-pore silica-based ether-bonded phases for separation of proteins by high-performance hydrophobic-interaction and size-exclusion chromatography. *J. Chromatogr.* **316**: 519-536.

20.   Chang, J.-P., Z. El Rassi, and C. Horváth (1985). Silica-bound polyethyleneglycol as stationary phase for separation of proteins by high-performance liquid chromatography. *J. Chromatogr.* **319**: 396-399.

21.   Alpert, A.J. (1986). High-performance hydrophobic-interaction chromatography of proteins on a series of poly(alkyl aspartamide)-silicas. *J. Chromatogr.* **359**: 85-97.

22.   Kato, Y., T. Kitamura, and T. Hashimoto (1984). New support for hydrophobic interaction chromatography of proteins. *J. Chromatogr.* **292**: 418-426.

23.   Kato, Y., T. Kitamura, and T. Hashimoto (1984). Operational variables in high-performance hydrophobic interaction chromatography of proteins on TSK-gel Phenyl-5PW. *J. Chromatogr.* **298**: 407-418.

24.   Kato, Y., T. Kitamura, and T. Hashimoto (1985). Preparative high-performance hydrophobic interaction chromatography of proteins on TSK-gel Phenyl-5PW. *J. Chromatogr.* **333**: 202-210.

25.   Kato, Y., T. Kitamura, and T. Hashimoto (1986). New resin-based hydrophilic support for high-performance hydrophobic interaction chromatography. *J. Chromatogr.* **360**: 260-265.

26.   Janzen, R., K.K. Unger, H. Giesche, J.N. Kinkel, and M.T.W. Hearn (1987). Evaluation of advanced silica packings for the separation of biopolymers by high-performance liquid chromatography. V. Performance of non-porous monodisperse 1.5 mm bonded silicas in the separation of proteins by hydrophobic-interaction chromatography. *J. Chromatogr.* **397**: 91-97.

27.   Kato, Y., T. Kitamura, S. Nakatani, and T. Hashimoto (1989). High-performance hydrophobic interaction chromatography of proteins on a pellicular support based on hydrophilic resin. *J. Chromatogr.* **483**: 401-405.

28.   Kalghatgi, K. (1990). Micropellicular stationary phases for rapid protein analysis by high-performance liquid chromatography. *J. Chromatogr.* **499**: 267-278.

29.   Sadler, A.J., R. Micanovic, G.E. Katzenstein, R.V. Lewis, and C.R. Middaugh (1984). Protein conformation and reversed-phase high-performance liquid chromatography. *J. Chromatogr.* **317**: 93-101.

30.   Drake, A.F., M.A. Fung, and C.F. Simpson (1989). Protein conformation changes as the result of binding to reversed-phase chromatography column materials. *J. Chromatogr.* **476**: 159-163.

31.   Vogel, H.J., L. Lindahl, and E. Thulin (1983). Calcium-dependent hydrophobic interaction chromatography of calmodulin, troponin C and their proteolytic fragments. *FEBS Lett.* **157**: 241-246.

32.   Lee, B. and F.M. Richards (1971). The interpretation of protein structures: Estimation of static accessibility. *J. Mol. Biol.* **55**: 379-400.

33.   Miller, S., J. Janin, A.M. Lesk, and C. Chothia (1987). Interior and surface of monomeric proteins. *J. Mol. Biol.* **196**: 641-656.

34.   Fausnaugh, J.L. and F.E. Regnier (1986). Solute and mobile phase contributions to retention in hydrophobic interaction chromatography of proteins. *J. Chromatogr.* **359**: 131-146.

35.   Jennissen, H.P. (1976). Evidence for negative cooperativity in the adsorption of phosphorylase b on hydrophobic agaroses. *Biochemistry* **15**: 5683-5692.

36.   Wu, S.-L., W.S. Hancock, B. Pavlu, and P. Gellerfors (1990). Application of high-performance hydrophobic-interaction chromatography to the characterization of recombinant DNA-derived human growth hormone. *J. Chromatogr.* **500**: 595-606.

37.   Gellerfors, P., K. Axelsson, A. Helander, S. Johansson, L. Kenne, S. Lindqvist, B. Pavlu, A. Skottner, and L. Fryklund (1989). Isolation and characterization of a glycosylated form of human insulin-like growth factor I produced in Saccharomyces cerevisiae. *J. Biol. Chem.* **264**: 11444-11449.

38.   Lin, L.S. and R. Yamamoto (1987). Purification of native and recombinant tumor necrosis factor, in *Protein Purification: Micro to Macro*, R. Burgess, Editor. A. R. Liss, New York: 409-419.

39.   Miller, N.T., B. Feibush, K. Corina, S. Powers-Lee, and B.L. Karger (1985). High-performance hydrophobic interaction chromatography: Purification of rat liver carbamoylphosphate synthetase I and ornithine transcarbamoylase. *Anal. Biochem.* **148**: 510-517.

40.   Kato, Y., T. Kitamura, and T. Hashimoto (1986). High-performance hydrophobic interaction chromatography of proteins on TSKgel Phenyl-5PW preparative column. *J. Liquid Chromatogr.* **9**: 3209-3244.

41.   Hyder, S.M., R.D. Wiehle, D.W. Brandt, and J.L. Wittliff (1985). High-performance hydrophobic-interaction chromatography of steroid hormone receptors. *J. Chromatogr.* **327**: 237-246.

42.   Hyder, S.M., N. Sato, and J.L. Wittliff (1987). Characterization of estrogen receptors and associated protein kinase activity by high-performance hydrophobic-interaction chromatography. *J. Chromatogr.* **397**: 251-267.

43.   Hyder, S.M. and J.L. Wittliff (1988). High-performance hydrophobic-interaction chromatography as a means of identifying estrogen receptors expressing different binding domains. *J. Chromatogr.* **444**: 225-237.

44.   Pavlu, B., U. Johansson, C. Nyhlén, and A. Wichman (1986). Rapid purification of monoclonal antibodies by high-performance liquid chromatography. *J. Chromatogr.* **359**: 449-460.

476

45.   Berkowitz, S.A. and M.P. Henry (1987). Use of high-performance hydrophobic-interaction chromatography for the determination of salting-out conditions of proteins. *J. Chromatogr.* **389**: 317-321.

46.   Alpert, A.J. (1988). Hydrophobic interaction chromatography of peptides as an alternative to reversed-phase chromatography. *J. Chromatogr.* **444**: 269-274.

47.   Tanford, C. (1980). *The Hydrophobic Effect: Formation of Micelles and Biological Membranes.* New York, Wiley-Interscience.

48.   Ha, J.-H., R.S. Spolar, and M.T. Record Jr. (1989). Role of the hydrophobic effect in stability of site-specific protein-DNA complexes. *J. Mol. Biol.* **209**: 801-816.

49.   Kauzmann, W. (1959). Some factors in the interpretation of protein denaturation. *Adv. Protein Chem.* **14**: 1-63.

50.   Tanford, C. (1978). The hydrophobic effect and the organization of living matter. *Science* **200**: 1012-1018.

51.   Baldwin, R.L. (1986). Temperature dependence of the hydrophobic interaction in protein folding. *Proc. Natl. Acad. Sci. U.S.A.* **83**: 8069-8072.

52.   Privalov, P.L. and S.J. Gill (1988). Stability of protein structure and hydrophobic interaction. *Adv. Protein Chem.* **39**: 191-234.

53.   Dill, K.A. (1990). Dominant forces in protein folding. *Biochemistry* **29**: 7133-7155.

54.   Frank, H.S. and M.W. Evans (1945). Free volume and entropy in condensed systems. III. Entropy in binary liquid mixtures; partial molal entropy in dilute solutions; structure and thermodynamics in aqueous electrolytes. *J. Chem. Phys.* **13**: 507-532.

55.   Edsall, J.T. (1935). Apparent molal heat capacities of amino acids and other organic compounds. *J. Am. Chem. Soc.* **57**: 1506-1507.

56.   Nemethy, G. and H.A. Scheraga (1962). Structure of water and hydrophobic bonding in proteins. I. A model for the thermodynamic properties of liquid water. *J. Chem. Phys.* **36**: 3382-3400.

57.   van Oss, C.J., D.R. Absolom, and A.W. Neumann (1980). The "Hydrophobic effect": Essentially a van der Waals interaction. *Colloid Polym. Sci.* **258**: 424-427.

58.   Visser, J. (1981). The concept of negative Hamaker coefficients. I. History and present status. *Adv. Colloid Interface Sci.* **15**: 157-169.

59.   van Oss, C.J., J. Visser, D.R. Absolom, S.N. Omenyi, and A.W. Neumann (1983). The concept of negative Hamaker coefficients. II. Thermodynamics, experimental evidence and applications. *Adv. Colloid Interface Sci.* **18**: 133-148.

60.   van Oss, C.J., D.R. Absolom, and A.W. Neumann (1979). Repulsive van der Waals forces. II. Mechanism of hydrophobic chromatography. *Sep. Sci. Technol.* **14**: 305-317.

61.   Hamaker, H.C. (1937). *Physica* **4**: 1058-1070.

62.   Neumann, A.W., S.N. Omenyi, and C.J. van Oss (1979). Negative Hamaker coefficients I. Particle engulfment or rejection at solidification fronts. *Colloid Polym. Sci.* **257**: 413-419.

63.   van Oss, C.J., S.N. Omenyi, and A.W. Neumann (1979). Negative Hamaker coefficients II. Phase separation of polymer solutions. *Colloid Polym. Sci.* **257**: 737-744.

64.   van Oss, C.J., R.J. Good, and M.K. Chaudhury (1986). The role of van der Waals forces and hydrogen bonds in "hydrophobic interactions" between biopolymers and low energy surfaces. *J. Colloid Interface Sci.* **111**: 378-390.

65.   van Oss, C.J., R.J. Good, and M.K. Chaudhury (1987). Determination of the hydrophobic interaction energy - Application to separation processes. *Sep. Sci. Technol.* **22**: 1-24.

66. van Oss, C.J. and R.J. Good (1988). On the mechanism of "hydrophobic interactions". *J. Disp. Sci. Technol.* **9**: 355-362.

67. Ben-Naim, A. (1980). *Hydrophobic Interactions.* New York, Plenum Press.

68. Elkoshi, Z. and E. Grushka (1981). Hydrophobic interactions and chromatographic processes. *J. Phys. Chem.* **85**: 2980-2986.

69. Martire, D.E. and R.E. Boehm (1983). Unified theory of retention and selectivity in liquid chromatography. 2. Reversed-phase liquid chromatography with chemically bonded phases. *J. Phys. Chem.* **87**: 1045-1062.

70. Martire, D.E. (1988). Unified theory of adsorption chromatography: Gas, liquid and supercritical fluid mobile phases. *J. Chromatogr.* **452**: 17-30.

71. Dill, K.A. (1985). Theory for the folding and stability of globular proteins. *Biochemistry* **24**: 1501-1509.

72. Dill, K.A., D.O.V. Alonso, and K. Hutchinson (1989). Thermal stabilities of globular proteins. *Biochemistry* **28**: 5439-5449.

73. Sinanoglu, O. and S. Abdulnur (1965). *Fed. Proc.* **24**: 12.

74. Sinanoglu, O. (1968), in *Molecular Associations in Biology*, B. Pullman, Editor. Academic Press, New York: 427-445.

75. Horváth, C., W.R. Melander, and I. Molnár (1976). Solvophobic interactions in liquid chromatography with nonpolar stationary phases. *J. Chromatogr.* **125**: 129-156.

76. Melander, W.R. and C. Horváth (1977). Salt effects on hydrophobic interactions in precipitation and chromatography of proteins: an interpretation of the lyotropic series. *Arch. Biochem. Biophys.* **183**: 200-215.

77. Absolom, D.R., W. Zingg, and A.W. Neumann (1985). Surface thermodynamics of cellular and protein interactions, in *Comprehensive Biotechnology*, C.W. Robinson and J.A. Howell, Editors. Vol. 4, Pergamon Press, New York: 433-446.

78. Neumann, A.W., R.J. Good, C.J. Hope, and M. Sejpal (1974). An equation-of-state approach to determine surface tensions of low-energy solids from contact angles. *J. Colloid Interface Sci.* **49**: 291-304.

79. Neuman, A.W., D.R. Absolom, D.W. Frances, and C.J. van Oss (1980). Conversion tables of contact angles to surface tensions. For use in determining the contribution of the van der Waals attraction or repulsion to various separation processes. *Sep. Purif. Methods* **9**: 69-163.

80. van Oss, C.J., D.R. Absolom, A.W. Neumann, and W. Zingg (1981). Determination of the surface tension of proteins I. Surface tension of native serum proteins in aqueous media. *Biochim. Biophys. Acta* **670**: 64-73.

81. Absolom, D.R. and R.A. Barford (1988). Determination of surface tension of packings for high-performance liquid chromatography. *Anal. Chem.* **60**: 210-212.

82. Barford, R.A., B.J. Sliwinski, A.C. Breyer, and H.L. Rothbart (1982). Mechanism of protein retention in reversed-phase high performance liquid chromatography. *J. Chromatogr.* **235**: 281-288.

83. Melander, W.R., D. Corradini, and C. Horváth (1984). Salt-mediated retention of proteins in hydrophobic-interaction chromatography. Application of solvophobic theory. *J. Chromatogr.* **317**: 67-85.

84. Melander, W.R. and C. Horváth (1977). Effect of neutral salts on the formation and dissociation of protein aggregates. *J. Solid-Phase Biochem.* **2**: 141-161.

85. Dill, K.A. (1987). The mechanism of solute retention in reversed-phase liquid chromatography. *J. Phys. Chem.* **91**: 1980-1988.

478

86.   Dorsey, J.G. and K.A. Dill (1989). The molecular mechanism of retention in reversed-phase liquid chromatography. *Chem. Rev.* **89**: 331-346.

87.   Hofmeister, F. (1888). *Arch. Exp. Pathol. Pharmakol.* **24**: 247-263.

88.   El Rassi, Z. and C. Horváth (1986). Hydrophobic interaction chromatography of t-RNA's and proteins. *J. Liquid Chromatogr.* **9**: 3245-3268.

89.   Katti, A., Y.-F. Maa, and C. Horváth (1987). Protein surface area and retention in hydrophobic interaction chromatography. *Chromatographia* **24**: 646-650.

90.   Melander, W.R., Z. El Rassi, and C. Horváth (1989). Interplay of hydrophobic and electrostatic interactions in biopolymer chromatography. Effect of salts on the retention of proteins. *J. Chromatogr.* **469**: 3-27.

91.   Miller, N.T. and B.L. Karger (1985). High-performance hydrophobic-interaction chromatography on ether-bonded phases: Chromatographic characteristics and gradient optimization. *J. Chromatogr.* **326**: 45-61.

92.   Szepesy, L. and C. Horváth (1988). Specific salt effects in hydrophobic interaction chromatography. *Chromatographia* **26**: 13-18.

93.   Arakawa, T. and S.N. Timasheff (1984). Mechanism of protein salting in and salting out by divalent cation salts: Balance between hydration and salt binding. *Biochemistry* **23**: 5912-5923.

94.   Arakawa, T. and S.N. Timasheff (1982). Preferential interaction of proteins with salts in concentrated solutions. *Biochemistry* **21**: 6545-6552.

95.   Wu, S.-L., K. Benedek, and B.L. Karger (1986). Thermal behavior of proteins in high-performance hydrophobic-interaction chromatography: On-line spectroscopic and chromatographic characterization. *J. Chromatogr.* **359**: 3-17.

96.   Wu, S.-L., A. Figueroa, and B.L. Karger (1986). Protein conformational effects in hydrophobic interaction chromatography: Retention characterization and the role of mobile phase additives and stationary phase hydrophobicity. *J. Chromatogr.* **371**: 3-27.

97.   Roettger, B.F., J.A. Myers, M.R. Ladisch, and F.E. Regnier (1989). Adsorption phenomena in hydrophobic interaction chromatography. *Biotechnology Progress* **5**: 79-88.

98.   Wyman, J. (1964). Linked functions and reciprocal effects in hemoglobin: a second look. *Adv. Protein Chem.* **19**: 223-286.

99.   Arakawa, T. and S.N. Timasheff (1982). Stabilization of protein structure by sugars. *Biochemistry* **21**: 6536-6544.

100.   Lee, J.C., K. Gekko, and S.N. Timasheff (1979). Measurements of preferential solvent interactions by densimetric techniques. *Methods Enzymol.* **61**: 26-49.

101.   Lee, J.C. and S.N. Timasheff (1981). The stabilization of proteins by sucrose. *J. Biol. Chem.* **256**: 7193-7201.

102.   Arakawa, T. (1986). Thermodynamic analysis of the effect of concentrated salts on protein interaction with hydrophobic and polysaccharide columns. *Arch. Biochem. Biophys.* **248**: 101-105.

103.   Barford, R.A., T.F. Kumosinski, N. Parris, and A.E. White (1988). Salt-binding effects in hydrophobic-interaction chromatography. *J. Chromatogr.* **458**: 57-66.

104.   Melander, W.R., D.E. Campbell, and C. Horváth (1978). Enthalpy-entropy compensation in reversed-phase chromatography. *J. Chromatogr.* **158**: 215-225.

105.   Melander, W.R. and C. Horváth (1980). Reversed-phase chromatography, in *High Performance Liquid Chromatography - Advances and Perspectives*, C. Horváth, Editor. Vol. 2, Academic Press, New York: 113-319.

106. Melander, W.R., A. Nahum, and C. Horváth (1979). Mobile phase effects in reversed-phase chromatography III. Changes in conformation and retention of oligo(ethylene glycol) derivatives with temperature and eluent composition. *J. Chromatogr.* **185**: 129-152.

107. Nahum, A. and C. Horváth (1981). Surface silanols in silica-bonded hydrocarbonaceous stationary phases I. Dual retention mechanism in reversed-phase chromatography. *J. Chromatogr.* **203**: 53-63.

108. Kalghatgi, K. and C. Horváth (1987). Rapid analysis of proteins and peptides by reversed-phase chromatography. *J. Chromatogr.* **398**: 335-339.

109. Goheen, S.C. and S.C. Engelhorn (1984). Hydrophobic interaction high-performance liquid chromatography of proteins. *J. Chromatogr.* **317**: 55-65.

110. Ingraham, R.H., S.Y.M. Lau, A.K. Taneja, and R.S. Hodges (1985). Denaturation and the effects of temperature on hydrophobic-interaction and reversed-phase high-performance liquid chromatography of proteins: Bio-Gel TSK-Phenyl-5-PW column. *J. Chromatogr.* **327**: 77-92.

111. Gehas, J. and D.B. Wetlaufer (1990). Isocratic hydrophobic interaction chromatography of dansyl amino acids: Correlation of hydrophobicity and retention parameters. *J. Chromatogr.* **511**: 123-130.

112. Gelsema, W.J., P.M. Brandts, C.L. de Ligny, A.G.M. Theeuwes, and A.M.P. Roozen (1984). Hydrophobic interaction chromatography of aliphatic alcohols and carboxylic acids on octyl-Sepharose CL-4B: mechanism and thermodynamics. *J. Chromatogr.* **295**: 13-29.

113. Brandts, P.M., W.J. Gelsema, and C.L. de Ligny (1986). Influence of additives to the eluent on hydrophobic interaction chromatography of simple compounds III. Thermodynamics of interaction of n-alcohols with octyl-agarose in mixtures of water with methanol, urea or ethylene glycol. *J. Chromatogr.* **354**: 19-36.

114. Murphy, K.P., P.L. Privalov, and S.J. Gill (1990). Common features of protein unfolding and dissolution of hydrophobic compounds. *Science* **247**: 559-561.

115. Sturtevant, J.M. (1977). Heat capacity and entropy changes in processes involving proteins. *Proc. Natl. Acad. Sci. U.S.A.* **74**: 2236-2240.

116. Privalov, P.L. (1979). Stability of proteins: Small globular proteins. *Adv. Protein Chem.* **33**: 167-241.

117. Lee, B. (1991). Isoenthalpic and isoentropic temperatures and the thermodynamics of protein denaturation. *Proc. Natl. Acad. Sci. U.S.A.* **88**: 5154-5158.

118. Gill, S.J., S.F. Dec, G. Olofsson, and I. Wadsö (1985). Anomalous heat capacity of hydrophobic solvation. *J. Phys. Chem.* **89**: 3758-3761.

119. Spolar, R.S., J.-H. Ha, and M.T. Record Jr. (1989). Hydrophobic effect in protein folding and other noncovalent processes involving proteins. *Proc. Natl. Acad. Sci. U.S.A.* **86**: 8382-8385.

120. Richards, F.M. (1991). The protein folding problem. *Scientific American* **1**: 54-63.

121. Jones, G. and W.A. Ray (1941). The surface tension of solutions of electrolytes as a function of the concentration II. *J. Am. Chem. Soc.* **63**: 288-294.

122. *International Critical Tables of Numerical Data, Physics, Chemistry and Technology* (1929). Vol. 4. New York, McGraw-Hill.

123. Livingston, J., R. Morgan, and W.W. McKirahan (1913). The weight of a falling drop and the laws of tate. XIV. The drop weights of aqueous solutions of the salts of organic acids. *J. Am. Chem. Soc.* **35**: 1759-1767.

124. Livingston, J., R. Morgan, and G.A. Bole (1913). The weight of a falling drop and the laws of tate. XIII. The drop weights of aqueous solutions and the surface tensions calculated from them. *J. Am. Chem. Soc.* **35**: 1750-1759.
125. Schäfer, K., A.P. Masia, and H. Jüntgen (1955). Untersuchungen über Grenzflächeneffekte kapillar-inaktiver wäßriger Lösungen. *Z. Elektrochem.* **59**: 425-434.
126. Jones, G. and W.A. Ray (1937). The surface tension of solutions of electrolytes as a function of the concentration. I. A differential method for measuring relative surface tension. *J. Am. Chem. Soc.* **59**: 187-198.
127. Murphy, K.P. and S.J. Gill (1989). Thermodynamics of dissolution of solid cyclic dipeptides containing hydrophobic side groups. *J. Chem. Thermodyn.* **21**: 903-913.

THEORETICAL ASPECTS OF QUANTITATIVE AFFINITY CHROMATOGRAPHY

C. VIDAL-MADJAR and A. JAULMES
Laboratoire de Physico-Chimie des Biopolymères, CNRS
2-8 rue Henry Dunant
94320 Thiais
France

ABSTRACT. Affinity chromatography, a technique based on the unique properties of selective binding between biological macromolecules, is generally used as a preparative tool for protein purification. The quantitative method mainly differs from the preparative technique because isocratic elution is used, the experimental conditions allowing solute reversible binding to the affinity matrix. The method is ideal for studying molecular interactions between biological species.

Frontal analysis is advantageous because the chromatograhic data allows the direct determination of equilibrium isotherms for complex interacting systems. Because of theoretical difficulties, zonal analysis is mainly restricted to infinite solute dilution. In the competitive elution mode, zonal analysis is useful for measuring association equilibrium constants in solution.

The methodology of equilibrium and kinetic measurements is described for monovalent interacting systems. Plate height methods enable chemical rate constant measurements if other causes for slow mass transfer are negligible or are accounted for. The analysis of the unusual chromatographic behavior, the "split-peak" effect, offers interesting perspectives.

1. INTRODUCTION

Unlike the other chromatographic methods based on physico-chemical differences between the interacting species, affinity chromatography exploits the unique properties of selective binding between a ligand and a biopolymer to isolate and purify biological macromolecules such as proteins, enzymes and antibodies.

The principles of separation by preparative affinity chromatography are described in Figure 1. The purification is accomplished by passing the crude extract containing the protein of interest through a column packed with a support on which a ligand has been covalently bonded. The molecules which do not exhibit any specificity for the immobilized ligand will pass through the column, while the molecules having an affinity for the ligand will be retained. After a washing step, the solvent conditions will be changed in order to elute the bound species.

481

F. Dondi and G. Guiochon (eds.),
Theoretical Advancement in Chromatography and Related Separation Techniques, 481–511.
© 1992 Kluwer Academic Publishers.

482

This may be achieved by modifying ionic strength, pH, temperature or by adding a competitive inhibitor or a substrate to the mobile phase.

Affinity chromatography is not only a technique for isolating bio-logically active substances, but is also an ideal method for studying molecular interactions between biological species, for examining the mechanisms of enzymatic processes or for elucidating molecular structures. Quantitative analysis in affinity chromatography is also useful to study the various factors affecting chromatographic separations and to give guide-lines based on the physico-chemical properties of the solute, the eluent and the affinity support. Several reviews [1-5] describe in details the theoretical basis and the applications of quantitative affinity chromatography.

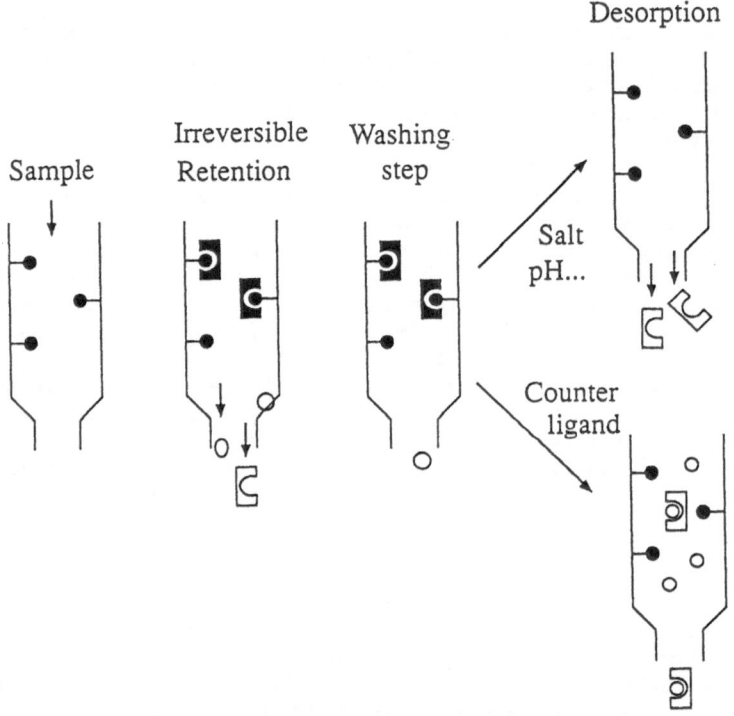

Figure 1. Illustration of the various steps for preparative affinity chromatography

Analytical affinity chromatography differs mainly from the preparative technique by the elution mode: isocratic elution is used under conditions generally allowing the reversible biospecific binding to the affinity matrix. A low molecular mass ligand is immobilized to the affinity matrix, and a competing ligand or inhibitor must generally be added to the mobile phase to decrease the retention of the macro-molecule.

Steric reasons may affect the formation of a complex between the immobilized ligand and the macromolecule. Macroporous supports should be selected for the matrix. Moreover, to increase the flexibility and the mobility of the ligand, one should bind the ligand to the matrix through a spacer arm. If the spacer is too short, no complex formation can take place. If the spacer chain is too long, non-specific effects may inter-fere, such as hydrophobic interactions that may affect the mechanisms of biospecific recognition.

For measuring protein-ligand or protein-protein interactions the macromolecule is sometimes immobilized on the matrix and the ligand or another macromolecule is injected as a solute. In this case, in order to decrease retention, the inhibitor is added to the eluent and will interact with the immobilized biopolymer. Again the immobilization procedure may alter the biological properties of the macromolecule and therefore the values of affinity constants with the immobilized protein may differ considerably from those in solution.

2. BASIC CHROMATOGRAPHIC METHODS

Two classical elution techniques are used which differ by the input concentration signal, while the output signal is measured as a function of the time elapsed from the injection moment:

1) Frontal elution, in which solute concentration is suddenly raised at the moment chosen for the injection, and then maintained at a constant value (Dirac step function).

2) Zonal elution, in which solute concentration is raised at a high value during a short time and is then suddenly lowered back to zero value (Dirac rectangular pulse function).

2.1. Frontal Elution

In frontal elution, a sufficient volume of solution is applied in order to obtain a plateau corresponding to the concentration C of the injected solution. The concentration of the outlet signal will rise from 0 to C and a "front" will be generated. With a convex adsorption isotherm, the sheerness of this front will depend on the speed of attai-ning the equilibrium state (Figure 2). Before performing another experiment, a desorption step is then obtained by applying a solution of pure buffer. A trailing profile is observed in which the concentration decreases from the plateau to zero.

If one integrates the volumes on the concentration domain of the front (0 to $\bar{C}$), one obtains the quantity of solute A retained on the column added to the amount contained in the whole mobile phase. This result may be translated into an average retention volume according to the classical calculation [2] :

$$V_R = \frac{1}{\bar{C}} \int_0^{\bar{C}} V(\bar{C}) \ d\bar{C} \tag{1}$$

Fortunately this average retention volume is independent of the slowness of adsorption of the chemical processes. The volume of mobile phase having flown before the front is detected, includes the elution volume needed for the mobile phase to reach the column outlet, that for the solute to enter the pores, that corresponding to a non-specific adsorption and that for the specific binding to the affinity matrix.

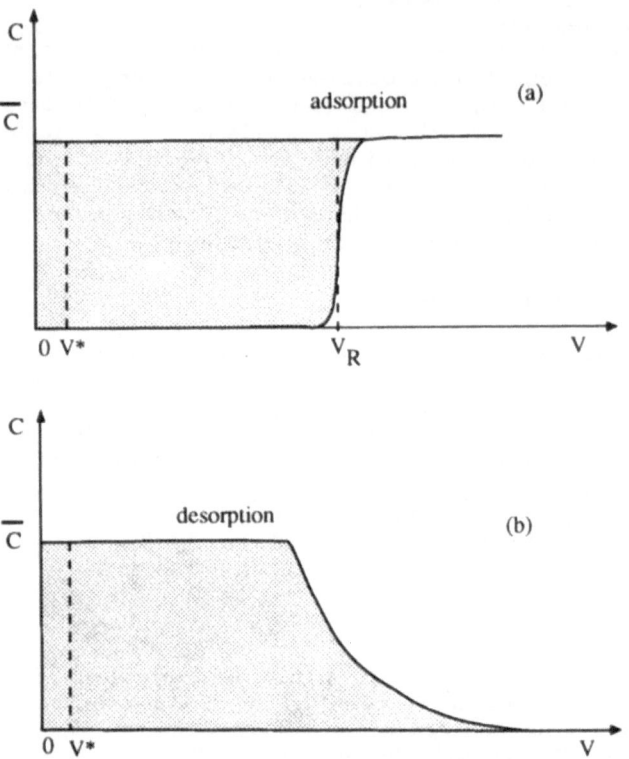

Figure 2. Schematic representation of frontal elution with a convex partition isotherm. a) Adsorption step: change from buffer to a solution containing the solute at a concentration $\bar{C}$; b) Desorption step: change from solution to pure buffer.

The solute amount which has been retained on the solid phase only by affinity $\bar{Q}_A$ is measured (Figure 2) from the difference between this average retention volume V_R and the one under conditions where there is no retention on the solid phase V^*:

$$\bar{Q}_A = \bar{C} \ (V_R - V^*) = f(\bar{C}) \tag{2}$$

Therefore, frontal experiments at various eluent concentrations yield directly the amount of interacting molecules as a function of the total solute concentration $\bar{C}$ in the mobile phase. It is the global partition isotherm, which depends on the type of binding of solute A to the affinity matrix and on the temperature. The method permits its determination without any hypothesis concerning the chemical equilibria. On a practical point of view, in the presence of several chemical equilibria, one will essentially know the total concentration [A], here equal to $\bar{C}$. There is a direct chemical relationship between [A] and the free solute concentration [A].

In equation (2), the retention volume V^* for no adsorption on the support, depends on the solute used since, with porous particles, the permeation of the solute is a function of its molecular size. Such a dependence may be represented by the equation:

$$V^* = V_o + \sigma_A V_i$$

or

$$= V_o(1 + k_o) \qquad \qquad \Big\} \quad (3)$$

where V_o is the liquid external volume.
V_i the internal pore volume.
σ_A the permeation coefficient of solute A.
k_o the ratio: $\sigma_A V_i / V_o$

2.2. Zonal Elution

In zonal elution a small volume of solute is injected into the column. The method is generally preferred to frontal analysis because of its simplicity and the small amount of solute required. The analyses by zonal elution are generally performed by extrapolating the retention data to zero sample size. Since the total solute concentration varies continuously during the elution, the problems for analyzing zonal experiments are complex and require approaches based on the non-linear theories of chromatography (Figure 3). The analyses by zonal elution are generally restricted to the infinite dilution domain in which the global partition isotherm can be considered as linear.

The mass balance equations characterizing the solute migration through the column is given by:

$$\frac{\partial \bar{C}}{\partial t} + u \cdot \frac{\partial \bar{C}}{\partial z} - D \cdot \frac{\partial^2 \bar{C}}{\partial z^2} = - \frac{1}{V_o} \cdot \frac{\partial (\bar{Q}_A + \bar{Q}'_A)}{\partial t} \qquad (4)$$

where $\bar{C}$ denotes the solute concentration in the bulk mobile phase; $\bar{Q}'_A$ and $\bar{Q}_A$ are respectively the total amount of solute (moles) in the pores and the one of solute fixed on the support by affinity. The time elapsed from the injection is t; z is the abscissa along the column (length = L), u the interstitial mobile phase velocity and D the dispersion coefficient or global diffusion parameter.

Taking into account the kinetic effects, a relationship exists between the sum $(\bar{Q}'_A + \bar{Q}_A)$ and $(\bar{Q}'_A{}^* + \bar{Q}_A^*)$, the value which would have been reached at equilibrium:

$$\frac{\partial (\bar{Q}_A + \bar{Q}'_A)}{\partial t} = \alpha \left[(\bar{Q}_A + \bar{Q}'_A) - (\bar{Q}'_A{}^* + \bar{Q}_A^*) \right] \qquad (5)$$

486

where α is the rate constant, which may be a complicated function of the solute concentration in the bulk.

The mathematical solution of the mass balance expressions (equations 4 and 5) involves considerable difficulties for the general treatment with a non-linear adsorption isotherm, when non-equilibrium effects are included. Taking into consideration the diffusion kinetics and the chemical reactions involved, numerical methods could be used, in principle, to model the elution profile. Then the physico-chemical constants implied in the chromatographic phenomena can be extracted from curve fitting of the model onto the experimental profile.

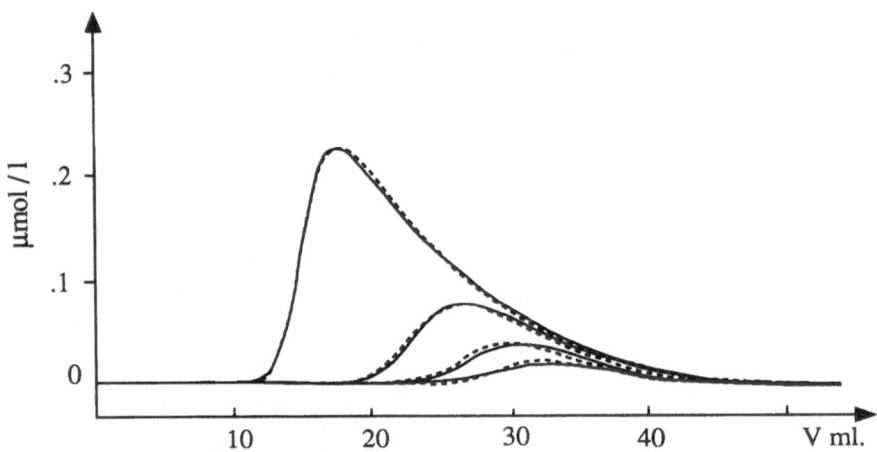

Figure 3. Influence of sample size on zonal elution. Chromatography of phenylbutazone on immobilized human serum albumin. (····) Experimental data, (———) numerical simulation. Figure 4 of Reference 6, reproduced with permission of the authors and Elsevier Science Publishers.

A solution of the above system of differential equations exists in the Laplace domain, in the case of a linear partition isotherm, with first order kinetic laws and assuming a Dirac pulse function for the injection signal. In zonal elution, theoretical expressions of the central statistical moments can be found that characterize the peak profile. The zeroth order moment or peak area is defined as:

$$\mu_o = \int_o^\infty \overline{C}(t) \cdot dt \tag{6}$$

where $\overline{C}(t)$ is the solute concentration $[\overline{A}]$ at the column outlet.
The first order moment, or average retention time, is:

$$\mu_1' = \frac{\int_o^\infty \overline{C}(t) \cdot t \cdot dt}{\mu_o} \tag{7}$$

The second order central moment, or peak variance, is defined by the equation:

$$\mu_2 = \frac{\int_o^\infty \overline{C}(t)(t-\mu_1')^2 dt}{\mu_o} \tag{8}$$

It is related through the plate height H according to

$$H = L\mu_2/\mu_1'^2 \tag{9}$$

A numerical integration will give the moments of the experimental signal, but when the peak is gaussian μ_1' coincide with peak summit and μ_2, which is equal to the variance σ^2, may be determined from the peak width at the fraction $e^{-0.6}$ of the peak height.
The theoretical expression for the first moment is given here as a function of the column capacity factor k' and the column permeation ratio $k_o = \sigma_A V_i / V_o$:

$$\mu_1' = \left(\frac{L}{u} + \frac{2D}{u^2} \right) \cdot (1 + k_o) \cdot (1 + k') \tag{10}$$

k' is related to the slope at origin of the equilibrium partition isotherm:

$$k' = \frac{1}{V*} \cdot \left(\frac{d\overline{Q_A}}{d\overline{C}} \right)_{\overline{C}=0} = \frac{V_z - V*}{V*} \tag{11}$$

where V_z is the retention volume in zonal elution. Generally the term $2D/u^2$ can be neglected and V_z is measured from the average retention time extrapolated to zero sample size:

$$V_z = \mu_1' \cdot F = V* (1 + k') \tag{12}$$

The results for the second moment will be given later when discussing the plate height method for kinetic measurements.
Let us first compare the retention volumes measured by the frontal and zonal elution methods. If one looks at the two functions:

$$\text{a)} \quad V_R - V* = \frac{\overline{Q_A}}{\overline{C}} . \qquad \text{b)} \quad V_z - V* = \left(\frac{d\overline{Q_A}}{d\overline{C}} \right)_{\overline{C}=0} \tag{13}$$

One can observe and keep in mind that they will be different in any non-linear case, except if $\overline{C}$ is zero! The zonal retention volume is related to the slope of the tangent to the isotherm at origin while the retention volume from frontal measurements is equal to the slope of the segment connecting the origin to a point of the isotherm at a given C concentration.

3. EQUILIBRIUM MEASUREMENTS

Quantitative affinity chromatography is now a well recognized method for measuring ligand-protein and protein-protein interactions. The dependence of the measured retention volumes is based on a set of equilibrium equations which describe the interactions between a solute A, the immobilized ligand X and the soluble ligand L. The theories have considered only a limited number of cases involving the interactions with monovalent monomeric solutes, with multivalent monomeric biopolymers and those between self-associating proteins. In this chapter, we shall limit the study to monovalent systems, but the description given here can help the approaches to more complex systems [5] since the methodologies used to examine the various chemical equilibria involved are close.

The theory was first developed [7-9] to describe situations where a monomeric solute having a single site interacts with a single-site immobilized ligand. The solute is univalent or, if it is multivalent, should interact with only one site because of steric requirements. The quantitative treatment of affinity chromatography was developed by Nichol et al. [7] and general relationships are given that include all possible interactions between A, L and X.

3.1. Interaction of a monovalent monomeric macromolecule with a single-site immobilized ligand

The simplest equilibria are those involving interactions of the macromolecule with the mobile and immobilized ligands. There is no interaction between the mobile ligand or the ligand-protein complex and the matrix.

Two equilibria define the interactions of the protein:

$$A + X \rightleftharpoons AX \qquad \{AX\} = \{X\}[A] \cdot K_{AX} \qquad (14)$$
$$A + L \rightleftharpoons AL \qquad \{AL\} = [L][A] \cdot K_{AL} \qquad (15)$$

where K_{AX} and K_{AL} are the binding constants and the braces { } represent the surface concentrations in mol/m^2. In this case there is no interaction of the mobile ligand L with the immobilized one X at the surface of the matrix.

The total concentration of protein $[\bar{A}]$ in the liquid phase is related to that of the free species by:

$$[\bar{A}] = [A] \cdot (1 + [L] \cdot K_{AL}) \qquad (16)$$

The total amount of immobilized protein $\bar{Q}_A$ is:

$$\bar{Q}_A = \{AX\} \cdot S = \bar{Q}_X \frac{[A] \cdot K_{AX}}{1 + [A] \cdot K_{AX}} \qquad (17)$$

where S is the surface area of the matrix and $\bar{Q}_X$ the total amount of immobilized ligand. The combination of equations 16 and 17 yields the partition equilibrium isotherm:

$$\bar{Q}_A = \frac{\bar{Q}_X \cdot [\bar{A}] \cdot K_{AX}}{1 + [\bar{A}] \cdot K_{AX} + [L] \cdot K_{AL}} \qquad (18)$$

This equation, where $\bar{Q}_A$ is a function of the total concentration of the protein $[\bar{A}]$, can be directly used in frontal experiments. Since $\bar{Q}_A/[\bar{A}] = V_R - V^*$, therefore:

$$V_R - V^* = \frac{\bar{Q}_X \cdot K_{AX}}{1 + [\bar{A}] \cdot K_{AX} + [L] \cdot K_{AL}} \qquad (19)$$

The experiments are performed at given protein or ligand concentrations and the constants K_{AL}, K_{AX} and $\bar{Q}_X$ are determined by a linear least-square fit method or any graphic one. It must be noticed however that the concentration of free ligand [L] is generally not directly measurable [10] and the only directly available magnitude is the total concentration of the ligand $[\bar{L}]$ which could be calculated by combining equations 15 and 16 with:

$$[\bar{L}] = [L] + [LA] \qquad (20)$$

Generally, as a first approximation, in the presence of a large concentration of free ligand in comparison to the protein, [L] is nearly equal to $[\bar{L}]$.

The main advantage of using frontal analysis instead of the zonal elution mode is that one need not make the hypothesis of linear chromatography; the equilibrium is obtained when the steady state is reached (plateau on the chromatogram). The total number of ligand moles $\bar{Q}_X$ actually active towards the protein is measurable from these experiments. For large amounts of protein, as compared with free ligand concentration, we have:

$$[\bar{A}] \cdot (V_R - V^*) \longrightarrow \bar{Q}_X \qquad (21)$$

Equation 18 is conveniently rearranged to:

$$\frac{1}{[\bar{A}] \cdot (V_R - V^*)} = \frac{1}{\bar{Q}_X} + \frac{1 + [L]K_{AL}}{\bar{Q}_X \cdot K_{AX} \cdot [\bar{A}]} \qquad (22)$$

The reciprocal of $[\bar{A}] \cdot (V_R - V^*)$, (the adsorbed amount $\bar{Q}_A$) is a linear function of the reciprocal of $[\bar{A}]$. The intercept on the ordinate axis will yield $1/\bar{Q}_X$. Figure 4 shows the elution profiles obtained in frontal affinity chromatography of guanosine on a carboxy-methylated RN-ase T-Sepharose column [11] in the absence of a competitive ligand ([L]=0). The dissociation constant is calculated from the intercept on the abscissa axis ($K_D = 1/K_{AX}$).

The interaction of trypsin with various L-arginine-terminated oligopeptides immobilized on agarose gels was studied by frontal affinity chromatography [3,11]. The effect of the addition of a counter-ligand was then studied and the dissociation constant for the enzyme-soluble competitive inhibitor was determined ($K_i = 1/K_{AL}$). The values obtained are in good agreement with those from kinetic analyses. Further the interaction of β-trypsin immobilized on Sepharose with competitive inhibitors was studied. Frontal chromatography gives two important parameters: the dissociation constant ($K_D = 1/K_{AX}$) and the total amount Q_X of active trypsin immobilized on Sepharose. The K_D value for the immobilized trypsin-benzamidine complex is similar to the K_i value. These results show that trypsin activity is hardly affected upon immobilization. The chromatographic method can be superior to enzyme kinetic measurements because it is applicable even when the enzyme is no longer active.

The main drawback of frontal chromatography is the large amount of protein required for every measurement, specially with long retention times. Therefore the zonal method is generally preferred. However this method assumes that local equilibrium is reached throughout the column and is independent of the amount injected. To check linear elution chromatography, the retention volume is to be independent of sample size, while the assumption of local equilibrium is valid if no variation of the retention volume is observed with a change in flow rate.

It is equal to the slope of the equilibrium isotherm at origin (equation 17):

$$V_z - V^* = \frac{K_{AX}\bar{Q}_X}{1 + [\bar{L}] \cdot K_{AL}} \tag{23}$$

A linear relationship is obtained when $1/(V_z - V^*)$ is plotted *vs.* the total ligand concentration $[\bar{L}]$ (figure 5) and K_{LX} is deduced from the slope to ordinate intercept ratio [9]:

$$\frac{1}{V_z - V^*} = \frac{1}{K_{AX}\bar{Q}_X} + \frac{K_{AL}}{K_{AX}Q_X} [\bar{L}] \tag{24}$$

This theoretical relationship will not apply if the retention volume depends on the amount of protein injected. It is necessary to inject small sample sizes near the limit of the detector sensitivity. Chaiken and Taylor [12], when studying the ribonuclease-nucleotide interactions, have shown that the retention volume of the ribonuclease peak decreases with increasing sample size. The procedure generally adopted is to extrapolate retention volumes to zero sample size.

Another drawback of the zonal elution method is that the value of K_{AX}, defining the interaction between the protein and the immobilized ligand, is not directly measurable: the effective amount of immobilized ligand active towards the protein has to be known. If measured from an independent titration method, it may be found considerably lower than the amount originally immobilized on the support. Q_X may be determined by frontal chromatography by measuring the maximum amount of mobile component that binds with the derivatized support [2].

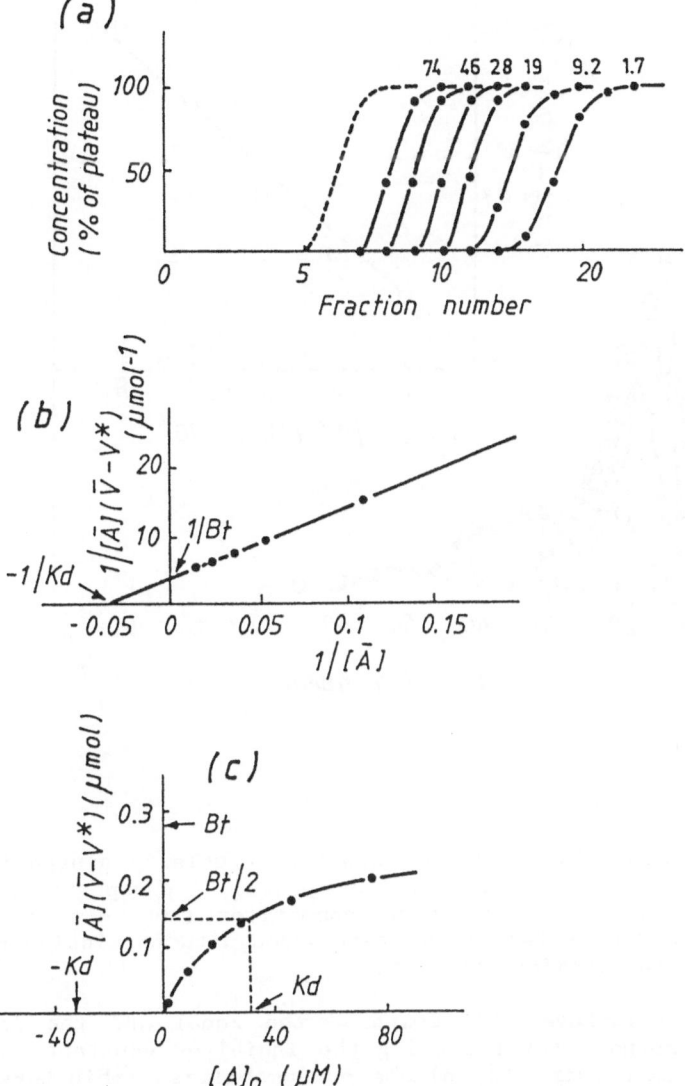

Figure 4. Frontal affinity chromatography for various concentrations of guanosine on a carboxy-methylated RNase T_1 Sepharose column in the absence of a competitive ligand (a). Analysis of the above data (b). Equilibrium isotherm (c). Figure 3 of Reference 11, reproduced with permission of the authors and Elsevier Science Publishers.

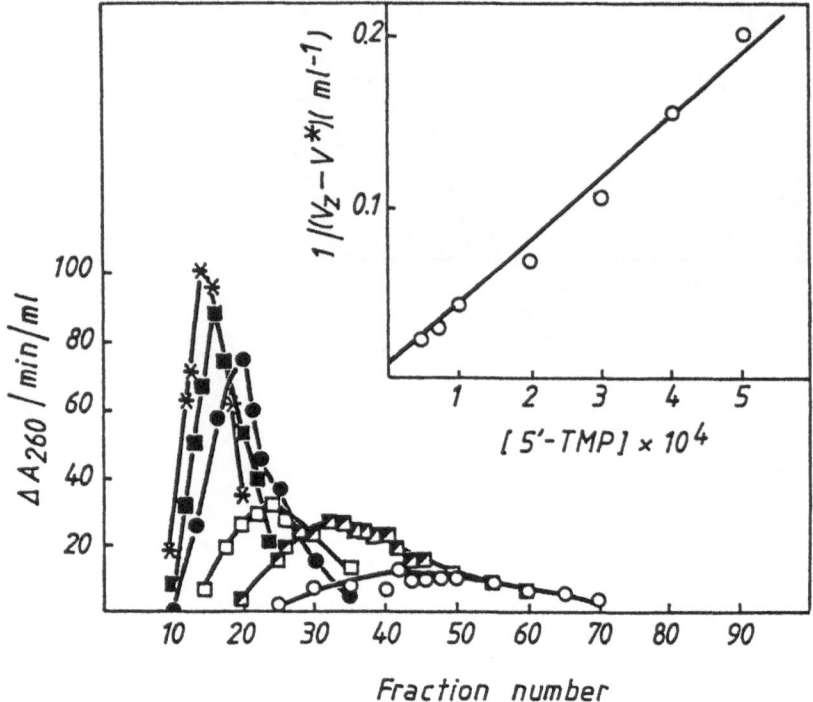

Figure 5. Competition zonal elution profiles: nuclease eluted from a thymidine 3'-(p-Sepharose-aminophenyl phosphate) 5'-phosphate affinity column with varying concentration of competitor (thymidine 5'-monophosphate. Figure 3 of Reference 9, reproduced with permission of the authors and American Chemical Society.

Malanikova and Turková [13] compared the zonal and the frontal chromatographic methods for measuring the inhibitor constants of the complexes of trypsin with its soluble or immobilized inhibitors. The elution behavior of trypsin on p-aminobenzamidine immobilized on hydroxyalkylmethacrylate gel (Spheron) in the presence of mobile ligands (benzylamine, benzoyl-L-arginine, N-butylamine, benzamidine and p-aminobenzamidine). The values of the dissociation constants $(K_i = 1/K_{AL})$ obtained by affinity chromatography with zonal or frontal method are in good agreement with the data measured kinetically. The dissociation constant of the complex of trypsin with immobilized p-aminobenzamidine $(1.6$ to $3.7 \cdot 10^{-6} M)$ is lower than the dissociation constant of the complex of trypsin with free p-aminobenzamidine $(1.9 \cdot 10^{-5} M)$.

High performance liquid affinity chromatography (HPLAC) was successfully used [14] to determine the dissociation constants by a zonal elution method. This method combines the advantages of conventional affinity chromatography and high performance liquid chromatography (HPLC) *i.e.* high selectivity with a short analysis time and an increased sensitive detection. The various aspects of HPLAC were discussed by Larsson *et al.* [4]. The high sensitivity of detection permits to work with small concentrations of solute and therefore to reach the conditions for linear chromatography, where the formulae established for frontal elution are approximately valid even in zonal elution.

3.2. Interaction of a solute with single site immobilized macromolecules

Anderson and Walters [15] studied another type of system where the macromolecule X is immobilized and the solute A is eluted with a competing ligand L. The equilibria are:

$$X \; + \; A \; \rightleftharpoons \; XA \qquad \{XA\} \; = \; \{X\}[A] \cdot K_{AX} \qquad (25)$$
$$X \; + \; L \; \rightleftharpoons \; XL \qquad \{XL\} \; = \; \{X\}[L] \cdot K_{LX} \qquad (26)$$

The total amount of immobilized solute, $\bar{Q}_A$ is:

$$\bar{Q}_A \; = \; S \, \{XA\} \; = \; \frac{\bar{Q}_X \cdot [A] \cdot K_{AX}}{1 \; + \; [A] \cdot K_{AX} \; + [L] \cdot K_{LX}} \qquad (27)$$

Since the total concentrations $[\bar{A}]$ and $[\bar{L}]$ are equal to those of the free species, they are replaced here by [A] and [L] and the equilibrium constants K_{AX} and K_{LX} can be measured by frontal chromatography from the equilibrium partition isotherm expression (equation 27):

$$\frac{1}{V_R - V^*} \; = \; \frac{1}{K_{AX} \bar{Q}_X} \; + \; \frac{[A]}{\bar{Q}_X} \; + \; \frac{K_{LX}}{K_{AX} \bar{Q}_X} \cdot [L] \qquad (28)$$

For small solute concentrations from a linear plot of $1/(V_z - V^*)$ *vs.* [L] the slope to intercept ratio allows the determination of K_{AX}.

The equilibrium constants for the binding of various sugars to immobilized concanavalin A were determined by HPLAC [15]. The values obtained by zonal and frontal analysis on columns of variable concanavalin A coverage were in close agreement but twice as large as literature values from solution studies.

3.3 Interaction of a solute with multiple-site immobilized macromolecules

Lagercrantz *et al.* [16] have described a system involving the interaction of a low molecular mass ligand A in solution and an immobilized protein carrying several sites. The equilibria are described as follows:

$$\begin{aligned} X \; + \; A \; &\rightleftharpoons \; XA \qquad & \{XA\} \; &= \; K_1 \cdot \{X\} \cdot [A] \\ \cdots\cdots\cdots\cdots & \qquad & \cdots\cdots\cdots\cdots\cdots & \\ XA_{m-1} \; + \; A \; &\rightleftharpoons \; XA_m \qquad & \{XA_m\} \; &= \; K_1 K_2 \; \cdots \; K_m \{X\} \cdot [A]^m \end{aligned} \qquad (29)$$

where K_1, K_2, ... K_m are the successive association constants and m the number of sites.

The expression of the equilibrium isotherm deduced from the above equilibria is:

$$\bar{Q}_A \ - \ \bar{Q}_X [A] \ \frac{K_1 \ + \ 2K_1 K_2 [A] \ + \ \cdots \ + \ mK_1 K_2 \ \cdots \ K_m [A]^{m-1}}{1 \ + \ K_1 [A] \ + \ K_1 K_2 [A]^2 \ + \ K_1 K_2 \ \cdots \ K_m [A]^m} \ - \ \bar{Q}_X Y \quad (30)$$

where $\bar{Q}_A$ is the total amount of ligand bound to the matrix and Y is the molar ratio between bound ligand and serum albumin. The Scatchard model involves n apparent binding constants β_i, with:

$$Y \ - \ \sum_{i=1}^{n} N_i \beta_i [A] / \left(1 \ + \ \beta_i [A]\right) \quad (31)$$

The retention volume in frontal chromatography is deduced from the above expression since:

$$V_R \ - \ V^* \ - \ \bar{Q}_A / [A] \ - \ \bar{Q}_X \sum_{i=1}^{n} N_i \beta_i / \left(1 \ + \ \beta_i [A]\right) \quad (32)$$

The constants β_i and N_i can be determined by identification to the stoichiometric binding constants K_i, with

$$K_1 \ - \ \sum_{i=1}^{n} N_i \beta_i . \quad (33)$$

The equilibrium constant K_1 for the interaction of some fatty acids, steroids and drugs with bovine serum albumin immobilized on Sepharose was obtained with zonal chromatographic experiments while frontal chromatography was used to evaluate the complete set of binding constants for salicylic acid. The plot of $1/(V_R - V^*)$ vs. the concentration of ligand [A] is not linear, indicating the presence of multiple binding sites with different association constants (figure 6a). The data points of Y obtained by frontal chromatography were fitted to the Scatchard model according to equation 31 (figure 6b).

Numerical simulations of the chromatographic process were applied by Vidal-Madjar et al. [6] to the determination of the equilibrium isotherm of a drug (phenylbutazone) with human serum albumin immobilized on diol-silica using zonal HPLC. It was shown that a three parameter isotherm equation is adequate to define the system (Figure 3). These measurements reveal two types of site on the immobilized protein, with specific and non-specific interactions. The association constant with the high affinity sites is about three times as large as that for the drug with the free protein in solution.

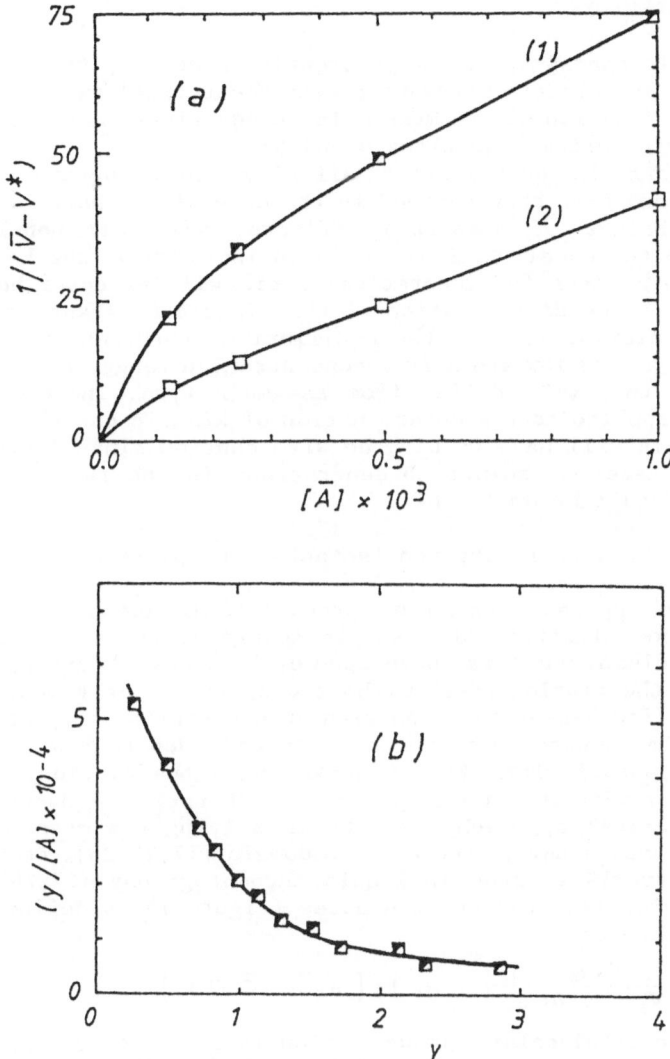

Figure 6. Frontal chromatographic analysis of the interaction of ligands with immobilized bovine serum albumine. (a) Plot of $1/(\bar{V} - V*)$ vs. ligand concentration: (1) [^{14}C]-octanoate-octanoate; (2) [^{14}C]-decanoate-decanoate. (b) Scatchard diagram for [^{14}C]-salicylic acid-salicylic acid. Figure 3 and 4 of Reference 16, reproduced with permission of the authors and Academic Press.

4. KINETIC MEASUREMENTS

The elution profile contains all the information necessary to determine the thermodynamic and kinetic parameters. Its theoretical expression is obtained from the solution of the mass balance equations characterizing the chromatographic system (equations 4 and 5).

For applications in quantitative affinity chromatography exact analytical solutions have been derived in the case of a linear isotherm [17,18] or in absence of dispersive effects, with the non-linear sorption rate limited model which leads to an isotherm of the Langmuir type [19]. For both cases the theoretical basis will be described and the applications to the determination of the kinetic constants will be reviewed. We saw previously that the equilibrium isotherms are seldom linear, except in a very low and narrow concentration range, furthermore for many systems they will differ from Langmuir type. Therefore the theoretical laws applied to the determination of kinetic constants from peak shape analysis will have to be used with caution since large peak distortions originate for solute concentrations in the range of non-linearity of the equilibrium isotherm.

4.1. Kinetics with a linear sorption isotherm: the plate height method

The stochastic approach, based on probabilistic ideas [20] which consider the random migration of a single solute molecule through the chromatographic column, was first developed by Giddings and Eyring [21]. It is shown that the elution peak "without dispersion" is given by the product of a modified Bessel function with an exponential step and with the inverse of time square root, which is to be added to a Dirac step function. Its asymptotic form is a Gaussian when the elution time is large in comparison with the time required for adsorption or desorption.

. Another theoretical approach consists in solving the mass balance equations of chromatography in the Laplace domain [17,22-26]. According to the model generally adopted in liquid chromatography [27,28], the solute mass transfer flow through the pores (Figure 7) is defined by a linear kinetic process:

$$\frac{1}{V_o} \cdot \frac{\partial(\bar{Q}_A + \bar{Q}'_A)}{\partial t} = k_2 \left(\sigma_A \bar{C} - \bar{C}'(R) \right) \tag{34}$$

where $\bar{C}'(R)$ is the total solute concentration in the pores at the outer surface of every spherical particle (when r, the distance from the particle center is equal to R, the particle radius).

The diffusion law for the solute inside the porous spherical particle is represented by the equation:

$$\frac{1}{V_i} \frac{\partial \bar{Q}_A(r)}{\partial t} = \frac{D'}{r^2} \cdot \frac{\partial}{\partial r} \left[r^2 \frac{\partial \bar{C}'(r)}{\partial r} \right] - \frac{\partial \bar{C}'(r)}{\partial t} \tag{35}$$

with the boundary condition:

$$\frac{\partial \bar{C}'(0)}{\partial r} = 0$$

where r is the radial position in the grain and D' the global diffusion coefficient inside the porous particle.

The affinity process is assumed, as a first approximation, to follow a linear kinetic law:

$$\frac{\partial \overline{Q}_A (r)}{\partial t} = k_a V_i \overline{C}'(r) - k_d \overline{Q}_A (r) \tag{36}$$

At equilibrium, we have:

$$\frac{\overline{Q}_A^*}{V_i \overline{C}'} = \frac{k_a}{k_d} = K_A \tag{37}$$

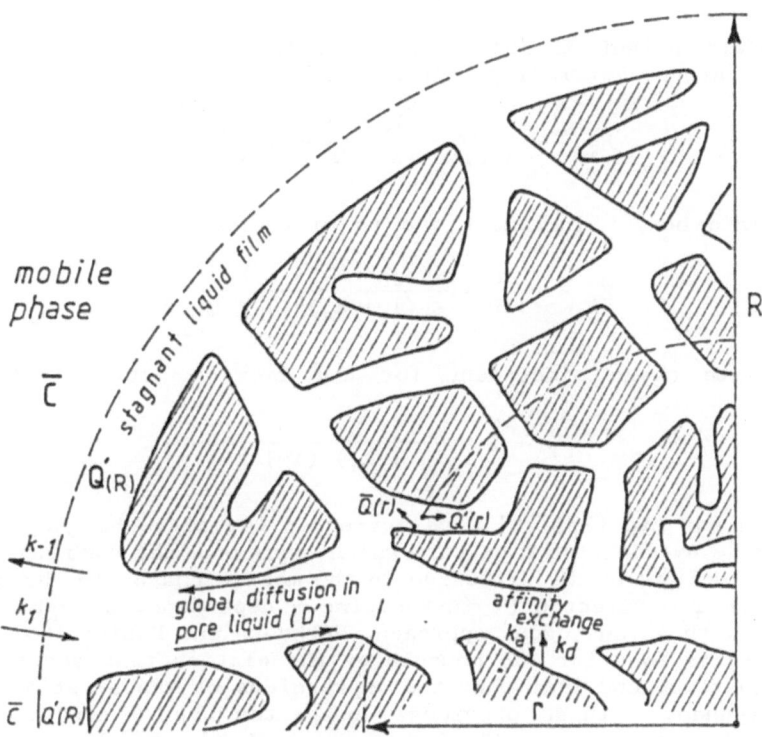

Figure 7. Schematic illustration of the mass transfer kinetics into porous spherical particles. Figure 11 of Reference 5, reproduced with permission of the authors and Marcel Dekker Inc.

Combining equations 3 and 36 with the column capacity factor referred to the retention volume V^* with no affinity (equation 11), we have:

$$k'(0) = \frac{\partial \overline{Q}_A}{v^* \cdot \partial \overline{C}} = \sigma_A \cdot K_A \frac{V_i}{V^*} = \frac{k_o K_A}{1 + k_o} \tag{38}$$

The set of differential equations describing the chromatographic system with a linear sorption isotherm can be solved using a Laplace transformation. The method gives the expression of the statistical moments characterizing the retention and the band broadening of the elution peak. The mass center μ_i' is given by equation 7.

The second moment is related to the plate height (equation 9) which is generally given as a sum of increments [27,28]:

$$H = H_D + H_E + H_I + H_K \qquad (39)$$

The first term H_D expresses the plate height increment due to axial dispersion:

$$H_D = 2D/u \qquad (40)$$

The plate height increment H_E is related to the mass transfer resistance at the external particle boundary:

$$H_E = \frac{2}{\sigma_A k_2} \cdot \frac{(k_o + k' + k'k_o)^2}{(1 + k_o)^2(1 + k')^2} \cdot u \qquad (41)$$

The plate height increment due to diffusion inside the pores is:

$$H_I = \frac{2R^2}{15D'} \cdot \frac{(k_o + k' + k'k_o)^2}{k_o(1 + k_o)^2(1 + k')^2} \cdot u \qquad (42)$$

The plate height increment for the mass transfer kinetics is:

$$H_K = 2 \cdot \frac{k'}{k_d(1 + k_o) \cdot (1 + k')^2} \cdot u \qquad (43)$$

Horváth and Lin [27, 28] have corrected the plate height diffusion increment H_D by addition of a term which expresses the combined effects of flow velocity and the diffusion in the mobile phase. In the same way the H_E term is corrected by introducing a dependence of $\sigma_A k_2$ on flow velocity. With an analogous approach, Hethcote and DeLisi [25] derived similar expressions for the moments of the elution peak and the plate height equation. and show that the expressions of the first and second moments reduce to those previously presented when there is no free ligand in the mobile phase. The kinetic contribution to the plate height is then similar to that described by equation 43.

The plate height method can be used to determine the rate constants from the slope of the straight line relating the plate height to the linear velocity (Figure 8). The other contributions to band broadening must then be negligible or subtracted. When the solute is a protein, large increase in plate height will be observed if the pore diameter is close to that of the protein molecule, because of the restricted diffusion into the pores.

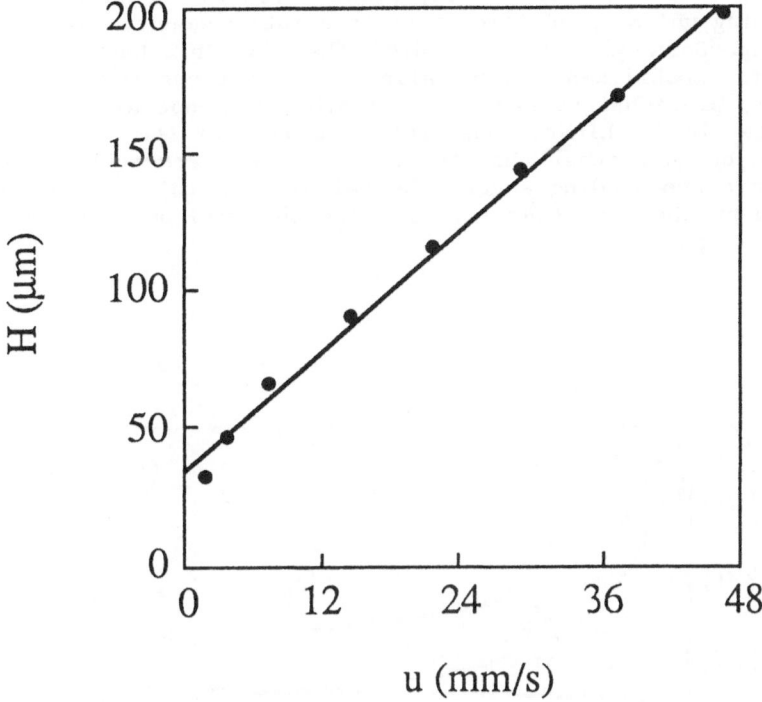

Figure 8. Plate height as a function of linear velocity, for α-D-manno-pyranoside on Lichrospher SI 500 diol. Figure 5 of Reference 15, reproduced with permission of the authors and Elsevier Science Publishers.

Anderson and Walters [15] have applied the plate height method based on Horváth and Lin's approach [28] to study the kinetics of the interaction between several sugar derivatives and silica bound concanavalin A. The equilibrium constants obtained by zonal and frontal analysis were approximately twice as large as literature values from solution studies. Under linear elution conditions, assuming immobilized concanavalin A to be homogeneous in nature, it was shown that the contribution to plate height is inadequately described as a function of k' (Figure 9). The method used to determine the dissociation rates was

500

quite inaccurate because of the important diffusion contributions to band broadening which are difficult to take into account.

Muller and Carr [29] have studied the same biochemical system consisting in immobilized concanavalin A with various sugars, noting that the only plate height contribution which does not depend on particle diameter is the kinetic one. They assumed that the contributions to the plate height, other than the kinetic ones, are equal to those measured for a non-binding sugar. The validity of this correction is demonstrated by the fact that the k_d value obtained on $10\mu m$ and $50\mu m$ particles are equal.

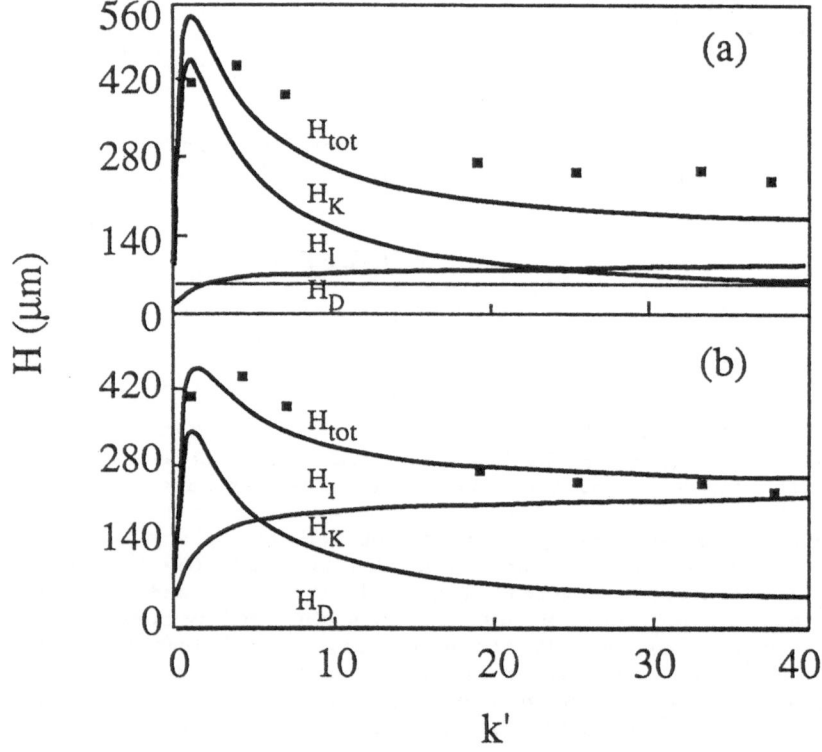

Figure 9. Plate height vs. k' for p-nitrophenyl α-D-mannopyranoside on a support with a high density of binding sites (Concanavalin A immobilized on a HPLC silica support). (a) Calculated H_D and H_I values. (b) H_I allowed to increase to better account for band broadening at high k' values. Figure 8 of Reference 15, reproduced with permission of the authors and Elsevier Science Publishers.

4.2. Second order kinetics with a "Langmuir" type isotherm

Assuming that there is instantaneous equilibrium between pore and bulk liquid phase, the interaction between the solute A with the immobilized binding sites X on the affinity adsorbent can be described by an equilibrium equation of the form:

$$A + X \xrightleftharpoons[k_r]{k_f} A\!-\!X \tag{44}$$

where k_f is the second order rate constant governing the forward reaction and k_r the first order rate constant governing the reverse one. This corresponds to a Langmuir type equilibrium isotherm. The ratio k_f/k_r is equal to the association constant K_{AX} relative to the chemical equilibrium.

The rate of formation of the affinity complex is described by the relationship:

$$\frac{\partial \bar{Q}_A}{\partial t} = k_f \bar{C} \cdot (\bar{Q}_X - \bar{Q}_A) - k_r \bar{Q}_A \tag{45}$$

with:

$$\bar{Q}'_A - \bar{Q}'_A{}^* = V_i \bar{C} \cdot \sigma_A \tag{46}$$

At equilibrium the amount of adsorbed material is related to its concentration, with $K_{AX} = k_f/k_r$:

$$\bar{Q}^*_A = \frac{K_{AX} \bar{C} \cdot \bar{Q}_X}{1 + K_{AX} \bar{C}} \tag{47}$$

The material balance equation of the chromatographic system (equation 4) writes now:

$$(1 + k_o)\frac{\partial \bar{C}}{\partial t} + u \cdot \frac{\partial \bar{C}}{\partial z} - D \cdot \frac{\partial^2 \bar{C}}{\partial z^2} = -\frac{1}{V_o} \cdot \frac{\partial \bar{Q}_A}{\partial t} \tag{48}$$

Assuming that the effect of axial dispersion is negligible, Thomas [30] gave a solution for the system (equations 45 and 48) which was later adapted by Chase [31] for affinity chromatography. For frontal elution, the concentration at the column outlet is given by:

$$C = C_0 \; P/E \tag{49}$$

where C_0 is the plateau concentration.

P and E are given by:

$$P = J(\nu w, \nu T) \tag{49a}$$

$$E = J(\nu w, \nu T) + \left[1 - J(\nu, w\nu T)\right] \cdot \exp\left[(1-w) \cdot (\nu - \nu T)\right] \tag{49b}$$

where J function is given by:

$$J(x,y) = 1 - e^{-y} \int_0^x I_0 (2\sqrt{\tau y}) \cdot e^{-\tau} d\tau \tag{50}$$

where I_0 is the Bessel function of zeroth order and F the flow rate.

502

The expressions of the dimensionless parameters ν, w and T are :

$$\nu = Q_x \, k_f \, / \, F \qquad w = 1 \, / \, (1 + K_{Ax} \, C_0)$$
$$\left. \begin{array}{c} \\ \\ \end{array} \right\} \qquad (51)$$
$$T = (tF - V*) \, /(\, K_{Ax} \, Q_x \, w)$$

ν is the number of transfer units and is related to the mass transfer kinetic contribution according to

$$\nu = \frac{2L \cdot k'^2}{H_K (1+k')^2} \qquad (52)$$

In fact, the constants k_f and k_r are not simply the rate constants for the chemical exchange process. They include also the resistances to mass transfer from the mobile phase to the adsorbent, i.e. the mass transfer at the particle boundary and the diffusion in the stagnant fluid into the pores or between the particles.

The experimental system studied by Chase was the binding of proteins to Cibacron Blue Sepharose and that of E. coli β-galactosidase to an immobilized monoclonal antibody. The model for frontal elution (equations 49-51) agrees well with the experimental data as shown in Figure 10 for the elution of lysozyme from a column packed with immobilized Cibacron blue.

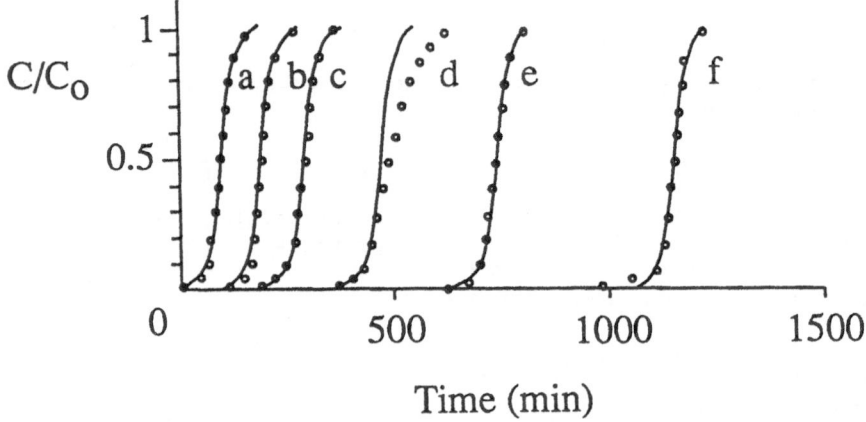

Figure 10. Variation of shape and position of breakthrough curve with column length. Elution of lysozyme on Cibacron Blue Sepharose. The figure shows the experimental determinations of C/C_0 (o) and the solid lines are the theoretical predictions determined by use of eqn. 50. Figure 5 of Reference 31, reproduced with permission of the authors and Elsevier Science Publishers.

Arnold et al. [19,32] extended the theoretical model to include the dispersion effects through the column. They discussed the applications of the theoretical model to various types of chromatographic operations: adsorption, wash and elution processes. It was shown that the dispersion contributions (axial dispersion, diffusion inside the pores) are to be added to the mass transfer kinetics via a dimensionless number, the number of transfer units ν.

Arnold and Blanch [19] have shown that the method for investigating the binding kinetics from breakthrough curves can be used if the extra-contributions to peak spreading do not overwhelm the kinetic one. These extra-contributions can be determined from the value of the mass transfer units ν, measured for non-adsorbed tracers.

On the basis of the same hypotheses as those used for the frontal elution model, Wade et al. [33] gave an analytical solution of the output signal in zonal elution assuming a Dirac injection function. The retention behavior of p-nitrophenyl-α-D-mannopyranoside on immobilized concanavalin A was examined. The adsorption-desorption rate constants and the binding site density were determined for the more populous binding sites. The non-linear model reveals the heterogeneous adsorption on immobilized concanavalin A.

4.3. Study of the "split-peak" effect

Assuming a linear isotherm equilibrium and first order kinetics, Giddings and Eyring [20] have pointed out that the probability for solute molecules to pass through the column without adsorption is: $\exp(-h_1 L/u)$. A linear kinetic law is then assumed for the global mass transfer of the solutes between the mobile phase and the particles:

$$A \quad \underset{h_{-1}}{\overset{h_1}{\rightleftarrows}} \quad A_g \qquad (53)$$

where A is the solute in the bulk liquid phase, A_g the solute inside the grains and h_1 and h_{-1} are global rate constants for entrance into, and exit from the beads.

The solute mass transfer flow from the grains is:

$$\frac{\partial (\overline{Q}_A' + \overline{Q}_A)}{\partial t} = h_1 V_o \overline{C} - h_{-1} (\overline{Q}_P' + \overline{Q}_P) \qquad (54)$$

At equilibrium:

$$h_1/h_{-1} = (1 + K_A)k_o \qquad (55)$$

Equation 54 combined with the mass balance equation of chromatography (equation 4) can be solved in the Laplace domain [22-26]. The shape of the theoretical peak presents a tailing and with some particular experimental conditions (slow adsorption kinetics, high flow rate and short columns) a bimodal peak is generated (Figure 11).

The theory was further developed by Walters et al. [34,35] to combine the mass transfer at the particle boundary:

$$A \quad \underset{k_{-1}}{\overset{k_1}{\rightleftarrows}} \quad A_p \qquad (56)$$

504

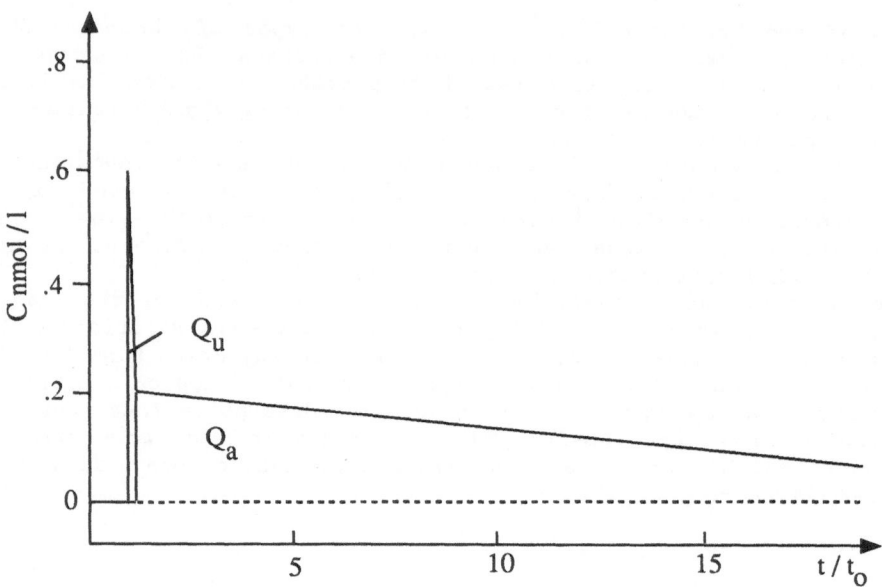

Figure 11. Theoretical bimodal peak in linear elution chromatography. L= 5 cm; t_0= 30 s; h_{-1}= 0.005 s^{-1}; k'= 10. Q_u — unretained amount, Q_a = adsorbed amount ($f = Q_u / Q_a$).

and the surface binding process:

$$A_p + X \underset{k_{-3}}{\overset{k_3}{\rightleftarrows}} AX \qquad (57)$$

With the partial differential equations:

$$\frac{\partial(\overline{Q'_A} + \overline{Q_A})}{\partial t} = k_1 V_o \overline{C} - k_{-1} \overline{Q'_A} \qquad (58)$$

$$\frac{\partial \overline{Q_A}}{\partial t} = k_3 \cdot \frac{\overline{Q_X}}{V_i \sigma_A} \cdot \overline{Q'_A} - k_{-3} \overline{Q_A} \qquad (59)$$

$k_a/k_d = K_{AX}$, where K_{AX} is the equilibrium constant for a monomeric solute interacting with a single-site immobilized ligand. The rate constant k_3 (M$^{-1}\cdot$s^{-1}) is relative to a second order process and the others k_1, k_{-1} and k_{-3} (s^{-1}) to first order ones.

With our previous notations (equation 34):

$$\left.\begin{array}{c} k_1/k_{-1} = \sigma_A V_i/V_o = k_o \\ k_1 = \sigma_A k_2 \quad ; \quad k_{-1} = k_2 V_o/V_i \\ k_3/k_{-3} = K_A \sigma_A V_i/Q_X \end{array}\right\} \qquad (60)$$

When the adsorption is quasi-irreversible towards the time scale of the experiment (k_{-3} very small), the following scheme is obtained:

$$A \underset{k_{-1}}{\overset{k_1}{\rightleftharpoons}} A_P + X \overset{k_3}{\longrightarrow} AX \tag{61}$$

and for the partial kinetic differential equations:

$$\frac{\partial \overline{Q_A'}}{\partial t} = V_o k_1 \overline{C} - \left(k_{-1} + k_3 \cdot \frac{\overline{Q_x}}{\sigma_A V_i} \right) \overline{Q_A'} \tag{62}$$

$$\frac{\partial \overline{Q_A}}{\partial t} = k_a \cdot \frac{\overline{Q_x}}{\sigma_A V_i} \cdot \overline{Q_A'} \tag{63}$$

where the amount of free immobilized ligand $\overline{Q_x}$ is assumed to be independent of the solute concentration (hypothesis of linear chromatography).

It may be shown by solving the system of differential equations (equations 4, 62 and 63) using a Laplace transform that the fraction f of solute which has not been retained by affinity is given by the equation:

$$f = \exp \left(-\frac{L}{u} / \left(\frac{1}{k_1} + \frac{V_o}{k_3 Q_x} \right) \right) \tag{64}$$

or, as a function of flow rate F:

$$-1/\ln(f) = F \left(\frac{1}{k_1 V_o} + \frac{1}{k_3 Q_x} \right) \tag{65}$$

The value f = 0 is valid for a normal chromatographic behavior, without the appearance of an unretained peak. Equation 65 shows that the "split-peak" phenomenon will increase with increasing flow rate.

The split-peak behavior was studied [34] for a chromatographic system consisting of immobilized protein A and immunoglobulin G by varying the flow rate through short protein columns. When the inverse of the logarithm of the fraction of non-adsorbed protein was plotted as a function of flow rate, straight lines were obtained in agreement with equation 65. The activity of immobilized protein A was determined from break-through curves and an independent estimate of k_1 was obtained from the plate height method (equation 39), using a matrix without immobilized protein. For an adsorption limited process, a reasonable value for the association rate constant of protein A with IgG, $k_3 = 1.2 \cdot 10^5$ $M^{-1} \cdot s^{-1}$, was measured when protein A was immobilized on a diol-bonded support using the "Schiff-base" method.

It was found that the kinetic properties of the protein depend on the immobilization method. The split-peak method is particularly useful for studying antibody-antigen reaction when the retention is strong. The derivation of equation 65 is based on the assumptions of linear elution chromatography. Since a dependence of the retention volume with the amount injected was observed, an extrapolation to zero sample size was necessary to extract kinetic parameters.

It is therefore useful to have an expression for the split-peak effect valid in the non-linear range of the adsorption isotherm. On the basis of the same hypotheses used by Chase [31] and Arnold et al. [19,32], Jaulmes and Vidal-Madjar [36] have developed a model to

506

characterize the split-peak effect in mass-overload conditions. Its expression is easily obtained by integrating equation 50 between 0 and t_i, the time corresponding to the duration of injection of a rectangular pulse. In the case of irreversible adsorption, the non-retained fraction f is a function of the amount injected Q_i, the capacity of the column Q_x and the number of transfer units ν:

$$ f = \frac{Q_x}{\nu\, Q_i}\, \text{Ln}\left[1 + \left(e^{\nu Q_i/Q_x} - 1\right)e^{-\nu} \right] \tag{66} $$

The adsorption yield ($\rho = 1 - f$) which is a useful quantity in preparative affinity chromatography may be predicted from the above expression. Its variation with the sample size is shown in Figure 12. The lower the number of transfer unit ν is, the larger the split-peak effect. Therefore, the non-retained fraction will increase with the flow rate or with the adsorption rate constant.

For given ν values, the characteristic curves of the adsorption yield are only function of the ratio Q_i/Q_x. For $Q_i \gg Q_x$ or for zero flow rate, the characteristics reach the asymptotic law (dotted line) corresponding to the trivial expression of the adsorption yield in batch experiments: $\rho = 1 - Q_x/Q_i$ if $Q_i > Q_x$ and $\rho = 1$ if $Q_i < Q_x$.

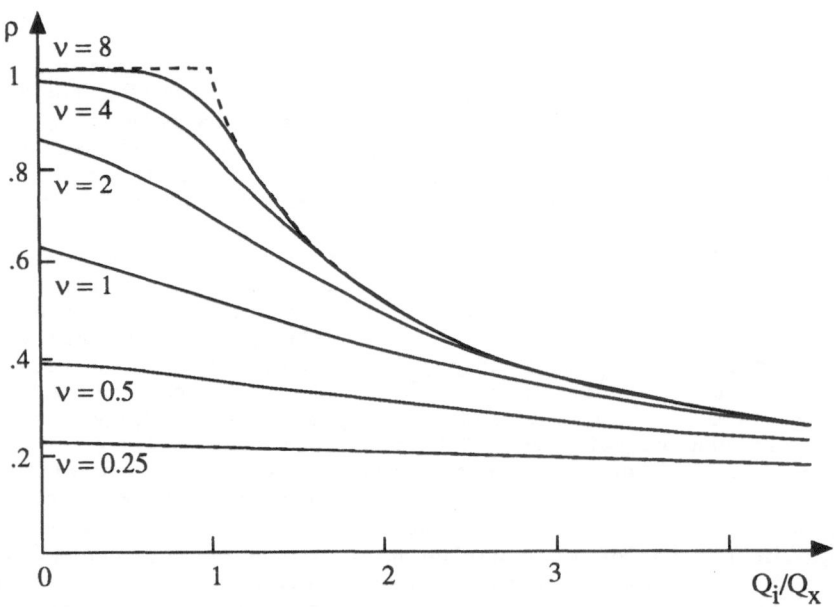

Figure 12. Variation of the adsorption yield with the sample size, assuming a Langmuir type adsorption isotherm . Figure 5 from Reference 36, reproduced with permission of the authors and American Chemical Society.

5. CONCLUSION

Affinity chromatography is a reliable technique for measuring the interactions betweeen biological species in solution. The equilibrium constants are then generally in good agreement with those determined by other techniques. Larger differences exist, however, with the constants characterizing the interaction with the immobilized species. These discrepancies are due to the difficulty in determining precisely the effective amount of the active immobilized ligand. This is particularly true in zonal analysis. Moreover, the properties of the immobilized species will depend on the point of attachment to the matrix and the immobilized molecule may loose its properties of biological recognition. The trend is now to produce supports with immobilized ligands having affinity properties as close as possible to those of the free ones, to decrease the non-specific interactions of the support itself and to increase the mass transfer rates by using non-porous particles or particles of extremely large porosity.

An important development of quantitative affinity chromatography is to be expected since the method is now combined with HPLC [4], a technique which offers high sensitivity and speed of operation. The injection of a pulse at the column inlet is generally preferred to the more rigorous frontal chromatographic method, because the amount of solute required can be very small. However, several difficulties arise in the treatment of the chromatographic results because the chromatographic system may be out of equilibrium and a non-linear system involved. The shorter the analysis time, the more important the kinetic phenomena. An increasing attention is now given to kinetic studies.

6. SYMBOLS

A	:	solute studied (bulk liquid phase)
A_g	:	solute studied (in the grains)
A_p	:	solute studied (pore liquid phase)
$[\underline{A}]$	:	free solute concentration (M)
$[\bar{A}]$	:	total solute concentration (M)
C	:	free solute concentration in the mobile phase $C = [A]$ (M)
$\bar{C}$	:	global solute concentration in the mobile phase $\bar{C} = [\bar{A}]$ (M)
$\bar{C}'$	:	global solute concentration in the pores (M) (eqn. 34)
C_0	:	plateau concentration (M) (eqn. 49)
D	:	dispersion coefficient (cm^2/s) (eqn. 4)
D'	:	diffusion coefficient inside the porous particle (cm^2/s) (eqn. 35)
E	:	auxiliary variable (eqn. 49b)
F	:	flow rate of the mobile phase (cm^3 s^{-1}) (eqn. 65)
f	:	unretained fraction of solute (eqn. 64)
H	:	total plate height related to the second moment (cm) (eqn. 9)
H_C	:	sum of H_E, H_I and H_K (cm)

H_D : plate height increment due to axial dispersion (cm) (eqn. 40)

H_E, H_I, H_K : plate height increments due to mass transfers (cm) (eqn. 41-43)

h_1, h_{-1} : global rate constants (s^{-1}) (eqn. 53)

K_1, K_2..K_n : successive association constants (M^{-1}) (eqn. 29)

K_A : partition ratio (eqn. 37)

K_D : dissociation constant ($K_D = 1/K_{AX}$) (M)

K_i : dissociation constant for solute-ligand in the liquid phase ($K_i = 1/K_{AL}$) (M)

K_{IJ} : association constant for $I + J \rightleftharpoons IJ$ (M^{-1})

k_o : column permeation ratio (eqn. 3)

k_1, k_{-1} : rate constants relative to pore inlet (s^{-1})(eqn. 56)

k_2 : rate constant relative to the global exchange at pore inlet (s^{-1})(eqn. 60)

k_3 : second order adsorption rate constant relative to the binding process ($M^{-1} \cdot s^{-1}$) (eqn. 57 and 60)

k_{-3} : desorption rate constant relative to the binding process (s^{-1}) (eqn. 57 and 60)

k_a, k_d : adsorption and desorption rate constants in the first order binding model (s^{-1}) (eqn. 36)

k_f : second order apparent forward rate constant ($M^{-1} \cdot s^{-1}$) (eqn. 44)

k_r : apparent reverse rate constant for second order kinetics (s^{-1}) (eqn. 44)

k' : generalized apparent capacity factor (eqn. 11)

L : ligand in the mobile phase

$[\underline{L}]$: free ligand concentration (M)

$[\overline{L}]$: total ligand concentration (M)

N_i : coefficients in Scatchard expansion (eqn. 31)

n : number of binding sites in Scatchard model (eqn. 32)

m : number of successive association constants (eqn. 29)

$\underline{P}$: auxiliary variable (eqn. 49a)

Q_A : total amount of A fixed in the column (moles) (eqn. 4)

Q'_A : total amount of A in the pores of the column (moles) (eqn. 4)

$\overline{Q}_A^*$: value of $\overline{Q}_A$ at equilibrium (moles) (eqn. 5)

$Q'_A{}^*$: value of Q'_A at equilibrium (moles) (eqn. 5)

Q_i : amount injected (moles) (eqn. 66)

Q_X : total amount of immobilized ligand (moles) (eqn. 14)

R : particle radius ($0<r<R$) (μm)

r : distance to the particle center (μm) (eqn. 35)

S : solid phase surface area (m^2)

t : time elapsed from the injection instant (s) (eqn. 4)

t_i : duration of the injection pulse (s)

T : auxiliary variable (eqn. 51)

u : interstitial mobile phase velocity (cm s^{-1}) (eqn. 4)

V : retention volume (cm^3) (eqn. 1)

V^* : retention volume of a non-retained solute (cm^3) (eqn. 3)

V_o	:	liquid external volume (cm^3) (eqn. 3)
V_i	:	internal pore volume (cm^3) (eqn. 3)
V_R	:	average retention volume (cm^3) (eqn. 1)
V_z	:	limit value of V_R; $V_z = V_R(0)$
x	:	auxiliary variable (eqn. 50)
X	:	immobilized ligand
$\{X\}$	:	surface concentration of immobilized ligand (mole cm^{-2})
y	:	auxiliary variable (eqn. 50)
Y	:	binding number $Y = Q_A/Q_X$ (eqn. 31)
w	:	auxiliary variable (eqn. 51)
z	:	abscissa along the column length (eqn. 4)
α	:	rate constant (s^{-1})(eqn. 5)
β_i	:	binding constants in Scatchard expansion (eqn. 56)
μ_o, μ_1', μ_2	:	statistical moments of the output peak (eqns. 6-9)
ν	:	number of transfer units (eqn. 51)
σ_A	:	permeation coefficient of solute A (eqn. 3)
τ	:	integration variable (eqn. 50)

7. REFERENCES

1 - Turková, J. (1978) "Affinity Chromatography", Elsevier, Amsterdam.

2 - Winzor, D.J. (1985) "Quantitative characterisation of interactions by affinity chromatography" in P.D.G. Dean, W.S. Johnson and F.A. Middle (eds.), Affinity Chromatography. A practical approach, IRL Press, Oxford, pp. 149-168.

3 - Chaiken, I.M. (1986) "Analytical affinity chromatography in studies of molecular recognition in biology: a review", J. Chromatogr. 376, 11-32.

4 - Larsson, P.O., Glad, M., Hansson, L., Mansson, M.O., Ohlson, S. and Mosbach, X. (1983) "High-performance liquid affinity chromatography", Adv. Chromatogr. 21, 41-85.

5 - Jaulmes, A. and Vidal-Madjar, C. (1989) "Theoretical aspects of quantitative affinity chromatography. An overview", Adv. Chromatogr. 28, 1-64.

6 - Vidal-Madjar, C., Jaulmes, A., Racine, M. and Sébille, B. (1988) "Determination of binding equilibrium constants by numerical simulation in zonal high-performance affinity chromatography", J. Chromatogr., 458, 13-25.

7 - Nichol, L.W., Ogston, Q.G., Winzor, D.J. and Sawyer, W.H. (1974) "Evaluation of equilibrium constants by affinity chromatography", Biochem. J., 143, 435-443.

8 - Dunn, B.M. and Chaiken, I.M. (1974) "Quantitative affinity chromatography. Determination of binding constants by elution with competitive inhibitors", Proc. Natl. Acad. Sci. U.S.A., 71, 2382-2385.

9 - Dunn, B.M. and Chaiken, I.M. (1975) "Evaluation of quantitative affinity chromatography by comparison with kinetic and equilibrium dialysis methods for the analysis of nucleotide binding to Staphylococcal Nuclease", Biochemistry 14, 2343- 2349.

10 - Brinkworth, R.I., Masters, C.J. and Winzor, D.J. (1975)
     "Evaluation of equilibrium constants for the interaction of
     lactate dehydrogenase isoenzymes with reduced nicotinamide-
     adenine dinucleotide by affinity chromatography", Biochem. J.,
     151, 631-636.
11 - Kasai, K., Oda, Y., Nishikata, M. and Ishii, S. (1986) "Frontal
     affinity chromatography : Theory for its application to studies
     on specific interactions of biomolecules.", J. Chromatogr. 376,
     33-47.
12 - Chaiken, I.M. and Taylor, H.C. (1976) "Analysis of ribonuclease-
     nucleotide interactions by quantitative affinity chromatogra-
     phy", J. Biol. Chem., 251, 2044-2048.
13 - Malanikova, M. and Turkova, J. (1977) "Determination of disso-
     ciation constants of complexes of trypsin and its low molecular
     weight inhibitors by affinity chromatography in zonal and fron-
     tal analysis arrangement", J. Solid-Phase Biochem., 2, 237-249.
14 - Nilsson, K. and Larsson, P.O. (1983) "High-performance liquid
     affinity chromatography on silica-bound alcohol dehydrogenase"
     Anal. Biochem., 134, 60-72.
15 - Anderson, D.J. and Walters, R.R. (1986) "Equilibrium and rate
     constants of immobilized concanavalin A determined by high-per-
     formance affinity chromatography.", J. Chromatogr. 376, 69-85.
16 - Lagercrantz, C., Larsson, T. and Karlsson, H. (1979) "Binding
     of some fatty acids and drugs to immobilized bovine serum
     albumine studied by column affinity chromatography", Anal.
     Biochem. 99, 352- 364.
17 - Hethcote, H.W. and DeLisi, C. (1982) "Determination of
     equilibrium and rate constants by affinity chromatography",
     J. Chromatogr., 248, 183-202.
18 - Arnold, F.H., Schofield, S.A., and Blanch, H.W. (1986) "Analy-
     tical affinity chromatography. I. Local equilibrium theory and
     the measurement of association and inhibition constants", J.
     Chromatogr., 355, 1-12.
19 - Arnold, F.H. and Blanch, H.W. (1986) "Analytical affinity
     chromatography. II. Rate theory and the mesurement of biological
     binding kinetics", J. Chromatogr., 355, 13-27.
20 - Giddings, J.C. (1965) "Dynamics of Chromatography, Part I,
     Principles and Theory", Marcel Dekker, New-York, N.Y.
21 - Giddings J.C. and Eyring, H. (1955) "A molecular dynamic
     theory of chromatography", J. Phys. Chem., 59, 416-421.
22 - Kučera, E. (1965) "Contribution to the theory of chromato-
     graphy. Linear non-equilibrium elution chromatography", J.
     Chromatogr., 19, 237-248.
23 - Grushka, E. (1972) "Chromatographic peak shapes. Their origin
     and dependence on the experimental parameters", J. Phys. Chem.,
     76, 2586-2593.
24 - Villermaux, J. (1974) "Deformation of chromatographic peaks
     under the influence of mass transfer phenomena", J. Chromatogr.
     Science, 12, 822-831.

25 - Hethcote, H.W. and De Lisi, C. (1982) "Non-equilibrium model of liquid column chromatography. I. Exact expressions for elution profile moments and relation to plate height theory", J. Chromatogr., 240, 269-281.

26 - De Lisi, C., Hethcote, H.W. and Brettler, J.W. (1982) "Non-equilibrium model of liquid column chromatography. II. Explicit solutions and non-ideal conditions", J. Chromatogr., 240, 283-295.

27 - Horváth, Cs. and Lin, H.J. (1976) "Movement and band spreading of unsorbed solutes in liquid chromatography", J. Chromatogr., 126, 401-420.

28 - Horváth, Cs. and Lin, H.J. (1978) "Band spreading in liquid chromatography. General plate height equation and a method for the evaluation of the individual plate height contributions", J. Chromatogr., 149, 43-70.

29 - Muller, A.J. and Carr, P.W. (1984) "Chromatographic study of the thermodynamic and kinetic characteristics of silica-bound concanavalin A", J. Chromatogr., 284, 33-51.

30 - Thomas, H. (1944) "Heterogeneous ion exchange in a flowing system", J. Amer. Chem. Soc., 66, 1664-1666.

31 - Chase, H.A., (1984) "Prediction of the performance of preparative affinity chromatography", J. Chromatogr., 297, 179-202.

32 - Arnold, F.H., Blanch, H.W. and Wilke, C.R. (1985) "Analysis of affinity separations. I. Predicting the performance of affinity adsorbers", Chem. Eng. J., 30, B9-B23.

33 - Wade, J.L., Bergold, A.F. and Carr, P.W. (1987) "Theoretical description of nonlinear chromatography, with applications to physicochemical measurements in affinity chromatography and implications for preparative-scale separations", Anal. Chem., 59, 1286-1295.

34 - Hage, D.S., Walters, R.R. and Hethcote H.W. (1986) "Split-peak affinity chromatographic studies of the immobilization-dependent adsorption kinetics of protein A", Anal. Chem., 58, 274-279.

35 - Larew, L.A. and Walters, R.R. (1987) "A kinetic, chromatographic method for studying protein hydrodynamic behavior", Anal. Biochem., 164, 537-546.

36 - Jaulmes, A. and Vidal-Madjar, C. (1991) "Split-peak phenomenon in non-linear chromatography. 1. A theoretical model for irreversible adsorption", Anal. Chem., 63, 1165-1174.

THEORETICAL BASIS OF FIELD-FLOW FRACTIONATION.

Michel MARTIN
Ecole Supérieure de Physique et Chimie Industrielles
Laboratoire de Physique et Mécanique des Milieux Hétérogènes (URA CNRS 857)
10, rue Vauquelin. 75231 Paris Cedex 05. France

and

P. Stephen WILLIAMS
Field-Flow Fractionation Research Center
Chemistry Department, Henry Eyring Building
University of Utah
Salt Lake City, Utah 84112, USA

ABSTRACT. In field-flow fractionation (FFF), macromolecular or particulate species (including colloids and particles up to about 50 μm) are separated in thin flow channels under the influence of a transverse external field. The separation relies on the non-uniform distribution of the species molecules or particles in the channel cross-section. In the ideal case corresponding to molecules or particles of negligible size and a uniform applied field, the migration toward one of the channel walls is countered by Brownian motion resulting in an exponential transverse concentration profile. This characterizes the Brownian, or normal, mode of operation in FFF. Interactions of species with nonuniform fields, transverse gradients, the flow of carrier fluid, and/or the channel walls may result in quite different transverse concentration profiles. These profiles distinguish the non-Brownian modes of operation.

This review of the theory of FFF begins with a discussion of the formation and distribution of colloidal zones under the influence of the field and this is followed by a discussion of the velocity profiles for the various FFF channel types. The basic theory of retention is presented for the Brownian mode and alternative modes of operation. The various FFF subtechniques, distinguished according to the nature of the applied field, are described together with their application domain. This is followed by a discussion of selectivity and sample characterization. Zone broadening in the Brownian mode, separation optimization, and programming in FFF are also discussed. Finally, the capabilities of separation methods based on split flow in thin channels (SPLITT cells) are presented.

CONTENTS

F. Dondi and G. Guiochon (eds.),
Theoretical Advancement in Chromatography and Related Separation Techniques, 513–580.
© 1992 *Kluwer Academic Publishers.*

514

1. Introduction

The term "field-flow fractionation" (FFF) encompasses a relatively large family of separation methods for macromolecular, colloidal and particulate materials which are all characterized by the combination of a nonuniform axial flow and different transverse concentration profiles induced by an applied field. FFF was conceived twenty-five years ago by J.C. Giddings (1966). A year later, the concept was successfully demonstrated by Thompson et al. (1967) and, independently, by Berg et al. (1967b,c). Separation occurs in one phase and the method has sometimes been termed "one-phase chromatography" (Giddings 1976a). Various types of fields can be used, leading to various FFF subtechniques, such as a thermal gradient (thermal FFF), an electrical or magnetic field (electrical FFF or magnetic FFF), a gravitational or centrifugal field (sedimentation FFF) or a transverse flow through semi-permeable membranes (flow FFF).

The technique has been mostly developed in Professor Giddings' laboratory and is now used in an increasing number of laboratories in nearly all continents. The literature on FFF includes up to mid-1991 about 350 scientific articles covering the theoretical, instrumental and applications aspects of the method.

1.1. CONFIGURATION OF THE FFF CHANNEL

The FFF separation system is generally a thin ribbon-like channel, as shown in the schematic diagram of Figure 1 together with the coordinate system used in the following text. The main part of the channel has the form of a parallelepiped with a typical length, L, of about 50 cm, a breadth, b, of about 1 or 2 cm and a thickness, w, of about 0.25 mm. Because of the very large cross-sectional aspect ratio of the channel, infinite parallel plate theories are often used to describe the behavior of the sample components. The extremities of the channel are usually tapered to allow for the flow of a carrier fluid.

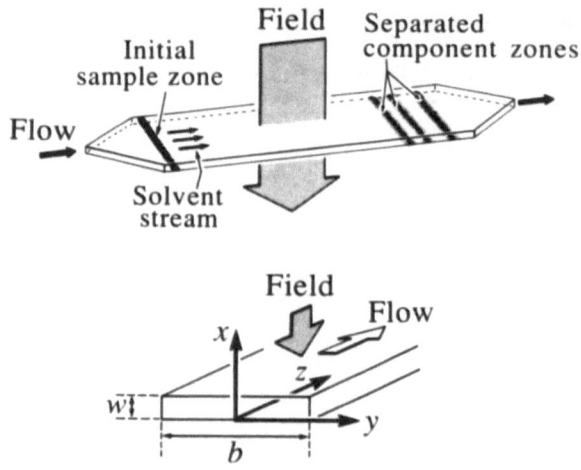

Figure 1. Schematic diagram of the FFF channel and coordinate system.

The sample is introduced at the inlet of the channel by some appropriate injection device. Generally, the injection volume and time are small in comparison with the channel volume and residence time, respectively. The different components of the sample are displaced at various velocities along the channel and elute as more or less narrow and symmetrical zones which are observed at the channel outlet by a detector. The external force field is applied all along the channel perpendicularly to the plates (or, it has at least a significant component along this x-direction).

1.2. BASIS OF FFF SEPARATION

The force field induces a nonuniform distribution of a sample component in the channel. For instance, component A may be more concentrated than component B in the channel centre where, due to the viscous character of the carrier fluid flow, the flow velocity is larger than near the walls.

Consequently, component A is displaced along the channel at a faster velocity and elutes earlier than component B. Eventually, if the broadening of the zones can be kept small enough, the two components are fully separated at the outlet of the channel. The fractogram, which is the output signal of the concentration-sensitive detector versus time, appears therefore, like an analytical chromatogram, as a succession of more or less completely separated peaks, each peak corresponding to a sample component.

There are various ways of inducing concentration nonuniformities along the field axis in the channel. They lead to various possible modes of retention. The Brownian mode, in which diffusion tends to oppose the component accumulation near one wall, is the mode for which theoretical developments are the most advanced as well as the mode in which most applications have been performed. It is first described in the following. Then alternative modes of retention are considered.

2. Concentration Distribution in the Thickness of the FFF Channel

2.1. EQUILIBRIUM CONCENTRATION PROFILE

Let us first consider the effect of the field on the distribution of a given species in the channel thickness. Although FFF is applicable to both macromolecular and particulate species, one uses, here, the term 'particles' to refer to the individual, identical constituents of the species, but it does not matter, for the present purpose, whether the species is made of suspended particles or of dissolved macromolecules. The force field is assumed to be directed in the direction of the negative x-coordinate. Under the influence of the field, the particles move towards one channel wall where their accumulation is counteracted by the diffusion process. Let J_x be the particle flux, i.e the number of particles passing per unit time through a flat surface of unit area located at distance x from the accumulation wall. This flux is made of a convective part, U c, and of a diffusive part, expressed by the first Fick law defining the diffusion coefficient, D :

$$J_x = U c - D \frac{d c}{d x} \tag{1}$$

where U is the velocity impelled by the field on the particles in the negative x-direction, c the number concentration of the particles at distance x from the accumulation wall.

Eventually, when the field is allowed to act on the particles for a sufficiently long time, a steady-state is reached where the field-induced convective flux and the diffusive flux exactly equilibrate each other and the overall flux J_x is zero at all distances from the walls. Then, one has :

$$\frac{c(x)}{c_0} = \exp\left(\int_0^x \frac{U}{D} d x \right) \tag{2}$$

where c_0 is the particle concentration at the accumulation wall.

2.2. BASIC EXPONENTIAL CONCENTRATION PROFILE

In the case where the ratio U/D does not depend on the position in the channel thickness, which happens when both U and D are constant, equation 2 becomes :

$$\frac{c(x)}{c_0} = \exp\left(-\frac{|U|}{D}x\right) \tag{3}$$

or :

$$\frac{c(x)}{c_0} = \exp\left(-\frac{x}{\ell}\right) \tag{4}$$

with :

$$\ell = \frac{D}{|U|} \tag{5}$$

In equations 3 and 5, the absolute value of U is used because U is negative with the direction chosen for the x-axis. In these conditions, the particle concentration decreases exponentially from the accumulation wall. The space constant, ℓ, is a characteristic of the species under consideration. Species strongly affected by the field and/or slowly diffusing have a low ℓ value and are concentrated in a narrow zone near the accumulation wall. The particle cloud of species with large D and/or low U extends relatively far away from the wall. At the limit, for species unaffected by the field, for which U=0 and ℓ is infinite, equation 3 indicates that the concentration is uniform through the channel thickness, as expected. The space constant is conveniently expressed in dimensionless coordinate as :

$$\lambda = \frac{\ell}{w} = \frac{D}{|U|\, w} \tag{6}$$

λ is the characteristic parameter of the species which is directly derived from retention data in this classical FFF retention mode where field-induced convection and diffusion compete with each other to form exponentially distributed particle layers near the accumulation wall.

2.3. BROWNIAN MODE OF OPERATION

This mode is refered to as the Brownian mode of operation. It is clear that the corresponding distribution profile derived above is a limiting profile. Implicitly, when considering U and D constant, one assumes that the particles are sufficiently small to be considered as points in the flux expression especially near the walls (no steric exclusion effects) and sufficiently diluted for avoiding particle-particle interactions and non-linear effects. It is also assumed that there are no particle-wall interactions. In spite of these limiting assumptions, the Brownian mode of operation provides the primary basis on which the particle behavior in the channel can be gauged.

2.4. EINSTEIN EQUATION FOR THE DIFFUSION COEFFICIENT

A deeper insight into the process of distribution of the particles in the channel thickness and into the expression of the diffusion coefficient is provided by Einstein's investigation into Brownian motion. Let us consider that particles at a given concentration are contained in a vessel with walls permeable to the solvent or suspending fluid but impermeable to the particles. As a result, a pressure excess is exerted on the particle-containing side of the vessel. This is the osmotic pressure. It depends on the particle concentration. Accordingly, whether a semi-permeable partition is present or not, particles at a given concentration can potentially exert an osmotic pressure on the surrounding. If the semi-permeable partition is removed, the particles will diffuse into the external region. Therefore, the osmotic pressure difference between two regions is the driving force for the diffusion process. In order for a steady concentration gradient to exist, the corresponding force of osmotic pressure must be balanced by an external force. In the FFF channel, this is provided by the external field.

2.4.1. *Diffusion Coefficient and Osmotic Pressure Gradient.* Let us follow Einstein (1905) to compute the particle flux resulting from an osmotic pressure gradient. If an osmotic pressure difference $d\,\Pi$ acts in a region of a cylinder with unit cross-sectional area and height $d\,x$, the corresponding force per unit volume is $-\,d\,\Pi/d\,x$. The force per particle is then $-\,(d\,\Pi/d\,x)\,/\,c$. The velocity, υ, imparted by this force to one particle is the ratio of the force to the particle friction coefficient, f, i.e. $-\,(d\,\Pi/d\,x)\,/\,(fc)$. The particle flux is then :

$$\upsilon\,c \;=\; -\,\frac{1}{f}\,\frac{d\,\Pi}{d\,x} \;=\; -\,\frac{1}{f}\,\frac{d\,\Pi}{d\,c}\,\frac{d\,c}{d\,x} \tag{7}$$

This flux corresponds to the diffusive Fickian flux, $-\,D\,(d\,c/d\,x)$, which gives the expression of the diffusion coefficient :

$$D \;=\; \frac{1}{f}\,\frac{d\,\Pi}{d\,c} \tag{8}$$

2.4.2. *Osmotic Pressure of a Diluted Suspension or Solution.* When the suspension or solution is diluted enough, the behavior of particles in a vessel of volume V with semi-permeable walls can be regarded, on the point of view of the pressure (osmotic pressure, in that case) exerted by the thermal motion of the particles on the vessel walls, as equivalent to that of an ideal gas. This gives for the equation of state of the particle cloud at temperature T :

$$\Pi\,V \;=\; n'\,\mathcal{R}T \tag{9}$$

where n' is the number of moles of particles contained in volume V and $\mathcal{R}$ the ideal gas constant. In terms of the number concentration of particles, c, one gets :

$$\Pi \;=\; \frac{c\,\mathcal{R}\,T}{\mathcal{N}} \tag{10}$$

and :

$$\frac{d\,\Pi}{d\,c} \;=\; \frac{\mathcal{R}T}{\mathcal{N}} \;=\; k\,T \tag{11}$$

where $\mathcal{N}$ is the Avogadro number and k the Boltzmann constant.

2.4.3. *Einstein and Stokes-Einstein Equations.* Combining equations 8 and 11, one gets the Einstein relation for the diffusion coefficient :

$$D \;=\; \frac{k\,T}{f} \tag{12}$$

In the case where the particles are solid spheres of diameter d moving at low velocity in a fluid of viscosity η, the friction coefficient is, according to Stokes, equal to :

$$f \;=\; 3\,\pi\,\eta\,d \tag{13}$$

which, combined with equation 12, gives :

$$D = \frac{k\,T}{3\,\pi\,\eta\,d} \tag{14}$$

This is the well-known Stokes-Einstein expression of the diffusion coefficient. It is worth noting that the validity of this equation is restricted to diluted suspensions of slowly moving, solid spherical particles with a size much larger than the molecular size of the suspending fluid.

It can be noted that equation 12 for the diffusion coefficient may be obtained as well by considering the concentration dependence of the chemical potential of a diluted solution (Giddings 1991a).

2.5. EXPRESSION OF THE BASIC FFF PARAMETER λ

If F is the force exerted by the field on an individual particle, the field-induced velocity is simply :

$$U = F/f \tag{15}$$

It can be seen that the assumption of a constant U which leads to an exponential concentration profile corresponds to that of a constant force applied on the particle at any position in the channel. Using equations 6, 12 and 15, one gets :

$$\lambda = \frac{k\,T}{|F|\,w} \tag{16}$$

λ is seen to be equal to the ratio of the thermal energy, $k\,T$, to the work performed for displacing a particle through the channel thickness w. With the help of equations 4, 6 and 16, the concentration profile is given by :

$$\frac{c}{c_0} = \exp\left(-\frac{x/w}{\lambda}\right) = \exp\left(-\frac{|F|\,x}{k\,T}\right) \tag{17}$$

2.6. ALTERNATIVE DERIVATION OF THE CONCENTRATION PROFILE

This equation can be alternatively derived from the Boltzmann distribution law :

$$\frac{c}{c_0} = \exp\left(-\frac{W(x) - W(0)}{k\,T}\right) \tag{18}$$

where $W(x)$ and $W(0)$ are the potential energies of a particle at distance x from the accumulation wall and at this wall, respectively. Since the force exerted on a particle derives from this potential :

$$F = -\frac{d\,W}{d\,x} \tag{19}$$

one gets when this force does not depend on the distance from the wall :

$$W(x) - W(0) = -F\,x = |F|\,x \tag{20}$$

which gives, as expected for a constant applied force, the profile of equation 17 when combined with equation 18.

2.7. CROSS-SECTIONAL AVERAGE CONCENTRATION AND CENTRE-OF-GRAVITY

The above concentration profiles are expressed in terms of the particle concentration at the accumulation wall. They can also be expressed in terms of the average concentration in the channel thickness, $<c>$, by means of the following relationship :

$$\frac{<c>}{c_0} = \lambda \left[1 - \exp(-1/\lambda) \right] \tag{21}$$

It can be noted that, because of the finite thickness of the channel, the distance of the centre-of-gravity, x_{cg}, of the particle cloud to the wall is not exactly equal to ℓ. One has indeed :

$$\frac{x_{cg}}{w} = \frac{<c(x)\, x>}{<c(x)>\, w} = \lambda - \frac{\exp(-1/\lambda)}{1 - \exp(-1/\lambda)} \tag{22}$$

where the symbol "$<\ >$" indicates a cross-sectional average (or, presently, an average in the x-direction) :

$$<f(x)> = \frac{1}{w} \int_0^w f(x)\, d\, x \tag{23}$$

For low and high λ values, one gets, respectively :

$$\lim_{\lambda \to 0} \frac{x_{cg}}{w} = \lambda \tag{24}$$

and :

$$\lim_{\lambda \to \infty} \frac{x_{cg}}{w} = \frac{1}{2} \tag{25}$$

According to equation 24, the centre-of-gravity of a particle cloud sufficiently compressed near the wall is equal to ℓ.

3.   Velocity Distribution in the Channel

The determination of the velocity profile in the channel is essential for computing the species retention as the particles are transported along the channel by the flow of the carrier fluid, which, in nearly all applications of FFF, is a liquid. The basic equation for deriving the flow profile is given by the Navier-Stokes equation of motion, which is a momentum balance equation, in combination with the continuity equation, which expresses the conservation of the mass of a fluid element. Because of its highly non-linear character, this differential equation can be solved analytically only in quite restricted situations.

3.1. BASIC PARABOLIC FLOW PROFILE IN INFINITE PARALLEL PLATE CHANNELS

The basic FFF retention theory considers the steady-state, fully developed, laminar flow of an isothermal, incompressible fluid between two stationary infinite parallel plates, so that the velocity vectors are in the z-direction and depend only on x. The Navier-Stokes equation becomes then simply (Happel & Brenner 1973) :

$$\eta \frac{d^2v}{dx^2} = \frac{dP}{dz} \tag{26}$$

where v is the velocity and P the pressure. A double integration of this equation, using the two boundary conditions v(x=0) = v(x=w) = 0 (no slip at the walls), in terms of the pressure drop ΔP across the column of length L gives the classical (Poiseuille) parabolic profile :

$$v(x) = \frac{w^2 \Delta P}{2 \eta L} \left[\left(\frac{x}{w}\right) - \left(\frac{x}{w}\right)^2 \right] \tag{27}$$

The cross-sectional average velocity, <v>, is then :

$$<v> = \frac{w^2 \Delta P}{12 \eta L} \tag{28}$$

which gives :

$$v(x) = 6 <v> \left[\left(\frac{x}{w}\right) - \left(\frac{x}{w}\right)^2 \right] \tag{29}$$

Because of the planar geometry of the system, the maximum velocity is 1.5 times larger than the average velocity (instead of 2 for a circular geometry). It can be noted that the pressure drop in the FFF channel itself is very small and rarely exceeds 1 mbar, therefore, pressure-induced non-linear effects can be neglected in FFF.

3.2. FLOW PROFILE IN RECTANGULAR CHANNELS

In practice, obviously, the FFF channel cannot be an infinite parallel plate system but has a rectangular cross-section with breadth, b, which is usually much larger than w. Because of the presence of these edges, the system loses its two-dimensional geometry and becomes three-dimensional. For a straight channel, there are still no transverse velocity components, but the axial velocity depends not only on x but also on y, in the vicinity of the edges, extending to a distance of about w from the edges. The basic equation of motion is then :

$$\eta \left(\frac{\partial^2 v}{\partial x^2} + \frac{\partial^2 v}{\partial y^2} \right) = \frac{dP}{dz} \tag{30}$$

This equation has been solved analytically for the boundary conditions corresponding to the FFF channel (Cornish 1928). The resulting equation is given as an infinite series. However, a good approximation can be obtained by multiplying the x-dependence of v given in equation 29 by the function h(y) given by Takahashi and Gill (1980), when setting the origin of the x and y axes at the center of the accumulation wall :

$$h(y) = 1 - \frac{\cosh (2\sqrt{3} \, y / w)}{\cosh (\sqrt{3} \, b / w)} \tag{31}$$

The correction brought by this equation becomes insignificant for very large aspect ratios of the channel cross-section.

3.3. END EFFECTS

It should be noted that the above flow equations apply for a fully developed, steady flow. The entrance length of a rectangular channel with large aspect ratio is of the order of 10^{-1} Re w, where Re is the axial Reynolds number :

$$Re = \frac{\rho <v> w}{\eta} \tag{32}$$

and ρ is the eluent density (Han 1960). With eluent average velocities of about 1cm/s, it is seen that the entrance length is about a few percents of the channel thickness, which is extremely small. However, because of the tapered-end configuration of most FFF channels, the effective length required to a have steady-state velocity profile is of the order of the magnitude of the channel breadth, b (Williams et al. 1986).

3.4. DEVIATIONS FROM THE PARABOLIC PROFILE IN TWO-DIMENSIONAL FLOWS

Although the parabolic profile of equation 28 will be used below to compute the particle retention and serves as a basis for gauging the true velocity profile in the FFF channel, there are situations, however, where, apart from taking into account the presence of edges, the assumptions behind equations 26 or 30 are not correct.

3.4.1. *Non-Constant Viscosity.* This is the case, for instance, when the fluid viscosity is not constant but depends on x. This happens in thermal FFF because of the temperature dependence of the viscosity or in FFF subtechniques where the carrier fluid is itself a non-uniformly distributed multicomponent system, for instance in focusing FFF (see below), because of the composition dependence of the viscosity. It would also happen if a non-Newtonian eluent were used. In these cases, equation 26 must be written as (Bird et al. 1960) :

$$\frac{d}{d\,x}\left(\eta\,\frac{d\,v}{d\,x} \right) = \frac{d\,P}{d\,z} \tag{33}$$

which gives (Westermann-Clark 1978, Shah et al. 1979) :

$$v(x) = \left(-\frac{\Delta P}{L} \right) \left(\int_0^x \frac{x}{\eta(x)}\,d\,x - C' \int_0^x \frac{d\,x}{\eta(x)} \right) \tag{34}$$

with :

$$C' = \int_0^w \frac{x}{\eta(x)}\,d\,x \,/\, \int_0^w \frac{d\,x}{\eta(x)} \tag{35}$$

These equations have been solved analytically in thermal FFF by using appropriate relationships between η and T and between T and x (Gunderson et al. 1984) and numerically for sedimentation FFF in the focusing mode (Schure et al. 1986). The case of the flow of a non-Newtonian fluid, for which the viscosity depends on the local shear stress belongs to this category. It has been treated in terms of the shear stress profile with some approximations (Janča & Giddings 1981). Obviously equation 27 is retrieved by setting η constant in equation 34.

3.4.2. *Transverse Flow.* Another situation where the simple equation of motion given by equation 26 has to be modified is encountered in flow FFF in which a cross-flow of eluent through two semi-permeable membranes in the x-direction forces the particles which cannot leak through the membrane to accumulate at the downward wall. When the cross-flow velocity, U, is constant all along the channel, the equation of motion becomes (Berman 1953) :

$$\eta \frac{\partial^2 v}{\partial x^2} = \frac{\partial P}{\partial z} + \rho U \frac{\partial v}{\partial x} \qquad (36)$$

The last term represents a convective acceleration. The analytical solution for the infinite parallel plate channel has been given (Berman 1958) and depends on the cross-flow Reynolds number, Re_c, equal to :

$$Re_c = \frac{\rho |U| w}{\eta} \qquad (37)$$

The velocity profile is no more symmetrical and the position of the maximum velocity appears somewhat shifted towards the accumulation wall.

3.4.3. *Flow in Curved Channels.*

A third case where deviation from parabolic flow occurs is found in curved channels, like those used in sedimentation (centrifugation) FFF, simply because of the non-planar geometry of the flow system. The axial flow profile is found by solving an equation of motion similar to equation 26 but expressed in cylindrical coordinates (Chandrasekhar 1961).

It is interesting to note that the velocity of a fluid flowing in a curved round pipe has always a transversal component. This secondary flow contributes to equalize the particle transverse distribution and, thus, prohibits significant retention in FFF. When using a parallel plate channel configuration, no secondary flow occurs below a flow threshold, corresponding to a Dean number, Dn, equal to 35.92, with :

$$Dn = Re \sqrt{\beta} = \frac{\rho <v> w}{\eta} \sqrt{\frac{w}{r_0}} \qquad (38)$$

where β is the curvature ratio and r_0 the curvature radius of the channel centerline. When the channel is rotating, as in sedimentation FFF, the threshold Dean number decreases or increases, depending on the interplay of the centrifugal and Coriolis effects (Mutabazi et al. 1991). In practical FFF conditions with low curvature ratios, this threshold value is never reached.

3.4.4. *Channels with Different Geometries*

Channels with a non-parallelepipedal geometry have sometimes been advocated for some applications. Annular channels with axial flow between the gap of two coaxial cylinders may ease the application of some fields and avoid the disturbance of the edges of rectangular channels (Davis & Giddings 1985a). The velocity distribution can be exactly determined by solving the equation of motion in cylindrical coordinates (Bird et al. 1960). In addition, one of the cylinders can be made to rotate as proposed for implementing shear FFF (Giddings & Brantley 1984a).

In the case of asymmetrical flow FFF, i.e. flow FFF without cross-flow but with one semi-permeable wall (Litzén & Wahlund 1991a), the full description of the flow profile includes $\partial v/\partial z$, and $\partial^2 v/\partial z^2$ terms in the equation of motion (Granger et al. 1986). A similar situation is found for channels with two permeable walls (Doshi & Gill 1979) and for flow FFF in hollow fiber channels (Kozinski et al. 1970, Doshi et al. 1975, Jönsson & Carlshaf 1989).

The case of unidirectional flow in channels with a quadrilateral cross-section for which at least two opposing faces are not parallel (for instance when w depends on y) has to be solved by means of equation 30 with appropriate boundary conditions. These channels have been advocated for isopycnic focusing in FFF (Janča & Jahnová 1983).

3.5. THIRD-DEGREE VELOCITY PROFILE APPROXIMATION

The velocity profile expressions obtained in cases where equation 26 does not apply are more or less complex although they represent, in practice, a rather small deviation from the parabolic equation 28. One can get a much simpler representation of these profiles by approximating them to a third-degree velocity profile which allows one free parameter to adjust to the departure from the parabolic flow. This flow distortion parameter, v, is conveniently selected in such a way that :

$$v(x/w) / <v> = (v / <v>)_2 + v \; (v / <v>)_3 \tag{39}$$

where $(v/<v>)_2$ is the parabolic profile of equation 29 and $(v/<v>)_3$ a third-degree profile equal to :

$$\left(\frac{v(x/w)}{<v>}\right)_3 = 6 \frac{x}{w}\left(1 - \frac{x}{w}\right)\left(1 - 2\frac{x}{w}\right) \tag{40}$$

In these conditions, the velocity profile is not symmetrical. The maximum velocity, v_{max}, is then :

$$\frac{v_{max}}{<v>} = \frac{3}{2}\left(1 + \frac{v^2}{4}\right) \tag{41}$$

and appears at a position, $(x/w)_{max}$ equal to :

$$\left(\frac{x}{w}\right)_{max} = \frac{1}{2}\left(1 - \frac{v}{2}\right) \tag{42}$$

It is then closer to the accumulation wall than to the other if v is positive. It is interesting to note the slope of the velocity profile near the accumulation wall :

$$\left(\frac{d\,(v/<v>)}{d\,(x/w)}\right)_{x/w=0} = 6\,(1+v) \tag{43}$$

The axial velocity of the particles near this wall is then amplified by a factor $(1+v)$.

In thermal FFF, the value of v depends on the nature of the eluent and on the temperature profile. In typical experiments, when taking the origin of the x-axis at the cold wall, v is around -0.1 or -0.2 (Martin et al. 1979, Martin & Reynaud 1980). In flow FFF, one can easily show (Andreev et al. 1987) that v is equal to $Re_c/6$, which is typically smaller than 10^{-3} (Davis 1991). In sedimentation FFF, it has been shown (Mattson & Alfredsson 1990) that, if the origin of the x-axis is taken at the outer wall, v is equal to $-w/R_c$ which, in typical instruments, is also of the order of 10^{-3}. Therefore, in practice, only in thermal FFF will the correction for the deviation of the flow profile from the parabolic one have to be taken into consideration. It should be noted that this third-degree profile describes also the free convection, gravity-induced, velocity profile obtained when orienting the thermal FFF channel vertically. The corresponding v parameter has to be added to that described above for the temperature dependence of the viscosity in forced flow (Giddings et al. 1979a).

4.   Basic Theory of Retention in the Brownian Mode

The characterization of FFF, a zonal separation method, can be done in the same way and using the same parameters as used in analytical chromatography.

4.1. RETENTION RATIO

The retention ratio, R, is defined as the ratio of the zone migration velocity, $\mathcal{V}$, to the average carrier liquid velocity, $<v>$:

$$R = \frac{\mathcal{V}}{<v>} \qquad (44)$$

Since $<v> = L / t^0$ and $\mathcal{V} = L / t_r$, where t^0 is the average residence time of the eluent and t_r the average residence time of the species, called retention time, one gets :

$$R = \frac{t^0}{t_r} \qquad (45)$$

This equation is used for the experimental determination of the retention ratio. Alternatively, as noting the volumes of carrier which flow through the channel during the times t^0 and t_r as V^0 and V_r, respectively, one has, when the flow-rate is constant :

$$R = \frac{V^0}{V_r} \qquad (46)$$

V^0 is the void volume of the channel and V_r is called the retention volume of the species.

4.2. BASIC RETENTION THEORY

A given particle, located at some distance from the walls, is axially displaced instantaneously by the flow at the velocity of the carrier at the particle position (in fact, a correction must be brought to the particle velocity relative to the fluid velocity when the particle is relatively large, but this effect is neglected here). But because of the thermal motion and of the field-induced motion, the particle does not stay at the same transverse position for long, but rather experiences some erratic transverse motion. In fact, during a sufficiently long time spent in the channel, a given particle will sample all distances from the accumulation wall with a frequency which is, in an average, given by the concentration distribution profile. This profile, given by equations 2 or 3, is indeed a dynamic, rather than static, equilibrium profile.

A simplified expression of the mean particle migration velocity can be obtained by averaging the instantaneous axial velocities of all particles assuming they are distributed according to the equilibrium concentration profile (Giddings 1968) :

$$\mathcal{V} = <\frac{c}{<c>} v > = \frac{< c v >}{<c>} \qquad (47)$$

which gives for R :

$$R = \frac{< c v >}{<c> <v>} \qquad (48)$$

Equation 48 is the basic equation for computing the retention ratio in various experimental situations, for various modes of retention.

4.3. RETENTION RATIO IN THE BROWNIAN MODE WITH A PARABOLIC FLOW

For the exponential lateral concentration profile (equation 17) and the parabolic velocity profile obtained in the infinite parallel plate channel configuration (equation 29), equation 48 gives the basic retention equation in the Brownian mode :

$$R = 6\lambda \left(\coth \left(\frac{1}{2\lambda} \right) - 2\lambda \right) \tag{49}$$

The retention ratio depends only on the basic FFF parameter λ. This equation allows to determine λ from the measurement of the retention time and to extract the basic field-sensitive physico-chemical parameter characterizing the particles by means of the dependence of λ on this parameter (see section 5).

It becomes clear that a necessary condition for the successful separation of two species is that their λ values are different, i.e. they must have either different field-induced transverse velocities or different diffusivities or both.

When λ is relatively large (low retention), the particles move on average at the carrier velocity and R approaches 1 :

$$\lim_{\lambda \to \infty} R = 1 - \frac{1}{60 \lambda^2} \tag{50}$$

In this situation, the separation becomes difficult as all species tend to move at the same eluent velocity. In the high retention limit, one has :

$$\lim_{\lambda \to 0} R = 6\lambda - 12\lambda^2 \tag{51}$$

The error in R brought by this equation is less than 1% in the most useful range where $R < 0.67$.

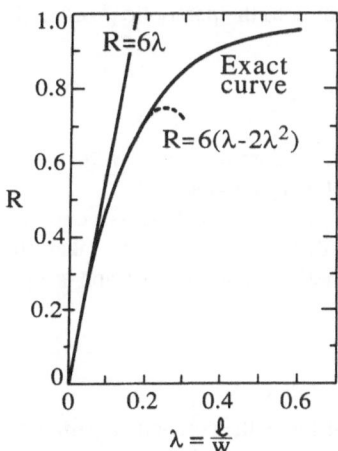

Figure 2. Retention ratio versus the basic FFF parameter $\lambda = \ell / w$. Exact retention curve based on the assumption of an exponential concentration profile and a parabolic velocity profile together with high retention (low R) approximations.

One can get some simplified view of the retention process when λ is small, i.e when the particle cloud is compressed in a narrow zone near the accumulation wall where the velocity profile is nearly linear and equal to $6 <v> x/w$. One can imagine the zone located at a single thin layer at distance ℓ from the wall. This layer is displaced at a velocity of $6 <v> \ell/w$, giving $R = 6 \lambda$, which is the limiting expression of equation 49 for very small λ (Giddings 1973a). The curve $R(\lambda)$ is shown in Figure 2 together with the high retention approximations.

4.4. RETENTION RATIO IN THE BROWNIAN MODE WITH A NON-PARABOLIC FLOW

The validity of equation 49 relies on that of the concentration profile and of the flow velocity profile. When the concentration is still exponentially distributed but the velocity profile no longer parabolic, the expression of R becomes quite complicated. It has been computed by means of equation 48 for various situations for which an exact flow profile can be obtained. This is the case for thermal FFF with an appropriate expression for the temperature dependence of viscosity (Gunderson et al. 1984), for flow FFF with a uniform cross-flow (Davis 1991), sedimentation FFF with a curved channel (Davis 1986a) or when the viscosity in the zone depends on the particle concentration (Shah et al. 1979). The resulting expressions of R depend on some instrumental factor causing the deviation from the parabolic flow as well as on λ. (The calculation has also been performed for annular channels in the case where the field-induced velocity is not constant but varies with w according to a power law (Davis & Giddings 1985a)).

Because the deviation of these flow profiles from the parabolic one is generally small, a retention expression based on the third degree velocity profile of equations 39 and 40 is likely to provide a good approximation of R. Combining them with equations 17 and 48, one gets R in terms of λ and v (Martin & Giddings 1981a) :

$$R = R_p + 6\lambda v (1 - R_p) \tag{52}$$

where R_p is the retention ratio corresponding to the parabolic profile, given by equation 49. The high retention (low R) limit is :

$$R = 6\lambda (1 + v) - 12\lambda^2 (1 + 3v) \tag{53}$$

which shows that for a particle cloud closely compressed near the wall, the migration velocity is amplified by a factor $(1 + v)$ relative to that in a parabolic flow.

The influence of the finite aspect ratio of the channel cross-section on R has been investigated (Giddings & Schure 1987a). It is found that this effect can be taken into account by applying a correction factor, $(1 - jw/b)$, to $\mathcal{V}$ where j depends slightly on λ and is equal to about 0.55.

5. The FFF Subtechniques

Any externally applied field that can interact with the component particles or molecules of a sample, to drive them towards the accumulation wall, may be used to bring about retention in FFF. Differences in the strength of interaction with the applied field, and/or differences in rates of back diffusion (for the Brownian mode) or strength of opposing forces (for lift and focusing modes as described in section 9) result in differential retention for the sample components, and hence, in sample fractionation.

In the Brownian mode of retention, as explained in section 2, the field induced migration is opposed by diffusion. This results in an exponential concentration profile adjacent to the accumulation wall characterized by the retention parameter λ given by :

$$\lambda = \frac{D}{|U|w} = \frac{kT}{|F|w} \tag{54}$$

In order to characterize a sample in terms of some property of interest, such as particle diameter or molecular weight, we require a monotonic variation of λ, and consequently of elution time, with that property (see section 9.1.2). Since different field types interact with different sample properties, the nature of the required information will determine the field that must be used. It is the type of field employed that characterizes the FFF subtechnique. The most commonly used subtechniques to date make use of sedimentation, cross flow, thermal gradient and electrical fields.

Each of the subtechniques may also be operated in the steric mode (Koch & Giddings 1986) or in the lift or steric-hyperlayer mode (Ratanathanawongs & Giddings 1989, Giddings et al. 1988a), the mechanisms of which will be discussed in detail in section 9. Briefly, sample components are of such a size that Brownian motion is suppressed (generally true when $d > 1$ μm) and transport by back diffusion from the accumulation wall is replaced by some position-dependent force opposing the applied field. The suppression of Brownian motion results in each component being held by the opposing forces at some fixed height above the accumulation wall. Migration of a component is therefore coherent and nonequilibrium band spreading greatly reduced. Generally the larger components migrate at higher positions above the wall and elute before smaller components. The requirement for monotonic variation of retention with a sample parameter of interest is equally necessary for these modes of elution. Since the order of elution for Brownian and steric modes are reversed it is therefore of the utmost importance that the sample be fully eluted in a single mode. Fortunately, this may often be achieved through proper selection of experimental conditions (Myers & Giddings 1982, Lee & Giddings 1988a).

Some of the subtechniques may also be operated in focusing, or hyperlayer, mode (see section 9). We make the distinction here that focusing is achieved through the combined effect of the generally uniform applied field and some gradient in carrier property across the channel thickness (Giddings 1977, Janča 1982, Giddings 1983a). We are also at liberty to apply two or more fields simultaneously, either in the same direction or opposed, and these methods are described below as hybrid techniques.

5.1. SEDIMENTATION FFF

A gravitational or centrifugal field provides the driving force towards the accumulation wall. In the latter case the channel is placed around the inner wall of a centrifuge basket, the inlet and outlet flow lines being carried to the channel via rotating seals (Giddings et al. 1974, Giddings et al. 1980, Giddings & Caldwell 1989a).

For the Brownian mode, the retention parameter λ is given by :

$$\lambda = \frac{\mathcal{R}T}{M(1 - \bar{v}_s\rho)Sw} = \frac{\mathcal{R}T}{M(\Delta\rho/\rho_s)Sw} = \frac{6kT}{\pi d^3 \Delta\rho Sw} \tag{55}$$

where S is the acceleration due to gravity or centrifugation (in which case $S = \omega^2 r_0$ where ω is the angular velocity and r_0 is the radius of the channel), M is the component relative molecular mass, $\bar{v}_s$

its partial specific volume and ρ_s its density, ρ is the carrier density, $\Delta\rho$ is the difference in density between particle and carrier, and d is the particle spherical (or equivalent spherical) diameter. Material is seen to be retained according to its effective, or buoyant, mass.

The potential applications for sedimentation FFF are many and varied. The Brownian mode is restricted to the colloidal range of a few tens of nm up to about 1μm. Samples analysed include latex materials, viruses, subcellular particles, protein aggregates, and emulsions (reviewed by Giddings & Caldwell 1989a). Fractionation of DNA materials has been achieved (Kirkland et al. 1983a, Schallinger et al. 1984, 1985). Natural colloids of river water have been examined (Karaiskakis et al. 1982, Beckett et al. 1988a). Other submicron materials fractionated include metal hydrosols (Oppenheimer & Smith 1989), metal powders, Teflon particles, zirconia, clays (Giddings et al. 1991b), pigments (Koch et al. 1990), and both colloidal and fumed silicas (Giddings et al. submitted). The high resolution and gentle nature of the separation process have been exploited in the study of the formation and breakup of colloidal latex aggregates (Jones et al. 1988, Giddings et al. 1989b, Barman & Giddings 1991). Silica particle aggregates have also been studied in organic systems by FFF (Yonker et al. 1987). The gentleness of separation by FFF has also been shown to be ideally suited to the study of emulsions and their stability (Caldwell et al. 1981, Kirkland et al. 1982b, Yang et al. 1983, 1984, Hansen & Short 1990). The dependence of retention on the product of $\Delta\rho$ and particle volume allows the determination of both particle diameter and density via the measurement of retention for a range of different carrier densities (Giddings et al. 1981, Kirkland & Yau 1983b, Nagy 1989, Koliadima et al. 1990, Kirkland et al. 1990).

The steric or lift modes are restricted to the size range of about 0.3μm to 100μm. Polystyrene particles (Koch & Giddings 1986), glass beads (Giddings et al. 1991c, Williams et al. 1992b), metal powders (Giddings et al. 1991b), porous silicas (Giddings & Moon 1991d) and biological cells (Caldwell et al. 1984, Cardot et al. 1991) have been fractionated. The dependence of sedimentation force on the product of field strength S and $\Delta\rho$, and the apparent insensitivity of lift force to any particle parameter other than size (Williams et al. 1992a) has led to the development of a density compensation procedure (which entails keeping the product $S\Delta\rho$ constant) that allows the use of calibration runs with polystyrene standards, relating retention to particle size, to be used for any particulate sample of known density (Giddings et al. 1991c, Williams et al. 1992b).

A focusing mode of elution is possible when a density gradient is set up across the channel thickness by the action of the primary sedimentation field on a suspended or dissolved density modifier in the carrier (Giddings 1977a, 1983a, Janča 1982, Schure et al. 1986). Focusing takes place by isopycnic sedimentation under stopped-flow conditions at a point removed from the channel inlet (see Schure et al. 1986). Elution of the components is then effected at velocities corresponding to their isopycnic points. Separation is therefore based on density differences in sample components. Such a focusing mode of elution has also been proposed for trapezoidal cross-section channels (Janča & Jahnová 1983, Janča & Chmelík 1984, Chmelík & Janča 1986). The use of step density gradients in such systems has also been considered (Janča 1987a, Janča & Nováková 1987b, 1988). Note however that the proper accounting for viscosity differences in the immiscible carrier components has been detailed by Wičar (1988). Although gradients are likely to be more easily set up in such trapezoidal cross-section channels, these systems suffer from the much greater time required to achieve isopycnic equilibrium due to the much greater cross channel migration distances. Also, bandspreading contributions due to dispersion across velocity gradients in two coordinates will be significant (Janča & Chmelík 1984).

5.2. FLOW FFF

In the conventional flow FFF systems (see, for example, Giddings et al. 1980, Ratanathanawongs & Giddings 1991) the parallel plane walls are composed of a permeable frit material. The driving force towards the accumulation wall is provided by a flow of carrier fluid pumped from a chamber behind one frit and withdrawn at the same rate through the opposite frit. A semipermeable membrane (commonly an ultrafiltration membrane) is laid over the frit through which the outflow passes. This serves the purpose of preventing loss of sample through the frit and defines the accumulation wall. A channel flow along the length of the system elutes material in the usual way. For even the most extreme conditions the velocity profile in the direction of elution is predicted to deviate only slightly from parabolic (Davis 1991) so that the standard retention ratio equations may be used without modification.

In flow FFF the field induced velocity is equal to the velocity of the cross flow. All sample components are transported towards the accumulation wall at the same rate, and differences in λ, and hence in retention, are due solely to differences in rates of back diffusion. We have (Giddings et al. 1976c, Giddings et al. 1976d) :

$$\lambda = \frac{D}{|U| w} = \frac{D V^o}{\dot{V}_c w^2} = \frac{k T V^o}{3 \pi \eta \dot{V}_c w^2 d} \tag{56}$$

where V^o is the channel void volume and $\dot{V}_c$ is the cross flow rate. In the third form for λ in equation 56 the Stokes-Einstein equation has been assumed for diffusion coefficient and d is therefore the Stokes diameter of the component. Significantly retained components elute in times that are inversely proportional to diffusion coefficient, or directly proportional to Stokes diameter. Flow FFF in the Brownian mode results effectively in a size-based separation. In the lift, or steric-hyperlayer, mode the separation is also size-based if lift forces are dependent only upon the particle size (Giddings et al. 1987b, Williams et al. 1992a).

A variant of conventional flow FFF has been proposed (Granger et al. 1986, Wahlund & Giddings 1987). In these systems that were briefly mentioned in section 3.4.4, the upper frit (the depletion wall) is replaced by an impermeable material such as a glass plate. The channel inlet flow then contributes to both the flow through the membrane forming the accumulation wall, and the flow along the length of the system. This configuration was termed asymmetrical flow FFF due to the velocity gradients that are set up in both longitudinal and transverse directions. The velocity gradient in the transverse direction was shown to have little effect at high retention, and that in the longitudinal direction is easily accounted for. These systems are technically simpler to construct and the glass upper wall allows for examination of flow paths by the injection of colored dyes. A modification of these systems was later proposed (Litzén & Wahlund 1991a, 1991b) where the channel breadth is reduced linearly along the system length. These trapezoidal channels may be operated with more uniform channel flow velocities and are predicted to offer advantages in allowing greater loadability and reduced dilution of sample during elution.

The viscous drag of the cross flow is exerted equally on any sample type which means that, of all the subtechniques, flow FFF has the potential for the widest range of application. The materials that have been fractionated include viruses (Giddings et al. 1977b, Litzén & Wahlund 1989, 1991a), proteins (Giddings et al. 1977c, Wahlund & Giddings 1987, Wahlund & Litzén 1989, Litzén & Wahlund 1989), DNA materials (Wahlund & Litzén 1989, Litzén & Wahlund 1989), blood cells (Giddings et al. 1991e), pollens and spores (Ratanathanawongs & Giddings 1991), organic- (Brimhall et al. 1984) and water-soluble polymers (Giddings et al. 1978a, Wahlund et al. 1986, Wahlund & Litzén 1989, Benincasa & Giddings 1992), naturally occurring fulvic and humic acids (Beckett et al. 1987, Beckett et al. 1988b), polystyrene latexes (Giddings et al. 1987b, Ratanathanawongs & Giddings 1991), colloidal, fumed and chromatographic support silicas (Giddings et al. 1978b, Ratanathanawongs & Giddings 1989, 1991, Giddings et al. submitted), and various ground minerals (Barman et al. 1989).

5.3. THERMAL FFF

In thermal FFF the driving force is provided by a temperature gradient across the channel thickness maintained by electrical heating of a polished metal block (usually a chrome-plated copper block) forming one wall and cooling of a similar block forming the opposite wall. The cooling is usually accomplished by passage of cold water through holes bored in the block (Thompson et al. 1967, 1969). Macromolecules in solution are driven by thermal diffusion usually towards the cold wall at a velocity equal to $D_T \, dT/dx$, where D_T is the coefficient of thermal diffusion and dT/dx is the local thermal gradient. The resulting retention parameter has been shown to be given approximately by (Hovingh et al. 1970) :

$$\lambda \approx \left[w\left(\frac{\alpha}{T}+\gamma\right)\frac{dT}{dx} \right]^{-1} \tag{57}$$

where γ is the coefficient of expansion for the local solution and α is the thermal diffusion factor defined by (Tyrrell 1961) :

$$\alpha = D_T T/D \tag{58}$$

where D is the ordinary diffusion coefficient. Both α and dT/dx are functions of x, the former through its dependence on T and the latter due to the variation of thermal conductivity of the solution with temperature. Consequently the concentration profile deviates slightly from the exponential form. However at high retention the sample cloud is restricted to a thin layer close to the cold wall and an acceptably accurate λ may be obtained using the value of dT/dx at the cold wall and the value of α at the cold wall temperature (Giddings et al. 1976b). At high retention γ also becomes insignificant (Hovingh et al. 1970) and λ may be approximated by :

$$\lambda \approx \frac{T_c}{\alpha \, w \, (dT/dx)_c} = \frac{D}{D_T \, w \, (dT/dx)_c} \tag{59}$$

where α, D and D_T are here the values of the thermal diffusion factor, the ordinary diffusion coefficient and the thermal diffusion coefficient at the cold wall temperature T_c, respectively, and $(dT/dx)_c$ is the temperature gradient at the cold wall. A center of gravity approach for approximation of λ, α, and dT/dx has also been suggested (Giddings et al. 1976b). It has been found that for given solute/solvent systems D_T is independent of molecular weight (Schimpf & Giddings 1987a, 1989a), and components are separated according to their differences in D as for flow FFF.

The mechanism for thermal diffusion of polymers in solution is not well understood and the thermal diffusion factor and coefficient remain empirical quantities. The measurement of retention of narrow polymer standards in thermal FFF systems is an effective means of determining these factors (Giddings et al. 1970, 1976b, Martin & Reynaud 1980, Brimhall et al. 1985, Song et al. 1986, Schimpf & Giddings 1987a, 1989a, 1990a). Once the thermal diffusion factor is known for a given solute/solvent system then thermal FFF may be used to determine molecular weights and molecular weight distributions of unknown samples (Brimhall et al. 1981, Gao et al. 1985, Schimpf et al. 1989b, Kirkland & Yau 1985, 1986, Kirkland et al. 1988). The range of application for thermal FFF in the field of polymer analysis has been reviewed by Giddings & Caldwell (1989a) and more recently by Schimpf (1990b).

As for the other subtechniques discused up to this point, elution of large components in a lift or steric-hyperlayer mode has been observed (Giddings et al. 1988a) although the lift forces operate via a mechanism specific to polymer chains in sheared flow.

Finally, thermal diffusion of particles has recently been observed (Liu & Giddings 1991a). Submicron polystyrene latex standards and silica particles were separated by thermal FFF with acetonitrile as carrier.

5.4. ELECTRICAL FFF

In this subtechnique charged species are driven towards one or other wall by the influence of an electric field acting across the channel thickness. The field induced velocity is given by the product of the electrophoretic mobility of the species μ and the field strength E so that :

$$\lambda = \frac{D}{\mu |E| w} \tag{60}$$

Retention is therefore governed by the ratio μ/D. The force exerted on a molecule of some species is given by :

$$F = z e E \tag{61}$$

where z is the effective charge on the species and e is the charge on the electron. The retention parameter may therefore be expressed as :

$$\lambda = \frac{kT}{z |E| e w} = \frac{kT}{z |\Delta V| e} \tag{62}$$

where ΔV is the potential difference across the channel thickness. Retention is therefore seen to be dependent only upon the magnitude of the effective charge on the species (Davis et al. 1987, Giddings 1989c).

Conventional planar electrical FFF systems have been constructed with walls defined by membranes, preferably supported by frits much like in flow FFF systems. The membranes prevent the loss of sample while allowing the passage of electric charge. Platinum wire electrodes, lying external to the channel, provide the electric field. The carrier solution is necessarily buffered, and a solution of identical composition is circulated through the chambers housing the electrodes in order to remove electrolysis products and reduce polarization (Caldwell et al. 1972, Kesner et al. 1976, Giddings et al. 1976e, 1980). Other geometries such as the hollow fibre systems described by Lightfoot and co-workers (Reis & Lightfoot 1976, Shah et al. 1979, Chiang et al. 1979, Lightfoot et al. 1981) and an annular porous glass channel described by Davis et al. (1987) have been constructed.

The theoretical potential of the technique has been discussed by Giddings (1989c), but in practice performance has thus far fallen far short of prediction. The technique has been applied almost exclusively to protein separation, the exceptions being a poly(styrene sulfonate) and a large anionic species reported by Davis et al. (1987).

A focusing (or hyperlayer) mode of operation that utilized isoelectric focusing in a pH gradient across the channel has been reported by Thormann et al. (1989). Such utilization of isoelectric focusing has also been described by Chmelík et al. (1989a), Chmelík & Janča (1989b), and Chmelík (1991a, 1991b), but this application was in a channel of trapezoidal cross section.

534

5.5. OTHER SUBTECHNIQUES

Every field that is applied to a sample in solution or suspension has the potential to interact with a different sample property. Each of the various subtechniques of FFF consequently provides information on sample distribution according to a specific sample parameter. Several field types, in addition to the four already discused, have been proposed.

A concentration gradient in the components of a mixed carrier solution is predicted to drive sample components to the wall corresponding to lower chemical potential (Giddings et al. 1977d). Selectivity would therefore be based on chemical rather than physical properties. Lateral forces due to sheared flow have been considered (Giddings & Brantley 1984a). Dielectrical FFF has been the topic of a feasibility study by Davis & Giddings (1986b).

Nickel complexes with bovine serum albumin (Vickrey & Garcia-Ramirez 1980, Mori 1986) and with egg albumin and EDTA (Mori 1986) have been shown to be selectively retained by a magnetic field. In each of these cases the channel was constructed of Teflon capillary tubing. A parallel plate system was constructed by Schunk et al. (1984) and used to separate iron oxide particles into singlets, doublets and triplets.

Acoustic FFF has been demonstrated (Semyonov & Maslow 1988a). When the wavelength of the standing acoustic wave field is optionally set at four times the channel thickness, the retention parameter for strongly retained spherical particles may be shown to be given by :

$$\lambda = \frac{48\,kT}{\pi^2 \rho \, u^2 d^3 \, |f|} \tag{63}$$

where u is the ultrasonic oscillating velocity amplitude, d is the particle diameter and f is given by :

$$f = 1 - \frac{\beta_s}{\beta} + \frac{3\,(\rho_s/\rho - 1)}{2\,\rho_s/\rho + 1} \tag{64}$$

where β_s and β are the adiabatic volume compressibilities of the particle and the carrier, respectively, and ρ_s and ρ are the particle and carrier densities, respectively. Depending on the symmetry of the harmonics only those particles with either positive or negative f-values will be retained; those of opposite sign are driven towards the channel center. Separation of particles differing in sign of f should therefore be very effective.

We might add that the possibilities for field types in FFF have not yet been exhausted. Even photophoresis has been proposed as a mechanism for retention (Giddings 1988b).

5.6. HYBRID TECHNIQUES

Generally one might expect the application of two fields in the same direction to display only subtle differences from the application of the dominant field alone. However, if the two fields dominate at opposite extremes of some sample parameter under study (e.g., at small and large particle size) then simultaneous use may be justified. Gravity-augmented flow/steric FFF (Chen et al. 1988) is such an example. For cross and channel flow rates conducive to elution of small particles (>1μm), the additional force due to gravity was usefully found to counter the strong hydrodynamic lift on the larger particles. Another example may be found in the case of thermal FFF of submicron particles where the superimposition of an electrical field has been used to increase retention via an interaction with particle surface charge (Liu & Giddings 1991a). The resulting hybrid is known as thermal-electrical FFF.

Many combinations of opposed fields may be used to effect focusing, or hyperlayer, modes of FFF. The principles of operation of an elutriation form of hyperlayer FFF were given by Giddings (1986c). In this system a uniform gravitational or centrifugal force is opposed by a nonuniform cross flow. The latter is achieved simply by arranging a higher input cross flow than is drawn off at the upper frit. The excess flow exits with the channel flow at the channel outlet. The focusing effect of the opposed forces is expected to be of far higher resolution than found in conventional elutriation systems where the flow velocity profile is inherently nonuniform. In the thin channel the cross flow velocity profile will be essentially uniform with deviation only at the edges. Elutriation focusing FFF has been implemented in trapezoidal cross-section channels (Janča 1987c, Urbánková & Janča 1990). Here the nonuniform cross flow velocity is due to a widening of the channel (the input and output volumetric cross flow rates being equal). Once again this is a less satisfactory arrangement due to the high cross channel migration distances and the mixing effect of the nonuniform cross channel velocity profile. A second example concerns acoustic FFF where some advantage has been proposed for opposing the acoustic field by gravitation (Semyonov & Maslow 1988a).

We might also mention in this discussion of hybrid techniques the combination of FFF with surface effects at the accumulation wall as utilized in potential barrier FFF (Karaiskakis & Koliadima 1989) and the hybrid FFF / adhesion chromatography technique (Bigelow et al. 1989, 1991). The mechanisms for these will be discused later in sections 9.6.2 and 9.6.3, respectively.

The considerable advantages to be expected of two-dimensional FFF have been discused by Giddings (1990a). Here the fields operate at right angles to one another as well as to the direction of flow.

6. Dispersion (Zone Broadening) Theory

A necessary condition for the resolution of two species is that their transverse concentration distribution are different. However complete separation requires that the widths of the corresponding zones are kept sufficiently small for minimizing their overlap. The control of the peak broadening is, like in chromatography, of paramount importance in FFF.

6.1. DEFINITION OF THE PLATE HEIGHT

Because of the non-uniformity of the velocity profile and of the ubiquitous thermal motion of the particles, zones are spread to some extent along the channel length. At a given time, one can characterize the distribution of the distances covered by the particles of the same species by a variance, σ_z^2. Some time later, when the centre-of-gravity of the zone has covered a distance δz, this variance has increased by an amount $\delta \sigma_z^2$. The plate height, H, characterizing this zone broadening process, is defined, like in chromatography, as (Giddings 1965) :

$$H = \lim_{\delta z \to 0} \frac{\delta \sigma_z^2}{\delta z} = \frac{d \sigma_z^2}{d z} \qquad (65)$$

H, according to this definition, is a local parameter and the total variance of the zone when the peak has traveled through the column can be obtained by integration of equation 65 over the channel length. However, as long as the physical parameters controlling the zone dispersion do not change their value when the peak is moving, H is constant and can be determined experimentally from the fractogram. This is true for dilute concentrations when the separation conditions are kept constant (no change in the temperature, force field or flow rate). In that case, indeed, provided that the variance of the particle zone as it enters the channel is negligible in comparison with the outlet variance, one can easily show that :

$$H = \frac{\sigma_L^2}{L} = L \left(\frac{\sigma_t}{t_r}\right)^2 \tag{66}$$

where σ_L^2 is the variance of the distribution of the distances traveled by the particles, $\sigma_z^2(t_r)$, at the time when the centre-of-gravity of the zone leaves the channel and σ_t the standard deviation of the distribution of the residence times of the particles of the species in the channel, which is determined from the peak on the fractogram. H is then related to the commonly used number of theoretical plates, N, which in these conditions is equal to :

$$N = \frac{L}{H} = \left(\frac{t_r}{\sigma_t}\right)^2 = \left(\frac{L}{\sigma_L}\right)^2 \tag{67}$$

It is noticeable that H has the dimension of length (it is twice the dispersion length commonly used by statistical physicists) and that it is expressed as a variance per unit channel length. Accordingly, the various independent processes of band broadening additively contribute to H.

6.2. VARIOUS CONTRIBUTIONS TO THE PLATE HEIGHT

Among the various contributions to the plate height, one can distinguish those arising from : the longitudinal diffusion, H_{ld}, the nonuniformity of the flow profile, H_n, the relaxation process, H_r, the sample polydispersity, H_p, and several instrumental contributions, ΣH_i, which, since independent, are additive :

$$H = H_{ld} + H_n + H_r + H_p + \Sigma H_i \tag{68}$$

6.2.1. *Longitudinal Diffusion.* This contribution arises from the diffusion process associated with the presence of concentration gradients, $\partial c/\partial z$, on each side of the particle zone. According to Einstein (1905), after a time t_r spend in the channel, the variance due to this process alone is :

$$\sigma_L^2 = 2 D t_r \tag{69}$$

which gives, with equation 66 :

$$H_{ld} = \frac{2 D}{\mathcal{V}} = \frac{2 D}{R <v>} \tag{70}$$

This contribution is inversely proportional to the eluent velocity and proportional to the particle diffusion coefficient. However, due to the very low diffusion coefficient of macromolecular and particulate species, this contribution to H is generally, in practice, negligibly small.

6.2.2. *Nonuniformity of the Flow Profile.* This contribution, which arises from the fact that particles at different transverse locations are displaced at different velocities by the nonuniform flow and that a finite time is needed for them to travel transversally from one velocity region to another has been termed the nonequilibrium contribution (Giddings 1965). This is because the nonuniform flow disturbs the equilibrium concentration profile and that the departure from this equilibrium profile cannot be instantaneously relaxed due to the finite transverse mass-transfer time. This contribution is in essence equivalent to the flow-induced tracer dispersion in tubular flow (Taylor dispersion) and is also called hydrodynamic dispersion. The determination of this contribution to H is relatively complex and is discussed in section 6.3.

6.2.3. *Relaxation*. When a species is introduced in the channel, it is generally uniformly distributed in the channel cross-section. A finite time, the relaxation time, is required before the equilibrium concentration profile is established. During that time, a relatively large zone broadening occurs. An approximate theory of the contribution of this process to H has been developed by computing the trajectories of the particles, initially uniformly distributed along x, assuming they are displaced transversally without diffusion by the constant field-induced velocity. Then, the variance of the distribution of the distances traveled by the particles when they all have reached the accumulation wall (Hovingh et al. 1970) is calculated. This gives for H_r :

$$H_r = \frac{17}{140} \frac{w^2 <v>^2}{L \, |U|^2} = \frac{17}{140} \frac{1}{L} \left(\frac{\lambda \, w^2 \, <v>}{D} \right)^2 \tag{71}$$

This contribution to H can be relatively important, especially when operating at high flow-rates and for slowly diffusing species, as it is proportional to $(<v>/D)^2$. This is particularly true in sedimentation FFF or flow FFF when working with relatively large colloidal particles, while it is of a lesser importance in thermal FFF with polymers of moderately high molecular weights. However, there are fortunately some experimental possibilities for considerably reducing this relaxation contribution to H.

First of all, the flow can be stopped immediately after the sample introduction for allowing the concentration equilibration to take place before the flow is restarted. The relaxation time for which the flow should be stopped can be estimated as the time, t_{rel}, required for the centre-of-gravity of the initial particle distribution (w/2) to move to its final transverse position (x_{cg}), that is :

$$t_{rel} = \frac{1}{|U|} \left(\frac{w}{2} - x_{cg} \right) = \frac{w^2 \lambda}{D} \left(\frac{1}{2} - \lambda + \frac{\exp(-1/\lambda)}{1 - \exp(-1/\lambda)} \right) \tag{72}$$

which has for limits $w^2\lambda/2D$ and $w^2/12D$ for low and high λ values, respectively (Hovingh et al. 1970). An upper limit of the relaxation time can simply be estimated as $w/|U|$ or $w^2\lambda/D$. However, a more detailed analysis indicates that, in order to closely reach the equilibrium profile, this time should be multiplied by a factor of about 2 (Janča et al. 1985).

Secondly, relaxation can be greatly accelerated by various stopless flow procedures, which avoid the disadvantages of the stopflow procedure (increased run time, baseline instabilities, risk of adhesion at the accumulation wall). For instance, the channel thickness can be reduced by means of a blocking element in the inlet region of the FFF channel (Giddings 1989d). Another possibility is to use a thin flow splitter in the inlet region separating two flow ports located at the two walls. The sample is introduced in the port near the accumulation wall, the carrier in the other one. The effective thickness of the sample stream is then controlled by the ratio of these two inlet flows (Giddings 1985a). A third stopless, hydrodynamic relaxation procedure uses a small piece of permeable wall material (a frit element) imbedded in the depletion wall (opposite to the accumulation wall), which allows an independent flow stream permeating into the channel through this element to compress the sample near the accumulation wall (Giddings 1990b).

6.2.4. *Sample Polydispersity*. The individual elements (molecules or particles) of a macromolecular or particulate species are not always identical (this is generally true for synthetic polymers but not for biopolymers) but differ to some extent in molecular weight or size from the average value for the ensemble of the species elements. The distribution of masses or sizes can generally be considered as continuous (this is not rigorously exact for synthetic polymers but, because the selectivity of separation methods does not generally allow to distinguish two macromolecules differing by one monomeric unit, they can in practice be considered as so). As a result, their fractogram appears as a single peak which is broadened by the incomplete separation of the various elements within the envelope. The corresponding contribution to H, H_p, which is not an undesirable effect, as it corresponds to a true separation and allows to characterize the distribution of the molecular weights or particle sizes, is related to the variance of this distribution. This contribution is, according to equation 66, given for a molecular weight (MW) distribution as :

$$H_p = \frac{1}{z}\left(\frac{d\,z}{d\,M}\right)^2 \sigma_w^2 \tag{73}$$

where z is the distance migrated by the species of molecular weight M and σ_w^2 the variance of the weight MW distribution (the weight, rather than number, MW distribution is considered since detectors usually sense masses rather than numbers of moles). Assuming that σ_w^2/M_w^2 is equal to $(\mu-1)/\mu$ where μ is the polydispersity index, $\mu=M_w/M_n$, M_w and M_n the weight- and number-average molecular weights, respectively, H_p becomes (Smith et al. 1977, Hovingh et al. 1970) :

$$H_p = L\ S_M^2\ \frac{\mu-1}{\mu} \tag{74}$$

It is seen to increase greatly with the molecular weight selectivity, S_M, of the separation system given by :

$$S_M = |\frac{d\ln V_r}{d\ln M}| = |\frac{d\ln R}{d\ln M}| \tag{75}$$

Similar expressions can be obtained for particle size distributions (Giddings et al. 1974).

6.2.5. *Instrumental Contributions.* The ΣH_i term in equation 68 includes instrumental effects, such as the finite volume and time of introduction of the sample in the channel, irregularities in the channel geometries (longitudinal and/or lateral variations of w due to a lack of parallelism between the plates and surface roughness), edge and end effects, finite connecting volumes to the injector and to the detector, finite detector response time, etc. With the present typical configuration of the FFF channels, these contributions can be considered as small and, more or less, negligible. However, they constitute a technological barrier for separation improvements by means, for instance, of a significant reduction of the channel thickness.

6.3. THEORY OF THE HYDRODYNAMIC DISPERSION

The H_n hydrodynamic dispersion term due to the transverse nonuniformity of the velocity profile in equation 68 is generally the factor dominating the peak broadening process. Various models allow its determination.

6.3.1. *Simplified Random Walk Model.* This model assumes that a particle in a cloud with mean thickness ℓ can be found in only two velocity states, a high one with velocity $2\mathcal{V}$ at distance 2ℓ from the accumulation wall, a low one at the wall with velocity 0 (Giddings 1973a). A molecule stays in a state for the time t_d required to diffuse from one state to the other ($t_d = (2\ell)^2/2D$ according to the Einstein diffusion law). During that time, it gains or loses a distance $\mathcal{L}$ over the zone as a whole which is moving at velocity $\mathcal{V}(L = (2\mathcal{V}\text{-}\mathcal{V})t_d = \mathcal{V}t_d)$. For a large number, n, of random steps, with $n = L/\mathcal{L}$, the variance of the zone becomes, according to the random walk model, $\sigma_z^2 = \mathcal{L}^2 n$, which gives $H_n = \sigma_z^2/L = 2\mathcal{L}^2 \mathcal{V}/D$. It is remarkable that this crude model, providing a simple conceptual view of the band broadening process, gives the exact functional form of H_n and is only a factor 2 smaller than the limiting value for small $\mathcal{L}$ given by more rigorous theories.

6.3.2. *Nonequilibrium Theory.* The nonequilibrium theory of FFF has been developed by Giddings (1968) in the same line as similar theories developed for chromatography (Giddings 1965). It starts from the convection-diffusion mass transfer equation and expresses the departure of the actual concentration, c, from the equilibrium concentration, c*, in terms of an equilibrium departure term ε such that $c = c^* (1+\varepsilon)$. It assumes that the separation occurs in near-equilibrium conditions, i.e. that the zone moves forward under conditions of total lateral equilibrium. Then, a differential equation is found for the departure term in the case of constant field-induced velocity and constant lateral diffusivity (however, the lateral diffusion coefficient may be different from the axial diffusion coefficient). The axial flux equation is then expressed in terms of ε and assumed to obey the Fick law which provides an effective dispersion coefficient, $\mathcal{D}$, related to the plate height as :

$$H_n = \frac{2 \mathcal{D}}{\mathcal{V}} \tag{76}$$

The differential equation for a departure term, ϕ, proportional to ε is given in the two-dimensional geometry (no variation along y) by :

$$\frac{d^2\phi}{d\zeta^2} - \frac{d\phi}{d\zeta} = \frac{v(\zeta)}{\mathcal{V}} - 1 \tag{77}$$

where ζ is equal to $x/\mathcal{L}$. Then, $\mathcal{D}$ is equal to :

$$\mathcal{D} = - \frac{\mathcal{V} \mathcal{L}^2}{D_l} \frac{<c\ v\ \phi>}{<c>} \tag{78}$$

In this equation D_l is the lateral diffusion coefficient which in many situations can be taken as equal to D. Accordingly, H_n can be expressed as :

$$H_n = \psi \frac{\mathcal{V} \mathcal{L}^2}{D} = \chi \frac{w^2 <v>}{D} \tag{79}$$

with :

$$\psi = -2 \frac{<c\ v\ \phi>}{<c> <v>} \tag{80}$$

and :

$$\chi = \psi \lambda^2 R \tag{81}$$

These equations have been solved in the case of a parabolic flow profile (Giddings et al. 1975) which gives for χ :

$$\chi = \frac{24 \lambda^3 A}{1 + \exp(-1/\lambda) - 2\lambda (1 - \exp(-1/\lambda))}$$

$$A = (28\lambda^2 + 1)(1 - \exp(-1/\lambda)) - 10\lambda(1 + \exp(-1/\lambda)) - \frac{1}{3\lambda^2} - \frac{2}{\lambda} + 4$$

$$- \frac{1/\lambda}{(1 \exp(-1/\lambda))}(4\lambda - \frac{1}{3\lambda} - 6) - \frac{4/\lambda}{(1 - \exp(-1/\lambda))^2} \tag{82}$$

It is instructive to look at the limiting expressions of χ and ψ for high and low retention. One gets from equation 82 :

$$\lim_{\lambda \to 0} \chi = 24\lambda^3 (1 - 8\lambda + 12\lambda^2 + 24\lambda^3 + 48\lambda^4) \tag{83}$$

$$\lim_{\lambda \to 0} \psi = 4 (1 - 6\lambda + 24\lambda^3) \tag{84}$$

$$\lim_{\lambda \to \infty} \chi = \frac{1}{105} \tag{85}$$

$$\lim_{\lambda \to \infty} \psi = 0 \tag{86}$$

The curve χ versus λ is plotted on Figure 3.

It appears that the nonequilibrium contribution to H increases linearly as a function of the average flow velocity and of the reciprocal of the diffusion coefficient, as expected from the behavior encountered in chromatography. When λ and, hence, χ are kept constant, H is proportional to the square of the channel thickness, i.e. of the distance over which flow nonuniformities occur. This provides the rationale for using channels as thin as possible in FFF (analogous to using packing particles as small as possible in liquid chromatography).

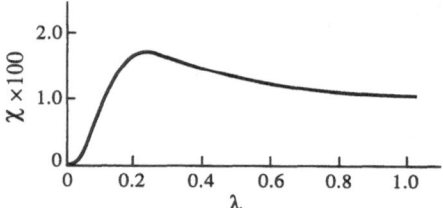

Figure 3. Variation of the nonequilibrium band broadening parameter, χ, (equation 79) versus λ for a parabolic flow profile and an exponential transverse equilibrium concentration profile

An unique characteristic of FFF, however, lies in the dependence of peak broadening on retention. Equation 83 shows that H_n decreases very quickly as retention increases (λ decreases). This can be seen more clearly by expressing χ in terms of R :

$$\lim_{R \to 0} \chi = \frac{R^3}{9} (1 - \frac{R}{3} - \frac{4 R^2}{9} - \frac{10 R^3}{27}) \tag{87}$$

This is basically due to the fact that the effective distance over which particles have to diffuse to relax the flow nonuniformities is of the order of magnitude of the particle cloud thickness, ℓ, which decreases with increasing retention (in liquid chromatography, this distance remains scaled to the particle size whatever the level of retention). The error resulting from the use of the limiting equation 87 is less than 1% and 3% when R is smaller than 0.4 and 0.7, respectively, which is quite satisfying taking into account the precision of the experimental determination of χ.

6.3.3. *Special Case of Thermal FFF*. The nonequilibrium theory developed by Giddings assumes a constant transverse diffusion coefficient. It is therefore not strictly applicable to thermal FFF since the diffusion coefficient depends on the temperature. Nevertheless, because the variation of D with x is not very rapid, the above nonequilibrium equation can be used in that case by taking the value of D at the temperature of the centre-of-gravity, x_{cg}, of the zone. However the diffusion coefficient appearing in the longitudinal diffusion term (equation 70) should be the cross-sectional average coefficient, $<D\,c> / <c>$ (Hovingh et al. 1970).

6.3.4. *Discrete Random Walk Model*. Instead of assuming that a particle can be found in only two velocity states as in the simplified random walk model of section 6.3.1., a relatively large number of states can be allowed by dividing the channel thickness into a number of equal intervals each moving at a given velocity. The probability for a particle to move upward or downward is biased according to the concentration distribution profile of equation 17. The statistical analysis of this multi-state random walk process in the long time asymptotic domain allows Berg and Purcell (1967a) to determine a band broadening parameter, f, as a function of λ for a parabolic or half-parabolic velocity profile, such that :

$$\sigma_z^2 = \frac{f^2}{2} <z>^2 \frac{x_{cg}^2}{D\,t} + 2\,D\,t \tag{88}$$

where $<z>$ is the average distance traveled by the particles at time t. Obviously the last term represents the longitudinal diffusion. Accordingly, one should have :

$$\chi = \frac{f^2\,R}{2} \left(\lambda - \frac{\exp(-1/\lambda)}{1 - \exp(-1/\lambda)} \right) \tag{89}$$

Unfortunately, the graphical form of the presentation of the results does not allow an easy verification of this relationship.

6.3.5. *Continuous Random Walk Model*. A generalized random walk model has been developed (Van den Broeck & Maes 1987) which can allow any flow or concentration profile as well as adsorption in a retentive layer near one or two walls. General expressions of the long time dispersion coefficient are given for both discontinuous and continuous models. An analytical expression of this coefficient is given for the case of a parabolic velocity profile and of the exponential concentration distribution, with and without adsorption at the walls. Unfortunately, the complicated forms of the expression and of equation 82 do not allow their comparison in a simple way.

6.4. HYDRODYNAMIC DISPERSION FOR A GENERALIZED VELOCITY PROFILE

The basic nonequilibrium equation for an exponential concentration profile has been solved in the case of a generalized n-degree polynomial velocity profile (Martin & Giddings 1981a). The calculation of χ and ψ has been explicitly performed as a function of λ and of the third-degree velocity profile parameter ν of equation 39. The resulting expression of χ is equal to :

$$\chi = \frac{2\,\lambda^2\,A'}{R\,(1 - \exp(-1/\lambda))}$$

with :

$$A' = 2A'' [6(1+v) - (1/\lambda) - (A''/\lambda) + 36v\lambda^2 - 6\lambda(1+6v) + 18\lambda(1-10v\lambda)exp(-1/\lambda)]$$
$$+ 72\lambda^2 [(1+v)^2 - 10\lambda(1+4v+3v^2) + 4\lambda^2(7+69v+90v^2) - 672v\lambda^3(1+3v)$$
$$+ 4464v^2\lambda^4]$$
$$- 72\lambda^2(exp(-1/\lambda)) [(7-2v+v^2) + 2\lambda(5-68v+15v^2) + 4\lambda^2(7-69v+180v^2)$$
$$- 672v\lambda^3(1-3v) + 4464v^2\lambda^4]$$

and :

$$A'' = \frac{12\lambda (6v\lambda - 1) exp(-1/\lambda)}{1 - exp(-1/\lambda)} \tag{90}$$

which gives in the high retention limit (for $v \neq -1$) :

$$\lim_{\lambda \to 0} \chi = 24\lambda^3 [(1+v) - 8\lambda (1+3v) + 12\lambda^2 (1+14v+17v^2) / (1+v)$$
$$+ 24\lambda^3 (1+3v) (1-10v-7v^2) / (1+v)^2$$
$$+ 48\lambda^4 (1-7v-10v^2-39v^3-21v^4) / (1+v)^3] \tag{91}$$

Obviously, when putting $v = 0$ in equations 90 and 91, one finds again the equations 82 and 83 for the parabolic flow. The calculation of the peak broadening parameter χ for the non-parabolic velocity profiles discussed in section 3.4. is relatively complex and has not been attempted except in the case of annular channels (Davis 1989, Ugrozov et al. 1989). The third-degree profile approximation has been used to calculate χ in thermal FFF (Gunderson et al. 1984) by fitting the more exact velocity to equation 39 for adjusting the parameter v. It is noticeable that in this case where v is negative, the peak broadening in the high retention limit is lower than for a parabolic profile. This is because of the smaller value of the slope of the velocity profile near the accumulation wall.

7. Other Theoretical Models in the Brownian Mode

A number of more or less complex theoretical models have been developed to describe the behavior of a particle cloud subjected to a lateral field force in a unidirectional flow. These models generally start from the convective diffusion equation in the two-dimensional space (x,z) :

$$\frac{\partial c}{\partial t} + v(x)\frac{\partial c}{\partial z} + U\frac{\partial c}{\partial x} = D \left(\frac{\partial^2 c}{\partial z^2} + \frac{\partial^2 c}{\partial x^2} \right) \tag{92}$$

This equation assumes a fully developed laminar flow and a constant diffusion coefficient. A solution is searched for the evolution in time and space of the cross-sectional average concentration, $<c(t,z)>$. This is given by the generalized dispersion theory of Gill and Sankarasubramanian (1970, 1971) in reduced variables as :

$$\frac{\partial \theta}{\partial \tau} = \sum_{i=1}^{\infty} K_i(\tau) \frac{\partial^i \theta}{\partial \xi^i} \tag{93}$$

where the reduced cross-sectional average concentration, θ, the reduced time, τ, and the reduced axial coordinate, ξ, are equal to :

$$\theta = <c> / c_{ref} \tag{94}$$

$$\tau = \frac{4 D t}{w^2} \tag{95}$$

$$\xi = \frac{8}{3} \frac{D z}{w^2 <v>} \tag{96}$$

c_{ref} is a reference concentration related to the total number of particles. Clearly, from equation 93, K_1 is a reduced form of the velocity of the particle zone and K_2 a reduced dispersion coefficient. Higher-order K terms describe the evolution of the shape of the cross-sectional particle concentration in the channel. All these K terms are functions of the time. Therefore the various theoretical models based on this analysis can be divided in transient models describing the time dependence of the K factors and asymptotic models focusing on their long time asymptotic values.

7.1. TRANSIENT MODELS

Lee and Lightfoot (1976) as well as Krishnamurthy and Subramanian (1977) applied this model to the case of a parabolic flow and for a uniform initial particle distribution. They neglected terms of order higher than 2 in equation 93 which then becomes a one-dimensional diffusion equation with time-dependent convection and dispersion such that, with the above dimensionless variables :

$$K_1(t) = -\frac{2}{3} R(t) \tag{97}$$

and :

$$K_2(t) = \frac{8}{9} R(t) \chi(t) \tag{98}$$

In the asymptotic long time limit, they found the same R expression versus λ as equation 49 and an expression for $\chi(\lambda)$ in terms of K_2 which has the same asymptoptic values as equations 83 and 85 for the high and low retention limits, respectively.

The time dependence of K_1 and K_2 for different λ values is plotted in Figures 4 and 5.

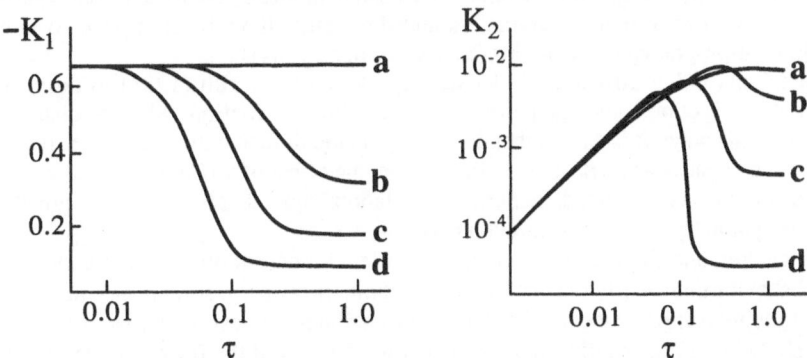

Figure 4. Variation of the convective coefficient K_1 vs. dimensionless time τ for different λ values : (a) $\lambda = \infty$; (b) $\lambda = 0.1$; (c) $\lambda = 0.05$; (d) $\lambda = 0.025$. Adapted from Krishnamurthy & Subramanian 1977, courtesy of Marcel Dekker, Inc.

Figure 5. Variation of the nonequilibrium part of dispersion coefficient K_2 vs. dimensionless time τ for different λ values: (a) $\lambda = \infty$; (b) $\lambda = 0.1$; (c) $\lambda = 0.05$; (d) $\lambda = 0.025$. Adapted from Krishnamurthy & Subramanian 1977, courtesy of Marcel Dekker, Inc.

Note that, in equation 98 and in Figure 5, the longitudinal diffusion contribution to the dispersion, which is part of K_2 in equation 93, has been removed. Figure 4 indicates that, at very early times, all particulate species move at the eluent velocity. Then they reach their asymptotic R values for a time of the order of 0.025 w^2/D to 0.25 w^2/D, depending on λ. This time decreases with increasing λ as expected from the discussion in section 6.2.3. A similar behavior is observed for the dispersion coefficient.

The validity of the truncation of equation 93 after the second order has been checked by Jayaraj and Subramanian (1978) who solved equation 92 for a uniform slug injection by means of a finite difference numerical method (the alternating direction implicit technique). Their study indicates that the axial concentration profile <c(z)> can be strongly asymmetric and even bimodal in the early times of the particle migration. (A bimodal fractogram for a single monodisperse species has been observed in some experimental conditions (Yang et al. 1977)). The axial concentration distribution obtained by this analysis is seen numerically to approach the Gaussian shape predicted by the truncated model for time of the order of $w^2/2D$.

The calculation of the time-dependent values of K_1 and K_2 has been performed for a rectangular channel with adjustable aspect ratio, b/w, by means of the correction of equation 31 to the parabolic profile (Kim & Chung 1986) and of the Gill-Sankarasubramanian model truncated after the second order. There is only a slight change in K_1 when b/w increases from 10 to 100 and no change above 100. The K_2 vs. τ curve is close to the two-dimensional curve (corresponding to b/w $\rightarrow \infty$) up to τ equal to 1 to 10 (depending on λ), then increases gradually to the value given by the asymptotic model of Takahashi and Gill (1980). However, the large values of K_2 obtained in this situation are, in practice, unrealistic because FFF elution times are usually much shorter than the diffusion times in the y-direction. A similar calculation for the rectangular channel geometry shows that an initial stopflow step considerably increases the resolution (Andreev & Khidekel 1990).

A digital simulation of the FFF migration process has been performed by means of a random walk Monte-Carlo simulation (Schure 1988). It indicates a close correspondence with the asymptotic band broadening theory at medium retention levels but discrepancies are found in the low retention range due to the non-Gaussian peak shapes.

Transient average particle migration velocities have been numerically computed from equation 92 as a function of some experimental parameters and the results have been applied to interpret retention data in field-programmed conditions (Yau & Kirkland 1984).

A one-dimensional delay-diffusion model was developed by Smith (1983) to describe the evolution of the cross-sectional average particle concentration distribution. Skewed distributions can be observed in this way. It is shown that depending on the time and the value of λ, positively or negatively skewed peaks can be observed. In addition, the effect of an initial relaxation step can be included in this model. As expected, numerical calculations show that this improves greatly both the width and the symmetry of the concentration band.

Generally speaking, one should note that, while transient models allow a deeper understanding of the particle behavior in the early times of their migration, their practical relevance may be small since, by one of the relaxation procedures discussed in section 5.2.3., especially stopflow, one can closely approach the long time equilibrium profile before the start of the flow process. A detailed analysis of the two first moments distribution of the particle zone along the channel as a function of time and of the duration of the initial relaxation step has been performed by Ugrozov (1990).

7.2. ASYMPTOTIC MODELS

The Takahashi and Gill model discussed above for channels with a finite aspect ratio belongs to this category of asymptotic models, although it is generally of little relevance for typical FFF experiments.

A general theory of FFF has been developed by Gajdos and Brenner (1978) for non-spherical particles when the external field force produces not only displacement but also reorientation of the individual particles. It gives coefficients analogous to R and χ in the long time limit since their time-dependent parts have been ignored. The general frame of the theory corresponds then to an anisotropic convective diffusion equation in which diffusion coefficients in the direction parallel and perpendicular to the field may be different. The effects of finite particle size and wall potential are also considered. The calculation of R and χ from their numerically computed analogs for some selected values of λ, for small spheres in a parabolic flow, gives results identical to those of equations 49 and 82, respectively.

General expressions of $\mathcal{V}$ and $\mathcal{D}$ have been given in the long time limit for general diffusion coefficient, potential and velocity profiles, D(x), W(x) and v(x), respectively, in the two-dimensional geometry for non-adsorbed particles (Prieve 1982) :

$$\mathcal{V} = \int_a^{w-a} v(x) \exp(-W(x))\, d\,x \quad / \quad \int_a^{w-a} \exp(-W(x))\, d\,x \tag{99}$$

and :

$$\mathcal{D} = \int_a^{w-a} \frac{I^2(x)\, d\,x}{D(x)\exp(-W(x))} \quad / \quad \int_a^{w-a} \exp(-W(x))\, d\,x$$

$$I(x) = \int_a^x [\, v(x') - \mathcal{V}\,] \exp(-W(x'))\, d\,x' \tag{100}$$

where a is the particle radius. These equations are particularly interesting in the case where interaction potential have to be added to the field-induced potential.

The variation of the particle friction coefficient in the vicinity of the wall (lubrication effect) influences the diffusivity near the wall as well as the field-induced velocity (except in the case of flow FFF where U rather than F is imposed). This does not change λ as these influences cancel out in the D/U ratio, except for flow FFF. The retention in flow FFF was calculated to account for this effect (Davis & Giddings 1985b). However, it is seen to be small. The Giddings nonequilibrium theory has also been modified to account for this influence in the expression of ψ.

The retention and plate height long time asymptotic expressions for annular channels have been computed using the method of the generalized dispersion theory for a combined free convection (thermogravitational effect) and forced convection profile (Zolotarev et al. 1988, Ugrozov et al. 1989).

8. Separation Optimization

The plate height H of a monodisperse species in ideal conditions, i.e. such that $H_r = \Sigma H_i = 0$, has the same velocity-dependence as in capillary chromatography :

$$H = \frac{2\,D}{\mathcal{V}} + \frac{4\,l^2\,\mathcal{V}}{D} = \frac{B}{\mathcal{V}} + C\,\mathcal{V} \tag{101}$$

The plate height curve has a minimum which is given, in the high retention limit where $R = 6 \lambda$, by (Giddings 1973b) :

and :

$$H_{min} = \frac{\sqrt{32}}{6} R w = 0.94 R w \approx R w \qquad (102)$$

or:

$$\mathcal{V}_{opt} = \sqrt{18} \frac{D}{R w} \qquad (103)$$

$$<v>_{opt} = \sqrt{18} \frac{D}{R^2 w} \qquad (104)$$

This analysis shows that, for a given R, H is the smallest and $<v>_{opt}$ the greatest for channels of narrow thickness, w. Unlike in chromatography, the minimum plate height is seen to decrease and the optimum velocity to increase with increasing retention (decreasing R). Indeed, writing $R = 6D/|U|w$, one gets $\mathcal{V}_{opt} = |U|/\sqrt{2}$. However, this velocity is very low. It is generally impractical to operate at the optimum velocity.

By manipulating the basic high retention equations to obtain the maximum rate of generation of plates, $\dot{N}_{max}$, in FFF, one gets (Giddings 1973b) :

$$\dot{N}_{max} = \frac{9 D}{R^2 w^2} \qquad (105)$$

This maximum rate increases dramatically with decreasing both the retention ratio and the channel thickness. This is realized when operating at a velocity much larger than the optimal one.

The resolution, R_s, between two particle zones can be expressed, as in chromatography, by :

$$R_s = \frac{\Delta t_r}{4 \sigma_t} \qquad (106)$$

where Δt_r is the difference of their retention times, σ_t their average time standard deviation. If δM is the difference of the masses of the individual particles of the two species (or the molecular weight difference of the two macromolecular substances) and M the average mass, then one can show that, using average values for the diffusion coefficients, retention times and retention ratios of the two species, in the high retention domain, t_r is given by (Giddings et al. 1978c) :

$$t_r = \frac{16 R^2 w^2}{9 \omega_M^2 D} \left(\frac{R_s M}{\delta M} \right)^2 \qquad (107)$$

with :

$$\omega_M = \frac{d \ln \lambda}{d \ln M} \qquad (108)$$

ω_M is a constant depending on the field type. The last group of terms in equation 107 expresses the requirement for the separation in terms of resolution and relative mass difference. It shows that a trade-off must be found between analysis time and resolution. The Rw factor in the numerator, proportional to the particle layer thickness, ℓ, indicates that time optimization in FFF requires using the highest possible retention conditions and the lowest possible channel thickness. An equation similar to equation 107 can be written in terms of the relative particle size difference, $\delta d/d$, instead of $\delta M/M$, by replacing ω_M by ω_d with :

$$\omega_d = \frac{d \ln \lambda}{d \ln d} \qquad (109)$$

In spite of the technological limitation in reducing w, separations of multicomponent species have been performed in less than one minute (Giddings et al. 1978c, Giddings et al. 1987b, Ratanathanawongs & Giddings 1989).

The multi-component separation capability of the FFF techniques can be compared to that of other separation methods for macromolecules and particles in terms of the peak capacity, defined as the maximum number of components resolvable at unit resolution. The influence of various operational parameters on the peak capacity has been described (Martin & Jaulmes 1981b).

9.   Alternative modes of operation in FFF

The primary mode of retention in FFF is the Brownian mode, also called "normal" mode, for which the equilibrium transverse particle concentration profile decreases exponentially from the accumulation wall. Departure from this profile can occur if the finite average particle concentration induces non-linearities into the transport properties D and U or particle-particle interactions. Some of these effects have been studied such as the influence of the average concentration on D (Caldwell et al.1988), on the transverse concentration profile (Hansen et al. 1989a, Mori et al. 1990) and on R in the high retention limit by means a virial approach (Inagaki & Tanaka 1980). However, in the alternative modes of retention discussed below, the form of the transverse concentration profile deviates from the exponential one even for infinite dilution particle suspensions because of the non-linearity of the potential profile independently of concentration overloading effects.

9.1. STERIC MODE OF RETENTION

When the particles become bigger and bigger, they reach a size for which they can no longer be considered as point-like and, eventually, which becomes of the same order of magnitude, or even larger, than the cloud thickness ℓ calculated from equation 6. Due to the steric exclusion effect, the particle center cannot have access to the low velocity region near the accumulation wall. Large particles are thus rejected in the fast velocity streamlines of the accessible core in the center part of the channel.

9.1.1. *Retention Model in the Steric Mode.* A theoretical model for retention and dispersion in this steric mode of migration has been developed by Giddings (1978d). A solid sphere of radius a is assumed to execute Brownian displacements in the accessible region of the channel defined by a $\leq$ x $\leq$ w-a, as shown in Figure 6. The Brownian model is applied to the channel constituted by this accessible core moving relatively to the laboratory frame of reference with velocity v(x=a) = $6<v>(\alpha-\alpha^2)$ where α is:

$$\alpha = \frac{a}{w} \tag{110}$$

Then the retention ratio is given by :

$$R = 6(\alpha - \alpha^2) + 6\lambda(1 - 2\alpha)\left[\coth\left(\frac{1 - 2\alpha}{2\lambda}\right) - \frac{2\lambda}{1 - 2\alpha}\right] \tag{111}$$

When both α and λ are small, one gets :

$$\lim_{\lambda \text{ and } \alpha \to 0} R = 6\alpha + 6\lambda \tag{112}$$

548

Equation 111 corresponds to a hard-sphere potential with $W = \infty$ for $0 \leq x \leq a$ and $w-a \leq x \leq w$ and with $W(x) - W(a) = |F| (x-a)$ for $a \leq x \leq w-a$. It gives a R value identical to that resulting from the model of Gajdos and Brenner (1978) applied to spherical particles. Equation 112 describes the limiting behavior of R as a function of the particle mass or size. For small particles with $\alpha \ll 1$, equation 111 gives the classical result for the Brownian mode (equation 49). Then, the λ terms in equations 111 and 112 are predominant and R decreases with increasing particle size since ω_d, the relative coefficient of variation of λ with the particle size in equation 109, is generally negative. Ultimately, with further increase in the particle size, the α terms in equations 111 and 112 become overwhelmingly predominant and R increases with particle size.

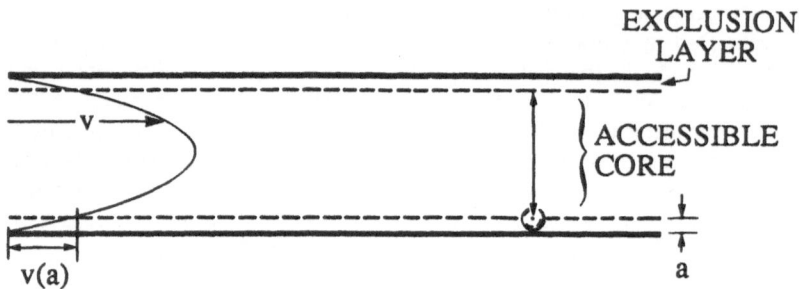

EXCLUSION LAYER

ACCESSIBLE CORE

v

v(a)

a

Figure 6. Schematic representation of the steric model of migration. Reprinted from Giddings 1978d, courtesy of Marcel Dekker, Inc.

9.1.2. *Inversion of the Order of Elution.* A minimum is observed in the R vs. particle size curve for some critical particle size, a_{crit}, as shown in Figure 7. To a first approximation, one can consider that particles with $a < a_{crit}$ are migrating in the Brownian mode while particles with $a > a_{crit}$ are eluted in the steric mode of retention. (In fact, the steric effect appears slightly before the minimum in the curve of Figure 7 and the Brownian motion has not completely disappeared just after this minimum). The exact value of a_{crit} depends on the experimental conditions but is of the order of 1 µm (Martin & Jaulmes 1981b, Myers & Giddings 1982, Lee & Giddings 1988a). Such an inversion in the order of elution vs. particle size poses problems for the analytical characterization of a distribution of particle sizes since particles of two different sizes can elute simultaneously, one with d smaller, the other with d larger than the critical value. Accordingly, sample particle size distributions should be such that all sample particles are either smaller or larger than the critical size.

In equations 111 and 112, the particle is assumed to move at the velocity of the unperturbed fluid at the particle centre-of-gravity. This neglects some hydrodynamic effects occurring when the particle is moving near the wall (particle rotation, viscous drag). Accordingly, equation 112 has been modified by a velocity correction factor, γ, which gives when $\lambda \ll \alpha$ (Caldwell et al. 1979) :

$$R = 6\gamma\alpha \tag{113}$$

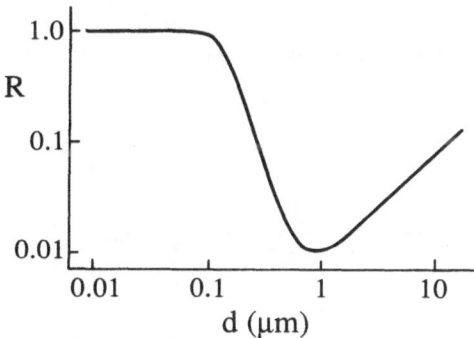

Figure 7. Variation of the retention with the particle size (relative value) showing the inversion in the elution order after a critical size, due to the onset of the steric mode of migration. Adapted from Myers & Giddings 1982, courtesy of American Chemical Society.

9.2. HYDRODYNAMIC LIFT MODE OF OPERATION

According to equations 111 to 113, R should not depend on the carrier flow-rate. This is well verified in the Brownian domain of retention. However, experimental data show that, for particles larger than a_{crit}, R increases significantly with $<v>$ (Caldwell et al. 1979). This suggests that the particles drift away from the wall into faster moving streamlines under the influence of velocity-dependent lift forces. This behavior seems to be linked to the "tubular pinch effect" observed by Segré & Silberberg (1961) who found that particles of finite size moving in a tube tend to focus at a radial position equal to about 0.6 tube radius from the axis (Segré & Silberberg 1962). Accordingly, the transverse lift force depends on both the average carrier velocity and on the lateral position in the channel, x/w. The particle is thus repulsed from the axis and from the walls. At some transverse position, the lift force is zero. As no sidewise lift forces can arise in the low velocity Stokes flow regime, this effect is necessarily inertial in origin (Brenner 1966).

9.2.1. *Expression of the Inertial Lift Force and Equilibrium Position.* The calculation of this force is a relatively difficult fluid mechanical problem. This calculation was performed with some limiting approximations (particle relatively far away from the wall, small particle Reynolds number, Re_p, with $Re_p = \alpha^2 Re$, and small channel Reynolds number, Re) by Ho and Leal (1974) and Vasseur and Cox (1976). The two models give similar results. The variation of the inertial lift force with x/w is antisymmetrical about x/w = 0.5. The force is repulsive near the walls and near the axis and becomes zero at lateral positions x/w = 0.2 and 0.8. These equilibrium positions are somewhat shifted toward the walls when Re increases above 15 (Schonberg & Hinch 1989).

These positions do not depend on the particle size although the amplitude of the force does. In the FFF channel, the lift force combines with the field force near the depletion wall while it competes with the field force near the accumulation wall. As a result, the equilibrium position where the net force, F_n, is zero (since it corresponds to the minimum of the potential profile, i.e. to $d\,W/d\,x = -F_n = 0$) is shifted toward the accumulation wall and the amount of this shift depends on the particle size. This provides the basis of the particle size selectivity in this lift (or inertial) mode of retention. In practice, only the lower of the two equilibrium positions is significant because the particles are driven by the field below the mid-plane at x/w=0.5. This is seen in Figure 8.

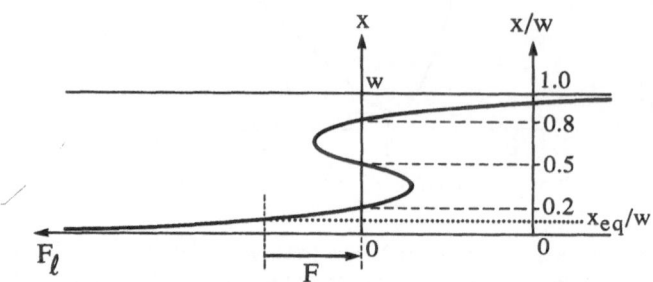

Figure 8. Variation of the lift force, F_l, (horizontal axis) vs. the lateral position, x/w. The force is an odd function of position across the centerline. The equilibrium positions are at x/w = 0.2 (stable), 0.5 (unstable) and 0.8 (stable). When the field force, F, competes with the lift force, the resulting net force is obtained by substracting |F| from F_l. The new equilibrium position is where the net force is zero.

The lift force, F_l, computed by Ho and Leal is equal to :

$$F_l = \frac{\rho \ a^4 \ <v>^2}{w^2} \ G(x/w) \tag{114}$$

where G(x/w) is a dimensionless function of the lateral position which has the shape of curve (a) in Figure 8. It is noticeable that this force increases with the square of the carrier velocity and the fourth power of the particle size. This explains why lift effects are generally not effective for sub-micron particles.

The equilibrium position is found in the region $a \le x \le 0.2$ where G(x/w) is positive and decreasing with increasing x (dG/dx <0). This position, x_{eq}, where $F + F_l = 0$, corresponds to :

$$G(x_{eq}/w) = \frac{w^2 \ |F|}{\rho \ a^4 \ <v>^2} \tag{115}$$

The size selectivity of this lift mode of separation depends on the field type through ω_d (d lnG/d lnd = - ω_d - 4). If the field force increases as a^4, then x_{eq} does not depend on size. Generally $|\omega_d|$ is smaller than 4 and, therefore, G decreases and x_{eq} increases with increasing particle size. This effect is larger for flow FFF (ω_d = -1) than for sedimentation FFF (ω_d = -3). As $\mathcal{V}$ increases with x_{eq} in the near-wall domain, large particles elute before small ones. The elution order, vs. particle size, of the lift mode is thus the same as that of the steric mode and the problems associated with the inversion in the order of elution vs. particle size are the same in the two modes.

When particles are forced at high field strength in a region very near the wall where x is lower than about 2a, experimental evidence suggests an additional contribution to the lift force given by equation 114 (Williams et al. 1992a).

9.2.2. *Steric vs. Lift Mode of Retention.* In practice, there does not seem to be a clear-cut frontier between the steric and lift modes of migration which both become significant for particles larger than about 1 μm. In principle, one could note from the above equations that, when the flow velocity is sufficiently low for x_{eq} calculated from equation 115 to be smaller than a, the steric mode should be operating and that then R should be independent of <v>. However, in this situation, the viscous drag influence on $\mathcal{V}$ results in a dependence of R on <v> through the γ factor of equation 113 (Peterson II et al. 1984). It has been suggested to arbitrarily assign to the steric mode situations for which γ ≤ 2 while particles for which γ > 2 are considered to migrate in the lift mode (Barman et al. 1989). Due to the fact that particles tend to be concentrated in a zone of a quite narrow thickness at the transverse equilibrium position where they form a so-called hyperlayer, the lift mode is also described in the literature as the hyperlayer mode of operation. Because of the inertial origin of the lift force, this lift mode has also been called inertial mode (Martin et al. 1982).

9.2.3. *Retention Model in the Lift Mode* With the help of equation 115 and the tabulated values of G(x/w), a theoretical model of retention in this mode, taking into account the lift and field forces, the steric effect and the Brownian motion of the particles, has been developed (Martin et al. 1982). A R vs. <v> curve resulting from this model is plotted in Figure 9 together with some experimental data obtained under the natural gravitational field. It is seen that the direction and amplitude of variation of R with <v> is relatively satisfyingly described by the model.

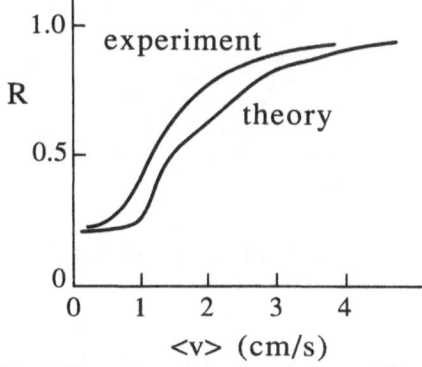

Figure 9. Variation of experimental and theoretical curves of variation of R with the carrier velocity for 7μm particles. The important variation of R with <v> is a manifestation of the influence of a flow-dependent lift force. The theoretical curve was computed by means of equation 114 (Martin & al. 1982).

The model can similarly describe the influence of various experimental parameters on R (Martin 1988). However some discrepancies remain between the absolute experimental and theoretical values. It is believed that they arise, first, because of the inaccuracy of the values of G(x/w) tabulated by Ho and Leal in the near-wall region of interest (in their calculations, they assumed (x-a)/a >> 1) and, second, because the finite time of relaxation of the particles to their equilibrium position. This is a general problem with the lift mode of migration since the equilibrium position (or concentration profile) depends on the flow conditions and cannot be reached before starting the flow by means of the stopflow relaxation method. Still the experimental R values are observed to depend on the extent of the relaxation step (Cardot & Martin 1990). Due to the lack of a sufficiently accurate theoretical model of retention, the characterization of particles eluted in the lift mode must rely on appropriate calibration. The size dependence of the duration of the relaxation period has been suggested to furnish the basis for separating particles during the transient regime even without force field (Kononenko & Shimkus 1990). The separation of buoyant particles by means of the lift force in horizontal channels of circular cross-section was modeled, neglecting the Brownian motion, by a similar analysis (Afromowitz & Samaras 1989).

9.3. FOCUSING FFF

9.3.1. *Basis of Focusing FFF.* The technique of focusing FFF has been proposed by Giddings (1977a, 1983a) and Janča (1982) as an extension to FFF of equilibrium-gradient separation methods. The basic principle of this FFF mode is illustrated in Figure 10 in the case of a density gradient.

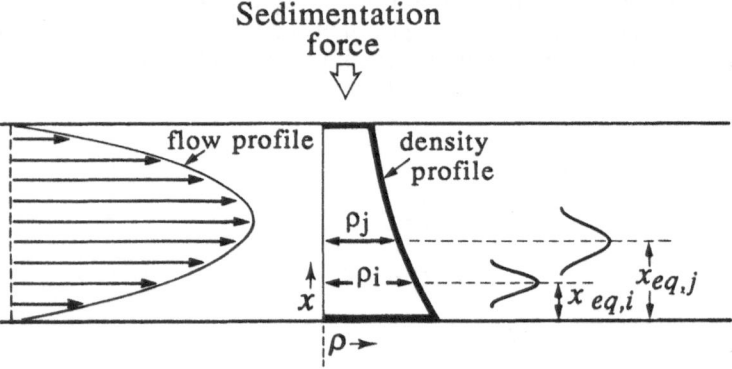

Figure 10. Schematic illustration of the principle of focusing FFF in the case of a density gradient. Reprinted from Giddings 1983a, courtesy of Marcel Dekker, Inc.

The carrier liquid is not a single component fluid but instead a multi-component solution or suspension. Let us assume that a multi-gravitational field (centrifugal) is applied and that the carrier is a solution or suspension of small colloidal buoyant particles or macromolecules. After a relaxation period, a transverse nonuniform concentration profile is established for these particles. As their average concentration in the incoming fluid is constant, this equilibrium concentration profile is stationary. Accordingly, a steady carrier density profile is present all along the channel. When relatively large sample particles are injected in this carrier, they are transported by the gravitational force up to the position of their isopycnic (isodensity) point where the force is zero. As soon as particles, due to the thermal motion, tend to escape from this position, they experience a Hookian force restoring them to this equilibrium focusing position. They form there narrow layers (hyperlayers), the thickness of which depends on their diffusion coefficient as well as slope of the density-gradient. Sample particles are transported axially at the fluid velocity corresponding to the equilibrium position and R is then equal, to a first approximation, to $v(x_{eq})/<v>$.

9.3.2. *Focusing vs. Lift Mode of Operation.* The focusing mode of separation bears some similarity to the lift mode as particles form narrow zones focused at some lateral equilibrium position, x_{eq}, hence the name "hyperlayer" sometimes used by Giddings and co-workers to characterize the lift mode. However, they differ basically in that the potential well at x_{eq} results, in the lift mode, from ubiquitous hydrodynamic forces, and in the focusing mode, from a force gradient intentionally applied by means of a multi-component carrier and the external primary field. Accordingly, the basis of separation is, in principle, different. Particles of the same material but of various sizes can be separated in the lift mode. However, such particles all move to the same lateral equilibrium position and, thus, cannot be separated in the isopycnic focusing mode, which separates particles on the basis of density differences. In fact, both modes can cooperate to provide a mixed size and density separation if sample particles are large enough to be influenced by the lift effect.

9.3.3. *Retention and Dispersion Models in the Focusing Mode.* An approximate theory of the focusing mode has been developed, assuming the zones have Gaussian lateral concentrations profiles, by means of a two-state random walk model (Giddings 1983a). A more detailed calculation takes into account the polydispersity of the density-modifying particles, the exact sample particle concentration profile, the velocity profile resulting from the dependence of viscosity on the carrier particle concentration (Schure et al. 1986). The theory predicts that the resulting plate height should be very small giving more than 10^6 plates per meter. A dispersion model based on the long time asymptotic solution of the convective diffusion equation indicates that, in typical conditions, the nonequilibrium parameter χ of equation 79 can be as low as 10^{-7} (Semenov & Kuznetsov 1986).

9.3.4. *Focusing FFF in channels with modulated permeability.* As the zones are focused in a narrow layer in the field direction, the use of thin channels is no longer an absolute requirement in the focusing mode of FFF. Accordingly, Janča & Jahnová (1983) suggest to use channels with the field applied in the large cross-sectional dimension (y) with w increasing continuously with y, w(y), in order to have a linear or slightly parabolic axial velocity profile, v(y), in the mid-plane which is then parallel to the field. The retention theory with these geometries has been given (Janča & Chmelík 1984). In this case, apart from the axial dispersion resulting from the particle distribution around their equilibrium position in the field direction (y), an additional peak broadening contribution arises from the nonuniform, nearly parabolic, velocity profile in the lateral direction perpendicular to the field (x).

9.3.5. *Various Forms of Focusing FFF.* The principle of focusing FFF, which has been demonstrated in a gravitational field with either a continuous (Chmelík & Janča 1986) or a step density gradient (Janča & Nováková 1987b), is not restricted to the case of isopycnic focusing with a sedimentation field. The separation effect of a pH gradient in the direction of the electrical field in a channel with a variable thickness by means of isoelectric focusing FFF has been demonstrated (Chmelík & Janča 1989b). Elutriation focusing FFF has also been proposed in such a channel with semi-permeable horizontal membranes and an upward cross-flow, with a varying velocity due to the trapezoidal shape of the channel (Janča 1987c). Particles are then focused at the position where the altitude-dependent upward cross-flow velocity exactly balances their constant downward field-induced velocity. In principle, this kind of counter-current method, which has been demonstrated in the case of the gravitational field (Urbánková & Janča 1990), is not restricted to sedimentation but could be applied with different types of force field.

9.4. CYCLICAL-FIELD FFF

In all FFF methods described above, the field strength was maintained constant during the course of the separation. Field programming techniques are very useful to extend the range of particle sizes or masses analyzed during a single run and to optimize the separation (see section 11.). In these programming techniques, the direction of application of the field and the sign of the rate of variation of the field are maintained unchanged. Such is not the case for the cyclical-field mode of FFF operation proposed by Giddings (1986a). In this method, the field strength is periodic with a cycle time much smaller than the particle residence time in the channel. In addition, the period should be smaller than the relaxation time of the particles so that they will be migrating laterally in a transient state for most of the time they spend in the channel. The lateral concentration is then a function of both x and t and the particle migration velocity becomes a time average over the cycle time. Accordingly, R should be dependent on transport rates (generalized mobilities) rather than on equilibrium parameters. This provides a new selectivity basis for FFF.

The principle of the method has been described in some details for a square-wave form of the field strength assuming that particles behave coherently, i.e. that identical particles follow the same trajectory (the concentration profile at a given time is then a Dirac function). The influence of the cycle shape on retention of micron-size particles with a gravitational field has been demonstrated (Lee et al. 1988b). There can be nearly an unlimited number of variations of the method according to the shape and relative duration of the field strength cycle.

9.5. FFF WITH SECONDARY CHEMICAL EQUILIBRIA

The range of application of FFF can be extended to small molecules by using secondary chemical equilibria. In this method, multi-phase carriers are used and the solute molecules distribute themselves between the continuous phase and the discontinuous phase which is retarded by its interaction with the external field. Then separations of solutes are possible according to their distribution coefficients between the two phases. The retention theory of this mode of separation has been established when the "support" particles of the discontinuous phase are strongly compressed by the field near the accumulation wall (Berthod & Armstrong 1987). Then the average "support" concentration is larger than its concentration in the carrier entering the channel and the centre-of-gravity of the solute is given by :

$$\frac{x_{cg}}{w} = \frac{3 + \phi_s (K-1)}{6 + \phi_s (K-1) / \lambda_s} \tag{116}$$

where ϕ_s is the "support" volume fraction in the entering carrier, λ_s the reduced thickness of the "support" particle cloud and K the solute distribution coefficient between the "support" particles and the continuous phase. The separation of small molecules in this mode has been demonstrated by means of micro-emulsion "support" particles in sedimentation FFF (Berthod et al. 1988).

9.6. FFF WITH WALL INTERACTIONS

In the basic retention theory of the Brownian mode, it was implicitly assumed there were no particle interactions with the walls of the channel. However, as particles are pushed near the accumulation wall by the force field, this assumption is no longer correct. The potential energy of the particle then includes the particle-wall interaction contribution, which modifies the concentration profile and, consequently, the retention ratio. They are calculated by means of equations 18 and 48, respectively, provided that the potential energy profile W(x) can be known.

9.6.1. Electrostatic Particle-Wall Interactions.

9.6.1. Electrostatic Particle-Wall Interactions. Two kinds of electrostatic interactions are most commonly encountered : surface charge interactions and van der Waals interactions. Colloidal materials are electrically charged by nature. So are the channels walls. The particles and the walls generally bear surface charges of the same sign and this charge-charge interaction is repulsive. It is strongly influenced by the ionic strength of the carrier solution. Indeed, high ionic strength solutions compress the electrical double layers on each surface and thus reduce the distance over which this repulsive interaction is active.

The van der Waals interactions or London dispersion forces are associated with electrostatic interactions between instantaneous multipoles resulting from the non-coordinated displacements of the electrons around the atomic nuclei of the two bodies. They are always attractive and independent of the solution ionic strength but depend on the polarizabilities of the materials. For macroscopic bodies they are characterized by the Hamaker constants of the materials.

The DLVO (Derjaguin-Landau-Verwey-Overbeek) theory characterizes the net potential energy resulting from the combination of these two kinds of electrostatic interactions. At low ionic strengths, the charge repulsion dominates the dispersive interactions and the particle seeks an equilibrium position, corresponding to a minimum in the potential energy, relatively far from the wall. At high ionic strengths, the coulombic screening reduces the repulsive potential so that the net potential may become positive, leading to deposition of the particle at the wall.

These interactions modify the concentration profile. For this reason, one can consider they lead to another mode of FFF, although they are, in practice, present in nearly all FFF systems where particles are in the vicinity of the walls. The equations for the repulsive and attractive potentials are known but their dependence on the particle position is too complex to provide analytical expressions for the concentration profile and for the retention ratio. They have been computed numerically for typical experimental conditions (Hansen & Giddings 1989b, Mori et al. 1990). It is found that, instead of continuously decreasing from the accumulation wall, the concentration profile has a peak close to this wall with a depletion layer in the immediate vicinity of the wall. The amplitude of the concentration peak increases and the thickness of the depletion layer decreases with increasing solution ionic strength. The influence of these interactions on retention and band broadening with and without applied field have been calculated by considering the region where these interactions are active as a chromatographic-like "stationary phase" (Semyonov 1988b).

9.6.2. Potential Barrier FFF.

9.6.2. Potential Barrier FFF. By similarity to the technique of potential barrier chromatography which relies solely on particle-wall interactions, the technique of potential barrier FFF has been proposed to separate particles in an FFF channel with a force field according to the strength of their interactions with the wall (Hansen & Giddings 1989b, Karaiskakis & Koliadima 1989). In this technique, particles are adsorbed on the channel wall when the ionic strength of the carrier liquid is larger that some critical value for which the peak in the potential energy profile near the wall is lower than the activation energy of adsorption. Desorption and elution of the particles is performed by changing the carrier composition to one of lower ionic strength. This provides a means for separating particles according to the corresponding critical ionic strength for desorption as well as for studying adsorption of particles on solid surfaces.

9.6.3. Hybrid FFF/chromatography.

9.6.3. Hybrid FFF/chromatography. The idea of combining FFF with chromatography originated from the conception of FFF (Giddings 1966). Due to the large varieties of techniques available for both chromatography and FFF, the number of possible combinations of the two methods is quite large. One possibility, combining affinity chromatography and gravitational sedimentation FFF has been recently explored (Bigelow et al. 1989). The accumulation wall is coated with an affinity material on which cells are allowed to adhere. The particles are selectively detached from the wall by the shear forces which are constant all along the channel length. Flow programming allows the recovery and separation of different cell populations according to their binding forces to the surface.

10. Selectivity and Sample Characterization

10.1. SELECTIVITY

The ability of a system to separate material according to some parameter of interest, such as particle diameter or molecular weight, may be described in terms of its selectivity (Giddings 1979b) which is defined by

$$S_\phi = \left| \frac{d \ln t_r}{d \ln \phi} \right| = \frac{\phi}{t_r} \left| \frac{d t_r}{d \phi} \right| \tag{117}$$

where ϕ is the parameter of interest and S_ϕ is the ϕ-based selectivity. It is simply the slope of a plot of the logarithm of elution time *versus* the logarithm of ϕ. The selectivity is always given as a positive quantity. For significant retention in Brownian mode FFF the selectivity is equal to the inverse power dependence of λ on the parameter of interest, e.g., $S_d = 3$ and $S_m = 1$ for sedimentation FFF, $S_d = 1$ and $S_m = 1/3$ for spherical particles in flow FFF. The logarithmic plots of t_r *versus* ϕ also happen to be linear over wide retention ranges in steric and lift modes of FFF. In sedimentation/steric FFF S_d is generally a little less than unity (Giddings et al. 1991c, Williams et al. 1992b), whereas in lift mode flow FFF (i.e., flow/hyperlayer FFF) S_d is usually greater than unity (Giddings et al. 1987b, Ratanathanawongs & Giddings 1989). Thermal FFF typically displays a molecular weight-based selectivity S_M of around 0.6. This compares to S_M of around 0.1 for size exclusion chromatography (Gunderson & Giddings 1986).

10.2. TRANSFORMATION OF FRACTOGRAM TO SAMPLE DISTRIBUTION

The requirement for a monotonic variation of elution time with the sample parameter of interest ϕ has already been pointed out. Consequently it is important that conditions are adjusted such that all sample components are eluted either in Brownian or in steric or lift modes (i.e., in either the so-called normal or steric order). The transformation from the time scale of the fractogram to the ϕ scale of the desired distribution first requires the association of ϕ values with a set of discrete elution times. If these discrete times have a small constant interval δt_r then the mass eluted during one such interval at time t_r will be equal to $c(t_r) \dot{V} \delta t_r$, where $c(t_r)$ represents the mass concentration in the outlet stream at time t_r, and $\dot{V}$ is the volumetric channel flow rate. It is assumed for the moment that a component of discrete ϕ is associated with each discrete t_r (i.e., system band spreading is negligible). A mass distribution with respect to ϕ will indicate the abundance of material for the interval $\delta\phi$ corresponding to that particular δt_r. Integrating the mass distribution $m(\phi)$ over the interval $\delta\phi$ and equating with the known mass eluted in interval δt_r results in

$$m(\phi) \, \delta\phi = c(t_r) \, \dot{V} \, \delta t_r \tag{118}$$

which, in the limit of $\delta t_r \to 0$, may be transformed to

$$m(\phi) = c(t_r) \dot{V} \left| \frac{dt_r}{d\phi} \right| \tag{119}$$

The modulus of the differential $dt_r/d\phi$ is considered so that the transformation is equally valid for the steric and lift modes.

Brownian mode elution time is predictable from first principles but steric or lift mode elution generally requires a calibration procedure involving the construction of a plot of $\ln t_r$ versus $\ln \phi$ for a set of narrow standards. From equations 117 and 119 it may be shown (Giddings et al. 1991c) that for these modes

$$m(\phi) = c(t_r) \dot{V} S_\phi t_{r1} \left(\frac{t_r}{t_{r1}} \right)^{(S_\phi+1)/S_\phi} \tag{120}$$

where t_{r1} is the elution time extrapolated to a component of unit ϕ. Until recently for sedimentation FFF, calibrations were carried out using standards of the same density as the unknown and under the same experimental conditions. It has been shown (Giddings et al. 1991c), as mentioned earlier in section 5.1, that standards do not have to be of the same density provided field strength is adjusted to compensate for the difference, i.e., such that the field-induced force is the same for particles of equal size but different density. This allows greater freedom in selecting a suitable range of standard materials, and allows analysis of unknowns having densities that do not correspond to any standard materials.

If the detector responds linearly to mass concentration then the relative mass distribution with respect to ϕ is obtainable directly through application of equation 119 to the recorded fractogram. If the response is not linear (as in the case of light scattering by particles) a correction must be applied. The response from a detector that senses the number concentration of particles may be transformed to a mass concentration by multiplying by the particle mass eluting at each discrete time.

10.3. LIGHT SCATTERING CORRECTION

The detectors used in FFF are commonly of the UV/visible type employed in HPLC. Particulate materials are detected via a combination of light scattering and absorption. When particles are of the same order of magnitude as the detector wavelength, or smaller, the beam is attenuated principally by the scattering process. The scattered intensity is highly dependent on the particle size relative to the wavelength, its refractive index and the angle to the incident beam as described by Mie theory (see Kerker 1969 and Bohren & Huffman 1983). If mass concentration is required the signal has to be corrected accordingly (Schure et al. 1989).

When particles are large relative to the wavelength, as is usually the case in the steric modes of FFF, the angular distribution of scattered light is strongly weighted in the forward direction and the acceptance angle of many common HPLC detectors is such that light is effectively unattenuated by scattering. The dominant mechanism for attenuation is then absorption which is proportional to the summed area of cross section of particles in the beam. The result is that the detector gives a surface area response. This is easily transformed to a mass or number distribution as desired (Giddings et al. 1991c).

10.4. CORRECTION FOR SYSTEM BAND SPREADING

It was mentioned in section 6.3 that the dominant contribution to band spreading in Brownian mode FFF is due to nonequilibrium effects. This constitutes a non-selective band spreading process. The system selectivity also causes broadening of a sample band that is not absolutely monodisperse. All other contributions can be made negligible in practice. The overall apparent plate height for a narrow particulate or soluble polymer standard is then given by

$$H_{app} = \frac{\chi w^2}{D} <v> + H_p \tag{121}$$

where the first term represents the contribution due to nonequilibrium effects and H_p is a contribution due to small but finite polydispersity. A plot of H_{app} *versus* $<v>$ should then result in a straight line with intercept close to H_p. Note that this approach makes no assumptions regarding the veracity of the expressions for χ and D.

For polymeric solutes equation 74 relates H_p to the polydispersity index μ. This provides a method of determining values for μ for very narrow standards since the intercept is magnified by the relatively high selectivity of the FFF subtechniques (Schimpf et al. 1987b).

For particulate materials the following expression for H_p has been derived (Ratanathanawongs & Giddings 1989)

$$H_p = L S_d^2 \left(\frac{\sigma_d}{d} \right)^2 = L S_d^2 (C.V.)^2 \tag{122}$$

where C.V. represents the coefficient of variation, or the relative standard deviation in particle diameter.

For the more typical polydisperse samples this approach will not work because 1) the nonequilibrium band spreading contribution to H, being such a strong function of λ, will vary across the width of the eluted band, and 2) very often an accurate particle size or molecular weight distribution is required rather than merely an estimate of polydispersity. In such cases we need to turn to techniques of deconvolution.

The convolution of the ideal distribution on the time axis $W(t_r)$, for example, with a function describing the system band spreading $G(t, t_r)$, results in the observed elution profile $F(t)$, as described by the convolution integral

$$F(t) = \int_0^\infty W(t_r) \, G(t, t_r) \, dt_r \tag{123}$$

The reverse process of removing the effects of $G(t, t_r)$ from $F(t)$ to obtain $W(t_r)$ is known as deconvolution. Various methods have been developed to achieve this. Fourier methods are very successful when the band spreading function does not vary. Iterative methods are required when the function is not constant across the response function $F(t)$ (see, for example, Jansson 1984). These iterative methods entail the convolution of $G(t, t_r)$ with an estimate to $W(t_r)$, comparison of the result with $F(t)$, followed by adjustment of the estimate $W(t_r)$ to minimize the difference between successive results and $F(t)$. Gold's ratio method (Gold 1964) has been successfully applied to data treatment in size exclusion chromatography (Ishige et al. 1971), sedimentation FFF (Jahnová et al. 1987) and thermal FFF (Schimpf et al. 1989b). In these procedures FFF holds the advantage over SEC in that the dominant contribution to band spreading is predictable from first principles. This is true even in the case of thermal FFF where an elution time is characterized by some value of λ (taking into account the deviations from parabolic flow profile found in this system). The bandspreading function may then be obtained using this λ together with values for various system parameters (Schimpf et al. 1989b) (again accounting for the deviation from the parabolic flow profile as explained in section 6.4 (Martin & Giddings 1981a)). More sophisticated methods such as that due to Jansson (1984) have also been employed (Schure et al. 1989). This work involved isocratic sedimentation FFF of particulate samples and included a light scattering correction at each iteration. These methods will hopefully allow the reduction of analysis times by the use of high flow rates in conjunction with field programming, with almost full recovery of information on particle size distribution.

11. Programming in FFF

The need for programming was quickly realized for normal mode elution of polydisperse samples (Yang et al. 1974). The relative displacement velocities of species differing slightly in some selective quantity f (where f is commonly diameter d, molecular weight M, etc.) is described by the selectivity S_ϕ given by $| d \ln t_r / d \ln \phi |$ (see section 10.1). Significantly retained material in Brownian mode flow FFF exhibits diameter-based selectivity S_d of unity but for Brownian mode sedimentation FFF $S_d = 3$. A polydisperse sample covering a 10:1 ratio of diameters may be fractionated using flow FFF in a reasonable time whereas sedimentation FFF would elute the material over a 1000:1 retention time range. The conditions must be set to retain the smallest components sufficiently for their resolution. The largest components would therefore be excessively retained and would be relatively difficult to detect. The difficulty in detection is due to the high selectivity of the technique which results in the spreading of the continuous distribution of material across the elution time scale. Equation 119 indicates that mass concentration of the eluent will be proportional to $m(\phi) / | d t_r / d \phi |$. In the case of Brownian mode sedimentation FFF at constant field and flow, it follows that $c(t_r) \propto m(d)/d^2$. It is the influence of field programming on $| d t_r/d d |$ as a function of d that results in improved detection of the larger components. We note that in theory the elution of absolutely monodisperse materials under constant field conditions will not suffer from this detectability problem. In fact, since the major contribution to band spreading decreases with increasing retention (see equations 83 and 128), detectability would tend to increase with increase of retention. The programmed reduction of field strength or increase of channel flow rate, or their simultaneous programming (dual programmed FFF), may be used to ensure sufficient retention of the smallest components while speeding the elution of larger components. Note that the programming of channel flow rate alone may bring about the desired reduction in analysis time (Giddings et al. 1979c) but will not alleviate the problem of detectability of the later components. Other system parameters having an influence on retention, such as the carrier fluid density in sedimentation FFF, may also be programmed (Yang et al. 1974, Giddings & Caldwell 1984b). The following discussion will be restricted to field programmed FFF.

11.1. FIELD PROGRAMMING FOR BROWNIAN MODE

Field decay programs were initially selected in a fairly arbitrary manner. The principal objective was to elute all components of a sample within some acceptable time having first retained and fractionated the smallest components of interest at some initial high field strength. Gradual reduction of the field strength over some period of time to a very low level, or even to zero, ensured the elution of even the largest components in a comparable time. The tools for selecting an optional functional form for the programmed decay, i.e., the function describing the variation of field strength with time, were not available at this time. For the arbitrary programs examined particle diameter or molecular weight could be related to retention time through solution of the perfectly general integral equation (Yang et al. 1974)

$$L = \int_0^{t_r} R <v> dt \tag{124}$$

where L is the length of the channel, t_r is the retention time for a species whose retention ratio R varies with time due to the changing field strength, and $<v>$ is the mean channel flow velocity, which may also be programmed but for this discussion of field programming will be considered constant.

11.1.1. *Fractionating Power.* It was the introduction of the concept of fractionating power that allowed the determination of how well components of a polydisperse sample are resolved under isocratic or programmed conditions (Giddings et al. 1987c). Only then did it become possible to compare program types in terms of their ability to fractionate broad samples. Fractionating power is a measure of relative resolution and is defined as the resolution for two closely eluting species divided by their relative difference in the selective quantity of interest ϕ. The diameter-based fractionating power F_d is therefore given by

$$F_d = \frac{R_s}{\delta d/d} = \frac{d}{4\sigma_t} \frac{\delta t_r}{\delta d} \tag{125}$$

where R_s is the resolution for components differing by δd in diameter (having mean diameter d) and consequently differing by δt_r in retention, and σ_t is the standard deviation in retention. In the limit of $\delta d/d \to 0$ we have

$$F_d = \frac{d}{4\sigma_t} \frac{dt_r}{dd} = \frac{t_r}{4\sigma_t} \frac{d\ln t_r}{d\ln d} = \frac{t_r}{4\sigma_t} S_d \tag{126}$$

The molecular weight-based fractionating power F_M is defined in an equivalent manner (Giddings et al. 1990c). In fact a fractionating power may be defined for any selective quantity.

For the Brownian mode of elution in FFF the dominant contribution to nonselective band spreading is due to the nonequilibrium effect which in terms of plate height is given by equation 79 (Giddings et al. 1975, Karaiskakis et al. 1981). Substituting equation 79 for H in equation 65 and integrating over the length of the channel results in

$$\sigma_L^2 = \int_0^L \frac{\chi w^2 <v>}{D} dz = \frac{w^2}{D} \int_0^{t_r} \chi R <v>^2 dt \tag{127}$$

The standard deviation in retention σ_t is obtained from σ_L by dividing by the velocity of the band as it elutes from the system, i.e.,

$$\sigma_t = \frac{\sigma_L}{R_r <v>} = \frac{w}{D^{1/2} R_r} \left(\int_0^{t_r} \chi R \, dt \right)^{1/2} \tag{128}$$

where R_r is the retention ratio of a component at its elution time t_r. For systems where the flow profile is effectively parabolic, χ is a function of λ alone (see equation 82). In thermal FFF the velocity profile is dependent on the instantaneous temperature drop across the channel (Gunderson et al. 1984) and χ is therefore dependent on both λ and the third-degree velocity profile parameter ν

(see equation 90). In either case χ is a known function of time for each component, and equation 128 may be evaluated for any species for which equation 124 has been solved. The solution of equations 124 and 128 allow calculation of F_d or F_M via equation 125 for any species eluting under field programmed conditions.

Under isocratic conditions F_d was shown to increase with $(D/\lambda^3)^{1/2}$, i.e., with d raised to the power $(3n-1)/2$ (assuming D is given by the Stokes-Einstein equation) where n = 1 for flow FFF and 3 for sedimentation FFF. We see that F_d increases linearly with d in flow FFF, but with d^4 in sedimentation FFF. All programmed runs show such an initial rise in F_d for those components that are just retained at the initial field strength. The variation of F_d with d deviates from this as the field starts to decay. For thermal FFF under conditions of constant temperature drop the relationship between λ and D shown in equation 59, and the lack of molecular weight dependence of D_T, reveals that F_M increases only with the reciprocal of D. The advantages of programming, including the enhanced detectability and faster elution of the larger components, are however equally desirable for thermal FFF.

11.1.2. *Linear and Parabolic Field Decay.* Linear and parabolic decay programs were shown by Williams & Giddings (1987a) to result in the initial rise in F_d described above followed by a rapid decrease in F_d as field strength approached zero. Fractionating power was predicted to attain very high levels but over a very restricted range of particle diameter. Such behavior is expected for any non-asymptotic decay of field strength to zero. This is due to the fact that relative change in field strength rapidly increases as the field decays and all components that have not already been eluted are quickly swept out of the channel.

Simple linear and parabolic field decay programs (the latter corresponding to linear decay of rotation rate in sedimentation FFF) were the first to be investigated (Yang et al. 1974, Giddings et al. 1976f, Wahlund et al. 1986). The decay followed an optional period during which the field was held constant at the initial high level. Although the conditions were not optimized the results showed a great improvement over isocratic operation.

11.1.3. *Exponential Field Decay.* Yau & Kirkland (1981) and Kirkland & Yau (1981, 1982a) introduced the exponential decay program which yielded a simple linear relationship between the elution time and the logarithm of particle diameter. The field decay follows the equation

$$S(t) = S_0 \exp\left(-\left(\frac{t-t_1}{\tau'}\right)\right) \tag{129}$$

where S(t) is the field strength at time t, S_0 is the initial field strength, t_1 is the period preceding decay during which the field is held constant, and τ' is the decay constant. The linear relationship mentioned above was shown to be valid for a wider range of particle size when the initial period t_1 was set equal to the decay constant τ', and this special case was refered to as the *time-delayed exponential (TDE) program.*

This program form proved to be a great improvement over linear and parabolic field decay and has been widely used for sedimentation FFF (Kirkland et al. 1980, Kirkland & Yau 1981, Yau & Kirkland 1981), thermal FFF (Kirkland et al. 1988) and flow FFF (Wahlund et al. 1986). The detectability problem for the more strongly retained components of polydisperse samples was solved. In the case of sedimentation FFF it may be shown that for the later eluting components $c(t_r) \propto d\,m(d)$, so that in fact detectability is seen to be enhanced.

The exponential decay program was shown by Giddings et al. (1987c) to result in a level of F_d that, following an initial increase, gradually fell with increasing particle diameter. Fractionating power was predicted to fall with the reciprocal of $d^{1/2}$. Predicted fractionating powers have been confirmed experimentally for narrow particle standards eluted under exponential field decay in sedimentation FFF (Williams et al. 1988).

11.1.4. *Power Programmed Field Decay.* With the object of avoiding the decline in F_d given by the exponential program a new program function was proposed by Williams & Giddings (1987b), having the form

$$S(t) = S_0 \left(\frac{t_1 - t_a}{t - t_a} \right)^p \tag{130}$$

where t_1 is again a period of constant initial field strength S_0, t_a is a second time parameter, and p is a real power. It is required that $t > t_1 > t_a$ and $p > 0$ for programmed field decay. It is refered to as the *power program* due to the power-law dependence of the field decay. As for exponential programming of field strength, the detectability of the later eluting components of a polydisperse sample is enhanced, with $c(t_r) \propto d^{2/3} m(d)$ in the case of sedimentation FFF. Such a program was shown to result in constant F_d over a wide range of d when $p = 3n-1$, i.e., $p = 8$ for sedimentation FFF, and $p = 2$ for flow FFF. In addition, the range is extended towards smaller d when $t_a = -pt_1$. For polydisperse samples this behavior is of course highly desirable. The level of F_d and the lower limit on the range of d eluted at this F_d are easily fixed by adjustment of S_0 and the ratio t_1/t^0 where t^0 is the channel void time. The differing features of the power and exponential programs have been discussed recently by Williams & Giddings (1991).

For thermal FFF, a power programmed decay of the temperature drop ΔT has been shown to result in constant molecular weight-based fractionating power F_M when p is close to 2 (Giddings et al. 1990c).

11.2. SECONDARY RELAXATION EFFECT

The quasi-equilibrium exponential concentration profile for a band migrating under steady-state conditions was described in section 2.2. When field strength decays according to some program there will be a continuous change in the concentration profile. The concentration profile will tend to lag behind that steady-state distribution corresponding to the instantaneous field strength due to the finite time required for the relaxation of the band. There will also be an accompanying distortion in the concentration profile. The change in concentration profile to accommodate a change in field strength is termed *secondary relaxation* to distinguish it from the *primary relaxation* that occurs (often under stopped-flow conditions) at the head of the channel.

Yau and Kirkland (1984) recognized this potential problem and derived expressions based on nested sums of infinite series for predicting the effect on the migration velocity of a band. Giddings (1986b) was later able to derive a first order correction to the concentration profile to account for the perturbation due to secondary relaxation, and from this a first order correction for the instantaneous retention ratio of a band. This approach allows the calculation of a corrected retention time for a band eluted under any set of programmed conditions, provided deviation from steady-state is not too great. An expression for the corrected retention time under exponential field decay has been obtained (Hansen et al. 1988), together with an expression for the relative error in particle diameter that would be expected should this effect be ignored.

11.3. PROGRAMMING FOR LIFT MODE FLOW FFF

The simultaneous programming of both cross flow and channel flow (dual programming) in steric-hyperlayer flow FFF has been shown to extend the range of particle size that may be analyzed in a single run (Ratanathanawongs & Giddings 1992). The equilibrium position of a hyperlayer for a given particle size is determined by the balance between the drag of the cross flow and the hydrodynamic lift forces. The drag is proportional to the cross flow rate whereas the lift is a function of the channel flow rate, among other things. Without the use of programming there is a tendency for the smaller particles to become immobilized. The gradual reduction in cross flow and increase in channel flow raises these particles to higher elevations and ensures their elution.

12. Split Flow in Thin Channels

12.1. FLOW SPLITTING

The success of the techniques described here depends on the predictable laminar flow that takes place within thin parallel plate systems. As explained earlier, when the ratio of width to thickness of the channel is high, a uniform parabolic flow velocity profile is very quickly set up across the thickness with only small perturbations near the channel edges where the flow velocity approaches zero. The bulk of the flow is therefore simply described by a two-dimensional parabolic velocity profile that is uniform across the third dimension (the width).

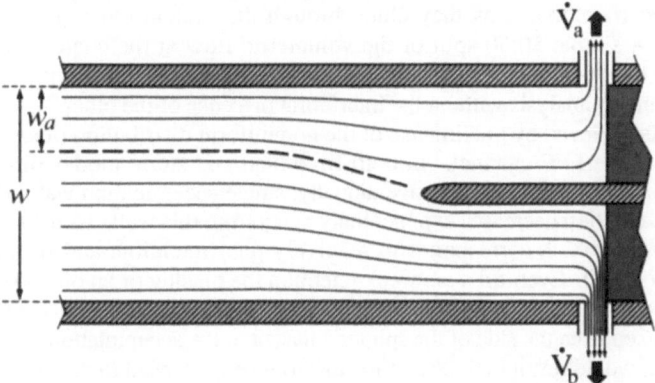

Figure 11. Edge view of split outlet flow pattern. The splitting plane is shown as a broken line at a distance w_a from the upper wall.

If the flow is divided by a thin rigid knife-edge, or "flow-splitter", set parallel to the major walls, then all fluid elements flowing below a so-called splitting plane within the main body of the system will be swept below the splitter while those elements flowing above the plane will be swept above the splitter. The splitter may be placed near the channel outlet and the fluid flowing to either side conveyed to two separate outlets. It is the ratio of the volumetric outlet flows that determines the position of the splitting plane relative to the plane channel walls. Generally, for practical facility, a splitter is placed at the mid-point of the channel thickness, but the splitting plane for the outlet streams (the outlet splitting plane) may be positioned at any desired point by adjusting the ratio of the outlet flows.

A schematic edge view of a split outlet flow pattern is shown in Figure 11. The outlet flow rates $\dot{V}_a$ and $\dot{V}_b$, which together make up the total channel flow rate $\dot{V}$, fix the outlet splitting plane at a distance w_a from the upper wall. The position of the splitting plane is given by (Williams et al. accepted)

$$\frac{w_a}{w} = \sin\left(\frac{\theta}{3}\right) + \frac{1}{2} \qquad (131)$$

where

$$\sin\theta = (2\dot{V}_a/\dot{V}) - 1 \qquad (132)$$

Fluid may be introduced to the system on each side of a splitter at the channel inlet. The ratio of the inlet flow rates determines in the same way the position of an inlet splitting plane.

12.2. SPLIT FLOW FFF

The splitting of flow as described above was first utilized at the outlet of FFF channels for obtaining 1) a binary separation over and above the fractionation brought about by FFF (Jones et al. 1987), and 2) a concentration of the outlet stream carried to the detector (Giddings et al. 1983b). The first effect may be realized, for example, in the case of sedimentation FFF of a complex sample consisting of two materials differing in density. If the carrier is adjusted to some intermediate density then the two types of material will be driven to opposite walls by the applied field. The two materials will then be fractionated as they elute through the system close to their respective accumulation walls. A simple 50:50 split of the volumetric flow at the channel outlet may be employed to carry the two fractionated materials to individual detectors. Fractions of each material may be collected for further analysis without the interfering presence of the other.

The second effect is achieved by making use of the nonuniform distribution of sample material across the thickness of the FFF channel. In both Brownian and steric modes of operation the concentration is greatest at or close to the, conventionally, single accumulation wall. For Brownian mode elution the concentration decreases exponentially away from this wall. Consider for example a retention ratio of 0.480 which corresponds to relatively poor retention, and for the Brownian mode is given by a λ of 0.10. Even for such weak retention the placing of an outlet splitting plane at $x/w = 0.25$ (obtained by setting the ratio $\dot{V}_a/\dot{V} = 0.156$ (Williams et al. accepted), where $\dot{V}_a$ is the volumetric flow rate taken from the side of the splitter adjacent to the accumulation wall and $\dot{V}$ is the total channel flow rate) would result in 91.8% of the material being carried to the detector by 15.6% of the channel flow. An enrichment factor of 5.9 is obtained with loss of only 8.2% of the sample. More strongly retained material would have enrichment factors approaching 6.4 with smaller losses at this split ratio, as would all sterically migrating species which are confined to very narrow regions close to the accumulation wall.

Split flow at the FFF channel inlet was later used for the hydrodynamic relaxation of sample towards the accumulation wall (Lee et al. 1989). The procedure was briefly mentioned in section 6.2.3. The sample stream (introduced adjacent to the accumulation wall) is merged with a comparatively large flow of pure carrier. This has the effect of confining the sample to a relatively thin region at the head of the channel and little further relaxation is required to attain an equilibrium concentration profile. The relatively slow flow of the finite sample volume into the channel gives rise to a band spreading effect when it merges with the other inlet stream. This effect increases with increase of inlet split ratio whereas the contribution from residual relaxation decreases. An expression for the optimum inlet split ratio has been derived (Liu et al. 1991b).

12.3. SPLITT CELLS

Split-flow thin, or SPLITT, cells are dimensionally similar to FFF channels, and operate with orthogonal field and flow vectors (Giddings 1985b). Their operation differs fundamentally from FFF in that separation across the thickness of the channel is exploited while differential displacement in the flow direction is not. The latter property allows for separations to take place in continuous mode. The fact that separations take place across the thin dimension of the system limits the degree of separation (generally binary separations are carried out) but the separation achieved can be extremely fast. Greater than binary separation could be obtained by splitting the outlet flow into three or more streams with the use of two or more splitters, but this would involve practical difficulties in maintaining integrity and exact parallel placement of the splitters and would inevitably lead to undesireable increase in thickness of the SPLITT cell. An alternative and recommended approach is to carry out a binary separation in a simple binary SPLITT cell and then feed one or both of the outlet streams to additional cells for further separation. The outlet streams of these can be more finely fractionated by the addition of further successive SPLITT cells.

Two modes of operation for SPLITT cells are possible, namely the *equilibrium mode* and the *transport mode*. These are described below.

12.3.1. *Equilibrium Mode.*
In these systems the species to be separated are driven towards opposite walls of the cell (Levin et al. 1989). If sufficient time elapses for the species to be driven to each side of an outlet splitting plane then separation will be complete, the species exiting via opposite outlet streams. This mode of operation therefore strictly requires an outlet splitter only. Conventional SPLITT cells with both inlet and outlet splitters are commonly used however. One or other of the inlets may optionally be closed. The required time for separation may be obtained by adjustment of the system flow rate and/or the strength of the field. A conventional electrical SPLITT cell used in equilibrium mode is shown in an edge view in Figure 12.

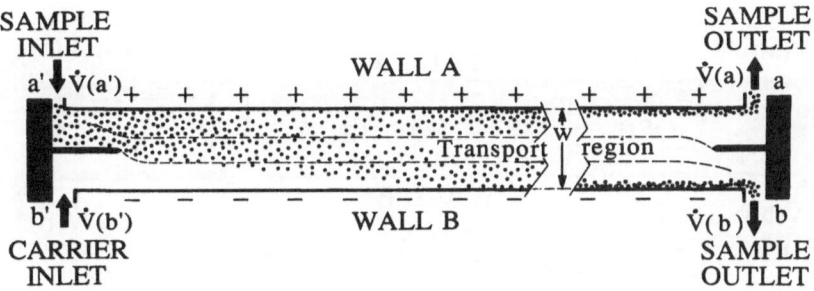

Figure 12. Conventional electrical SPLITT cell used in equilibrium mode. The inlet and outlet splitting planes are shown as dashed lines. Reprinted from Levin et al. 1989, courtesy of Marcel Dekker, Inc.

566

In this figure the two species achieve equilibrium distributions at opposite walls of the cell, but as mentioned above it is sufficient that the distributions fall exclusively to each side of the outlet splitting plane during transport through the cell.

The opposed action of sedimentation and an upwards cross flow has been proposed for SPLITT cell operation (Giddings 1986c). This is a form of elutriation which, because of the relatively uniform upward flow velocity in comparison with conventional elutriation systems, and the very small transport distances, should result in very rapid separations with high resolution. A uniform cross flow obtained by maintaining equal input and output of cross flow rate would result in an equilibrium mode separation where all particles larger than a critical size would fall to the lower outlet and those smaller than the critical size would be swept to the upper outlet. A nonuniform cross flow velocity may be obtained for example by maintaining a higher cross flow rate into the cell than is withdrawn. In this case various particle sizes would be carried to hyperlayers within the channel thickness. The opposition of sedimentation and hydrodynamic lift forces resulting in nonuniform net forces on the particles and consequent hyperlayer formation has also been proposed (Giddings 1988c).

12.3.2. *Transport Mode.* For this mode of operation use is made of differing transport rates of species from one wall towards the other (Springston et al. 1987). This requires that the sample stream be introduced to the system via a split inlet so that all species are initially confined to a thin region close to one wall. This is achieved by merging a relatively slow sample stream $\dot{V}_a$, with a flow of pure carrier $\dot{V}_b$, introduced at the opposing inlet. During transport through the cell, species are driven towards the opposite wall either by the influence of some applied external field, or by simple diffusion from the initial region of high concentration into regions of lower concentration. In the case of diffusional transport, only partial separations are possible. Such a system is shown again in edge view in Figure 13. This shows a special case where a macromolecular species is driven by diffusion across the cell thickness at a slow enough rate that the majority does not migrate beyond the outlet splitting plane (OSP). At the same time a particulate species falls under the influence of gravity through the transport region and is carried to outlet b. A continuous mode separation is achieved (Levin & Giddings 1991).

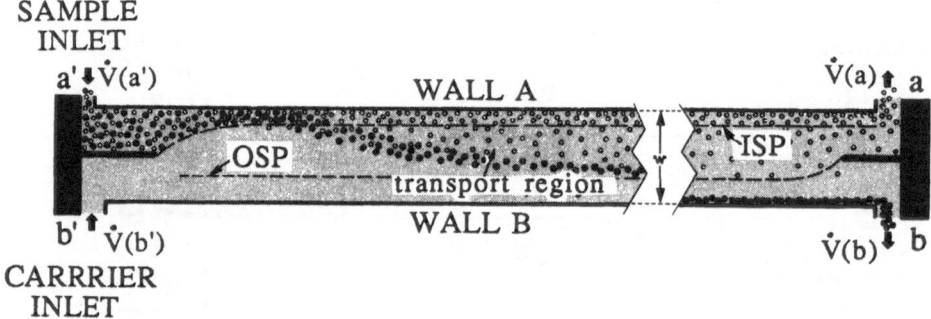

Figure 13. SPLITT cell operated in transport mode. The two components are driven by diffusion and by gravitation, respectively. Reprinted from Levin & Giddings 1991, courtesy of Society of Chemical Industry.

The sample is introduced as flowstream $\dot{V}_{a'}$ and initially occupies the zone above the inlet splitting plane (Zone 1). Only material that traverses the transport region between splitting planes during migration through the cell will exit in flowstream $\dot{V}_{b}$. All other material will exit at outlet a. It is a simple matter to determine the fraction of a given material migrating under the influence of an externally applied field that will exit each outlet (Springston et al. 1987). The fraction of a material undergoing diffusional transport exiting each outlet is obtainable by numerically solving the diffusion-convection equations (Williams et al. accepted).

12.3.3. *Applications*. To date particle separations have been reported for sedimentation under one gravity in the transport mode (Springston et al. 1987) and in the equilibrium mode (Barman - unpublished data, and see Giddings 1988d). Particles have been separated from macromolecules in solution using the transport mode as described above (Levin & Giddings 1991). Electric fields have been used to separate proteins (Levin et al. 1989) where the equilibrium mode was achieved by the use of a carrier buffered to a pH between the pI's of the two proteins.

13. Acknowledgements

The material contributed by PSW is based upon work supported by the National Science Foundation under Grant No. CHE-9102321.

14. List of symbols

a	radius of a spherical particle
A	intermediary term in equation 82
A'	intermediary term in equation 90
A"	intermediary term in equation 90
b	channel breadth
B	proportionality coefficient of $1/\mathcal{V}$ in equation 101
c	number concentration of particles
c*	equilibrium number concentration of particles
c_0	equilibrium number concentration of particles at the accumulation wall
c_{ref}	reference concentration (equation 94)
$c(t_r)$	mass concentration of sample in outlet stream at time t_r
<c>	cross-sectional average number concentration of particles
C	proportionality coefficient of $\mathcal{V}$ in equation 101
C'	constant defined by equation 35
C.V.	coefficient of variation $(=\sigma_d/d)$
d	particle diameter
δd	small interval in d
$(dT/dx)_c$	temperature gradient at cold wall in thermal FFF
$\mathcal{D}$	effective axial dispersion coefficient
D	particle diffusion coefficient
Dn	Dean number
D_T	thermal diffusion coefficient
e	charge on the electron

E	electrical field strength
f	function of adiabatic volume compressibilities and densities of particle and carrier (equation 63)
f	particle friction coefficient
f	band broadening factor (equation 88)
F	force exerted by the field on the particle
F_d	diameter-based fractionating power (equation 125)
F_l	lift force
F_M	molecular weight-based fractionating power
F_n	net force
F(t)	observed fractogram (equation 123)
G(x/w)	functional dependence of the lift force with the lateral position (equation 113)
$G(t, t_r)$	system band spreading function (equation 123)
h	velocity correction function (equation 31)
H	plate height
H_{app}	apparent plate height
H_i	instrumental contribution to the plate height
H_{ld}	longitudinal diffusion plate height
H_{min}	minimum plate height
H_n	nonequilibrium (or hydrodynamic) plate height
H_p	polydispersity contribution to the plate height
H_r	relaxation contribution to the plate height
I	intermediary term in equation 100
j	edge-effect correction factor
J_x	flux of particles in the x-direction (equation 1)
k	Boltzmann constant
K	dimensionless distribution coefficient
K_i	coefficients in equation 93
ℓ	space constant of the particle exponential concentration profile (equation 4)
$\mathcal{L}$	distance traveled during one random walk step
L	channel length
m(ϕ)	mass distribution with respect to ϕ
M	molecular weight or mass
δM	difference in mass or molecular weight
n	number of steps of a random walk
n	inverse power dependence of λ on selective parameter ϕ
	e.g., $n = 3$ for $\phi = d$ in sedimentation FFF
	$n = 1$ for $\phi = d$ in flow FFF
	$n \approx 0.6$ for $\phi = M$ in thermal FFF
n'	number of moles
N	number of theoretical plates
$\mathcal{N}$	Avogadro number
$\dot{N}$	rate of generation of plates
$\dot{N}_{max}$	maximum rate of generation of plates
p	power parameter for power programmed field decay (equation 130)
P	pressure
ΔP	pressure drop

r_0	radius of sedimentation FFF channel = curvature radius of the channel centerline
$\mathcal{R}$	ideal gas constant
R	retention ratio
Re	axial Reynolds number
Re_c	cross-flow Reynolds number
Re_p	particle Reynolds number
R_p	retention ratio for parabolic flow profile
R_r	retention ratio of component at time of elution
R_s	resolution
S	field strength (acceleration due to gravity or centrifugation)
S_d	diameter-based selectivity
S_m	mass selectivity
S_M	molecular weight selectivity
S_0	initial field strength (centrifugal acceleration)
$S(t)$	field strength as function of time in programmed FFF
S_ϕ	selectivity based on sample parameter ϕ
t	time
t^0	average eluent residence time
t_a	time parameter for power programmed field decay (equation 130)
t_d	duration of one random walk step
t_r	particle retention time
δt_r	interval in retention time
t_{r1}	retention time extrapolated to component of unit ϕ (equation 120)
t_1	period preceeding field decay in programmed FFF
Δt_r	difference in retention time
t_{rel}	relaxation time
T	absolute temperature
T_c	cold wall temperature in thermal FFF
u	ultrasonic oscillating velocity amplitude in acoustic FFF (equation 63)
U	field-induced velocity (in the x-direction)
v	carrier fluid velocity
$\bar{v}_s$	partial specific volume of sample component
$<v>$	cross-sectional average carrier fluid velocity
$<v>_{opt}$	optimum eluent velocity
$\mathcal{V}$	zone migration velocity
$\mathcal{V}_{opt}$	optimum zone migration velocity
V	volume
ΔV	potential difference across channel thickness in electrical FFF
V^0	channel void volume
V_r	particle retention volume
$\dot{V}$	volumetric channel flow rate
$\dot{V}_a$	volumetric flow rate at outlet a (sample side) of SPLITT cell
$\dot{V}_{a'}$	volumetric flow rate of sample stream at inlet a' in SPLITT cell
$\dot{V}_b$	volumetric flow rate at outlet b (opposite to sample side) of SPLITT cell
$\dot{V}_{b'}$	volumetric flow rate of carrier stream at inlet b' in SPLITT cell
$\dot{V}_c$	volumetric cross flow rate in flow FFF
w	channel thickness

w_a	distance of outlet splitting plane from wall a (sample side) of SPLITT cell
W	potential energy of the particle
W_o	potential energy of the particle at the wall
$W(t_r)$	ideal fractogram (equation 123)
x	transverse coordinate parallel to the field, distance to the accumulation wall
x_{cg}	distance of the centre-of-gravity of the particle cloud to the accumulation wall
x_{eq}	equilibrium lateral position
y	transverse coordinate perpendicular to the field
z	effective electric charge on sample component (section 5.4.)
z	axial coordinate
δz	incremental axial distance
α	thermal diffusion factor (equation 58, section 5.3.)
α	ratio of particle radius to channel thickness (equation 110, section 9.1.)
β	curvature ratio (section 3.4.3.)
β	adiabatic volume compressibility of carrier (section 5.5.)
β_s	adiabatic volume compressibility of particle (section 5.5.)
γ	coefficient of thermal expansion of sample solution in thermal FFF (equation 57, section 5.3.)
γ	drag correction factor (equation 113, sections 9.1. and 9.2.)
ϵ	equilibrium departure term
ζ	reduced transverse coordinate equal to x / ℓ
η	carrier fluid viscosity
θ	reduced concentration (equation 94, section 7.),
θ	intermediary term used in equations 131 and 132 (section 12.1.)
λ	basic FFF parameter, reduced space constant of the exponential concentration profile
λs	reduced thickness of the "support" particle cloud in FFF with secondary equilibrium
μ	molecular weight polydispersity
μ	electrophoretic mobility of sample component (equation 60, section 5.4.)
ν	third-degree velocity profile parameter
ξ	reduced axial coordinate (equation 96)
Π	osmotic pressure
ρ	carrier fluid density
$\Delta\rho$	density difference between sample component and carrier
ρ_s	density of sample component
σ_d	standard deviation in particle diameter
σ_L	standard deviation of the distribution of distances traveled by the particles at time t_r
σ_t	standard deviation of the distribution of particle residence times
σ_w	standard deviation of the weight molecular weight distribution
σ_z	standard deviation of the distribution of distances traveled by the particles
$\delta\sigma_z$	incremental standard deviation of distances traveled by the particles
τ	reduced time (equation 95)
τ'	exponential field decay constant (equation 129)
υ	particle velocity resulting from osmotic pressure gradient (equation 7)
ϕ	equilibrium departure term (section 6.3.2.)
ϕ	selective sample parameter (sections 10. and 11.)
$\delta\phi$	interval in ϕ corresponding to δt_r (equation 118, section 10.2.)
ϕ_s	volume fraction of "support" particles in carrier in FFF with secondary equilibrium

χ nonequilibrium plate height parameter (equation 79)
ψ nonequilibrium plate height parameter (equation 79)
ω angular velocity of centrifuge
ω_d size-sensitive field-type parameter (equation 109)
ω_M mass-sensitive field-type parameter (equation 108)
$<>$ cross-sectional averaging symbol (equation 23)

15. References

Afromowitz, M.A., Samaras, J.E. (1989) 'Pinch field-flow fractionation using flow injection techniques', Sep. Sci. Technol. 24, 325-339.

Andreev, V.P., Semenov, S.N., Kuznetsov, A.A., Reifsman, L.S. (1987) 'The flow fractionation methods in fields of different physical nature. II. Theoretical models of the flow fractionation process', Zh. Fiz. Khim. 61, 1-12.

Andreev, V.P., Khidekel, M.I. (1989) 'Flow fractionation in the rectangular section channel: three-dimensional model', Nauch. Apparat. 4, 123-137.

Barman, B.N., Myers, M.N., Giddings, J.C. (1989) 'Rapid particle size analysis of ground mineral by flow/hyperlayer field-flow fractionation', Powder Technol. 59, 53-63.

Barman, B.N., Giddings, J.C. (1991) 'Overview of colloid aggregation by sedimentation field-flow fractionation', in T. Provder (ed.), *Particle Size Distribution II*, ACS Symp. Series No. 472, Chapter 14.

Beckett, R., Jue, Z., Giddings, J.C. (1987) 'Determination of molecular weight distributions of fulvic and humic acids using flow field-flow fractionation', Environ. Sci. Technol. 21, 289-295.

Beckett, R., Nicholson, G., Hart, B.T., Hansen, M. (1988a) 'Separation and size characterization of colloidal particles in river water by sedimentation field-flow fractionation', Wat. Res. 22, 1535-1545.

Beckett, R., Bigelow, J.C., Zhang, J., Giddings, J.C. (1988b) 'Analysis of humic substances using flow field-flow fractionation', in P. MacCarthy and I.H. Suffet (eds.), *Influence of Aquatic Humic Substances on Fate and Treatment of Pollutants*, ACS Advances in Chemistry Series No. 219, Chapter 5.

Benincasa, M.A., Giddings, J.C. (1992) 'Separation and molecular weight distribution of anionic and cationic water soluble polymers by flow field-flow fractionation', Anal. Chem. 64, 790-798.

Berg, H.C., Purcell, E.M. (1967a) 'A method for separating according to mass a mixture of macromolecules or small particles suspended in a fluid. I. Theory', Proc. Nat. Acad. Sci. 58, 862-869.

Berg, H.C., Purcell, E.M., Stewart, W.W. (1967b) 'A method for separating according to mass a mixture of macromolecules or small particles suspended in a fluid. II. Experiments in a gravitational field', Proc. Nat. Acad. Sci. 58, 1286-1291.

Berg, H.C., Purcell, E.M. (1967c) 'A method for separating according to mass a mixture of macromolecules or small particles suspended in a fluid. III. Experiments in a centrifugal field', Proc. Nat. Acad. Sci. 58, 1821-1828.

Berman, A.S. (1953) 'Laminar flow in channels with porous walls', J. Appl. Phys. 24, 1232-1235.

Berman, A.S. (1958) 'Laminar flow in an annulus with porous walls', J. Appl. Phys. 29, 71-75.

Berthod, A., Armstrong, D.W. (1987) 'Theoretical study on the use of secondary equilibria for the separation of small solutes by field flow fractionation', Anal. Chem. 59, 2410-2413.

Berthod, A., Armstrong, D.W., Myers, M.N., Giddings, J.C. (1988) 'Use of secondary equilibria for the separation of small solutes by field-flow fractionation', Anal. Chem. 60, 2138-2141.

Bigelow, J.C., Giddings, J.C., Nabeshima, Y., Tsuruta, T., Kataoka, K., Okano, T., Yui, N., Sakurai, Y. (1989) 'Separation of B and T lymphocytes by a hybrid field-flow fractionation/ adhesion chromatography technique', J. Immunol. Meth. 117, 289-293.

Bigelow, J.C., Nabeshima, Y., Kataoka, K., Giddings, J.C. (1991) 'Separation of cells and measurement of surface adhesion forces using a hybrid of field-flow fractionation and adhesion chromatography', in D.S. Kompala and P. Todd (eds.), *Cell Separation Science and Technology*, ACS, Washington.

Bird, R.B., Stewart, W.E., Lightfoot, E.N. (1960) Transport phenomena, John Wiley, New York, Chap. 3.

Bohren, C.F., Huffman, D.R. (1983) Absorption and scattering of light by small particles, Wiley, New York.

Brenner, H. (1966) 'Hydrodynamic resistance of particles at small Reynolds numbers', Adv. Chem. Eng. 6, 287-438.

Brimhall, S.L., Myers, M.N., Caldwell, K.D., Giddings, J.C. (1981) 'High temperature thermal field-flow fractionation for the characterization of polyethylene', Sep. Sci. Technol. 16, 671-689.

Brimhall, S.L., Myers, M.N., Caldwell, K.D., Giddings, J.C. (1984) 'Separation of polymers by field-flow fractionation', J. Polym. Sci., Polym. Lett. Ed. 22, 339-345.

Brimhall, S.L., Myers, M.N., Caldwell, K.D., Giddings, J.C. (1985) 'Study of temperature dependence of thermal diffusion in polystyrene/ethylbenzene by thermal field-flow fractionation', J. Polym. Sci., Polym. Phys. Ed. 23, 2443-2456.

Caldwell, K.D., Kesner, L.F., Myers, M.N., Giddings, J.C. (1972) 'Electrical field-flow fractionation of proteins', Science 176, 296-298.

Caldwell, K.D., Nguyen, T.T., Myers, M.N., Giddings, J.C. (1979) 'Observations on anomalous retention in steric field-flow fractionation', Sep. Sci. Technol. 14, 935-946.

Caldwell, K.D., Karaiskakis, G., Giddings, J.C. (1981) 'Characterization of liposomes by sedimentation field-flow fractionation', Colloids Surfaces, 3, 233-238.

Caldwell, K.D., Cheng, Z.-Q., Hradecky, P., Giddings, J.C. (1984) 'Separation of human and animal cells by steric field-flow fractionation', Cell Biophys. 6, 233-251.

Caldwell, K.D., Brimhall, S.L., Gao, Y., Giddings, J.C. (1988) 'Sample overloading effects in polymer characterization by field-flow fractionation', J. Appl. Polym. Sci. 36, 703-719.

Cardot, P., Martin, M. (1990) 'Influence of relaxation effects on the elution behavior of particles in gravitational field-flow fractionation', Poster presented at the 18th International Symposium on Chromatography, Amsterdam, September 23-28.

Cardot, P.J.P., Gerota, J., Martin, M. (1991) 'Separation of living red blood cells by gravitational field-flow fractionation', J. Chromatogr. 568, 93-103.

Chandrasekhar, S. (1961) Hydrodynamic and hydromagnetic stability, Oxford University Press, Chap. 8.

Chen, X., Wahlund, K.-G., Giddings, J.C. (1988) 'Gravity-augmented high-speed flow/steric field-flow fractionation: simultaneous use of two fields', Anal. Chem. 60, 362-365.

Chiang, A.S., Kmiotek, E.H., Langan, S.M., Noble, P.T., Reis, J.F.G., Lightfoot, E.N. (1979) 'Preliminary experimental survey of hollow-fiber electropolarization chromatography (electrical field-flow fractionation) for protein fractionation', Sep. Sci. Technol. 14, 453-474.

Chmelík, J., Janča J. (1986) 'Sedimentation-flotation focusing field-flow fractionation in channels with modulated cross-sectional permeability. II. Experimental implementation', J. Liq. Chromatogr. 9, 55-66.

Chmelík, J., Deml, M., Janča, J. (1989a) 'Separation of two components of horse myoglobin by isoelectric focusing field-flow fractionation', Anal. Chem. 61, 912-914.

Chmelík, J., Janča, J. (1989b) 'Isoelectric focusing field-flow fractionation - A new method of separation of amphoteric compounds', Chem. Listy 83, 321-325.

Chmelík, J. (1991a) 'Isoelectric focusing field-flow fractionation. Experimental study of the generation of pH gradient', J. Chromatogr. 539, 111-121.

Chmelík, J. (1991b) 'Isoelectric focusing field-flow fractionation. II. Experimental study of focusing of methyl red in the trapezoidal cross-section channel', J. Chromatogr. 545, 349-358.

Cornish, R.J. (1928) 'Flow in a pipe of rectangular cross-section', Proc. R. Soc. 120, 691-700.

Davis, J.M., Giddings, J.C. (1985a) 'Retention theory for field-flow fractionation in annular channels', J. Phys. Chem. 89, 3398-3405.

Davis, J.M., Giddings, J.C. (1985b) 'Influence of wall-retarded transport on retention and plate height in field-flow fractionation', Sep. Sci. Technol. 20, 699-724.

Davis, J.M. (1986a) 'General retention theory for sedimentation field-flow fractionation', Anal. Chem. 58, 161-164.

Davis, J.M., Giddings, J.C. (1986b) 'Feasibility study of dielectrical field-flow fractionation', Sep. Sci. Technol. 21, 969-989.

Davis, J.M., Fan, F.-R.F., Bard, A.J. (1987) 'Retention by electrical field-flow fractionation of anions in a new apparatus with annular porous glass channels', Anal. Chem. 59, 1339-1348.

Davis, J.M. (1989) 'Nonequilibrium theory for field-flow fractionation in annular channels', Sep. Sci. Technol. 24, 219-245.

Davis, J.M. (1991) 'Influence of cross-flow hydrodynamics on retention ratio in flow field-flow fractionation', Anal. Chim. Acta 246, 161-169.

Doshi, M.R., Gill, W.N., Subramanian, R.S. (1975) 'Unsteady reverse osmosis or ultrafiltration in a tube', Chem. Eng. Sci. 30, 1467-1476.

Doshi, M.R., Gill, W.N. (1979) 'Pressure field-flow fractionation or polarization chromatography', Chem. Eng. Sci. 34, 725-731.

Einstein, A. (1905) Investigations on the theory of the Brownian movement, Dover Publications, New York, 1956, Chap. 5.

Gao, Y.S., Caldwell, K.D., Myers, M.N., Giddings, J.C. (1985) 'Extension of thermal field-flow fractionation to ultrahigh (20×10^6) molecular weight polystyrenes', Macromolecules 18, 1272-1277.

Gajdos, L.J., Brenner, H. (1978) 'Field-flow fractionation: extension to nonspherical particles and wall effects', Sep. Sci. Technol. 13, 215-240.

Giddings, J.C. (1965) Dynamics of chromatography. Part I. Principles and theory, Edward Arnold Ltd., London; Marcel Dekker, Inc., New York.

Giddings, J.C. (1966) 'A new separation concept based on a coupling of concentration and flow nonuniformities', Sep. Sci. 1, 123-125.

Giddings, J.C. (1968) 'Nonequilibrium theory of field-flow fractionation', J. Chem. Phys. 49, 81-85

Giddings, J.C., Hovingh, M.E., Thompson, G.H. (1970) 'Measurement of thermal diffusion factors by thermal field-flow fractionation', J. Phys. Chem. 74, 4291-4294.

Giddings, J.C. (1973a) 'The conceptual basis of field-flow fractionation', J. Chem. Educ. 50, 667-669.

Giddings, J.C. (1973b) 'Parameters for optimum separations in field-flow fractionation', Sep. Sci. 8, 567-575.

Giddings, J.C., Yang, F.J.F., Myers, M.N. (1974) 'Sedimentation field-flow fractionation', Anal. Chem. 46, 1917-1924.

Giddings, J.C., Yoon, Y.H., Caldwell, K.D., Myers, M.N., Hovingh, M.E. (1975) 'Nonequilibrium plate height for field-flow fractionation in ideal parallel plate columns', Sep. Sci. 10, 447-460.

Giddings, J.C. (1976a) 'Field-flow fractionation. Extending the molecular weight range of liquid chromatography to one trillion (10^{12})', J. Chromatogr. 125, 3-16.

Giddings, J.C., Caldwell, K.D., Myers, M.N. (1976b) 'Thermal diffusion of polystyrene in eight solvents by an improved thermal field-flow fractionation methodology', Macromolecules 9, 106-112.

Giddings, J.C., Yang, F.J., Myers, M.N. (1976c) 'Flow field-flow fractionation: a versatile new separation method', Science 193, 1244-1245.

Giddings, J.C., Yang, F.J., Myers, M.N. (1976d) 'Theoretical and experimental characterization of flow field-flow fractionation', Anal. Chem. 48, 1126-1132.

Giddings, J.C., Lin, G.-C., Myers, M.N. (1976e) 'Electrical field-flow fractionation in a rigid membrane channel', Sep. Sci. 11, 553-568.

Giddings, J.C., Smith, L.K., Myers, M.N. (1976f) 'Programmed thermal field-flow fractionation', Anal. Chem. 48, 1587-1592.

Giddings, J.C. (1977a) 'FFF with double gradients', Lecture presented at a private group seminar, Salt Lake City, February 9; 'Hyperlayer field-flow fractionation', University of Utah Patent Office file no. U-728, February 1977.

Giddings, J.C., Yang, F.J., Myers, M.N. (1977b) 'Flow field-flow fractionation: a new method for separating, purifying, and characterizing the diffusivity of viruses', J. Virol. 21, 131-138.

Giddings, J.C., Yang, F.J., Myers, M.N. (1977c) 'Flow field-flow fractionation as a methodology for protein separation and characterization', Anal. Biochem. 81, 395-407.

Giddings, J.C., Yang, F.J., Myers, M.N. (1977d) 'Criteria for concentration field-flow fractionation', Sep. Sci. 12, 381-393.

Giddings, J.C., Lin, G.C., Myers, M.N. (1978a) 'Fractionation and size distribution of water soluble polymers by flow field-flow fractionation', J. Liq. Chromatogr. 1, 1-20.

Giddings, J.C., Lin, G.C., Myers, M.N. (1978b) 'Fractionation and size analysis of colloidal silica by flow field-flow fractionation', J. Colloid Interface Sci. 65, 67-78.

Giddings, J.C., Martin, M., Myers, M.N. (1978c) 'High-speed polymer separations by thermal field-flow fractionation', J. Chromatogr. 158, 419-435.

Giddings, J.C. (1978d) 'Displacement and dispersion of particles of finite size in flow channels with lateral forces. Field-flow fractionation and hydrodynamic chromatography', Sep. Sci. Technol. 13,241-254 (Corrigendum: Sep. Sci. Technol. 14 (1979) 869-870).

Giddings, J.C., Martin, M., Myers, M.N. (1979a) 'Thermogravitational field-flow fractionation: an elution thermogravitational column', Sep. Sci. Technol. 14, 611-643.

Giddings, J.C. (1979b) 'Field-flow fractionation of polymers: one-phase chromatography', Pure Appl. Chem. 51, 1459-1471.

Giddings, J.C., Caldwell, K.D., Moellmer, J.F., Dickinson, T.H., Myers, M.N., Martin, M. (1979c) 'Flow programmed field-flow fractionation', Anal. Chem. 51, 30-33.

Giddings, J.C., Myers, M.N., Caldwell, K.D., Fisher, S.R. (1980) 'Analysis of biological macromolecules and particles by field-flow fractionation', in D. Glick (ed.), Methods of Biochemical Analysis, Vol. 26, John Wiley, New York, pp. 79-136.

Giddings, J.C., Karaiskakis, G., Caldwell, K.D. (1981) 'Density and particle size of colloidal materials measured by carrier density variations in sedimentation field-flow fractionation', Sep. Sci. Technol. 16, 607-618.

Giddings, J.C. (1983a) 'Hyperlayer field-flow fractionation', Sep. Sci. Technol. 18, 765-773.

Giddings, J.C., Lin, H.C., Caldwell, K.D., Myers, M.N. (1983b) 'Outlet stream splitting for sample concentration in field-flow fractionation', Sep. Sci. Technol. 18, 293-306.

Giddings, J.C., Brantley, S.L. (1984a) 'Shear field-flow fractionation: theoretical basis of a new, highly selective technique', Sep. Sci. Technol. 19, 631-651.

Giddings, J.C., Caldwell, K.D. (1984b) 'Field-flow fractionation: choices in programmed and nonprogrammed operation', Anal. Chem. 56, 2093-2099.

Giddings, J.C. (1985a) 'Optimized field-flow fractionation system based on dual stream splitters', Anal. Chem. 57, 945-947.

Giddings, J.C. (1985b) 'A system based on split-flow lateral-transport thin (SPLITT) separation cells for rapid and continuous particle fractionation', Sep. Sci. Technol. 20, 749-768.

Giddings, J.C. (1986a) 'Cyclical-field field-flow fractionation: a new method based on transport rates', Anal. Chem. 58, 2052-2056.

Giddings, J.C. (1986b) 'Simplified nonequilibrium theory of secondary relaxation effects in programmed field-flow fractionation', Anal. Chem. 58, 735-740.

Giddings, J.C. (1986c) 'Crossflow gradients in thin channels for separation by hyperlayer FFF, SPLITT cells, elutriation, and related methods', Sep. Sci. Technol. 21, 831-843.

Giddings, J.C., Schure, M.R. (1987a) 'Theoretical analysis of edge effects in field-flow fractionation', Chem. Eng. Sci. 42, 1471-1479.

Giddings, J.C., Chen, X., Wahlund, K.-G., Myers, M.N. (1987b) 'Fast particle separation by flow/steric field-flow fractionation', Anal. Chem. 59, 1957-1962.

Giddings, J.C., Williams, P.S., Beckett, R. (1987c) 'Fractionating power in programmed field-flow fractionation: exponential sedimentation field decay', Anal. Chem. 59, 28-37.

Giddings, J.C., Li, S., Williams, P.S., Schimpf, M.E. (1988a) 'High speed separation of ultra-high-molecular-weight polymers by thermal/hyperlayer field-flow fractionation', Makromol. Chem., Rapid Commun. 9, 817-823.

Giddings, J.C. (1988b) 'Field-flow fractionation', C & E News 66, 34-45.

Giddings, J.C. (1988c) 'Continuous particle separation in split-flow thin (SPLITT) cells using hydrodynamic lift forces', Sep. Sci. Technol. 23, 119-131.

Giddings, J.C. (1988d) 'Continuous separation in split-flow thin (SPLITT) cells: potential applications to biological materials', Sep. Sci. Technol. 23, 931-943.

Giddings, J.C., Caldwell, K.D. (1989a) 'Field-flow fractionation', in B.W. Rossiter and J.F. Hamilton (eds.), Physical Methods of Chemistry, Vol. 3B, John Wiley, New York, pp. 867-938.

Giddings, J.C., Barman, B.N., Li, H. (1989b) 'Colloid characterization by sedimentation field-flow fractionation. VII. Colloidal aggregates', J. Colloid Interface Sci. 132, 554-565.

Giddings, J.C. (1989c) 'Harnessing electrical forces for separation. Capillary zone electrophoresis, isoelectric focusing, field-flow fractionation, split-flow thin-cell continuous-separation and other techniques', J. Chromatogr. 480, 21-33.

Giddings, J.C. (1989d) 'A pinched inlet system for reducing relaxation effects and stopless flow injection in field-flow fractionation', Sep. Sci. Technol. 24, 755-768.

Giddings, J.C. (1990a) 'Two-dimensional field-flow fractionation', J. Chromatogr. 504, 247-258.

Giddings, J.C. (1990b) 'Hydrodynamic relaxation and sample concentration in field-flow fractionation using permeable wall elements', Anal. Chem. 62, 2306-2312.

Giddings, J.C., Kumar, V., Williams, P.S., Myers, M.N. (1990c) 'Polymer separation by thermal field-flow fractionation: high-speed power programming', C.D. Craver and T. Provder (eds.), *Polymer Characterization by Interdisciplinary Methods*, ACS Advances in Chemistry Series 227, American Chemical Society, Washington, DC, pp. 3-21.

Giddings, J.C. (1991a) Unified separation science, John Wiley & Sons, Inc., New York, p. 44-45.

Giddings, J.C., Myers, M.N., Moon, M.H., Barman, B.N. (1991b) 'Particle separation and size characterization by sedimentation field-flow fractionation', in T. Provder (ed.), *Particle Size Distribution II*, ACS Symp. Series No. 427, Chapter 13.

Giddings, J.C., Moon, M.H., Williams, P.S., Myers, M.N. (1991c) 'Particle size distribution by sedimentation/steric field-flow fractionation: development of a calibration procedure based on density compensation', Anal. Chem. 63, 1366-1372.

Giddings, J.C., Moon, M.H. (1991d) 'Measurement of particle density, porosity, and size distributions by sedimentation/steric field-flow fractionation: application to chromatographic supports', Anal. Chem. 63, 2869-2877.

Giddings, J.C., Barman, B.N., Liu, M.-K. (1991e) 'Separation of cells by field-flow fractionation and related methods', in D. Kompala and P. Todd (eds.), *Cell Separation Science and Technology*, ACS Symp. Series No. 464, Chapter 9.

Giddings, J.C., Ratanathanawongs, S.K., Liu, G., Tjelta, B.L., Moon, M.H., Hansen, M.E., Barman, B.N. 'Characterization of colloidal and particulate silica by field-flow fractionation', *Ralph K. Iler Memorial Symposium Volume*, ACS Books, submitted.

Gill, W.N., Sankarasubramanian, R. (1970) 'Exact analysis of unsteady convective diffusion', Proc. Roy. Soc. Lond. A. 316, 341-350.

Gill, W.N., Sankarasubramanian, R. (1971) 'Dispersion of a non-uniform slug in time-dependent flow', Proc. Roy. Soc. Lond. A. 322, 101-117.

Gold, R. (1964) AEC Research and Development Report ANL-6984, Argonne National Laboratory, Argonne, Illinois.

Granger, J., Dodds, J., Leclerc, D., Midoux, N. (1986) 'Flow and diffusion of particles in a channel with one porous wall: polarization chromatography', Chem. Eng. Sci. 41, 3119-3128.

Gunderson, J.J., Caldwell, K.D., Giddings, J.C. (1984) 'Influence of temperature gradients on velocity profiles and separation parameters in thermal field-flow fractionation', Sep. Sci. Technol. 19, 667-683.

Gunderson, J.J., Giddings, J.C. (1986) 'Comparison of polymer resolution in thermal field-flow fractionation and size-exclusion chromatography', Anal. Chim. Acta 189, 1-15.

Han, L.S. (1960) 'Hydrodynamic entrance lengths for incompressible laminar flow in rectangular ducts', J. Appl. Mech., Sept., 403-409.

Hansen, M.E., Giddings, J.C., Schure, M.R., Beckett, R. (1988) 'Corrections for secondary relaxation in exponentially programmed field-flow fractionation', Anal. Chem. 60, 1434-1442.

Hansen, M.E., Giddings, J.C., Beckett, R. (1989a) 'Colloid characterization by sedimentation field-flow fractionation. VI. Perturbations due to overloading and electrostatic repulsion', J. Colloid Interface Sci. 132, 300-312.

Hansen, M.E., Giddings, J.C. (1989b) 'Retention perturbations due to particle-wall interactions in sedimentation field-flow fractionation', Anal. Chem. 61, 811-819.

Hansen, M.E., Short, D.C. (1990) 'Optimization study of octane-in-water emulsions by sedimentation field-flow fractionation', J. Chromatogr. 517, 333-344.

576

Happel, J., Brenner, H. (1973) Low Reynolds number hydrodynamics, Noordhoff Internat. Publ., Leyden, Chap. 2.

Ho, B.P., Leal, L.G. (1974) 'Inertial migration of rigid spheres in two-dimensional unidirectional flows', J. Fluid Mech. 65, 365-400.

Hovingh, M.E., Thompson, G.H., Giddings, J.C. (1970) 'Column parameters in thermal field-flow fractionation', Anal. Chem. 42, 195-203.

Inagaki, H., Tanaka, T. (1980) 'Sedimentation field-flow fractionation in macromolecule characterization', Anal. Chem. 52, 201-203.

Ishige, T., Lee, S.-I., Hamielec, A.E. (1971) 'Solution of Tung's axial dispersion equation by numerical techniques', J. Appl. Polym. Sci. 15, 1607-1622.

Jahnová, V., Matulík, F., Janča, J. (1987) 'Correction for zone spreading in sedimentation field-flow fractionation', Anal. Chem. 59, 1039-1043.

Janča J., Giddings, J.C. (1981) 'Non-Newtonian flow for retention control in field-flow fractionation', Sep. Sci. Technol. 16, 805-815.

Janča J. (1982) 'Sedimentation-flotation focusing field-flow fractionation', Makromol. Chem., Rapid Commun. 3, 887-890 (Corrigendum: Makromol. Chem., Rapid Commun. 4 (1983) 267).

Janča J., Jahnová, V. (1983) 'Sedimentation-flotation focusing field-flow frationation in channels with modulated cross-sectional permeability. I. Theoretical analysis', J. Liq. Chromatogr. 6, 1559-1576.

Janča J., Chmelík, J. (1984) 'Focusing in field-flow fractionation', Anal. Chem. 56, 2481-2484.

Janča J., Chmelík, J., Přibylová, D. (1985) 'Optimization of field-flow fractionation with respect to relaxation and use of stop-flow technique at constant field operation', J. Liq. Chromatogr. 8, 2343-2368.

Janča, J. (1987a) 'Flow velocity profile in channels with modulated cross-sectional permeability for sedimentation-flotation focusing field-flow fractionation using step density gradients', J. Chromatogr. 404, 23-32.

Janča J., Nováková, N. (1987b) 'Sedimentation-flotation focusing field-flow fractionation in channels with modulated cross-sectional permeability. III. Application of step density gradient', J. Liq. Chromatogr. 10, 2869-2876.

Janča J. (1987c) 'Elutriation focusing field-flow fractionation', Makromol. Chem., Rapid Commun. 8, 233-236.

Janča, J., Nováková, N. (1988) 'Retention in sedimentation-flotation focusing field-flow fractionation using a step density gradient', J. Chromatogr. 452, 549-562.

Jansson, P.A. (1984) Deconvolution with applications in spectroscopy, Academic Press, New York.

Jarayaj, K., Subramanian, R.S. (1978) 'On relaxation phenomena in field-flow fractionation', Sep. Sci. Technol. 13, 791-817.

Jones, H.K., Phelan, K., Myers, M.N., Giddings, J.C. (1987) 'Colloid characterization by sedimentation field-flow fractionation. V. Split outlet system for complex colloids of mixed density', J. Colloid Interface Sci. 120, 140-152.

Jones, H.K., Barman, B.N., Giddings, J.C. (1988) 'Resolution of colloidal latex aggregates by sedimentation field-flow fractionation', J. Chromatogr. 455, 1-15.

Jönsson, J.A., Carlshaf, A. (1989) 'Flow field flow fractionation in hollow cylindrical fibers', Anal. Chem. 61, 11-18.

Karaiskakis, G., Myers, M.N., Caldwell, K.D., Giddings, J.C. (1981) 'Verification of retention and zone spreading equations in sedimentation field-flow fractionation', Anal. Chem. 53, 1314-1317.

Karaiskakis, G., Graff, K.A., Caldwell, K.D., Giddings, J.C. (1982) 'Sedimentation field-flow fractionation of colloidal particles in river water', Intern. J. Environ. Anal. Chem. 12, 1-15.

Karaiskakis, G., Koliadima, A. (1989) 'Potential barrier field-flow fractionation for the separation and characterization of colloidal materials', Chromatographia 28, 31-32.

Kerker, M. (1969) The scattering of light, Academic Press, New York.

Kesner, L.F., Caldwell, K.D., Myers, M.N., Giddings, J.C. (1976) 'Performance characteristics of electrical field-flow fractionation in a flexible membrane channel', Anal. Chem. 48, 1834-1839.

Kim, E.-K., Chung, I.J. (1986) 'Transient convective mass transfer in rectangular field-flow fractionation channels', Chem. Eng. Commun. 42, 349-365.

Kirkland, J.J., Yau, W.W., Doerner, W.A., Grant, J.W. (1980) 'Sedimentation field flow fractionation of macromolecules and colloids', Anal. Chem. 52, 1944-1954.

Kirkland, J.J., Yau, W.W. (1981) 'Time-delayed exponential field-programmed sedimentation field flow fractionation for particle-size-distribution analyses', Anal. Chem. 53, 1730-1736.

Kirkland, J.J., Yau, W.W. (1982a) 'Sedimentation field flow fractionation: applications', Science 218, 121-127.

Kirkland, J.J., Yau, W.W., Szoka, F.C. (1982b) 'Sedimentation field flow fractionation of liposomes', Science 215, 296-298.

Kirkland, J.J., Dilks, Jr., C.H., Yau, W.W. (1983a) 'Sedimentation field flow fractionation at high force fields', J. Chromatogr. 255, 255-271.

Kirkland, J.J., Yau, W.W. (1983b) 'Simultaneous determination of particle size and density by sedimentation field flow fractionation', Anal. Chem. 55, 2165-2170.

Kirkland, J.J., Yau, W.W. (1985) 'Thermal field-flow fractionation of polymers with exponential temperature programming', Macromolecules 18, 2305-2311.

Kirkland, J.J., Yau, W.W. (1986) 'Thermal field-flow fractionation of water-soluble macromolecules', J. Chromatogr. 353, 95-107.

Kirkland, J.J., Rementer, S.W., Yau, W.W. (1988) 'Molecular-weight distributions of polymers by thermal field flow fractionation with exponential temperature programming', Anal. Chem. 60, 610-616.

Kirkland, J.J., Liebald, W., Unger, K.K. (1990) 'Characterization of diesel soot by sedimentation field flow fractionation', J. Chromatogr. Sci. 28, 374-378.

Koch, L., Koch, T., Widmer, H.M. (1990) 'Sedimentation field-flow fractionation for pigment quality assessment', J. Chromatogr. 517, 395-403.

Koch, T., Giddings, J.C. (1986) 'High-speed separation of large (>1 μm) particles by steric field-flow fractionation', Anal. Chem. 58, 994-997.

Koliadima, A., Dalas, E., Karaiskakis, G. (1990) 'Simultaneous determination of particle size and density in polydisperse colloidal samples by sedimentation field-flow fractionation', J. High Resolut. Chromatogr. Chromatogr. Commun. 13, 338-342.

Kononenko, V.L., Shimkus, J.K. (1990) 'Non-equilibrium analytical focusing field-flow fractionation using hydrodynamic force and integral Doppler anemometry', J. Chromatogr. 520, 271-285.

Kozinski, A.A., Schmidt, F.P., Lightfoot, E.N. (1970) 'Velocity profiles in porous-walled ducts', Ind. Eng. Chem. Fundam. 9, 502-505.

Krishnamurthy, S., Subramanian, R.S. (1977) 'Exact analysis of field-flow fractionation', Sep. Sci., 12, 347-379.

Lee, H.L., Lightfoot, E.N. (1976) 'Preliminary report on ultrafiltration-induced polarization chromatography - An analog of field-flow fractionation', Sep. Sci. 11, 417-440.

Lee, S., Giddings, J.C. (1988a) 'Experimental observation of steric transition phenomena in sedimentation field-flow fractionation', Anal. Chem. 60, 2328-2333.

Lee, S., Myers, M.N., Beckett, Giddings, J.C. (1988b) 'Particle separation and characterization by sedimentation/cyclical-field field-flow fractionation', Anal. Chem. 60, 1129-1135.

Lee, S., Myers, M.N., Giddings, J.C. (1989) 'Hydrodynamic relaxation using stopless flow injection in split inlet sedimentation field-flow fractionation', Anal. Chem. 61, 2439-2444.

Levin, S., Myers, M.N., Giddings, J.C. (1989) 'Continuous separation of proteins in electrical split-flow thin (SPLITT) cell with equilibrium operation', Sep. Sci. Technol. 24, 1245-1259.

Levin, S., Giddings, J.C. (1991) 'Continuous separation of particles from macromolecules in split-flow thin (SPLITT) cells', J. Chem. Tech. Biotechnol. 50, 43-56.

Lightfoot, E.N., Noble, P.T., Chiang, A.S., Ugulini, T.A. (1981) 'Characterization of an improved electropolarization chromatographic system using homogenous proteins', Sep. Sci. Technol. 16, 619-656.

Litzén, A., Wahlund, K.-G. (1989) 'Improved separation speed and efficiency for proteins, nucleic acids and viruses in asymmetrical flow field-flow fractionation', J. Chromatogr. 476, 413-421.

Litzén, A., Wahlund, K.-G. (1991a) 'Zone broadening and dilution in rectangular and trapezoidal asymmetrical flow field-flow fractionation channels', Anal. Chem. 63, 1001-1007.

Litzén, A., Wahlund, K.-G. (1991b) 'Effects of temperature, carrier composition and sample load in asymmetrical flow field-flow fractionation', J. Chromatogr. 548, 393-406.

Liu, G., Giddings, J.C. (1991a) 'Separation of particles in nonaqueous suspensions by thermal-electrical field-flow fractionation', Anal. Chem. 63, 296-299.

Liu, M.-K., Williams, P.S., Myers, M.N., Giddings, J.C. (1991b) 'Hydrodynamic relaxation in flow field-flow fractionation using both split and frit inlets', Anal. Chem. 63, 2115-2122.

Martin, M., Myers, M.N., Giddings, J.C. (1979) 'Nonequilibrium and polydispersity peak broadening in thermal field-flow fractionation', J. Liq. Chromatogr. 2, 147-164.

Martin, M., Reynaud, R. (1980) 'Polymer analysis by thermal field-flow fractionation', Anal. Chem. 52, 2293-2298.

Martin, M., Giddings, J.C. (1981a) 'Retention and nonequilibrium peak broadening for a generalized flow profile in field-flow fractionation', J. Phys. Chem. 85, 727-733.

Martin, M., Jaulmes, A. (1981b) 'Peak capacity in field-flow fractionation', Sep. Sci. Technol. 16, 691-724.

Martin, M., Gril, J., Besançon, O., Reynaud, M. (1982) 'Influence of inertia effects on particle retention in field-flow fractionation', Lecture presented at the 6th International Symposium on Column Liquid Chromatography, Cherry Hill, June 7-11.

Martin, M. (1988) 'Influence of operational parameters on particle retention in inertial field-flow fractionation', Lecture presented at the 12th International Symposium on Column Liquid Chromatography, Washington, June 19-24.

Mattson, O.J.E., Alfredsson, P.H. (1990) 'Curvature- and rotation-induced instabilities in channel flow', J. Fluid Mech. 210, 537-563.

Mori, S. (1986) 'Magnetic field-flow fractionation using capillary tubing', Chromatographia 11, 642-644.

Mori, Y., Kimura, K., Tanigaki, M. (1990) 'Influence of particle-wall and particle-particle interactions on retention behavior in sedimentation field-flow fractionation', Anal. Chem. 62, 2668-2672.

Mutabazi, I., Wesfreid, J.E., Martin, M. (1991) 'Avoidance of hydrodynamic centrifugal instabilities in sedimentation FFF', Lecture presented at the 2nd International Symposium on Field-Flow Fractionation, Salt Lake City, February 2-4.

Myers, M.N., Giddings, J.C. (1982) 'Properties of the transition from normal to steric field-flow fractionation', Anal. Chem. 54, 2284-2289.

Nagy, D.J. (1989) 'Density determination of low-density polymer latexes by sedimentation field-flow fractionation', Anal. Chem. 61, 1934-1937.

Oppenheimer, L.E., Smith, G.A. (1989) 'Sedimentation field-flow fractionation of colloidal metal hydrosols', J. Chromatogr. 461, 103-110.

Peterson II, R.E., Myers, M.N., Giddings, J.C. (1984) 'Characterization of steric field-flow fractionation using particles to 100μm diameter', Sep. Sci. Technol. 19, 307-319.

Prieve, D.C. (1982) 'Axial dispersion of sedimented colloids', Sep. Sci. Technol. 17, 1587-1607.

Ratanathanawongs, S.K., Giddings, J.C. (1989) 'High-speed size characterization of chromatographic silica by flow/hyperlayer field-flow fractionation', J. Chromatogr. 467, 341-356.

Ratanathanawongs, S.K., Giddings, J.C. (1991) 'Separation and characterization of 0.01-50 μm particles using flow field-flow fractionation', in T. Provder (ed.), Particle Size Distribution II, ACS Symp. Series No. 472, Chapter 15.

Ratanathanawongs, S.K. and Giddings, J.C. (1992) 'Dual field and flow programmed lift hyperlayer field-flow fractionation', Anal. Chem., 64, 6-15.

Reis, J.F.G., Lightfoot, E.N. (1976) 'Electropolarization chromatography', AIChE. J. 22, 779-785.

Schallinger, L.E., Yau, W.W., Kirkland, J.J. (1984) 'Sedimentation field flow fractionation of DNA's', Science 225, 434-437.

Schallinger, L.E., Gray, J.E., Wagner, L.W., Knowlton, S., Kirkland, J.J. (1985) 'Preparative isolation of plasmid DNA with sedimentation field flow fractionation', J. Chromatogr. 342, 67-77.

Schimpf, M.E., Giddings, J.C. (1987a) 'Characterization of thermal diffusion in polymer solutions by thermal field-flow fractionation: effects of molecular weight and branching', Macromolecules 20, 1561-1563.

Schimpf, M.E., Myers, M.N., Giddings, J.C. (1987b) 'Measurement of polydispersity of ultra-narrow polymer fractions by thermal field-flow fractionation', J. Appl. Polym. Sci. 33, 117-135.

Schimpf, M.E., Giddings, J.C. (1989a) 'Characterization of thermal diffusion in polymer solutions by thermal field-flow fractionation: dependence on polymer and solvent parameters', J. Polym. Sci., Part B: Polym. Phys. 27, 1317-1332.

Schimpf, M.E., Williams, P.S., Giddings, J.C. (1989b) 'Accurate molecular weight distribution of polymers using thermal field-flow fractionation with deconvolution to remove system dispersion', J. Appl. Polym. Sci. 37, 2059-2076.

Schimpf, M.E., Giddings, J.C. (1990a) 'Characterization of thermal diffusion of copolymers in solution by thermal field-flow fractionation', J. Polym. Sci., Part B: Polym. Phys. 28, 2673-2680.

Schimpf, M.E. (1990b) 'Characterization of polymers by thermal field-flow fractionation', J. Chromatogr. 517, 405-421.

Schonberg, J.A., Hinch E.J. (1989) 'Inertial migration of a sphere in Poiseuille flow', J. Fluid Mech. 203, 517-524.

Schunk, T.C., Gorse, J., Burke, M.F. (1984) 'Parameters affecting magnetic field-flow fractionation of metal oxide particles', Sep. Sci. Technol. 19, 653-666.

Schure, M.R., Caldwell, K.D., Giddings, J.C. (1986) 'Theory of sedimentation hyperlayer field-flow fractionation', Anal. Chem. 58, 1509-1516.

Schure, M.R. (1988) 'Digital simulation of sedimentation field-flow fractionation', Anal. Chem. 60, 1109-1119.

Schure, M.R., Barman, B.N., Giddings, J.C. (1989) 'Deconvolution of nonequilibrium band broadening effects for accurate particle size distributions by sedimentation field-flow fractionation', Anal. Chem. 61, 2735-2743.

Segré, G., Silberberg, A. (1961) 'Radial particle displacements in Poiseuille flow of suspensions', Nature 189, 209-210.

Segré, G., Silberberg, A. (1962) 'Behaviour of macroscopic rigid spheres in Poiseuille flow. Part 2. Experimental results and interpretation', J. Fluid Mech. 14, 136-157.

Semenov, S.N., Kuznetsov, A.A. (1986) 'Possible application of the focusing flow fractionation of particles in a transverse field', Zh. Fiz. Khim. 60, 439-443.

Semyonov, S.N., Maslow, K.I. (1988a) 'Acoustic field-flow fractionation', J. Chromatogr. 446, 151-156.

Semyonov, S.N. (1988b) 'Near-wall field-flow fractionation. Application of surface forces and transverse voltage to fractionation in a laminar flow', J. Chromatogr. 446, 131-139.

Shah, A.B., Reis, J.F.G., Lightfoot, E.N., Moore, R.E. (1979) 'Modeling electroretention of proteins during electropolarization chromatography', Sep. Sci. Technol. 14, 475-497.

Smith, L.K., Myers, M.N., Giddings, J.C. (1977) 'Peak broadening factors in thermal field-flow fractionation', Anal. Chem. 49, 1750-1756.

Smith, R. (1983) 'Field-flow fractionation', J. Fluid Mech. 129, 347-364.

Song, K.-C., Kim, E.-K., Chung, I.J. (1986) 'Measurement of thermal diffusion parameters of polystyrene using a thermal field flow fractionation method', Korean J. of Chem. Eng. 3, 171-175.

Springston, S.R., Myers, M.N., Giddings, J.C. (1987) 'Continuous particle fractionation based on gravitational sedimentation in split thin cells', Anal. Chem. 59, 344-350.

Takahashi, T., Gill, W.N. (1980) 'Hydrodynamic chromatography: three dimensional laminar dispersion in rectangular conduits with transverse flow', Chem. Eng. Commun. 5, 367-385.

Thompson, G.H., Myers, M.N., Giddings, J.C. (1967) 'An observation of a field-flow fractionation effect with polystyrene samples', Sep. Sci. 2, 797-800.

Thompson, G.H., Myers, M.N., Giddings, J.C. (1969) 'Thermal field-flow fractionation of polystyrene samples', Anal. Chem. 41, 1219-1222.

Thormann, W., Firestone, M.A., Dietz, M.L., Cecconie, T., Mosher, R.A. (1989) 'Focusing counterparts of electrical field flow fractionation and capillary zone electrophoresis. Electrical hyperlayer field flow fractionation and capillary isoelectric focusing', J. Chromatogr. 461, 95-101.

Tyrrell, H.J.V. (1961) Diffusion and heat flow in liquids, Butterworths, London, p 45.

Ugrozov, V.V., Maksimycheva, M.A., Zolotarev, P.P. (1989) 'Theory of flow fractionation in a coaxial channel', Zh. Fiz. Khim. 63, 1048-1053.

Ugrozov, V.V. (1990) 'The method of flow fractionation with finite analysis times', Zh. Fiz. Khim. 64, 1006-1011.

Urbánková, E., Janča J. (1990) 'An attempt at experimental elutriation focusing field-flow fractionation', J. Liq. Chromatogr. 13, 1877-1895.

Van den Broeck, C., Maes, D. (1987) 'Retention and peak width in field-flow fractionation with wall effects', Sep. Sci. Technol. 22, 1269-1280.

Vasseur, P., Cox, R.G. (1976) 'The lateral migration of a spherical particle in two-dimensional shear flows', J. Fluid Mech. 78, 385-413.

Vickrey, T.M., Garcia-Ramirez, J.A. (1980) 'Magnetic field-flow fractionation: theoretical basis', Sep. Sci. Technol. 15, 1297-1304.

Wahlund, K.-G., Winegarner, H.S., Caldwell, K.D., Giddings, J.C. (1986) 'Improved flow field-flow fractionation system applied to water-soluble polymers: programming, outlet stream splitting, and flow optimization', Anal. Chem. 58, 573-578.

Wahlund, K.-G., Giddings, J.C. (1987) 'Properties of an asymmetrical flow field-flow fractionation channel having one permeable wall', Anal. Chem. 59, 1332-1339.

Wahlund, K.-G., Litzén, A. (1989) 'Application of an asymmetrical flow field-flow fractionation channel to the separation and characterization of proteins, plasmids, plasmid fragments, polysaccharides and unicellular algae', J. Chromatogr. 461, 73-87.

Westermann-Clark, G. (1978) 'Note on nonisothermal flow in field-flow fractionation', Sep. Sci. Technol. 13, 819-822.

Wičar, S. (1988) 'Flow velocity profiles in rectangular channels containing two liquid layers of different densities and viscosities for sedimentation field flow fractionation', J. Chromatogr. 448, 456-457.

Williams, P.S., Giddings, S.B., Giddings, J.C. (1986) 'Calculation of the flow properties and end effects in field-flow fractionation channels by a conformal mapping procedure', Anal. Chem. 58, 2397-2403.

Williams, P.S., Giddings, J.C. (1987a) 'Fractionating power in sedimentation field-flow fractionation with linear and parabolic field decay programming', J. Liq. Chromatogr. 10, 1961-1998.

Williams, P.S., Giddings, J.C. (1987b) 'Power programmed field-flow fractionation: a new program form for improved uniformity of fractionating power', Anal. Chem. 59, 2038-2044.

Williams, P.S., Kellner, L., Beckett, R., Giddings, J.C. (1988) 'Comparison of experimental and theoretical fractionating power for exponential field decay sedimentation field-flow fractionation', Analyst 113, 1253-1259.

Williams, P.S., Giddings, J.C. (1991) 'Comparison of power and exponential field programming in field-flow fractionation', J. Chromatogr. 550, 787-797.

Williams, P.S., Koch, T., Giddings, J.C. (1992a) 'Characterization of near-wall hydrodynamic lift forces using sedimentation field-flow fractionation', Chem. Eng. Commun. 111, 121-147.

Williams, P.S., Moon, M.H., Giddings, J.C. (1992b) 'Fast separation and characterization of micron size particles by sedimentation/steric field-flow fractionation: role of lift forces', in N. Stanley-Wood and R. Lines (eds.), Proceedings of 25th Anniversary Conference on Particle Size Analysis (PSA91), Royal Society of Chemistry, Cambridge, U.K.

Williams, P.S., Levin, S., Lenczycki, T., Giddings, J.C. 'Continuous fractionation by diffusion in split-flow thin (SPLITT) separation cells', Ind. Eng. Chem. Res. accepted.

Yang, F.J.F., Myers, M.N., Giddings, J.C. (1974) 'Programmed sedimentation field-flow fractionation', Anal. Chem. 46, 1924-1930.

Yang, F.J., Myers, M.N., Giddings, J.C. (1977) 'Peak shifts and distortion due to solute relaxation in flow field-flow fractionation', Anal. Chem. 49, 659-662.

Yang, F.-S., Caldwell, K.D., Myers, M.N., Giddings, J.C. (1983) 'Colloid characterization by sedimentation field-flow fractionation. III. Emulsions', J. Colloid Interface Sci. 93, 115-125.

Yang, F.-S., Caldwell, K.D., Giddings, J.C., Astle, L. (1984) 'Stability of perfluorocarbon blood substitutes determined by sedimentation field-flow fractionation', Anal. Biochem. 138, 488-494.

Yau, W.W., Kirkland, J.J. (1981) 'Retention characteristics of time-delayed exponential field-programmed sedimentation field-flow fractionation', Sep. Sci. Technol. 16, 577-605.

Yau, W.W., Kirkland, J.J. (1984) 'Nonequilibrium effects in sedimentation field flow fractionation', Anal. Chem. 56, 1461-1466.

Yonker, C.R., Jones, H.K., Robertson, D.M. (1987) 'Nonaqueous sedimentation field flow fractionation', Anal. Chem. 59, 2574-2579.

Zolotarev, P.P., Ugrozov, V.V., Skornyakov, E.P. (1988) 'Theory of separation of mixtures in a coaxial column with a transverse temperature gradient', Zh. Fiz. Khim. 62, 1896-1903.

CONVENTIONAL ISOELECTRIC FOCUSING AND IMMOBILIZED pH GRADIENTS

P. G. RIGHETTI and C. TONANI
Dept. Biomedical Sciences and Technologies
University of Milano, Via Celoria 2
Milano 20133, Italy.

ABSTRACT. We review here some mathematical models pertaining to separations by isoelectric focusing (IEF) in carrier ampholyte (CA) buffers and in immobilized pH gradients (IPG). In CA-IEF, the steady-state distribution profile of amphoteric compounds predicts gaussian distributions only if and when the background conductivity remains constant within the focused zone. This is in practice only achieved when focusing across neutrality. In all other pH regions, the tendency is to produce asymmetric peaks, with negatively (in acidic regions) or positively (in basic gradients) skewed profiles. The transient state, in uniformly loaded samples, is seen as two discernible peaks migrating from the two ends of the column and merging into one at the pI position. The decay state seems to be generated by an isotachophoretic phenomenon of anodic, cathodic, or symmetric loss of carrier ampholytes. IPGs are characterized by indefinitely stable pH gradients and by an extremely high resolution. For producing wide pH gradients, computer algorithms are described which optimize any mixture of Immobiline chemicals for generating linear or, in general, concave exponential pH gradients.

1. Conventional Isoelectric Focusing

Isoelectric focusing (IEF) in carrier ampholyte (CA) buffers was born through the stubborn efforts of Svensson (1961, 1962), who first understood that, in order to create and maintain a pH gradient (*i.e.* a proton concentration gradient) in an electric field, one had to resort to a set of peculiar buffers which would have to be, at the very least, a) amphoteric and b) carrier (of buffering power and conductivity) at the isoelectric point (pI). Such a gradient, generated by electrolysis of an initially homogeneous mixture of the CA-buffers, is called a natural pH gradient, as opposed to 'artificial' pH gradients obtained by diffusion of two limiting solutions in the absence of an electric field (Kolin, 1976). Svensson first perceived the fact, which now appears obvious, that an ampholyte in the isoelectric state may have a conductivity despite its zero mobility and the fact that the overall charge is close to zero (let us recall here that the values of the isoelectric and isoionic points differ slightly). Svensson also first used the concept of an ampholyte with two dissociation stages close to the isoelectric point (*conditio sine qua non* for buffering and conducting at the pI!).

F. Dondi and G. Guiochon (eds.),
Theoretical Advancement in Chromatography and Related Separation Techniques, 581–605.

582

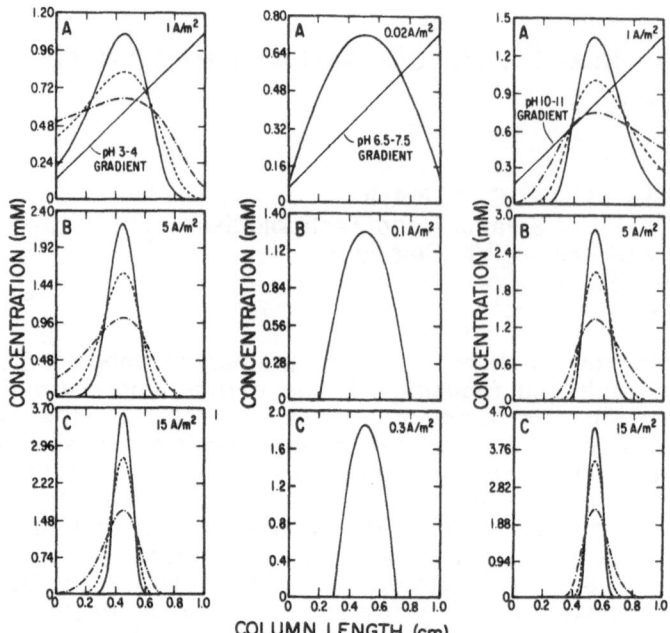

COLUMN LENGTH (cm)

Figure 1. Computer simulation of focusing ampholytes at neutral and pH extremes. Left 3 panels: focusing in the pH 3-4 interval. Three pI 3.5 ampholytes, with ΔpKs 1 (solid curve), 2 (broken) and 3 (broken and dotted curve) are focused in a pH 3-4 range formed by linear concentration gradients of an imaginary acidic Immobiline (pK 3.0) and a basic one (pK 4.0) ranging from 10 to 5 mM. The steady state profiles are reported for current densities of 1 (A), 5 (B) and 15 (C) A/m², with a fourth curve reporting the pH gradient. Central 3 panels: focusing in the pH 6.5-7.5 interval. All conditions as in the pH 3-4 range, except that the three ampholytes with ΔpKs 1, 2 and 3 have a pI of 7.0 and that the two background Immobilines have a pK of 6.5 (acidic species) and of 7.5 (basic component). Here the three current densities are 0.02 (A), 0.1 (B) and 0.3 (C) A/m². Right 3 panels: focusing in the pH 10-11 range. All conditions as in the pH 3-4 interval, except that the three ampholytes with ΔpKs 1, 2 and 3 have a pI of 10.5 and that the two background Immobilines have a pK of 10.0 (acidic species) and of 11.0 (basic component) (modified from Mosher *et al.*, 1986).

1.1 THE STEADY STATE.

1.1.1 *Distribution Profile of an Ampholyte About Its Isoelectric Point.* Under steady-state conditions (as obtained by balancing the simultaneous electrophoretic and diffusional mass transports) Svensson (1961) has derived the following differential eqn. describing the concentration profile of a focused zone:

$$Cui/qk = D(dC/dx) \qquad (1)$$

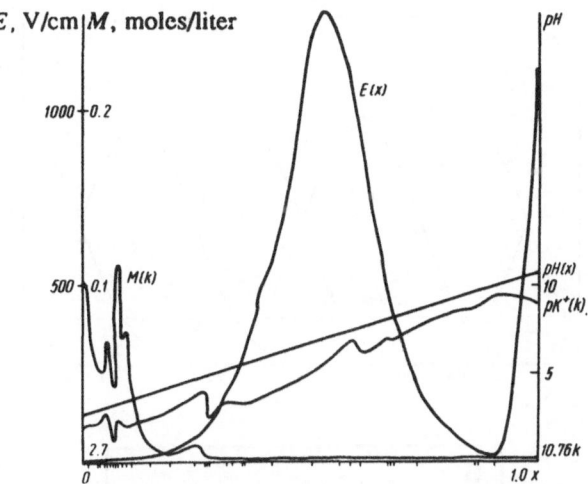

Figure 2. Results of the calculation of the total concentration M(k) and field intensity E(x) for the set of substances given in Table I, with which a linear pH profile is created in the range $2.77 \leq pH \leq 10.76$ (reprinted with permission from Babskii *et al.*, 1989).

where:
C = concentration of a component in arbitrary mass units per arbitrary volume unit;
u = electric mobility in $cm^2\ V^{-1}\ sec^{-1}$ of ion constituent except H^+ and OH^-, with positive sign for cationic and negative sign for anionic migration;
i = electric current in A;
q = cross-sectional area in cm^2 of electrolytic medium, measured perpendicularly to the direction of current;
k = conductance of medium, in $ohm^{-1}\ cm^{-1}$;
D = diffusion coefficient in $cm^2\ sec^{-1}$ of component corresponding to the ion constituent with mobility u;
x = coordinate along the direction of current, x increasing from 0 to the anode towards the cathode.

Each term in eqn. (1) expresses the mass flow per second and cm^2 of the cross-section, that to the left being the electric, that to the right the diffusional mass flows. If eqn. (1) is written in the form:

$$(iu/q)\ (dx/k) = D\ (dC/C) \qquad (2)$$

it is seen that it is possible to integrate it if u is known as a function of pH and D as a function of C. Specifically, if the conductance, the diffusion coefficient, and the derivative:

$$p = -du/dx = -[du/d(pH)\ [d(pH)/dx] \qquad (3)$$

(where p is the ratio between the protein titration curve and the slope of the pH gradient over the separation axis) can be regarded as constant within the focused zone, then $u = -px$ and one obtains the following analytical solution:

584

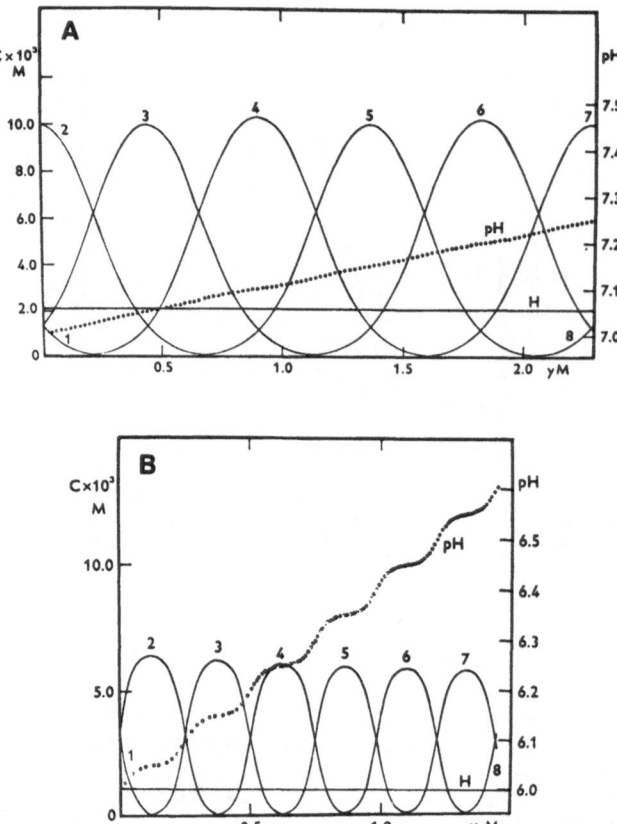

Figure 3. Computer simulation of focusing of ideal, mono-monovalent ampholytes with equal diffusion coefficients and electric mobilities. The abscissa (yM) has the dimensions of moles/liter; the left ordinate is in moles, H is the conductivity and the dotted line is the pH gradient. A: focusing of 8 carrier ampholytes over 0.25 pH units, having $\Delta pK = 2$ and spaced at $\Delta pI = 0.05$. B: focusing of 8 carrier ampholytes over 0.5 pH units, having $\Delta pK = 2$ and spaced at $\Delta pI = 0.1$ (reprinted with permission from Almgren, 1971).

$$C = C_o \exp[-(pix^2)/(2qkD)] \qquad (4)$$

where x is now defined as being $= 0$ at the concentration maximum C_o. This is a Gaussian concentration distribution with the inflection points at:

$$x_i = \pm\sqrt{(qkD)/(pi)} \qquad (5)$$

where x_i denotes the width of the Gaussian distribution of the focused zone measured from the top of the distribution of the focused ampholyte to the inflection point (one standard deviation). The course of the pH gradient is d(pH)/dx and du/d(pH) represents the titration curve of the ampholyte. It should be borne in mind that this Gaussian profile holds only if and as long as the conductivity of the bulk solution comprised within the zone is constant. Constant conductivity along a pH gradient is quite difficult to maintain, especially as one approaches pH extremes (below pH 4 and above pH 10), if for no other reason, because the non-negligible concentration of H^+ and OH^- present in the bulk liquid begin to strongly contribute. We have simulated such focusing conditions in two extreme pH intervals (pH 3-4 and pH 10-11) and across neutrality (pH 6.5-7.5) (Mosher et al., 1986). For simplicity, the gradients have been assumed to be formed by immobilized buffering species, which would contribute minimal conductivity to the liquid phase, and whose diffusion coefficients, transport numbers and mobilities would be zero. The gradients within the state pH intervals are assumed to be linear. In such simulations we have computed the steady-state distribution of three different ampholytes, all having pI in the middle of each gradient (thus pI 3.5, pI 7.0 and pI 10.5, respectively) but with different ΔpKs of 1, 2 or 3 (thus, in the example of the pH 3-4 interval, $\Delta pK = 1$ is an ampholyte with $pK_1 = 3.0$ and $pK_2 = 4.0$; $\Delta pK = 2$ has $pK_1 = 2.5$ and $pK_2 = 4.5$ and $\Delta pK = 3$ has $pK_1 = 2.0$ and $pK_2 = 5.0$; the same applies to the pI 7.0 and 10.5 ampholytes in the other two pH ranges). All ampholytes selected for simulation are simple biprotic species. The steady-state profiles have been computes for 3 different current densities of 1, 5 and 15 A/m^2, except for simulations across neutrality (pH 6.5-7.5) where much lower values had to be used (respectively 0.02, 0.1 and 0.3 A/m^2). The results are summarized in Fig. 1 (A to C). The steady-state profiles for the three different ΔpK species are shown: a) in the pH 3-4 gradient (left panel); b) in the pH 6.5-7.5 interval (central section) and c) in the pH 10-11 span (right panel). As shown in the left and right panels, at low current densities (1 A/m^2) the three ampholytes exhibit negatively (pH 3-4) and positively (pH 10-11) skewed profiles, the degree of skewness increasing substantially from the $\Delta pK = 1$ to the $\Delta pK = 3$ species. As a consequence, their kurtosis varies from leptokurtic ($\Delta pK = 1$) to mesokurtic ($\Delta pK = 2$) to platykurtic ($\Delta pK = 3$). As the current densities are increased to 5 (B panels) or to 15 (C panels) A/m^2, two phenomena are apparent: the kurtosis is reduced for the species with larger ΔpKs (i.e. the zones sharpen up into taller and narrower bands) and also the skewness is quite decreased. However, neither is completely eliminated and it is seen that even at the highest current densities the $\Delta pK = 1$ (considered to be a good carrier ampholyte) is still distinctly skewed. It should be noted that, in the pH 10-11 interval, both phenomena, skewness and kurtosis, are considerably reduced as compared with the pH 3-4 interval. This is due to the fact that the conductivity profile is not a steep exponential as in the acidic gradient (in general, the ratio of background conductivities is ca. 2:1). In turn, this is linked to the mobility of H^+ and OH^- ions, the former being 3.625×10^{-4}, the latter 1.985×10^{-4} (cm^2 V^{-1} sec^{-1} at 25°C in free solution). As a result of this, at any current density, the peaks focused in the pH 10-11 gradient are sharper than the corresponding ones in the pH 3-4 span; e.g., by measuring the fourth moment of the $\Delta pI = 1$ peaks at pI 10.5 and 3.5, respectively, at 5 A/m^2, it is found that the former is spread over 0.38 pH units, while the latter spans a pH 0.52 interval. Conversely, only focusing across neutrality generates symmetric steady-state profiles (see middle panels of Fig. 1) and this occurs at any current density and for the three pI 7.0 species, irrespective of their ΔpKs.

This does not pose a major problem for proteins, which usually have a rather small diffusion coefficient and remarkably good titration curves (*i.e.* with steep slopes about the pI value) (although one can appreciate that protein zones focusing at pH extremes have band-widths 2-3 times greater than in the pH 5-9 interval) but it could severely decrease resolution in case of peptides or small amphoteric molecules, such as amino acids.

1.1.2 Description of Infinite-Component Mixtures. In this section we propose a model for a mixture of chemical subsystems when the number of mixture components is sufficiently high (Babskii *et al.*, 1989). We can consider a mixture composed of a solvent (water), its ions, and an infinite number of closed chemical subsystems. We will characterize the individual closed chemical subsystems by the sorting parameter s. In order to describe an infinite component mixture, we will introduce the concentration distribution function for the components of the mixture, $X(s,x,t)$, depending on the multidimensional sorting parameter s. The composition of the mixture occupying the volume at the initial instant of time is specified by the global distribution function

$$M(s) = \int R\, X(s,x,t)dv \qquad (6)$$

where M is the mass concentration in a zone (moles/liter) and dv is an infinitesimal incremental volume. The physico chemical properties of such a mixture will be described using the distribution functions of the diffusion coefficient $D[s, \psi(x,t)]$, the mobilities $\mu[s, \psi(x,t)]$, the conductivities $\sigma(s, \psi(x,t))$, the diffusional conductivity $\sigma^0(s, \psi(x,t))$ and the molar charge $e[s, \psi(x,t)]$. We will recall that, for closed chemical subsystems, D, μ, σ, σ^0 and e depend only on the acidity of the mixture $\psi(x,t)$. We shall restrict ourselves to the case in which the mixture as a whole is at rest, its density is constant, there are no barodiffusion and thermal diffusion effects, the dielectric constant of the mixture equals that of the solvent and the D and μ values of individual components do not depend on temperature. We shall additionally assume that the mixture of chemical subsystems is electrically neutral. For a specified monotonic acidity profile $\psi(x)$, for a known set of amphoteric substances characterized by the quantities D, μ, σ, σ^0 and e above, it is possible to calculate the steady-state distribution profile M(s) if the potential V or the electric current density j are specified.

Let us now proceed with a practical example. In 1962, Svensson, upon a six-month search in chemical catalogues, listed a series (Table I) of amphoteric substances potentially useful in IEF. The quality mark of such substances is the | pI -pK | value, which should be as small as possible (ideally around 0.5; note that only 3 chemicals in Table I fulfill this requirement, aspartylaspartic acid, histidylhistidine and lysillysine). By simulating the steady-state distribution profile in the pH 2.7-10.7 range (j = 102 A/cm²; V = 3400) one obtains the results of Fig. 2: it is seen that it is possible to obtain a fairly linear pH gradient, followed by a more hectic pK variation. The M(s) profile (in moles/L) is clearly uneven, and it follows the relative distribution of pIs of the 37 compounds in Table I (note that there are 12 amphoteres in the pH 3-4, 5 in the pH 4-5 and 5 in the pH 5-6 ranges). Since a quite linear pH gradient has been obtained, one might think that such a mixtures of 37 amphoteric buffers is sufficient for good

Table I[*]

Possible carrier ampholytes arranged in order of pI values

Ampholyte	pI	pI - pK
Aspartic acid	2.77	0.89
Glutathione	2.82	0.70
Aspartyl-tyrosine	2.85	0.72
o-Aminophenylarsonic acid	3.00	0.77
Aspartyl-aspartic acid	3.04	0.34
p-Aminophenylarsonic acid	3.15	0.92
Picolinic acid	3.16	2.15
L-Glutamic acid	3.22	1.03
ß-Hydroxyglutamic acid	3.29	0.96
Aspartyl-glycine	3.31	1.21
Isonicotinic acid	3.35	1.51
Nicotinic acid	3.44	1.37
Anthranilic acid	3.51	1.47
p-Amino benzoic acid	3.62	1.30
Glycyl-aspartic acid	3.63	0.82
m-Amino benzoic acid	3.93	0.81
Diiodotyrosine	4.29	2.17
Cystinyl-diglycine	4.74	1.62
a-Hydroxyasparagine	4.74	2.43
a-Aspartyl-histidine	4.92	1.90
ß-Aspartyl-histidine	4.94	2.00
Cysteinyl-cysteine	4.96	2.31
Pentaglycine	5.32	2.27
Tetraglycine	5.40	2.35
Triglycine	5.59	2.32
Tyrosyl-tyrosine	5.60	2.08
Isoglutamine	5.85	2.04
Lysyl-glutamic acid	6.10	1.65
Histidyl-glycine	6.81	1.00
Histidyl-histidine	7.30	0.50
Histidine	7.47	1.50
L-Methylhistidine	7.67	1.19
Carnosine	8.17	1.34
a,ß-Diaminopropionic acid	8.20	1.40
Anserine	8.27	1.23
Tyrosyl-arginine	8.38	1.00
L-Ornithine	9.70	1.05
Lysine	9.74	0.79
Lysyl-lysine	10.04	0.59
Arginine	10.76	1.72

*From Svensson (1962).

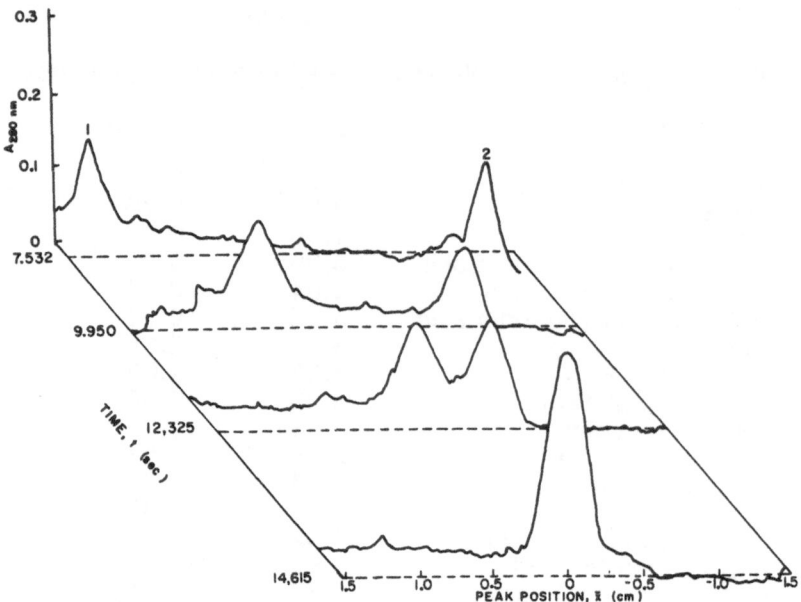

Figure 4. Scanning electrophoretic patterns of histidyl-tyrosine peaks migrating toward the pI position. Peak 1 migrates away from the negative electrode and peak 2 from the positive. The pI position is arbitrarily set at x = 0. The original sample distribution was uniform and the Ampholine concentration 2% (reprinted with permission from Catsimpoolas, 1976).

focusing. In fact, it is not, since the field intensity (E, in V/cm) is highly irregular, varying from as low as a few volts in regions of high σ up to >1200 V/cm across neutrality. Such an irregular V profile will never allow for proper sample focusing: in addition, due to σ minima around pH 7, with production of a tremendous joule heat, the liquid could literally boil over in such region (Rilbe, 1976). It will be noted, in fact, that there are huge pI gaps among tyrosyltyrosine (pI 5.85), isoglutamine (pI 6.10) lysylglutamic acid (pI 6.81) and histidylglycine (pI 7.30). Nature has been quite parsimonious in creating amphoteric species buffering around neutrality. Even protein lysates, used as background electrolytes for IEF, did not help much: very few peptides focus and buffer around neutrality. The problem was solved by Vesterberg (1969), who advanced the brilliant idea of producing a highly heterogeneous mixture of amphoteres by reacting oligoamines (up to six nitrogen long) with acrylic acid: a 'chaotic' organic chemistry, able to produce several hundred amphoteres with the required property of closing the gaps around neutrality and producing a fairly even buffering power all along the pH scale. It might be asked, then, how many different species of carrier ampholytes should be present per pH unit for generation of a stepless pH course. This question can be answered by formulating a condition for completely unresolved double zones of adjacent carrier ampholyte peaks. As demonstrated by Svensson (1966), the sum of two similar Gaussian curves has one flat maximum and no minimum at all for a peak to peak distance equal or less than two standard deviations. If all CA

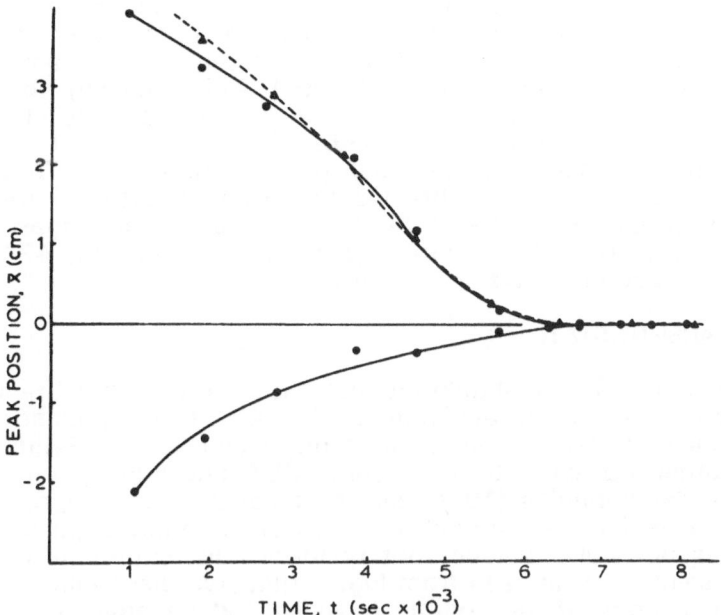

Figure 5. Plot of peak position difference (Δx) versus time. Histidyltyrosine, 2% Ampholine concentration (reprinted with permission from Catsimpoolas, 1976).

buffers lie as closely spaced as that, all of them are completely unresolved from their neighbours and no pH plateau can be expected. Therefore, the greatest allowable pI difference between adjacent CA buffers compatible with a smooth pH course is given by the inequality:

$$\Delta pI < 2 \sqrt{[D \ d(pH/dx) \ (1 + s)]}/Eu \ln 10 \qquad (7)$$

where the lower is the s value of a CA species, the better is its conductivity and buffering capacity.

Could one actually calculate how many different species of carrier ampholytes should be present per pH unit for generation of a stepless pH course? In a real system, this might turn out to be an impossible task. However, for an ideal system, comprising monovalent ampholytes with equal diffusion coefficients and electric mobilities, equal relative concentrations and evenly spaced pI values on the pH scale and exhibiting the same difference between their two pK values, Almgren (1971) has been able to simulate the steady-state profiles. In order to solve this problem, we have to know the following parameters: ΔpI and ΔpK and the width of the Gaussian distribution curve at the steady-state (expressed in the abscissa of Fig. 3A,B as yM: because it is a column coordinate, but it has the dimensions of moles/liter). As shown in Fig. 3A, a system comprising 8 carrier ampholytes, distributed over 0.25 pH units (from pH 7.0 to 7.25), in which each individual species has a $\Delta pK = 2$, and spaced at $\Delta pI = 0.05$, is able to generate smooth pH and conductivity (H) courses. According to these data, it would appear that at least 30 carrier ampholytes/pH unit are needed for production of a stepless pH gradient. Conversely, is the same CA species ($\Delta pK = 2$) are now spaced at $\Delta pI = 0.1$ (thus halving the number of amphoteres to 15/pH unit) the pH gradient will increase by steps (Fig. 3B). These plateaus are centred in the middle of each peak, while the pH grows linearly in the boundary between each pair of ampholytes. Thus, the notion that good focusing can be obtained by using an artificial mixture of 47 buffers, as proposed in 1982 by Cuono and Chapo, goes against the elementary rules of IEF. Needles to say, this idea of the 47 buffers was abandoned as soon as it was proposed.

1.2 THE TRANSIENT STATE

It sounds odd that one should mention the transient after the steady-state, but this is historically the evolution of the field. Nobody elaborated on this topic, having the steady-state so elegantly discussed by Svensson (as well as, as it turned out later, by Kaumann, 1957, and Schumaker, 1957). Only much later Catsimpoolas (1976) started measuring the approach to the pI position. He devised a protocol in which an ampholyte (histidyltyrosine in Fig. 4) was not pulse loaded, but uniformly distributed between the two electrodic solution. It is seen from Fig. 4 that two discernible peaks migrate from the two ends of the column (positive and negative) until they merge into one at the pI position. These data illustrate one of the fundamental aspects of the transient state in IEF: the changes which occur begin at the boundaries (electrodes) and propagate inward. If one plots the peak position difference (Δx) versus time (Fig. 5) one can appreciate another phenomenon: at any given time the cation (the positively-charged species moving away from the anode) move faster toward the pI position than anions (*i.e.* the species repelled from the cathode). For simple amphoteric species, this might be due to the well known fact that anions are in general more hydrated that cations (Edward and Waldron-Edward, 1965), so that their radius is somewhat more larger and thus the velocity (which depends on the charge/radius ratio) correspondingly smaller. For proteins, there is usually another added cause: the two branches of the titration curve about the pI value might have quite different slopes, so that the net positive or negative charges away from the pI value are unequal: this calls for markedly different rates of approach to the pI position. We can recall here at least one glamorous example (Righetti, 1983): bovine carbonic anhydrase presents a family of 6-8 isoforms, all of which focus as sharp bands if applied on the anodic side, whereas, upon cathodic sample loading, the two most basic isoforms never seem to be able to reach the pI position. Upon performing a two-dimensional electrophoresis procedure displaying the 'titration curves' of all species, it was found that most of these isoforms had very steep pH/mobility curves on both sides of the pI value, and thus would focus sharply independently from the anodic or cathodic application point in IEF. However, the two most basic isozymes had very flat titration curves above their pI values (anionic species), while presenting a steeper slope in the branch below the pI (cationic species). Consequently, these two bands focus sharply when driven to equilibrium from their anodic side, but present diffuse zones, unable to reach their pI position, if applied at the cathode. These

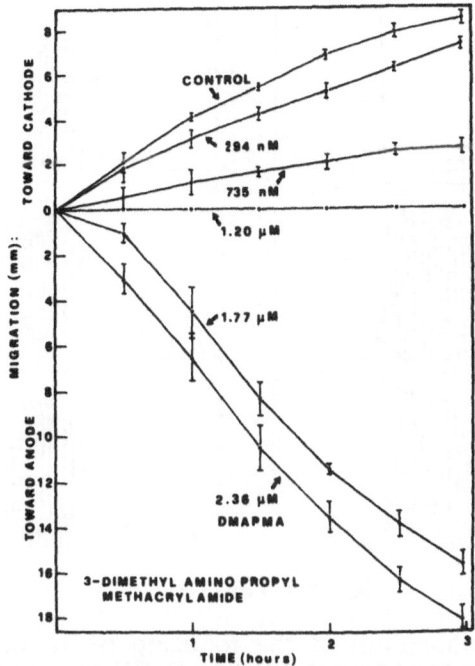

Figure 6. Effect of a tertiary base on the cathodic drift. The gel was polymerized either as such (control, no DMAPMA added) or in presence of increasing amounts of 3-dimethylaminopropylmethacrylamide (DMAPMA), from a minimum of 294 nM up to 2.36 μM. Notice that, above the equivalence point (1.20 μM DMAPMA added) the focused hemoglobin zones start drifting towards the anode (drift reversal). The gel was 6%T, 4%C Bis matrix, 360 μm thick, and contained 2% Ampholine pH 3.5-10. Focusing was at 10°C, 20 min pre-run, followed by 1 h sample run (normal human adult hemoglobin). The time period marked on the abscissa begins after this total time of 80 min. All cathodic (or anodic) drift measurements were made at a constant voltage of 1000 V (100 V/cm) (reprinted with permission from Righetti and Macelloni, 1982).

phenomena should be borne in mind when focusing proteins.

1.3 THE 'DECAY' STATE

Equilibria are the most difficult thing to maintain here in earth (and perhaps in the universe). It is thus no surprise that IEF, guarantied by mathematical formulae to be indefinitely stable, would be subject to decaying in real life. But the causes of this degradation have been quite obscure for a long time. Murel *et al.* (1979) by computer simulation, have suggested that pH gradient stability may depend on a balance between mobilities and concentrations of protons and hydroxyl ions migrating in opposite directions. In other words, according to these authors, the 'proton could' generated at the anode travels towards the cathode faster than the ' -OH⁻ cloud' moving in the

592

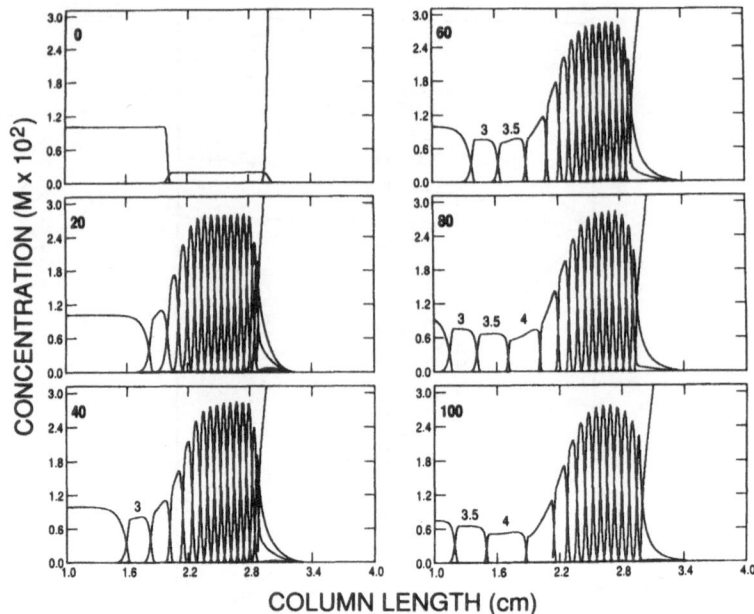

Figure 7. Computer simulation of the focusing of 15 ampholytes with 10 mM phosphoric acid as anolyte and 90 mM NaOH as catholyte. The anode is to the left. The ampholytes are initially distributed uniformly, 2 mM each, at time zero (upper left-hand panel) over the central 1 cm. Time points presented are after 20, 40, 60, 80 and 100 min of current flow. At this current density the ampholytes are focused by 20 min. The migration of phosphoric acid toward the anode causes the acidic ampholytes to form an isotachophoretic stack which results in those ampholytes being lost from the focusing space. The ampholytes zones are identified by their pIs (reprinted with permission from Mosher and Thormann, 1990).

opposite direction, hence a net drift towards the cathode. They suggest a mechanism of 'pressure' of the proton cloud, just like in a thermoionic valve: such 'cloud' has the effect of protecting the anodic end and pushing the cathodic end of the pH gradient out of the electrophoretic space. If this hypothesis were correct, equalization of the product (H^+ molarity x H^+ mobility) with the product (-OH^- molarity x OH^- mobility) should stop the cathodic drift, but this does not seem to be the case (Delincée, 1980). Probably, the 'cathodic drift' (so called because the pH gradient decay is most often at the cathodic end) has a multifactorial origin. One of the possible causes could be ingrained in the anticonvective support itself. The most popular matrix used today (a polyacrylamide gel), while believed to be devoid of ionizable groups, is indeed not quite neutral. We have found that, on the contrary, polyacrylamides are negatively charged due to: (a) trace impurities of acrylic acid in the gel; (b) covalent incorporation of catalysts (persulphate) as terminal groups in the matrix chains; and (c) hydrolysis of amide groups to acrylic acid in the gel layer close to the cathodic gel end (Righetti and Macelloni, 1982). Thus, the only way we found effective in

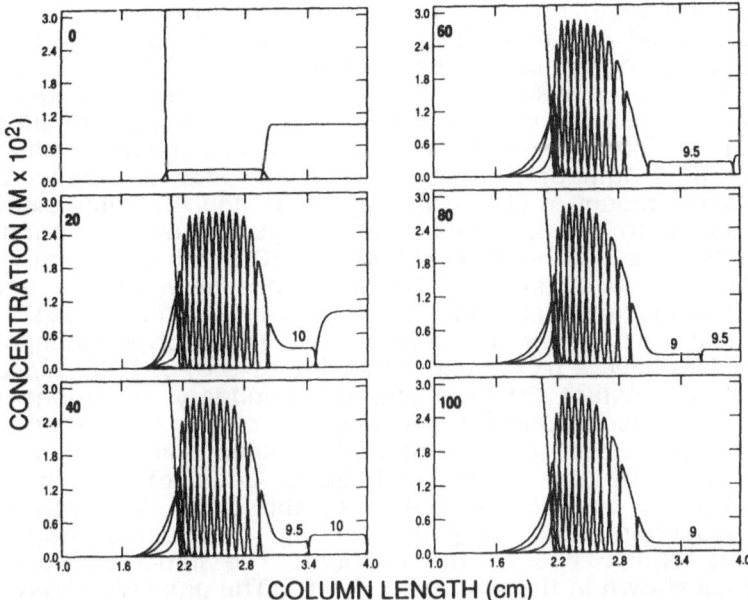

Figure 8. Computer simulation of the focusing of 15 ampholytes with 100 mM H₃PO₄ and 10 mM NaOH as anolyte and catholyte, respectively. The anode is to the left. All other conditions as in Fig. 7. The NaOH acts as a leader in the isotachophoretic sense for the basic ampholytes, causing their progressive loss from the focusing space (the central 1 cm of the column). The ampholyte zones are identified by their pIs. None of the acidic ampholytes have been lost from the focusing space by 100 min (reprinted with permission from Mosher and Thormann, 1990).

blocking pH gradient decay in IEF was the production of balanced matrices', *i.e.*, the covalent incorporation of tertiary (or better quaternary) amines in stoichiometric amounts as compared with the gel negative charges. To this end we have utilized 3-dimethylamino propyl methacrylamide (DMAPMA) or methacrylamido propyltrimethyl ammonium chloride (MAPTAC), at such levels (in general μM) as to balance the autochthonous negative charges. As shown in Fig. 6, incorporation of DMAPMA in increasing amounts in polyacrylamide gels progressively quenches the cathodic decay down to 'zero drift' conditions at the equivalence point. When an excess of DMAPMA is incorporated into the gel, the drift reverses its polarity and becomes anodic. Thus, the fact that, according to the net charge incorporated into the gel, a cathodic drift, zero drift (balanced gels) or anodic drift can be generated, speaks in favour of electrosmosis being the major cause of gradient decay in polyacrylamide gels. Yet, even this might not be the ultimate cause for pH gradient decay. Svendsen and Schafer-Nielsen (1983) have suggested that pH gradients in IEF might decay by an isotachophoretic (ITP) mechanism. They additionally postulated that the nature of the drift could be controlled by appropriate choice of electrolyte type and concentration as well as the volume of the electrode reservoir. This model was further expanded, recently, by Mosher and Thormann (1990). These authors have used for their simulation work the generalized model for transient electrophoretic processes of Bier *et al.* (1983). This model is isothermal and one-dimensional and assumes the absence of liquid flows. It predicts the evolution of concentration, pH and conductivity profiles as a function of time. Inputs required include the pK and mobility values of each component, the length of the separation space and its segmentation, the current density, amount of electrophoresis time, and the initial distribution of each component. In the particular case of Figs. 7 and 8, a mixture of 15 biprotic ampholytes with equally spaced pIs (0.5 pH units) between pH 3 and 10 was used for simulation. The difference between pK_1 and pK_2 for each was 2 units, and each ionic mobility was 3×10^{-8} m²/Vs. The ionic mobility of Na$^+$ was 5.19×10^{-8} m²/Vs, while the anolyte (phosphoric acid) was assumed to have a pK = 2 and a mobility of 3.67×10^{-8} m²/Vs. Current density was a constant 20 A/m² and the boundary conditions employed allowed for free mass transport into and out of the separation space. By these simulations it was found that both, anodic and cathodic drifts (*i.e.* progressive loss of the acidic and basic ends of the pH gradient, respectively) were possible, according to the initial concentration in the electrolyte reservoirs. The six panels of Fig. 7 illustrate the progressive decay of the anodic end of the gradient. The initial distribution of all components is shown in the first panel (T = 0). The ampholytes are uniformly distributed at a concentration of 2 mM across the central 1 cm of the separation axis (the focusing space). The anolyte (phosphoric acid, 10 mM) occupies the left side (anodic portion) of the separation space. The catholyte (NaOH, 90 mM) is located in the right side. The additional 5 panels present the concentration distributions of all components at 20 min intervals up to 100 min. In the second panel (T = 20), it is seen that the ampholytes have focused and the H_3PO_4 boundary has been displaced towards the anode, whereas the Na$^+$ boundary remains stationary. The 40 min profile shows ampholyte pI = 3 (the most acidic species) clearly forming an isotachophoretic zone, with a characteristic plateau shape, behind the phosphate boundary. In the 60 and 80 min profiles it is seen that the adjacent species (pI = 3.5 and pI = 4) are also mobilized by forming an isotachophoretic train; in addition, the phosphoric acid zone has nearly left the electrophoretic space. In the last panel (T = 100 min) a stack of ampholytes has formed and the most acidic species (pI = 3) has now vacated the column too. Note that, according to the Kohlrausch autoregulating function, the adjusted concentrations decrease from the pI = 2 to the pI = 4 zones, as typical of an isotachophoretic stack. Conversely, at this current density and time interval, the NaOH boundary has hardly moved and, as a consequence, no ampholytes are lost from the cathodic end of the gradient. If now one reverses the molarity of the two end electrolytes, one should then observe a cathodic drift. This situation is simulated in Fig. 8: here the anolyte is 100 mM H_3PO_4 and the catholyte is reduced to 10 mM NaOH. The same mixture of ampholytes of Fig. 7 is used, with the same initial distribution (T = 0, upper left panel). At T = 20 min, the ampholytes are focused and the Na$^+$ boundary has migrated a substantial distance toward the cathode. This boundary is now acting as a leading ion for ampholyte pI = 10, which is forming an isotachophoretic zone immediately behind. At T = 40 and 60, additional ampholytes are seen stacking behind the leading ion, and again adjusting their plateau concentration according to their mobility. At T=60, the entire Na⁺ zone has vacated the electrophoretic space and also the pl = 10 ampholyte has almost completely disappeared. The decay progresses as time goes by and new stacked zones begin to form and move out. According to these data, it would appear that, if one were to equalize the product of the concentration x mobility of the two end species, one might hope to block such drift. In fact, equalization of the velocities of the two boundaries (which occurs when the NaOH concentration is 2.25 fold the H_3PO_4 molarity) has merely the effect of producing a symmetrical drift, *i.e.* a pH gradient which decays at both extremes.

2. Immobilized pH Gradients

Pédant: 'L'animal seul, monsieur, qu'Aristophane
Appelle Hippocampelephantocamélos
Dut avoir sous le front tant de chair sur tant d'os!'
La tirade du nez, from Cyrano de Bergerac of Edmond Rostand

We have seen, in the first part of this 'tirade', how much 'flesh' there was in conventional IEF; now, in this second part, let us examine the 'bone' underneath it. Conventional IEF, in addition to the decay problems outlined above, suffers from several drawbacks, such as: a) uneven buffering capacity; b) uneven conductivity; c) unknown chemical environment; d) very low and unknown ionic strength; e) cathodic drift (or pH gradient instability) and f) unwieldy to pH gradient engineering. Most of these vexatious problems have never been solved in 25 years of focusing with the carrier ampholyte buffers. For all these reasons, as a result of an intensive collaboration with Dr. Bjellqvist group in Stockholm and Dr. Görg's group in Munich, in 1982 we launched the technique of immobilized pH gradients (IPG) (Bjellqvist *et al.*, 1982). IPGs are based on the principle that the pH gradient, which exists prior to the IEF run itself, is copolymerized, and thus insolubilized, within the fibres of a polyacrylamide matrix. This is achieved by using, as buffers, a set of non-amphoteric, weak acids and bases, having the following general chemical composition: $CH_2 = CH-CO-NH-R$, where R denotes either weak carboxyl or weak tertiary amino groups. Six of these chemicals are available, from Pharmacia-LKB, under the trade-name Immobiline; however, we have described a total of 16 of these compounds, whose formulas and general properties can be found in a recent survey (Chiari *et al.*, 1991). During gel polymerization, these buffering species are efficiently incorporated into the gel (84-86% conversion efficiency at 50°C for 1 h). Immobiline-based pH gradients can be cast in the same way as conventional polyacrylamide gradient gels, using a density gradient to stabilize the Immobiline concentration gradient, with the aid of a standard, two-vessel gradient mixer. As shown by their chemical formulae, these buffers are not any longer amphoteric, as in conventional IEF, but are bifunctional. At one end of the molecule is located the buffering (or titrant) group, and at the other end is an acrylic double bond which disappears during immobilization of the buffer in the gel matrix. IPGs have proven most successful in protein fractionation, and today are believed to represent the electrokinetic technique with the highest resolving power. The technique is

596

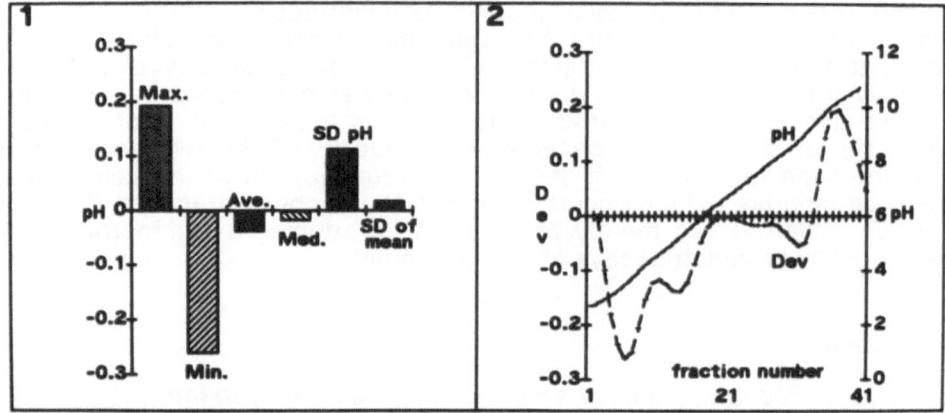

Figure 9. pH profile and deviation from linearity of the pH 2.5-11 recipe of Table II. Max. and Min.: maximum and minimum deviation; Ave.: average; SD: standard deviation.

straighforward in its use; however, the complexity of the system came when formulating extended pH gradients. It is thus our pleasure to recollect here the history of development of our computational approach to IPGs and to give new examples of this method. Soon after publishing the fundamentals of the technique, it became apparent to us that IPGs would not survive if we could not find a way to compute recipes for extended pH gradients (we recall here that IPGs were born as a continuation of conventional IEF in CA buffers and originally were only meant to extend the fractionation capability of the conventional technique to extremely narrow pH ranges). In fact, in the cornerstone publication (Bjellqvist *et al.*, 1982), only modified Handerson-Hasselbalch's equations were given for computing narrow to ultranarrow pH gradients, extending to a maximum of 1 pH unit. Such equations were based on the 'tandem' principle, *i.e.* on the use of only two Immobiline species, one buffering and the other titrating within the desired pH interval. However, when several buffering species had to be mixed and titrated to form an extended pH gradient, the problem became quite complex. A starting point and a model for the calculations were also missing, when one day we realized that we could further develop the basic concepts of gradient elution as first proposed in 1959 by Peterson and Sober. We were so fond of this idea, that we even built a multichambered mixing device, which soon became a museum piece (Dossi *et al.*, 1983). In our first computer simulation of extended pH gradients, we introduced:
a) the dissociations constants of acidic and basic Immobilines;
b) the law of electroneutrality;
c) a polynomial calculating the value of $[H^+]$ (and thus pH) along the titration as a function of all dissociated species present;
d) the generalized Peterson-Sober equation (1959), for solving the above polynomial by calculating the actual concentration of each Immobiline in the

output flow;

e) an equation for calculating the ionic strength along the pH gradient;

f) an equation for computing the buffering power (ß) during titration.

In practice, this procedure proved too laborious: we had to prepare 5 different solutions (as we used a 5 chamber mixer) pre-titrated to given pH values, *in crescendo*, from anode to cathode. This approach was thus abandoned in favour of a two-vessel gradient mixer (Gianazza *et al.*, 1983): the same set of equations was used; in addition we introduced an optimization algorithm (based on Cauchy's minimum slope criterion) which had as a goal the linearization of the pH gradient, by keeping the ß power constant along the titration. This version proved to be quite popular, and allowed us to publish a series of wide pH gradients (spanning 2-6 pH units) optimized in terms of linearity of pH gradient and constant ß profile (Gianazza *et al.*, 1984). The set goal: within each formulation, the deviation from linearity had to be maintained within 1% of the stated pH interval. This meant that, in a 2-pH unit gradient, the deviation at no point could be greater than 0.02 pH unit, or 0.03 pH unit in a 3-pH unit gradient and so on. In this paper, we described two approaches to the generation of extended pH gradients: a) in one case, the 'same concentration' of buffering Immobilines would be placed in the two vessels of the mixer; b) in the other, 'different concentrations' of buffering groups could be present in each chamber. The latter case (a sort of unorthodox titration) allows the shift of the apparent pK of a given buffer, thus filling-in 'holes' of buffering power produced by too wide ΔpK values.

Even this last approach would not have been optimal in quite a few cases: e.g., in developing two-dimensional (2-D) maps, best resolution in the focusing dimension would clearly be obtained by non-linear pH gradients, following the relative abundance of isoelectric proteins along the pH scale. Thus, we soon recalculated wide, non-linear IPG recipes for use in 2-D maps and in all cases requiring analysis of highly heterogeneous samples (Gianazza *et al.*, 1985a). This approach was particularly interesting as it proved, for the first time, the possibility of having in the gel cassette a non-linear pH gradient supported by a linear density gradient (the latter is always present when pouring IPGs, as the liquid elements, during delivery, have to be stabilized against convection and remixing). As at that time we had given extended recipes (especially the 5 and 6 pH unit intervals) utilizing, in addition to the set of 6 weak acidic and basic Immobilines, also two titrants, one strong acid and a strong base (not commercially available) we were forced by colleagues to publish a new set of formulations, optimized in the absence of such titrants (Gianazza *et al.*, 1985b). During that period, other types of focusing in non-amphoteric buffers were introduced: Chrambach's 'arrested stack' (Chrambach and Hjelmeland, 1984), Rilbe's (1978) 'steady-state rheoelectrolysis' and Bier's 'physically' immobilized pH gradients (Bier *et al.*, 1984). We were able to use our IPG computer program to re-simulate their data at the light of our acquired knowledge and find the merits and limits of each alternative route to 'non-amphoteric buffer focusing' (Righetti and Gianazza, 1985).

At the beginning of 1986, we started thinking of expanding the fractionation capability of IPGs: up to this moment, the most extended pH interval described was a pH 4-10. For this reason, we had not included the

598

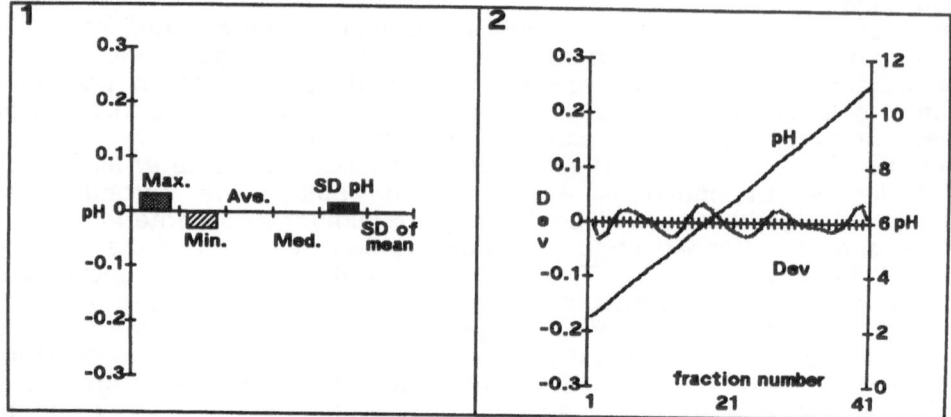

Figure 10. pH profile and deviation from linearity of the pH 2.5-11 recipe of Table III. Max. and Min.: maximum and minimum deviation; Ave.: average; SD: standard deviation.

dissociation products of water (H⁺ and OH⁻) in our simulations, since within the pH 4-10 range their concentration is negligible. At that time, we started focusing dansylated amino acids (which exhibited pIs in the pH 3-4 interval) (Bianchi-Bosisio *et al.*, 1986) and we realized that there was a strong divergence between calculated and experimental pH gradients: thus our computer program was expanded to include the effects of H⁺ and OH⁻ on ß, ionic strength and pH profile (Righetti *et al.*, 1986). As a provocative concept, we introduced the idea of water as two 'Immobiline' species, one with pK = -1.74 [H⁺], the other with pK = 15.75 [OH⁻]. In fact, simulations were not only limited to acidic, but included also quite basic (pH 10-11) intervals: in this latter case, we were able to model IPG behaviour by modifying the more general program for steady-state IEF, as developed by Bier *et al.* (1983). All the work produced up to the end of 1986 was summarized in an article (Celentano *et al.*, 1987) which gave the overall view of the computational method and described the architecture of the program.

In 1988, we started a long-range program on the characterization of existing Immobilines and on the synthesis of new species. We wanted in fact to expand to the utmost the capabilities of IPGs, so new chemicals were sorely needed. In a series of papers, we could in fact:
a) describe the synthesis and purification of existing Immobilines (Chiari *et al.*, 1989a, b);
b) produce a new, more acidic acrylamido buffer with a pK of 3.1 (Righetti *et al.*, 1988) for fractionations extending down to pH 2.5;
c) prepare two analogues of the two morpholino species (pK 6.2 and 7.0) having slightly increased pK values (6.6 and 7.4, respectively) by substituting the morpholino- with a thio-morpholino-ring (Chiari *et al.*, 1990a);
d) synthesize a new hydrophilic Immobiline with a pK = 8.05 (Chiari *et al.*, 1990b).

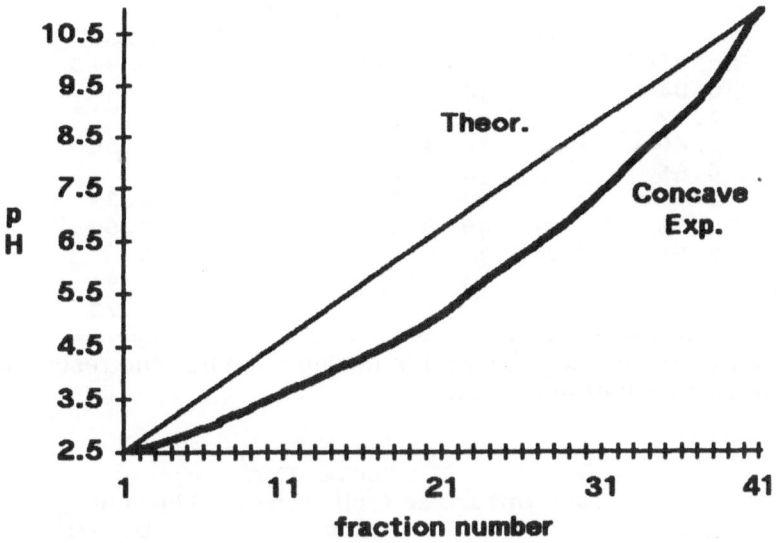

Figure 11. Concave exponential (Exp.) gradient obtained with the recipe of Table IV. Theor.: theoretical linear curve.

At this point the family of 'Immobilines' had considerably expanded and our former program (which was limited to mixtures of no more than 10 species, including buffers and titrants) could not handle the new generation any longer. In addition to this, some authors (Charlionet *et al.*, 1984) had described oligoprotic acrylamido buffers, and these species too were not contemplated in our computational approach. Thus, we changed completely our pH simulation program and we described a new, powerful system, having the following characteristics:
a) it is under MS-DOS, thus it is fully compatible with IBM hardware;
b) the modules are written in FORTRAN 77 subset implemented by the Microsoft FORTRAN 3.2 compiler, with the graphic functions in GW-BASIC;
c) it allows computation of pH gradient, ß and ionic strength not only of monoprotic, but also of polyprotic (polyacids, polybases, zwitterions) buffers each with up to 10 protolytic groups;
d) it can handle a much more complex mixture of up to 48 polyprotic buffers plus two titrants (as opposed to 10 monoprotic species in the former program).
e) it includes automatically in the computation the two 'Immobilines' generated by water dissociation (inserted in the formulations as a base, pK = 0, and an acid, pK = 14, taken at unit activity, *i.e.* 1 Mole/L).

As the accompanying equations describing the system are quite complex, we are just quoting here this recent development and refer the interested readers to the original articles (Celentano *et al.*, 1988a, b; Righetti *et al.*, 1988). As a result of this new program, we have recently given new formulations for IPGs including pH extremes: the widest possible pH range that could be formulated is a pH 2.5-11 interval, spanning 8.5 pH units (Gianazza *et al.*, 1989).

Table II[*]
Recipe for a pH 2.5-11 IPG gradient

acidic[1]	Immobiline	basic[1]
20.04	pK 0.8	0
2.97	pK 3.1	0
3.60	pK 4.6	9.56
5.86	pK 6.2	0
0	pK 7.0	6.33
10.38	pK 8.5	2.69
2.25	pK 9.3	0
0	pK 10.3	6.94
0	pK 14	8.73

[1]Buffer concentrations (mM/L) in mixing chamber and reservoir.
[*]from Gianazza et al., 1989.

Table III[*]
New optimized pH 2.5-11 IPG recipe

acidic[1]	Immobiline	basic[1]
24.85	pK 0.8	0
1.20	pK 3.1	39.90
0.57	pK 3.6	2.29
0.73	pK 4.6	8.75
4.94	pK 6.2	0.18
0.06	pK 7.0	4.49
4.08	pK 8.5	2.14
3.84	pK 9.3	1.57
8.59	pK 10.3	1.77
0	pK 14	52.11

[1]Buffer concentrations (mM/L) in mixing chamber and reservoir.
[*]from Tonani and Righetti, unpublished.

Table IV[*]
Concave exponential pH 2.5-11 IPG recipe

acidic[1]	Immobiline	basic[1]
23.12	pK 0.8	0
0.31	pK 3.1	5.36
1.99	pK 3.6	3.09
4.02	pK 4.6	3.41
3.44	pK 6.2	2.65
0.12	pK 7.0	1.85
3.06	pK 8.5	1.91
6.57	pK 9.3	0.82
6.57	pK 10.3	0.21
0	pK 14	12.91

[1]Buffer concentrations (mM/L) in mixing chamber and reservoir.
[*]from Tonani and Righetti, unpublished.

More recently, we have modified our monoelectrolyte gradient simulator (MGS) program, by introducing new optimization algorithms, with better target functions (Tonani and Righetti, 1991; Righetti and Tonani, 1991a). We shall give here an example on the performance of this new program. As just stated above, we were able to calculate, with the Celentano et al. (1988a) program, new extended recipes, spanning as much as 8.5 pH units (Gianazzα et al., 1989). Table II gives such a recipe, spanning a pH 2.5-11 range, as calculated and optimized using as a target function the smoothing of buffering power (ß). Fig. 9 gives the profile and the statistics on this formulation: first of all, it is seen that, while reasonably linear in the pH 3.5-10 interval, it deviates substantially from linearity outside these boundaries with a maximum negative deviation of -0.25 units (at ca. pH 3) and a maximum positive of +0.2 units (at ca. pH 10.2). The reasons: smoothing of ß works at its best only when, for each buffering Immobiline, the same molarity is used in each of the two vessels; in addition, such a target cannot function outside a pH 4-10 interval, due to the buffering power of water, represented by two branches of an hyperbole. We have resimulated the pH 2.5-11 interval with the new program MGS-2: this program has a new target function to be minimized, namely the standard deviation of the residues from linearity, taking as ideal shape a straight pH line joining the two pH extremes in the limiting solutions of the two starting recipes (in the mixing chamber and reservoir). Table III gives the new recipe, and Fig. 10 offers the 'pedigree' of such an interval. It is immediately apparent how, by using the same set of 9 chemicals (plus the pK 3.6), the new optimized gradient is remarkably linear. In the points of maximum deflection, the positive and negative deviations from linearity are never >0.03 pH units (compare with 0.2 and 0.25, respectively, in the previous recipe). Thus, it is seen that this new optimization algorithm can perform almost 10 times better than the previous one. The 'philosophy' of the new algorithm can be appreciated on the profile of the deviation curve in Fig. 10A: it is seen that such deviations oscillate constantly around zero, in an almost sinusoidal way. In other words, each positive deviation is counteracted immediately by a negative one, the sum of the areas of positive and negative deviations tending to zero (see the bar SD of mean in Fig. 10B). We believe that this is the most promising 'philosophy' by which an approach to linear gradient optimization (utilizing different molarities of the same buffering ion in the two chambers) can properly work. However, our new program can do even better than that: in fact, when working with 2-D maps or with very complex samples, it would not pay to use a linear wide gradient, since 70% of the proteins have pIs below pH 7 and only 30% are alkaline. Thus, ideally, the pH 2.5-11 IPG described above should be non-linear, in fact should be a concave exponential, with a pH shape following the relative abundance of proteins along the pH scale. Table IV and Fig. 11 give, respectively, the recipe and the profile for non-linear pH 2.5-11 interval, as calculated by our new program. This new recipe has exactly the required properties, since it gives 2/3 of the separation space to acidic proteins and only 1/3 to basic species. It should be the ultimate in 2-D maps of highly heterogeneous samples (work in progress).

Conclusions
We have reviewed here 30 years of isoelectric focusing, from an early,

confused stage up to present-day highly sophisticated techniques. With the latest variant, immobilized pH gradients (IPG), and the well-known chemistry of the buffering and titrant compounds, coupled to highly evolved expert systems simulating the most complex pH gradient courses, it is felt that the technique has reached a plateau and is currently trouble-free. What can be envisaged in the future? A big challenge is represented by capillary zone electrophoresis, which couples the high resolving power typical of electrokinetic methodologies with an instrumental approach, offering full automation and precise peak quantitation. The IPG technique still has much to offer: on the preparative scale, with the revolutionary approach of multicompartment electrolyzers with isoelectric membranes (Righetti *et al.*, 1989, Righetti *et al.*, 1990), coupling a very high resolving power with a very high load ability. Outside the realm of electrophoresis, with the recent invention of 'macroreticulate buffers', based on the Immobiline technology, able to support cellular growth even under the most adverse conditions, by offering an extremely high buffering power with no ionic strength increments of the bulk culture liquid (Righetti *et al.*, 1991b, Pompei *et al.*, 1991).

Acknowledgements
Supported in part by Agenzia Spaziale Italiana (ASI, Roma) and by Progetto Finalizzato Biotecnologia e Biostrumentazione (CNR, Roma).

3. References

Almgren, M. (1971) 'Isoelectric fractionation, analysis and characterization of ampholytes in natural pH gradients. X: Conditions for linear pH courses', Chem. Scripta 1, 69-75.

Babskii, V.G., Zhukov, M.Y. and Yudovich, V.I. (1989) Mathematical Theory of Electrophoresis, Plenum Press, New York.

Bianchi-Bosisio, A., Righetti, P.G., Egen, N.B. and Bier, M. (1986) 'Isoelectric focusing of dansylated amino acids in immobilized pH gradients', Electrophoresis 7, 128-133.

Bier, M., Palusinski, O., Mosher, R.A. and Saville, D.A. (1983) 'Electrophoresis: mathematical modelling and computer simulations', Science 219, 1281-1287.

Bier, M., Mosher, R.A., Thormann, W. and Graham, A. (1984) 'Isoelectric focusing in stable preformed buffer pH gradients', in H. Hirai (ed.), Electrophoresis '83, de Gruyter, Berlin, pp. 99-107.

Bjellqvist, B., Ek, K., Righetti, P.G., Gianazza, E., Görg, A., Westermeier, R. and Postel, W. (1982) 'Isoelectric focusing in immobilized pH gradients: principle, methodology and some applications', J. Biochem. Biophys. Methods 6 317-339.

Catsimpoolas, N. (1976) 'Transient state isoelectric focusing', in N. Catsimpoolas (ed.) Isoelectric Focusing, Academic Press, New York, pp. 1-11.

Celentano, F., Gianazza, E., Dossi, G. and Righetti, P.G. (1987) 'Buffer systems and pH gradient simulation', Chemometr. Intel. Lab. Systems 1, 349-358.

Celentano, F.C., Tonani, C., Fazio, M., Gianazza, E. and Righetti, P.G. (1988a) 'pH gradients generated by polyprotic buffers. I: theory and computer simulation', J. Biochem. Biophys. Methods 16, 109-128.

Celentano, C.F., Gianazza, E., Tonani, C. and Righetti, P.G. (1988b) 'Polyelectrolyte gradients simulation: theoretical and computational aspects', in C.Schafer-Nielsen (ed.) *Electrophoresis '88*, VCH, Weinheim, pp. 15-27.

Charlionet , R., Sesboüé, R. and Davrinche, C. (1984) 'Easy home made immobilized pH gradients for isoelectric focusing', Electrophoresis 5, 176-178.

Chiari, M., Casale, E., Santaniello, E. and Righetti, P.G.(1989a) 'Synthesis of buffers for generating immobilized pH gradients. I: acidic acrylamido buffers', Theor. Applied Electr. 1, 99-102.

Chiari, M., Casale, E., Santaniello, E. and Righetti, P.G. (1989b) 'Synthesis of buffers for generating immobilized pH gradients. II: basic acrylamido buffers', Theor. Applied Electr. 1, 103-107.

Chiari, M., Righetti, P.G., Ferraboschi, P., Jain, T. and Shorr, R. (1990a) 'Synthesis of thiomorpholino buffers for isoelectric focusing in immobilized pH gradients', Electrophoresis 11, 617-620.

Chiari, M., Pagani, L., Righetti, P.G., Jain, T., Shorr, R. & Rabilloud, T. (1990b) 'Synthesis of an hydrophilic, pK 8.05 buffer for isoelectric focusing in immobilized pH gradients', J. Biochem. Biophys. Methods 21, 165-172.

Chiari, M., Ettori, C. and Righetti, P.G. (1991) 'Capillary zone electrophoresis analysis of acrylamido buffers for isoelectric focusing in immobilized pH gradients', J. Chromatogr. 559, 119-131.

Chrambach, A. and Hjelmeland, L.M. (1984) 'Some recent conceptual advances in moving boundary electrophoresis and their practical implications' in H. Hirai (ed.) Electrophoresis '83, de Gruyter, Berlin, pp. 81-97.

Cuono, C.B. and Chapo, G.A. (1982) 'Gel electrofocusing in a natural pH gradient of pH 3-10 generated by a 47-component buffer mixture', Electrophoresis 3, 65-70.

Delincée, H. (1980) 'The effect of electrode solutions on the pH drift in thin-layer isoelectric focusing', in B.J. Radola (ed.) Electrophoresis '89, de Gruyter, Berlin, pp. 165-171.

Dossi, G., Celentano, F., Gianazza, E. and Righetti, P.G. (1983) 'Isoelectric focusing in immobilized pH gradients: generation of extended pH intervals', J. Biochem. Biophys. Methods 7, 123-142.

Edward, J.T. and Waldron-Edward, D (1965) J. Chromatogr. 20, 563-573.

Gianazza, E., Dossi, G., Celentano, F. and Righetti, P.G. (1983) 'Isoelectric focusing in immobilized pH gradients: generation and optimization of wide pH intervals with two-chamber mixers', J. Biochem.Biophys. Methods 8, 109-133.

Gianazza, E., Celentano, F., Dossi, G., Bjellqvist, B. and Righetti, P.G. (1984) 'Preparation of immobilized pH gradients spanning 2-6 pH units with two chamber mixers: evaluation of two experimental approaches', Electrophoresis 5, 88-97.

Gianazza, E., Giacon, P., Sahlin, B. and Righetti, P.G. (1985a) 'Non-linear pH courses with immobilized pH gradients', Electrophoresis 6, 53-56.

Gianazza, E., Astrua-Testori, S. and Righetti, P.G. (1985b) 'Some more formulations for immobilized pH gradients', Electrophoresis 6, 113-117.

Gianazza, E., Celentano, F., Magenes, S., Ettori, C. and Righetti, P.G. (1989) 'Formulations for immobilized pH gradients including pH extremes', Electrophoresis 10, 806-808.

Kauman, W.G. (1957) 'On the electrophoretic separation of ampholytes in a medium of non-uniform pH', Classe Sciences Academie Royale Belgique 43, 854-868.

Kolin, A. (1976) 'Isoelectric focusing', in N. Catsimpoolas (ed.) Isoelectric Focusing, Academic Press, New York, pp. 1-11.

Mosher, R.A., Bier, M. and Righetti, P.G. (1986) 'Computer simulation of immobilized pH gradients at acidic and alkaline extremes: a quest for extended pH intervals', Electrophoresis 7, 59-66.

Mosher, R. and Thormann, W. (1990) 'Experimental and theoretical dynamics of isoelectric focusing. IV: cathodic, anodic and symmetrical drifts of the pH gradient', Electrophoresis 11, 717-723.

Murel, A., Kirjanen, I. and Kirret, O. (1979) 'Instability and non-linearity of the pH gradient formed in isoelectric focusing', J. Chromatogr. 174, 1-11.

Pompei, R., Lampis, G., Chiari, M. and Righetti, P.G. (1991) 'Culture of eukaryotic cells with macroreticulate buffers: fermentation of cellulolytic fungi', BioTechniques 11, 701-706.

Righetti, P.G. and Macelloni, C. (1982) 'New polyacrylamide matrices for drift-free isoelectric focusing', J. Biochem. Biophys. Methods 6, 1-15.

Righetti, P.G. (1983) Isoelectric Focusing: Theory, Methodology and Applications', Elsevier, Amsterdam.

Righetti, P.G. and Gianazza, E. (1985) 'Isoelectric focusing in non-amphoteric buffers: catastrophe and non-catastrophe theories', J. Chromatogr. 334, 71-82.

Righetti, P.G., Gianazza, E. and Celentano, F. (1986) 'A recipe for a pH 3-4 immobilized gradient for isoelectric focusing', J. Chromatogr. 356, 9-14.

Righetti, P.G., Fazio, M., Tonani, C., Gianazza, E. and Celentano, F.C. (1988a) 'pH gradients generated by polyprotic buffers. II: experimental validation', J. Biochem. Biophys. Methods 16, 129-140.

Righetti, P.G., Chiari, M., Sinha, P.K. and Santaniello, E. (1988b) 'Focusing of pepsin in strongly acidic immobilized pH gradients', J. Biochem. Biophys. Methods 16, 185-192.

Righetti, P.G., Wenisch, E. and Faupel, M. (1989) 'Preparative protein purification in a multi-compartment electrolyzer with Immobiline membranes', J. Chromatogr. 475, 293-309.

Righetti, P.G. (1990a) Immobilized pH Gradients: Theory and Methodology, Elsevier, Amsterdam.

Righetti, P.G., Wenisch, E., Jungbauer, A., Katinger, H. and Faupel. M. (1990b) 'Preparative purification of human monoclonal antibody isoforms in a multicompartment electrolyzer with Immobiline membranes', J. Chromatogr. 500, 681-696.

Righetti, P.G. and Tonani, C. (1991a) 'Immobilized pH gradient simulator: an additional step in pH gradient engineering. II: nonlinear pH gradients', Electrophoresis 12, 1021-1027.

Righetti, P.G., Chiari, M. and Crippa, L. (1991b) 'Macroreticulate buffers: a novel approach to pH control in living systems', J. Biotechnol. 17 169-176.

Rilbe, H. (1976) 'Theoretical aspects of steady-state isoelectric focusing', in N. Catsimpoolas (ed.) Isoelectric Focusing, Academic Press, New York, pp. 14-52.

Rilbe, H. (1978) 'Steady-state rheoelectrolysis' J. Chromatogr. 159, 193-205.

Schumacher, E. (1957) 'Über fokussierenden ionenaustausch IV. Zur theorie des fokussierungseffektes', Helv. Chim. Acta 40, 2322-2340.

Svendsen, P.J. and Schafer-Nielsen, C. (1983) 'Computer simulation of model isoelectric focusing experiments', in D. Stathakos, (ed.) Electrophoresis '82, de Gruyter, Berlin, pp.83-89.

Svensson, H. (1961) 'Isoelectric fractionation, analysis and characterization of ampholytes in natural pH gradients. I. The differential equation of solute concentration at a steady-state and its solution for simple cases', Acta Chem. Scand. 15, 325-341.

Svensson, H. (1962) 'Isoelectric fractionation, analysis and characterization of ampholytes in natural pH gradients. II. Buffering capacity and conductance of isoionic ampholytes', Acta Chem. Scand. 16, 456-466.

Svensson, H. (1966) 'A suggestion for the definition of zone (or boundary) resolution in physico-chemical separation techniques', J. Chromatogr. 25, 266-273.

Tonani. C. and Righetti, P.G. (1991) 'Immobilized pH gradient simulator: an additional step in pH gradient engineering. I: linear pH gradients', Electrophoresis 12, 1011-1021.

Vesterberg, O. (1969) 'Synthesis and isoelectric fractionation of carrier ampholytes', Acta Chem. Scand. 23, 2653-2666.

SOLUTE BAND SPREADING IN CAPILLARY ELECTROPHORESIS

ELI GRUSHKA
Department of Inorganic and analytical Chemistry
The Hebrew University
Jerusalem, ISRAEL

ABSTRACT. Classification of electrophoresis as a separation technique and its relationship to chromatography is discussed. Special emphasis is placed on capillary electrophoresis (CE). The separation powers of both techniques will be analyzed. The mass balance approach to solute behavior in CE is discussed. Initially, analytical solutions to the mass balance equations are given. Four cases are examined: (1) theoretical behavior where molecular diffusion is the sole broadening mechanism, (2) effects of temperature gradients, (3) the effects of hydrostatic flow and (4) the effects of wall adsorption. In each case, an expression for the plate height is given and the theoretical importance and experimental ramification will be discussed.

1. Introduction

Electrophoresis is a technique where interaction with external electric field is responsible for differential migration leading to separation. Using the classification of Giddings [1], classical electrophoresis is a category Sc separation technique, which means a static (no flow) system with a continuous potential driving force. Some variants of electrophoresis can be classified as category $F(=)c$, which indicates a separation system with flow parallel to the continuous force field. Capillary electrophoresis (CE) also fall into one of the above two categories. If only electrophoretic velocity is responsible for the transport of the solutes then the system is Sc. On the other hand, if electroosmotic flow component is present and non ionic solutes are being separated, the $F(=)c$ classification is more appropriate.

Electrophoresis seems quite removed from chromatography. Using the Giddings scheme [1], chromatography belongs to category $F(+)d$; flow is perpendicular to a discontinuous potential field. This category is quite different than Sc or $F(=)c$. In chromatography, equilibrium processes control the separation. In electrophoresis kinetic considerations are important. In chromatography the driving force for the separation is internal - the difference in chemical potential between two phases. In electrophoresis the driving force is external to the system - an electric field. In chromatography the potential field is discontinuous. In electrophoresis the potential field is continuous. Why, then, include electrophoresis in a meeting discussing chromatography and related techniques?

Electrophoresis is, perhaps, not related to chromatography, but there are many similarities between the two methods. In both techniques differential migration leads to the separation. The solutes in both techniques obey basically the same mass balance equations which means

607

F. Dondi and G. Guiochon (eds.),
Theoretical Advancement in Chromatography and Related Separation Techniques, 607–632.
© 1992 *Kluwer Academic Publishers.*

that solute zone shape is similar (Gaussian) in chromatography and in electrophoresis. Moreover, zone spreading is controlled by similar mechanisms in both methods of separation. Most chromatographic methods are elution techniques. Some of the electrophoresis variants, especially capillary electrophoresis, are also elution techniques.

The similarities between the two techniques allow us to use a common language to characterize the behavior of solutes undergoing chromatographic or electrophoretic separations. Terms such as selectivity, resolution, plate numbers, plate height and peak capacity are applicable to both methods. Moreover, some experimental considerations are identical to both methods i.e. the need to keep extra-column effects to a minimum. As a result, many chromatographers are doing research in electrophoresis, especially in CE, and sessions dealing with the latter technique are now featured regularly in symposia ostensibly devoted to chromatography.

2. The Mass Balance Equation

2.1 THEORETICAL BEHAVIOR OF SOLUTES IN CE.

As discussed above, the solutes' behavior in chromatography or electrophoresis can be expressed in identical mass balance equations. This paper will discuss solute behavior only in capillary electrophoresis. The equations describing that behavior will be written and analyzed for several cases which will take into account some experimental limitations.

In theory, if solute migration is due only to electrophoretic and/or electroosmotic mobility, then the solute's velocity profile is flat; i.e., constant across the cross-section of the capillary, and the mass balance equation can be written as:

$$\frac{\partial C}{\partial t} = D\left[\frac{1}{r}\frac{\partial}{\partial r}(r\frac{\partial C}{\partial r})\right] + D\frac{\partial^2 C}{\partial x^2} - u\frac{\partial C}{\partial x} \tag{1}$$

C is the solute concentration, D is the solute's diffusion coefficient, t is analysis time and x is longitudinal direction, r is the radial position and u is the solute's velocity which can include electrophoretic and/or electroosmotic component. Equation (1) further assumes that the diffusion coefficient is independent of the radial position. Equation (1), which is also known as equation of continuity, is a result of conservation of matter; the rate of concentration change with respect to time is equal to the total transport of the solute in and out of a unit volume element in the capillary. The solute is transported by diffusion and by convective; i.e. electrophoretic and/or electroosmotic, means.

Equation (1) can be further simplified by assuming that in the presence of only electrophoretic and/or electroosmotic flow the radial distribution of the solute is uniform.

$$\frac{\partial C}{\partial t} = D\frac{\partial^2 C}{\partial x^2} - u\frac{\partial C}{\partial x} \tag{2}$$

Equation (2) will fail if the capillary radius is less than about 7 times the double layer thickness, which can happens if the buffer in the capillary is extremely diluted or if the capillary diameter is very small.

To solve Equation (2) we need boundary and initial conditions. Suitable conditions are

$$C(t, \infty) = 0$$
$$C(0,0) = C_o$$
$$C(t,0) = 0 \text{ for } t > 0$$
$$u = u_{ef} \pm u_{eo}$$

u_{ef} and u_{eo} indicate electrophoretic and electroosmotic velocity respectively.

Equation (2) is quite easy to solve by several different techniques. Although the chromatographically experienced reader will have little difficulties in anticipating the correct solution, we will give, for the sake of completion, a detailed explanation of one approach to the solution of Equation (2).

We will solve Equation (2) using the method of Laplace transform [2]. The Laplace transform, with respect to t, of Equation (2) is:

$$s\bar{C} - C_o = D \frac{d^2\bar{C}}{dx^2} - u \frac{d\bar{C}}{dx} \tag{3}$$

s is the Laplace variable and $\bar{C}$ indicates the Laplace transform of C. In the Laplace domain, the boundary and initial conditions are:

$$\bar{C}(s, \infty) = 0$$
$$\bar{C}(0,0) = C_o$$
$$\bar{C}(s,0) = 0 \text{ for } s > 0$$

The solution of Equation (3) is [2]:

$$\bar{C}(x,s) = \frac{1}{2D\sqrt{(\frac{u}{2D})^2 + \frac{s}{D}}} \int_{-\infty}^{\infty} C_i \exp[\frac{u}{2D} - \sqrt{(\frac{u}{2D})^2 + \frac{s}{D}}](x - x')dx' \tag{4}$$

If the injection plug is of the form of a delta function, then Equation (4) can be integrated to yield:

$$\bar{C}(l,s) = \frac{A}{2D\sqrt{(\frac{u}{2D})^2 + \frac{s}{D}}} \exp[\frac{ul}{2D} - l\sqrt{(\frac{u}{2D})^2 + \frac{s}{D}}] \tag{5}$$

where A is a constant related to the concentration of the solute and l is the separation length. The inverse of Equation (5) is the solution of the mass balance equation in Equation (2). While it may be difficult to invert this expression, Equation (5) can give some insight as to the chromatographic behavior of the solute. Use will be made here of the following property of the Laplace transform:

$$\int_0^\infty t^n f(t)dt = (-1)^n \lim_{s \to 0} \frac{d^n \bar{f}(s)}{ds^n} \tag{6}$$

The left-hand side of the above equation is recognized as the definition of the moments of the function $f(t)$. When that function is the concentration profile of the solute, t he moments have the following physical significance. The zeroth moment ($n=0$ in the above equation) is the peak area. The first moment ($n=1$) is the center of gravity of the concentration profile or the retention time of the solute. The second moment ($n=2$), when calculated around the first moment, yields the variance, and hence the plate height, of the concentration profile. Moments taken around the first moment are called central moments. Moments higher than the second characterize the peak shape. These higher moments will not be discussed in the present work.

The zeroth moment is simply A/u. The first moment is given by:

$$m_1 = \frac{l}{u} + \frac{2D}{u^2} \tag{7}$$

The last term in the right-hand side of the above equation is usually negligibly small and the first moment becomes the well known equation for the migration time, l/u. It should be reminded here that u is the electrophoretic and/or the electroosmotic velocity.

The second central moment is given by:

$$m_2 = \frac{2Dl}{u} + \frac{8D^2}{u^4} \tag{8}$$

Again, the second term on the right-hand side of the equation is negligible and the second central moment, which is the variance, becomes:

$$m_2 = \frac{2Dl}{u} \tag{9}$$

Using the definition of the height equivalent to theoretical plate, we get the following plate height equation:

$$H \equiv \frac{m_2 u^2}{l} = \frac{2D}{u} \tag{10}$$

Equation (10) is the theoretical HETP equation for capillary electrophoresis. The equation assume the ideal case where factors not directly related to the separation process do not contribute to zone broadening. The direct result of the lack of radial gradient in the velocity is that molecular diffusion is the sole contributor to zone broadening in CE. Equation (10) indicates that the efficiency of a CE system will improve with increasing electrophoretic and/or electroosmotic velocities. The behavior of H in CE is unlike that in chromatography, where there is an optimum velocity below or above which the efficiency deteriorates. Figure 1 shows a comparison between the HETP behavior for CE and for LC for a 25 μm radius capillary. The figure is somewhat complicated since both axes are broken in order to cover the wide range of H and of the velocity. However, The left half of the figure shows the minimum in H in the case of chromatography which does not exist in the case of CE. The divergence of the two plots is particularly noteworthy at high velocity; i.e. the right hand side of the figure.

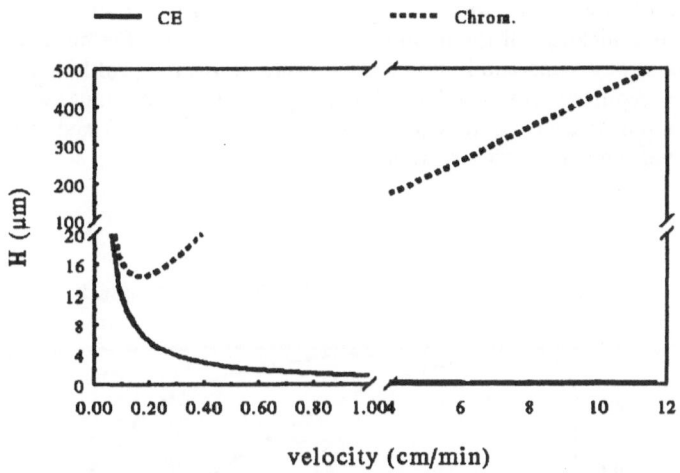

Figure 1. Comparison of HETP curves in chromatography and in capillary electrophoresis. Capillary radius: 25 μm. Solute's Diff. Coef.: 1×10^{-9} m^2/s

2.2. EFFECT OF PARABOLIC FLOW PROFILE NOT DUE TO TEMPERATURE OR HYDROSTATIC FLOW.

The discussion up to now assumed a flat electrophoretic and/or electroosmotic velocity profile. When electroosmotic flow is an important component of the separation, complication might arise. Rice and Whitehead [3] showed that the electroosmotic flow profile is given by:

$$u_{eo} = \frac{\epsilon_0 \epsilon_r \zeta E}{4 \pi \eta} [1 - \frac{I_o(\kappa r)}{I_o(\kappa a)}] \tag{11}$$

ϵ_0 is the permittivity of a vacuum, ϵ_r is the dielectric constant of the buffer, ζ is the average zeta potential of the solutes, E is the field strength (applied voltage divided by capillary length), η is the buffer viscosity, κ is the reciprocal of the double layer thickness, a is the capillary radius, r is the radial position and $I_o(x)$ is zero order modified Bessel function of argument x. κ is a function of the ionic strength of the buffer, I [4]:

$$\kappa = 3.288 \sqrt{I} \tag{12}$$

κ in the above equation has the units of nm^{-1}.

Typically, the second term in the right-hand side of Equation (11) is negligible and the expression for the electroosmotic velocity reduces to the classical flat profile. However, when using very dilute buffers or very narrow capillaries, the second term in Equation (11) becomes important and the velocity profile becomes more parabolic in nature. Figure 2 shows a parabolic profile as well as the Rice-Whitehead profile for $\kappa a = 7$ and 70. When the radius is seven times the thickness of the double layer, the velocity profile has a strong curvature, although not as strong as parabolic flow. At a value of 70, the profile is flat over the great majority of the capillary cross section. Under typical CE conditions, with 25 μm radius capillary and a buffer concentration of .05 M, κa is about 18,000 and the flow profile is virtually flat over the whole cross section.

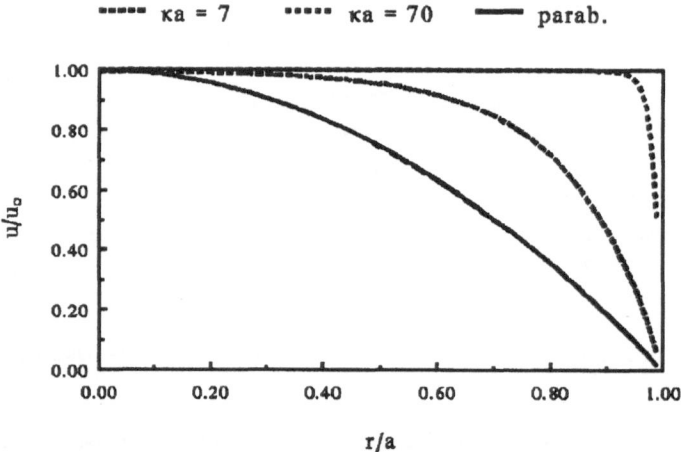

Figure 2. Rice-Whitehead Velocity profiles for several κa values.

When the velocity profile is not flat, Equation (1) should be modified to:

$$\frac{\partial C}{\partial t} = D[\frac{1}{r}\frac{\partial}{\partial r}(r\frac{\partial C}{\partial r})] + D\frac{\partial^2 C}{\partial x^2} - u(r)\frac{\partial C}{\partial x} \tag{13}$$

where the radial gradient in the velocity is indicated in the last term of the right-hand side of the equation. If electroosmotic flow is the major flow component, then the expression in Equation (11) is substituted for u(r) in Equation (13). Because of the velocity profile the radial component of diffusion must be retained in the mass balance equation.

The solution of Equation (13) is more complicated. There are several ways to solve the equation. One such solution is that of Aris [5] as applied by Martin and co-workers [6,7] to elctroosmotically driven open tubular liquid chromatographic system. Their approach can be summarized as follows. The plate height equation is the summation of the terms contributing to zone broadening:

$$H = \frac{2D}{u} + \frac{C_m a^2 u}{D} + C_s u \tag{14}$$

The first term is contribution due to molecular diffusion of the solute in the mobile phase. The second term is the contribution of the resistance to mass transfer in the mobile phase. This term occurs whenever the flow profile is not flat. The third term is due to the resistance to mass transfer in the stationary phase. C_m is the quantity which we would like to know since it is directly related to the shape of the flow profile. According to the Aris treatment, C_m, which results from the solution of the mass balance equation, is given by the following integral:

$$C_m = \int_0^1 [P(y) - Ry^2]^2 \frac{dy}{y} \tag{15}$$

where y is reduced radius r/a, R is solute relative velocity and P(y) is a radial average of reduced velocity given by:

$$P(y) = \int_0^{y'} 2y' \frac{u(y')}{\bar{u}} dy' \tag{16}$$

$\bar{u}$ is the cross sectional averaged velocity. To obtain the exact expression for C_m the velocity expression is needed, which, for electroosmotic flow is given by Equation (11). However, that equation presents difficulties due to the Bessel function part. Martin and colleagues [6,7] used two approximations to overcome that difficulty. The most successful approach [7] was to approximate the velocity profile by the equation:

614

$$u(r) = \bar{u}[1 - (\frac{r}{a})^n]$$ (17)

Figure 3 shows several flow profiles for several n values. It is seen that large n values lead to a flattened profile.

To make Equation (17) useful, a connection between the parameter n and the physically significant κa quantity needs to be accomplished. Martin and co-workers [7] show that n can be estimated as:

$$n = \kappa a - \frac{3}{2} + \frac{3}{8\kappa a} (1 + \frac{1}{\kappa a} + ...)$$ (18)

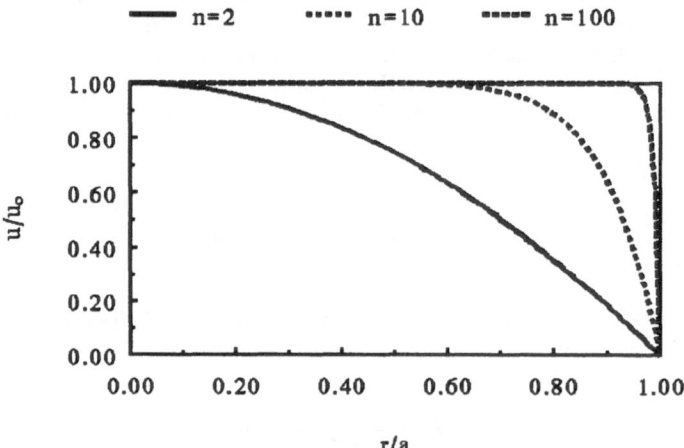

Figure 3. Martin's velocity profiles for several n values.

Using Equation (17), the Martin et al. approach gives the following plate height equation:

$$H = \frac{2D}{u} + \frac{R^2[(n+2)(n+4)] - 2R[(n+2)(n+6)]+n^2+10n+20}{4(n+2)(n+4)} \frac{a^2u}{D}$$ (19)

If R = 1 and n = 2 (that is; a parabolic flow), Equation (19) reduces to the Taylor-Aris dispersion expression in open tubes. Figure 4 shows three H plots: for conventional parabolic flow, for n = 5.55 (κa about 7) and for n = 98.5 (κa about 100). As expected, as n, and therefore as κa, becomes large, the efficiency improves at high velocities.

Under typical operating conditions of CE, broadening due to electroosmotic flow profile is not expected to be significant. However, it is possible that in the case of electroosmotically driven chromatography in packed columns, the distance between support particles is small enough to make the curvature of the electroosmotic flow near the walls of the channel become an important broadening factor. Several research groups, most notably that of Knox [8], are investigating currently electroosmotic flow in chromatography. It would be of interest to examine the observed versus the expected efficiency since the difference between the two could be traced to velocity profiles which are not flat

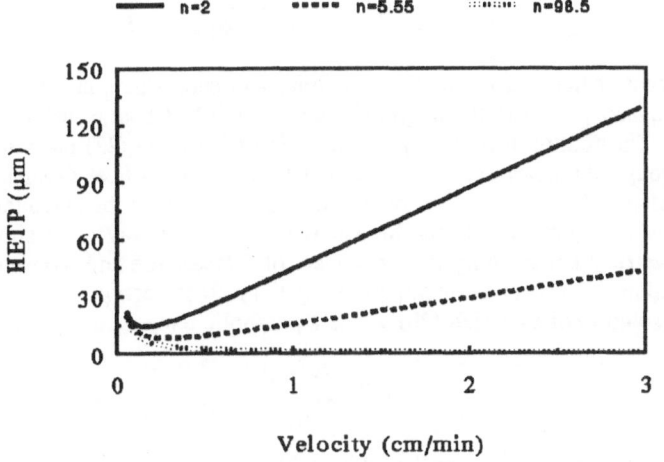

Figure 4. HETP curves for several Martin velocity profiles.

2.3. EFFECT OF TEMPERATURE

Passage of electric current is accompanied with generation of heat, the so called Joule heating. Temperature can affect the electrophoretic separation in several ways. In the context of this work, the most important thermal effect is on the electrophoretic velocity profile through changes in the buffer viscosity and in the ionic mobilities. There have been several studies on temperature profiles in open tubing due to the passage of current. For example, Hinkley [9] and Coxon and Binder [10] have examined the development of temperature profiles in cylindrical tubes. Jones and Grushka [11] and Gobie and Ivory [12] investigated temperature gradients in capillaries, which included polyimide coating, suited for CE. The temperature profile is obtained by solving a heat balance differential equation, which is much similar to the mass balance equation:

$$\frac{1}{r}\frac{d}{dr}(r\frac{dT}{dr}) = -\frac{G}{k_1} \tag{20}$$

T is temperature, G is heat generated per unit volume and k_1 is the thermal conductivity of the buffer. A suitable boundary condition is:

$$-ak_1\frac{dT}{dr} = Ua(T_1-T_s) \tag{21}$$

T_1 and T_s are the temperature at inner wall of the capillary and it's surroundings respectively, and U is the overall heat transfer coefficient. The product aU is a measure of the heat dissipation through the capillary wall and it is given by:

$$\frac{1}{aU} = \frac{1}{k_2}\ln(\frac{a_1}{a}) + \frac{1}{k_c}\ln(\frac{a_c}{a_1}) + \frac{1}{a_ch} \tag{22}$$

a and k represent radius and thermal conductivity and subscript 1 and c indicate quantities related to the quartz glass and the polyimide coating and h is heat transfer coefficient to the surroundings. The first term in the right-hand side of Equation (22) takes into account heat dissipation through the quartz glass, the second term is due to heat dissipation through the polyimide coating and the last term is related to heat removal by the surroundings.

The solution of Equation (20) can be acheived in several ways. In simplified case we assume that electrical conductivity is independent of temperature and, consequently, the rate of heat generation is constant throughout the capillary cross section. Under the constant G constrain, the solution of Equation (20) yields a parabolic temperature profile:

$$T = T_1 + \frac{Ga^2}{4k_1}(1 - \frac{r^2}{a^2}) \tag{23}$$

T_1 is the temperature at the inner wall of the capillary, given by:

$$T_1 = T_s + \frac{Ga}{2U} \tag{24}$$

where U can be calculated from Equation (22).

The more rigorous solution of Equation (20) takes into account the temperature dependence of the electrical conductivity. In such cases, the solution of the heat balance equation is more complicated, being dependent on Bessel functions. Jones and Grushka [11] showed that, under typical experimental conditions, the difference between the parabolic temperature profile and the Bessel function related one is very small and in the following treatment the parabolic profile will be used. However, when the exact temperature profile is needed, the equations can be modified to approximate the Bessel function; see, for example, the work of Gobie and Ivory [12].

Before moving into the analysis of band broadening due to temperature gradients, it is important to stress a point which is often neglected. The temperature drop across the radial position is a function of the power input and it is essentially independent of the surroundings' temperature. Thus cooling by thermostating the capillary will affect the average temperature of the capillary but will not affect ΔT. Still, because the electrophoretic mobility is temperature dependent, thermostating is advantageous as it will ensure better precision in the data. I addition, cooling down will reduce the solute's diffusion coefficient, which may help to improve the system's efficiency [13].

Another point to note is that radial temperature gradients in the capillary perturb mostly electrophoretic flow and to a much lesser extent electroosmotic flow. Electroosmotic mobility is controlled by the magnitude of the relevant parameters at a distance $1/\kappa$ from the capillary wall, which is negligibly small in conventional CE. Therefore, temperature will change the magnitude of the electroosmotic flow, but temperature gradients across the capillary cross section will have a minimal effect on it. However, in electroosmotically driven chromatography with packed columns, flow channels may be sufficiently narrow so as to make the temperature gradient effect electroosmotic flow as well.

Temperature affects the electrophoretic velocity via the viscosity. The electrophoretic velocity can be expressed as:

$$u_{ef} = \epsilon_o \epsilon_r \zeta \frac{E}{\eta} \tag{25}$$

The temperature dependence of the viscosity is given by:

$$\eta = A \exp\left(\frac{B}{T}\right) \tag{26}$$

where A and B are constants depending on the nature of the buffer. When Equation (26) is substituted in Equation (25) we get the dependence of the velocity on the temperature gradient:

$$u_{ef}(r) = \frac{1}{A}\epsilon_o \epsilon_r \zeta E \exp\left(-\frac{B}{T}\right) \tag{27}$$

Note the use of the symbol $u_{ef}(r)$ to indicate the radial distribution of T. Equation (27) can be handled best if we expand the exponential part using Taylor series around T_1, the temperature at the wall, and if we retain only the first two terms of that series:

$$u(r) = \frac{\epsilon_o \epsilon_r \zeta E \exp(-B/T_1)}{A}\left[1 + \frac{B}{T_1^2}(T - T_1)\right] \tag{28}$$

The term in front of the square brackets is the velocity, u_1, at T_1; that is, at the wall. For $T - T_1$ we can substitute the expression in Equation (23) to get:

$$u(r) = u_1 [1 + \frac{GBa^2}{4k_1 T_1^2} (1 - \frac{r^2}{a^2})] \tag{29}$$

The second term in the squared brackets, to a good approximation, represents the perturbation of the velocity caused by temperature gradients.

The heat generated G can be expressed in terms of measurable quantities:

$$G = \frac{EI}{\pi a^2} = \Lambda C_b E^2 \tag{30}$$

I is the current, Λ is the equivalent conductance of the buffer and C_b is the buffer concentration. Using Equation (30), the expression for the electrophoretic velocity becomes:

$$u(r) = u_1 [1 + \frac{\Lambda C_b E^2 Ba^2}{4k_1 T_1^2} (1 - \frac{r^2}{a^2})] \tag{31}$$

Equation (31) indicates how experimental variables, such as buffer concentration, field strength and capillary radius perturb the normally flat profile of electrophoretic velocity.

If we assume that only radial temperature gradient effects the efficiency, then Equation (31) can be substituted for u(r) in the mass balance equation, i.e. Equation (13). The method of Aris [5] can be used to derive the contribution of temperature gradient to the plate height; see for example the work of Knox and Grant [8] and of Gobie and Ivory [12]. However, we will employ here the approach of Gill and Co-workers [14], which better suits our purpose of obtaining a total effective dispersion coefficient. Moreover, the Gill approach will enable us to demonstrate one additional method to solve the mass balance equation in CE. The following discussion is based on our recent work on the effect of temperature gradients in CE [15]

The approach of Gill and colleagues [14] assumes that the solution of the mass balance equation is in the form of an expansion series and that, to a very good approximation, the first two terms of the series are sufficient to give an accurate description of the concentration profile. The solution is of the form:

$$C = C_m + \frac{\partial C_m}{\partial x_i} g(r) \tag{32}$$

where C_m is cross-sectional average of the concentration and g(r) is an expansion coefficient which needs to be determined. We are interested in the variance of the concentration distribution since it is directly related to the plate height. For that purpose Equation (32) is substituted back into the mass balance equation. Integration with respect to r yields the following equation:

$$\frac{\partial C_m}{\partial t} = [D - \frac{2}{a^2}\int_0^a u(r)\,g(r)\,rdr]\,\frac{\partial^2 C_m}{\partial x^2}$$ (33)

The integral in Equation (33) represents a weighted cross-sectional average velocity. Equation (33) is a modified Fick's second law, with the terms in the squared bracket being the dispersion coefficient, K:

$$K = D - \frac{2}{a^2}\int_0^a u(r)\,g(r)\,rdr$$ (34)

the units of K are distance^2/time. K is the quantity we are after since it is related to the variance of the concentration distribution, and therefore to H.

Gill and co-workers showed that g(r) is related to the average radial deviation of the velocity from its radial average, u_m:

$$g(y) - g(0) = \frac{a^2}{D}\int_0^y \frac{1}{y'}\{\int_0^y [u(r) - u_m]\,y''dy''\}\,dy'$$ (35)

where y is the reduced velocity r/a. To evaluate g(r) we need to find an expression for the cross-sectional averaged velocity u_m:

$$u_m = \frac{2}{a^2}\int_0^a u(r)\,rdr$$ (36)

$$u_m = \frac{2u}{a^2}\int_0^a \{1 + \frac{\lambda C_b E^2 B a^2}{4k_1 T_1^2}[1 - (\frac{r}{a})^2]\}\,rdr$$ (37)

the integration of Equation (37) is straight forward, and the average velocity is:

$$u_m = u_1 \frac{2 + q}{2}$$ (38)

where

$$q = \frac{\Delta C_b E^2 B a^2}{4k_1 T_1^2} \tag{39}$$

When the expression for the average velocity is inserted in Equation (31), the electrophoretic velocity becomes:

$$u(r) = \frac{2u_m}{2 + q}[1 + q(1 - \frac{r^2}{a^2})] \tag{40}$$

Therefore, the inner integrant in Equation (35) can be written as:

$$u(r) - u_m = u_m[\frac{2}{2 + q} + \frac{2q}{2 + q}(1 - y^2) - 1] \tag{41}$$

Performing the double integration in Equation (35) yields:

$$g(r) - g(0) = \frac{a^2 u_m}{D}(\frac{2y^2}{4(2 + q)} + \frac{2qy^2}{4(2 + q)} - \frac{2qy^4}{16(2 + q)} - \frac{y^2}{4}) \tag{42}$$

To solve for g(r) we need to evaluate g(0). Gill *et al.* showed that the cross-sectional average of g(y) is zero:

$$\int_0^1 g(y)y\,dy = 0 \tag{43}$$

Thus, from Equation (42) we write the following condition:

$$\int_0^1 \{\frac{a^2 u_m}{D}[\frac{2y^2}{4(2 + q)} + \frac{2qy^2}{4(2 + q)} - \frac{2qy^4}{16(2 + q)} - \frac{y^2}{4}] + g(0)\}dy = 0 \tag{44}$$

Integration and rearrangement allows to solve for g(0):

$$g(0) = \frac{2a^2 u_m}{D}[\frac{1}{16} - \frac{5q}{8(2 + q)} - \frac{1}{8(2 + q)}] \tag{45}$$

Substitution of the above expression for g(0) in Equation (42) gives g(y):

$$g(y) = \frac{a^2 u_m}{D}[\frac{y^2}{2(2+q)} + \frac{qy^2}{2(2+q)} - \frac{qy^4}{8(2+q)} - \frac{y^2}{4} - \frac{1}{4(2+q)} - \frac{5q}{24(2+q)} + \frac{1}{8}] \tag{46}$$

When Equation (46) is entered in Equation (34), the integration can be easily accomplished, and the expression for the dispersion coefficient becomes:

$$K = D + \frac{u_m^2 \Lambda^2 C_b^2 E^4 B^2 a^6}{48 D (8 k_1 T_1^2 - \Lambda C_b E^2 B a^2)^2} \tag{47}$$

Once we have the dispersion coefficient we can get the expression of the plate height. Using the relationships:

$$\sigma^2 = 2Kt \tag{48}$$

and

$$H = \frac{\sigma^2}{l} \tag{49}$$

we get the following equation for the plate height:

$$H = \frac{2D}{u_m} + \frac{\Lambda^2 C_b^2 E^4 B^2 a^6 u_m}{24 D (8 k_1 T_1^2 - \Lambda C_b E^2 B a^2)^2} \tag{50}$$

Variations of this equation were derived by Cox et al. [16] and by Knox and Grant [8].

The above plate height equation shows the effect of temperature gradients on the efficiency in CE when electrophoretic velocity contributes, either wholly or partially, to the migration time. The second term in the right-hand side of the equation represents temperature gradients contribution to zone broadening. Temperature induced parabolic flow profile gives rise to a resistance to mass transfer term. The behavior of the plate height as a function of the velocity is similar to that in chromatography. A previous paper of ours [15] analyzed in details temperature effects in CE. What follows is a summery of the important ramification of Equation (50).

(a) The sixth power dependence on the capillary radius shows the limitation of conventional electrophoresis and explains the excitement generated by capillary electrophoresis. Figure 5 shows five HETP plots for five radii. It is seen that the wider the capillary the greater is the effect of temperature. Figure 6 is similar to Figure 5 but for a larger solute ($D = 1 \times 10^{-10}$ m^2/s as compared to 1×10^{-9} m^2/sec). Due to the smaller diffusion coefficient, the temperature effect is much more noticeable.

(b) The square dependence in the second term in the plate height equation on the buffer concentration and on the field strength points the way to minimizing temperature effects in CE. Figure 7 shows the HETP behavior for a large solute in several capillaries with more dilute buffer than in Figure 6. The marked decrease in the temperature effect is the prominent feature of Figure 7.

622

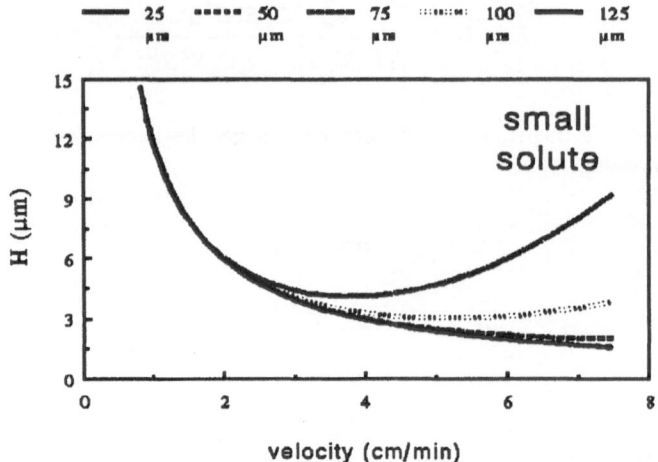

Figure 5. HETP curves for small solute with several capillary radii. Parameters used for the calculations: C_b = 100 mol/m³; V = 30kV; a_2 = 345 μm; a_c = μm; T_s = 298 K; h = 10,000 W/(m² K); k_1 = 0.605 W/(m K); k_2 = 1.5 W/(m K); k_c = 0.155 W/(m K), Λ = 0.015 m² mol⁻¹ Ω⁻¹; B = 2400 K; D = 1X10⁻⁹ m²/s

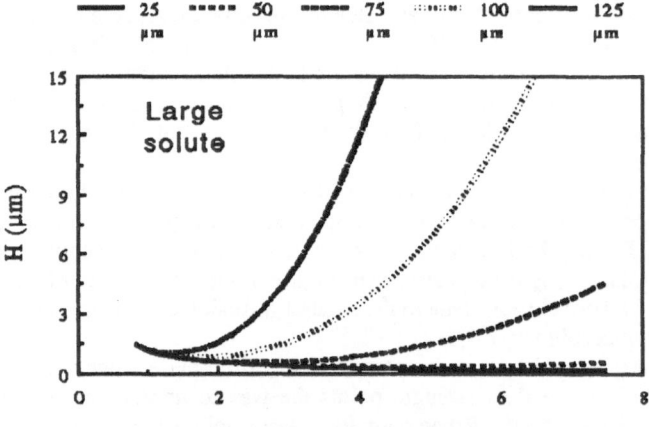

Figure 6. HETP curves for a large solute with several capillary radii. Parameters used for calculation same as in Figure 5 except for D = 1X10⁻¹⁰ m²/s

(c) Equation (50) allows the calculation of the upper capillary radius limit for a given loss in the plate height. If, for a large solute, a 20% loss in the efficiency is acceptable and if a fast analysis time is desirable, then a very narrow capillary is needed. The demand for narrow capillary can be alleviated somewhat by the use of dilute buffers. However, the penalty for using dilute buffers is smaller concentration of the injected solute.

It should be pointed out again that Equation (50) does not take into account the temperature dependence of the diffusion coefficient. Davis [17] has examined in details the plate height behavior when the temperature dependence of the diffusion coefficient is taken into account. His conclusion is that, most often, Equation (50) is sufficiently accurate. On the other hand, Knox [13] found some evidence that the temperature dependence of D can further aggravate the deterioration of efficiency.

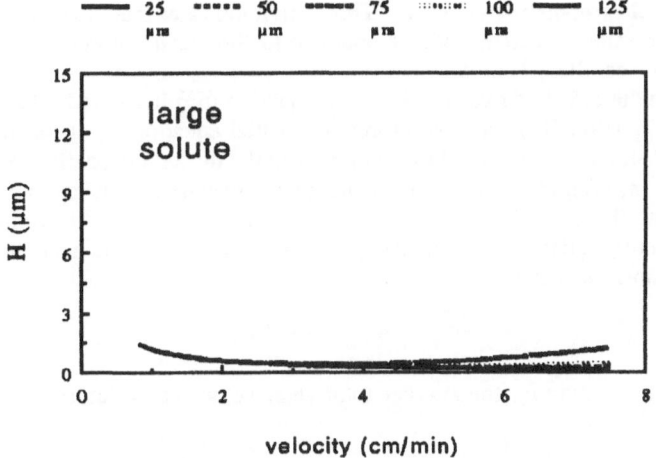

Figure 7. HETP curves for a large solute with several capillary radii. Parameters used for calculation same as in Figure 6 except for $C_b = 10$ mol/m³ (0.01 M)

2.4. EFFECT OF HYDROSTATIC FLOW

The discussion to this point assumed identical buffer levels at the two ends of the capillary. When the two buffers are not at the same height, hydrostatic flow results. Since hydrostatic flow in pressure-driven, it has the typical parabolic flow profile:

$$u_{hs} = 2u_s\left(1 - \frac{r^2}{a^2}\right) \tag{51}$$

u_{hs} is hydrostatic velocity and u_s is the cross-sectional average of that velocity, which is given by:

$$u_s = \frac{\Delta h \rho g a^2}{8\eta L} \tag{52}$$

Δh is the height difference between the buffer levels, ρ is the buffer density, g is the gravitational acceleration, η is the buffer viscosity and L is the capillary length. Assuming no temperature effects and no electroosmotic flow, then the migration velocity is:

$$u(r) = u_{ef} \pm 2u_s\left(1 - \frac{r^2}{a^2}\right) \tag{53}$$

where u_{ef} is the electrophoretic velocity. The $\pm$ sign indicates that the hydrostatic component can flow in the same direction with or opposite to the electrophoretic flow, depending on which end of the capillary is higher.

To ascertain the effect of hydrostatic flow, Equation (53) is entered into the mass balance expression in Equation (13). The resulting differential equation is, again, best solved using the method of Gill and co-workers [13], which yield the dispersion coefficient; Equation (34). The velocity term, u(r), in the dispersion equation is given here by Equation (53). Again, we need to evaluate the coefficient of the expansion series which is the solution of the mass balance differential equation. To evaluate g(r), we need to know the average velocity, which can be easily shown to be:

$$u_m = u_{ef} \pm u_s \tag{54}$$

Using the above equation for the average total velocity, we can calculate the difference g(y)-g(0):

$$g(y) - g(0) = \frac{a^2 u_s}{2D}\left[\frac{y^2}{2} - \frac{y^4}{4}\right] \tag{55}$$

Using the same arguments employed when discussing Equations (43) and (44), we can show that g(0) is:

$$g(0) = -\frac{a^2 u_s}{12D} \tag{56}$$

Thus, g(y) is equal to:

$$g(y) = \frac{a^2 u_s}{2D} \left(\frac{y^2}{2} - \frac{y^4}{4} - \frac{1}{6} \right) \tag{57}$$

Remembering that y is r/a, we now have an expression for the velocity and for g(r), which allows us to get the expression for the dispersion coefficient K:

$$K = D + \frac{a^2 u_s^2}{48D} \tag{58}$$

Using the relationship between the variance and K as well as between the variance and the plate height, Equations (48) and (49), we get the expression for the plate height in the presence of hydrostatic flow:

$$H + \frac{2D}{u_{ef} \pm u_s} + \frac{a^2 u_s^2}{24D(u_{ef} \pm u_s)} \tag{59}$$

We will assume here that both velocity components are in the same direction (the extension to cases where the two components are in the opposite direction is quite easy to carry out). Therefore, only the summation of the two velocities will be considered in the present discussion.

The above plate height equation shows that the presence of an hydrostatic component causes an additional zone broadening. In addition to the molecular diffusion term, there is also a mass transfer term. Implicitly, H depends on Δh via u_s. Equation (59) seems very similar to the plate height equation in chromatography. However, there is one major difference between the two HETP expressions for the two separation techniques: In chromatography, the resistance to mass transfer term is a linear function of the mobile phase velocity. In CE, provided the Δh is constant from run to run, the mass transfer term is proportional to the inverse of the electrophoretic velocity. Thus, H versus u plot will still decrease continuously with the velocity, albeit at a higher value than the theoretical curve. Figure 8 shows a comparison between theoretical [i.e. Equation (10)] H plot and several Δh values for a large solute and a 50 μm radius capillary. We see that a 5 mm height difference increases H by a substantial factor.

The hydrostatic flow contribution to the plate height is an explicit function of the capillary radius and the solute's diffusion coefficient and, as already mentioned, an implicit function of the height difference between the two ends of the reservoir. The diffusion coefficient dependence is particularly important since it indicates that HF problems are more pronounced with large molecules. Figure 9 shows the hydrostatic flow effect for large and small solutes in a 75 μm radius capillary. The solid lines give the theoretical HETP plots while the broken lines give the efficiency of the two solutes when Δh is 5 mm. We see that while a 5 mm height difference between the capillary ends produces a small increase in H with small solutes, with large solutes this Δh is sufficient to completely obliterate the separation.

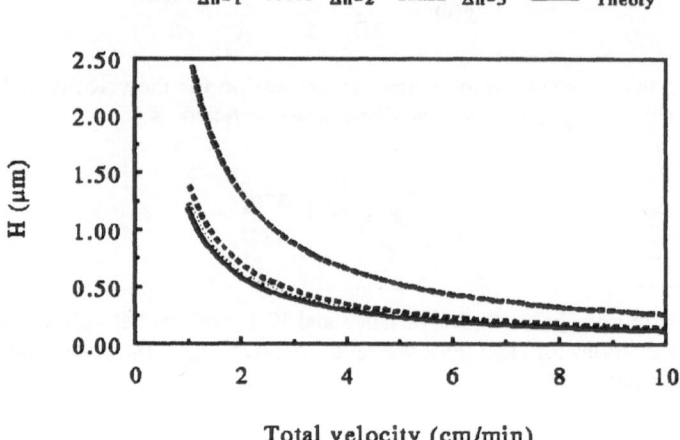

Figure 8. Effect of hydrostatic flow on HETP. Parameters used in the calculations: a = 4.0X10^{-5} m; V = 30 kV; D = 1X10^{-10} m^2/s; L = 1 m; g = 9.807 m/s^2; ρ = 1000 kg/m^3; η = 0.001 kg/(m s)

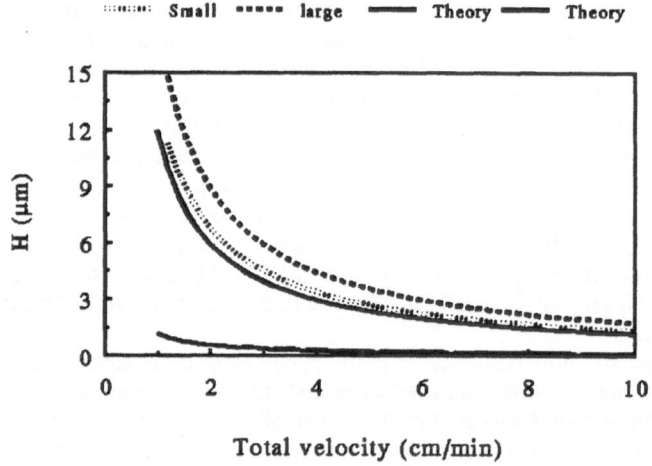

Figure 9. Dependence of hydrostatic effect on diffusion coefficient. Parameters used in calculation as in Figure 8 except: a = 7.5X10^{-5} m; Δh = 0.005 m; small solute D = 1X10^{-9} m^2/s

According to Equation (59), reducing the radius of the capillary will minimize the hydrostatic flow effect. In Figure 8 the radius was 50 μm. If we were to use a 25 μm radius, then even a 5 mm height difference could have been tolerated with a minimum effect on the efficiency. On the other hand, with a 75 μm radius capillary, a 2 mm height difference would triple H!

Equation (59) allows us to calculate maximum permissible Δh values for a given loss in the capillary efficiency. Table I gives some Δh values for several capillaries, for small and large solutes for give losses in H.

Table I: Maximum allowed height difference, in mm, for several capillary radii and several acceptable losses in H. Shown are values for large solutes (small diffusion coefficient) and small solute (larger diffusion coefficient).

Radius (μm)	Percent loss in H		
	10%	20%	40%
Δh_{max} for large solute (D = 1X10^{-10} m^2/s)			
25	11.4	16.2	22.9
50	1.43	2.01	2.86
75	0.42	0.60	0.85
Δh_{max} for small solute (D = 1X10^{-9} m^2/s)			
25	114.4	161.8	228.8
50	14.3	20.2	28.6
75	4.24	5.99	8.47

If the hydrostatic flow component is in the same direction as the electrophoretic flow, then we have an "aliasing" effect in which the actual velocity of the solute is faster than expected. At times, this could actually lead to efficiencies which, apparently, are better than theoretical prediction. The reason for the apparent decrease in H is the faster velocity due to hydrostatic flow. While this increase in H is misleading, at times it can be used advantageously. For large molecules, narrow capillaries and slow electrophoretic velocities, Δh can be increased to about 5 mm and the efficiencies will improve. Larger height differences will give higher H values (lower efficiencies) than operation without hydrostatic flow.

The existence of hydrostatic flow in capillary zone electrophoresis can cause an additional zone broadening, which lowers the expected efficiency of the method. The hydrostatic flow effect is particularly important with large solutes and wide capillaries. In such cases, theory shows that the height levels of the buffer solutions, at both ends of the capillary, should be controlled to better than 1 or 2 mm. The situation is less critical with narrow capillaries, or with small molecules. Thus, narrow capillaries are beneficial not only because they minimize Joule heating effect, but also because they reduce the influence of hydrostatic flow.

2.5. EFFECTS OF WALL ABSORPTION

Certain solutes can adsorb onto the wall of the capillary. If the adsorption is reversible on the time scale of the electrophoretic run, then we have an additional contribution to zone broadening. In essence, we have a possible combination of electrophoretic and chromatographic separations. The chromatographic component will introduce a resistance to mass transfer term in the plate height equation.

The presence of wall adsorption alters the mass balance equation, Equation(2), to the following differential equation:

$$\frac{\partial C}{\partial t} = D\frac{\partial^2 C}{\partial x^2} - u\frac{\partial C}{\partial x} + k_f(KC - C_s|_{z=a}) \tag{60}$$

where k_f is mass transfer coefficient between the liquid and the capillary wall, K is adsorption coefficient and C_s is the adsorbed solute concentration. Equation (60), and variations thereof, were discussed in chromatography by various workers such as Kucera [18], Grubner [19], Horn [20] and Grushka [21]. The equation is best solved using the Laplace transform technique as discussed in an earlier section of this paper. Using the Laplace transform method, the equation for the migration time (first moment) becomes:

$$t_r = [\frac{l}{u_{ef} \pm u_{eo}} + \frac{2D^2}{(u_{ef} \pm u_{eo})^2}](1 + k) \tag{61}$$

where l is separation length and k is the chromatographic capacity factor. Normally, the second term in the squared brackets is negligibly small and the equation reduces to its chromatographic analog:

$$t_r = [\frac{l}{u_{ef} \pm u_{eo}}](1 + k) \tag{62}$$

The time based variance, τ^2 (second central moment) is given by:

$$\tau^2 = [\frac{2Dl}{(u_{ef}\pm u_{eo})^3} + \frac{8D^2}{(u_{ef}\pm u_{eo})^4}](1+k)^2 + [\frac{2l}{u_{ef}\pm u_{eo}} + \frac{4D}{(u_{ef}\pm u_{eo})^2}]k(\frac{A}{V_m k_f}) \tag{63}$$

where V_m indicates the buffer volume and A is the capillary surface area. Again, the second term in the squared brackets is negligible. Using the following working definition of the plate height:

$$H = \frac{l\tau^2}{t_r^2} \qquad (64)$$

we get the plate equation for CE with wall adsorption:

$$H = \frac{2D}{u_{ef} \pm u_{eo}} + \frac{Ak(u_{ef} \pm u_{eo})}{V_m(1 + k)k_f} \qquad (65)$$

This plate height equation is similar to H expressions in chromatography. While the presence of wall adsorption may improve the electrophoretic selectivity, it most certainly will decrease the efficiency of the system.

Equation (65) indicates the way to minimize the effect of wall adsorption. Changing the diameter of the capillary will alter the ratio A/V_m, thus affecting the relative importance of the second term in Equation (65). Changing the mass transfer coefficient, k_f, by varying the chemical nature of the wall or by manipulating the temperature, also will influence the importance of the resistance to mass transfer term in Equation (65).

3. Conclusions

The aim of this chapter is two fold: (a) to demonstrate some of the mathematical tools available to describe more accurately solute zone spreading in capillary electrophoresis and (b) to discuss some processes which can contribute excessively to zone broadening. It was demonstrated that distorted electrophoretic flow profile, temperature, hydrostatic flow and wall adsorption can all have an impact on the efficiency of the CE system. The first three processes are similar in their nature since they all modify an otherwise flat electrophoretic and/or electroosmotic flow profile. The last contribution to plate height is different since it does not alter the flow profile. Rather, wall adsorption introduces a radial mass transfer component due solely to chemical partition between two phases. The equations which were developed to describe the plate behavior for each broadening mechanism allow to design CE systems with minimal interference from extraneous factors not directly related to electrophoresis.

4. Acknowledgements

This research was supported by grant No. 88-00021 from the United States-Israel Binational Science Foundation (BSF), Jerusalem, Israel.

5. References

1.   Giddings, J.C. (1991) Unified Separation Science, John Wiley & Sons, Inc. N.Y.

2.   Boyce,W.E. and DiPrima, R.C. (1986) Elementary Differential Equations and Boundary Value Problems, 4th Ed., John Wiley & Sons, Inc., N.Y.

3.   Rice, C.L. and Whitehead, R. (1965) J. Phys. Chem., *69*, 4017

4.   Hunter, R.H. (1981) Zeta Potential in Colloid Science, Academic Press, London, p. 27

5.   Aris, A. (1959) Proc. R. Soc: London, Part A, *A252*, 528

6.   Martin, M. and Guiochon G. (1984), Anal. Chem., *56*, 614

7.   Martin, M., Guiochon, G., Walbroehl, Y. and Jorgenson, J.W. (1985) Anal. Chem., *57*, 559

8.   Knox, J.H. and Grant, I.H (1987) Chromatographia, *24*, 135

9.   Hinckly, J.O.N. (1975) J. Chromatogr., *109*, 209

10.   Coxon, M. and Binder, M.J. (1974) J. Chromatogr., *101*, 1

11.   Jones, A.E. and Grushka, E. (1989) J. Chromatogr., *466*, 219

12.   Gobie, W.A. and Ivory, C.F (1990) J. Chromatogr., *516*, 191

13.   Knox, J.H. 13th International Symposium on Column Liquid Chromatography, June

14.   Reejhsinghani, N.S., Barduhn, A.J. and Gill, W.N (1968) AIChE, *14*, 100

15.   Grushka. W. McCormick, R.M. and Kirkland, J.J (1989) Anal. Chem., *61*, 241

16.   Cox, H.C., J.K. Hessels, and J.M.G. Teven, J.M.G. (1972) J. Chromatogr., *66*, 19

17.   Davis, J.M. (1990) J. Chromatogr., *517*, 521

18.   Kucera, E. J. (1965) Chromatogr., *19*, 237

19.   Grubner, O. (1968) in J.C. Giddings and R.A. Keller, (Eds) Advances In Chromatography Vol. 6,., Marcel Dekker Inc, N.Y., p 173

20.   Horn, F.J.M. (1971) AIChE J., *17*, 613

21.   Grushka, E. (1972) J. Phys. Chem., *76*, 2586

Glossary

Some terms have two meanings. The exact meaning is clear in the context of the paper.

a	Radius	m
a_1	Outside radius of capillary	m
a_c	Radius including polyimide coating	m
A	Constant relating to viscosity of buffer	
A	Peak area	
B	Constant relating to viscosity of buffer	K
C	Solute concentration	mole/m^3
C_b	Buffer concentration	"
C_s	Concentration of adsorbed solute	mole/m^2
D	Diffusion Coefficient	m^2/s
E	Electric field strength	V/m
g	gravitational constant	m/s^2
g(r)	Series coefficient in Gill's solution	
G	Heat generation rate	W/m^3
h	Heat transfer coef.	W/(m^2 K)
H	PLate height	m
I	Current	A
k	Capacity factor	
k_f	Mass transfer coefficient	
k_1	Buffer's thermal conductivity	W/(m K)
k_2	Capillary wall's Thermal conductivity	"
K	Dispersion Coefficient	m^2/s
K	Adsorption particion Coef.	m
l	Separation length	m
L	Capillary length	m
m_i	ith moment	s^i
n	constant in Martin's velocity equation	
q	Constant	
r	Radial position	m
t	time	s
T	Temperature	K
T_1	Temperature at inside wall	K
T_s	Temperature of surroundings	K
u	velocity	m/s
u_{ef}	Electrophoretic velocity	"
u_{eo}	Electroosmotic velocity	"
u_{hs}	Hydrostatic velocity	"
u_m	Cross sectional average velocity	"

632

u_s	Cross sectional average hydrostatic velocity	"
u_1	Velocity at inner capillary wall	"
V	Applied field	V
V_m	Volume of buffer	m^3
x	Longtitudinal distance	m
y	Reduced radius	
Δh	Buffer height difference	m
ϵ_0	Permittivity of vacuum	
ϵ_r	Dielectric constant	
η	Buffer viscosity	$N\ s\ m^2$
κ	inverse of double layer thickness	m^{-1}
ρ	Buffer density	kg/m^3
Λ	Equivalent conductance of buffer	$m^2/(\Omega\ mol)$
σ^2	Variance	m^2
ζ	Zeta potential	V
τ^2	variance	s^2

Index